Radio-Television-Cable Management

THIRD EDITION

Radio-Television-Cable Management

JAMES A. BROWN
University of Alabama

WARD L. QUAAL
The Ward L. Quaal Company

Boston, Massachusetts   Burr Ridge, Illinois   Dubuque, Iowa
Madison, Wisconsin   New York, New York   San Francisco, California   St. Louis, Missouri

McGraw-Hill
   A Division of the McGraw-Hill Companies

Radio-Television-Cable Management

 This book is printed on recycled, acid-free paper containing 10% postconsumer waste.

1 2 3 4 5 6 7 8 9 0 QPF 9 0 9 8 7

ISBN 0-697-13237-4

Publisher: *Phil Butcher*
Sponsoring Editor: *Marjorie Byers*
Developmental Editor: *Valerie Raymond*
Marketing Manager: *Carl Leonard*
Project Manager: *Jayne Klein*
Production Supervisor: *Deborah Donner*
Designer: *Kathy Theis*
Cover designer: *Elise Lansdon*
Credit: Excerpt from *Television: The Business Behind the Box,* copyright © 1971 by
       Lester L. Brown, reprinted by permission of Harcourt Brace & Company.
Compositor: *Electronic Publishing Services, Inc.*
Typeface: *10/12 Times*
Printer: *Quebecor Printing Book Group/Fairfield*

Library of Congress Number 97-70794

www.mhhe.com.

CONTENTS

LIST OF TABLES

PREFACE

—◦◦◦—

We began the Preface to the second edition (1976) of our *Broadcast Management* text: "In an era of social change, often unprecedented in nature, an institution as dynamic as broadcasting cannot afford complacency." Between 1976 and 1997 the field of radio-television-cable-telecommunications offered everything but complacency, rushing through convolutions in technology, economics, and regulation that not even the experts anticipated. This accelerated rate of change has thrust American mass media into that vortex characterized in 1970 by Alvin Toffler as "future shock." Approaching the twenty-first century, the American mass media systems continued to be reshaped and redefined into the 1990s.

Despite the cycle of changes transforming broadcasting and cable, it was imperative that the topics in that book be updated. This text, including tables and endnote references, offers current statistical data and other factual information. Parts were heavily revised and major portions rewritten to reflect changes in the past two decades.

We wish to serve the countless readers and users of the previous book who have repeatedly urged us to adapt it to reflect the contemporary scene. The result is the text you now have in your hands—in somewhat different form and with a new publisher.

Both "the professional and the professor" who again collaborated on the present book still stand by our observations offered in the Preface to the second edition. Those comments seem almost as apt today as in 1976, once again demonstrating that the more things change, the more they stay the same:

■   Above all, broadcasters need to realize that, whether they like it or not, the public expects them to behave responsibly as members of a profession. The commercial station exists to offer community service while making a profit for its shareholders. It is, however, an overemphasis on profit return that causes some commercial managers to lose sight of the many other rewards to be achieved through radio and television broadcasting operations. When everything is subordinated to the dollar, standards of good broadcasting are sacrificed. Managing a radio or television station then holds no more meaning or challenge than operating a lemonade stand. To be penny-wise is in many instances to be socially foolish. Neither men nor institutions grow in stature through huckster interests in a narrow field.

Lee DeForest, pioneer in radio broadcasting, lamented towards the end of his life:

As I look back today over the entire history of radio broadcasting since [1907] . . . I . . . am filled with a heartsickness. Throughout my long career I have lost no opportunity to cry out in earnest against the crass commercialism, the etheric vandalism of the vulgar hucksters, agencies, advertisers, station owners—all who, lacking awareness of their grand opportunities and moral responsibilities to make of radio an uplifting influence, continue to enslave and sell for quick cash the grandest medium that has yet been given to man to help upward his struggling spirit.[1]

While that may be a harsh evaluation by an inventor who met ill success as an entrepreneur of radio, echoes of his criticism are voiced today by broadcasters as well as critics.[2] The late Edward R. Murrow, in his last speech (October, 1964), noted the need for proper human harnessing of these man-made media:

> The speed of communications is wondrous to behold. It is also true that speed can multiply the distribution of information that we know to be untrue. The most sophisticated satellite has no conscience. The newest computer can merely compound, at speed, the oldest problem in the relations between human beings, and in the end the communicator will be confronted with the old problem, of what to say and how to say it.[3]

Perhaps DeForest himself provided appropriately balanced commentary on the shared responsibility inherent in any kind of communication between sender and receiver who together make the process happen: "To me the quality of radio broadcasting is the index of the mental or moral qualities of the people who formulate its policies, or who continue faithfully to listen to its output."[4]

This book is directed to those who formulate those policies and to those who aspire to that role, the station managers and administrators of broadcast companies, as well as students engaged in formal study of telecommunications management.

Therefore, while this book offers representative patterns of practical management procedures, it does not attempt merely to set down minutely detailed and standardized practices for managing radio or television stations. Such a volume would be obsolete before publication and useless to those managers who want their stations to have distinct images or personalities that make them stand out because they are importantly different. Thus, we believe it appropriate to suggest areas for improved managerial leadership, to point out the values of change in those areas, and to inspire individual managers to action on the basis of their particular present and future local resources.

We offer considerable statistical documentation as well as excerpts from and references to other commentators about media management. Hopefully these data, together with the authors' syntheses and conclusions, will challenge the reader to analyze more fully the facts and principles involved, before accepting or rejecting either the status quo or alternate options in broadcast administrative practices. Regularly cited in footnotes and end-of-chapter references are related publications about specialized aspects of management; secondary sources are cited when they are more readily available than the original publications or reports. Our intent has been to offer to broadcasters, teachers, and students of broadcast management a range of generally available sources that provide fuller detail and differing viewpoints. We urge readers not to overlook those further discussions of controverted issues that are in the end-chapter notes. ■

The authors for the most part present and interpret factual data, offering descriptive and predictive analyses—what *is* and also what *will* most probably occur as a result of certain factors. We also take the liberty of making value judgments, offering prescriptive statements about what *ought* to be. Readers are reminded to seek information from all kinds of sources—books, journals, magazines, conversations with broadcast managers and staffs, and through firsthand experience—to build a personal synthesis from which to form their own judgment about such matters. We urge readers to keep an open mind, not accepting as the final word what an author writes (including the present ones), a teacher asserts, a broadcaster claims, or a critic concludes. Readers should check for themselves whether such judgments are corroborated by other knowledgeable sources and also confirmed by their own experience.

We thank many colleagues and friends who encouraged and assisted along the way, providing data and further perspective. Any lapses or errors, of course, are our own. This text has been strengthened by media professionals who readily granted permission to quote them and their published work. Particularly noteworthy is the contribution made by *Broadcasting & Cable* magazine's editors and staff whose thorough reporting over the decades provided a contemporary record of industry developments. We have relied heavily on that publication for specific factual data, including information provided with the permission of Nielsen Media Research, Arbitron, and other companies. A phalanx of successive graduate research assistants at The University of Alabama pored through and indexed literally cartons of notes representing decades of trade press coverage, scholarly journal articles, and industry and government reports. We thank reviewers of the manuscript draft for their reactions and suggestions, including Steven Adams, Cameron University; Robert L. Clark, Missouri Southern State College; John E. Craft, Arizona State University; Lynne S. Gross, Pepperdine University; Michael J. Havice, Marquette University; Jeff McCall, DePauw University; Regis Tucci, Mississippi Valley State University; and W. Joseph Oliver, Stephen F. Austin State University. Bringing this project to fruition with practical guidance and support at all stages were editors Stan Stoga, Eric Ziegler, and Kassi Radmoski, plus project manager Jayne Klein and copy editor Cynthia Cechota. Our thanks to all.

James A. Brown
Tuscaloosa, Alabama

Ward L. Quaal
Chicago, Illinois

ᨆ Preface Notes ᨆ

1. Lee DeForest, *Father of Radio* (Chicago: Wilcox & Follett, 1950), 442–443.

2. See appraisals by CBS alumni—not only disaffected ones terminated during takeover threats and severe budget cutbacks in the 1980s and 1990s, but also those who departed the company even in preceding decades; for example, Fred W. Friendly, *Due to Circumstances Beyond Our Control...* (New York: Random House, 1967); Alexander Kendrick, *Prime Time: The Life of Edward R. Murrow* (New York: Little, Brown & Co., 1970).

3. Kendrick, "Prime Time," 5.

4. DeForest, *Father of Radio,* 446.

INTRODUCTION

———∽∾∽———

This book presents a theory of management drawn from classic works and applies it to broadcasting (radio-television stations and networks), cable (local/group systems and program services), and noncommercial public broadcasting.

After surveying the recent and current status of the field, the book explores principles behind managing. Then it looks at each of the major components of the broadcast industry by successive treatment of personnel, audiences, programming, sales, finance, regulation, and engineering.

The goal is to provide a theoretical base, a value-oriented perspective, and concrete specifics where those principles are applied. Comprehensive notes are offered after each chapter to reflect the wide-ranging sources relevant to analysis. Quantitative data and statistics also profile the "realities" of the broadcast industry. The intent is to offer a comprehensive, descriptive, and discursive study that provides a range of source material, references, and points of analysis (including unresolved, controverted issues). The professional, teacher, or student can emphasize those aspects most suited to their own context.

This text is not a handbook for learning the daily mechanics of how to run a local station; such manuals already exist. Nor is it oriented to teach technical equipment or program production skills, although the crucial role of technology in broadcast/cable operations is discussed at many points.

The book is intended to be read and discussed by students with their professors. It is also directed to professional broadcasters seeking to further understand electronic mass media as a profession with social impact in a government-regulated environment within competitive free enterprise. While giving emphasis to career considerations, the text also looks beyond—for the benefit of professional practitioners as well as the informed citizen and consumer—to broadcast/cable media as a social phenomenon, a force in commerce, and a sometime art form. Because the book is intended for professionals as well as students, it purposely avoids a "syllabus" approach with outlined summaries, study questions, and other activities.

The text fuses the perspectives of a broadcast executive who provides "realistic idealism" about professional considerations with an academic who offers a scholarly viewpoint mingled with professional experience. Other sources add a range of evaluative viewpoints and factual data. Cited to document data and assertions in the text are philosophical treatises, legal studies, social and literary sources, professional and scholarly journals, magazines, trade newsletters, station and network files, and personal interviews.

The great sea-swell of change in media structures—including financial and regulatory—during the past decade continue to date even recent books. We have stressed principles and patterns over specific practices, partly to avoid outdating the text.

This successor to the second edition of *Broadcast Management* is recast to retain the perspective and major portions of content, but is presented in a more readable manner. This text adds analysis of cable and satellite systems. Factual data, references to media personnel, companies, and programming, have all been intensively updated. Topics addressed more fully include those recommended by previous adopters of the text: more coverage of financial aspects of the broadcasting/cable business; fuller treatment of sales and marketing; and extended analysis of program acquisition and long-term depreciation/amortization in smaller markets as well as major ones. Public noncommercial broadcasting is treated without adding unduly to the scope of the book.

The sequence of chapters follows this rationale: Chapter 1 offers a perspective on recent patterns of change in the broadcast/cable industry and the social impact of these changes. "Theories of Managing" (chapter 2) emphasize personnel in media structures and processes. The next three chapters look to the "persons" equation in broadcast media: managers (chapter 3) and the ones managed—middle-management staffs and personnel (chapters 4 and 5). The efficiency of reaching vast audiences (chapter 6) with local and national programming (chapters 7 and 8), coupled with marketing, determines the success of sales (chapter 9), which pays for the entire enterprise (chapter 10). These electronic mass media operate in a uniquely regulated environment (chapter 11), and are built on the foundation of continually evolving technology (chapter 12).

The 1960s and 1970s were a period of social upheaval in the United States (the earlier edition reflects those trends). The decades of the 1980s and 1990s brought economic, regulatory, and technological evolution—so swift as to be almost revolutionary. The present book reflects those massive shifts central to contemporary broadcast management.

A NOTE ABOUT THIS BOOK

—⟋⟍⟋—

Different readers will have different purposes and goals as they read this book. Many will be professional broadcasters or cable operators. Others will be media critics or public interest activists concerned about social implications of media structures and processes. Still others will be teachers and their advanced students exploring how radio-TV stations and networks and cable systems operate.

Each group mentioned above will approach the book differently. Some will skim over or omit the chapter on management theories (although they under-lie the very concept of what managing is all about in broadcast/cable or anywhere else). Some may ignore references in chapter endnotes, while others will check them for sources of comments, factual data, and occasional commentary about points made in the text.

Those who find some sections of chapters not wholly relevant to their needs will skim or read only selected parts. To aid the reader, this text introduces variations in format to highlight three levels of material: (a) standard typeface for most content; (b) a light grey background "screen" behind paragraphs of histor-ical background material; and (c) slightly smaller size type (with a square bullet ■ at start and end of paragraph or section) for heavily documented numerical data such as ratings and financial figures. That way the one continuous text addresses itself to three levels of interest or to various kinds of readers among profession-als, professors, students, and general readers—each with differing purposes and seeking varied kinds of topic development.

Some may find portions of chapters such as national programming (chap-ter 8) daunting at times with the plethora of factual information and statistical data. They may well glide across paragraphs studded with particulars to spot major themes involved. However, patterns do emerge out of closely observed data. Managers must be fact-based as well as instinct oriented; they must base creative decisions on demonstrable facts. Rather than write admittedly more flowing prose, the authors of this book chose to support statements throughout with hard data from a wide range of sources. This will enable the reader to test the validity of statements from the data provided in the text and referenced in chapter endnotes.

Much of the factual data will be evident to those already in management, including middle managers, and perhaps to those aspiring to such posts. For the majority of readers who do not intend to manage broadcast/cable properties, but who work in operations overseen by managers, this book should offer insight into the what, why, and how of radio-TV-cable management.

1

Managing Electronic Mass Media Systems

Perfection of means and confusion of goals seem in my opinion to characterize our age.

ALBERT EINSTEIN

This instrument [television] can teach, it can illuminate; yes, and it can even inspire. But it can do so only to the extent that humans are determined to use it to those ends. Otherwise it is merely wires and lights in a box.

EDWARD R. MURROW[1]

THE SCOPE OF THIS BOOK

"Broadcasting" includes transmission of electromagnetic energy intended to be received by the public.[2] This book concentrates on local commercial radio and television stations. It discusses network administration insofar as local stations are affected by network decisions about programming and sales. It treats public (noncommercial or educational) broadcasting explicitly where it differs from commercial operations. This book reviews cable television as a competitive consideration for local station managers and also as part of management's planning for expansion or collaboration.

This text is concerned with analyzing electronic mass media systems engaged in producing, distributing, and exhibiting information and entertainment. Beyond eliminating motion pictures and print, this description also excludes telecommunication forms such as computers and telephone, videotex and teletext, and other limited but significant nonmass or nonsystem media technology.[3] Also excluded (to put some limits on the book) are the important areas of audio and video recording industries, corporate mass communications—within a company's sprawling divisions, and from companies to the public—and governmental media operations, including the military. All have impact on society, and media careers including management can be found in each; but they are beyond the scope of the present work.

We specifically study electronic distribution of information and entertainment programming to mass audiences through radio, television, and cable. This includes (a) *local* entities at the heart of the distribution service—radio and TV stations, cable franchises with head-ends wired to households; (b) *national* distributors—networks, syndicators, cable program services, and direct broadcast satellite services; and only indirectly (c) program suppliers (producers and studios) and creators of commercials (advertising agencies). It also connotes *regional* ownership clusters: group-owned broadcast (over-the-air) stations and multiple-system cable operators (MSOs). These include commercial and noncommercial as well as pay/subscription services.

EVOLVING MEDIA SYSTEMS AND THE MARKETPLACE

THE 1990s

The final two decades of the twentieth century saw rapidly changing forms of telecommunications and mass media systems. Management reflected those changes. While essential purposes and functions remain similar to the past, the role of the manager—especially in major media corporations—expanded from coordinating people and creative planning to specializing in sophisticated macro-budgeting. National and regional managers became engaged in high finance. Their decisions were shaped by corporate mergers, leveraged buyouts, acquisitions, and resale of massive holdings based on stock market fluctuations partly reflecting government's shifting regulatory patterns.

The frenzy of "merger mania" undercut predictability, as did changing technology, international forces, national debt, lowered value of the dollar, and consumer shifts. While warning that "no firm can take anything in its market for granted," management consultant Tom Peters underscored the need for continual improvement amid constant change, counseling success by building on rampant chaos rather than trying to avoid it.[4] In the first half of the 1980s, Fortune 500 companies eliminated 2.8 million jobs, including many in middle management. Prior to buying out RCA and NBC, General Electric in the first six years of the 1980s had acquired 326 other businesses (paying out $12 billion) and divested itself of more than 225 companies (for $8 billion). Mergers and "downsizing" in the 1990s brought another wave of widespread layoffs and lost jobs. Media corporations reflected trends in U.S. business generally: economies of scale involved cutting back staffs. Between 1980 and 1994 the nation's five hundred largest manufacturing companies reduced jobs 25 percent—4.7 million fewer employees on their payrolls. IBM alone cut back employment by more than 100,000 from 1991 to 1993 (while losing $5 billion in 1992). In early 1996, AT&T announced it would split into three separate companies, resulting in dropping forty thousand of its 302,000 jobs—partly by attrition plus offering buyout packages to 77,800 managers if they left within six weeks).[5]

As the corporate squeeze became tighter and tighter, managers less and less enjoyed the luxury of mingling with their staffs—the MBWA concept of "managing by wandering around." Instead they tended to be engulfed in budget paperwork. In larger companies managers were obligated to mix with the power elite in a metropolitan area, particularly in financial and governmental centers. Media executives in the 1980s and 1990s became players in the corporation's strategy of manipulating massive finances—or at least paper representing equity value.

This shift of management's focus affects staffs, procedures, and even products of manufacturing and service companies. It has significant impact on the operation of broadcast/cable companies whose business is creating and distributing programming for millions of human beings.

Historical Background   It all began with the inventions of the telegraph and telephone in the late 1800s; then wireless telephony by 1920 expanded into the early mass medium of radio broadcasting.[6] Electronic communication progressed from wireless inventor Guglielmo Marconi to communications satellite Telstar within the span of a single lifetime: only forty years lay between the beginnings of mass radio broadcasting and the first international television by satellite. Within that short time the major technological developments of radio and television were introduced, refined, and expanded. From the crystal set grew the marvels of high-powered transmitters, FM radio, stereo, videotape, transistorized receivers, color television, videocassettes and cartridges, computerized station operation, laser and holograph, and satellite transmission. In the quarter century since that first U.S. communications satellite in 1962, compact disks, digital audio, interactive TV, high definition television and direct broadcast satellites as well as other refinements of broadcast and cable signals further enhanced electronic communications.

As wireless broadcast transmission matured it was hotly pursued by competing wired systems of coaxial cable, which in turn faced hybrid "wireless cable" operations along with direct broadcast satellite (DBS) transmission. The ageless baseball pitcher Satchel Paige advised: "Never look back because something may be gaining on you!" That is certainly true in the dynamic field of electronic innovation. When president of the CBS Television Network, Robert D. Wood noted: "In this business, when you slam the door a window goes up!" The very *rate* of accelerating change in technology—labeled as "future shock" by Alvin Toffler in 1970—is a vexing challenge to media managers.

Superimposed on technical achievements of hardware and electronic systems are the art and skills of producing programs—made financially possible by techniques of sales and advertising. In the public mind producers, programmers, salespersons, and advertisers are more closely associated with broadcasting than the scientists and engineers who made it all possible.

Engineering of course remains a vital part of the broadcast service. Without the contributions of electronic engineers, producers and salesmen could not keep radio-TV stations or cable systems operating. But the importance of engineers' technical contributions to mass media often is overshadowed by the social, cultural, and commercial applications of their creations.

Broadcast and cable managers—given incentive, freedom of enterprise, inspiration, and a sense of direction—can achieve results in programming and sales that parallel and even exceed the technical accomplishments of the industry.

INDUSTRY GROWTH: MANAGEMENT OPPORTUNITIES

Radio   Opportunities for managers in the emerging radio industry peaked in the late 1940s. When World War II ended in 1945 most of the 900 radio stations on the air were managed by men who had been in radio since its infancy two decades earlier, often as entertainers or salesmen. But the glamour of broadcasting and the thousands of authorized but unused frequency assignments made local radio an attractive (if unusual) form of investment, especially for many returning war veterans desiring to go into business for themselves.

The real growth of local radio began in the late 1940s. During the previous five years of war, experimental development of frequency modulation (FM) radio and television had been put aside while inventors, engineers, and technical resources were dedicated exclusively to the nation's defense effort. They developed elaborate communications systems and applied electronics to highly sophisticated weaponry. When they returned to civilian life, they brought those new technologies that speeded progress of FM and TV.

Most AM stations then on the air were substantially financed operations affiliated with one of the four national radio networks (NBC, CBS, Blue/ABC, Mutual). From 1946 through 1950, some 1,800 new radio stations came on the air. In 1948 alone, 533 stations began operations, many of them using the new FM frequencies for which few people had receivers. Those were the most authorizations for new stations in any single year since broadcasting began.

Despite television's surge to popularity in the post-war decades, FM stations gradually succeeded. They eventually dominated radio markets by the 1980s—initially because of their few commercials, but later for their specialized programming formats and high-quality transmission of stereophonic music.

Radios installed as factory equipment in cars, and bulky but portable radios, first brought out-of-home listening to millions. The transistor, printed circuits and miniaturization, made possible pocket radios and Sony "Walkman"-type headsets; out-of-home listening grew exponentially. By 1987, 95% of all cars had radios and 18 million people had "walk-along" receivers.

Radio's proliferation as well as its social importance were demonstrated most dramatically on November 9, 1965, when a massive power failure plunged the northeast portion of the nation into darkness for many hours. With studio and transmitting equipment powered by emergency generators and miniature receivers powered by batteries, radio kept millions informed, averting the panic of noninformation, until electric power was restored. No other event, no educational campaign, could have convinced so many people so quickly of the importance of keeping handy a transistorized, battery-powered radio, and of radio's reliability in a major emergency.

Multiplying local radio outlets led to various formats and "sound" for selected kinds of audiences. Syndication companies distributed custom programming matching these formats. Meanwhile, the original radio networks, unable to compete with TV's national programming attracting mass audiences, restructured after the 1960s into more than twenty-two commercial (plus two public, noncommercial) networks supplying specialized program services to local stations around the country.

By the mid-1960s, the number of radio stations had more than quadrupled. The rush to seek station licenses was so great that the Federal Communications Commission twice put a "freeze" on new authorizations (one lasting half a decade). By 1973, the FCC had authorized more than 7,000 commercial AM and FM radio stations. That year more than one-third of all commercial radio stations in the United States were FM; over 33% of portable radios had FM tuners (up from only 2% in 1961); and four out of five homes had sets capable of receiving FM.

By 1996, more than half of all stations were FM which attracted 75 percent of all radio listening. So the original AM radio service was first outflanked by television and then overtaken by FM. Local managers and their staffs operated

12,000 commercial radio stations across the country—5,285 FM and 4,906 AM. Another 1,810 noncommercial FM stations offered alternative radio service.[7]

The sale of radio receivers, the number of people listening, and station profits generally kept pace with population growth and the increasing number of stations. Virtually every household in the United States (99%) owns radio receivers, averaging almost six working sets each. Over one-third (36%) of the nation's 585 million radio receivers are out of homes—in cars and places of business.

Television   Radio was surpassed as a national mass medium after television became well established in 1950. UHF stations were slow to develop because at first few sets were built to receive UHF ("ultra high frequency") transmission. With the government's order for all television sets manufactured after May 1, 1964 to include both VHF (channels 1–13) and UHF (channels 14–83), potential audiences increased as did applications for UHF licenses. To markets already served by VHF/UHF stations affiliated with networks came new UHF stations as independent channels; their coverage was later extended when carried over regional cable systems.

By 1996, of 1,181 commercial TV stations, just over half (622) were UHF. Two-thirds of 363 noncommercial educational TV stations were UHF.

The high degree of mobility associated with radio became increasingly true for television; not only receivers but color cameras and transmitting units (including remote units for up-linking to satellites) became lightweight, compact, and highly portable. By 1995, over 98% of the nation's 95 million households owned television sets; almost all owned color sets, and 68% had two or more sets. Most homes could choose from five or more TV stations; two-thirds could receive signals from nine or more stations. In addition, almost two out of three homes in the country could watch stations from distant markets, including "superstations," whose programs were carried by most of the nation's 11,660 cable systems. Three out of four homes owned videocassette recorders. The number of prerecorded tapes sold to dealers (for resale or multiple rentals) quadrupled in four years to 200 million cassettes by 1989; 280 million blank tapes were also sold that year.

The "Big-Three" networks' national audiences dwindled annually, dropping from 90% of all prime-time viewers in the 1970s to 61% in 1993. Their audiences leveled off in 1994, then dropped to 55% in 1995. More than one third of the nation's homes in the evening were tuned to nonnetwork sources. They included local independent stations; the parttime Fox network (which in 1990 increased its program schedule from two nights to five nights, and subsequently to all seven nights); parttime networks of Warner Brothers and United Paramount (UPN); public noncommercial programs; cable services offering dozens of channels on local systems; programs on over 100 channels broadcast directly from satellites to home-receiving dishes; and rented movies on videotape (4.6 billion videos were rented each year, one out of five from Blockbuster stores whose 1994 revenue totaled $4 billion).

Cable TV   Cable television is almost as old as commercial over-the-air TV. It began in the late 1940s as community antenna television (CATV), designed to bring the service of urban TV stations to outlying areas where terrain made reception inadequate or nonexistent. Entrepreneurs hoped to profit by investing in local cable operations by collecting monthly subscription fees from users. Consequently, regions considered likely prospects for new UHF stations suddenly became high risks because of the competitive threat of CATV systems importing television programs from distant stations. In April, 1965, the FCC asserted jurisdiction over some 400 CATV systems that used microwaves for transmission; the following March the agency brought under its control 1,200 systems delivering service by means of telephone lines. The FCC's actions for the first time brought all CATV under government oversight, imposing certain conditions for their operation. Those regulations were relaxed in the following decade to permit less fettered development of cable TV, which subsequently grew into a sprawling giant in the mass media mix. In 1975, 3,240 CATV systems in 6,980 communities served 8 million subscribers.

During the twenty years after 1975 the number of local cable systems more than tripled to 11,660 in 1996; communities served quadrupled to over 25,000; and the number of subscribers multiplied seven-fold to 62 million households—nearly two thirds of all TV homes. More than fifty "basic" and multiple "pay" cable services provided national programming. On the heels of the FCC's deregulation of broadcasting in the 1980s came Congressional concern for reimposing restraints on the now vast cable industry. Federal debates continued into the 1990s about respective roles, rights, and restrictions of over-air stations and local cable systems. The Telecommunications Act of 1996 removed many limitations on cable operations while also authorizing entry of telephone companies and other competitors into the cable arena.

Other Forms of Broadcast Distribution   Not treated directly in this text are technically limited forms of distributing broadcast signals, each attracting small portions of total audiences available to mass media. In 1996 they included 1,772 low-power television transmitters (LPTV)—two-thirds of which were on UHF frequencies—serving local areas with highly specialized service. More than 7,000 translators extended the reach of regular local stations to distant locales by retransmitting 2,189 FM, 2,263 VHF, and 2,562 UHF signals. These add to the media mix vying for attention of potential listeners and viewers. Increased competition for audiences forces media companies to put a priority on marketing strategies merely to survive, much less to succeed.

MEDIA CHANGES AS CHALLENGES

As the number of TV and radio stations and cable services grew, competition became the top priority for most managers. Stations that previously enjoyed loyal audiences found new competitors attracting their listeners and viewers. Small-market broadcasters had to be aggressive even if theirs was one of the few stations in town. They were excluded from advertising budgets of many giant national advertisers who allocated spot dollars only to the largest 50 or 100 markets, or only to the top three radio stations in any market. And for limited advertising by local businesses they also had to compete with hometown newspapers, radio stations, and direct mail services.

While large and successfully managed TV and radio stations in major markets generate enormous revenue, the majority of AM and FM radio and UHF television stations earn modest profits, if any (only one out of three radio stations made money in 1993 and 1994). The federal government considers the broadcast industry a "small business" field because FM and low-wattage AM stations commonly employ staffs of fewer than a dozen, and UHF TV stations operate with staffs of only 25 or 30. Stations rely primarily on time sales to local retailers in their community. Average profits (before taxes) for TV stations in most markets were 15%-20% of revenues. Some in the top 25 markets averaged 24% profitability while network owned-and-operated (O&O) stations earned as much as 45%; but TV stations in small markets averaged from 5% down to annual losses ranging from $50,000 to half a million dollars. Radio stations generally profited 5% to 10% of annual revenue; profits exceeded 20% at the high end, down to losses equal to 22% of yearly revenue.

As the broadcast industry had grown, problems began to multiply regarding: sales and advertising practices; "orphan" independent stations vs. network affiliates; clear channel radio vs. local reduced-schedule stations; government regulation; and program formats and content. Some problems were so widespread and complicated that broadcasters formed industry associations to deal with them, such as INTV/Association of Independent Television Stations (renamed Association of Local Television Stations in 1995) and the Television Information Office, a public relations operation of the National Association of Broadcasters (after several decades of service, it was closed by the 1990s).

Paralleling radio-TV's growing impact on daily lives, criticism of programming and practices came from intellectual critics and other opinion leaders, political figures, consumer organizers, and the FCC. Criticism mounted early with quiz show scandal revelations in 1959 and social turmoil throughout the 1960s which spawned assassinations, civil rights and anti-war activists, urban riots, and consumer movements. TV's menu of alleged sex and violence along with radio's music lyrics about drugs and sex were blamed as significant factors in the rising tide of crime, violence, and sexual permissiveness in American society. The phrase "vast wasteland," coined in 1961 by FCC chairman Newton Minow to describe his perception of broadcasting's schedule of programs and commercials, has since haunted station managers and network executives.[8] Critics through the decades have seized upon that phrase when excoriating broadcasting in general, often with charges not fully justified. Broadcasters soon became hypersensitive to criticism and devoted more time to defensive reactions than to positive adjustments in significant problem areas.

Men and women in administrative positions usually are more aware of such problems in their own industry than outside critics think. Broadcast and cable executives have usually risen to their positions because they understand the industry's complexity and have proved able to guide personnel and procedures properly as well as profitably. But unlike many other businesses, no matter how effective their efforts are on making good business decisions, management's judgments affect program service that reaches out day and night into the community of citizens. The public's reactions and critics' appraisals often are valid on grounds other than business and economics. A manager cannot ignore this "social reach" of broadcast and cable media.

COMPETING CREATIVELY

Preoccupation with merely defensive strategy has plagued broadcasters throughout the history of radio and television. A positive approach is needed. The station manager must exercise initiative and leadership (see chapters 2 and 3). "Playing it safe" is seldom really playing it safe. Imitating other stations' practices and clinging to trends already established by competing operations is to abdicate creativity and does not gain audiences. With trained and experienced personnel widely available, as well as professional consultants, a station should initiate fresh programming and promotion to attract public support.

A predominant tendency among managers of radio and television stations is to imitate successful program formats and business practices of other stations. Even in the early years of television's development, comedian Fred Allen quipped that "Imitation is the sincerest form of television." Evidence of originality or true creativity is rare. Station management often disproportionately respects ratings services (discussed in chapter 6), sometimes ignoring other measures of impact more significant for the long-range future. People given too much of the same kind of programming grow bored and turn to other media and other forms of recreation. Broadcasting's greatest asset has always been its popularity. But as people acquire a blasé attitude toward local stations and network TV, the industry falls on hard times. Witness viewers' attraction to alternative program sources—cable and satellite services, and rented videocassettes—as well as their restless "grazing" among TV channels, plus zapping away from commercials (or zipping through them at high speed on VCRs). Successful managers are those able to inspire originality in staffs to hold onto and build audiences.

An observer and prominent force in broadcasting throughout its history, the late Sol Taishoff as chairman and editor of *Broadcasting* magazine received in 1966 the Distinguished Service Award of the National Association of Broadcasters. On that occasion he challenged broadcasters to establish "a proper heritage of imagination and boldness" for those people of the new generation who would some day be in charge of stations. "The current climate of the broadcasting business," said Mr. Taishoff, "is not especially conducive to imaginative ventures and risky progress."[9] Rather, he said, the prevailing attitude is "conservative," with suggestions not to "rock the boat." Two decades later, another pioneer sounded a similar note when he accepted the same NAB award in 1987. Martin Umansky's professional career paralleled the successful history of station KAKE-TV in Wichita, Kansas, which he managed and then chaired until retirement. He urged his colleagues in the industry:

> [G]ood, serious broadcasters who do not live in a vacuum . . . know that serving the public well is good broadcasting and that's good business. . . . It's the key to success.
>
> And when you develop a Number 1 news department . . . when your local programming hits the mark . . . when you get deeply into community activities and you are recognized as a committed, effective broadcaster . . . that's when you become the Number 1 'favorite station' in the market. . . .
>
> That bottom line may be better served in the long term, if investment is made in investigative reporters and public affairs people. Money spent on developing a sound and respected relationship with the community and becoming the trusted, relied-upon communicator . . . will come back many times over.[10]

The solid advice offered by both Taishoff and Umansky was heard less and less in the 1980s and 1990s during the extended period of deregulation, corporate financial stratagems, and formal FCC challenge of the "trusteeship" concept of American broadcasting.

Broadcasters sometimes separate themselves from their listener-viewers by what advertising executive Gene Accas once called "an idea-tight, communications-proof wall constructed of ego, misplaced self-importance and mis-understandings."[11] The broadcaster cannot afford such wasteful luxury. Empathy with the audience and qualitative analysis can reveal much-needed information that rating services cannot provide.

This is particularly important as the twentieth century concludes its final decade because, in addition to economic and regulatory changes, the American population is also undergoing significant demographic developments. The concentration of "baby boomers" is moving into middle-age; smaller families (almost half with one-parent households) with few children means smaller youthful audiences in coming years; and the elderly and retired populations—with considerable disposable income—are expanding. The nation's homes in previous decades were made up of three to five people or more; now they average 2.3 persons. Millions of households consist of a single individual. Television networks, radio stations, and cable services must assess those shifts to alter programs and formats to attract audiences that meet advertisers' target demographics.

BROADCAST/CABLE MANAGEMENT IN PERSPECTIVE

FOUNDATION PRINCIPLES

A fundamental problem lies in differing philosophies of "what it's all about" as reflected in the perennial question of who owns the airwaves. With true ownership comes control, and the right to determine what goes out over those airwaves. Often enough the licensee who has invested risk capital in an uncertain venture in a highly competitive market is impatient with the question. He is acutely conscious of the enormous amount of money paid for legal and technical fees, equipment, personnel, and programming. He knows that potential listeners or viewers will pay nothing more or less whether his station transmits or not. So the broadcast investor-owner finds the question of "ownership" academic at best, and a point of law at worst. But that point of law is also a point of fact.*

As for the law that structured U.S. broadcasting for almost two-thirds of the century (1934-1996), The Communications Act of 1934 (section §301) explicitly stated that no licensee possessed property rights over channels or frequencies. The Act provided "for the use of such channels, but not the ownership thereof, by persons for limited periods of time, under licenses granted by federal authority, and no such license shall be construed to create any right beyond the terms, conditions, and periods of the license."[12] The Communications Act further reaffirmed that any license is granted conditionally by the FCC, because it "shall not vest in the licensee any right to operate the station nor any right in the use of the frequencies designated in the license beyond the term thereof nor in any other manner than authorized therein." The Telecommunications Act of

---

*The co-authors of this book disagree on this controverted point. Professional broadcast executive Ward Quaal firmly holds that risk-taking entrepreneurs made it all possible and sustain the service by massively investing in developing technologies and creative programming, deserving a hard-won ownership claim. Professor James Brown affirms that the valuable resource is the heritage of the citizenry at large whom media affect and who also spend enormous amounts to make the sending/receiving broadcast service economically feasible. Both agree that, in any interpretation, electronically transmitted communication to the nation's homes demands conscientious owners and managers who are responsible professionals; proper media use likewise demands selective viewing-listening.

1996 is less emphatic on the point but preserves the notion of serving the public as a touchstone of broadcast and cable policy.

Apart from legal technicalities, there is the pragmatic fact of capital investment in facilities as a criterion of who really "owns" the airwaves. The billions of dollars invested by owners of station facilities and property are exceeded by the greater billions of dollars invested by purchasers of radio and television receivers—without which stations would have no audience to reach or airtime worth selling to advertisers. In 1981, for example, the total property and capital equipment of TV stations and networks was worth $3.6 billion. However, in that same year consumers purchased 18,479,000 color and black-and-white sets valued at $5.2 billion, while the previous five years averaged $4 billion each. Those annual totals increased in following years, to $11.5 billion spent for TV receivers in 1993, plus $2.1 billion spent on new radio sets (home, personal, and auto).[13] Thus, dollar figures alone by licensees do not imply "ownership" of the electromagnetic spectrum for electronic media which depend as much on continued reception by an audience as they do on continued transmission by stations in order to exist at all.[14] But if one determines "who owns the airwaves" according to dollars invested, then the public clearly would have greater proprietary rights over those airwaves.

So the issue must be resolved more in the arena of philosophy. To whom does the phenomenon of the "natural resource" of the electromagnetic spectrum rightly belong: to individual broadcast licensees or to the public at large? The viewpoint about this single pivotal question affects the professional and pragmatic, as well as philosophical, policies of the broadcast manager and staff. If the airwaves belong to the public, broadcasters are licensed to harness them on that public's behalf, serving as trustees of the public resource. But if the airwaves are owned by broadcasters, they have full right to program whatever they wish that is profitable without consideration of "public service" imposed on them in using their private resource. The latter view reasons that nothing is effectively "owned" until it is used. The airwaves—a portion of the electromagnetic spectrum of frequencies—were always there. (Former secretary of state Dean Rusk asserted that "the people do not own the airwaves any more than they own the North Star or gravity.") It took dedicated individuals, with courage, money, and faith in themselves and the potential of radio and television, to develop the present system of American broadcasting.

This issue is reflected in the government's successively shifting stance about first regulating broadcasting, then cable. The pendulum swings both ways, as witnessed by the FCC's sustained oversight in the 1960s. Less restrictive "reregulation" was introduced by Republican FCC chairman Richard Wiley in the mid-1970s; that was expanded by his Democratic successor Charles Ferris, and pushed to its ultimate of permissive "deregulation" by Republican Mark Fowler during the 1980s. But it was later pushed back in the opposite direction by Congress and assenting members of the Commission as the decade turned into the 1990s. During his six years as FCC chairman until 1987, Fowler strove effectively to deregulate broadcasters by eliminating dozens of major restrictions and minor procedures. His often stated premise was that the "public trustee" model was no longer relevant because the vast number of stations eliminated the "scarcity" premise for regulating; instead, competitive "marketplace forces" would determine the public's interest (that is, what interested the public most). The FCC itself waffled between the two views in its 53rd Annual Report (1987): "Overall, however, the Commission concluded that the approach that best achieves the First Amendment principles underlying the doctrine and its own public interest objectives, would be reliance on an unregulated marketplace of *ideas*" (emphasis added to highlight the concept whose forebears include Socrates and Jefferson).[15] But Chairman Fowler's marketplace was one primarily of *economics*—based on the criterion of business profitability among competitive station owners. After Fowler left the FCC, Congress pressured the Commission to reestablish the "public interest" standard as the central criterion for granting and renewing broadcast licenses (as in the 1934 Act's benchmark phrase "public interest, convenience, and necessity"). Democratic chair Reed Hundt in the mid-1990s vigorously reasserted social responsibility as a factor in licensing broadcast stations.

Over the years similar challenges were raised about the proper place of cable in a community and its obligations to provide moderately priced quality service when it was exclusively franchised by local government. The Cable Communications Policy Act of 1984 freed local systems from much governmental oversight; but criticism of shoddy service and price increases around the country prompted Congressional enactment of a new cable law mandating the FCC to analyze and reduce local rates judged excessive (see chapter 11). The Telecommunications Act of 1996 restored more autonomy to cable systems while authorizing telephone companies and others to compete with established cable systems.

Involvement in Media Issues

Some media administrators have been reluctant to take positions on issues of public importance, concerned that such stands might limit future business from advertisers who view these issues differently. But strong stands on vital issues can gain as much business as they lose, while increasing the stature of broadcast managers who state their minds, earning new respect for themselves and their stations. The late Edgar Kobak, former executive of ABC, NBC, and the Mutual Broadcasting System, claimed that in his professional lifetime he lost significant business only twice for refusing to compromise his standards in operating those broadcast chains. One instance personally involved the late Henry Ford, who eventually returned his advertising to the network because he respected Kobak's firmness and integrity. Broadcasters have a special role of such leadership because their industry not only generates interest in the social environment but is itself a key part of it.

Whenever significant issues arise within the industry, a few national media leaders devote great amounts of time toward their solution; the individual station manager's voice is seldom heard. Local managers should participate actively in their state associations and with committees of the National Association of Broadcasters and other organizations. Each manager must supplement the work of state, regional, and national professional organizations by continually analyzing her own station, audience, and market.* The owner and manager should determine where they want the station to be at the end of another decade and then concentrate on achieving that goal. Their goals might include greater professional practice among staff, expanded station services, program development, wider public acceptance, and more productive time sales—all resulting in increased profitability along with enhanced service to the community. Managers cannot run only a "holding" operation, sandbagged and dug in, and expect to succeed.

---

*Female and male pronouns will be randomly alternated throughout the text to avoid clumsy forms such as "he/she" while yet reflecting the major role of women in media—one out of three employees, a large percentage of news directors and sales managers, significant numbers of station and network executives.

Conversely, they cannot jump back and forth among radio formats, air personalities, TV syndication scheduling, news anchors, and news or program directors based merely on a few rating books.

CBS executive Tom Leahy reaffirmed goals and principles when he spoke to the network's annual conference of affiliated station managers in 1986. His remarks bridge nicely into the next section of this chapter.

> Some worry that if what we see today continues to accelerate, broadcast properties could become mere commodities to be traded as dispassionately and as fungibly as soybeans or cotton futures.
>
> And somehow, that notion is at great variance with what we have always felt about the essential character of broadcasting. That is not the kind of business that you and I signed on for. It is not the kind of business that we have been in. It is not the kind of business that we believe the public wishes us to be in. . . .
>
> Above all, everyone working in radio or television developed the keen awareness of being rooted in the community. We knew we were a community resource. The public interest has been the standard by which the industry operated, and we believe that we have thrived as commercial enterprises because we have served the public interest.[16]

FCC commissioner Ervin Duggan underscored the point when he spoke to the Association of Independent Television Stations convention in 1993:

> The best defense against the dead hand of government is the lively imagination of the industry—imagination to work voluntarily to improve service to children; to illuminate public issues; to improve news coverage—and to insure that this nation's system of publicly licensed broadcast stations does not descend into a mire of lowest-common-denominator programming. If your voluntary concern for the public interest is present and visible, you have little to fear from government.[17]

Broadcast manager Yale Roe cited John D. Rockefeller's view that true justification for economic power lies in service to the public.[18]

MANAGERS AND SOCIAL VALUES

Most managers in broadcasting have been socially responsible. In past decades, the FCC's careful screening of license applications and station transfers protected stations from being operated by people of questionable character (granted that as licensees the owners are scrutinized, not their hired management and staffs). Reacting to oversight loosened through deregulation in the 1980s, the governmental pendulum began to swing back again. Federal regulators reconsidered the need for "character qualifications" when determining suitable owner-licensees.

The public nature of electronic media regulated by government generates an unending series of crises for an industry operating under the spur of risks and rewards of competitive free enterprise. Despite more than a decade of deregulation, station managers still are not as free to conduct their affairs as most other businesspersons—including those in print media. As an individual, an owner or manager is virtually powerless to effect any changes in the overall system. As a group, broadcasters have seldom mobilized anything approaching unanimity of opinion on any issues except those that would penalize them individually or collectively.[19] As a result, they have become easy targets for negative criticism, some of it valid and needed, some of it vitriolic and without foundation.

While preoccupation with governmental pressures and constant assault by critics and opinion leaders relaxed somewhat with deregulation, broadcast managers still expend too much time and energy on merely defensive reaction. They could manage more efficiently, with substantial advances, by more constructive attention to station and community needs. A few leaders in the industry keep busy defending the actions of some fellow broadcasters and arguing for conditions benefiting all broadcasters. Often, however, a hard-won right is not used by enough broadcasters after it has been granted, such as the right to editorialize and the right of access to report electronically from state legislatures.

The broadcast executive should not be a faceless proprietor of an electronic money-making machine, but a person responsible and responsive to urgent issues of the times, especially local ones. The broadcaster will not be given a place in the community unless he or she earns it. That place is not one of a servant of surveys and profit statements, but of a powerful participant in a free society building for the future.

MANAGING MEDIA DISTRIBUTION SYSTEMS

Once king of the broadcast hill, AM radio in the 1980s struggled to survive by introducing stereo transmission (dismissed by Westwood One's chairman Norm Pattiz as merely "offering static on *two* channels"), to be enhanced by digital audio engineering for superior quality. Both AM stations and the dominant FM service not only proliferated formats but kept modifying and even changing formats to match the shifting fancy and fads of audiences. Companies syndicated nationally by satellite to local stations, offering complete formats as well as specialized feature services (such as a wide range of music forms, talk, sports, financial news). The "classic" radio networks permuted into multiple special-service niche networks.

MEDIA CONGLOMERATES

The ten-year span between 1985 and 1995 saw massive restructuring of the radio-TV-cable industry; enormous changes in ownership configurations affected the structure and processes of giant broadcast media. The traditional "Big Three" networks had already been faced with a serious alternative service by the Atlanta-based multiple operations of entrepreneur Ted Turner; his WTBS "superstation" served the nation's cable systems by satellite, later joined by the Cable News Network (CNN) and the TNT movie channel. A surge in the number of independent television stations created a need for more programming than off-network reruns could provide, so syndication of first-run product accelerated. The traditional movie studios were joined by smaller companies, often formed by former broadcast executives, which quickly moved into satellite distribution of their programming to local stations around the country.

CABLE SYSTEMS AND MSOs

Cable service leaped forward in the mid-1970s when Time Inc. first distributed HBO to local systems by satellite (1975), followed by Ted Turner's WTBS national feeds (1976). By 1985 broadcast interests owned almost half of the nation's top-50 multiple system operators (MSOs) in cable; newspaper and magazine publishers owned another 34%; film production and distribution companies owned 12%, while only a third of the largest MSOs were owned by independent non-media interests.[20] Tele-Communications Incorporated (TCI) grew into the nation's largest multiple system operator (MSO) by building, developing, or buying out local cable systems around the country; by 1996 TCI distributed cable signals to 15 million homes in 48 states—almost one of every four cable households.

■　　Further, John Malone's TCI partly owned many of the twenty top basic cable program networks, including CNN, TNT, 20% of Black Entertainment Television, half each of the Discovery Channel and of American Movie Classics, 60% of the Prime Network group of regional sports channels, and 50% of Viacom International's two pay services, Showtime and The Movie Channel. Just those partial ownerships together totaled assets worth an estimated $3 billion. With TCI and his Liberty Media company, Malone's investments involve 29 cable services, including the QVC shopping channel and Court TV. TCI's size was evident by comparing its estimated 1990 "operating cash flow" (that is, income before taxes, interest, depreciation, and amortization) at $1.4 billion, against the 1990 operating cash flow of Capital Cities/ABC ($1.1 billion) and CBS ($322 million). Yet, in 1995 TCI's annual revenue was still less than one-third of those generated by mega-merged Disney/CapCities-ABC or by Viacom/Blockbuster/Paramount. In the two decades since Malone joined TCI in 1973, the firm purchased or became a partner or investor in 650 companies.[21] TCI's assets were valued at $20 billion in 1995.　　■

INTRA-MEDIA DEALS

Investment strategies spawned corporate mergers and buyouts. The aggressive competition of Turner and others such as Rupert Murdoch (News Corporation of Australia and Fox) created a flurry of business "mega-deals" involving billions of dollars, as newly formed conglomerates bought out broadcast properties of pioneer owners. Classic names associated with the earlier growth of American radio and television, such as Storer and Metromedia, turned over their broadcast holdings—for up to hundreds of millions of dollars per station, in some instances (the price for all Storer's properties exceeded $1 billion)—to other corporations or to newly formed entities in the telecommunications field.

■　　General Electric sold its classic pioneering station in Schenectady, New York, and its other properties, only to reemerge a few years later as purchaser of RCA and NBC for $6.4 billion. It subsequently sold off the NBC radio network to Westwood One radio syndication company which earlier had acquired the Mutual Broadcasting System. In a $3.5 billion merger, Capital Cities Communications bought out the four-times larger American Broadcasting Companies, Inc.; then within a decade was itself taken over in 1996 by the Disney conglomerate for $18.5 billion. Ted Turner tried various tactics to purchase controlling interest in CBS Inc. While unsuccessful, his efforts forced that corporation to restructure its financial posture—affecting all its corporate divisions by drastic reductions in personnel (especially in broadcast news); organizational structure (selling off its book publishing and record divisions, among others); and broadcast holdings (sale of KMOX-TV, its company-owned station in St. Louis, temporarily reducing its owned-and-operated TV stations from five to four). In 1993 Bell Atlantic Corporation announced a merger with TCI involving $33 billion (but it subsequently withdrew partly because Congress forced cable systems to lower their subscriber rates). In 1994 Viacom paid $9.5 billion for the Paramount media empire. In 1995 Westinghouse Electric bought CBS for $5.4 billion, despite its own current debt of $6.9 billion (from 1985 to 1987 alone Westinghouse had sold off 70 generally unprofitable enterprises and acquired 55 new ones). That same year Time-Warner and Turner Broadcasting announced an $8 billion merger deal. In 1996 Westinghouse/CBS paid $5 billion (mostly in stock) to acquire Infinity's group of 44 large radio stations, bringing its total to 82 in fifteen markets, most in the top-10 markets—generating over $1 billion in sales revenue in 1996 (triple the revenue of 112 radio stations acquired by Clear Channel Communications by then). Yet these figures were dwarfed by the announced merger of regional telephone companies Bell Atlantic and NYNEX—each with estimated assets of $25 billion—in a deal valued at $22 billion; their alliance would expedite entry into household video service by cable and fiber optic. ■

In 1995 alone, 65 radio and TV deals involved $26.3 billion, plus another 84 cable and broadcast-related deals were valued at $9.5 billion. Trade magazine articles attempted to provide a scorecard so you could tell the players. Top management/ownership of media corporations remained highly visible, but their lineups kept changing overnight and all systems were in a state of flux.

INTERNATIONAL/FOREIGN OWNERSHIP

New deals for program production and distribution to the $3 billion syndication market scrambled the lines of doing business, with foreign co-production and syndication/satellite leasing growing in importance to U.S. media corporations. The goal of many deals was *vertical integration* into a single company of: (a) hardware systems, (b) software content and "talent," and (c) distribution of media products.

■   Rupert Murdoch's Australian News Corporation purchased 20th Century-Fox studios, then a number of TV stations in the United States, including the Metromedia group. Sony Corporation purchased CBS Records in 1988 ($2 billion) and Columbia Pictures Studios the following year ($3.4 billion). In 1989 Warner Communications (including Warner Brothers Studio and extensive local cable holdings) merged with Time, Inc. (owner of HBO/Home Box Office and local cable systems); the previous year Warner Communications had bought Lorimar Telepictures, later selling it to British interests. In 1990 the French media company Pathé bought out MGM/United Artists (except for the extensive film library already bought by Ted Turner for his Turner Broadcasting System and TNT/Turner Network Television on cable). In 1991 Matsushita (manufacturer of Panasonic, Quasar, and Technics electronics) bought massive MCA for $6.1 billion—including Universal Television, Universal Pictures studios and film library, Motown Records, and a Florida theme park; four year later Seagrams of Canada took control when it bought 80 percent of MCA-Universal.   ■

NETWORKS/CABLE COLLABORATION

Meanwhile, the traditional networks hedged their bets on their own future by collaborating with rival domestic satellite and cable systems.

■   CapCities/ABC owned 80% of ESPN sports cable network and a portion of the Arts & Entertainment cable channel. NBC created CNBC—Consumer News and Business Channel for cable (it also created the new executive position of president of NBC Cable and Business Development). CBS launched an unsuccessful fine arts cable program service, as did Rockefeller Center—later reemerging as the A&E/Arts and Entertainment network jointly owned by CapCities/ABC, General Electric (NBC), and Hearst. In 1988–89 ABC-TV promoted its massive mini-series *War and Remembrance* by coordinating advertising with the Sony manufacturing company—a joint national ad campaign urged the public to buy Sony brand videotapes to record ABC program episodes they could not watch during consecutive evenings scheduled. Adding to its

CNBC in 1996, NBC joined forces with computer software giant Microsoft to mount MSNBC channel, a news and information program service by satellite-cable.   ■

NEW TECHNOLOGIES AND AUDIENCES

During the 1980s, federal deregulation was spawned as much by the reality of proliferating, competing mass media as by the sociopolitical philosophy of the Republican administration. The influx of independent over-air stations—linked to a growing fourth network (Fox) and to Paramount's and Warner Brothers' emerging UPN and WB networks—competed for traditional over-air media audiences, as did multiple-channel cable systems, satellite programmers, and videocassette rental services. The resulting media mix fragmented not only audiences but also advertisers and their purchases of commercial time. Veteran networks and broadcast corporations found themselves in a massively shifting marketplace.

"Alternate" technologies expanded at a swift pace (see chapter 12). Satellite distribution at the sending end of the communications process was matched by videocassette recordings (VCR) at the receiving end. Within several years VCR units programmed TV sets in two-thirds of the nation's households, by "time-shifting"—delayed playback of programs previously taped off the air or from cable—or by renting or purchasing pre-recorded presentations to play at home at their convenience. (The audience was becoming the scheduler as well, moving electronic media consumption away from the passive, inexorable "temporal treadmill" of the over-air broadcast model and towards the print model's highly selective consumer determination of time and place for using media.) Into that mix were added distribution by direct broadcast satellite (DBS) and the approach of high-definition television (HDTV). Other innovations moved forward at an accelerated pace, assisted by computerized technology that was already a major component not only of technical operations and equipment in broadcast studios and control rooms—but also in the offices of sales, billing, traffic, programming, and other departments.

WHAT NEXT?

It is an understatement to call the decade of 1985-1995 a time of major transition in the broadcast industry. Probably the most dramatic, visible manifestation was the fading of traditional TV network dominance of airwaves and audiences in the United States, coupled with the rise of cable service to almost two-thirds of the nation's homes (over 90% of households have cable lines available if they want to subscribe). Symptomatic of the shift in network strength was the unprecedented full-page ad late in 1990, signed with the three competing network logos, captioned "The Picture Sure Has Changed in 20 Years." The ad claimed that multibillion dollar purchases of media companies by international corporations were unfair competition against networks hampered by regulatory constraints against investing in or syndicating program series they carried. The networks' role may lead to a new stage of international dominance as other countries liberalize their state systems, opening up media to competitive capital and marketplace forces. More probably, rather than remaining full-service broadcast companies, the networks will reconfigure themselves as mass delivery systems of specialized services, similar to radio syndicators or cable program services. In a word, TV will follow the route of radio when it responded to video's inroads in the 1950s: abandon a broad-based network programming service to become suppliers to local formats with niche programming and satellite-syndicated special features services such as news or sports.

But before announcing the demise of TV networks, one must note that even with "only" 55%–60% of the prime-time audience, the three networks far exceed in nationwide mass reach—socially as well as with advertising impact—the fragmented 40% of U.S. audience scattered among dozens of cable services and satellite systems, and hundreds of independent local outlets and public noncommercial stations. (When cable entrepreneur Ted Turner labeled the major networks "dinosaurs," CBS executive Tony Malara responded that the allusion encouraged him: dinosaurs had roamed the planet for thousands of years, so the networks—only around for over half a century—had prospects for a long future.)

Cable forges ahead, with digital technology and optical fiber replacing coaxial cable, offering capacity for up to 500 cable channels including high definition TV (HDTV). Direct broadcast satellites (DBS) literally hover in competition, beaming directly to 18-inch receiving dishes on individual homes many feature services now carried by traditional networks and cable systems.

Business in the mid-1990s was booming, but whose business was it? Multimedia ventures and vertical integration mergers and buyouts reconfigured classic lines formerly separating communication industries. The Telecommunications Act of 1996 opened the door to cross-industry competition. That year saw local phone business (primarily seven regional "Baby Bells") of $98 billion, long distance service (mostly AT&T, MCI, and Sprint) at $67 billion, along with television's $30 billion, radio's $11.6 billion, and cable business at $23 billion.[22] New legislation spawned further deregulation, now permitting each to move into one another's territory.

American mass media systems continue to change as the new century approaches. Economic reconfiguration of major media companies is compounded by proliferating alternative communication systems. Audience patterns of using and reacting to mass media keep shifting. The government's role vis-à-vis broadcasting continues to be modified by the FCC and reevaluated by Congress and media critics with passage of the Telecommunications Act of 1996. Competing sectors of the industry even parlay trade-offs of concessions to regulation in return for stabilizing their established businesses against competing systems—such as license renewal guarantees; protection from cable or DBS or massive telephone companies; limiting children's advertising and requiring a minimum three hours weekly of educational-oriented TV for kids; reinstating the "fairness doctrine"; auctioning off spectrum space; and computerized "V-chip" technology to restrict TV set reception of violent programs.

In this context of a continuing "churn" in mass media, the following chapters explore the complex role of broadcast and cable managers.

⸺ CHAPTER 1 NOTES ⸺

1. A. M. Sperber, *Murrow: His Life and Times* (New York: Freundlich Books, 1986), 540.

2. The operative word, "intended," in the Communications Act of 1934, section §3(o), distinguished mass-oriented programming via AM-FM-TV radiated signals from specialized person-to-person and private services such as citizens band, amateur, maritime, police, mobile, etc.

3. For apt definitions and distinctions, see the Broadcast Education Association's symposium of articles on "The Place of Telecommunications," ed. Ray Carroll, *Feedback* 27:3

(Winter 1985): 2–31. One example of synergism brought to new technologies by multiple corporations was the early joint effort in videotex—electronic publishing—by IBM (hardware manufacturer), CBS (software/content producer), and Sears (distributor). CBS eventually dropped out while the other two developed Prodigy on-line computer service. In 1966 half-owner Sears announced its departure and IBM prepared to sell off the service.

4. Tom Peters, *Thriving on Chaos: Handbook for a Management Revolution* (New York: Alfred A. Knopf, 1987), 11; other data in this paragraph are from ibid., 3, 5.

5. From Mortimer B. Zuckerman, "The Glass is Half Full," *U.S. News & World Report,* 27 February 1995, 80; and Associated Press articles in *Tuscaloosa News,* 31 March 1993, 7C, and 16 November 1995, 5C.

6. For a definitive history of broadcasting's historical development, current structure and processes, see Sydney W. Head and Christopher H. Sterling, *Broadcasting in America: A Survey of Electronic Media,* 6th ed. (Boston: Houghton Mifflin Company, 1990). A similar analysis with less exhaustive detail is Joseph Dominick, Barry L. Sherman, and Gary Copeland, *Broadcasting/Cable and Beyond: An Introduction to Modern Electronic Media* (New York: McGraw-Hill, 1990). Another readable historical account, with 33 pages tabulating comparative historical statistics, is Christopher H. Sterling and John M. Kittross, *Stay Tuned: A Concise History of American Broadcasting,* 2nd ed. (Belmont, Calif.: Wadsworth, 1990).

The authors recommend that, to understand antecedents and the current context of mass media management, readers familiarize themselves with Head and Sterling's excellent survey of broadcasting and cable in the United States. Their lucid and precise style is supported by painstaking documentation in 615 pages—including 177 tables, graphs, figures, and illustrations, as well as 53 pages of annotated bibliographies. For advanced students and professionals it probably offers the best analysis of major structures and patterns of American electronic media.

7. These and following data are from "By the Numbers," *Broadcasting & Cable,* 18 March 1996, 71; and from *Broadcasting & Cable Yearbook '95* (Washington, D.C: Broadcasting & Cable Publications, 1995).

8. Minow's speech to the National Association of Broadcasters convention might well be reread today, to compare with contemporary media criticism three decades later. This was his first speech as FCC chairman, appointed by President John F. Kennedy; he left the Commission after only 27 months but his "vast wasteland" phrase became part of the literature characterizing television programming. Newton N. Minow, *Equal Time: The Private Broadcaster and the Public Interest,* ed. Lawrence Laurent (New York: Atheneum, 1974), 48–64.

9. Sol Taishoff, "Penalty of Success," *Broadcasting,* 6 June 1966, 102.

10. Martin Umansky, "Viewpoints: Fragmentation of Business Makes Local Commitment More Important Than Ever," *Television/Radio Age,* 13 April 1987, 57.

11. Gene Accas, "An Open Letter to the NAB's Future President," *Broadcasting,* 21 December 1964, 18.

12. *Communications Act of 1934,* 73rd Cong., section §309 (n. 2), Public Law 416, 19 June 1934.

13. Data from *Broadcasting,* 10 August 1981; *Broadcasting/Cable Yearbook '89* (Washington, D.C.: Broadcasting Publications, 1989), H-72; and *Broadcasting/Cable Yearbook 1995* (Washington, D.C.: Broadcasting Publications, 1995), B-654, 656. In 1993 consumers spent over $6.5 billion on 21.8 million portable and table color TV sets ($299 average cost); $700 million for 1.2 million console color TVs ($583 average); over $4.2 billion for 9.8 million stereo color TVs ($429 each); and $40 million for 550,000 mono TVs ($80). Consumers also spent $307 million on 20 million table, clock, and portable radios (average cost $15) and almost $17 billion for 15.8 million auto radios ($107 each).

14. This weighing of respective financial investments by consumers and broadcasters has been compared over half a century. The Rand Corporation's study of two decades (1946–1966) showed that, for each dollar the TV industry spent on physical facilities to prepare and send out programming, $40 were spent by consumers for equipment needed to receive those programs—a ratio of 40 to 1. Americans were estimated to have spent a total of $35 billion for TV sets. (Source: WBAT-FM and The Network Project, "Feedback 2: The Television Industry," *Performance* 3 [July/August 1972]: 42–43.) A report to the Federal Government in 1963 estimated the public's share of capital investment in broadcasting as 96% compared with the radio-TV industry's 4%—a ratio of 24 to 1. (Source: Sydney W. Head, *Broadcasting in America,* 2nd ed. (Boston: Houghton Mifflin Company, 1972), 250–251, and note 6, quoting testimony in government hearings.)

Not included in those estimates were consumers' costs for electricity and repairs, nor the percentage of each consumer dollar spent on retail purchases which companies allocate to advertising those products through television. But also not included were broadcasters' enormous continuing costs of personnel salaries and programming. On the other hand, stations derive massive income from advertising, whereas consumers receive program service but no fiscal return on their investment. (Also recall comments above about widespread annual losses by station operations; see chapter 10, "Financial Management.")

15. Federal Communications Commission, *53rd Annual Report/Fiscal Year 1987* (Washington, D.C.: U.S. Government Printing Office, 1988), 30.

16. Thomas F. Leahy, "Viewpoints: Transfer of TV Properties Went from $750 Million to $6 Billion in Four Years," *TV/Radio Age,* 9 June 1986, 61. Then the executive vice-president of CBS/Broadcast Group, Leahy had been general manager of WCBS-TV in New York City; subsequently he was appointed president of CBS Marketing for all divisions of the corporation.

17. He spoke at the 20th annual convention of the Association of Independent Television Stations, San Francisco, January 1993; quoted in *Broadcasting,* 1 February 1993, 27.

18. Yale Roe, *The Television Dilemma* (New York: Hastings House, 1962), 130–136. A former Federal Communications Commissioner recounted various recommendations through ten years to create some sort of national citizens' committee to appraise the general performance of broadcast media; he listed ten activities such a commission ought to analyze and publicly evaluate. This approach attempts to avoid the twin problem of industry favoritism and of governmental intrusion into free enterprise broadcasting. Of course there are strong objections to such efforts as unwarranted meddling; thus even these recommendations constituted part of the problem or "dilemma" of broadcasting. See Nicholas Johnson, *How to Talk Back to Your Television Set* (New York: Little, Brown, 1970); paperback edition (New York: Bantam, 1970), 175–183.

19. Reflecting the disparate, fragmented, multifocused self-interests of various subgroups among members of the National Association of Broadcasters, Vincent Wasilewski quipped in his convention speech upon retiring as NAB president that his greatest accomplishment during two decades heading the organization was "keeping all you contentious bastards in the same boat!"

20. Herbert H. Howard, "An Update on Cable TV Ownership: 1985," *Journalism Quarterly* 63(4) (Winter 1986): 706–709, 781.

21. See Ken Auletta, "John Malone: Flying Solo," *New Yorker,* 7 February 1994, 52; also L. J. Davis, "Television's Real-Life Cable Baron," *The New York Times,* 2 December 1990; 16, 38.

22. Warren Cohen and Katia Hetter, "Racing into the Future," *U.S. News & World Report,* 12 February 1996, 49–51.

Theories of Managing

Scientific management [is] not a collection of techniques to increase efficiency, but a philosophy of management and a way of thinking.

FREDERICK W. TAYLOR (THE "FATHER
OF SCIENTIFIC MANAGEMENT")[1]

In colleges and universities, seniors and graduate students as well as experienced media people study station management along with other aspects of broadcasting and cable. Professional associations offer seminars in management for those actively engaged in running radio and television stations. In these contexts the topic of "management" transcends narrow concepts and routine tasks.

Such study seems to presuppose a common set of principles and methods suitable for any station. Is this true? Can the manager of a 500-watt, daytime-only radio station serving a rural small town follow the same procedures as at a 50,000-watt, clear-channel metropolitan radio station? Are procedures and decision making at independent television stations similar to those at network affiliates? Does managing a broadcast property differ significantly from managing other kinds of enterprises? Obviously there are differences in all managerial situations, but successful principles and methods should apply universally if ingenuity is used in adapting them.

After 80 years of radio and some 50 years of television, broadcasters generally realize two broad areas of need. First needed is a genuine *philosophy* concerning broadcasting. A positive philosophy is one that defines the purposes of broadcasting, its place in contemporary society, and the principles or ethics to which broadcasters subscribe. Those considerations, partly noted in chapter 1, underlie all the chapters of this book.

Second, the special nature of radio and television as commercial enterprises makes it imperative to organize *principles* of station management. The increasing complexity of managerial assignments and ongoing significant developments call for defining basic duties and responsibilities as well as challenges confronting the broadcast manager. Foundation principles serve as a valuable yardstick for evaluating the manager's own accomplishments (or lack of them) against others engaged in similar activity. This chapter and the next specifically address theories and principles of managing.

After defining the term "management" (including its etymology), this chapter analyzes the phenomenon of management and formulates principles for a coherent theory. While abstract, a theoretical analysis is not irrelevant; it clarifies essential characteristics of the management process. Subsequent chapters explore implications of that theory for practical broadcast management.

TERMINOLOGY

"Management" means many different things, partly because the roots of the word are ambiguous and its range of meanings has grown over the years. The academic and professional field of management itself recognizes no universally accepted semantic content to the term "management."[2] Here lies the problem of developing a theory of management with clear structural and functional elements. A theory's validity is based in the integrity of its terms; but with the word "manage" there is considerable etymological ambiguity.[3]

A further clue to the fuller meaning of the term "management" is derived from usage—what people have meant when they used it. That is how words get their operative meaning; those meanings can change and evolve over decades and centuries until they vary widely from their original denotations. This is useful to keep in mind when considering Douglas McGregor's observation that conventional, classical principles of theories of organization and management are derived from the military and the Catholic church.[4] He cautions that "new theory, changed assumptions, and more understanding of the nature of human behavior in organizational settings" are needed before applying to contemporary business the models drawn from far different political, social, and economic contexts. Yet, by reflecting back through the centuries of humanity's social groupings, we realize that without sophisticated weaponry, transportation, or communication even the military as well as the church depended on face-to-face transactions (and hand-to-hand battle!). In our present era, military and ecclesiastical management includes extensive material resources—administrative paperwork and files, buildings, vehicles, equipment, budgeted finances, and elaborate procedures—as well as personnel. But the root of the word "manage" came from the more directly personal phenomenon of influencing and directing individual persons in their consciousness and conscience as decision makers and doers, including physical contact.[5] Indeed, major modes of earliest transportation and communication were animals and human propelled boats (by oars or by wind-assisted hand-hauled sails); early vocabulary such as *manège* bespoke the personal handling and directing of animals to collaborate with men in achieving some physical goal. Management thus has everything to do with people and actions, and only by historical accretion refers to mere physical resources of things.

This analysis of the etymology of "management" leads to our distinctive theoretical statement of management, after first clarifying some related terms.

A common contemporary meaning of "management" is the spectrum of all those who have any authority or jurisdiction over other workers or who participate in executive decision making and planning, thus distinguishing them from all members of the labor force. In this book we do not necessarily refer to that *class* of "management" as differentiated from union-organized "labor" because at some levels broadcast managers may well be members of unions and guilds (for example, chief engineers, news directors, program directors).

The rungs of the management ladder can be divided roughly into executives at the higher end, through middle managers, to supervisors at the lower end. Executives (e.g., regional or plant managers or, in broadcasting, general managers) chart the course or initiate entrepreneurial risks and set creative directions for an enterprise, but they generally do not work directly with those carrying out specific tasks. At the other end of the spectrum, supervisors (in a shop or office, or a production manager over TV studio crews) oversee detailed activities that they appraise and either report back or reward/punish with predetermined sanctions. But they do not creatively modify or strategically manipulate policies regarding movement of material or relationships among people. That is the role of middle managers (e.g., shop supervisors, news and program directors) who neither establish major policy (as do executives) nor merely oversee the activities involving things (appraising them and reporting back, as do supervisors); they "handle" or "direct" or "manipulate" (in the good sense) people and policies.[6]

Finally, a distinction may be made about persons identified as professionals. While not necessarily among management as such, they still differ from the laboring workforce. Professionals are often hired as consultants and experts from specialized fields in education or research; they may be on staff as resource people. In the broader sense of "professionalism"—implying specialized training, commitment to excellence, sense of ethics and community service, and standards of practice—all broadcasters are expected to be professionals, as emphasized throughout this book.

In the following chapters, we use "management" as a composite term sometimes including several of the categories above, at other times specifying a particular management position or role (e.g., network executive, station manager, program director).

BACKGROUND: THEORIES OF MANAGEMENT

Most authors agree that management is an art more than a science of objectively measured phenomena. And business management is the art of accomplishing desired results through and with the members of the organization. Management is defined by Filley and House as "a process, mental and physical, whereby subordinates are brought to execute prescribed formal duties and to accomplish certain given objectives."[7] Terry defines management as "a distinct process consisting of planning, organizing, actuating, and controlling, performed to determine and accomplish the objectives."[8] Henri Fayol, considered the father of modern management theory, included as functions of management: planning, organizing, commanding, coordinating, and controlling.[9]

Eleven authors of eight books (including Fayol, Koontz & O'Donnell, and Terry) vary somewhat in listing roles or functions of the management process: all eight include *planning, organizing,* and *controlling;* three add *staffing;* three include *leading;* one each adds *commanding, directing, influencing, actuating,* and *coordinating.*[10]

Different authors offer varying categories by which to analyze the phenomenon of management. McCavitt and Pringle succinctly trace patterns among those authors in contexts of history, theories, leadership styles, motivation, and organizational change and development.[11] Koontz classifies schools of management according to the principles that each theory emphasizes. The "management process" school emphasizes managers' functions in getting things done by people who act in groups. The "empirical" school stresses case study to derive experience and principles from the success and failures of others. The "human behavioral" school focuses on interpersonal relations and on understanding what people expect from work situations and how they perform in groups. The "social system" school views management as a cooperative, social system of cultural relationships; it emphasizes communication and personal behavior.[12] Koontz somewhat anticipates our own effort to formulate a theory when he notes that, being a logical process (which can thus be applied to varied situations), management can be expressed in terms of mathematical symbols and relationships; this highly scientific study can be considered the "mathematics" school of management. To these categories, Terry adds schools of decision theory and of economic analysis and accounting.

THEORIES X AND Y

Douglas McGregor, in *The Human Side of Enterprise,* characterized broad patterns of management styles for motivating others to work. He divided them into two contrasting trends, in effect authoritarian and humanitarian—the former focusing on physical resources and reward/punishment while the latter looks to human resources, especially collaborative interaction among people engaged in an enterprise.

He cautioned that every practical action by management grows out of at least implied theory in the form of assumptions, generalizations, and hypotheses and that acting without explicitly examining those theoretical bases results in widely inconsistent managerial behavior. McGregor emphasized that "the theoretical assumptions management holds about controlling its human resources determine the whole character of the enterprise. They determine also the quality of its successive generations of management."[13]

He characterized previous literature on organization and management practice as supporting an absolute concept of authority in which authoritarian structures and procedures bring about the goals of an enterprise. Labeling that conventional analysis as "Theory X," he described it as attempting to manage people by motivating them in earlier times through physical force and power and more recently by moral authority coupled with monetary compensation. In opposition, he posited "Theory Y" whose assumptions are dynamic rather than static because they emphasize the need for selective adaptation and flexibility in forms of control, based on the essential possibility of continued human growth and development. Thus his Theory Y describes management's improved ability to control through influencing others not by the amount of authority they can exert but by appropriately selecting different means of influence according to given circumstances. Those "means of influence" are drawn from the underlying assumptions of Theory Y which look upon human persons as having wants and needs. McGregor's six key assumptions are:

1. The expenditure of physical and menial effort in work is as natural as play or rest.
2. External control and the threat of punishment are not the only means for bringing about effort toward organizational objectives. Man will exercise self-direction and self-control in the service of objectives to which he is committed.
3. Commitment to objectives is a function of the rewards associated with their achievement.
4. The average human being learns, under proper conditions, not only to accept but to seek responsibility.
5. The capacity to exercise a relatively high degree of imagination, ingenuity, and creativity in the solution of organizational problems is widely, not narrowly, distributed in the population.

6. Under the conditions of modern industrial life, the intellectual poten-
tialities of the average human being are only partially utilized.[14]

Precisely by responding to those needs managers most positively influence or control and dynamically motivate subordinates. McGregor outlined the progressive
importance of human wants as: physiological needs; safety and security needs
(protection against danger, threat, deprivation, and "for the fairest possible
break"); social needs (for belonging, for association, for acceptance by one's fellows, for giving and receiving friendship and love); and finally two kinds of ego-
istic needs, which are most significant to the person—and thus to managers
attempting to motivate others:

1. Those that relate to one's self-esteem: needs for self-respect and self-confidence, for autonomy, for achievement, for competence, for knowledge.
2. Those that relate to one's reputation: needs for status, for recognition,
for appreciation, for the deserved respect of one's fellows.[15]

These are similar to Maslow's hierarchy of needs.[16]

Herzberg's two-factor theory in employee motivation looked to "maintenance
factors" (pay, work conditions, relationships, supervisors, company policy) and
"motivational factors" (achievement, recognition, responsibility, advancement
and growth, and the work itself).[17] McGregor claimed that any manager's successful
social influence and control depends on helping other persons achieve their personal goals and satisfy their human needs. He stresses *integration* as the central
principle derived from Theory Y: creating conditions so employees best achieve
their own personal goals by applying their efforts toward the success of the organization. His theory looks to the manager's role as complex and flexible in the
superior-subordinate relationship. The manager must sometimes be a leader of the
group, but in other circumstances must act as a member of a peer group. He or
she must variously deal with other departmental managers, with an immediate superior, or with administrators and executives at various higher levels in the company.

Therefore Theory Y emphasizes relationships among people in the organization. Guided by Theory Y assumptions for a theoretical base, the manager seeks
to create "an environment which will provide opportunities for the maximum exercise of initiative, ingenuity, and self-direction in achieving them." For McGregor,
human collaboration in the changing circumstances of an organizational setting
depends not on human nature or a static assessment of line-and-staff people's past
record of productivity. Rather it results from management's ingenuity in discovering ways to tap the potential commitment and achievement (and thus personal
fulfillment) of employees, who constitute the organization's human resources. At
the same time he noted that management which is creative, flexible, and socially
responsible to employees is not the same as merely permissive management.

Translating those principles into more concrete terms, by following Theory
Y assumptions, managers try to motivate a person to find within the enterprise a
way to live as fully free a human being as possible in the specific context of an
organization and job. That employee is prompted to look to the job as an integral
part of personal daily living. The place of employment and the job should contribute to what an employee is becoming as a person. It should not just be a place
where employees make money in order to go home or on vacation to really "be"
and grow as individual persons. Management's role is to find the right people for
whom working in that specific enterprise with its given goals and procedures will
be attractive and rewarding. Precisely by joining the enterprise to help achieve the

goals of that organization, the individual at the same time contributes to personal growth and fulfillment.

This brief background provides the context for our own theoretical formulation. Reflecting McGregor's Theory Y assumptions, it is more closely aligned with the highest of Maslow's hierarchy of needs—self-actualization (surpassing next-ranked self-esteem and status)—and Herzberg's factors of responsibility and the work itself. But we analyze more precisely the very process of managing, emphasizing the dynamics of managerial relationships in action.

The "V Theory" of Managing

"Management," as commonly used today, may refer to either (1) that complex of actions constituting the process of managing, or (2) that complex of persons who possess managing authority. What these two common uses of "management" lack is not a general sense of what management involves (indeed, the generalness is so expansive that confusing and widely divergent definitions are employed to describe this phenomenon), but a structurally specific and comprehensive explanation of what managing must entail if it is to be management at all.

MANAGING IS THIS BI-RELATIONAL PROCESS:
THE INTENTION OF THE MANAGER IS DIRECTED
TO AND ACTUALIZED BY THE MANAGEE*

The managing process is bi-relational in that it absolutely depends on both manager and managee; without one there is not the other. If a manager gives directives that are not carried out by the managee, then properly speaking that person is not managing (just as when the manager does not initiate directives or oversee in some way their being acted on, that person is not managing). If the managee acts on matters unrelated to or different from the manager's directives and intent, then that person is not acting as a managee. Thus "managing" is a process involving persons relating to one another through directives in the context of communicated intention. It is this dynamic interrelation of person to person through directives whose purpose or intention is understood, and their being actualized through execution, which constitutes the essence of the process of managing. Management is thus triadic and may be investigated according to any one of, or any combination of, its three components: manager, managing, managee.

Furthermore, the dynamism of the manager's purposeful intention in those directives may be considered as either micro-intention (i) or macro-intention (I). The micro-intention consists of specific directives that establish, maintain, or further the relationship between the manager and the managee. The macro-intention consists of the stated purposes and policies of the total enterprise which establish, maintain, or further the relationship between one unit or level of management and another unit or level of management.

Summing up: the dynamism of management's interacting vectors (V) is a function of the manager as subject (M) directing (D) the managee as object (m) to actualize (a) those directives according to the enterprise's and manager's intentions (Ii), or

$$V = Ii(M{\rightarrow}D)(a{\leftarrow}m)$$

The formulation attempts to express the dynamics of management as trivalent (thus the "V Theory"), with three interacting vectors consisting of "the one who" (*M*), "'the one whom" *(m),* and "the what/how"—actualizing *(a)* of directives *(D).* Strictly speaking, without the central dynamism of those directives being

---

*As the word "employee" relates to the word "employer," so the word "managee" relates to the word "manager."

actualized *(D, a)*, there is no management relationship operative, and to that extent neither manager *(M)* nor managee *(m)* is properly acting in their respective roles within the enterprise. Intentionality becomes operative precisely in the process of directing/actualizing.[18]

Therefore the "V Theory" focuses on the very *act* of managing as the heart of all management theory and practice. The V Theory underscores the dynamic relation of the manager as person to the managee as person, which relation is constituted by the managee's actualizing the manager's directives in the context of management's intentions. This emphasis has immediate implications for establishing criteria to appraise the success of management situations.

INTENTIONALITY

The central element of intentionality—the intention or purposefulness of the enterprise—is fused into every part of the V Theory. As such it is the primary criterion by which the entire bi-relational *process* of managing can be evaluated. To ask what is the primary and abiding intentionality of management is to address oneself to that enterprise's management philosophy.

Intentionality includes both "micro-intention" and "macro-intention." Micro-intention refers to the manager's specific objectives and purposes in directing a managee to execute some action. The macro-intention refers to goals of the total enterprise as determined by all the stakeholders—primarily the owners and stockholders—but also by employees, by government, and by the public (as actual or potential customers, consumers in the public market generally, critics, and voting citizens). The macro-intention refers to all those public, governmental, and economic factors that shape the whole enterprise—its product or service, policies, and procedures.[19] Some refer to this as the "overarching" goal, others as the "superordinate" goal or mission of the organization.[20] The entire corporate purpose is embodied in the corporate mission from which flows corporate strategy in setting goals. Secretan says

> Long- and short-range goals are the milestones along the path to fulfillment of the corporate mission. . . . They are dreams with deadlines. They define the specific performance, results, and standards that the activities of an organization are designed to produce.[21]

The key managerial function of planning involves selecting organizational objectives and policies, together with methods for achieving them through programs and procedures. Intentionality expressed in planning provides a framework for integrated decision making vital to all collaborative activities involving people.[22]

The macro-intention provides criteria against which to appraise the manager's success as manager—in one sense as advancing company goals by interpreting them and forming directives, in another sense as effectively motivating managees to respond to those directives. To the extent that these central vectors in the management process—directives *(D)* and actualization *(a)*—manifest disparity or imbalance or are at cross-purposes, there is inefficient and ineffective organizational process. Managerial actions are to that extent not fulfilling their purpose or macro-/micro-intentionality *(I/i)*. Translated, that means poor management and eventually even chaos and failure in the enterprise.

The Manager (M)

While the manager's knowledge of intended product and processes is important in order to determine directives, it is not managers' detailed factual knowledge that this theory looks to so much as their ability to effectively communicate directives (always in the context of intentionality) in a way that motivates managees to actualize them. "Communicating" refers not merely to adroit interaction—written, verbal, or nonverbal—between manager and managees; it embraces the entire complex dynamic discussed in McGregor's Theory Y and Japanese Theory Z and further embodied in our formula of

$$V = Ii(M \rightarrow D)(a \leftarrow m).$$

The very relationship between the manager and those managed is undercut without at least some knowledge and experience of processes and product with which managees are engaged. The more managers know, the better they can make judgments about the quality of managees' work, precisely in the context of the organization's goals (macro-intention). But a person who knows only factual details and specific procedures may be an ineffective manager if he cannot also motivate subordinates. On the other hand, a person who does not fully know detailed processes and product but who understands people and how to motivate and direct them may be an effective manager.[23]

Therefore the manager must know primarily the structures and relationships involved in the company's activity and in the activity of the personnel producing the product. If the manager has solid knowledge of these relationships and knows how to "power" them, he can call upon resource people to contribute more particularized expertise about details—which he can then assess for achieving the organization's goal.

This makes logically supportable and consistent the practice of advancing a person good in managing one area to a new and broader sphere of responsibility where she may well be just as effective, despite her less detailed knowledge and experience of the expanded area of responsibility. She can be truly effective as a manager if, while endeavoring to learn about the new activities now overseen, she continues to apply her previous ability to manage *per se*. Certainly one cannot know personally all that there is to know about an enterprise—especially when a person rises to top management involving several corporations or clusters of companies in distantly related fields, as happens in mergers.

For example, a successful program director at a station who has background in performance and production may have learned a good bit along the way about sales; then he is made station manager with responsibilities over not only programming and sales but also engineering; eventually he may become an executive of a group of broadcast stations in charge of several radio and television stations and even of cable and more distantly related subsidiaries; finally he may become an executive administrator within a media conglomerate. The point is that a successful manager knows how to get the right people to do the job well, and has ability to select subordinates who themselves also know how to get other apt people to do the job well. Thus the manager—despite changing work procedures and factual data—continually deals with the same context: collaborating with people and motivating those who are competent while accurately assessing them and their work in light of the company's goals and objectives.[24] This is the manager's essential role in the enterprise.

Directives (D)

The manager's role in the bi-relational process is to communicate micro-intentionality directly, and indirectly the undergirding macro-intention of the enterprise, through directives. The operative note here is *communicated* intent, selected from the cluster of intentions that constitute the policy and planning process of management and which are not communicated similarly to all managees. The power of this bi-relational process (as expressed by the V Theory of managing) lies in getting directives actualized. Therefore the more fully managers can convey their intention through directives, the more perfectly directives can be actualized. A managee who knows the directive in a way that is translucent of the manager's full intention—the "why" of the directive—better understands and can better actualize that directive. This is paramount when changing conditions prompt the managee to adapt actualization (in effect, modifying the directive) so as to serve the micro-/macro-intention behind the original directive.

Actualizing (a)

The word chosen for the V Theory formula is not "execute" but "actualize" because the former suggests a literal, one-for-one carrying out of what is directed, where the directive *(D)* alone is considered and the execution is precise.[25] (That might be desirable in some instances, but it reflects Theory X's authoritarian emphasis of mere subservience to management.) Theory Y would support the term "actualize" because it connotes more flexibility for the managee to interpret and apply at the point of action not merely literal directives but the manager's micro-intention (and even underlying aspects of the enterprise's macro-intention), thus allowing the managee to participate in the total purposefulness of the manager's directives. Here is where managees truly serve the enterprise by initiative in adjusting—engaging, as it were, in on-the-spot replanning or refining directives to match micro-intentions. This is far from extrinsic benefits of merely fulfilling one's human needs or achieving a measure of personal satisfaction (Theory Y). One becomes substantively engaged in the very process of managing *(M/m),* thereby exercising abilities by truly participating in the company's operation.

Of course it is the manager's further role to determine ways to assess how well directives are actualized, to ensure (by correction, variant motivation, or further guiding directives) a growing equation between directives *(D)* and managees' actualizing *(a).*

The Managee (m)

A manager's directive role in the dynamic relationship is correlative to the managee's actualizing those directives. To the extent that a managee does put them into action, she is then acting precisely as managee of this given manager in a dynamic, effective management relationship. To the extent that she is unable to actualize the directives because of competence, attitude, or the like, she is less a managee. To the extent that $D = a$, and I/i are being realized, there is an ideally efficient management process and perfectly correlative roles of manager and managee. To the extent that there is some discrepancy between D and a (directives and actualizing), in the context of intentionality *(I/i),* there is a less efficient and proper relationship between the two.

It is that dynamic relationship between directives and the managee's actualizing (so that $D = a$) which the manager is primarily concerned with—not so much the details of that actualizing, but the relationship of the actualizing to the directives and thus to the overall "why?" of the micro- and macro-intentionality in the management process. The manager's role is to select the managees who can best bring about this equation. The manager's further responsibility is to plan, control, and discern through feedback, testing, and other techniques to what extent there is a favorable equation between D and a, and then to modify or correct elements (D, a, m) in the management process, including replacing the managees with others who will better actualize management's directives. This is the essential role of managing.

In summary, a theory of management must include, regardless of the particular content or concrete circumstances to which applied, the dynamic relationship (the vectors of the V Theory) among people (M, m) and the directed collaborative activity (D, a) of those people to produce the product or service (the micro- and macro-intentionality of the enterprise). Thus the formulation of the V Theory of management:

$$V = Ii(M\rightarrow D)(a\leftarrow m).$$

Theory Z

The present authors first advanced their V Theory formulation in 1976. Half a decade later William Ouchi described American versions of Japanese interactive management processes with employees, labeling it Theory Z.[26] Many characteristics of Theory Z reflect Theory Y, and also some of our own earlier analysis under title of the V Theory. Japanese management is highly participatory, with workers engaged in planning goals, objectives, and procedures through consensus—featuring "quality control circles" of employee team discussion. Marked by trust, intimacy, and subtlety, relationships of employees to the company are enhanced by lifetime employment, evaluation by peer group rather than by superiors, and collective decision making—infusing morale and collective values and sense of responsibility for company philosophy and objectives (our *"I/i"*). But hierarchical aspects in Theory Z, including elements of sexism and racism, also reflect some of Theory X.

Entrepreneurial Partnership Strategy

Reinhard Mohn, recently retired as chairman of multinational media conglomerate Bertlesmann AG (based in Germany for 150 years, approaching $10 billion annual sales, with 42,000 employees worldwide in 1991). He wrote about his company's "partnership" approach to managing. His 1986 book, translated into English in 1988, reflects in many ways our own V Theory concepts in practice among Bertlesmann companies.[27] He repeatedly says to "awaken the creativity which is absolutely essential for development of the work process, employees must identify themselves with their tasks" and with their jobs and the company. Echoing the employee as "managee" who can aptly interpret management's intentions and directives, Mohn notes the pace of change demands flexibility and adaptability, adding that without initiative and participation in decision making, employees "do things simply 'by the book,' without any personal commitment." He acknowledges that "perhaps fewer mistakes are made if things are done 'by the book.' But this also eliminates necessary entrepreneurial decision making" that ultimately results in more profitability for owners and investors. Again and again he comes back to the central focus of what we have called intentionality: "the decentralized management system presumes that the managers identify with their tasks and with the goals and conduct of the company as a whole." This strengthens the entire enterprise: "the internal structure of the company is stable thanks to its social policy, which minimizes conflicts and motivates great creativity and commitment on the part of the employees, because they identify with the company" and its purposes, as well as with their job. He alludes to the mission or macro-intentionality of the corporation that strives "to implement the principles of a people-oriented and socially responsible method of operation also in its foreign subsidiaries." He summarizes the overall goals of this "V Theory"-type company:

> The self-concept of a company based on partnership is characterized by the fact that all the persons involved in it, from stockholders to management to employees, consider themselves to be working together with the same goals and responsibility. The primary objective of all companies, that of making a contribution to society, can only be achieved by making the company secure and by helping it to develop successfully.... The second goal of a partnership company is giving all the persons working in the organization the opportunity for self-fulfillment.[28]

But "making a contribution to society is seen as being the supreme corporate goal." As for micro-intentions and managees' comprehension of directives to be actualized, "the employees understand the purpose of the orders they are given, rather than following them blindly. Accordingly, the manager must give reasons for his orders with explanations and dialogue, and must make them comprehensible. In a partnership company, all employees must desire success and contribute to it. They can only do this if they understand the corporate policy and the instructions of their superiors." Mohn reverses our semantic equation of manager/managee *(M/m)* but to the same effect as the V Theory formulation: "In a partnership company, the manager must understand himself to be one employee among many others." The Bertlesmann corporation—managers as well as employees—codified their management ideals in 1985 by drawing up a corporate constitution and by-laws that list the elements of this entrepreneurial partnership.[29] They closely parallel the V Theory concept of managing. Mohn concludes his book by reaffirming idealistic priorities among macro-intentions of corporate management in today's world:

> In my opinion, an incorrect understanding of objectives is the main reason for the failure of a corporation. Only when we have reached agreement concerning the fact that the prime purpose of a company is to make a contribution to society and that the task of management must be understood as a mandate will we be able to draw the correct conclusions for maintaining the continuity of a company. . . .
>   The tasks of the future will demand cooperation, not tough confrontation. The purpose of business activity is no longer maximizing profits but making the greatest possible contribution to our society.[30]

All three theories (X, Y, and Z) are designed to characterize dominant styles of management; they are built on descriptive analyses and case studies of concrete instances of management situations. Barry Sherman notes how broadcasting's pressure of time and competitive sales keeps Theory X "alive and well" to some extent in most media operations, along with more benevolent Theory Y and creative Theory Z approaches.[31] The "Contingency Theory" draws from all forms, adapting them to various contexts of "task-organizations-people fit."[32] Only the V Theory attempts to dissect the very process itself to get at its constitutive essence. Perhaps closest to V in applying theory of motivation is the so-called Participative Management by Objectives (PMBO) hybrid, linking the traditional Management by Objectives (MBO) with Z.[33]

Tannenbaum and Schmidt plot gradual stages along their "Continuum of Manager-Nonmanager Behavior" in the decision-making process, where the latter (managees) increasingly participate in power and influence.[34] In terms of the discussion above, the extreme left of their model diagram characterizes Theory X style, while the extreme right represents Theory Z and our own V Theory, including some of Theory Y. Concentric ellipses ("organizational" and "societal environment") surrounding their diagram represent important aspects of a company's macro-intentionality. Rensis Likert's classification of management styles would progress from left to right on their model: successively as exploitive authoritative, benevolent authoritative, consultative, and participative group.[35]

These concepts can be found paralleled in Secretan's sketch of three management models, drawing on Miles' "leadership attitudes": the *traditional* manager (X/authoritarian), the *human relations* manager (Y/humanistic), and the *human resources* manager (Z and V/collaborative, empowering).[36]

—◁∿▷—

We provided a definition of managing which includes the essential structural elements of that dynamic social phenomenon. We have further suggested various sources by which a person can verify major propositions of the V Theory of managing: linguistic analysis, etymology and semantics; common sense meanings of executive, manager, and supervisor; people's own experience of what motivates them regarding their job and the place of that job in their life; and specialists' writings about management theories, including McGregor's Theory X and Y and Japanese Theory Z.

This book emphasizes this relationship of persons in action within the common enterprise of a broadcasting company (chapters 3-5), and then explores more detailed applications of management knowledge and decision making to specific areas of broadcasting (chapters 6-12). We approach radio-TV broadcasting and cable from the perspective of the manager rather than from the viewpoint of an expert in each of the areas of programming, sales, engineering, and regulation. The burden of comprehensive knowledge of each of these specialized areas is placed neither on the authors of this book nor on the managers about whom they are writing. In later chapters, therefore, the primary emphasis will be on broader, structural kinds of knowledge that a manager should have in order to address problems arising in those areas, rather than focusing on the subordinate kind of detailed knowledge that no manager with even a fine memory could recall daily even if he were competent in many of the areas. Other books describe specific data and procedures involved in operating broadcast stations. We emphasize management's effective collaboration with middle managers or supervisors and their staffs. Management's proper concern is how to get the total staff producing creatively while fulfilling themselves as persons and professionals, and thereby achieving the purposes of the enterprise. Thus the distinctions between network and station management, or between local and group management, or independent and affiliate and owned-and-operated management, or even radio and television and cable management, are not that critical. To all those areas of broadcasting can be applied the broader theoretical analysis embodied in the V Theory of managing.

⟿ CHAPTER 2 NOTES ⟿

1. William J. McLarney and William M. Berliner, *Management Training: Cases and Principles*, 5th ed. (Homewood, Ill.: Richard D. Irwin, Inc., 1970), 391.

2. See George R. Terry, *Principles of Management*, 4th ed. (Homewood, Ill.: Richard D. Irwin, Inc., 1964), 14.

3. The word *manage* had a convoluted etymological odyssey, especially in its Low Latin root *manidiare*. While clearly derived from the Latin word *manus* (meaning *hand*), the semantic evolution of *manage* includes a plethora of meanings. The most apposite include: to handle, direct, govern, or control in action or use (a horse); to succeed in accomplishing; to handle, wield, make use of (a weapon, tool, implement, etc.); to control and direct the affairs of (a household, institution, state, and so forth); to operate upon, manipulate for a purpose; to bring (a person) to consent to one's wishes by artifice, flattery, or judicious suggestion of motives. Cf.: *Random House Dictionary of the English Language*, unabridged ed. (New York: Random House, 1967); *A New English Dictionary on Historical Principles*, ed. Sir James Murray, vol. VI, part II M-N (Oxford: Clarendon Press, 1908), 104–105; A. Ernout and A. Maillet, *Dictionaire Etymologique de la Langue Latinae, Histoire des Mots* (Paris: Librarie C. Klincksieck, 1949), 590–591; Alfred Hoare, *An Italian Dictionary*, 2nd ed. (Cambridge: University Press, 1925), 363; R. E. Latham, *Revised Medieval Latin Word List* (London: Oxford University Press, 1965), 287–288; Charles Grandgent, *From Latin to Italian* (Cambridge: Harvard University Press, 1940), 110.

4. Douglas McGregor, *The Human Side of Enterprise* (New York: McGraw-Hill Book Co., 1960), 18.

5. The early Christian community of people had little to do with institutionalized money and buildings; it had everything to do with an authority structure that grew out of a people united together under a God believed to be personal.

6. To the extent that executive administrators are such, they are not strictly speaking managing if they spend little time guiding people but almost all their time administering fiscal affairs, analyzing research data, planning, and other tasks. Similarly, at the lower end of the spectrum—while it is hard to delimit supervisory-type activities from purely management ones—insofar as a person is merely carrying out assigned duties, prescribed tasks, policy guidebooks, assessing, and passing the word back up for directives about further actions, he or she is then working and possibly supervising rather than managing.

7. Alan C. Filley and Robert J. House, *Managerial Process and Organizational Behavior* (Glenview, Ill.: Scott, Foresman, 1969), 391. See chapter 3 for analysis of principles of management.

8. Terry, *Principles of Management*, 56; pp. 18–23 treat material referenced below.

9. Henri Fayol, *Industrial and General Administration,* trans. J. A. Coubrough from French ed. (Geneva: International Management Institute, 1929), 20; see McLarney and Berliner, *Management Training,* 13–14. Cannon notes that twentieth century management began with Fayol who first developed and practiced management principles from the chief executive's perspective; Thomas Cannon, *Business Strategy and Policy* (New York: Harcourt, Brace and World, 1968), 347.

An entire book analyzes these five characteristics: Harold Koontz and Cyril O'Donnell, *Essentials of Management,* 2nd ed. (New York: McGraw-Hill, 1978); the 562-page work compresses their earlier *Management: A Systems and Contingency Analysis of Managerial Functions* in the McGraw-Hill management series. See also J. W. Haynes and J. Massie, *Management: Analysis, Concepts and Cases* (Englewood Cliffs, N.J.:

Prentice-Hall, 1961), 3. A practical workbook of analysis, cases, and self-tests centers on the five functions: Lester R. Bittel, *The McGraw-Hill 36-Hour Management Course* (New York: McGraw-Hill, 1989), 71–201.

10. Richard M. Hodgetts, *Management: Theory, Process and Practice* (Philadelphia: W. B. Saunders, 1979), 64.

11. William E. McCavitt and Peter K. Pringle, *Electronic Media Management* (Boston: Focal Press, 1986), 2–18.

12. See Harold Koontz, *Principles of Management* (New York: McGraw-Hill, 1959), 89–99; also Richard M. Hodgetts, *Management: Theory, Process and Practice,* 2nd ed. (Philadelphia: W. B. Saunders, 1979).

13. McGregor, *The Human Side,* vii; see also 6–7.

14. Ibid., 47–48. Reproduced with permission of the McGraw-Hill Companies.

15. Ibid., 38; two excerpts below are cited from pp. 49 and 132. Reproduced with permission of the McGraw-Hill Companies.

16. Abraham H. Maslow, *Motivation of Personality* (New York: Harper, 1954).

17. Frederick Herzberg, *Work and the Nature of Man* (New York: World Publishing, 1966); cited by John M. Lavine and Daniel B. Wackman, *Managing Media Organizations: Effective Leadership of the Media* (New York: Longman, 1988), 190–191.

18. The centrality of the dynamic verbal element, which gives content to the nouns precisely as subject of that verbal action or as object, is clear in the literary paradigm. This formula and definition of "managing" was derived from the SVO theory of communication of Terrance A. Sweeney, first presented in May 1973, as part of his doctoral presentation for the Graduate Theological Union, Berkeley, California. The SVO Theory is a detailed language paradigm that investigates philosophically the structure of cinema and kerygma as communication events. The theory—so named because the foundations of its paradigm are Subject (S), Verb (V), Object (O)—has wider implications and applications than filmic and kerygmatic communication events: it presents an explanation of the structure of all human communication.

19. Terry, *Principles of Management,* 174, diagrams the tripartite clusters of "wants" which management must attempt to unify. (1) *Owners* want: the enterprise kept solvent and efficiently operated; a fair return on their invested capital; current information about the status and prospects of the enterprise; greatest possible efficient utilization of facilities; and long-range planning for stability and growth. (2) *Employees* want: protection from accidents, and security in sickness and old age; steady employment; fair wages; "completeness of daily living—satisfaction from their jobs, a feeling that they are making a contribution"; information about what is going on. (3) The *public* wants: goods and services available at fair prices; greater usefulness by improvements in goods and services; "harmonious relations among owners, employees, and managers"; and more fulfilling lifestyle provided by these goods and services. All of these "wants" are unified by management as macro- and micro-intentions.

20. David L. Bradford and Allan R. Cohen, *Managing for Excellence: The Guide to Developing High Performance in Contemporary Organizations* (New York: John Wiley & Sons, 1984), Chapter 4, especially pp. 100–101. R. T. Pascale and A. G. Athos, *The Art of Japanese Management* (New York: Simon and Schuster, 1981). See Lance H. K. Secretan, *Managerial Moxie: The 8 Proven Steps to Empowering Employees and Supercharging Your Company* (Rocklin, Calif.: Prima Publishing, 1993); he nicely develops in a chapter on "Corporate Purpose" this concept as four-fold, including corporate mission, statement of values, statement of vision, and the "ALDO audit" of the company's internal *assets/liabilities* (the micro environment of stockholders, employees, suppliers, customers, competitors, government, interest groups, professional associations), and the external *dangers/opportunities* in the societal environment of economic, sociocultural, technological, and political/legal forces and constraints.

21. Secretan, 60–61.

22. See Richard A. Johnson, Fremont E. Kast, and James E. Rosenweig, *The Theory and Management of Systems* (New York: McGraw-Hill Book Co., Inc., 1967), 4.

23. Malcolm P. McNair cautioned against business and industry's over-emphasizing human relations, causing executives to lose perspective about their primary responsibilities in achieving results, especially if such emphasis was merely intended to develop skills in manipulating people to achieve pragmatic goals. "Thinking Ahead; What Price Human Relations?" *Harvard Business Review,* 32:2 (March/April 1957):15–23; cited by McLarney and Berliner, *Management Training,* 405. The latter authors note (pp. 529–530) that lower-management supervisors can make good decisions about work and lead their people to successful performance if they have adequate knowledge of details of that work; their own competence generates confidence that communicates to their people and engenders respect. On the other hand, Terry, *Principles of Management,* p. 15, emphasizes that "Management, by its very nature, implies social and ethical considerations. In the final analysis, the whole being for most management exists for the betterment of human beings."

24. University administration reflects these necessary qualities of management; vice presidents and deans do not themselves possess knowledge about the details of every course of study in the institution, nor even about the precise manner of teaching by each faculty member. Yet those administrators must oversee, appraise, plan, and activate policies that affect students and faculty as well as curricular structures and daily operation of the total enterprise. To the extent that administrators themselves analyze and revise curriculum content and classroom procedures without collaborating with those involved first-hand, those administrators are no longer managing but actually mismanaging by assuming lesser (or at least different) nonadministrative roles. The same may be said for the family-owned factory, office, or broadcast company.

25. We do not use the term *actualize* in the sense that Terry refers to "actuating" as the methods which a manager uses to put a group into action (such as by leadership, communication, instruction and discipline); Terry, *Principles of Management,* 51, and Part V, "Actuating," 463–587.

26. William G. Ouchi, *Theory Z: How American Business Can Meet the Japanese Challenge* (Reading, Mass.: Addison-Wesley, 1981; also New York: Avon Books, 1981). See also Keitaro Hasegawa, *Japanese-Style Management: An Insider's Analysis* (Tokyo: Kodansha International Ltd., 1986), especially concerning *nemawashi* or peer discussion leading to *ringi* system of forwarding written recommendations by stages for ultimate approval by top management (pp. 28–37).

This concept, popularized in the 1980s, differs from the mix of variations under that title in preceding decades, which sought variations or blends of Theories X and Y: Rosenfield and Smith (1965), Morse and Lorsch (1970), and Colin (1971). Colin most nearly anticipated elements of the V Theory, and partly reflected participatory management. See Phillip V. Lewis, *Organizational Communications: The Essence of Effective Management* (Columbus, Ohio: Grid, Inc., 1975), 131–133.

27. Reinhard Mohn, *Success Through Partnership: An Entrepreneurial Strategy* (New York: Doubleday, 1988); English translation of *Erfolg durch Partnerschaft* (Berlin: Wolf Jobst Siedler Verlag GmbH, 1986). The following excerpts are from pages 5, 7, 28, 79, 80–82, 84, and 86 of the 1988 English translation. From *Success Through Partnership* by Reinhard Mohn, translation copyright © 1988 by Doubleday, a division of BANTAM DOUBLEDAY DELL PUBLISHING GROUP, INC. Used by permission of Doubleday, a division of Bantam Doubleday Dell Publishing Group, Inc.

28. Ibid., 81–82.

29. Excerpts are from "The Corporate Constitution of Bertelsmann AG" reproduced in Part Three of Mohn's book, pp. 120–127: "It is important for the employees to be able to identify with their tasks, the objectives, and the conduct of the company, within the framework of critical loyalty. This can only be achieved if they are kept informed about the work and the development of the company and have the opportunity to contribute their expertise and opinions to the decision-making process." Regarding "Management Conduct":

2. Managers should develop initiative, creativity, and the ability to implement decisions, and the actions they take in their areas must be results-oriented and must demonstrate social concern. Risks must be responsibly weighed. . . .

6. Managers should allow responsible employees to act and make decisions autonomously within their functions, in order to stimulate initiative and a sense of responsibility, and to promote identification with the job.

7. As a prerequisite for successful work, employees must be kept fully and promptly informed. They must be given an opportunity to voice their opinions. Suggestions and criticism should be noted and supported. . . .

9. The superior must discuss with his subordinates the objectives of their work, explain the connection with overall goals to them, and give reasons for his decisions.

From Success Through Partnership by Reinhard Mohn, translation copyright © 1988 by Doubleday, a division of BANTAM DOUBLEDAY DELL PUBLISHING GROUP, INC. Used by permission of Doubleday, a division of Bantam Doubleday Dell Publishing Group, Inc.

30. Mohn, 140–141, 174.

31. Barry L. Sherman, *Telecommunications Management: The Broadcast and Cable Industries* (New York: McGraw-Hill, 1987), 56–58.

32. John H. Morse and Jay W. Lorsch, "Beyond Theory Y," *Harvard Business Review* (May/June 1970); reprinted in Michael T. Matteson and John M. Ivancevich, *Management Classics,* 2nd ed. (Santa Monica, Calif.: Goodyear Publishing, 1981), 396–407. Also see Fred E. Fiedler, "The Contingency Model—New Directions for Leadership Utilization," *Journal of Contemporary Business* 3(4) (1974): 65–80; reprinted in Matteson and Ivancevich, 408-418.

33. William F. Christopher, *Management for the 1980's* [rev. ed. of *The Achieving Enterprise* (1974)] (New York: AMACOM/American Management Association, 1980), 124–129.

34. Robert Tannenbaum and Warren H. Schmidt, "How to Choose a Leadership Pattern," *Harvard Business Review* (May/June 1973), in Michael T. Matteson and John M. Ivancevich, eds., *Management Classics,* 2nd ed. (Santa Monica, Calif.: Goodyear Publishing, 1981), 265–280; figure on 278.

35. Rensis Likert, *The Achieving Society* (Princeton, N.J.: Van Nostrand, 1961); in William F. Christopher, *Management for the 1980s* (New York: Amacom/American Management Association, 1980), 121.

36. Secretan, *Managerial Moxie*, 153; he reproduces as table 5-1 a two-page outline of parallel assumptions, policies, and expectations of the three patterns from Raymond E. Miles, Lyman W. Porter, and James A. Craft, "Leadership Attitudes Among Public Health Officers," *American Journal of Public Health,* 56:(December 1966), 1990–2005.

Principles of Managing and the Manager's Role

Don't be clever, be conscientious.

PETER F. DRUCKER[1]

The V Theory of managing outlined in Chapter 2 relates to management in general. In this chapter the theory's principles are applied to broadcast and cable management, particularly the station manager's role.

A European Broadcasting Union conference once analyzed "the possibility of applying modern management techniques to cultural organizations like radio and television."[2] It produced forty-one reports about broadcasting's internal organization, personnel, budgets, and computers as a management tool. They assessed the problem of increasingly complex managing as equally important to engineering, programming, and legal problems.

Budget allocations confirm the importance of management-related activities. Typical AM, FM, and network-affiliated TV stations devote fully one-third of annual budgets to administration (and "general" expenses); they divide the remaining two-thirds of expenses among programming/production, news, sales, engineering, and advertising/promotion.[3] By 1990 most of a network's costs for entertainment and sports programming were for contractual license fees; sports expenses included 80% to 90% for negotiated rights, with less than 20% for talent, production, and other costs.[4] Managers at stations as well as networks strive to contain expenses by cutting back news expenses (fewer regional news bureaus, smaller staffs, less coverage of events) or, alternately, by replacing costly entertainment shows with expanded news programming.

Many people throughout an organization exercise management responsibilities and skills. Executives or "top management"—such as heads of networks and group-owned clusters of stations, and owners and general managers of local radio-TV stations and cable systems—are concerned with the organization's long-range goals, overall planning to achieve those goals, and the company's relationships with the community, industry (including advertisers), and government. "Middle managers" reporting to those executives are responsible for the station staff's efficient productivity and for facilities and materials.[5]

The manager is responsible for achieving objectives planned by company policy and strategy—the "intentionality" or goals and purposes of the enterprise and its top management. She does this by effectively directing people and their equipment and materials. This all relates to the manager's shared responsibility for capital invested in the enterprise and cash flow in the company's daily business. A middle manager spends time planning, organizing, and coordinating the work by department heads and others reporting to her. But she leaves to department heads the details of accomplishing objectives she has formulated and communicated to them. She develops policies for those areas, monitors costs, and compiles and evaluates data to forward to upper management.

This description of management's role reflects the classic theory of administration outlined by Henri Fayol.[6] His listed principles emphasize components of an authoritarian centered system of management: division of work, authority and responsibility, discipline, unity of command, unity of direction, subordination of individual interests to the general interests, remuneration of personnel, centralization, line authority, order equity, stability of personnel, initiative, and esprit de corps or morale.[7] In contrast, McGregor's Theory Y, Japanese Theory Z, and our own V Theory of managing look not to static structures imposed on a laboring force but rather to dynamic collaborative relationships among all personnel—executives, middle managers, and staff and line workers.

As discussed in the previous chapter, various criteria shape the manager's role. Peter Drucker puts it:

> Who is a manager can be defined only by a man's function and the contribution he is expected to make. And the function which distinguishes the manager above all others is his *educational* one. The one contribution he is uniquely expected to make is to give others vision and ability to perform. It is vision and moral responsibility that, in the last analysis, define the manager.[8]

Some commentators equate management with leadership, describing it as getting the job done through other people. But managers at various levels of most organizations are also occupied with tasks they do themselves rather than delegate to others. Those kinds of activity, however, are not strictly managerial; they are usually undertaken by default of available competent personnel or perhaps by the manager's personal choice for the satisfaction he derives from them.

While primary aspects of management include planning, staffing, and controlling, V Theory emphasizes the central role of leadership in effective managing. Management is an art more than a science precisely because much of the manager's work is solving problems among people. These kinds of problems require the manager to acquire not merely detailed knowledge and experience of an enterprise's operation but more especially skills of social interaction. Those skills include supervisory methods in leadership, communicating, counseling, brainstorming and group dynamics, using staff, and delegating.[9] The manager's role is to motivate people to work for the company's and management's objectives. They will do this effectively only when the manager understands their abilities and needs. By understanding them, he must skillfully communicate with them so they become participants by sharing macro- and micro-intentions behind directives.

Management is even more an art in the context of radio and television because, beyond executive ability and administrative skills, the broadcast manager must be proficient in a field that is "a unique blending of public utility, private business, and showmanship."[10]

Few who listen to radio and view television programs know who manage those stations. People identify stations by call letters, positions on the dial or channel numbers, air personalities and programs. People not engaged in broadcasting are seldom aware that a station "image" is developed by carefully planned policies determined and implemented by management. Public response to a station's service is largely shaped by the ability, beliefs, attitudes, character, personality, and philosophy of the manager.

A position of such impact and responsibility requires certain qualifications (part A). The range of duties (part B) can be better understood by analyzing people who manage stations.

A. Qualifications of Station Managers

Extensive broadcast experience is usually prerequisite to executive status in the industry. But other factors may not be so apparent.

In earlier decades, station administration was the exclusive domain of white males. But today there are many highly successful women managing stations and in executive positions at networks, cable systems, production companies, advertising agencies, and federal regulatory agencies. Approaching the mid-1990s, women were news directors at 16% of TV stations and at one out of four radio stations (29%).[11] Similarly, persons of minority groups have gradually been moving into management positions, serving as news directors at 8.5% of TV stations and 7.2% of radio stations. Jennifer Lawson, an African American woman served as chief program executive for PBS noncommercial television network (see the next chapter about personnel).

Because the manager's role includes motivating others to work for management objectives, the ability to deal with people is critical to successful managing. A manager's attitude toward staff persons is the key to fostering a sense of purpose and guiding them in carrying out their duties. He must therefore have some understanding of their abilities, wants and feelings, and their behavior patterns. He must be able to communicate with them effectively. In order to fulfill managerial responsibilities, managers must possess—or swiftly develop—certain qualities that can be communicated to their subordinates. Haimann lists "direction, enthusiasm, friendliness, affection, integrity, technical mastery, decisiveness, intelligence, teaching skill, and faith."[12]

From among many possible qualities, seven personal characteristics and attitudes stand out as vital to success in broadcast management. This subjective listing was confirmed by checking colleagues who successfully manage media companies. Each of them is strong in: leadership, intelligence and knowledge, judgment, personal integrity, sense of responsibility, and attitude toward work.

Leadership

Most agree with the premise that an effective manager should be a leader. But opinions differ about what human factors tend to produce media managers who can provide leadership. Some define leadership as a process by which a person exerts social influence over members of a group in order to influence their behavior.[13] Consonant with the V Theory of managing, Filley and House caution against looking merely to the leader's personal characteristics for a high level of group effectiveness, because that effectiveness depends on interaction between leader and group and also between individual members of the group. McGregor insists that leadership involves a relationship among at least four major variables: (a) a leader's own characteristics (discussed later); (b) the attitudes, needs, and other personal characteristics of followers (discussed later and in chapter 5 on personnel); (c) characteristics of the organization, including its purpose, structure, process, and kinds of tasks (chapters 6–9, plus 12); and (d) the wider social, economic, and political milieu (chapters 10 and 11). McGregor concludes that, depending on those factors, personal characteristics needed by a leader to perform effectively may vary. "It means that leadership is not a property of the individual, but a complex relationship among these variables."[14] The relationship between leader, colleagues, and context is essentially circular. This is the focus of "situational leadership" and "contingency" theories of managing.[15]

This circular process is subject to adjustment, application, and interpretation. Therefore, supervisory leadership must be flexible, with the enterprise's macro-intention—its "overarching" or "superordinate" goal or mission—providing overall direction and coherence.[16] To achieve the overriding purpose of the enterprise, a manager must retain adaptability along with creative decisiveness in the context of understanding the organization's larger intentions and policies. A manager may have to modify intermediate micro-intentionality and directives, at times, by either passing along some duties to subordinates or not passing them along. While presenting many plans to managees for their reactions and advice, she makes other decisions without consulting subordinates—depending on how she discerns the company's operation taken as a totality. This suggests that managees cannot participate excessively in decision making. They must realize that in some instances flexibility may prompt their manager either to override their recommendations or to anticipate by prior decision some deliberation she might otherwise let them share. The manager-managee relationship can be kept clear and effective if managees know the implications of flexible decision making, and as long as the manager communicates about such actions soon afterwards so managees can know what was done and the reasons for it. A manager who provides managee-centered participatory planning and decision making, but then usurps it occasionally without explanation, will destroy credibility and eventually the effectiveness of this shared process. So management must account for this flexibility when exercised. Further, as McLarney and Berliner note:

> [I]t is important that the manager be honest in letting the group know how much authority he is keeping for himself and how much he is giving to the subordinates. If the boss intends to make the decision himself, he should not try to fool the subordinates into thinking it was their idea in the first place. Using participation as a technique to manipulate employees is a shortsighted policy that is apt to backfire.[17]

Following the V and Z theories, leadership is successful when it accomplishes the company's planned objectives while involving employees in deliberations about how best to achieve them. The ideal superior-subordinate relationship is one of social collaboration instead of command-over-people. Leadership functions primarily through personality and administrative technique. Personality-oriented leadership emphasizes social contexts and individual persons' attitudes and responses in order to achieve corporate work goals; it is effective in organizations of fewer than fifty people (thus including staffs of most radio stations and local cable operations, and one-third of TV stations). But with more then fifty employees, interrelationships are too many and complex for leadership to be exercised adequately without administrative techniques.[18] The latter seek to instill high morale by setting goals ("management by objectives" or MBO)—preferably in collaboration with affected employees ("participative management by objectives" or PMBO)[19]—and then helping employees achieve those goals. This is particularly apt for the inner-directed employee who is goal-oriented, fulfilled directly by work and by achieving work-oriented goals. Aware of management's macro- and micro-intentionality, the employee's understanding of the job helps him to self-correct and to grow with his achievements.[20]

Psychological factors affect how well one assumes the role of management leadership.[21] A person will strive to lead if thereby she can fulfill herself (just as other persons will follow leadership that offers them self-fulfillment). Although a manager's self-actualization within an organization depends essentially on a "people-centered" approach, it derives also from freedom in the job, status, and directing subordinates. A manager wants to be a leader for various reasons: because it offers power, or opportunity for growth in wisdom, or even respect and affection, or because it offers something of the role of a teacher communicating to subordinates what the manager believes in and finds rewarding, or because the person has a different approach or a distinctive way of thinking about administration. In terms of problem solving, the manager must be able to motivate people for whom she is responsible so as to gain their "emotional participation" not only in her specific directives but in the total enterprise.

The manager's leadership role demands a high degree of confidence. Normally, his training and experience qualify him to meet unexpected problems. As a realist and aware that some problems must be settled without benefit of precedent, he is willing to stake his future on his ability to face issues squarely. He learns to thrive under pressure. In a volatile industry that is partly art, science, and business in shifting social and technological contexts, a manager must be able to manage ambiguity and uncertainty, change and even seeming chaos.[22] Peter Drucker emphasized the theme of his *Managing in Turbulent Times:*

> There is one overall 'text' on which this book preaches. It is *'Don't be clever, be conscientious.'* Predicting the future can only get you into trouble. The task is to manage what there is and to work to create what could and should be.[23]

Among classic leaders with vision that motivated entire organizations were David Sarnoff (RCA/NBC), William Paley and Frank Stanton (CBS), and others such as Edward R. Murrow (CBS News), Donald McGannon (Westingouse/Group W), and Joan Ganz Cooney (Children's Television Workshop/PBS). Entrepreneurs with vision for creating new media combinations include Ted Turner (Turner Broadcasting System/WTBS, CNN, Turner Network Television [cable]), Tom Murphy (Capital Cities/ABC), Stanley S. Hubbard (Hubbard Broadcasting (CONUS news cooperative, and USSB direct satellite broadcasting), and Norm Pattiz (Westwood One/Mutual Broadcasting/NBC Radio networks).

Ted Turner was fond of the aphorism: "Lead, follow, or get out of the way."[24] William S. Paley, founder of CBS in 1928 and chairman until his death in 1990, reflected on his leadership role through half a century:

> As the chief executive I also had to take the time to pause, to step back and take the long view of CBS. Instead of looking at tomorrow, I would try to look years ahead, and ask myself: are we conducting our business properly for the long run? Is the balance right between broadcasting and other activities? Is the company itself properly organized? There is a certain excitement in looking at a company this way, for decisions on this level are designed for long-lasting effects. It is often years before one can say if he has been right or wrong. At CBS, many of our most important decisions have been made after pausing to take this long-range corporate view, which is uniquely the responsibility of the chief executive.[25]

Leadership is the keystone of effective management. Other characteristics discussed later reflect this quality or are constitutive of it. This is reflected in the National Association of Broadcasters' list of "Twenty Guidelines for Leadership" (adapted from Perry Smith's analysis):

1. Trust is vital [mutually between superior and subordinates].
2. A leader should be a good teacher and communicator.
3. A leader should rarely be a problem solver. [Instead, facilitate managees' solving most problems, building participation and self-esteem.]
4. A leader must have stamina.
5. A leader must manage time well and use it effectively.
6. A leader must have technical experience.
7. Leaders must not condone incompetence.
8. Leaders must take care of their people.
9. Leaders must provide vision.
10. Leaders must subordinate their ambitions and egos to the goals of the unit or the institution that they lead.
11. Leaders must know how to run meetings.
12. A leader must be a motivator.
13. Leaders must be visible and approachable.
14. Leaders should have a sense of humor.
15. Leaders must be decisive, but patiently decisive.
16. Leaders should be introspective. [Able to review their own mistakes objectively.]
17. Leaders should be reliable.
18. Leaders should be open-minded.
19. Leaders should establish and maintain high standards of dignity.
20. Leaders should exude integrity.[26]

Several items in the list relate to qualities noted below.

Motivating

One aspect of motivating managees involves sharing "the vision" of the enterprise with them, inspiring them to help achieve the organization's goals and objectives, purposes and values. The effective manager (in Burns' terms)[27] provides *transforming* leadership that builds on a person's need for meaning and institutional purposefulness. This goes far beyond *transactional* leadership that encompasses the duties and activities making up most managers' days (see the second part of this chapter). The transforming leader engages others to transcend daily details to embrace a larger perspective and a personal commitment to the endeavor. Burns explains:

> Various names are used for such leadership: elevating, mobilizing, inspiring, exalting, uplifting, exhorting, evangelizing. The relationship can be moralistic, of course. But transforming leadership ultimately becomes *moral* in that it raises the level of human conduct and ethical aspiration of both the leader and the led, and thus has a transforming effect on both.[28]

A second aspect of motivation pertains to a manager's sincere interest in the welfare of those working at the station and living in the community. With genuine regard for people plus an understanding of human motivations and aspirations, she can lead them toward positive growth. Her concern for staff members' personal development will characterize her success or failure as a leader. While the successful manager develops skill in correcting mistakes by others, she is equally adept at showing appreciation for good work. Employees respond positively to a manager who is clearly interested in their personal growth and achievement.

People expect certain advantages or satisfactions from working in an organization. McLarney and Berliner note that what people want from jobs can be translated into what they expect their supervisors to be concerned with on their behalf:

> Respect for their personal dignity and worth. Concern for their success on the job, their safety and health. Acceptance of their limitations and appreciation of their abilities. Understanding of their needs for security, fair play, approval, belonging, importance and recognition. Willingness to listen to them, to try to understand them, to spare them from unnecessary unpleasantness and worry.[29]

A manager's attitude affects employees' self-image and satisfaction with work. To foster this unpressured yet productive attitude, the manager deemphasizes his role as commanding or directing subordinates, promoting instead a relationship of collaboration. This is important despite sometimes having to communicate with and motivate individuals who are bored or frustrated, anxious, dissatisfied, or even maladjusted.[30] Norman Maier notes that efficiency in work correlates with level of motivation: "When motivation is low, fatigue effects become apparent very early, but, when motivation is high, the evidence of fatigue may not be apparent until considerable physical exhaustion is manifest."[31] In terms of energy expenditure, the level of motivation generated by work—and by management's support—will determine the amount of energy available for that work.

Broadcast employees generally are designated "professionals" because they often contribute to social and legal, as well as economic sectors of society. For example, newspersons are concerned with such issues as journalistic responsibility and applying the First Amendment to press freedom; their associations promote professional standards for reporting news events. They, along with salespeople and on-air talent such as announcers and disc jockeys, participate in national organizations that establish guidelines for proper practices. Professional-oriented staff members often follow schedules suited to the station's larger needs (rather than the 9-to-5 routine) and they often exercise other socially responsible roles beyond their salary-producing jobs. Further, trade publications and the general press report media activities, including public impact of decision making. College departments offer academic courses of study in related areas. All of this distinguishes broadcasters from the rank-and-file national workforce. When attributes of the "professional" apply to certain staff members, a manager relates somewhat differently to them than to other personnel. The manager often finds it appropriate to deal with them virtually as peers.

Another distinctive trait of many media personnel is creativity. To support and inspire the many creative people working in radio and television, a manager needs a high degree of personal creativity or else a fine sensitivity to creative efforts. Managers must also effectively challenge them, motivating them to do their best creative work.

In short, the manager must stress building and maintaining excellent staff morale. Somehow, she causes all staff persons to feel their work is vitally important to the station's success. She is pleased when this feeling exists even among custodial employees. When genuine, the manager's own idealism, commitment, and enthusiasm are contagious; her motivation permeates the staff.

Delegating

This characteristic is central to leadership in media management precisely because it affects the level of creativity in a staff that includes professionally-oriented persons. A manager who really knows his managees' qualities and expertise and responsiveness to motivation will delegate to them responsibilities (along with accountability) proportioned to their knowledge. To restate once again: a manager exercises leadership by finding ways to motivate managees so they can fulfill themselves while also carrying out the macro-intention of the whole enterprise. This leads to the corollary that there is no "average" person and that an individual tends to rise to the level that challenges his full capabilities. This is apparent in McLarney and Berliner's enumeration of what the manager wants to achieve through leadership of managees:

1. Willing, sustained, and high-level job performance.
2. Readiness to accept change.
3. Acceptance of responsibility.
4. Involvement of people so they will use their brains, abilities, initiative and ideas.
5. Improvement in problem solving, in cooperation, and morale.
6. Development of people to be self-starting and self-controlling.
7. Reduction of turnover, absences, grievances, tardiness and waste.[32]

In this area of delegating—of decentralization or subsidiarity—the V Theory of managing offers descriptive criteria. The more a manager deals with other people (managees) rather than with informational data and specific procedures, the more properly he is managing. The less he deals with people and instead engages in material things and procedures, the less he is acting as a manager and the more he is performing functions of an administrator or staff member.

A paradox can arise in the sequence of delegating tasks to employees to the point of promoting them for work well done. Laurence J. Peter pursues this conflict to its logical excess, identifying it as "the Peter Principle": everyone rises to the level of their incompetence.[33] That is, a person demonstrating ability in some area is promoted to greater responsibilities; his competence in that higher position eventually warrants a further promotion. He continues to move up until reaching a position where he does *not* function effectively; so he is promoted no more, but left to malfunction in that area. He has reached the level of his incompetence. But, optimistically assuming he never reaches that plateau of ineffectiveness, along the way he successively learns each job and adds that experience to his previous competence. Thus he gains considerable ability in the several areas which he subsequently manages. Yet if he reverts to those earlier tasks, getting involved in details, he is forsaking his larger role as manager and is in fact doing the work of other people—his subordinates or managees. This underscores the repeated theme: managing is not doing the job itself, but rather motivating, directing, and guiding other people to do those jobs.

This is clear in broadcasting. Often a sales manager subsequently becomes station manager where she oversees not only sales (in which she is experienced) but also programming (where she may have some knowledge) and administration (where she has a little related experience) and engineering (where she may have no background whatsoever).

In terms of the V Theory of managing, the manager's key role is to interpret the enterprise's intentionality—its purpose, policies, procedures—through directives to the managees, who can then exercise appropriately creative responsibility in actualizing those directives, in light of the understood macro-intention of the company. This means the manager does not want subordinates to come to her *for* decisions, but rather *with* judgments they have already made for her to endorse. As long as they understand the larger "why" (micro-intention of directives by managers), they can responsibly exercise initiative and provide creative input by participating in the company's rationale and philosophy (macro-intentions). Often, too, they are in a better position than the manager to judge how to carry out specific directives most efficiently on the spot.

Because delegated responsibility includes actualizing and even interpreting management decisions, delegating involves real risk. Risk means there is genuine chance of mistake or failure—affecting both manager and managee. And only *true risk* permits *true creativity*. The manager must accept this context of real risk. If there is no possibility of mistake then she has overly routinized things, and no genuine creativity can be exercised.

This opens the door to considerable latitude of interpretation as managees actualize directives from managers. But this is the heart of the V Theory of managing as well as of Theories Y and Z: human beings collaborate in a common enterprise. There is not a "dictatorship of meaning"—assigning actions to be performed in one exclusive and unique way—which management imposes univocally on all directives and actions. Personal initiative, in the context of informed commitment to the company's purposes, is supported and encouraged. McGregor again and again stresses the need for this higher "unity of *purpose*" (or our terms of macro- and micro-intentionality) rather than the older philosophy ascribed to organization dynamics, the "unity of command," where one person in charge makes decisions and everyone else is responsible to that superior for their actions. McLarney and Berliner agree:

> There should be more emphasis on the purpose or *why* of routines, more dependence upon people and less upon the system. The trend in better management today is to build meaning into jobs—to design jobs in such a way as to promote individual responsibility. If a man feels that responsibility rests upon him rather than upon a system, he will exercise initiative and judgment. But if he feels that he has to buck the routine in order to protect the company's interests, he may choose to follow routine. As a result he will be of less value to the company and he will have less satisfaction in his job. Procedures can't operate without people, and the performance of people hinges upon their knowledge, ability, and attitude, the design of their jobs, and the quality of the supervision.[34]

This effective and perhaps bold form of delegating is reasonable—even essential—for a creative, productive, person-fulfilling enterprise. Theodore Roosevelt appraised proper management leadership in delegating: "the best executive is the one who has enough sense to pick good men to do what he wants done, and self-restraint enough to keep from meddling with them while they do it."[35] Peters and Austin assert that delegation is "the toughest issue in management."[36]

Communicating

Closely tied to delegating responsibility is communication, because it develops throughout a staff shared vision and commitment to the enterprise's purposes and goals. Some observers stress that executive management's real role is "to manage the *values* of the organization" and to promote and protect those values.[37]

Management must clearly communicate the degree of freedom managees have for implementing or adapting directives. By offering wide scope for freedom and interpretation, yet with clearly stated boundaries, managers elicit the highest quality of creative, productive collaboration from staff members.[38] Results become increasingly unsatisfactory as managers lessen the area of freedom and define boundaries vaguely. An employee can be responsive to a manager's directives, according to Milton Blum, when that employee understands the communication and sees it as consistent with the organization's purposes as well as compatible with his own interests as an employee, and when he is mentally and physically able to carry it out.[39]

Various forms of communication within a company are described in chapter 5. Briefly noted here is that meetings and conferences are ineffective if participants perceive the manager's intent is to win the group over to his position or else to manipulate their attitudes, even if he does not indicate specifically what they should think or decide. Managees will be aware of whether the sessions are for open discussion of ideas or merely to confirm the manager's judgments when communicated orally to the staff.[40]

While effective communication is an important element of leadership, it includes more than being highly articulate in both written and spoken communication, having the ability to say "yes" or "no" with firmness, and knowing how to ask the right questions at the right times. It also includes the skill, and virtue, of being able to listen and to attempt to understand the other person's statements, meaning, and implied attitudes. Effective listening by both manager and managee is critical for creative collaboration, because mutual communication is the instrument by which motivation is generated. Guidelines for better communicating, and for sharing new ideas and insights between management and staff, include: avoid making value judgments, listen to the full story, recognize feelings and emotions, restate the other's position to clarify your understanding of it, and question with care.[41] Effective listening includes attending to the main thought or idea, noting not only verbal but also nonverbal communication (posture, gestures, facial expression), and summarizing what you think the speaker has just said.[42] Rapid and honest communication through conversation, meetings, and formalized procedures such as memos, bulletin boards, or newsletter will dispel haphazard informal communicating through the grapevine of rumor and gossip which undermines personnel confidence in management.

Many commentators stress the need for MBWA—Management By Wandering Around. This is more than merely drifting about physically among departments and staff. Listening is a primary objective of MBWA, along with coaching, teaching, or counseling—informally. This exchange of ideas and needs between manager and managees leads to facilitating results; it does not end in mere rhetoric that is in fact hypocritical.[43]

Finally, as a leader who communicates, the manager does not confine her activities to her station or cable company. She seeks out further responsibilities in the community and in various professional associations. Not content to assume a passive role, she volunteers for active service on committees, runs for office within organizations, and makes speeches at local, regional, and national meetings. While fellow broadcasters recognize her as a leader in the industry, she has a similar image as a leader in her community. Surveys have found that radio, TV, and cable managers participate more in community activities than do other business leaders.[44]

Their high degree of activity in civic and charitable organizations is partly motivated by strong inclination for meeting a wide variety of people. Of course the pragmatic motive of making and renewing contacts for business also prompts managers to enroll in memberships, give speeches, conduct seminars and meetings, attend conferences, and maintain continuing dialogue with community and civic leaders.

INTELLIGENCE AND KNOWLEDGE

It is obvious that a manager's intelligence should be well above average. Whether a particularly superior I.Q. is an advantage to the manager is another question. Various authorities judge that a person can achieve success in a managerial position with intelligence less than superior but considerably above the norm. Beyond the basic college degree (often in communication-related majors), broadcast managers increasingly have done postgraduate work and earned master's degrees. Networks and station groups in the past decade have hired persons with doctoratal degrees, especially for key research and marketing areas. Company presidents with doctorates have included Dr. Frank Stanton at CBS and Dr. William Baker at Westinghouse Broadcasting/Group W. For years, Dr. Charles Steinberg headed the CBS public information department, and Dr. Joseph Klapper its social research office; Dr. Alan Wurtzle served as vice president of ABC's broadcast standards division. Dr. John Able was vice president of the National Association of Broadcasters, as was Dr. Charles (Chuck) Sherman who managed local television stations before joining the NAB. Dr. John Malone has led Tele-Communications, Inc. to the forefront of the cable industry. Many "Ph.D.s" have served in executive, staff, and consulting positions with major broadcast companies, professional trade associations, advertising agencies, and the FCC and congressional committees overseeing telecommunications.

Vital to successful media managing is the ability to acquire knowledge quickly, retain it, and apply it effectively to various new and continuing problems of broadcasting. In addition to obvious areas of news and public affairs, entertainment programming, and regulatory subjects, a premium is often placed on familiarity with technical aspects of economics, finance, marketing, research, data-processing, and statistics.[45]

A radio, TV, or cable manager is expected to have specialized knowledge of the industry, his own station, and its market. Not only are there continuing *changes* in the business world, in entertainment forms, in radio and television and cable, and in governmental regulation; there is continued acceleration in the *rate* of change (as noted in chapter 1). This exploded geometrically during the 1980s. A manager must keep abreast of fast-paced developments to remain competitive and to be a leader within his company and among his colleagues.

Recent years brought uncertainty and risk: How would cable and satellite reception affect over-air broadcast business (it fragmented mass audiences and siphoned away some advertising dollars). Would domestic satellites mean the entry of one, possibly two additional television networks (in 1990 Fox linked 150 local independents, followed by Warner Brothers and Paramount). Will the network structure change so they eventually sell programming to affiliates (compensation was decreasing, even dropped for some programs, but later rebounded as newly emerging networks aggressively foraged for stations to affiliate with). What about the future of the agency and rep business (dozens of companies merged into megafirms).[46] Similarly, a business newsletter back in 1972 (presciently reading like a 1990s' trade piece) cautioned local TV stations not to stagnate with the status quo in the face of encroaching competitive media.[47] It called for more imaginative executives, not relying on network formulae and stereotyped syndication for prime-time access programming but rather responding to each market's different needs with local news and community shows; expanding production departments to better serve local advertisers; syndicating their own shows to other stations; and even becoming program centers for cable TV (entering the 1990s several radio and TV stations in fact shared news programming with regional cable operators).

Related developments were happening in full force in the 1990s, such as the impact of high definition TV equipment on station budgets (variously estimated from $5 million to $15 million per station); foreign ownership of film studios supplying programs and movies; international marketing of syndicated shows; cooperative ventures with foreign producer-investors; and paying off enormous debt from leveraged buyouts and mergers, which dictated slashing expenses (translation: cutting back on network and station staffs and program service).

A manager continually makes decisions about day-to-day operations, such as inter-station cooperative news pooling; cross-over services between broadcast and cable systems; and adopting new technologies: upgraded news gathering equipment (ENG), satellite up- and down-link remote trucks (SNG), cellular mobile phones, facsimile (fax), office and studio computerization, automated master-control, and digital and stereo (AM as well as FM) transmission. Meanwhile, all of this goes on amid the legislative/judicial churn of media that in recent years have been successively regulated, deregulated, and reregulated by federal government, amended and nuanced by a succession of court cases. "Bottom-line" considerations include the role of timely and effective promotion, aggressive marketing and competitive time sales, and collaborating with ad agencies and station rep firms that themselves have undergone waves of expansion in research methodologies, instantaneous communication systems, and mergers.

An informed manager studies ways to achieve greater efficiencies in budgeting and learns to be intrigued by the operation's financial affairs.[48] Added to short-term expenses are long-term issues that round out the manager's sphere of major concerns: financing capital expenditures and syndication package contracts along with fees for wire services and music licensing; cyclic negotiations with unions and labor relations agencies; and participation in trade associations.

To stay abreast of all the changes occurring regularly in the industry, a manager must read extensively. She needs to read important trade publications, periodicals and books in the field of business, and more general magazines and newspapers. She can save time by having an able staff member prepare digests of important publications for her (this is feasible at least in large stations). But digests of information cannot substitute for an executive's reading original sources. Analyzing materials is as important as reading them; this activity is essential although time-consuming. Surveys over past decades repeatedly indicate that the vast majority of broadcast managers read *Broadcasting & Cable* magazine regularly, followed distantly by other trade journals and specialized media magazines;* a large percentage also read the *Wall Street Journal*.[49]

Constant study of changes within the broadcast/cable industry is a heavy assignment. At the same time, managers must also expand their knowledge of their own stations. As noted earlier, most station managers are well acquainted with either

---

*This is one reason why this text relies on *Broadcasting & Cable* as a key source that chronicles major developments and detailed data of the media industry.

sales or programming or engineering; few have personal experience with more than one of those operations. Yet, the manager must be prepared to discuss each department intelligently, analyze the operations and practices of each, make major decisions affecting them, and provide them with continuing challenges. It is the manager who must establish effective working relationships among departments.

The manager must also acquire a wealth of knowledge about the market, and be able to analyze shrewdly that market's business potential along with the interests and needs of people in the company's coverage area. A manager must be familiar with retailing strategies and problems, with specialized understanding of the role of advertising in the business economy.

When appointed to manage a local station or cable system, a person enters a new level of the working world with a new set of requirements and demands. She finds much to learn about the whole field of broadcasting which she did not need to know in previous media positions. More than in previous positions, the alert manager notices complaints by critics of broadcasting. She finds herself examining criteria and standards for programs and commercials. She becomes conscious of public relations and reassesses her own station's facilities for publicity, promotion, and merchandising. She becomes well versed in the problems and achievements of all mass media and related forms in the entertainment field. She becomes active in efforts to improve the community. Her mind is expanded and challenged by all this new knowledge and experience.

Beyond formal college studies, often including specialized experience in internships and co-operative education professional work, media managers continually update themselves not only by general and specialized reading but by participating regularly in professional seminars and industry conferences. Managers with a liberal education or background who maintain true intellectual curiosity should be able to converse intelligently on almost any subject. Their wide range of interests should make them popular and stimulating guests at social events important to their professional careers.

JUDGMENT

It might be relatively easy to manage a radio, TV, or cable operation by imitating the practices of successful managers elsewhere. But circumstances differ enough from company to company to make this unfeasible. Furthermore, that sort of managing hardly appeals to those who place a high premium on their ability to think for themselves. Take away the need for using wisdom and the management position loses its attraction for one who regards decision making as a personal challenge.

Every executive is required to make decisions regularly. In broadcasting, countless problems arise that have few, if any, precedents. They must often be solved in a matter of minutes. Yet, managers must not make hasty or arbitrary decisions.

When decisions must be made on long-range problems, the most effective managers know how to define those problems, gather all the important facts, determine possible solutions, arrive at decisions, and implement those decisions in a manner that gets employee support. Good managers will not settle for insufficient evidence nor will they be influenced by unimportant or trivial data.[50] The effectiveness of long-range thinking depends on how much time the executive has available for uninterrupted reflection. A wise manager blocks out periods of time for thinking about major issues and for planning.

Putting it simply, an executive administrator must avoid the extremes of either uninformed, seat-of-the-pants guesstimates ("Ready, FIRE! . . . Aim"), or caution to the point of paralysis from endless analysis ("Ready, aim, aim, aim, aim. . . !"). Harold Leavitt sees managing as an interactive process of *pathfinding* (involving initiative, experimentation, design drawn by intuition, including values and vision); *decision making* (rational, analytical, antiexperimentation, based on data and logic); and *implementing* (pragmatic, idiosyncratic, both people- and task-oriented).[51]

Loring and Wells summarize judgmental skills needed for management:

Having the ability to develop goals and objectives.

Being able to build a plan to implement these goals interrelating them with others.

Capable of effective communications, especially with people but also written and formal presentations.

Able to resolve or balance conflicts between work, interests and people.

Good at problem-solving in all its phases, with work processes and people problems.

Balanced decision-making, carefully weighing the important elements and generally using good judgment.

Able to determine priorities with flexibility, to change as needed, and stick with them when necessary.[52]

PERSONAL INTEGRITY

The broadcast or cable manager must be a person of strong character. This position is in the spotlight, influencing staff personnel and people living in the community. The position obligates a manager to command respect, not by mandating it but by earning it.

The manager should be sensitive to deeper values, reflecting depth of character. Integrity involves consistency, following through on stated ideals. Peters and Austin studied these characteristics of excellence:

> The superb business leaders we use as models in this book epitomize paradox. All are as tough as nails and uncompromising about their value systems, but at the same time they care deeply about and respect their people; their very respect leads them to *demand* (in the best sense of the word) that each person be an innovative contributor. . . . The best bosses—in school, hospital, factory—are neither exclusively tough nor exclusively tender. They are both: tough on the values, tender in support of people who would dare to take a risk and try something new in support of those values. They speak constantly of vision, of values, of integrity; they harbor the most soaring, lofty and abstract notions. At the same time they pay obsessive attention to detail. No item is too small to pursue if it serves to make the vision a little bit clearer.[53]

Those authors conclude their book on the note that excellence in leadership comes about "when high purpose and intense pragmatism meet."

"Integrity" means "wholeness"—a balanced composite of sensitivities and values based on insight and experience, which supports principled convictions about oneself and about one's fellow human beings. This sense of ethics or morality may derive from formal religious commitment or from a self-understanding of human respect for and responsibility to fellow human beings, to help others in society develop fully as individual persons. Such integrity goes far beyond mere personal piety or scattered efforts at "good works"; it is the fabric of a person's 1ife.[54] The keynote of an ethical sense and a healthy morality is freedom—freedom to innovate, freedom to select from among real alternatives, freedom to counter the prevailing climate of value-structures when they are inadequate or harmful. Haselden concludes that this freedom "is the habitat of authentic morality." Otherwise, these alternatives may eventually be provided by other sources. For example, alternative concepts in content, form, and scheduling can be supplied by default by cable TV and pay television and by videocassette distribution, if commercial broadcasters (owners/licensees, stockholders, managers, staffs) and advertising agencies and their clients restrictively stereotype programming patterns only for a broad level of mass audiences. Rivers and Schramm excoriated the preponderance of merely "kitsch" or mass culture as the popular standard of American media.[55] Even on the heels of TV's so-called Golden Age, newsman Edward R. Murrow lamented that "television in the main is being used to distract, delude, amuse and insulate us."

Lynch noted four major areas of problems in media, especially in the sector of artistic creativity, which challenge the broadcast manager of integrity who must ultimately answer to society for the role that his station, network, or cable system plays in the local or national community's daily living.[56] The problems include: (a) mass media's failure to distinguish between fantasy and reality, (b) weakening audiences' feeling and sensibility, (c) restricting freedom of imagination (not by morality but by techniques for fixating the imagination), and (d) massive projection of make-believe that loses true lines of human reality. Every person of integrity, including media managers, should be concerned with preserving and helping to create (rather than protesting against) freedom of imagination in mass media. Integrity raises a challenge against imitating repeated presentations of merely one or other partial aspect of the human condition (violence, physical relations, war, unfairness, glib banter, and superficial sense pleasure, among others). Integrity supports the attempt to portray more of the fullness of reality all around us—in entertainment programs and also in the range of cultural and informational programs, including news, documentaries and public affairs. What Lynch calls the "mass media diseases" of fantasy, flatness, fixation, and empty pretentious magnificence affect "the condition of human sensibility in our civilization." Against this mediocrity of craftsmanship with routine technique and stereotypical image-and-sound clichés, broadcast leaders of integrity strive to support media creativity that "knows how to move us from insight to insight."[57]

Although individual media managers cannot hope on their own to correct systemic media weaknesses or to change major aspects of media content and practices, each manager can exercise personal integrity of judgment by reflecting on the potential of mass media in today's world, specifically for their station personnel, program service, and local audiences.

International churches, as well as some secular professional organizations, have noted what has been referred to as "the centrality of the communications vocation in modern society." The U.S. Catholic bishops' Communications Committee noted:

> The newsman, the broadcaster, the playwright, the film maker cannot be considered or consider themselves merely as entertainers or technicians. As much as anyone in contemporary society, they form the world-view of modern man. They convey information and ideas that are essential to the functioning of society; even more important, they help to shape the very ethos of the world in which we live. Theirs is, then, a calling of high honor—of heavy responsibility.
>
> Communicators in our country have generally met this responsibility with a conscientiousness and dedication which does them deep credit. In some areas, however, there are problems.[58]

Although any change in mass communication involves a long process, change must come about in individual persons before it can affect the larger system.

Professionalism

Individual as well as institutional professionalism contributes to ethical integrity of media through executive managers and middle managers (department heads). Broadcast educator and historian Sydney Head commented on media gatekeepers' personal accountability for entertainment and information:

> In the practical terms of day-to-day operations, the private conscience and sense of responsibility of the individual worker—writer, salesman, air personality, control operator, editor—govern what goes out over the air. The multitude of their small decisions determines the actual quality of the broadcast service. Legal regulation and institutionalized self-regulation can govern only a small proportion of these decisions; most remain personal.[59]

Rivers and Schramm acknowledge that most practitioners of mass communication are employees whose employers/managers are ultimately responsible for their work and the quality of their service. Nevertheless, those authors stress individual rather than corporate sense of responsibility; they look to each communicator's duty "as a public servant and as a professional, quite apart from his obligations to the business that employs him."[60] Founded partly on each person's sense of ethics, professionalism implies high standards of personal conduct voluntarily adopted in a career involving social responsibility.

Young adults often perceive the ethics of those in business in a less than flattering light. They offer chilling appraisals of their perceptions of business executives' sense of integrity in the competitive world of free enterprise. Samples taken over three decades (1968 to 1996) of college students majoring in broadcasting and mass communication reflect pessimistic assumptions about what business executives might consider as norms for "what *ethical* means."[61] Most college students consistently presumed that for those in business "ethical" would mean "customary behavior in society" or "about the same as legal." But a national study of executives and three small samples of professional broadcasters found most of them selecting criteria similar to college-age persons: "what does the most good for the most people" and "what my feelings tell me is right"—plus two categories ignored by most students, "in accord with my religious beliefs" and "conforms to the Golden Rule." Almost no businesspersons and few students (3.8%) over the years equated "about the same as legal" with what they themselves meant by "ethical" (although 12.5% of broadcasters' responses did indicate "legal"). Surprisingly, a large percentage of businesspeople but very few college students chose "in accord with my religious beliefs" as what "ethical" meant. These discrepant judgments suggest that college students' perceptions of business criteria for integrity may be far off the mark, when in fact their own judgments about what constitutes ethical norms are quite similar to business executives' own judgments. (It is important to clarify that both businesspeople and college students responded only to what they *meant* by the term "ethical," not necessarily to what was the norm by which they acted—either in business dealings or on campus.) The surveyed businesspeople also reflected greater ethical concern over personnel problems than about questions directly pertaining to profit-making aspects of business.

When entrepreneur and major newspaper publisher Francis Dale had the same list put to him, he found merit in many of the nine selections; but pressed for his preference he chose "what my feelings tell me is right"—just as collegians have done over three decades—explaining that by "feelings" he meant the accumulated sense of perception, sensitivity, and values that become part of a person over time from their upbringing and social experience (not the same as merely what "feels good" emotionally):

> I'm assuming that my feelings have a moral base, and I believe that is true. If you've come through a long life of experience where you've been placed in situations that call for moral decision and if you've had either a religious background or moral training, you build up a body of experience and hopefully, when you're faced with a situation, there is something that clicks in your mind and says, "Yes, that's right, this is wrong." If you have that kind of basic training, I think you do react that way. Very few of us have the time to sit down and think it all through: one, two, three, four. So I would be comfortable with "what my feelings tell me" [as descriptive of what "ethical" means], on the assumption that my feelings are well grounded.[62]

In studies of 1,512 businesspeople (1961) and another 1,200 (1976), most indicated the biggest influence on making *ethical* decisions was an individual's own personal code of behavior, followed rather distantly by formal company policy and by how a person's superiors in the company acted. On the other hand, primarily influencing one to make *unethical* decisions was how company managers acted and the industry's "ethical climate."[63] Fully 80% of all respondents in the second study discounted highly relative or "situational" ethics; instead, they agreed that business people should try to live up to an absolute moral standard rather than to the moral standard of their peer group. In both studies, almost 98% agreed that "in the long run, sound ethics is good business"; but half in both surveys noted that business executives are "preoccupied chiefly with gain" rather than applying major ethical considerations at work—partly because integrity is not rewarded if profitability is risked, and partly because pressures of competition overshadow ethical considerations.

The same study concluded from 1,200 business executives' responses: (a) public disclosure, including media coverage, and increasing public concern about unethical practices most contributed to raising ethical standards in business; (b) "hedonism, individual greed, and the general decay of social standards" most affected declining ethical standards; and (c) shifts in ethical business standards were influenced by forces managers usually do not directly control.[64]

A survey of twelve chief executives of America's largest corporations found that more than knowledge, skills and even experience, the most critical factor for professional success is character, defined as "moral excellence" and integrity.[65] For them, that quality encompasses integrity of personal conduct, product, and policy; it includes honesty, sincerity, steadfastness, and courage.

The decade of the 1980s evidenced lack of integrity in business dealings and in government (the public's business): The Watergate scandals that brought down a president and sent executive staffers to prison; Iran-Contra meddling in unlawful bartering of arms with a hostile nation which led to indictments and trials of public officials; the savings and loan crisis involving financial executives and congressmen (and up to $500 billion tax dollars to bail out the badly invested funds); the Whitewater financial involvement of former governor then U.S. president Bill Clinton. Wrongdoing occurred in media affairs: Payola practices at radio stations involved drugs, dollars, and sex; sales departments double-billed advertising clients; clients received payoffs and kickbacks from the chairman of William B. Tanner Co. (buyer/seller of radio-TV airtime); the chairman of Advertising to Women of New York used company assets for personal expenses and was cited for conspiracy and income tax evasion; J. Walter Thompson ad agency's TV syndication division exaggerated its sales record by $22.8 million with faked accounting entries; and so on. These excesses, distortions, and even criminal practices occur when integrity lapses. Managers as well as individual employees are responsible for preventing or else spotting and correcting such abuses.

SENSE OF RESPONSIBILITY

Closely linked with personal integrity is the manager's attitude toward various responsibilities to the people his company serves. A media manager has several kinds of "customers" or "stakeholders" in the enterprise: owners (individual or stockholders), advertisers, audiences, and the company's employees. A manager is constantly reminded that he alone is ultimately accountable for the many decisions made in their interests. The manager, in the context of the company's macro- and micro-intentions, determines priority among those various customers.

The most effective managers, while fully alert to the critical importance of profitability, also recognize other social values as important to overall success as efficiently accumulating money. This is one reason why they become active in groups working to improve conditions in their communities. They know their activity usually results in material benefits for the station, but only if they participate sincerely and effectively.

"Responsibility" typifies a broadcast manager's ability to respond to the enormous social challenge and ethical burden of overseeing a medium of mass communication. This responsibility was appraised by a world-wide church conference:

> The channels of social communication, even though they are addressed to individuals, reach and affect the whole of society. They inform a vast public about what goes on in the world and about contemporary attitudes and they do it swiftly. That is why they are indispensable to the smooth functioning of modern society, with its complex and ever changing needs, and the continual and often close consultation that it involves. This exactly coincides with the Christian conception of how men should live together.
>
> These technical advances have the higher purpose of bringing men into closer contact with one another. By passing on knowledge of their common fears and hopes they help men to resolve them. . . .
>
> All men of good will, then, are impelled to work together to ensure that the media of communication do in fact contribute to that pursuit of truth and the speeding up of progress . . . which is the brotherhood of man under the fatherhood of God.[66]

Mass media play multiple roles in contemporary society. They help us anticipate significant events and trends in the arts as well as public affairs; they assist in correlating our responses to coming challenges and opportunities. Media help us reach consensus on social action, transmit our culture, entertain us, and sell goods and services. Thus a broadcaster's responsibility involves value judgments about people, society at large and events within it, and the content of media presentations.

Although broadcast stations, especially in radio, are usually small businesses, many local stations and cable systems are parts of larger companies. It is illuminating to apply to media management some data and appraisals available from larger corporate enterprises, since mass media share the impact of American business on the general social scene.

Joseph C. Wilson, president of Xerox Corporation, held firmly that the corporation is an integral part of society and has concomitant social responsibilities. He noted that

> inevitably the corporation is involved in economic, social, and political dynamics whether it wills it or not. . . . [I]t seems beyond belief that anyone could contest the view that the businessman, who wields a vast amount of economic, and therefore social power, must expand his vision beyond the limits of making maximum profit, the traditional definition of his primary function, to encompass many of society's harassing problems that wash the edges of his island.[67]

It is probably no mere coincidence that Xerox regularly sponsored major television dramas and documentaries on commercial networks and award-winning cultural/educational series through the Public Broadcasting Service.

In the survey cited earlier, 1,200 respondents were split almost evenly on whether a company's social responsibility would in fact lower corporate profits in the short run (41% agreed, 43% disagreed, 16% neutral) but would bring higher corporate profits in the long run (43% agreed, 36% disagreed, 22% neutral). But more than half of all respondents (58%) supported the statement "the socially aware executive must show convincingly a net short-term or long-term economic advantage to the corporation in order to gain acceptance for any socially responsible measure he might propose."[68]

Profit cycles can lead to dehumanization; a business system whose original function included serving society by helping people can shift priorities and make people a function of the system. So claimed John Sack in *The Man-Eating Machine,* criticizing business administration that pushes beyond technology to a level of technocracy.[69] Media companies that grow into massive corporations or merge with sprawling conglomerates can risk distorting human values—among employees as well as mass audiences. During the past decade, broadcast networks struggled to reposition themselves financially through new corporate alignments; they consolidated staffs by widespread terminations and early retirements, including large percentages of their news divisions. Network veterans characterized these shifts as traumatic; some claimed they significantly affected traditional processes of producing news and entertainment programming.[70] Intense pressure from new forms of competition did demand greater efficiency in operations if networks were to survive, much less prosper. Investors depended on astute fiscal stewardship. But the networks' manner of responding with abrupt cut-backs drew acerbic allegations of forsaking their media birthright.

Beyond specifically legal guidelines (described in chapter 11), the broadcast/cable manager's sense of responsibility helps assure that programming reflects the community's pluralistic population (including minorities) and viewpoints—with impartiality and fairness, striving for accuracy and reliability in presenting news and other programming. Personal ideals and company policy can be incorporated into internal station guidelines, coupled with professional associations' broad statements of principle (such as by the Radio and Television News Directors Association and the National Association of Broadcasters in its 1990 statement of standards). Those are vehicles for exercising responsibility in matters involving hard and patient management decision making, including problems of access, fair and unfair criticism, pressure groups, and less direct pressures by large businesses and government—all of which relate to program decisions, staff conduct, news coverage, editorials, and business practices generally, including sales and commercials.

A manager's sense of the community's tastes, cultural level, needs and aspirations contributes to a sound appraisal of the station's proper balance of service. Those kinds of concerns, expressed in responsibly supervising a station's policies and programming practices, will at the same time fulfill the manager's duty on behalf of the owner-licensee's legal obligations to serve the "public interest, convenience and necessity."

ATTITUDE TOWARD WORK

Some people visualize the manager's assignment as an easy one. They seldom see her when she is at the station, and she arrives and leaves at irregular hours. Her office is usually carpeted and well furnished. She enjoys personal and social amenities. Her secretary or other staff person often seems to be in command of the station when she dines with associates, plays golf, or goes out of town. Daily on-air operations do not require her direct involvement. On the surface, a manager may seem to devote less time to work than to what looks like leisure.

Most station managers *do* have "fun," but a large part of their enjoyment comes from making decisions and solving problems. Both fun and solving problems were emphasized by George Gillett, chairman and CEO of his own multistation company (which also owned the Vail ski resort and a meatpacking firm). Managers work longer hours than their employees and carry work with them when they go home at night or leave on business trips. The informality in radio, TV, or cable operations can be deceiving. Pressures from stockholders, owners, staff personnel, the audience, government agencies, and critics are so constant, and changes in the industry so continual, that the manager is preoccupied during most waking hours with thinking about the station and its problems.

Former Federal Communications commissioner Robert Wells, himself a radio broadcaster, asked broadcast educators to put these questions to college students considering careers in radio and television:

> Do you like to work peculiar hours? Do you enjoy being in the public eye and being criticized with some regularity—not just by your boss or fellow workers, but by every citizen in town? Do you enjoy holding a position which brings with it great responsibility? Can you wonder on some days whether anybody appreciates anything you do without getting upset? [71]

He noted that broadcasting's social, economic, and political importance makes those who go into the field forsake the comfortable category of "ordinary citizen" and assume responsibilities far greater than in most careers. He cautioned:

> If they don't feel some sort of dedication, they probably are in the wrong business. The job will become more and more demanding. That is the history of broadcasting in the 50 years since its inception. But the very fact that broadcasters have continued to accept these responsibilities and to improve their techniques and service has made them vulnerable to criticism, because people expect more and more and the pressures mount. There is no indication that this will change.

A general public impression of broadcasting is that it is a "glamour" business. Yet, any description of the manager's position could hardly include that term. The job is basically hard work. It is, in effect, a demanding way of life that has to be lived with few genuine breaks or relief from daily pressure. That is the way a successful manager wants it!

Related to attitude toward work, two other distinct assets for managers are an even temper and the ability to control extremes of emotion.[72] When successes come their way or when they suffer reverses, they must be able to react with an outward appearance of stability and professionalism. A sense of humor provides helpful balance for many problems encountered. By using humor with sensitivity, managers can often relieve stress in other people as well as themselves and can defuse a tense situation. Successful executives seem able to develop a remarkable facility for laughing at themselves and for being amused at their own mistakes. Amused, but not too self-forgiving, either.

B. Duties of the Manager

The varied duties and activities of managing radio, television, and cable properties are discussed throughout the chapters of this book. Preceding and accompanying all other duties is the essential managerial task of planning goals, objectives, and operational procedures.[73]

Policy Planning

All broadcast and cable operations need well defined plans, both short-term and long-range.[74] Because these immediate and ultimate targets apply to every phase of a station's operation, the manager's position offers the best perspective for evaluating them. Key staff people can supply important data. Certainly their opinions should be considered carefully (as stressed in Theory Z and the V Theory). But management should make all final decisions regarding policy.

Effective planning involves forecasting. Most station employees work in a "here and now" atmosphere; their assignments for the most part involve today's deadlines or those of tomorrow or next week. It is imperative that someone take the long view and make accurate predictions about the station's future. As the person most directly responsible to station owners, the manager is expected to provide answers concerning long-range business prospects and their probable effect on the station's future. He must be aware of trends and signs in program popularity, advertisers' expectations, market research, and new developments in technical equipment. He is expected to analyze and project this knowledge into future plans so the station can lead, rather than follow, its competition.

Wise planning takes into account all possible eventualities. Without such planning, unexpected or unforeseen events can cause temporary chaos. Valuable time is saved by considering in advance specific steps to be taken in major emergencies. Such crises might include a tornado's destruction of a transmitting tower or loss of a key staff member in an accident. "Pre-planning" may sound redundant, but it underscores a good concept. Pre-planned actions, while never entirely adequate in emergencies, are preferable to spontaneous decisions made while in a state of shock or under heavy emotional stress.

Management should make plans clear to employees. A media company functions as a closely interrelated team, making such understanding particularly necessary. Every staff member should believe in management's objectives and have keen interest in working toward achieving them. Objectives are clarified by policy guidelines for dealing with clients and competitors, the public, and in conducting business properly (e.g., no padded expense accounts, "payola" in cash or drugs, double billing, or unauthorized discounting of rate cards). Well conceived planning also precludes employees' misperceiving themselves as indispensable; it reduces personality conflicts, displays of temperament, and damaging internal politics in the organization because all share an objective vision and goal.

The principal purpose of planning is to produce maximum return on investment in the company, while serving the community that is the station's audience. Achieving that goal depends on a clear policy, a well organized plan of attack, and a dedicated staff.

Adequate planning is as essential for small radio-TV stations and local cable companies as for large stations and media groups, networks, and cable's multiple system owners (MSOs) and satellite program services. Stations of comparable size have basically the same type of facilities and equipment. The important differences between them result from the people staffing each and how their activities are coordinated, guided, and challenged. Consonant with the V Theory of managing, the manager must plan carefully how to select the persons on his staff and motivate them for effective activity. The general manager or station manager must select key people in middle management (department heads in news, programming, engineering, sales) who are competent in their specialities as well as with administrative duties.[75]

Furthermore, the manager must be able to plan, direct, and control his own use of time.[76] He must emphasize supervising personnel, not personally looking after procedures and physical matters. He should plan to spend most time in creative activity and a minimum on merely routine work, leaving a cushion of unallocated time for emergencies.[77] A good manager is able to manage his own affairs and himself. "The difference between good planning and poor planning is the difference between order and confusion, between things done on time and not being done on time, between cooperation and conflict, and between pleasant working conditions and a workday full of discord."[78]

Owners of stations and cable operations need to give their managers authority to make whatever changes are necessary whenever they are needed. In a business operating as fast as mass media, owners should not throttle managers by requiring them to obtain superiors' approval for decisions, even major ones. An owner—whether individual or a board of directors representing stockholders—should regard the manager as fully in charge, authorized to act in the owners' behalf, with only courtesy consultations or informing about decisions already made.

When managers change, the new executive should expect to draw his own plans for the station's future, within the context of the macro-intention of the company and its owners. This same privilege is accorded every new president, new governor, and new manager of a baseball team. That is why a person is elected or appointed: to take charge (without excluding Theory Y or Z or V contexts of participatory collaboration).

The process of planning calls for the most realistic thinking and projections possible. When appointed, a manager assumes responsibility for service and profit; so anything short of realistic planning is inexcusable. Some radio and television stations unfortunately operate with highly unrealistic goals; others lack clearly defined objectives. Those stations deserve the negative results they get in audiences, sales, and profitability.

Variables that require new decisions and even altering policy include: shifting radio formats, TV/cable schedules, or individual programs; changing department heads, news personnel, or on-air personalities; adjusting rate structure or selling methods to respond to shifting markets; revising office procedures;

improving facilities and equipment; responding to revised FCC regulations and court interpretations of communication law. Most such decisions must be made under the pressure of time.

The manager's first important obligation, then, is to determine goals, including policies and procedures, for this radio or TV station or cable system in this particular market. Some operational practices are standard throughout the industry and can be borrowed from other stations. (The National Association of Broadcasters recounted concrete examples in its 150-page book *Radio in Search of Excellence: Lessons from America's Best-Run Radio Stations*.) But the committed manager will want all important decisions about policies and practices to be custom made for this specific station.

THE STATION POLICY BOOK

Once formulated, station policy needs to be put in writing and copies given to each employee and to the owners. Some stations also circulate selected portions of their policy statements to various community groups that might be interested and affected.

A written station policy: (a) promotes greater efficiency; (b) reduces chances for misunderstandings; (c) eliminates repeated decisions about routine matters; (d) provides time for managers to concentrate on other significant and distinctive matters; (e) sets standards for treating all employees equitably; and (f) offers a guide that, even if not central to the total station operation, may be very important to employees as individuals.[79] Unless management promotes it honestly and discreetly, of course, such a manual can be misused. An unnamed top manager objected to policy manuals: "Let's call them what they are. They aren't rule books. They're *excuse* books. Nobody reads the damned things unless there's a screwup. And then they only read them to figure out who to blame the foul-up on."[80] Lists of rules rigidly enforced can also stifle creativity and innovative risk-taking crucial for successful media operations, not only in programming and sales but in less visible operations such as bookkeeping, engineering maintenance, switchboard, and reception. The Visa company's former chairman cautioned: "Substituting rules for judgment starts a self-defeating cycle, since judgment can only be developed by using it. You end up with an army of people who live by rote rather than reason, and where reason cannot be depended upon."[81]

But without station policy committed to writing, a manager devotes disproportionate time to decisions about relatively minor issues. With clearly defined policy, many of these instances need never arise. When they do occur, they can be settled quickly by referring to explicit station policy. Printed guidelines, consistently supported by management, offer an objective means of monitoring and assessing performance at all levels of the company. It reflects "management by objective" by stating clearly in advance the criteria by which personnel are evaluated.

For instance, in the absence of a policy on "moonlighting," plans for rotating announcers' shifts may be upset due to a second job that an announcer is holding away from the station. An employee may be unfairly reprimanded for taking excessive sick leave if it has never been made clear just how much sick leave is "excessive." Staffing the station on holidays can be a problem unless employees know which people are expected to work and why. Station clients may complain about improper treatment on the telephone by station employees if no guidelines of telephone etiquette have been formulated. Cleanliness can be a continual problem in various offices, restrooms, or the reception room if no rules exist. The manager may note that coffee breaks for office personnel tend to get longer, that conversations around the water cooler tend to increase, that smoking continues in restricted areas, that more and more employees request to use station equipment, that official company stationery is used for personal correspondence, that some people on the staff engage in long personal conversations on the telephone during working hours, that the only way to make sure letters replying to listeners/viewers are appropriate is to read them before they are mailed. In such cases, a manager can spend far too much time issuing memoranda that have been circulated before; the same problems recur over time involving different people. Instead, once employees are made aware of operating policies, most accept and follow established rules and practices—as long as they are consistent and apply to all equally.

Policy statements can cover a wide range of employee interests, including job descriptions, wages and salaries and the bases for raises and promotion, rules concerning vacations and other fringe benefits (see chapter 5 on personnel). Complete job descriptions define each staff position's limitations and responsibilities and explain the relationship of each position's work to that of other people on the staff as well as to other station functions. Job descriptions need to be clearly framed. It is more efficient to define functions and then find people to match them who will be good team members than it is to hire people and then attempt to adapt the jobs to them. Station positions should be so well defined that whenever an individual is replaced, the basic assignment for that post remains unchanged.

Written policy is particularly valuable for new personnel. Too often, new employees at radio and television stations learn their job routines and absorb the company's philosophy very inefficiently—by talking with other employees, by making mistakes and having to be corrected, or by asking questions. Partly because small staffs work closely in informal surroundings, broadcasters often neglect indoctrination procedures for new people. Minimally, a statement of station policy, including various job descriptions as well as station rules, should be given to each candidate for a position before being hired. This policy statement can be regarded as part of the employee's contract. A prospective member of a staff has every right to know all the conditions of employment before accepting a position.

It is impossible to compose a complete station policy book when first developing it. New situations will constantly arise and conditions will frequently change. Provision can be made for regular revisions. The most carefully defined rules cover most cases; but there will always be exceptional circumstances requiring separate decisions. Policies should be stated broadly as principles that permit a limited amount of flexibility. Any exceptions can then be explained without embarrassment or criticism. When overriding written policy, a manager must exercise extreme care to treat staff members fairly.

Some media companies compile a few typewritten pages outlining objectives and procedures; others prepare loose-leaf binders (facilitating insertion of memos updating selected provisions); still others provide printed booklets offering considerable detail. Preparing a satisfactory policy statement is a time consuming process. But the investment improves management efficiency as well as station morale.

IMPLEMENTING POLICY

A manager's human qualities are extremely important in implementing station policy successfully. The manager's administrative skills and clarity in communicating ideas will determine how they are accepted by owners, employees, advertisers, and the public (where applicable, as in changing program formats and on-air personalities or newspersons). Ineffective communication of the true intent and meaning of new plans, or an authoritarian approach to enforcing their acceptance, can result in employees' complying at best out of a sense of obligation instead of with enthusiastic commitment. They can react defensively or with antagonism. To achieve positive results, the manager's leadership and communication skills are put to the test. The manager must be a good salesperson and coach as well as a good listener.

In communicating new plans to the staff, it is important to explain why changes are being made, and how they will specifically benefit the staff and station.

As stressed in Theory Z and the V Theory of managing, implementation should never be exclusively a downward process with management initiating all action. An operation following that procedure loses many good ideas and weakens staff support for final decisions. All personnel should be encouraged to contribute suggestions for improving the station; every employee making a workable suggestion should be given full credit for that contribution. Staff members often know deficiencies or changes that might be necessary; but the manager may be unaware of them. Some employees feel reluctant to report irregularities; others may be inclined to give only a partial accounting of adverse facts. It is important for the manager to have the confidence of personnel, while making clear her desire to be kept informed of any weaknesses in station staffing or operation.

Often effective for pooling information and exploring decisions is the conference approach, whereby an administrator gathers together junior executives (department heads) to discuss problems and issues.[82] When several minds collaborate, possible alternatives can usually be considered more effectively. Nevertheless, although groups can often craft apt policy, only individuals can implement that policy. The station manager provides the insight that brings issues to the conference table, guides objective appraisal of the problems, and has authority to administer adopted policy measures.

When duties are delegated to others, the limits of authority—including responsibility and accountability—must be explicitly understood. Station policy should clearly state the expectations of all administrative and supervisory roles as well as their relationships to the duties and responsibilities of other station executives. An organization chart helps clarify lines of supervision and accountability (see chapter 4). General managers and station managers need to make special efforts to keep department heads supplied with all information and support resources related to areas of their jurisdiction.

Delegating authority is required, of course, for executive efficiency. Station managers must be freed from details so they can concentrate on major policy and on duties central to the managerial position. Those duties constitute the remaining chapters of this book: working within media systems and among middle managers (chapter 4); attending to motivation, rewards and working conditions of station *personnel,* including supervising office staff (chapter 5); continually studying the station's market and *audience* (chapter 6); working closely with the director of *programming* (chapter 7 plus chapter 8 from a national perspective); *sales* manager (chapter 9); and the chief *engineer* (chapter 12); preparing overall budget and maintaining control over income and expenses in order to earn *profits* (chapter 10); continually acquiring information about *government* concerning proposals, rules and regulations of the Federal Communications Commission—maintaining relationships with the FCC and other officials at federal, state, and local levels (chapter 11).

Some of these areas may be combined in a cable system or a national satellite cable program service, others are replaced in public noncommercial broadcasting (e.g., development and fund-raising in place of commercial time sales). Networks may differ from local stations in executive designations (vice-presidents or divisional presidents instead of directors or managers). But these essential categories persist in radio, TV, and cable media in the 1990s. Many of them also pertain to specialized corporate video operations.

These areas continually interact in any mass media company. By way of concrete example, McLarney and Berliner's seven steps of problem-solving technique might be applied to a typical broadcast context.[83] First, *clearly define the problem;* the sales manager may note difficulty in finding clients (directly or through their agencies) to buy spot commercial time available in the local schedule in early evening of a given night of the week. The next step is to *gather the information,* perhaps by having a researcher in programming or sales get demographic data on that market, coupled with comparative statistical data on competing stations' ratings for that time period; further, the program director or operations manager may have her people provide information about feasibility of scheduling local program production in their own studios if that is a possible option. Third, the manager and her staff must *interpret the information,* in light of the sales manager's initial definition of the problem, the program/operations manager's appraisal of production scheduling (if part of a potential solution), and the chief engineer's assessing feasibility of studio facilities and crews for mounting more programming or commercial production in the studios. Fourth, the manager collaborates with other personnel to *develop specific solutions,* providing alternate ways to meet the sales problem. Fifth, the manager must *select the best practical solution,* in the context of all the staff's input. The actual exercise of that decision, to *put the solution into operation,* is the responsibility of key people in departments affected by the decision. For example, if the solution chosen is to buy filmed/videotaped programming for the schedule, there is little problem for the chief engineer; but if the choice is to produce more local news or even entertainment programming, in addition to traffic and programming considerations, the engineering department must determine how best to supply staff, crews, and facilities. Finally, the seventh step is to *evaluate the effectiveness of the solution;* again, the manager is provided with data from the department heads so together they can appraise the results. In some instances, the first stages of this process might find data coming to the manager from a competent staff to be almost self-determining, so the appropriate decision is apparent. On the other hand, the option might be only one of several reasonable alternatives, so the staff may be asked to determine among themselves their composite recommendation to management. Thus a manager's primary role is to support the staff to work creatively on the problem. The manager's actual appraising of all information provided and then selecting one option from among possible alternatives depends on thorough staff input; the manager's decision may in fact simply endorse their recommendation.

IDEALISM AND REALITY

Theoretical analysis deals with essential characteristics, often abstract, but the pragmatics of real-life situations often modify those idealized concepts. Mintzberg emphasized the gulf between what he called "folklore" of classical theories and actual fact about managers' actual duties.[84] The day-to-day job of typical managers in countless studies of manufacturing and service industries found that managers

(a) have little time for reflection and systematic planning, with their intense pace of activity "characterized by brevity, variety, and discontinuity"; (b) have a lofty disengaged perspective through delegating replaced by daily handling of exceptions, routine negotiating, processing information, and ceremonial functions (including visitors); and (c) synthesize "soft" information through endless rounds of meetings and telephone calls, rather than preside over "hard" data processed by elaborate information systems (despite computer technology). Mintzberg concluded that management, instead of being a science or profession, depends more on elusive *judgment* and *intuition,* pressured by overwork often leading to superficial reaction to increasingly complex forces within and without the organization. One index of this load on middle and top managers is their typical workday of ten to twelve hours or longer.[85]

A study of managers in manufacturing reported they spent 28% of their time supervising and 20% in planning; they divided the other half of their time variously among coordinating, evaluating, investigating, negotiating, staffing, and representing.[86] Lavine and Wackman's investigations found media managers spend about 15% to 20% of their time "in true leadership and change," about 25% in supervising, and more than 55% proportioned roughly between planning and supervising as in the previous listing.

A major-market TV news director described more concretely "what has to happen *inside* you, if you are ever to succeed over the long haul as a manager in broadcasting."[87] Expect frustration, but continually correct unanticipated daily problems when they hit. Be convinced there are no failures, only results by staff members which should be acknowledged and rewarded. Rely not on high-tech equipment but on people; coach them to growth. Set reachable goals to progressively higher levels; don't merely seek to catch the competition (especially when #1 or even #2 may simply not be beatable under given circumstances) or merely to remain ahead of the others if you are on top. Have a passion and will to achieve, not merely a hope or wish. Keep focused on the top priority—not paperwork and meetings, but (for news) the content of the next broadcast, or sales goals, or profitability linked with substantive program service. Live and work in the present, rather than regretting the past or worrying about the future, because that's what you are immediately able to control. Build your own happiness by keeping perspective ("don't sweat the small stuff, and remember, it's all small stuff").

These useful observations do not negate theoretical constructs or undermine managers' optimistic efforts to move toward the ideal. They do caution against naive idealism that ignores real-world complexity. They prompt more sophisticated management theories reflecting realistic ambiguity, such as contingency theory with variables in situational, management, and performance criteria.[88]

—∿— CHAPTER 3 NOTES —∿—

1. Peter F. Drucker, *Managing in Turbulent Times* (New York: Harper Business edition, 1993), 4.

2. Robert Wangermee, "Towards Modern Management in Broadcasting," *EBU Review: Programmes, Administration, Law* 23:6 (November 1972): 30–38. See his summation in *Towards Modern Management in Radio & Television: International Colloquy* (Brussels: European Broadcasting Union, 1974), 459–471.

3. In the early 1990s TV station budgets averaged 33% for administration and general expenses, radio stations 41%—contrasted with programming/production on which TV spent 28% (plus another 20% for news), radio stations 24% (plus 3% for news). But independent TV stations devoted 51% to programming/production (plus 5% for news), and only 24% to adminstration/general expenses. Data from various sources are noted in ch. 9, Table 9.1, and ch. 10 at note 34.

4. "Network TV: Between a Rights Fee and a Hard Place," *Broadcasting,* 22 October 1990, 31.

5. See William J. McLarney and William M. Berliner, *Management Training: Cases and Principles,* 5th ed. (Homewood, Ill.: Richard D. Irwin, Inc., 1970). In outlining principles of effective management that apply to broadcasting, we draw heavily on this work. It is a synthesis of good sense and research findings drawn from a wide range of reliable sources in the field of management theory and science. See also Harold Koontz and Cyril O'Donnell, *Essentials of Management,* 2nd ed. (New York: McGraw-Hill, 1978), especially Part 2, pp. 55–172.

6. Henri Fayol, *General and Industrial Management,* trans. Constance Storrs (New York: Pitman Publishing, 1949); McLarney and Berliner, *Management Training,* 13–14, cite Fayol's five-fold division of management activities: planning, organizing, directing, coordinating, and controlling.

7. Listed by Howard F. Merrill, *Classics in Management* (New York: American Management Association, 1960), 217–223.

8. Peter Drucker, *The Practice of Management* (New York: Harper & Row, 1954), 350. See also: Peter F. Drucker, *Management: Tasks, Responsibilities, Practice* (New York: Harper & Row, 1974).

9. See Douglas McGregor, *The Human Side of Enterprise* (New York: McGraw-Hill, 1960), 216–221; cf. Warren Haynes and Joseph L. Massie, *Management: Analysis, Concepts, and Cases* (Englewood Cliffs, N.J.: Prentice-Hall, 1961), 9.

10. J. Leonard Reinsch and Elmo I. Ellis, *Radio Station Management,* 2nd rev. ed. (New York: Harper & Row, 1960), 9; cf. Gene F. Seehafer and Jack W. Laemmar, *Successful Television and Radio Advertising* (New York: McGraw-Hill, 1959), 542.

11. Vernon A. Stone, "Status Quo," *RTNDA Communicator* 48 (August 1994): 16–18. He reported a leveling off of the rate of increase; that tempered his projection in 1989 that within twenty years almost half of all news directors in broadcasting would be women; Vernon Stone, "Women Gain as News Directors," *RTNDA Communicator* 43 (December 1989): 18–20.

12. Theo Haimann, *Professional Management* (Boston: Houghton Mifflin, 1962), 446.

13. Alan C. Filley and Robert J. House, *Managerial Process and Organizational Behavior* (Glenview, Ill.: Scott, Foresman & Company, 1969); pp. 391 ff. provide extended treatment of theories of leadership.

14. McGregor, *The Human Side,* 182.

15. Fiedler's situational leadership theory is summarized by Lester R. Bittel, *The McGraw-Hill 36-Hour Management Course* (New York: McGraw-Hill, 1989), 173; see Fred E. Fiedler, "The Contingency Model," in Harold Proshansky and Bernard Seidenberg, eds., *Basic Studies in Social Psychology* (New York: Holt, Rinehart & Winston, 1965), 538-551; also Fred E. Fiedler, "Validation and Extension of the Contingency Model of Leadership Effectiveness," *Psychology Bulletin* 76 (August 1971): 128–148. See also Fred Luthans and Todd I. Stewart, "A General Contingency Theory of Management," *Academy of Management Review* (April 1977): 189; cited by Richard M. Hodgetts, 356–357.

16. David L. Bradford and Allan R. Cohen, *Managing for Excellence: The Guide to Developing High Performance in Contemporary Organizations* (New York: John Wiley & Sons, 1984), 100-101; they cite R. T. Pascale and A. G. Athos, *The Art of Japanese Management* (New York: Simon & Schuster, 1981). For a similar treatment of theories conjoined with practical exercises and worksheets to apply them, see Lance H. K. Secretan, *Managerial Moxie: The 8 Proven Steps to Empowering Employees and Supercharging Your Company* (Rocklin, Calif.: Prima Publishing, 1993).

17. McLarney and Berliner, *Management Training,* 526–527.

18. See Burleigh Gardner, *Human Relations in Industry* (Homewood, Ill.: Richard D. Irwin, Inc., 1955), 379. Cf. Elizabeth Marting and Dorothy MacDonald, eds., *Management and Its People* (New York: American Management Association, 1965), 236–237.

19. Phrasing introduced by William F. Christopher, *Management for the 1980s* (New York: AMACOM/American Management Association, 1980), 124–129.

20. William B. Eddy, *Behavioral Science and the Manager's Role* (NTL Institute for Applied Behavioral Science, 1969), 114.

21. Assertions in this paragraph are supported by: Harold Koontz and Cyril O'Donnell, *Principles of Management: An Analysis of Managerial Functions,* 4th ed. (New York: McGraw-Hill, 1968), 615; Henry L. Sisk, *Principles of Management* (New York: South-Western Publishing, 1969), 393, 404; David Ewing, *The Managerial Mind* (London: Collier-MacMillan, 1964), 31, 156; Haimann, *Professional Management,* 446; and Robert Dubin, *Human Relations in Administration* (New York: Prentice-Hall, 1951), 101.

22. Cf. Thomas J. Peters and Robert H. Waterman, Jr., *In Search of Excellence: Lessons from America's Best-Run Companies* (New York: Harper & Row, 1982), 89–118 (Chapter 4, "Managing Ambiguity and Paradox"). See also Tom Peters, *Thriving on Chaos: Handbook for a Management Revolution* (New York: Knopf [distributed by Random House], 1987).

23. Drucker, *Managing in Turbulent Times,* 4.

24. See Christian Williams, *Lead, Follow or Get Out of the Way: The Story of Ted Turner* (New York: Times Books, 1981).

25. William S. Paley, *As It Happened* (Garden City, New York: Doubleday, 1979), 351.

26. National Association of Broadcasters, "Twenty Guidelines for Leadership," *Info-Pak* (NAB Research & Planning newsletter/report), December 1989/January 1990, pp. 1–2; adapted from Perry M. Smith, *Taking Charge: Making the Right Choices* (Garden City Park, New York: Avery Publishing Group, Inc., 1988).

27. James MacGregor Burns, *Leadership* (New York: Harper & Row, 1978), 13–35; cited by Peters and Waterman, *In Search of Excellence,* 81–86.

28. Burns, 20; quoted by Peters and Waterman, 82.

29. McLarney and Berliner, *Management Training,* 12.

30. See Elizabeth Marting and Dorothy MacDonald, eds., *Management and Its People* (New York: American Management Association, 1965), 236–237; also Henry Clay Smith, *Psychology of Industrial Behavior* (New York: McGraw-Hill, 1955), 420–422.

31. Norman Maier, *Creative Management* (New York: John Wiley & Sons, 1962), 302–303.

32. McLarney and Berliner, *Management Training,* 519–520.

33. Laurence J. Peter, *The Peter Principle* (New York: William Morrow, 1969; Bantam edition, 1970).

34. McLarney and Berliner, 82; see Burns, 172.

35. Quoted by Filley and House, *Managerial Process,* 239.

36. Thomas J. Peters and Nancy Austin, *A Passion for Excellence: The Leadership Difference* (New York: Random House, 1985), 378.

37. Peters and Waterman, Jr., *In Search of Excellence,* 26, 85.

38. See Norman Maier, *Creative Management,* 41–42.

39. Milton L. Blum, *Industrial Psychology and Its Social Foundations* (New York: Harper & Brothers, 1949), 306.

40. See Maier, 43-64; also Harold Leavitt, *Managerial Psychology* (Chicago: University of Chicago Press, 1964), 249–250.

41. Sisk, *Principles of Management,* 440–441; see Ewing, *The Managerial Mind,* 36.

42. Eddy, *Behavioral Science,* 75.

43. Peters and Austin, *A Passion for Excellence,* 378–380; also Chapter 2, "MBWA: The Technology of the Obvious," 8–33.

44. This was particularly evident in the social-activist 1960s. See Charles E. Winick, "The Television Station Manager," *Advanced Management Journal* 31:1 (January 1966): 53–60. Winick's study reported the average TV station manager as "actively involved" with more than forty community social and welfare groups; one manager was cited as serving on the board of no fewer than seventeen different civic organizations.

45. George Frank, *The Menace of Management Obsolescence: The Job in a Changing World* (New York: American Management Association, 1964) listed representative topics for analyzing management recruits; cited by Ted W. Engstrom and R. Alec Mackenzie, *Managing Your Time: Practical Guidelines on the Effective Use of Time* (Grand Rapids, Mich.: Zondervan Publishing House, 1967), 41–43. Topics include: (economics) discounted capital, income statistics, economic indicators, marginal productivity; (finance) turnover ratios, retirement and replacement programs—including depreciation and amortization—cost of money, rate of return, current value of money, debt servicing, music and program licensing; (marketing) designing and evaluating questionnaires and surveys, consumer panels, sampling, market penetration analysis, sales forecasting; (organization) responsibility charting, organizational planning, management audit; (personnel) job evaluation, wage/salary survey, wage regression curve, executive compensation projection, programmed instruction, union negotiations; (operations research) linear programming, inventory models, economic order points, queuing; (data processing) computer language, random access storage, digital/analog computers, automated logging/traffic/billing; (statistics) probability theory, sampling, descriptive/inference statistics, analysis of variance, product-moment correlation, Chi-square analysis; (industrial engineering) work-sampling study, I. E. Schematic models, production cost analysis, PERT and CPM—program evaluation and review technique and critical path method, and others. This underscores the need to know far more than how to plan facilities and personnel for playing records and scheduling syndication packages or cable channel services.

46. These questions were listed back in 1973 by trade magazine editor Sol J. Paul, "Publisher's Letter," *Television/Radio Age* 3 September 1973, 10.

47. Bernard Gallagher, "The Gallagher Report," 23 February 1972 (a weekly sales and marketing newsletter published in New York).

48. Among overlooked areas of budgeting is the matter of efficient communication and bookkeeping. The cost of paperwork alone, including computerized files, contributes to office expenses and clerical salaries. It has been estimated that paperwork and computer processing for $1 million of ad billings costs an agency more than $40,000 (compared with less than $10,000 in the early 1960s). Extra paperwork that requires hiring an extra clerk at only $5 an hour ($200 a week salary, or $10,400 a year) forces the company to do $200,000 additional business at a 5% net profit to pay that salary. (Source: Day-Timers, Inc., citing studies by National Records Management Council, Stanford Research Institute, Dannell Corporation of Chicago, consultant Roben A. Shiff.) The cost for dictating and transcribing a letter was estimated at 30 cents in 1930 by the Dartnell Institute of Business Research; the Institute appraised the cost in 1984 at $8.10 per letter (based on annual salaries of $33,500 for the manager and $15,285 for the secretary); cited in *U.S. News & World Report,* 8 July 1984, 50. By 1990 the cost of getting out a dictated business letter was estimated at over $9.00. Management must be alert to cumulative costs from administrative paperwork, attempting to cut those costs through greater efficiency in communicating.

49. A detailed survey reported in 1987 that almost all general managers (161 of 174) and most program directors (140 of 181) and news directors (120 of 155) responding from all size markets read *Broadcasting;* half the general managers read *TV/Radio Age* (ceased publishing in 1990) and *Electronic Media;* about one third of them read *Variety, Broadcast Week,* and *Advertising Age.* Marjorie J. Fish and R. C. Adams, "Composite Profiles of Key Television Managers" paper presented at Western Speech Communication Association, Salt Lake City, Utah, February 1987), p. 32. An early survey (Winnick, 1966) reported that 88% of TV station managers read magazines about current events; only one-third read business magazines; and one-third read "intellectual" or conceptual magazines such as *Scientific*

American. Gallagher's survey (1971) found nine out of ten radio station managers read *Broadcasting* magazine weekly and one-third read the *Wall Street Journal* regularly; one in five read *Business Week* (*Time, Newsweek,* and *U.S. News & World Report* were not listed). Among small-market radio managers, Bohn and Clark (1972) also cited *Broadcasting* as the most read magazine and the *Wall Street Journal* as the "top newspaper choice."

50. For detailed theoretical and practical analysis of the process of management decision making, see McLarney and Berliner, *Management Training,* 20–26; also Filley and House, *Managerial Process,* 101–130.

51. Harold Leavitt of Stanford, cited by Peters and Waterman, Jr., *In Search of Excellence,* 49; see Chapter 2, "The Rational Model," 29-54.

52. Rosalind Loring and Theodora Wells, *Breakthrough: Women into Management* (New York: Van Nostrand Reinhold Company, 1972), 61.

53. Peters and Austin, *A Passion for Excellence*, xix–xx; see also Chapter 3, "Integrity and the Technology of the Obvious," 34–50.

54. Cf. Kyle Haselden, *Morality and the Mass Media* (Nashville, Tenn.: Broadman Press, 1968), Chapter 10, especially p. 185: "Authentic morality has its focus in people, its objective in the transforming of people into persons, its habitat in freedom, its criterion and energy in love, and its source in God." For a thoughtful, homey, sometimes simplistic, exhortative perspective—brief and readable—see Kenneth Blanchard and Norman Vincent Peale, *The Power of Ethical Management* (New York: William Morrow, 1988).

55. William L. Rivers and Wilbur Schramm, *Responsibility in Mass Communication,* rev. ed. (New York: Harper & Row, 1969), Chapter 7, 190–233.

56. William F. Lynch, S.J., *The Image Industries* (New York: Sheed and Ward, 1959), 20. Cf. Norman Corwin, *Trivializing America* (Secaucus, N.J.: Lyle Stuart, 1983).

57. Lynch, 150; cf. 113, 145. That this pertains to small-market radio and TV stations as well as national television production centers is attested to by Robert H. Coddington, *Modern Radio Broadcasting: Management & Operation in Small-to-Medium Markets* (Blue Ridge Summit, Pa.: Tab Books, 1969), 10–14.

58. Most Rev. John L. May, chair of Communications Committee, U.S. Catholic Conference, "Commentary on the Pastoral Instruction on Social Communication," June 3, 1971; *Communications: A Pastoral Instruction on the Media, Public Opinion, and Human Progress* (Washington, D.C.: U.S. Catholic Conference, 1971), p. ix. This booklet also reprinted the complete *Pastoral Instruction.* . . . by the Vatican's Pontifical Commission for the Means of Social Communication; see note 66.

59. Sydney W. Head, *Broadcasting in America,* 2nd ed. (Boston: Houghton Mifflin, 1972), 471.

60. Rivers and Schramm, *Responsibility,* 241, 242; see also 243–248.

61. Head, 472. Cf. McLarney and Berliner, *Management Training,* 672–676, 682–684.
    Illustrating the discrepancy between how younger people assess businesspersons' sense of ethics, and what the latter themselves think, are data from three decades (1968-1996) of college students studying mass media in the Midwest, Far West, Northeast, and South—compared with a national survey (1961) of business executives and four samplings (1977, 1978, 1981, 1993) of broadcasters. All were given the same list of phrases from which to select: (a) two phrases best representing "what ethical means" for them, and then (b) two choices most reflecting *what they presumed "ethical" means for businesspeople.* Students at the University of Detroit, Southern California and Loyola Marymount (Los Angeles), Colgate (Hamilton, N.Y.), and the University of Alabama (Tuscaloosa) responded similarly throughout the quarter-century. Of responses from 1,589 students studying broadcasting and mass communication at private and public colleges over that twenty-eight-year span, more than a quarter (27%) speculated that, for those in business, "ethical" would mean "about the same as legal" (another 26% put that as their second choice); one-fifth (19.4%) said businesspersons would consider ethical whatever was "customary in society" (another 22% put that as an alternate choice), while one-fifth (19.7%) presumed businesspeople's ethical norm would be "what does the most good for the most people" (15% added it as second choice). As for their own personal choices about the meaning of "ethical," students consistently gave

TABLE 3.1
"What Ethical Means"

BUSINESS-PERSONS' ACTUAL CHOICES		LIST OF POSSIBLE CHOICES ABOUT WHAT "ETHICAL" MEANS	STUDENTS' PERSONAL CHOICES		STUDENTS' ASSUMPTION OF WHAT BUSINESS PEOPLE WOULD CHOOSE	
1st	*2nd*		*1st*	*2nd*	*1st*	*2nd*
50	8	What my feelings tell me is right	566	348	113	131
25	14	In accord with my religious beliefs	149	121	3	7
18	15	Conforms to the "Golden Rule"	120	173	44	58
3	7	What does the most good for the most people	463	345	312	240
3	6	Customary behavior in society	125	253	308	349
1	1	Corresponds to my self-interest	24	40	178	165
0	1	About the same as legal	31	106	430	410
0	1	Contributes most to personal liberty	65	149	20	62
0	1	What I want in that particular situation	33	56	177	186
100	54	Total Number Responding*	1,587	1,591	1,586	1,592

*Not all participants responded to both choices.

highest priority to "what my feelings tell me is right" (35.6% put it as first choice, plus another 22% as second); next came "what does the most good for the most people" (29% and 21.7%). Fewer than 2% chose for themselves "the same as legal" (plus 6.6% choosing it second), while just under 8% selected "customary behavior in society" (16% second)—the categories students most assumed for those in business).

Those data can be compared with a study of businesspeople surveyed by a Harvard scholar in management and business administration, who reported in 1968 that "what my feelings tell me is right" determined what half (50) of those executives considered "ethical," with another one-quarter (25) judging as ethical what was "in accord with my religious beliefs," and one-fifth (18) what "conforms to the 'Golden Rule.'" Not a single executive selected "about the same as legal" as their first criterion of what was ethical and only three responded "customary behavior in society." Yet college students sampled through three decades consistently conjectured those two categories as major criteria for business ethics.

Table 3.1 provides comparative tabulations from the original study by Raymond Baumhart, S.J., *An Honest Profit* (New York: Holt, Rinehart & Winston, 1968) and from students in fifty-two media courses at five universities around the country over three decades (1968–1996).

In other informal samplings, Los Angeles broadcasters in two seminars (1977, 1978), professionals from broadcast stations around the country attending a media management institute (1982), and a cluster of broadcasters from five of six U.S. regions (none from the western mountain states) at the 1993 NAB convention identified what they meant by "ethical" in the same order (though weighted differently) as students' personal selections. Consistently over that time-frame the first preference for over one-third was "what my feelings tell me is right"; another one-quarter selected "what does the most good for the most people"; and one in ten chose "customary behavior in our society." Interestingly, those broadcasters also reflected students' more cynical assumptions about *other* businesspeople: one in four projected "same as legal" and one-fifth "customary behavior," another one-fifth "what I want in that particular situation."

A related study of 1,227 *Harvard Business Review* readers reported similarly favorable self-statements about ethical criteria and behavior, coupled with negative expectations projected towards peers in the business world: Steven N. Brenner and Earl A. Molander, "Is the Ethics of Business Changing?" *Harvard Business Review* 55 (January/February 1977): 57–71.

62. Francis L. Dale, publisher, *Los Angeles Herald-Examiner,* transcript of video-taped discussion on "Ethics, Business, and Journalism" (transcript series number 10, Center for the Study of Business in Society, School of Business and Economics, California State State University, Los Angeles, May 1980), 3–4.

63. See Raymond Baumhart, *Ethics in Business* (New York: Holt, Rinehart & Winston, 1968), 46–47. Influences to make ethical decisions, ranked by averages (1 most influential, 5 least influential) of 807 respondents:

1.5	A man's personal code of behavior
2.8	Formal company policy
2.8	The behavior of a man's superiors in the company
3.8	Ethical climate of the industry
4.0	The behavior of a man's equals in the company

Influences to make unethical decisions, ranked by averages of 705 other respondents matched to the previous group:

1.8	The behavior of a man's superiors in the company
2.6	Ethical climate of the industry
3.2	The behavior of a man's equals in the company
3.3	Lack of company policy
4.1	Personal financial needs

See also Steven N. Brenner and Earl A. Molander, "Is the Ethics of Business Changing?" *Harvard Business Review* 55 (January/February 1977): 57–71, who replicated the previous study. They found among 1,200 respondents almost the same ranking for unethical influences, with "lack of company policy" moving up to second position.

64. Brenner and Molander, 63.

65. Henry O. Golightly, "Success Depends on Character," *American Way,* April 1976, 33–37. The author was president of Golightly & Company International, a management consulting firm.

66. Pontifical Commission for the Means of Social Communication, *Pastoral Instruction for the Application of the Decree of the Second Vatican Ecumenical Council on the Means of Social Communication* (Washington, D.C.: U. S. Catholic Conference, 1971), 1, 3–4.

67. L. L. Golden, "Public Relations: The Why of Xerox," *Saturday Review,* 14 August 1971, 53. Milton Friedman vigorously disagrees in *Capitalism and Freedom* (Chicago: University of Chicago Press, 1962), 133: "The view has been gaining widespread acceptance that corporate officials and labor leaders have a 'social responsibility' that goes beyond serving the interest of their stockholders or their members. This view shows a fundamental misconception of the character and nature of a free economy. In such an economy, there is one and only one social responsibility of business—to use its resources and engage in activities designed to increase its profits so long as it stays within the rules of the game, that is to say, engages in open and free competition, without deception or fraud. . . . Few trends could so thoroughly undermine the very foundations of our free society as the acceptance by corporate officials of a social responsibility other than to make as much money for their stockholders as possible."

68. Brenner and Molander, 69.

69. John Sack, *The Man-Eating Machine* (New York: Farrar, Strauss, & Giroux, 1973).

70. Critics within media chronicled how news operations were affected by the disruptive fiscal restructuring of CBS. They included CBS newspersons Bill Leonard, *In the Storm of the Eye: A Lifetime at CBS* (New York: G. P. Putnam's Sons, 1987); David Schoenbrun, *On and Off the Air: An Informal History of CBS News* (New York: E. P. Dutton, 1989); and Fred Graham, *Happy Talk: Confessions of a TV Newsman* (New York: W. W. Norton, 1990); among critical observers from print media were Peter McCabe, *Bad News at Black Rock* (New York: Arbor House, 1987) and Peter J. Boyer, *Who Killed CBS: The Undoing of America's Number One News Network* (New York: Random House, 1988).

71. Excerpts of Commissioner Well's speech to the Association for Professional Broadcasting Education, March 27, 1971; quoted by John Pennybacker, ed., *Feedback* (APBE newsletter), May 1971, 17.

72. Offering practical guidelines for understanding and correcting excesses in attitude and behavior are Matthew J. Culligan and Keith Sedlacek, M.D., *How to Avoid Stress Before It Kills You* (New York: Grammercy Publishing, 1976), and Carol Travis, *Anger: The Misunderstood Emotion* (New York: Simon & Schuster, 1982).

73. Lester R. Bittel, The *McGraw-Hill 36-Hour Management Course* (New York: McGraw-Hill, 1989); Chapters 4–12 offer succinct analysis and pragmatic exercises, including brief case studies and examples, on these points. Paul M. Hammaker and Louis T. Rader candidly address these and broader principles covered earlier, in *Plain Talk to Young Executives* (Homewood, Ill.: Dow Jones-Irwin, 1977). Less traditional treatments of managers' characteristics and duties are: Robert N. McMurry, *The Maverick Executive* (New York: AMACOM/American Management Association, 1974) who profiles not mere nonconformists but hypereffective, dynamic, idiosyncratic executives who move and shake the static status quo; Mark H. McCormack, *What They Don't Teach You at Harvard Business School* (Toronto: Bantam Books, 1984) offers "a street-smart executive's" tightly written guidelines about major concepts and processes in managing; and Richard S. Sloma, *No-Nonsense Management: A General Manager's Primer* (New York: Macmillan, 1977) distills one- and two-page aphoristic observations about management's daily challenges. See also W. L. Laney, *How to Be Boss in a Hurry* (Indianapolis: Bobbs-Merrill, 1982); and Don Beveridge, Jr. and Jeffrey P. Davidson, *The Achievement Challenge: How to Be a 10 in Business* (Homewood, Ill.: Dow Jones-Irwin, 1988).

74. General concepts of managerial planning are well provided by McLarney and Berliner, *Management Training,* 14–21, 199–228; Filley and House, *Managerial Process,* 191-210. The latter authors emphasize that planning consists of searching for alternatives, and finally selecting the optimum alternative (194–196). Establishing and implementing "overarching goals" (macro-intentionality) is treated by David L. Bradford and Allan R. Cohen, *Managing for Excellence: The Guide to Developing High Performance in Contemporary Organizations* (New York: John Wiley & Sons, 1984).

75. See Milton D. Friedland, "The Network-Affiliated Station," 61-62, and Richard B. Rawls, "The Independent Station," in Yale Roe, ed., *Television Station Management— The Business of Broadcasting* (New York: Hastings House, 1962). 77–78. See the present text's next chapter on personnel.

76. Offering extremely practical details about paperwork, filing, workday schedules, eliminating timewasters, coordinating projects, overseeing secretaries and staff as well as specific communication equipment and supplies is Stephanie Winston, *The Organized Executive: New Ways to Manage Time, Paper, and People* (New York: Warner Books, 1983). See also Engstrom and Mackenzie, *Managing Your Time.*

77. McLarney and Berliner, *Management Training,* 203. They suggest indicators for prompting a manager to review how he is misusing time: "If he is just keeping up with his work—that is, taking care of one emergency after another—or if he is just getting or not quite getting his essential job done, if he needs to be in three places at the same time, if he has to put in excessive overtime, if he has not time for self-improvement, if he has to do everything himself, or, finally, if he dare not take a day off when he is ill; then it behooves him to make an evaluation of how he is spending his time" (204). By redistributing his time he can reduce strain and excessive fatigue at his work, thus remaining fresh for the creative responsibilities unique to him as manager.

See also some helpful cautionary advice from Clarence B. Randall, "The Myth of the Overworked Executive," *The Folklore of Management* (New York: Mentor edition, 1962), 80–85.

78. McLarney and Berliner, 216.

79. McLarney and Berliner outline the role of policies and procedures (pp. 68–89), listing characteristics of good policy statements: stable, flexible, compatible, understandable, sincere, realistic, and written (71). On the other hand, Robert Townsend vehemently opposes written policy manuals; he claims "don't bother: if they're general, they're useless. If they're specific, they're how-to manuals—expensive to prepare and revise." He cautions that manipulators on the staff can misuse such manuals "to confine, frustrate, punish, and eventually drive out of the organization every imaginative, creative, adventuresome woman and man"; *Up the Organization* (Greenwich, Conn.: Fawcett edition, 1970), 129.

80. Unidentified source quoted by Peters and Austin, *A Passion for Excellence*, 250.

81. Dee Hock, quoted by Peters and Austin, 250.

82. See Roe (ed.), *Television Station Management*, 61–62, 77-78. The value of meetings is questioned by Robert Townsend, *Up the Organization*, 89–91, but he advocates that "some meetings should be long and leisurely. Some should be mercifully brief. A good way to handle the latter is to hold the meeting with everybody standing up. The meetees won't believe you at first. Then they get very uncomfortable and can hardly wait to get the meeting over with" (p. 171).

83. McLarney and Berliner, 22.

84. Henry Mintzberg, "The Manager's Job: Folklore and Fact," in Michael T. Matteson and John M. Ivancevich, eds., *Management Classics,* 2nd ed. (Santa Monica, Calif.: Goodyear Publishing, 1981), 63–84.

85. Bradford and Cohen, *Managing for Excellence,* 4; Sydney W. Head and Christopher H. Sterling, *Broadcasting in America: A Survey of Electronic Media,* 6th ed. (Boston: Houghton Mifflin, 1990), 188.

86. Mahoney, Jerdee, and Carroll's 1965 data were cited by John M. Lavine and Daniel B. Wackman, *Managing Media Organizations: Effective Leadership of the Media* (New York: Longman, 1988), 87.

87. Bob Rowe (news director, WXYZ-TV, Detroit), "The Letter: 7 Deadly Management Sins and How to Avoid Them," *RTNDA Communicator*, December 1991, 26.

88. Hodgetts, 356–357, describes the model by Fred Luthans and Todd I. Stewart, "A General Contingency Theory of Management," *Academy of Management Review* (April 1977): 181–195. Cf. Matteson and Ivancevich (1981), for reprints of John J. Morse and Jay W. Lorsch, "Beyond Theory Y," 396–407, and Fred E. Fielder, "The Contingency Model— New Directions for Leadership Utilization," 408–418.

Profiles of Middle Managers

*Guiding an organization is not an exact science. . . . [N]o chart
can reveal all the complexities of the human relationships in an
organization and there is no such thing as a perfect plan for any
organization. The best plan reconciles the theoretical ideal with
the human resources available to do the job.*
          J. LEONARD REINSCH AND ELMO I. ELLIS, WSB ATLANTA

ADMINISTRATIVE STRUCTURE

A manager's role is reflected in the company's administrative structure, which
consists of executive colleagues and the clusters of personnel for whom he is responsible. All managers and managees (using our formulation's terminology) in giving directives and actualizing them, constitute the process that accomplishes the
organization's macro-intentions—its broad purposes, goals, and objectives. Therefore, process is more important than structure. *Process* gets the job done and achieves
the goals of the enterprise, especially in such a flexible and people-oriented business as broadcasting. *Structure* stabilizes the organization and clarifies efficient
relationships among personnel with varying responsibilities for that process.[1]

ORGANIZATION CHARTS

Structure clarifies lines of authority and accountability. An organization chart schem-
atically portrays how various units of management relate to one another. However,
two points must be kept in mind: (a) probably no two broadcast companies will
operate effectively with precisely the same organizational structure, and (b) any
company's organizational relationships must be flexible because human beings carry
out the process under conditions that are not static. Townsend urges managers to
assume that every man and woman is a human being, not a rectangle; Randall distinguishes between the usefulness of preparing an organization chart and the
mindlessness of abusing the chart by letting it replace team play among persons,
inhibiting flexibility and responsible enterprise on the company's behalf.[2]
     Figures 4.1–4.4 offer sample organization charts for small and large radio
and television stations plus a national network as part of a corporate broadcasting division. They demonstrate that, beyond the basic design of departmental and
personnel relationships, each station must develop its own line and staff structures.[3]

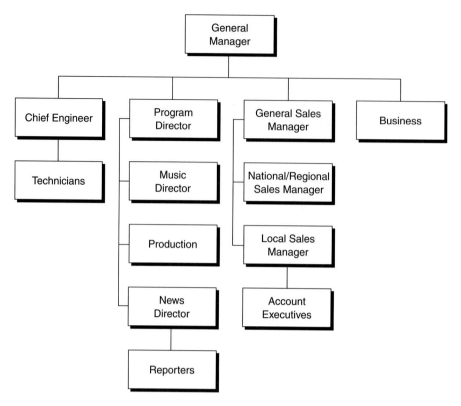

FIGURE 4.1 Organization chart of representative radio station

DEPARTMENT HEADS

Broadcasting's basic administration typically places three middle managers in charge of engineering, programming, and sales departments. The chief engineer, program director, and general sales manager report directly to the general manager; or, they report to the station manager, who in turn reports to a general manager if the local property includes an additional AM, FM, or TV operation—each with its own station manager.[4] Noncommercial stations replace sales with positions for "development" or fund-raising. Broadcast and cable both add business-related departments for general administrative functions which report to the manager. News directors at many TV stations and news/talk formatted radio stations oversee staffs independent of the programming department, and report directly to the manager instead of the program director (see figure 4.2). In fact, entering the 1990s, the position of program director was being phased out at many television stations. An operations manager or the general/station manager assume duties related to purchasing and scheduling syndicated properties, while the production manager and news director oversee most studio programming.

A medium market cable system de-emphasizes the programming position because its manager often determines contracted services for that system's available channels. Instead, cable stresses a customer service division for local subscribers, which is also reflected in a large technical staff to install and maintain residential cable connections. A cable sales department often emphasizes marketing and promotion.

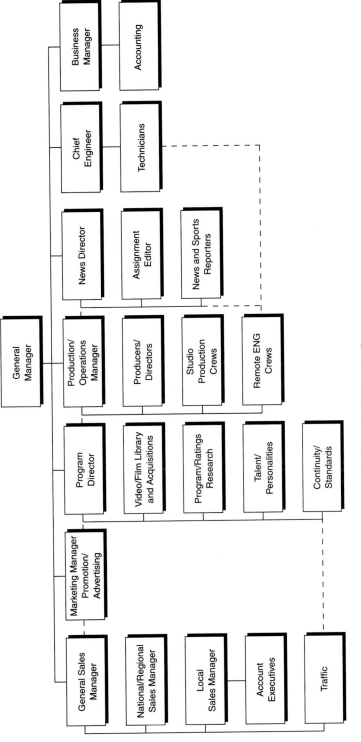

FIGURE 4.2 Organization chart of representative television station

73

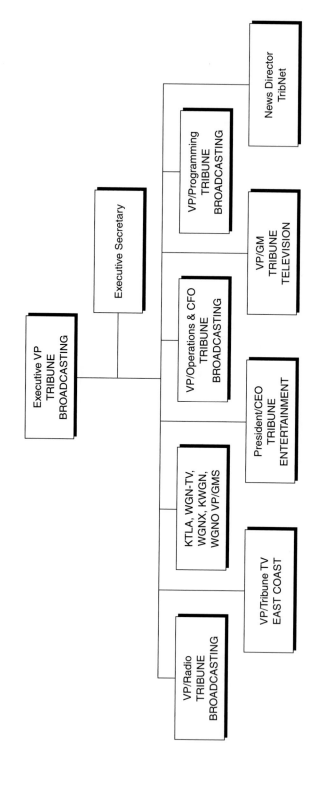

FIGURE 4.3 Organization chart of Tribune Broadcasting and WGN-AM-TV (September 1996)

74

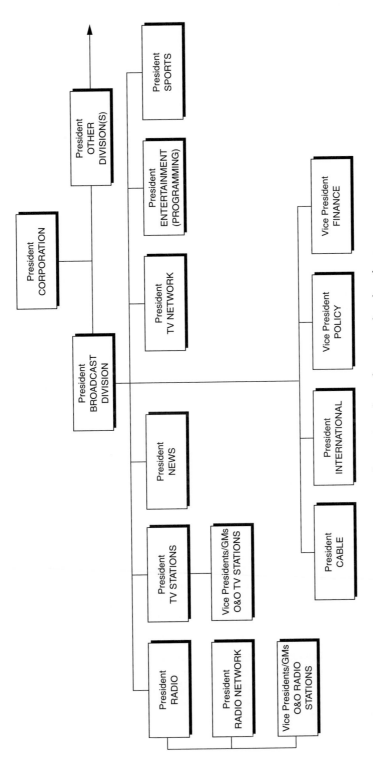

Figure 4.4 Organization chart of representative network and corporate structure

75

In some small radio stations, the manager doubles as sales manager or as program manager or both; in a few cases he is the station's chief engineer. He might also take a regular shift as an announcer of personality, and may write some of the commercial copy. The fulltime staff of some small radio stations total no more than five or six people. (This is a good training ground for young persons looking to a career in management, because they are forced to become engaged in all aspects of station operation.) At some TV stations in smaller markets, general managers are also designated "film buyer" or program director because they negotiate with syndicators of program series and movie packages that dominate the station's schedule and constitute a major portion of the annual budget. Small stations may have a technical "staff" of one fulltime engineer; others share an engineer with another station, or contract for a parttime or consulting engineer on an as-needed basis. Dependable, sophisticated electronic equipment as well as government deregulation in the past decade made this feasible.

At the opposite extreme are jointly owned radio and television stations located in large metropolitan markets (see figure 4.3). A general manager usually oversees both operations with two station managers, one for radio and the other for TV. Large stations with staffs of 150 to 300 may add assistant managers to carry part of the administrative load. Because it is important to a station's local identity and profitable service in a community, news operations are often split off from programming, with the news director reporting not to the program director but directly to the station or general manager. Stations with well budgeted, aggressive promotion and advertising departments may also operate independently of both programming and sales, reporting instead to the general or station manager.

GROUP/CORPORATE MANAGEMENT

Federal regulations since 1996 now permit a single company to own an almost unlimited number of AM, FM, and TV stations. That often means consolidating overlapping positions, especially when a company owns multiple stations in the same market.

■　　Deregulated ownership contrasts with the traditional limit of 7 each of AM, FM, and TV stations. Since the 1980s progressively 12, 18, then 20 each for AM and FM radio and 12 TV outlets; finally in the 1990s the number of radio or TV holdings were virtually unlimited. From five to eight radio stations may be owned in the same market, depending on the total number in that market; an owner's TV stations combined may reach a total potential audience of no more than 35% of the U.S. population—with UHF stations counting only half of their market size. While most of some 600 group owners each own half a dozen or so radio and/or TV stations, some largest groups own radio stations totaling 83 (Westinghouse/CBS including Infinity) and well over 100 (Clear Channel Communications). Depending on various market sizes, largest TV group owners reach: 34.8% of U.S. homes with 22 stations (Fox including New World Communications); 31% of U.S. homes with 14 stations (Westinghouse/CBS); 25% with 16 stations (Tribune including Renaissance) and also 25% with 11 stations (GE/NBC); or only 9% with 22 TV stations (Sinclair Broadcasting). By 1990 group owners already operated 90% of all VHF stations in the top 100 markets.[5]　■

Group-owned stations, in addition to their local administrative structures, have central headquarter offices such as Tribune Broadcasting in Chicago and Multimedia in Cincinnati. That cluster of stations may also be a division of a yet larger enterprise, not necessarily confined to media-related activities, such as Chris Craft Industries and Westinghouse Electric.

Corporate administrators in the group's central office often develop common operating policies for all its owned stations; many permit a more flexible arrangement whereby each station decides its local procedures. The central office coordinates financial data and legal matters for stations in the group. It may represent owned stations in union negotiations. The corporate office usually maintains a programming unit that leases program services—syndicated music/talk radio services or TV and film packages—at multiple purchase discounts for the group's stations.

NETWORKS

Corporate management at the network level is a subject for another book. But evidence of management challenges "at the top" is the expanded administrative complexity in broadcast divisions of media companies. Back in the 1970s a single network president oversaw all aspects of that operation, with a cluster of vice-presidents reporting to him. But a few years later the national broadcast corporations created layers of presidents (see figure 4.4).

■   For example, the president of the CBS/Broadcast Group reported to the president of CBS Inc. and its chairperson, and then in 1996 to the chairperson of new parent company Westinghouse Electric. That group president was assisted by separate presidents in charge of each of the following: CBS Television Network; CBS Entertainment (formerly, programming); CBS Sports (formerly part of programming); CBS Marketing (formerly sales); and CBS Affiliate Relations. Additionally, there were separate presidents of CBS Radio (the network and owned stations), CBS Television Stations (owned TV stations), and CBS News. Those eight second-level presidents were assisted by senior vice-presidents and executive vice-presidents for policy planning and financial matters, plus dozens of vice-presidents in those and other subdivisions of the Broadcast Group.

Similarly, NBC realigned its executive management structure in the 1980s. Reporting directly to NBC chairman Grant Tinker were NBC's executive vice-president and chief financial officer, an NBC vice-chairman overseeing staff functions (including business affairs, public information, corporate planning, law, personnel and labor relations, and research), and the president of NBC responsible for operations functions (news, sports, owned TV stations, the TV network, NBC Entertainment programming, and the radio divisions).[6]   ■

But the tight economy amid hostile takeover threats in the late 1980s and early 1990s prompted severe "downsizing" throughout all networks' administrative structures as well as at local stations everywhere. Middle managers and staff members at all levels were drastically reduced (see the next chapter on managing personnel).

STAFF SIZE

Local staffs generally reflect their market size, although very small radio and skeleton-staffed UHF independent stations are found in major metropolitan areas. Staffs employed by commercial radio and TV stations grew regularly until the mid-1980s, then began to decrease during the next five years (see tables 4.1 and 4.2).[7]

~~~ **TABLE 4.1** ~~~

AVERAGE STAFF SIZES,
COMMERCIAL RADIO STATIONS, 1974 AND 1989

AVERAGE NUMBER EMPLOYEES				MARKET SIZE BY POPULATION
Fulltime		Parttime		
1974	1989	1974	1989	
17		4		Over 2.5 million:
	47		11	Revenue $5,000,000+
	19		8	Revenue under $5,000,000
13	26	3	9	1 million-2.5 million
17	23	3	8	500,000-1 million
13	17	4	7	250,000-500,000
12	15	3	6	100,000-250,000
12		3		25,000-100,000
	11		6	50,000-100,000
	9		5	25,000-50,000
	7		5	0-25,000
9		3		10,000-25,000
6		3		Less than 10,000

Source: National Association of Broadcasters annual survey reports, 1974 and 1990; see endnote 7.

~~~ **TABLE 4.2** ~~~

AVERAGE STAFF SIZES,
COMMERCIAL TV STATIONS, 1974, 1985, AND 1989

AVERAGE NUMBER EMPLOYEES						MARKET SIZE BY ARB RANKING*		
Fulltime			Parttime					
1974	1985	1989	1974	1985	1989	1974	1985	1989
90	205	154	8	10	13	1-10	1-10	1-10
103	132	133	7	10.5	12	11-25	11-30	11-20
		106			11			21-30
88	97	93	6	6.5	8	26-50	31-50	31-40
		85			7			41-50
70	94	78	3	8	7	51-75	50-70	51-60
		77			8			61-70
55	69.5	73	5	8	8	76-100	71-100	71-80
		65			8			81-90
		67			7			91-100
42	62	61	5	7.5	6	101-125	101-130	100-110
		60			6			111-120
		56			10			121-130
37	53	52	6	7	8	126-150	131-150	131-150
30	44.5	51	4	6.5	8	150+	150+	150-175
		41			7			176+

*In successive survey reports, NAB broke out data in varying clusters of market sizes; they are grouped here to approximate previous years' categories.

Source: NAB survey reports, 1974, 1985, 1990; see note 7.

Over that decade and a half, the average staff size doubled at radio stations in major markets with populations over 1 million. Radio staffs in midsized markets increased slightly, but at stations in markets with populations less than 100,000, fulltime staffs decreased (while parttimers increased).

■   Between 1974 and 1985, average commercial television fulltime staffs grew by 25% (to 75 persons from 60 a decade earlier). Predictably, biggest growth in staffs came in the largest markets: up 127% in the top 10 markets. Average staff sizes for all markets grew by a quarter to a half. But by 1990, staffs were cut by 25% in the top 10 markets (down from 205 to 154 persons per station) and in the 51-70 markets (average dropping from 94 to 78 persons). Staff sizes dropped considerably in markets 11-30, somewhat in markets 30-50, but were roughly stable in markets 71-200.   ■

While network-owned TV stations[*] employ staffs of close to 300, UHF independents in the same markets lower the average with far smaller staffs. Similarly, the nation's markets have both large and small radio stations; the largest clear-channel stations in major markets might have staffs of 50 to 100 (for news/talk formats) while low-power stations in the same cities might employ only a dozen people.

The nation's 11,000 local cable systems employ anywhere from a dozen employees in the smallest operations (fewer than 5,000 households subscribing) to 35 in midsize systems, up to 70 fulltime persons in systems serving more than 20,000 households. The great majority of personnel are technicians and book-keepers, who install and maintain cable connections and process monthly billing and related office work.

STATION PHYSICAL LAYOUTS

Another way to conceptualize the size and variety of staffs necessary to operate a broadcast station is to study representative building floor plans depicting working areas for station personnel. Nonbroadcasters assume that most space in a TV station building is devoted to studios for producing programs. However, a local station is primarily a business involved in bookkeeping, sales research, and paperwork traffic to organize myriad details of commercial spots and record playlists, marketing and promotions, among others. A TV station depends on network and syndicated programming to fill most of its schedule; it is only secondarily a production site—usually several times a day for news, plus occasional public affairs taping for weekend playback. Typically, only one-tenth of space is allocated directly for studios; another 5% or 10% goes to production-related activity such as master control, engineering repair, and video editing. More than 80% of the building is for offices—administrative, financial, sales, newsroom, traffic and billing—and other support areas such as conference rooms, archival filing, lounges, and the reception lobby.

Organization charts and floor plans reflect the size and composition of station staffs that management oversees in the administrative structure. McLarney and Berliner aptly caution that whenever the number of employees in an organization is increasing (or the size and complexity of the physical plant is growing) at a faster rate than business, then top management should suspect that some of these people and facilities are there just to keep the "system" going and that the "system" and mere structure are becoming more important than the business.[8]

---

[*]Technically, a network is one division of a larger corporation which is licensee of local stations that make up a separate administrative division. Although the network neither owns nor operates those stations, the inaccurate designations "network-owned" and "network O&O" are shorthand used almost universally. CBS refers to them as company-owned stations that are locally operated by on-site managers. Similar to other networks, that corporation's stations are owned by CBS Inc. (now Westinghouse/CBS), not by the CBS-TV network.

The men and women represented by organization charts and floor plans are the personnel who operate the station or cable system. Managers and department heads are described below. The manager's concerns in overseeing staff personnel are explored in the next chapter.

WOMAN AND MINORITIES IN MANAGEMENT

WOMEN AS MANAGERS

The role of women in management has grown substantially in recent decades, that of minority persons less so.[*] Between 1965 and 1988, two-thirds of all new managers added to the nation's total work force were women. By the late 1980s, women held 38% of management and administrative positions in all U.S. businesses, although they held only 2%-3% of top executive posts.[9]

Studies have found women as effective as men in managing people, including initiating and organizing work in a group and attending to personnel needs and satisfaction. Nor do males and females differ in task-oriented or people-oriented emphases. But in handling poor performance by subordinates, males tend to underscore "equity" while females stress "equality" through training or by sanctions for behavior; male managers seek to influence subordinates' performance more than do female managers. While men's self-confidence and esteem appear higher, women are more accessible to subordinates. Men tend to motivate employees in terms of their self-interest by punishing poor work and rewarding superior performance (Theories X and Y); women lean more to inspiring employees to sublimate self-interest into broader goals (Theories V and Z). This "interactive leadership" by female managers seeks to "encourage participation, share power and information, enhance other people's self-worth, and get others excited about their work."[10] This management style involves what has been called "webs of inclusion" whereby women prefer horizontal organizations rather than hierarchical organizations, de-emphasizing memos and reports in favor of sharing information through frequent interaction among staff members (but so do 90% of all TV general managers, most of whom are male).

Similar for both genders is motivation as well as commitment to longer hours, extra work, and relocating. But women's commitment is more affected by family considerations than is men's. Both male and female managers cope with stress in job responsibilities about the same, despite women's added sources of stress in discrimination, gender stereotyping, social isolation as "tokens," and conflicting demands of family and job. An androgynous management style "blends behaviors previously deemed to belong exclusively to men or women"; it includes both feminine and masculine qualities that are adapted flexibly by both male and female managers.[11] Some specialists concluded early on:

> In conclusion, there is considerable evidence to support the fact that women managers psychologically are not significantly different from their male counterparts and that they may possess even superior attributes and skills in some areas related to managerial effectiveness. . . . It is recommended, therefore, that organizations begin treating women as equals, not because of moral obligations or pressures from outside interest groups to improve female-male ratios, but because they would effectively utilize valuable human resources.[12]

---

[*]See chapter 5 for further data about women and minorities on broadcast and cable staffs and in the general labor force; also see chapter 11 regarding the issue of equal employment opportunities (EEO).

FEMALE BROADCAST MANAGERS

In 1971 when the FCC began gathering female and minority employment data, women held 6% of the "top four classifications" of radio-TV positions. Their presence in these upper posts grew to 33% in 1993 (women were 40% of the total broadcast labor force). That year employees at cable systems were 41% women, filling 31% of "upper-level positions."[13]

■   Note that categories in the government reporting form were designed for all kinds of businesses and industries; they are not apt for mass media's mostly white-collar labor force. The "upper-four job categories" or "upper-level positions" can apply to most media positions, so do not identify true administrative-management posts; they are listed on federal forms as *officials/managers, professionals, technicians,* and *sales workers.* The FCC's *Annual Report* for 1994 reported 146,629 fulltime employees in radio and TV—down 30,000 from five years earlier. Fully 126,004 of them were designated in those top categories; but clearly 86% of all broadcast employees cannot be considered to hold managerial-related posts! With its army of installers and other technicians, cable's categories are more similar to other industries: listed in "upper-level" positions were 47,319 ("only" 43%) of all 109,230 fulltime employees—that total up almost 20% in half a decade.   ■

In 1986, 16.5% (235 of 1,428) middle and senior managers at the networks were women; at Cap Cities/ABC 20% (120 of 610); at NBC 15% (51 of 340); and at CBS 13% (64 of 478) of upper managers were female.[14]

■   By 1987, at NBC women held 21 of 174 executive positions at the level of vice-president or above.[*] In 1988 Nancy Widmann, one of many female vice-presidents at CBS, was appointed president of the CBS Radio Division; her female colleagues included vice-presidents of Current Programs in the CBS Entertainment division, of media relations, and of CBS radio stations.[15] By 1987, among four key management posts at each owned and operated (O&O) station around the country, women held 7 of 67 top positions at the 17 O&O TV stations and 20 of 162 positions at 43 O&O radio stations.[16] Among network news vice-presidents, women numbered two of 12 at ABC, three of 13 at NBC, and two of 11 at CBS—16.6% of all news vice-presidents in 1989. That same year at middle management levels, 40% of NBC's news division's "management/professional" posts were held by women, while *Variety* noted that "the CBS Washington [news] bureau is literally run by women, and one-half of that network's U.S. bureaus are headed by women."[17]   ■

By the 1990s women were managing key broadcast positions, such as senior vice-presidents of commercial and noncommercial television networks, and general managers of major TV stations. Whereas twenty years earlier no television station was managed by a woman, in 1991 a total of forty network affiliates had female general managers—although only four were in top-50 markets (two in San Francisco), with over two-thirds in markets ranked 100-and-smaller.[18] Almost one in ten broadcast station general managers (GMs) and owners were women (only 5% of TV GMs). Those female managers usually moved up from heading sales, program, or news departments in an industry where almost 40% of all employees

---

[*]Depending on one's perspective along the spectrum between traditionalist and feminist, the totals reported here could be annotated with either "as many as" or "only" (for example, "*as many as* 12 of 174" or *only* 12 out of 174"). Most major posts reported in this paragraph were 10% to 12% held by women, with some reaching 16% of all positions in a given category. Males, mostly white, thus held from 84% to 90% of all major posts in national broadcast companies almost three decades after legislation was enacted to support equitable (not privileged) advancement of women and minorities in the U.S. labor force.

were women. In 1992 Scripps-Howard Broadcasting appointed an African American woman vice-president and general manager of KSHB, Channel 41, in Kansas City, Missouri.

MINORITIES IN MEDIA MANAGEMENT

By 1994 minority persons, male and female, were 18% of all employees in broadcasting and one out of four in cable. Minority persons held 7% of the top four levels of radio-TV station positions back in 1971 (the first year the FCC compiled such data); that percentage more than doubled to 16% by 1994. Also in 1994 at cable systems minority persons held 20% of "upper-level positions" (again, these federal categories do not aptly reflect true management roles in media).

For 1990 the FCC reported minority persons held more than one in ten "official/managerial" positions (heading the four categories) at the nation's radio and TV stations; those jobs held by blacks and Hispanics were roughly split between men (5.5%) and women (4.9%), as listed in table 4.3. The Commission reported in 1994 that women held almost 33% of all four higher positions, and minorities 16% of them.[19]

■   In 1990 at four TV networks (the "Big Three" plus Fox) only six programming directors and two vice-presidents were African Americans, all but one of them female, among a total of 120 executives at that level. Examples of success include a black male who was president of a network TV station division (formerly a top-market TV station manager), a black male as vice-president of a major broadcast group while also serving as a board member of the National Association of Broadcasters, and a black female who was executive vice-president overseeing all programming at the noncommercial television network. Blacks serve in major positions at the Black Entertainment Television (BET) cable program service; radio's National Black Network and Sheridan Broadcasting Networks; at the 186 radio stations with black-oriented program formats; plus at the 182 radio stations and 17 TV stations licensed to 110 companies owned by blacks. The National Association of Black-Owned Broadcasters in 1990 included 109 members who were licensees of 185 radio and TV stations. In 1994 African Americans owned (half or more ownership) fewer than 3% of all broadcast stations—thirty-one of 1,155 TV stations and 292 of some 11,000 radio stations; they owned nine of 7,500 cable operations.[20]

Hispanics hold key posts as station owners and managers at Spanish Broadcasting System; Hispanic Radio Network; eight radio program services (UPI Spanish Radio Network, Spanish Information Service, SPM Radio Network, Lotus Satellite Network, Spanish International Marketing, Inc., Hola Amigos, Radio America, The World Radio Network); two Hispanic TV networks (Univision, Telemundo) and a cable programming service (Galavision); plus 237 radio stations and 15 television stations in eight states (plus 12 TV stations in Puerto Rico) featuring Spanish program service. Other companies not owned by minority persons but employing almost exclusively Hispanics, many of whom are also shareholders, include Tichenor Media Systems, Inc. of Dallas, with eight AM and four FM stations.   ■

From the FCC's formation in 1934 to late 1971, all Federal Communications commissioners were white males, except for Frieda Hennock (serving from 1948-1955). Following the civil rights movement, from October 1971 to late 1996, one Hispanic, three African American males, and eight women (one of Hispanic, another of Asian descent) have also been appointed to the agency. The fifty-four other commissioners since 1934, including all chairpersons, were Caucasian males.

TABLE 4.3
Percent of Minorities in "Top-Four" Categories, R-TV Stations, 1990

	WOMEN		MEN	
	Black	Hispanic	Black	Hispanic
Managerial	2.9%	2.0%	2.8%	2.7%
Professional	3.9	2.1	4.7	3.9
Technician	2.3	0.9	#	7.1
Sales	2.8	2.4	#	2.4

#Data not listed in secondary source; data from 1991 FCC report in *Variety*, 14 October 1991, 73 (see note 20).

SALARIES FOR BROADCAST/CABLE MANAGERS

Levels of compensation for station/cable managers and department heads vary with market sizes and kind of operation (AM, FM, TV, cable) along with efficient success in competitive programming and aggressive sales—all of which determine a company's annual revenues. Table 4.4 summarizes comparative salaries for major positions, extrapolating successive earlier data from several sources. The following text material offers further details about salaries for each management position.

■   Salary-levels are, of course, a function of station revenue which reflects market size. For example, total earnings—salary plus bonuses, incentive compensation, profit-sharing, (but not including other fringe benefits such as disability or health insurance, social security matching money)—by TV general managers in 1986 *averaged* $111,900 *for all U.S. stations*. But for independent TV stations, that average dropped to $103,000; for affiliates it rose slightly to $113,700. At those network-affiliated stations, over one-third (38.1%) of all TV GMs earned between $100,000 and $150,000 annually. Another quarter (23.8%) of GMs at affiliates earned $75,000-$100,000; 18.3% earned $50,000-$75,000; while 5% received less than $50,000 in total earnings annually. At the same time, topping the pay scale were 8.8% of TV-affiliate GMs who received $150,000-$200,000, and 6% over $200,000 in annual earnings.[21]   ■

MANAGERS AND DEPARTMENT HEADS

GENERAL/STATION MANAGER

Duties and Responsibilities

Major qualifications and duties for managing local radio or TV stations or cable operations were outlined in chapter 3 and in the preceding pages. General managers of *television* stations in one survey ranked these tasks as most important and least able to be delegated: budgeting and financial matters, followed by corporate planning, and matters pertaining to regulation and the station's license; but managers in medium and small markets were more closely involved with pricing and selling airtime along with planning and acquiring programming. Next came personnel matters, community relations, and general administration of the station. (Presumably, middle management department heads work more closely with day-to-day activities and concerns of personnel than do general managers.)

Issues concerning managers of small and medium market *radio* stations in 1990 were: sales and promotion, motivational techniques, compensation strategies for sales personnel (stressing not cutting costs but rather increasing revenue), retaining good people, strategic planning, using computers effectively, and ASCAP/BMI music licensing.[22]

Career Tracks

Top management in the 1930s and 1940s included former entertainers from show business and radio who knew how to attract audiences; in the 1950s and 1960s successful sales personnel moved into executive positions. In the 1970s—a decade of widespread criticism and social demands on broadcast licensees—those experienced in law and in business administration ("Harvard MBA types") began to move into top station management. By the 1980s and into the 1990s a mix of talents was found at higher levels of management: backgrounds in finance, sales, news and programming, and even from academic areas (a number of broadcast administrators have earned doctoral degrees). Budget cut-backs in the past decade dictated reducing staffs and closely monitoring expenditures in a tight economy amid take-overs, buyouts, and successive station sales; so financially skilled persons became a much sought commodity for the manager's office. Also, the key role of news for a local TV station's identity and its value for selling local airtime made news directors apt for managing a station's competitive position in the market.

Patterns in selecting managers reflect forces affecting broadcasting. Business entrepreneurs and financial investors (such as Laurence Tisch, chairman of CBS Inc., and Ted Turner, CEO of Turner Broadcasting) typically head major media corporations. But at times even social researchers and behavioral scientists have moved to lofty posts to collaborate in assessing and planning broadcasting's role in contemporary society as well as its efficiency in capturing audience attention and influencing consumer purchases of products and services.

Emphasis on financial aspects of media business led to bringing in managers from nonbroadcast companies. Bill James, a furniture salesperson in Grand Rapids, became successful general manager of clear-channel WJR in Detroit. At the corporate level, Arthur Taylor was a young executive previously with a Boston financial investment firm and a major paper manufacturer; he was appointed president of CBS Inc. in the 1970s. Laurence Tisch, the chairperson and co-CEO of Loews Corp. (insurance, tobacco, hotels, and theaters conglomerate), became president and chief executive officer of CBS Inc. in 1987 and also chairperson in late 1990.

But many local managers were previously employed in another position at the same station, usually radio or television sales, often programming (radio) and news (TV), rarely engineering. One study found just over half of TV general managers in the mid-1980s were appointed from within the company, while a third came from other stations outside; about 5% came from outside television, usually radio.[23] (Cox Cable has a long-term objective of hiring 80% of its general managers internally.) Half of television's GMs came from sales, and about 10% each from programming, production, and news. More than 7% came from management outside broadcasting, or from teaching, national rep firms, or legal, marketing, and engineering sectors. At Group W Broadcasting, while some general managers were appointed from finance, programming, and news departments, fifteen out of twenty appointments during one span came from sales.

Perhaps newspaper-owned TV stations are more alert for executives from news operations. For example, news directors became GMs at WHAS-TV in Louisville (then co-owned with the Louisville *Courier-Journal* by a local family) and at WVTM-TV in Birmingham, Alabama (owned by Los Angeles-based *Times-Mirror*). The latter station brought in a Boston TV news director as GM, subsequently appointing him manager at their Dallas station while selecting another news director from Albany, New York, to succeed him at WVTM. Another factor is proliferating TV sources vying for audiences: dozens of cable channel services, videotape playback, optical fiber service, and direct broadcast satellite with HDTV. The distinctive drawing power of over-air TV stations is news and other local programming. A former news director, Robert Morse, when GM of NBC-owned WMAQ in Chicago, asserted in 1991 that "the job is to compete. The one way you do this, particularly network affiliates, is to get entrenched in local news, information and programming. Differentiate yourself from all those other outside signals."[24] News directors are uniquely alert to this crucial role for a station's success in its community.

Management in local radio is oriented towards producing revenues in intensely competitive markets, with 68% of GMs appointed from sales. But radio GMs who were previously news directors doubled from 4% in 1981 to 8% in 1990 (despite the many radio operations devoted solely to music and syndicated material with little or no local news). Selected to manage radio stations, as in TV, are news directors who "run shops that contribute a large percentage to total station revenues" and "demonstrate an ability to use generally recognized management skills in [their] newsroom, from motivating and supporting people, to use of verbal praise, coming up with new ideas, and making quick decisions" as well as efficient budgeting.[25]

Today, station owners have the advantage of selecting managers with considerable experience in broadcasting who have successfully managed other radio and television properties. Many TV general managers previously managed radio stations, where leadership and experience can often be attained more readily than in TV. Local radio's volatile rivalry for market share makes a manager's decisions quickly evident; and there are more opportunities to be directly involved in all phases of the medium as an entertainment, business, and legal entity.

The volatile buying and selling of station properties in the 1990s accelerated churn in GM careers. By 1994 radio GMs in the top-25 markets averaged 5.3 years in their positions; those in markets 26-100 lasted only 4.1 years, while GMs in markets smaller than 100 averaged about 5 years at radio stations.[26]

Age

Early studies found local broadcast managers were typically in their early 40s, although small-market radio stations were often managed by those in their 30s. Approaching the 1990s, a survey reported one-third of TV general managers were 35-44 years old, more than one-fourth were 45-54, while another one-fourth were over 55 years old.

Education

Studies through the decades reflected the role of higher education for managerial positions in broadcasting. As early as the mid-1960s, three-fourths of all television managers had been to college and one-sixth had taken postgraduate study. Half the managers of radio and TV stations were college graduates. Of small-market radio managers, 85% had attended or graduated from college.[27] By 1987 two-thirds of general managers had received college degrees, another 15% had attended college but not graduated; and 15% of all GMs held graduate degrees.[28]

■   In 1990 *Channels* editors profiled the 74 top executives—chairpersons, presidents, CEOs—of publicly held media corporations with multiple holdings in TV, radio, cable, satellite transmission, and program production.[29] They ranged in age from 30 (heading Comcast Corporation with 2.5 million cable subscribers and $500 million annual revenue) to 80 (founder of Park Communications whose 7 TV and 19 radio stations plus dozens of newspapers earned annual revenues of $151 million); a third were in their 40s and another third were in their 50s, with 18 executives in their lower 60s. Among them, forty-six had earned bachelor of arts degrees and eighteen bachelor of science degrees. Thirty had added graduate degrees: six M.A.; two M.S.; eight M.B.A.; three Ph.D.; plus one medicine and ten law degrees; four others had taken some graduate studies. Six executives did not complete college degree work, while three had no college experience listed. In his late 40s, John C. Malone—president and CEO of TCI/Tele-Communications, Inc., the country's largest multiple system cable operator with 8 million subscribers—held four degrees: a B.S. from Yale, M.S. and Ph.D. from John Hopkins, and another M.S. from New York University. All these executives had moved up through radio, TV, cable, production, and electronics divisions, some reaching their top posts within a decade after college.   ■

Because the pool of college educated persons provides broadcasting with managers, it is worth noting that women have increasingly completed college degree work. Half or more of all bachelor's degrees in all major areas of study since 1982 have been awarded to women—contrasted with less than 25% in 1950 and 33% in 1965.[30] In 1987, a third of all M.B.A. degrees were earned by women (up from 12% in 1976 and only 2% in 1967). By the late 1980s, 58% of women in managerial/professional jobs in all U.S. businesses had completed four or more years of college.

Meanwhile, more than half—in some colleges up to two-thirds—of all college students majoring in broadcasting in the 1990s are women. Many aggressively seek out internships, work parttime at stations, and will move into important positions in broadcasting, especially news. Among them are many potential managers of stations, networks, and cable systems.

Similarly, the great majority of African Americans (86%) who in 1990 owned stations (106 AM, 63 FM, and 16 TV) had attended college. Among them 18% did not complete degree work, 68% had bachelor of arts degrees, 33% also completed master's programs (three-quarters of them in business administration) and 6% had earned doctorates; 18% earned professional degrees in law and medicine.[31]

■   More African Americans licensees had been in broadcasting (33%) than in any other business before becoming owners: 10% came from nonbroadcast sales; 10% from various services such as law or real estate; 7% each from banking, medicine, business, and other media; 6% each from education and government; and 3% from religion. Of those from a broadcast background, 73% had experience in administration (averaging over eight years), including 69% in programming and 15% in engineering (both groups averaging about ten years). Nine out of ten black broadcast owners were men.   ■

Areas of College Study

Surveys over three decades reflect steadily increasing support for academic study of broadcasting. This is due in part because more institutions in later years offered accredited programs of media study in departments of broadcasting (radio-telecommunications). Graduates entered the field in ever larger numbers from the 229 colleges offering bachelors, masters, and doctoral degrees in electronic mass communication studies in 1995.[32] In 1962 some 40% of broadcasters had majored in liberal arts, 24% in business administration, 10% in engineering, 8% in radio-TV, and 7.5% in journalism, among others (see the next chapter on personnel). A decade later a good share of small-market radio managers majored in business-related areas (20%), with 6% majoring in journalism and 2% in radio-television.

To those interested in broadcast management careers, managers most often suggest sales or promotion work, plus academic training in business, including study of sales, marketing, and accounting. In 1979 a majority of GMs (57%) recommended "a good, general education"; but 28% urged a marketing major in college, while 27.6% recommended a communications major; another 18% suggested further study for an M.B.A.[33] In 1987 two-thirds of general managers recommended media-related degrees in journalism, communication, and broadcasting; just under half also recommended a "well-rounded liberal arts background"; and one-third suggested business courses in economics, accounting, marketing, and management.[34]

Broadcast owners who were black earned undergraduate majors in arts and sciences (61%), education (14%), engineering (11%), journalism/mass communication (8%), and other areas (5.6%).

～～ TABLE 4.4 ～～
ESTIMATED ANNUAL COMPENSATION MANAGERS AND DEPARTMENT HEADS, 1996

Figures were computed by extrapolating Sherman's 1983 data and Head & Stirling's 1988 data, then projecting them proportioned to U.S. percentage raises for those five years plus the next three years, weighted to reflect media's previous 33% raises when the national private sector averaged 20%, and projected as 22% for media onto the next three years' national average of 15% increase (compounded).[35] See comparative data at bottom of table.

	RADIO	TELEVISION	CABLE
General Manager	$73,500	$163,000	$54,500
General Sales Manager	72,000	114,000	46,500
News Director	25,000*	77,000*	NA
Chief Engineer	34,000	54,000	62,500
Program Director	39,000	54,000	31,500
Business Manager	31,000	59,500	[no data]
Production/Operations Manager	40,000	47,500	43,000

NOTE: "National averages" or *means* are somewhat lower than typical compensation for these positions in *medium markets* (and often two-thirds or half of that in *large markets*), because of the number of positions at the great many stations in *small markets* where the figures are often one-third less than the national averages. A national *median*, of course, lies midway between the lowest and highest levels of compensation. See text for sample figures for general managers.

These estimates were roughly similar to actual compensation (salaries plus benefits and bonuses) for TV reported in the NAB/BCFMA's *1991 Television Employee Compensation and Fringe Benefits Report;* excerpts cited in *Broadcasting,* 25 November 1991, 50. TV general managers averaged $133,930; general sales managers $91,848; news directors $60,501; chief engineers $49,200; program directors $45,737; business managers $46,534; production managers $32,919; operations managers $46,605. Projected over five years to 1996 at 3% annual inflation rate, the general manager, general sales manager, and news director would total about $10,000 less in each position, the others within $2,000-$3,000 of the figures above. See table 5.4 in chapter 5.

*Ranges of 1996 median salaries, from smallest to largest markets, estimated by projecting 20% increases onto Stone's 1990 data: news directors in radio, $18,000-$38,500; in TV, $38,000-$96,000.[36]

Income

The income of television managers is often competitive with that earned by executives at a similar level of responsibility in other industries. In 1982, general managers' total compensation including salary and benefits (including the estimated value of bonuses, incentive compensation, profit-sharing, retirement benefits, stock-options) averaged almost $83,000; managers at stations with annual revenue less than $4 million earned $62,000 while those at stations with revenue over $8 million averaged $117,000.[37] By 1996 TV general managers' salaries had risen to about $160,000—more than twice what radio GMs earned. Both GMs' averages were double or triple the earnings of their program directors and news directors (see table 4.4).[38] At local cable systems, general managers in the mid-1990s averaged more than $50,000, while managers at MSO headquarters or regional offices earned twice that amount.[39]

Among small-market radio stations, more than two-thirds of the managers had part ownership. A quarter of TV general managers participated in ownership of their broadcast properties.

Personal Characteristics

In one survey, TV station general managers characterized themselves as generally informal rather than "hard-driving," but working long hours.[40] They much preferred business aspects to the programming side of TV—except for GMs whose station's annual revenue exceeded $5 million. They strongly favored face-to-face dealings with employees rather than memos and posted notices. They stressed intuition and personal judgment over applying scientific concepts of managing. They also somewhat favored leaving details to subordinates rather than being "a nut for detail" themselves. Given the choice between "my job requires long working hours" and "won't sacrifice family for job," fewer than one-third of managers selected the latter (25% of GMs at smaller stations, with revenues below $2 million; 34% where annual revenues were $2 million-$5 million; and 30% where station revenues were over $5 million).

■   Another national study found that almost all TV general managers agreed they "find real enjoyment working here," that "my present position is more satisfying than others I could get," the company they work for gave both them and their subordinates "a fair deal," and they were proud of the station where they worked. Most were satisfied with their present compensation package, although 20% replied "neutral" or "disagree." Only about 10% indicated they would have to change stations to receive the compensation they deserved, and 17% were neutral.[41]   ■

The survey reported by Fish and Adams profiled TV managers across the nation:

> The typical GM in broadcast television today would seem to be a male in the 35-44 age group who has come out of sales within the same company, has been continuously employed in broadcast television since entry shortly after completion of his baccalaureate in business or economics. He has about 20 years of experience in the industry and has occupied his present position for at least one but less than four years. He belongs to three national or regional organizations as a consequence of his employment, attends the convention of at least one of them a year, and reads three trade publications regularly—probably including *Broadcasting*. He devoted 50 to 60 hours per week to his work and work-related activities and does not hold an ownership interest in the station/company.[42]

PROGRAM DIRECTOR

Duties and Responsibilities

In radio's earlier days with station programming headed by a program director (PD), this was a key creative position. The assignment required a PD to plan and rehearse many programs broadcast live from the station's studios. Even with heavy network schedules, the program director had to fill many hours locally, considering it a challenge to match the quality of network production. This created an aura of excitement in radio stations.

With the advent of "formula" radio in the 1950s and 1960s—radio's first major response to the competition of television—there was little need for live production. At many stations, the program director served as a kind of chief office clerk. He arranged announcers' schedules, counted the number of times a particular record was played in a day, tried to maintain a measure of authority over the disc jockeys, and waited for direction from the station manager. He became a follower, overseeing mountains of paper. Today, with radio stations striving for identity amid competing signals, a premium is again placed on station creativity in developing a distinctive station "sound" and local personalities.

A similar scenario might be written for television stations. Creativity was once the hallmark of a program manager who could be inventive with meager materials in the dynamic chaos of early "live" television. But the increasing segments of the broadcast day supplied by national networking (two-thirds of the schedule) and syndicated materials (most of the rest except local news) narrowed the field for local program innovation. Except for local news and rare public affairs, program management centers on scheduling ready-made products—coordinating with network and syndication suppliers—rather than developing personalities and program forms in local studios. Still, the proliferation of independent and public noncommercial stations, especially on UHF channels, plus the growth of cable television brought the medium around somewhat to the early exploratory days of imaginative production with small staff, limited space and facilities, and minuscule budgets.

As noted earlier, at many stations the position has been merged with production into the title of operations manager, while the station manager oversees budgeting and scheduling of expensive syndication packages. This trend is reflected in annual conventions of NAPTE (National Association of Television Program Executives), where PDs gathered since 1963 to exchange program ideas and strategies; by 1989 the convention served as a venue more for station managers and sales staffs than for program directors. PDs were less involved in acquiring programming for their stations and more into interpreting ratings and demographic research, and marketing and promoting their station's program schedule.[43]

The program manager or operations director is ultimately responsible for the sound of the radio station and the look-and-sound of the television station which the public tunes in to. The PD must know her community, her audience, her own programming product and her competition's, her budget, her staff, her management, and herself (calling for personal objectivity coupled with creativity and self-reliance). Her goal is to produce the best program service possible—that is, most effectively drawing and holding an audience—given the staff, facilities, budget, and time availabilities in the schedule. She must be conscious of the mutual dependence between her work and the sales department because programming must provide a marketable product for the sales staff.[44]

The PD oversees the station's schedule and format (radio) and on-air "talent" (announcers, DJs, hosts, performers), including program standards regarding good taste and federal regulations. He maintains records of aired programs and coordinates with traffic and news departments in scheduling commercials and news. He is responsible for locally produced programs and for the overall look and sound of the station between network or syndication material. He may work closely with a production manager or operations director. The PD and his on-air staff collaborates with a promotion director to boost audience and advertiser awareness of the station's schedule. With the manager and accounting or finance assistants, he selects and budgets packages of TV syndication, including movie titles, or formatted music-talk-news distributed by satellite or tape (see chapter 7 on local programming); he negotiates contracts for those services with program suppliers and with music licensing organizations.

While program managers must be idea people they must also be competent businesspersons, combining creative judgment with practical judgment. The programmer must conscientiously gather and assess relevant data, while retaining independence and creative sensitivity, so his decisions achieve the goal of attracting and pleasing audiences and thus advertisers. This includes knowledge of the community and of ratings cycles and reports, including demographics and psychographics. A former PD offers a candid, concrete description of the many hats worn by the person in this position:

> He turns on the radio station and is treated to a mirror image of himself. . . . He had a hand in just about everything that comes out of the speaker. Whatever he is, however he sees himself, will be reflected in that mirror.
>
> Viewed from other perspectives, the program director's influence seems more limited than it is. A record promotion man sees the program director's most important responsibility as choosing the music the station will play. To the people on the air, the PD is seen as a combination boss, wailing wall, father confessor, and shrink. To the station manager, the PD is the guy who is always bucking the sales department, who overprotects his air people, and who is riding a fine line that may blow the license any minute.
>
> . . . Because the station is so much a reflection of himself, the program director takes rejection by the audience personally. This personal factor makes radio the program director's medium.[45]

A good PD is central to a station's success. A radio or TV station's local product is as strong as the person supervising its growth and development. Projecting the station's future programming strategy depends on the vision and wisdom of the person managing this department. Given a PD with these qualifications, combined with station management's support, a station can constantly develop and improve programming impact in a volatile local market of highly competitive stations.

Career Tracks

PDs generally advance through other positions in programming. But a national survey of TV program directors found that 30% were appointed PD from production posts and only 22% from programming; 7% had been in sales; 6% in promotion/business; 5% in administration; and 3% in other management posts, plus 19% scattered among traffic, news, engineering, teaching, or law.[46] Their earlier "specialty" before entering TV management had been 41% in production; 7% news; 6% promotion; 6% business; 4.4% operations; just under 4% in programming; 3.3% sales;

and 10% varied from traffic, national rep firm, advertising and marketing, and management elsewhere.

Women

The study cited above found males in almost 19 out of 20 positions as GMs and news directors in TV, but females were more than a quarter (28.2%) of program directors.

Education

Of the 60% of TV program directors with college degrees, one-third had gone on to further graduate work and another one-fifth completed graduate degrees.

■   When asked their advice for young people seeking to get into TV station programming or production, most (79%) said start at a small station, more than half (58%) urged "get a good, general education," while 49% recommended taking a major in communications.[47] One-third suggested a good background in broadcast equipment. From 3% to 6% for each cited experience in theater, a film or theater major, or the submission of films/tapes to TV stations as beneficial. In the 1987 Roper study, almost two-thirds recommended degrees in journalism, communication, and broadcasting, while 44% also supported broad liberal arts study, and only 26% suggested business courses.[48]   ■

Income

In 1996 program managers or PDs annually earned an average of over $39,000 in television and almost $27,000 in radio. This about equaled compensation to account executives (sales department) and promotion directors; it was about half the salary paid to general sales managers in the respective media (see table 4.4). Those managing programming at local cable systems earned about $24,000, and at MSO headquarters or regional offices about $46,000.[49] By 1991 PDs and operations managers in TV averaged almost $46,000, far more than production managers ($35,000) and promotion/publicity directors ($36,500) but still half of general sales manager's earnings ($91,500).[50]

Personal Characteristics

Program directors at television stations were surveyed on the qualifications they judged essential for their position.[51] More than three-quarters noted basic management qualities: executive skills, ability to deal with people, plus being well organized. Also widely picked were "innate feeling for what people like" (74%), coupled with "thorough knowledge of audience data for syndicated shows" (70%). About half included "tough negotiator on program prices" and also "ability to evaluate talent." Only one-third called for "thorough knowledge of what studio equipment can do"; one-fourth also added "solid background in theatrical and/or TV production."

■   Nine out of ten TV PDs surveyed found "real enjoyment" in their work, although 15% did not fully agree their position was more interesting than other ones and were unsure or unsatisfied with "my present job for the time being." Only about half were satisfied with their current compensation (6% were very dissatisfied), while more than one-fourth felt they would have to change stations "to receive the salary I deserve," with another one-quarter uncertain about that point. Similarly, while nine out of ten PDs were proud of the station where they worked, only 72% felt the station "gives me a fair deal."[52]   ■

Again, Fish and Adams provide a profile based on their national survey:

> The typical PD in broadcast television today would seem to be a male in the 25-34 age group who has come out of production, has been continuously employed in broadcast television since entry shortly after completion of his baccalaureate. He has 8 years of broadcast television experience, has occupied his present position for one to four years, was promoted to it from within the same company/station, and worked—just prior to promotion—within the same market. He typically belongs to two national or regional organizations as a consequence of his employment, attends the convention of at least one of them each year, and reads four trade publications—probably including *Broadcasting*—regularly. He devoted about 50 hours per week to his work or work-related activities and does not hold an ownership interest in the station or company.[53]

News Director

Duties and Responsibilities

To build strong news capability, management must attract and support a capable news director experienced in journalism and broadcasting.[54] Jay Hoffer quotes *Fortune* magazine's assessment of the integrity and independence needed in news media:

> The quality of journalism depends primarily on journalists—not on government and not on the legal owners of media. Publishers and executives of networks and broadcasting stations now have only a small fraction of the influence on news that owners used to exercise. As commercial bias diminishes, what counts now, for better or worse, is the bias of reporters, cameramen and editors. Their ideological bent is far less important than their artistic bias, the way they select and present what they regard as significant.[55]

The news director, with support of top management, must be able to gather a staff—whether it be one other fulltime person in a small radio station, or eighty to 150 in a major market television station—who are competent and dedicated to the principles of good journalism.

■ In 1972 three-fourths of radio news staffs had three or fewer fulltime employees (the national average was 2.8), while television or combined TV-radio news organizations had 10 or more fulltime staff members (10.5 average).[56] By 1986 radio news staffs had dropped to an average of 2.2 persons, while TV news staffs doubled to 20.1. Between 1988 and 1990, some 900 fulltime positions in radio news were reduced to parttime, with one of every ten major market radio stations dropping news operations entirely. Meanwhile, 120 independent TV stations originated (usually modest) local news by 1990, averaging 6 fulltime and 3 parttime staffs—up from 2 fulltime and 3 parttime in 1988.[57] ■

By 1991 half of all independent TV stations in the country (50.3%) did not program local news, while only 1.2% of network affiliated stations did not offer news; 3.4% of commercial radio stations were without local news, and one out of five noncommercial public radio stations (most with religious formats) had no local news.[58] By the deregulated and merger-swept mid-1990s staff cutbacks pared news coverage on almost a third of all radio stations to network feeds only—or to no news at all. Most stations without local news service assessed their markets as already served adequately by media news, considering themselves alternative programming; many cited expense as the main reason for dropping previous news programming, whereas very few cited deregulation as a factor.

■ By 1995 radio averages dropped further to 1.6 fulltime (plus 2.6 parttime) news staffers; the 10% of AM and FM stations with news or talk formats had largest radio news staffs with 6 to 8 fulltime newspersons plus almost 2 parttimers. Meanwhile at TV stations widespread shifts in network affiliations and the entry of Fox affiliates into news added some 2,000 new positions in the mid-1990s. Half the nation's TV stations added more local news in 1995 and 1996, usually about three more hours a week to their schedule. By 1995 TV staffs had grown to an average of 30 fulltime and 6 part-time (stations in large markets had 100 to 150 newsstaffers).[59] Stations formerly owned by Storer, then Gillett, and finally New World in Cleveland, Detroit, Atlanta, and Kansas City, switched from CBS to Fox but continued their six to seven hours of daily news staffed by 110 to 130 newspeople each; CBS had to affiliate with independents (usually UHF) which soon began to build news staffs of 40 to 50 persons each. ■

Managers best ensure competence and integrity in news service by (a) providing budgets for adequate staffs and (b) entrusting the news function to professional personnel free from interference or pressure. Effective news directors hire personnel with strong academic and professional credentials; they encourage them to update and grow intellectually by reading journals, by attending occasional meetings of colleagues, and even by advanced formal studies if possible. A well-trained, conscientious staff can then work effectively with the news director responsible for overseeing the news operation.

Career Tracks

Before becoming news directors, their special area of work predictably was news, with production a distant second (13%); but prior to entering management level posts 24% had specialized in production, and 6% variously in business, promotion, teaching, or sales. Unlike general/station managers and program directors, news directors are most often appointed not from within but from another broadcast company. This is a factor in their greater mobility among communities and companies than other department heads. During past decades, a typical news director remained about two to three years at a radio or TV station. In 1987 one-third of TV news directors and 42% of radio news directors (NDs) in a limited survey did not expect to be in the profession five years from then.[60] NDs typically spend a total of about seven years in news management, at approximately three different stations. After one news director's half-decade tenure, WCBS-TV in New York appointed a total of six successors within nine years (1977-1986).[61] Because a station's image and revenue is directly affected by the news operation, when local news ratings are unsatisfactory the ND tends to be replaced quickly with "new blood." This undercuts continuity and staff morale unless remaining news personnel include a strong executive producer, program producers, and assignment editor. Those support positions become increasingly important for a station's news service as news directors get more remote from day-to-day operations because of heavy paperwork and other management demands such as budgeting, union and talent negotiations, and legal matters.

Television NDs have become apt candidates to manage stations because they usually oversee the most employees and largest departmental budgets and are in charge of a station's profit center (strong local newscasts often bring in more than a third of a station's revenues).[62] One study found that of 35 news directors at ten top-50 markets in 1984, only 2 were still NDs in 1994; 6 had become GMs, 1 an assistant GM, and 2 were corporate vice-presidents for news.[63]

Women

The first national survey in 1972 found only 1% of TV stations and 4% of radio stations with female news directors. By 1988 women headed news operations at 18% of commercial TV stations and at one of every four radio stations—26% of commercial and 28% of public noncommercial ones.[64] In 1990, of 775 TV news directors, 125 (16%) were women; of 6,250 radio NDs, 1,590 (25.4%) were women.[65] If that rate of change (1972-1990) continued, women would hold the majority of news director positions in both media in two decades—by around 2010.

Minorities

In 1988 minority persons were news directors at 8% of TV stations surveyed (double the previous year), and at 4.5% of commercial radio and 13% of noncommercial public radio stations.[66] Minority news directors were most found in the top 25 TV markets: at one in ten of network O&Os and affiliates, and at one-third of independent stations. Almost half of those minorities were Hispanic males; about one-fourth were black women and men.

The 775 TV news directors in 1990 included 53 (6.7%) minority men and 25 (3.2%) minority women. They included 22 black men and 8 black women, 27 Hispanic men and 9 Hispanic women, 4 each of Asian men and women, and 4 American Indian women. Of 6,250 radio NDs, 385 (6.2%) were minority men and 175 (2.8%) minority women: 150 each of black men and women, 85 Hispanic men, 25 Asian women, and 150 Native American men.

Age

Like PDs, television news directors are relatively young, with one-third less than 35 years old. Females are significantly younger than their male counterparts in that position. In 1988 the median age (half younger, half older) was 38 for male TV news directors and 31.5 for females; for radio NDs it was almost 35 for males and 28 for women.

Education

By 1986 three-quarters of TV news directors and two-thirds in radio were college graduates. But only one in five NDs (similar to GMs, PDs, and sales directors) felt a graduate degree important for job applicants. In 1993 a survey of news directors at network affiliates across the country found that four out of five responding NDs had completed college degrees—70% had bachelor's degrees only, plus 13% held both bachelors and graduate degrees. Over half of both groups majored in journalism, broadcasting, or mass communications; the rest had majors scattered over twenty other areas mostly in humanities and social sciences.[67]

■ In the 1987 Roper survey, 70% of news directors recommended degrees in journalism and communication, compared with 64% of GMs and 63% of program directors so recommending. Similarly, more than half the NDs urged broad liberal arts background (55%, compared with 47% of GMs and 44% of PDs); but news directors were less supportive of business courses (21% vs. 35% and 26% respectively), despite their increasing role of budgeting and business-related responsibilities.

When asked in 1987 to appraise the importance of higher education for broadcast careers generally, far more news directors (28%) judged it "essential" than did general managers (19%), program directors (10%) or sales directors (6%).[68] Just over half of NDs considered higher education "important"—similar to other department heads. ∎

Income

The *average* or "mean" salary for news directors in 1992 was $50,500 in TV and about $22,000 in radio—lower than any other department head in radio except the traffic manager. (But recall that thousands of radio stations have news "staffs" of only one or two persons, so the news director is less a manager than an all-purpose news reporter-editor-announcer.) The *median* or mid-point weekly salary in TV was $908 or $47,250 a year. The *median* weekly salary for radio NDs was around $366 or $19,010 annually. These totals include neither benefit packages (health insurance, pension, or compensation contributions by the employer) nor funds withheld from paychecks for social security and income taxes. Table 4.4 listed estimated NDs' total compensation before taxes in 1996 as $77,000 at TV stations and $25,000 at radio stations.

The size market and the impact of a station—its audience and advertising effectiveness as measured by ratings and ratecard—determine the level of salary there. For example, 16% of TV news directors responding in 1992 earned a median (half less, half more) of $20,000 a year in ADI markets 151-210; while at the other end of the scale 17% of all NDs were compensated at a median of $92,500 in the top 25 markets—$107,500 at affiliates and half that, $52,900, at independents in the same markets.[69]

Personal Characteristics

In their mid-1980s study, Adams and Fish found TV news directors less positive about their current positions than were station managers and program directors.[70] Although the majority responded favorably to value judgments about their company and work, a large contingent was not so enthusiastic about some points.

∎ Almost 5% disagreed and over 8% were unsure that they found "real enjoyment working here." As many as 9% disagreed, and 14% were unsure, that their present position was "more satisfying than others I could get." Fully 16% disagreed, many of them strongly so, and 11% were not sure, that their company gave "my subordinates a fair deal." As to "satisfaction with my present position for the time being," 13% were unsure but another 11% were wholly dissatisfied. As to their current compensation package (for TV, much less radio), almost one out of four was dissatisfied and another 17% were not sure; 59% did feel they were compensated well. One-fourth thought they would have to change stations to get the compensation they deserved; another fourth were unsure. One in ten felt the station was not giving them a "fair deal," and another 17% were uncertain. Finally, while the large majority was proud of the stations where they worked, 10% disagreed—many strongly—and 17% were not so sure. ∎

A composite profile of television news directors was sketched by Adams and Fish:

> The typical ND from these data is a male in the 35-44 age group who has come out of a news background. He is just as likely to have worked in the same market as not prior to assuming his present position. However, of the three groups [including GMs and PDs], he is least likely to have been promoted to his present position within the same company. He has been continuously

employed in broadcast television since entry shortly after completion of his baccalaureate in journalism, has about eleven years of experience in the industry and has occupied his present position for about one year. He belongs to one national or regional organization as a consequence of his employment—probably RTNDA [Radio-Television News Directors Association], attends one convention with a frequency of at least once a year and reads three trade publications regularly—probably including *Broadcasting.* He devotes 50 to 60 hours per week to his work and work-related activities and does not hold an ownership interest in the company.[71]

SALES MANAGER

Heading the revenue-producing department of a commercial broadcast station or cable operation is the sales manager or director of sales. This position is usually designated general sales manager when assisted by a local sales manager and a national/regional sales manager (see chapter 9 for further details). A local sales manager works with the staff of salespersons—usually called account executives—who generate and service purchases of airtime by local clients. The national sales manager collaborates closely with the station rep organization selling that station's airtime to ad agencies in major markets. In smaller operations, the general sales manager (or, where none, the station manager) may also work on national sales while the sales manager exclusively handles local sales and staff. In cable systems, managers often combine sales and marketing responsibilities.

Duties and Responsibilities

The ideal sales manager has experience not only in sales but also in other phases of broadcasting. Since her function is to market the station's product, the more she knows about the total station operation the better she can design commercial campaigns and sales strategies to achieve maximum results.[72] Further, the head of sales collaborates closely with a station's key *accounting* and *financial* personnel, to forecast projected sales revenues—central to the entire station's budgeting for programming (including news), technical equipment, and salaries. The sales manager joins the *program director* in assessing for top management prospective purchase of syndicated programming to best serve advertisers' needs. The sales department sells airtime for local clients' commercials that are mounted by the *production* staff. And sales depends for many services provided by the *promotion* department.

The sales manager should know the strengths and weaknesses of all communication media—newspapers, magazines, supplements, direct mail, billboards and other forms of advertising. She must study competitive offerings by other media, including cable, as well as by other radio or television stations. That helps her determine how her station can outpace the competition by providing better service—effective audience impact and improved results for advertisers buying time on her station.

The ability to analyze media includes healthy respect for the value of research. A good sales manager raises significant questions for station research personnel to answer. By examining data from audience and marketing studies she can recommend to her staff a range of sales approaches.

The sales manager needs to be experienced in marketing and retailing. Understanding each advertiser's particular problems and objectives, she can make constructive suggestions toward building their business volume. Some of those recommendations may not even relate directly to an advertising campaign. Such broad business acumen generates confidence in merchandisers to place sizable amounts of their advertising budgets with her station.

As manager of the station's prime income-producing department, she must be aware of station expenses. A good sales manager is as proficient in controlling cost items as in producing income. She knows how to keep expense accounts and general costs in line without diminishing the quantity and quality of sales.

She must be adept at sales analysis, able to use tools and procedures of budgeting, projecting, evaluating performance, pricing, inventory control, and research. This results in the all-important rate card, setting airtime charges for commercials; it determines how much revenue a station's schedule can potentially generate. Those skills help set ambitious but realistic, attainable monthly and annual goals the sales staff can strive for. The sales manager must be willing to join her salespeople on the street and to work with them in conceiving and making presentations to agencies. She must have poise mingled with aggressiveness to gain true respect from colleagues who are themselves successful salespeople.

As an administrator, the sales manager needs to be a good organizer, with the ability and inclination to delegate most of the routine paper work to capable assistants. The time saved is devoted to the main functions of her position: (a) overseeing the sales staff; (b) directing the station's local sales, service, and sales development; (c) coordinating network and/or agency accounts carried on the station; and (d) maintaining contacts with the national/regional station representative.

With the perennial need for apt candidates for sales work, one of the sales manager's most difficult tasks is to find people for new or replacement positions on the staff. (One out of every three radio salespeople left their jobs in 1990, including 43% of those in the 100th and smaller markets.)[73] This is doubly important because wrong personnel choices can be extremely costly. Beyond the time, effort, and money needed to train new staff salespersons, station losses caused by dissatisfied clients or broken contracts can be just as large as they are difficult to estimate. Employing good salespeople results in excellent relationships with clients and higher revenue for the station.

The sales manager evaluates his staff's performance in selling the station's commercial time. By private conversation with salespeople, informal discussion with staff members, general sales department meetings, memoranda, and by forwarding printed materials (reports, statistical data, sales success stories, tips on selling techniques, and so forth), he can motivate and support their efforts to sell successfully.

The sales manager should never remove himself from selling. In addition to directing and motivating salespeople to specific targets and working with them for improvement, he needs to call on at least some agencies and clients regularly. Time, station circumstances, and market conditions permitting, it is important for him to be active personally with at least one or two agencies to avoid losing touch with the field. His calls also help him keep abreast of problems confronted by staff members. (A 1991 study found three out of four radio sales managers carried account lists.)

Career Tracks

Successful salespersons with strong track records and a talent for dealing with people are apt candidates for sales managers. The path might start in local radio or cable sales, or else in traffic, research, or sales service at a TV station. Another route is through a national sales rep firm or as an assistant buyer in an ad agency's media department. Yet another path is to begin in retail sales in non-media areas; becoming familiar with many aspects of promoting, selling, merchandising, and marketing that involve advertisers will be valuable later in effectively selling airtime to meet clients' needs.

Women

Key positions in sales are filled by women. Women are in the majority on sales staffs at an increasing number of TV stations. Warner's 1991 study reported that 27% of sales managers in radio and 7% in TV were women; almost half of radio sales staffs and more than a third of TV account executives were women.

Education

Sales managers' advice to those looking to TV sales: 57% recommended a general college education, 37% a marketing major, and 16% a broadcast major. One-third urged getting selling experience outside TV first. Equal numbers (about 13.5%) suggested reading books on selling or starting at a small station.

In the 1987 Roper survey, 60% of sales directors recommended degrees in journalism and communication, 40% added a liberal arts background, and 55% (far more than other department heads) also suggested courses in business.

Income

Sales managers earn about double what other department heads earn, almost equal to GMs' compensation; a very successful sales manager can earn much more than the station's top manager in total salary plus commission. In 1988 the national average for general sales managers in radio was $48,000 and in television $82,000.[74] Radio general sales managers in 1990 earned an average $66,000 nationally, and in the top 100 markets an average of $83,000. Late in 1991 compensation (salary plus benefits, commissions, bonuses) of TV GSMs averaged $92,000, while national sales managers earned $74,500 and local sales managers $70,500.[75] In 1996 GSMs averaged an estimated $114,000 in TV and $72,000 in radio (table 4.4).

Managers of sales and marketing for local cable systems had average salaries of $35,500 in 1988, and at MSO headquarters and regional offices $55,000.[76] Their role in cable grew in the 1990s as physical installations such as stringing cable began to plateau, and subscriber bases leveled off. Future revenues depended increasingly on attracting advertising to local cable channels and national systems. Effective sales and marketing personnel became more central to cable operations. Their compensation averaged an estimated $46,500 in 1996.

Personal Characteristics

An early survey found that, contrary to popular perception, general sales managers viewed themselves more as relaxed and informal (64%) than hard-driving (36%).[77] The vast majority (83%) believed "there are emotional factors than can be sold to buyers" more than analyzing programming for time-sales "strictly from audience size point of view" (17%). Similarly, for effective sales management they relied far more on intuition and personal judgment (87.4%) than "scientific, spelled-out management concepts" (12.6%). They strongly supported giving responsibility to others and letting them carry it out on their own (76%) rather than closely supervising sales personnel (24%). But they were about split between being "a nut for detail" and leaving details to subordinates. When forced to select between two characteristics (not necessarily contradictory ones), 40% said they "won't sacrifice family for the job" whereas 60% selected "my job requires long working hours." Only one-third of sales managers preferred managing to selling; two-thirds of them wished they could spend more time actually selling.

Chief Engineer

Completing the line-up of primary department heads in typical broadcast stations is the chief engineer, who oversees staff engineers and also at many stations production personnel. (At other stations, production crews report to an operations manager or production manager, or to the program director.)

Duties and Responsibilities

Regardless of his technical knowledge, the chief engineer can be truly effective only with training in the broadcast industry broad enough to make him an important factor in successful administration of the station. In addition to scheduling staff members with efficiency and cost control, the chief engineer's obligation to management is to seek every possible means of improving the radio or TV station's technical performance.

The head of the engineering department has varied functions involving personnel and equipment. He must maintain a balanced loyalty between management and the engineering staff. He sometimes assumes the duties of an executive and at other times of specialist in various phases of studio and station operation, master control, shortwave and microwave remote relays, videotape recording and broadcasts, antenna and transmitter work, satellite up- and down-linking, and maintaining and purchasing equipment. He takes full responsibility for all regular maintenance checks and usually is charged with supervising the building and its upkeep.

The chief engineer is expected by management to maintain proper liaison and provide assistance needed by other departments. He helps the station manager or GM better understand the engineering department's work, including its problems and achievements. His services are vital to the manager on FCC technical requirements. He must keep abreast of current trends and future developments in broadcast engineering, so he can recommend how the station can keep pace with those advances. He must be skilled in relationships with unions and he is often expected to participate in community affairs.

In short, the chief engineer, no less than the station manager, usually finds himself in an assignment somewhat different from the sort of career he anticipated when he started working in the industry. Generally, he finds that many adjustments are necessary. When a person with strong engineering background discovers the gratifications in exercising leadership and when the manager has appropriate intellectual curiosity about the engineering department, the station has achieved one of the most important working relationships in broadcasting. The ideal is a general/station manager and chief engineer who operate with mutual understanding and respect.

Balanced Loyalty

Historically, the engineer in broadcasting has not been inclined toward the business view. Yet, the chief engineer is expected to become a part of management's indispensable team with the program and sales managers. At the same time, engineering staff expect their chief to represent their interests fairly and firmly. The chief engineer is compelled to live a double existence, with his principal work in technology somewhat foreign to the manager, coupled with his duty of loyalty to corporate business and programming policy. This is a challenge to one who prefers finding satisfaction in his own creative way with electronic facilities, without being bothered by details expected of an administrator.

In a sense, engineers tend to form a closed fraternity. They can be jealous of their considerable skills and prerogatives. Management may seem remote and indifferent to those working on cameras or at control panels or at the maintenance bench. It is therefore doubly important that the engineering leader understand and communicate company policy to his staff. Shared knowledge fosters mutual support and cooperation. In a well run broadcast establishment, the chief engineer is conversant with the what and why of the overall operation—the policies and rationale (or macro- and micro-intentions) behind management directives.

Even then, conflicts are likely to arise. One problem is controlling the company's overtime costs at the expense of extra income for staff engineers. The question of priorities when engineers want more equipment or supplies than the manager judges the station can afford may create friction between management and the chief engineer and his staff. The need to sharpen engineers' awareness of a station's needs for cleanliness and good housekeeping, when individuals seem more satisfied with their own procedure for equipment care, creates minor irritations. Perhaps the hardest task for the chief engineer—and for any administrator— is to keep the flow of communication accurate and complete, from management to department personnel and from those personnel back to management.

It is no easy task to maintain the balance needed between the interests of technicians and the top manager's office, especially when so much may be misunderstood or misapplied because of unclear or incomplete communication. It applies both ways: managerial directives and motivations may be misconstrued, while the technical language and needs of engineering often must be translated into understandable terms for management.

Relationship with Other Station Departments

The chief engineer must regard his role and his staff as a service arm of the radio or television station or cable operation. While peak efficiency and specific cost control are demanded of engineering management, programming is controlled by the program department and not by engineering. The engineering department is there to furnish technical guidance and operational know-how to stage the station's programming and ensure dependable transmission of its daily schedule.

A good chief engineer works with respective departments to help develop programs and sales. He demonstrates what can be done with audio facilities and videotape in the studio and on remote locations. He assists sales and programming on ideas that are on the drawing board for future implementation. The chief engineer of any radio or TV station can be a willing colleague in this area, but needs to be invited to participate. Once he realizes his suggestions are wanted, he can make valuable contributions to other departments. Of course, the final command function lies in the front office of station management.

This section has stressed the interdependence of the chief engineer and his department with the general administrative routine of the station. While it may seem incongruous to find front office personnel delving into matters of engineering while electrical engineers become involved in finance and general business activity, this mutuality of shared concerns and competence is very important to a broadcast or cable operation's success—that is, to its efficiency and profitability as well as its effective programming and community impact.

Liaison with FCC

Every broadcast operation relies on its chief engineer for liaison with the FCC about technical operations. In fact, in many stations he is the only person truly qualified to handle many FCC matters. He must also know when he is not prepared to address technical legal questions, calling on Washington engineering consultants when necessary to protect the company's interests in matters before the Commission. Few engineering executives at the station level can handle all technical matters relating to federal agency regulations without needing outside engineering consultants. This competence in regulatory aspects of technical operations is critically important. It is difficult to add to or modify station holdings, and even to retain what one has, unless the chief engineer is able to keep management posted on regulatory and technical developments in engineering. Even in the climate of deregulation during the 1980s and 1990s—which reduced monitoring, logging, and reporting by certified engineers because of computerized, highly automated and dependable equipment—most forfeitures (fines) were still levied against stations for violations in technical operations.

Liaison with Unions

In larger markets, with formal labor contracts, the chief engineer carries major responsibility for liaison with union workers on his staff. This is true even of companies large enough to have their own directors of industrial relations. He must work closely with the union shop steward and with the president of the local, as well as with national and international offices of the union. Costly hours, days, and weeks of negotiation and work tie-ups can be prevented if the proper rapport exists.

The chief engineer also oversees the impact of computerized automation on personnel, along with budgetary effects of financial forecasting, change of station format, network affiliation, or ownership (see chapter 12).

Building Supervision

It is advisable to make the chief engineer responsible not only for facilities but for the entire physical plant, both in the studio-office building and at the transmitter. In a business based on electronics, almost every part of the building and all physical aspects of the broadcast operation benefit from the professional guidance and supervision of the engineering department head.

Equipment Purchase and Maintenance

Along with recommending strategic buying of updated equipment, one of the most important administrative areas for an engineering manager is maintenance—not only repair work but planned systematic preventative maintenance so equipment functions dependably (see chapter 12). That also involves either rotating technicians among routine monitoring tasks or assigning each one to specific duties, whichever is more efficient and responds to individuals' abilities and interests.

Future Planning

A good engineering executive plans long-term technical requirements. He stays ahead of developments in the industry, planning future steps for better performance and, where possible, at less cost than in the past. He plans now for eventualities, even for revolutionary changes. This may involve AM stereo, or digital audio broadcasting (DAB) that may replace both AM and FM terrestrial transmission; it can mean state-of-the-art ENG or SNG equipment, up-link satellite trucks, auxiliary services overlaid on transmission signals, and HDTV installation to replace all studio and transmission equipment at a TV station. For cable operations, it might involve shifting to higher capacity coaxial cable or fiber optic lines, and multi-use addressable "black boxes" in subscribers' homes.

The entire program and engineering operation of conventional TV and cable systems stands to be massively affected by direct broadcast satellite (DBS) service introduced in the mid-1990s, forcing terrestrial stations to concentrate on local and regional programming for survival, as occurred with radio when networks shifted over to TV in the 1950s. The engineering executive must plan sufficiently in advance so that when a change is anticipated, say in five years, he can arrange to absorb within his department, or in other phases of the station's operation, or by attrition in intervening years, personnel who would be relieved of their current duties due to the changes. Similarly, he must project reallocated budgeting for equipment and operations.

Career Tracks and Education

Predictably, electrical engineers specialize in that field in college. Major engineering colleges across the nation offer degree programs that prepare competent technicians and researchers who are also fully developed as persons. Their curricula require studies in humanities and social sciences so an engineering graduate can relate effectively to many sectors in the business and social world. Those broad studies in liberal arts and related areas prepare an engineering specialist particularly well for a career in broadcasting and cable—industries intimately linked to the lives of families, businesses, and politics in their communities and conducted under constant deadlines in close quarters with other department personnel. Co-operative education that alternates between semesters in the workplace and on campus, plus internships, parttime work, and initial technical jobs with local businesses, the telephone company, and other utilities, can lead to engineering positions in radio, television, and cable.

Rapid developments in new technologies require continually updating one's formal education and applied experience by reading trade magazines and journals, attending conventions and trade shows of newest equipment, and participating in conferences designed for professional engineers and broadcast-cable technical personnel. For example, the National Association of Broadcasters' annual convention features not only acres of sophisticated electronic gear and automated systems but also dozens of special panels and lectures by engineering specialists on late-breaking developments in technology and FCC regulations for operating standards.

Competent engineers keep abreast of federal regulations for the technical operation of licensed broadcast stations. The FCC has greatly reduced requirements for certifying radio and TV engineers. A first-class endorsement or license, granted after passing a series of examinations about technical matters, is no longer required of personnel overseeing station operation or running an audio board on the air. But top-quality, professionally run stations still place a premium on such training and certification when they hire engineering staff.

Income

Chief engineers at TV stations averaged an estimated $54,000 in 1996. That was about the same as what PDs earned, but half the compensation of general sales managers, and one-third of GMs' average income (table 4.4). At radio operations, chief engineers earned an estimated average of $34,000—$5,000 less than PDs, and about half the compensation paid to GMs and GSMs. But statistical salary averages do not necessarily reflect total annual compensation of professional engineers in radio because FCC deregulation no longer required stations to employ first-class licensed engineers on duty fulltime. Many radio stations hired only parttime engineers, sharing their services with other radio or television stations in the market.

——〰——

These four management positions form the nucleus of a broadcast or cable company's administrative staff: the general/station manager, program director, sales manager, and chief engineer. At large stations active in news/talk programming, a news director also plays a key role, reporting directly to top management; at smaller stations the ND often reports to the program director. Other positions reflected in organization charts variously serve in the top ranks or hold positions farther down the administrative structure: operations manager, financial affairs, business manager, and promotion director. Their responsibilities are treated in later chapters.

Network and corporate-group administration is far more complex but is built on the same foundation of programming, sales, and engineering, plus news, financial and business affairs, and often legal affairs. At national media corporations, duties and responsibilities, career tracks and education are similar to respective posts in individual stations. Further, there are layers of divisions and subdivisions, each with their own presidents and vice-presidents, as described earlier. Later chapters refer to some of their duties and decision making, although this book focuses on details of local station and cable management.

——〰—— CHAPTER 4 NOTES ——〰——

1. Descriptive analyses of management structures and organizations including discussions of "span of supervision," "chain of command," and "unity of command," as well as functions of line and staff personnel can be found in William J. McLarney and William M. Berliner, *Management Training: Cases and Principles,* 5th ed. (Homewood, Ill.: Richard D. Irwin, Inc., 1970), 38–56 and 100–119. See also Alan C. Filley and Robert J. House, *Managerial Process and Organizational Behavior* (Glenview, Ill.: Scott, Foresman, 1969), Chapter 4, "Management Process and Organizational Design," 69–100; Chapter 9, "Division of Labor," 212–238; and Chapter 11, "Line and Staff Relationships," 257–280; George R. Terry, *Principles of Management,* 4th ed. (Homewood, Ill.: Richard D. Irwin, Inc., 1964), Chapter 15, "Modern Organizational Concepts and Departmentalization," 303–331; Chapter 18, "Authority Relationships," 388–418; and Chapter 20, "Organization Charts and Manuals," 443–462; Richard M. Hodgetts, *Management: Theory, Process, and*

Practice, 2nd ed. (Philadelphia: W. B. Saunders, 1979), Chapter 6, "The Organizing Process," 109–134.

For organization charts of media companies, see John M. Lavine and Daniel B. Wachman, *Managing Media Organizations: Effective Leadership of the Media* (New York: Longman, 1988), Chapter 6, "Organizing Media Companies," 127–158. Barry L. Sherman, *Telecommunications Management: The Broadcast & Cable Industries* (New York: McGraw-Hill, 1987), 245–255, offers detailed tables of organization for electronic mass media.

2. Robert Townsend, *Up the Organization* (Greenwich, Conn.: Fawcett edition, 1970), 187; Clarence B. Randall, *The Folklore of Management* (New York: Mentor edition, 1962). Similar cautions were noted by J. Leonard Reinsch and Elmo I. Ellis, *Radio Station Management,* 2nd rev. ed. (New York: Harper & Row, 1960), 38–39; they were quoted at the start of this chapter.

3. Except for group-owner Tribune Broadcasting's WGN-AM-TV (Chicago), the charts are composites of several representative stations and networks.

4. For brief comments on broadcast station organization relationships, see Edd Routt, *The Business of Radio Broadcasting* (Blue Ridge Summit, Pa.: Tab Books, 1972), 167 ff. and 186–188 (itemized lists of on-air personnel needed for various kinds of radio formats); Robert H. Coddington, *Modern Radio Broadcasting: Management and Operation in Small-to-Medium Markets* (Blue Ridge Summit, Pa.: Tab Books, 1969), Chapter 9, "The Staff," 143–158; Sol Robinson, *Broadcast Station Operating Guide* (Blue Ridge Summit, Pa.; Tab Books, 1969), Chapter 5, "Staffing a Station," especially 103–104, 109–112, and 136–138; Barry L. Sherman, *Telecommunications Management: The Broadcast & Cable Industries* (New York: McGraw-Hill, 1987), 247–253. Organization charts for cable systems are in Norman Marcus, *Broadcast and Cable Management* (Englewood Cliffs, N.J.: Prentice- Hall, 1986), 156; and Thomas E. Baldwin and D. Stevens McVoy, *Cable Communication,* 2nd ed. (Englewood Cliffs, N.J.: Prentice-Hall, 1988), 229.

5. Sydney W. Head and Christopher H. Sterling, *Broadcasting in America: A Survey of Electronic Media,* 6th ed. (Boston: Houghton Mifflin, 1990), 187–189. Cf. listings in *Broadcasting & Cable,* 1 and 8 July 1996; see chapter 10 of this text for details.

6. "Tinker Lines Up NBC Execs," *Variety,* 26 August 1981, 51.

7. NAB Department of Broadcast Management, *Television 1974/Employee Wage & Salary Report* and *Radio 1974/Employee Wage & Salary Report* (Washington, D.C.: National Association of Broadcasters, 1974). These figures are similar to those reported by Lawrence W. Lichty and Joseph M. Ripley, "Size and Composition of Broadcasting Stations' Staffs," *Journal of Broadcasting* (Spring 1967): 139–151. More recent data are from National Association of Broadcasters and Broadcast Cable Financial Management Association, *1990: Radio Financial Report* and *1990: Television Financial Report* (Washington, D.C.: National Association of Broadcasters, 1990).

8. McLarney and Berliner, *Management Training,* 50. The EBU management conference in Brussels recognized that many broadcast organizations reach a growth threshold beyond which they can no longer be directed and controlled adequately without extensively decentralizing powers and responsibilities. They also pointed out that joint management of both radio and television operations may be feasible only with a staff of up to 100 people; "above this threshold, management becomes so complex that the [two] media must be separated if their specificity is to be respected"; Robert Wangermee, "Towards Modern Management in Broadcasting," *EBU* [European Broadcasting Union] *Review* (November 1972): 34.

9. Gary N. Powell, *Women and Men in Management* (Newbury Park, Calif.: Sage, 1988), 143. About 5% of all businesses have been started and are owned by women (p. 85); other data, pp. 13 and 152–153. In 1990 the U.S. Department of Labor analyzed ninety-four large corporations. It found that while one-third of all employees were women and 15.5% minorities, among managers only 17% were women and 6% minorities; top executives included 6.6% women and 2.6% minorities. "Business: Measuring the Glass Ceiling," *U.S. News & World Report,* 19 August 1991, 12.

10. Judy B. Rosener, professor of management at University of California at Irvine, reported her findings in "Ways Women Lead," *Harvard Business Review* (November/December 1990): 119–125, and also in "Why Not Give Women Credit?" *Harvard Business Review* (January/February 1991): 152–153. New York Times News Service, "Women's Role in Management Evolving Differently Than Men's," *Tuscaloosa News,* 5 May 1991, 7-E, citing Judy B. Rosener's article in *Harvard Business Review.* Also cited is the following description drawn from Chapter 2, "The Web of Inclusion" by Sally Helgesen, *The Female Advantage: Women's Ways of Leadership* (New York: Doubleday, 1990).

11. Alice Sargent, "Prologue" and Chapter 6, "New Models of Effective Managers in the 1980s," in *The Androgynous Manager* (New York: ANACOM, 1981), 2; cited by Powell, *Women and Men,* 168–169. Powell (p. 163) cites *Harvard Business Review* studies reflecting increasing acceptance of working for female managers. Responding to "I would feel comfortable working for a woman. . . ," male executives' acceptance almost doubled in two decades from 27% (1965) to 47% (1985), while female executives' acceptance rose from 75% to 82% in that period.

12. William E. Reif, John W. Newstrom, and Robert M. Monczka, "Exploding Some Myths About Women Managers," *California Management Review,* (Summer 1975): 79; cited by Richard M. Hodgetts, *Management: Theory, Process, and Practice,* 2nd ed. (Philadelphia: W. B. Saunders, 1979), 416.

13. Federal Communications Commission, *60th Annual Report/Fiscal Year 1994* (Washington, D.C.: U.S. Government Printing Office, 1994), 49.

14. "Job Profiles," *Variety,* 4 March 1987, 110. Even by the mid-1980s, 47% of middle management positions at CBS News were held by women, including bureau chiefs in Paris and Dallas, and three producers and the assignment manager in the Washington bureau; Edward M. Joyce, president, CBS News, in a speech to the annual convention of American Women in Radio and Television, Chicago, 30 May 1984; 5–7 of speech text.

15. Nancy C. Widmann joined CBS in 1972, became vice-president and general manager of CBS Radio Spot Sales in 1979, the following year vice-president and GM of WCBS-FM in New York City, subsequently vice-president of CBS-Owned AM Stations, and then of all CBS Owned Radio Stations in 1987 until her presidential move in 1988. "Nancy C. Widmann Named President, CBS Radio Division," press release from CBS/Broadcast Group, New York, 29 June 1988.

16. ABC had one female program director (PD) and no other female managers at 8 O&O TV stations, two women served as general sales managers, two as news directors, and one as program director at 17 O&O radio stations; CBS' 4 TV stations had women as two directors of broadcasting (PDs) and its 17 radio outlets had four as general sales managers, three news directors, and one PD; NBC's five TV stations had four PDs, while its 9 radio stations had three general sales managers, two PDs and two news directors who were women. Data compiled by NOW (National Organization of Women) Legal Defense and Education Fund and by *Variety* research; reported in "Job Profiles," *Variety,* 4 March 1987, 110.

17. Verne Gay, "Is There Room at the Top for the Women of TV News?" *Variety,* 10–16 [*sic,* the magazine over two periods of time was dated with the week's dates instead of a single publication day's date] May 1989, 86.

18. Women headed O&O stations for none of the three traditional networks but for two Fox networks. Female GMs ran 13 of ABC's 220 affiliates; 11 of NBC's 209, 3 of CBS' 220; plus 13 of Fox's 130 affiliated stations. "For Femmes, G.M. Means 'Generally Missing,'" *Variety,* 1 July 1991, 24. The only female general manager of a network O&O station had retired by 1990: Jeanne Findlater who headed ABC's WXYZ-TV in Detroit before new network owner Capital Cities Communications sold it to Scripps-Howard Broadcasting in 1986 to meet FCC limits. See Head and Sterling, *Broadcasting in America,* 213, citing a 1988 survey by American Women in Radio and Television (*Broadcasting,* 20 June 1988, 55); cf. Fish and Adams, 30 (see note 23).

19. Data from 1991 FCC report cited in "FCC Reports Minority B'cast Jobs Increasing," *Variety,* 14 October 1991, 73. Data for 1993 in Federal Communications

Commission, *60th Annual Report: Fiscal Year 1994* (Washington, D.C.: U.S. Government Printing Office, 1994), 49.

20. Sources for this paragraph: "Blacks' Gains on TV Aren't Reflected in Networks' Executive Suites," *Variety,* 30 May 1990, 1, 4; New York Times News Service, "Minority Broadcasting Makes Far Bigger Impact on Radio than TV," *Tuscaloosa News,* 3 August 1990, 7-H. Thomas H. Billingslea, Jr., "Monday Memo," *Broadcasting,* 28 May 1990, 25. Companies controlled 51% or more by minorities were licensees of 2.3% of all U.S. radio-TV stations in 1987, up from only 1% of broadcast stations in 1977. "More Minorities Buying B'cast Stations," *Variety,* 5–11 July 1989, 52. Dwight Ellis, vice-president, Human Resources Development, National Association of Broadcasters, January 31, 1991; in personal correspondence cited by James Phillip Jeter, "Black Broadcast Station Owners: A Profile," *Feedback,* 32:3 (Summer 1991): 14–15. Vice-President Al Gore's speech to National Association of Black Owned Broadcasters [*sic*], cited in *Broadcasting & Cable,* 19 September 1994, 6.

21. Alfred J. Jaffe, "Indie GM Earnings Show a Decline from '85 to '86," *Television/Radio Age,* 5 January 1987, 79–83.

22. The National Association of Broadcasters reported a roundtable conference of radio managers; *NAB News,* 3 December 1990, 9.

23. Marjorie J. Fish and R. C. Adams, "Composite Profiles of Key Television Managers" (unpublished report, Washington State University, Pullman, Washington, and California State University, Fresno, California; February 1987), 10, 29. Cf. "In Search of Executive Excellence," *IRTS News,* January 1988, 5.

24. Bob Morse, quoted by Lou Prato, "Bob Morse Interview: Management in the 1990s," *RTNDA Communicator,* April 1991, 13.

25. "Making the Jump from News Director to General Manager," *RadioWeek* (NAB: Washington, D.C.), 16 April 1990, 4.

26. *Radio Only,* 16 November 1994, 3; cited by "Broadcast Sales Training Executive Summary," 21 November 1994, 3.

27. Bohn and Clark, 213; but figures cited in their table on p. 207 indicate that only 38% of radio station managers had attended or completed college and 6% had done graduate studies. Previous data were by Winick and by the Association of Professional Broadcast Education and National Association of Broadcasters (APBE-NAB).

28. Fish and Adams, "Composite Profiles," 14–15.

29. "In Focus: Television Top Executives—The Powers That Run Television," *Channels,* 13 August 1990, 38-48.

30. Powell, *Women and Men in Management,* 86.

31. Data from survey conducted by James Phillip Jeter, "Black Broadcast Station Owners: A Profile," *Feedback* (Summer 1991): 14–15.

32. The Broadcast Education Association's 1990 directory listed major colleges that were institutional members: 144 of them offered baccalaureate degrees only (either Bachelor of Arts or Bachelor of Science); 74 others also offered masters degrees, and another 28 offered doctorates as well as B.A./B.S. and M.A. degrees. An additional 41 two-year community colleges presented broadcast study programs, and 39 four-year colleges provided courses but not full degree programs. Ray Carroll, ed., *BEA Directory of Membership, 1990: Broadcast and Electronic Media Programs in American and Canadian Colleges and Universities,* 17th Report (Washington, D.C.: Broadcast Education Association, 1990). In 1995, 229 colleges and universities, including community colleges, were institutional members of BEA. Barry S. Sapolski, ed., *BEA 1995-1996 Membership Directory* (Washington, D.C.: BEA, 1995).

33. *Television/Radio Age* mail survey, December 1978/January 1979; data reported in that magazine.

34. The Roper Organization, Inc., "Electronic Media Career Preparation Study; December 1987" (New York: Roper Organization, 1987), 19.

35. Cf. Sherman, *Telecommunications Management,* 94, 118, and 140, for 1983 data from NAB's *Employee Compensation Report* plus *Cable Television Business,* 15 June 1983,

43, and Head and Sterling, *Broadcasting in America* (1990 edition), 214, for 1988 data in NAB's *Radio. . . and Television Employee Compensation & Fringe Benefits Report.* Rates of increased compensation were computed for positions between 1983 and 1988 figures provided by those sources; those totals were extrapolated further to reflect private industry's raises of 4.1% in 1989; 4.5% in 1990; and 4.1% in 1991 (making U.S. 1991 averages about 15% higher than 1988 figures when raise percentages were compounded); source: *U.S. News & World Report,* 24 December 1990, 66, reporting data and projections from the Office of Personnel Management and Bureau of Labor Statistics. Projections were made from proportionate increases in positions from 1983 to 1988 in both radio and TV, which were respectively about 30% and 35%, and from 1988 to 1991—approximately 22% each. Computations were based on and resulted in:

U.S. AVERAGE SALARY: % INCREASES, 1983–1991

	1983–1988	*1988–1991*	*Composite, 8 Years*
Private-sector industries	20%	15%	35%
Radio stations	21%–45% [30%]	16%–32% [22%] (est.)	52%
Television stations	25%–45% [35%]	19%–32% [22%] (est.)	57%
Local cable systems	NA	NA	54%

36. For Vernon Stone's data on 1990 news salaries, see "News Salaries Lag Cost of Living," *RTNDA Intercom,* 1 February 1991, 1; and "News Salaries Lose Out to Cost of Living," *Broadcasting,* 11 February 1991, 32.

37. "TV Station Executive Earnings," *Television/Radio Age,* 8 February 1982, 54.

38. These and following data are projected from Head and Sterling, p. 214, citing National Association of Broadcasters, *Television. . . and Radio Employee Compensation & Fringe Benefits Reports* (Washington, D.C.: 1988); see table 4.4 for details. Estimates are approximately consistent with data reported in the mid-1990s from various sources.

39. Data extrapolated from figures cited by Sherman, *Telecommunications Management,* 140, from *Cable Television Business,* 15 June 1983, 43, and *Cable Marketing,* May 1986, 32; projections explained in notes to table 4.4.

40. Source: *Television/Radio Age* mail survey, December 1978-January 1979; data reported in that magazine.

41. R. C. Adams and Marjorie J. Fish, "A Descriptive Study of TV General Managers' Perceptions of Station Management Style and Selected Concomitant Variables" (unpublished paper), 29. A total of 174 general managers (64% of 274 representatively sampled) participated in the survey, from most kinds of stations and markets except for the disproportionate nonresponse from largest network-affiliated stations.

42. Marjorie J. Fish and R. C. Adams, "Composite Profiles of Key Television Managers" (unpublished paper presented at the Western Speech Communication Association, Salt Lake City, February 1987), 15.

43. "Program Directors: Getting Back in Decision-making Loop," *Broadcasting,* 30 January 1989, 39.

44. Succinct descriptions of the program director's role in radio stations are detailed by Bob Paiva, *The Program Director's Handbook* (Blue Ridge Summit, Pa.: Tab Books, 1983), Chapter 1, "The Program Director"; Jay Hoffer, *Organization & Operation of Broadcast Stations* (Blue Ridge Summit, Pa.: Tab Books, 1971), Chapters 2 and 3: "The Program Director" and "The PD's Staff & Public Relations"; and also in Edd Routt, *The Business of Radio Broadcasting* (Blue Ridge Summit, Pa.: Tab Books, 1972), 169–171. See also: Sol Robinson, *Broadcast Station Operating Guide* (Blue Ridge Summit, Pa.: Tab Books, 1969), 125–130; and Jay Hoffer, *Managing Today's Radio Station* (Blue Ridge Summit, Pa.: Tab Books, 1968), 105–107. A useful summary of the traditional TV program director's duties is provided by Edward A. Warren "Programming for the Commercial Station" in Yale Roe, ed., *Television Station Management: The Business of Broadcasting* (New York: Hastings House, 1964), 107–117.

A thoughtful analysis of concepts about audiences and radio programming that PDs could well study is David T. McFarland's *Contemporary Radio Programming Strategies* (Hillsdale, N.J.: Lawrence Erlbaum Associates, 1990).

See also programming texts by Carroll and Davis; Eastman, Head, and Klein; and related sections of broadcast management texts by Keith and Krause; Marcus; Pringle, Starr, and McCavitt; and Sherman.

45. Bob Paiva, *The Program Director's Handbook* (Blue Ridge Summit, Pa.: Tab Books, 1983), 1–2.

46. Marjorie J. Fish and R. C. Adams, "Descriptions of Personal Characteristics, and Coherence of Management Style and Job Satisfaction among TV GMs, PDs, and NDs" (unpublished paper presented to Western Speech Communication Association, Salt Lake City, February 1987), 18.

47. *Television/Radio Age* mail survey, December 1978-January 1979; data reported in that magazine.

48. The Roper Organization, "Career Preparation Study" p. 19.

49. Data extrapolated from figures cited by Sherman, *Telecommunications Management,* 140; from *Cable Television Business,* 15 June 1983, 43; and *Cable Marketing,* May 1986, 32. Cf. Head and Sterling, *Broadcasting in America,* 6th ed., 214.

50. Data from the NAB and Broadcast Cable Financial Management Association's *1991 Television Employee Compensation and Fringe Benefits Report* reported in "Most TV Pay Lags Inflation, Study Finds," *Broadcasting,* 25 November 1991, 50.

51. *Television/Radio Age* mail survey, December 1978-January 1979; data reported in that magazine.

52. Marjorie J. Fish and R. C. Adams, "A Descriptive Study of TV Program Directors' Perceptions of Station Management Style and Selected Concomitant Variables" (unpublished paper presented at the Speech Association Convention, Denver, Colorado, November 1985), 29.

53. Marjorie J. Fish and R. C. Adams, "Composite Profiles of Key Television Managers" (unpublished paper presented at Western Speech Communication Association, Salt Lake City, February 1987), 15.

54. Qualities of a competent news director are described in Hoffer, *Organization and Operation of Broadcast Stations,* Chapter 5, 91–107.

55. *Fortune,* November 1969, 161, cited by Hoffer, 91. For some of the extensive literature on bias and selectivity in broadcast media, see: Robert Cirino, *Don't Blame the People* (Los Angeles: Diversity Press, 1971); Cirino, *Power to Persuade: Mass Media and the News* (New York: Bantam Books, Inc., 1974); Edith Efron, *The News Twisters* (Los Angeles: Nash Publishing, 1971); Edward J. Epstein, *News from Nowhere: Television and the News* (New York: Vintage, 1973); Herbert Gans, *Deciding What's News* (New York: Pantheon, 1979); Gaye Tuchman, *Making News: A Study in the Construction of Reality* (New York: Free Press, 1978); Av Westin, *News Watch: How TV Decides the News* (New York: Simon and Schuster, 1982); John B. Robinson and Mark R. Levy, *The Main Source: Learning from Television News* (Beverly Hills, Calif.: Sage, 1986). More recent analyses include Martin A. Lee and Norman Solomon, *Unreliable Sources: A Guide to Detecting Bias in News Media* (New York: Carol Publishing Group, 1990, 1991); Robert Goldman and Arvind Rajagopal, *Mapping Hegemony: Television News Coverage and Industrial Conflict* (Norwood, New Jersey: Ablex, 1991); Michael Parenti, *Make Believe Media: The Politics of Entertainment* (New York: St. Martin's Press, 1991).

56. RTNDA survey responses from 752 broadcast news organizations, cited in "Profile of a News Director," *Broadcasting,* 4 December 1972, 47.

57. "RTNDA Radio-TV Staff Surveys," *Broadcasting,* 19 March 1990, 48.

58. Michael L. McKean and Vernon A. Stone, "RTNDA Research: Why Stations Don't Do News," *RTNDA Communicator,* June 1991, 22–24. See Papper, Gerhard, and Sharma, 26, for 1995 figures.

59. Vernon A. Stone, "Changing Profiles of News Directors of Radio and TV Stations, 1972-1986," *Journalism Quarterly* (Winter 1987): 745-749; Bob Papper, Michael

Gerhard, and Andrew Sharma, "More News, More Jobs," *RTNDA Communicator,* June 1996, 20–28; averages are based on the 679 TV stations (56.8% of all in the United States) and 449 radio stations (48.7% of 922 randomly surveyed) responding to the survey conducted in late 1995 by Radio Television News Directors Foundation and Ball State University.

60. Joseph Vitale, "The Newsroom's Revolving Door," *Channels,* September 1987, 46–48; data drawn from eighty-six radio and TV NDs reported by Tim Bock in "ND's Talk Back," p. 48.

61. See Kevin Goldman, "News Directors' Nomadic Life Impacts Local TV Journalism," *Variety,* 5 February 1986, 48, 136.

62. Joseph Vitale, "The Newsroom's Revolving Door," 48. He quotes Nick Lawler, seven years an ND before becoming a news consultant: "in market after market, the station with the number-one-rated [local] newscast is usually the leader overall" in program ratings for their entire schedule because early evening news leads into prime-time. Late evening news then leads into late-night programming which usually also determines where the TV channel is set when turned on the next day.

63. Andy Barton, "Ten Years After," *RTNDA Communicator,* February 1995, 11–12.

64. Vernon Stone, "RTNDA Research: Women Gain as News Directors," *RTNDA Communicator,* December 1989, 18–20.

65. Vernon A. Stone, "RTNDA Research: Minority Share of Work Force Grows for Third Year," *RTNDA Communicator,* May 1991, 20–22.

66. Vernon A. Stone, "Minorities Gain in TV News, Lose in Radio," *RTNDA Communicator,* August 1989, 32–35.

67. James W. Redmond, "A Case for Graduate Programs for Television News Directors, " *Journalism Educator* (Summer 1994): 33–42.

68. The Roper Organization, *Career Preparation Study,* p. 30.

69. Vernon A. Stone, "TV News Pay Lags Cost of Living, Radio News Directors Gain," *RTNDA Communicator,* March 1992, 12–14.

70. R. C. Adams and Marjorie J. Fish, "A Descriptive Study of TV News Directors' Perceptions of Station Management Style and Selected Concomitant Variables" (unpublished paper, California State University at Fresno and Washington State University, n.d.; internal data gathered in the early 1980s indicate this report was published sometime after 1984-1985).

71. Marjorie J. Fish and R. C. Adams, "Composite Profiles of Key Television Managers" (unpublished paper presented to the Western Speech Communication Association, Salt Lake City, February 1987), 16.

72. See Albert John Gillen, "Sales Management for the Network Affiliate," and Charles Young, "Sales Management for the Independent," Chapters 13 and 14 in Roe, ed., *Television Station Management,* 181–206. For sales management in radio, see Part III, "Sales," in Jay Hoffer, *Managing Today's Radio Station,* 201–288; also Chapter 9, "The Sales Manager," in Jay Hoffer, *Organization and Operation of Broadcast Stations,* 132–186. See also Jay Hoffer and John McRae, *The Complete Broadcast Sales Guide for Stations, Reps and Ad Agencies* (Blue Ridge Summit, Pa.: Tab Books, 1981); Charles Warner, *Broadcast and Cable Selling* (Belmont, Calif.: Wadsworth, 1986); Barton C. White and N. Doyle Satterthwaite, *But First, These Messages: The Selling of Broadcast Advertising* (Boston: Allyn and Bacon, 1989); and sections of other management books cited such as Sherman; and Pringle, Starr, and McCavitt.

73. Figures for a twelve-month period were reported early in 1991 from a survey conducted by Charles Warner for the Radio Advertising Bureau; "More Career Advancement for Women in Radio than TV, Study Shows," *Broadcasting,* 18 February 1991, 45. Following data are from this report and a similar one he conducted for the Television Bureau of Advertising.

74. National Association of Broadcasters, *Radio. . .* and *Television Employee Compensation & Fringe Benefits Report* (Washington, D.C.: NAB, 1988); cited by Head and Sterling, *Broadcasting in America,* 6th ed., 214. 1990 data are from Charles Warner's 1991 report (see preceding note).

75. NAB/BCFMA's data; Sukow, *Broadcasting,* 25 November 1991, 50.

76. Data extrapolated from figures cited by Sherman, *Telecommunications Management,* 140; from *Cable Television Business,* 15 June 1983, 43, and *Cable Marketing,* May 1986, 32. See Head and Sterling, 6th ed., 214.

77. *Television/Radio Age* mail survey, December 1978-January 1979; data reported in that magazine.

Managing Media Personnel

People and profits are inexorably linked. . . . Human beings will make or break a company.

RODNEY W. BARDY, PRESIDENT/CEO,
BONNEVILLE INTERNATIONAL

The number of people employed in American broadcasting is relatively small, compared with other major industries. In 1993 the FCC reported some 148,579 persons were employed fulltime in commercial and noncommercial radio and television stations, plus 108,280 in cable—mostly technicians and clerical staff.[1] In 1990 the U.S. Department of Labor estimated total employment in broadcasting (including networks) at 234,000, plus 124,600 more in cable systems and cable-related services such as DBS, multipoint distribution systems, and closed-circuit TV operations.[2] Single corporations employ more fulltime workers. IBM's world-wide staff in 1993 was 275,000 (down from 406,000 seven years earlier before massive downsizing began). In 1995 Boeing aircraft in Seattle employed a labor force of 105,500. It has been estimated that radio and TV stations, plus network headquarters, local cable systems, and related media companies employ almost 600,000 people fulltime—compared with 900,000 employed by all divisions of a single giant manufacturing corporation, General Motors.[3]

Men outnumber women in broadcasting by a ratio of only 1.5 to one—down from 1974's ratio of four to one. In 1995, 41% of radio-TV station employees were women and almost 20% were minorities.[4] Women increasingly held responsible positions in all phases of broadcasting, as noted in the previous chapter. In the 1990s several thousand women and minorities, male and female, manage or own stations and are executives with broadcast corporate owners including networks (see chapter 10 on "Financial Management" including broadcast ownership).

When we speak of managing personnel at broadcast stations, how many people are we talking about? In the 1990s, radio staffs ranged from 5 to 47 full-time employees, depending on market size; because most radio stations are in medium to small markets, the national statistical average is about 17 persons per station. Television stations employ between 41 and 155 (and in top 10 markets, up to 250 or more); the number of fulltime workers averages about 90 at affiliates and 60 at independent TV stations (see tables 4.1 and 4.2 in chapter 4).

A radio station with 15 fulltime employees typically employs 7 in programming and production, 5 in the sales department, 2 in administration offices, and 1 in engineering.[5] Of 94 fulltime employees at a typical TV station, 21 are in programming and production, 21 in the news department, 20 in engineering, 16 in sales and promotion, and 12 in administrative office positions.

No other element of the broadcasting enterprise can deliver as great a return on investment as its human resources. Superior physical plants and technical facilities depreciate in net worth over a period of years. But the value of people to a station should increase as their period of employment lengthens. Staff "turnover" is a two-edged sword: It eliminates nonproductive or troublesome employees, but it also loses talented and effective staff to competing stations or to larger markets, reducing continuity among those operating the company. Rodney H. Bardy, president/CEO of Bonneville International, refers to it as "a value-driven company, composed of values-driven people. . . . A broadcast organization is only as effective as its people."[6] Naisbitt and Aburdene put it more strongly, speaking of personnel in any kind of company:

> In the new information society, human capital has replaced dollar capital as the strategic resource. People and profits are inexorably linked. . . . [H]uman resources are any organization's competitive edge. . . . It is not a question of being nice to people. *It is simply a recognition that human beings will make or break a company.*[7]

The station manager's ultimate success depends on nurturing employees' interests and abilities to strengthen their competence, productivity, and satisfaction. A manager should give top priority to maintaining good human relations. An administrator who regards managing people as a major responsibility and devotes adequate attention to them finds that employee efficiency increases proportionately. Staff members with excellent morale and strong station loyalty respond with enhanced creativity and productivity. The manager must be continually vigilant in employee relations, attentive to matters vitally important to the working staff such as equitable wages and salaries, proper recognition for work achievement, and opportunities for advancing in the organization. Most employees want assurance that, if their assignments are accomplished satisfactorily, they need not be concerned about their job security. In recent years this sense of stability has been threatened by widespread restructuring and "shaking out" of media ownership, as well as cost efficiencies gained by "downsizing" staffs, as described in chapter 1.

Partly because it is looked on as a form of show business and has high visibility socially, broadcasting attracts many creative people into sales and promotion as well as to production and programming. If talented persons find work assignments merely routine with little chance for innovation, the most creative ones will seek positions elsewhere. Creative and competent people must have opportunity to use their talents. It is one achievement to hire the best qualified people available; it is another to keep them. Sustaining their top performance is a key duty of successful management.

Employee turnover is also expensive in money and time.[8] It has been estimated that a single resignation costs more than $400 to process. The total cost of replacing a TV station department head in the late 1980s went as high as $10,000; replacing a productive salesperson cost up to $12,000; less expensive was finding a TV news director ($1,500) or production manager ($1,200). These included expenses associated with moving and housing new hires, plus value of time spent by management in training them, plus revenue lost before a successor replaced a departed sales account executive. A search, interview, negotiation, and final hiring in each instance could take between two weeks to eight months for various positions, depending on market and station size. General managers typically needed two to three months to find TV department heads; news directors filled staff positions from one week to three months; production and sales managers took two to six weeks to hire new staffers. Meanwhile, work patterns of remaining employees are often disrupted until replacements are hired and become independently productive—usually after some in-house specific training.

"Personnel" Includes Middle Managers

This chapter looks to a general manager's responsibilities for staffing the station. Included in that total staff are the manager and department heads, who are the company's key personnel, along with all other employees. (Those middle management positions were profiled in chapter 4.) The general/station manager must work closely with her department heads to form a cohesive "coaching staff" who collaborate effectively in melding their respective staff members—in engineering, sales, production, programming, news, promotion, business affairs, traffic and billing—into a mutually supportive and productive station-wide team.

Together these middle and top managers plan how best to staff the total station operation. An organization chart (samples in chapter 4) outlines the division of labor, showing lines of authority and accountability between and within departments. Department heads develop job descriptions for staff positions, noting qualifications needed to carry out the purpose and responsibilities for each job. They assess current staff performance, retaining and promoting or terminating, and develop procedures for recruiting new employees.

Selecting Personnel

Because many broadcast companies follow the standard of two weeks' notice for resignation or for dismissal, that gives the station only a two-week period to locate replacements. This is hardly adequate for looking over the field of available personnel locally and in other markets and for making a satisfactory selection. Sometimes, as the time for securing a replacement nears an end without likely prospects, the first available person is hired.

Where do stations find short-notice replacement personnel? Sometimes they phone commercial and educational placement bureaus. They may contact other broadcasters. They fill a few positions with chance prospects who just happen by when there is a vacancy. Most stations advertise for help, especially to fulfill "equal opportunity" mandates. Alert managers keep a file of recent (past six months) inquiries about employment, including application forms filled out, when no positions were available.

Small stations often have to settle for people currently unemployed in broadcasting. Seldom can they make offers lucrative enough to interest employees of other stations. In some cases, they turn to people who are employed in other lines of work in the community. Large stations usually find people in smaller broadcast operations who have worked long enough to prove their abilities. Some stations even keep a marginally satisfactory employee on their payrolls rather than face the task of giving notice of termination and then searching for a replacement.

What about hiring personnel from other stations? People are motivated to change positions not only for more money, but often for greater security, better opportunity for advancement, and promise of greater satisfaction in the new position. Of course, larger stations and group-owned ones can seek replacements by advancing personnel already within the company, selecting on the basis of seniority, demonstrated ability, or even by competition. While enhancing staff morale, this efficiently utilizes veteran staffers' known abilities and familiarity with the operation. They recruit outside mostly for lower-level job replacement (as reflected in the CBS policy statement on internal promotion, described later). For distinctive positions such as news anchors or corporate management, major broadcast companies retain consulting firms that provide background data, résumés, air-check tapes, and other information about the continually changing field of apt candidates.

Neither recruitment nor replacement is an easy task. Both involve much searching and careful evaluation. A station committed to a long-range employment policy enjoys notable advantages over the station that merely fills vacancies with applicants who happen to be available.

Whatever the source of new employees, someone on the staff should be designated personnel director. In smaller stations, this may be the responsibility of the station manager or program director. Many large metropolitan stations employ a fulltime personnel director for all nonunion employees; some add a director of industrial relations responsible for liaison with union personnel. Occasionally it is a combined executive position. In addition to determining new or replacement staff people—collaborating with respective department heads involved—the personnel director oversees employee development. She confers with all staff personnel as the company's official representative in matters concerning individual and corporate welfare, benefits, and complaints.

Daily pressures and short-term crises often smother adequate concern for a station's future growth, resulting in some broadcast management's apathetic attitude toward planned recruitment. It is enormously wasteful not to screen employees adequately to select those with best prospects for long-term contribution to the station. Also, it is inefficient for large stations to depend on pirating people from smaller ones without developing at least a few new people on their own. It is unfair to expect only small stations to risk experimenting with inexperienced people, while training newcomers who move on as soon as competent, thereby absorbing the high costs of frequent turnovers. No one doubts the advantages of an informal "farm system," but an essential factor for long-term success even in a baseball club's system is a carefully conceived and executed recruiting plan.

Recruiting at Colleges and Universities

It can be useful even for small stations to conduct annual interview sessions at colleges nearest them whether they anticipate vacancies or not. This provides a list of prospects for the future when employees resign on short notice. Most colleges maintain facilities for interviewing graduating seniors; those facilities are usually used to capacity throughout the academic year by other industries and professions. But seniors interested in careers in broadcasting seldom have opportunities to talk with recruiters from broadcast companies. Consequently, many outstanding young prospects for the radio-television-cable industry are lost to other fields. This has been particularly true of young people interested in sales careers. It seems wasteful to disregard this resource of the college campus. Although major sectors of the broadcast industry were "downsizing" and jobs were scarce in the 1990s, media companies still need to identify and attract quality persons with intelligence, poise, and skill who are low- or no-risk employees destined to advance their employers' success as well as their own.

Most college students complete their degree work in May or June, others in December. Correspondence or phone calls to campus faculty can provide recommendations about top prospects for interviews. Obviously, a college degree in itself is no automatic guarantee of excellence; interviews with recommended students should be followed up by checking with faculty members in the students' major areas of study.

Minority persons can be recruited not only from colleges but also from local minority organizations that can identify good prospects for broadcast training and employment.

FINDING QUALIFIED EMPLOYEES

Even before the delicate issue of minority employment in broadcasting crested in the 1960s and 1970s, broadcasters noted difficulties in finding fully qualified personnel generally.[9] Most radio and TV station managers found it hardest to attract people skilled in writing and competent in sales or news. But they easily found persons interested in being announcers and disc jockeys, although too many of those interested were not adequately qualified. Recurring reasons for being "under-qualified" for all categories of jobs included: Applicants lacked knowledge of the industry, they desired to advance rapidly without commensurate expertise, and they were overly impressed by the glamour of broadcasting. Radio-TV executives have criticized entry-level applicants for "unrealistic expectations: they expect too high a starting salary, they expect to advance too quickly, and they come to the job with a misguided impression of the industry."[10] Three-fourths of executives cited "writing skills and style" among the most important qualities needed. Their stations offered best opportunities for recent college graduates in sales (cited by 49%), then news (32%) and production (28%).

Radio stations in small markets must particularly concern themselves with new employees' adaptability to perform various kinds of tasks and to work comfortably in a small physical area, close to the rest of the staff.[11] This need for versatility and compatibility is often coupled with a low wage scale that attracts new employees with only limited experience. Further, the small station repeatedly experiences turnover of personnel who, after developing their skills, are attracted away to larger operations. Small-market stations should seek employees who prefer the less hectic pace and wide local recognition and respect that small towns offer, where they can participate in civic activities as leaders. By attracting professionally qualified staff members not eager to move on, small stations can reduce the rate of turnover.

TRAINING FOR BROADCAST CAREERS

Professional broadcasters recommend a variety of education and prior work experience to qualify for broadcast employment. Media managers' recommendations reflect their own experience along with current trends at the time.

By 1971 more than half the TV news directors at 425 commercial TV stations preferred applicants with college training over those with experience but no college, even for entry-level positions.[12] They selected clusters of personal qualities desired in applicants:

energy, desire, eagerness, self-starter, enthusiasm (90 respondents)
positive attitudes, good manners, maturity, stable personality (53)
hard worker, not afraid to work long hours, a real work horse (22)
imagination, creativity, ingenuity, expressiveness (20)

good habits, clean and neat, clean-cut, no personal hang-ups (17)
ability to work with others, congenial, cooperative (16)
objective, open-minded, clear-headed (14)
dependable, reliable, dedicated (13)

In 1978 broadcast managers ranked specific media courses as more valuable for potential broadcasters than broader liberal arts courses.[13] Among forty course titles, they preferred commercial station operations, sales, internships, station management, news reporting, and the like over broadcast ethics, mass communication theory, media issues and effects, and mass media history; in addition, they strongly preferred those pragmatic courses over general economics, psychology, political science, sociology, survey of literature, and history of civilization courses.

However, in another study a year later, well over half of all television general managers (GMs), sales managers (SMs), and program directors (PDs) advised young people to "get a good, general education" to prepare for careers in broadcast business administration, sales, or programming at TV stations.[14] General managers recommended majoring in marketing (28%) and communications (27%), while sales managers advised majoring in marketing (38%) or broadcasting (16%), and PDs urged majors in communications (49%). About 18% of GMs also suggested an M.B.A (master's degree in business administration) while 1.3% noted "a Ph.D. would help" in the business end of TV. The three management groups variously supported starting at a small station (PDs 79%, GMs 27%, SMs 14%), with one-third of sales managers instead stressing prior experience in selling outside the field of TV.

On the other hand, backgrounds and college studies in areas far removed from mass media often serve well those who aspire to news-related positions on-air and as producers. Almost half the news directors (42% in TV and 48% in radio) with college degrees in 1976 had majored in English, social studies, business, and fields other than communications.[15]

Ted Koppel of ABC News reiterated at journalism conferences in the late 1980s his view that a broad liberal arts education, with a major concentration in a traditional academic subject rather than in journalism, best prepared one for broadcast news. While granting that there should be courses in journalism, he strongly objected to a major concentration in that area: "I believe that journalism is fundamentally a trade, and if our young journalists bring to that trade the real ability to read and write in their own native language, together with some experience in another subject or two or three, that the trade of journalism is something which is picked up very quickly."[16]

■   Reflecting that view in practice was Charles Rose, who began his TV career as an interviewer with WPIX-TV in New York in 1972, became managing editor and then executive producer of "Bill Moyers Journal" in the mid-1970s, followed by TV hosting successively in Chicago, Dallas/Forth Worth, and for NBC in Washington, D.C., until he became nighttime on-air interviewer for CBS-TV in 1983, and subsequently host-discussant on PBS' respected "Charlie Rose Show." His college preparation: a bachelor of arts degree in history and a law degree at Duke University, as well as attending the Graduate School of Business at New York University. Similarly, a vice-president/bureau chief of CBS News in Washington, D.C., had come up through print and local TV news reporting, after undergraduate studies in literature, history, and philosophy plus an M.B.A in finance; his successor in 1987 had a bachelor's degree in communications and an M.B.A degree prior to years of working in TV news.[17] Bryant Gumble rose to prominence in local and national sports and then news, including long-time co-host of

NBC's "Today" show, with an undergraduate degree major in Russian history; he has stressed to college students that their major study area need not determine their lifetime career.   ■

The Roper survey reported in 1987 that 72% of media executives judged an undergraduate college education as essential or important to entering the broadcast field; 25% thought a graduate degree was important. Two-thirds said a college degree with a major in mass communication—journalism, broadcasting—was "an important consideration" for evaluating applicants, while half said the same about a liberal arts degree. But for entry-level jobs, three-quarters of the managers gave more weight to nonacademic matters: "the general presentation of the individual, writing skills and style, information about or examples of previous work experience in the broadcast or cable industries, and the amount and kinds of 'hands on' experience in actual work situations."[18]

That survey did not inquire about qualities or abilities that broadcasters look for beyond initial employment, when considering personnel for advancement or even promotion to management-related positions. Professor Robert Blanchard commented on the Roper report's findings by recalling the broader skills required for a lifetime professional career:

> The obsession of professionals, by nature, is with the present or, more likely, with the immediate past. . . . What they do is based on what worked, or didn't work, last season. The university tradition reflects concern with identifying, assessing and transmitting enduring skills, principles and values, and understanding what they hold for the future. . . . The future for us is symbolized for our times with the advent of the 21st century, where today's students will be living and working most of their adult lives. It will be conceptual skills and principles and values, not last season's entry-level skills, that will guide them through our fast-changing and expanding information society.[19]

An informal survey found radio professionals urging college studies in applied areas directly related to the commercial field: research, audience ratings, computer text processing as well as program schedule logging and billing, legal aspects, writing skills, time sales, along with basic business knowledge including economics, finance, accounting, and marketing.[20] Some added the need for general education, plus being an "open thinker," creative, with common sense and practical experience.

EMPLOYING WOMEN AND MINORITIES

Larger broadcast companies may have, in addition to personnel directors, "affirmative action officers" who are responsible for attracting, and recommending for hiring and promotion, members of minority groups.[21] Some media companies provide "multicultural training" to help managers and other staff members learn how to respond to increasingly diverse backgrounds and characteristics of the employee labor pool. Among those programs are training in managing/valuing diversity, awareness, language, behavior modification, mentoring, and also processes for reducing bias and managing conflict.[22] Westinghouse Broadcasting sends selected station personnel to the parent corporation's training programs in New York. CBS provided human resources programs including multicultural training until 1986 when economic factors and new top executives eliminated them. Although public TV station KQED in San Francisco does not have a formal training program, it offers workshops conducted by consultants; and it reimburses employees for half the cost of participating in outside training. The Corporation for Public Broadcasting presents workshops in cultural diversity for public station supervisors and staffers. CPB's manager of EEO practices, Yoko Arthur, stressed that awareness and support for cultural and ethnic or racial diversity is more than an issue about handling personnel; it makes "good business sense" as part of a company's policy and strategy for doing business. Commitment to the principle must lead to positive programs engaging all management and staff in the process—with incentives plus procedures for evaluating each employee's responsibility and accountability for helping broadcasting mature into a multicultural industry. Everett sees this not only as "a matter of public policy or good will [but] a matter of business survival in the year 2000 and beyond."

Loring and Wells urge that, if at all possible, hiring and promoting minority persons should not be internally competitive, whereby one minority (such as black males or Hispanics) are given advantageous treatment over another minority (such as women or Puerto Ricans).

> To pit one group against the other is to undermine the objectives of both. One way to avoid such problems is to select an aware woman who can work with the minority man who is now in affirmative action work. Equal status between them eliminates possible conflicts of interest and makes visible management's equal commitment to both groups. This delicate balance in black/white, male/female advancement must be weighed in the final selection and placement of affirmative action officers as well as in implementing the program.[23]

Affirmative action officers—or personnel director or manager at smaller stations—should report directly to top management. Among other duties, they are responsible for reviewing goals and provisions of minority hiring practices. That includes announcing and advertising available jobs, listing job descriptions and qualifications, developing application forms and testing, plus training and development programs for further job progress. They also handle personnel records—including files on appraisals, promotion, discipline, and termination—which document data about minority persons, including women, on the staff.

Logic, as well as good business sense and an awareness of a station's role in its total community, can be coupled with humanity and a sense of ethics to support a broadcast company's efforts to attract minority persons. Of course federal law and agencies, including the FCC, strengthen such resolve, particularly Title VII of the Federal Civil Rights Act of 1964 and the Federal Equal Pay Act of 1963, plus enforcement powers given to the Equal Employment Opportunity Commission (EEOC) in 1972.[24] The 1990s' "post-Mark Fowler" FCC resumed close scrutiny of EEO matters, often resulting in forfeitures (fines) and short-term license renewals. For example, in one week late in 1990 the Commission levied two fines of $10,000 and one of $18,000 against radio stations in Florida and Georgia, plus short-term license renewals of two of them, for violating EEO rules—including failure "to undertake adequate EEO recruitment efforts, or to analyze its efforts to recruit, hire, promote and retain minorities." That policy of assessing noncompliant stations with forfeitures in the tens of thousands of dollars continued into the mid-1990s, with fines routinely of $10,000–$27,500 in 1995 to stations failing to recruit actively and hire adequate numbers of minorities.

■   The classic struggle for parity among white and black citizens became more complicated first as the Hispanic (or Latino) population multiplied, and then as Southeast Asians immigrated in great numbers, especially from Vietnam and Korea. They moved into low-paying jobs in the general labor force, sometimes displacing earlier ethnic and minority groups. Among 262 million U.S. citizens in 1993, 33 million (12.6%) were blacks, 27 million (10.3%) Hispanic or Spanish-surnamed, plus 12 million other non-Anglos including Asian-Oriental and Native Americans.[25] Thus more than one-quarter of the U.S. population were minorities—up from one-sixth in 1970. By 1991 whites were no longer in the majority in fifteen of twenty-eight U.S. cities with populations over 400,000 (within city limits—not counting suburbs and adjacent towns that are all part of local stations' coverage community). The city of Houston had 59.2% minorities, New York City 56.5%, Memphis 56.2%, San Francisco 53.2%, Dallas 52.2%, and Cleveland 52.1%; massive concentrations of various minorities lived in Los Angeles, Detroit, Miami, and other major metropolitan areas.[26] ▪

The working guideline for compliance with the FCC's EEO requirements was to employ minorities and females proportionate to one-half their percentage of each market's labor force. But since 1987 the Commission also held broadcasters to positive, systematic efforts to contact, evaluate, and attract minorities (see chapter 11).

A pragmatic advantage for stations is that staffs with minorities in responsible positions can be more alert to the varied publics in their signal area—responsive to their interests and needs in program content, including news coverage. With this awareness a station can build a broader local audience and is better able to serve the total community of license.

▪ Special attention must continue to be given to the complex problem of assimilating into the media workforce, including management roles, those not male Caucasians ("Anglo-Europeans"). Resources for guidelines and recommendations include professional and social organizations such as the National Association of Broadcasters, Radio and Television News Directors Association, Broadcast Education Association, Association for Education in Journalism and Mass Communication, Association of Black Communicators, National Black Media Coalition, National Association of Black Journalists, National Association of Hispanic Journalists, Native American Journalists Association, Asian American Journalists Association, American Women in Radio and Television, Inc., and Women in Communications, Inc.[27] ▪

The challenge lies in assisting apt minority persons to move into responsible positions so they can compete with and contribute to others on a basis of actual equality. Black leaders noted in the past broadcasters' limited efforts at developing capable minority persons to serve eventually in managerial capacities. Some identified those practices (or lack of them) as based on the traditional white-oriented structure, thus constituting "institutional racism" whether intentional or not. Those concerns remain today, despite recent years' accelerated progress in employing and promoting minorities in broadcasting and cable.

The issue is compounded: While substantial advances have been made over the past three decades, further imbalances remain to be righted, without at the same time causing alleged reverse discrimination to meet externally imposed percentage formulas or quotas. That vexing challenge was reflected in a weekly business magazine's editorial late in 1991:

> [F]or most Americans, especially for many young whites who have grown up thinking that the playing field is level, the special preferences and financial support that liberals say are required to achieve economic equity are no longer justified by the wrongs committed by previous generations of whites against previous generations of blacks. In overwhelming numbers, whites oppose minority preferences in hiring, promotion and college admissions; . . . many whites regard such programs as reverse racism.[28]

Executive management must take positive steps to ensure aggressive recruiting, training and promoting of all persons—including women and ethnic minorities—capable of productive employment and of advancing to media management. At the same time, all staff members should be held to performing their duties effectively. The NAACP's executive director (former FCC commissioner) Benjamin Hooks noted publicly that, while determined to assist minorities to enter broadcasting, he equally insisted they prepare themselves for employment and continue to advance their skills and understanding in order to merit promotion within broadcast companies.[29] Just prior to the deregulatory 1980s, experienced communications lawyer Erwin Krasnow, NAB general counsel, cautioned a statewide convention of broadcasters against overreacting or wrongly reacting to the FCC's EEO guidelines:

> Overreaction occurs when stations hire a minority or a female, and don't expect that person to perform well. In addition to the hiring being an unwise business decision, it also undermines the basic concept of affirmative action. . . .
> If you believe that someone won't perform well, it's usually a self-fulfilling prophecy. Hiring solely to meet what you perceive are FCC quotas is unfair to the person you hire and is totally contradictory to the spirit of affirmative action.[30]

He urged broadcasters to focus rather on "implementation of an aggressive affirmative action program" to help identify and train apt applicants among women and minority groups so they are equipped to compete successfully in media positions.

QUANTITATIVE PATTERNS OF EMPLOYMENT

Statistics and descriptive studies can lend perspective to the issue of minority employment. While the national scene continually changes, these data from representative points in time do reflect trends.

Fulltime and parttime minority employees at all levels of commercial TV stations numbered more than 4,000 (10% of total employment) in 1971. By 1989 some 12,600 minorities constituted over 18% of fulltime commercial TV station personnel (plus parttime for which data were not available). That same year another 1,700 minorities made up 19% of noncommercial TV station staffs, plus 7,900 (14%) of commercial radio employees, and another 484 (15%) at noncommercial radio stations. More than 4,500 (almost 20%) of fulltime employees at broadcast headquarters were minority persons, and almost 20,000 minority persons made up 21% of fulltime employees in cable by 1989.

The proportion of minorities among fulltime employees at *commercial TV stations* almost doubled from 10% in 1971 (the FCC's first year to report such data) to just over 18% by 1989. By 1995 minority persons, male and female, made up 19.7% of all employees in broadcasting and one out of four in cable. With minorities comprising nearly one-quarter of the U.S. population, minority employment thus stood at more than the 50% workforce-parity standard set by EEO and FCC guidelines.[31]

In 1989 women made up 45% of fulltime personnel in noncommercial TV, 39% in noncommercial radio, 39% at broadcast headquarters, 38% in commercial radio, and 36% in commercial TV. In 1995 total women—including those among minorities in figures cited above—had risen to almost 41% of fulltime employees at broadcast stations and 42% in cable systems. (See chapter 4 for details of women and minorities in upper management and executive positions in broadcasting and cable.)

Women in broadcasting served in the great majority of administrative and clerical positions, including business and data processing areas. One-third of news, programming, technical, and management posts were filled by women in radio and television in 1989 (Table 5.1).

A related issue is the "gender pay gap." American women in 1991 earned about 75 cents for every dollar earned by men, according to the U.S. Bureau of Labor Statistics; by 1994 that ratio was 77 cents to the "male dollar." The gap was slowly narrowing: back in 1979 it had been 62.5 cents and in 1988 it was 65 cents vs. $1 of male earnings. The gap is broader in this country than almost anywhere else in the world except Japan (in Australia women earn around 90% of parity with men, in Sweden and Norway 80%, but in Japan only 51% of parity with men's earnings).[32] Years of striving, despite federal guidelines, have not yet resulted in fully "equal pay for equal work" and for equal skills and qualifications among males and females. Because management controls compensation levels, it is up to them to implement equity in pay levels.

Sexual harassment has been a hidden problem for decades throughout the labor force generally; in recent years celebrated instances brought its pervasiveness to widespread public attention. Studies have uncovered wide-ranging abuses in government, the military, in corporate life and all kinds of business, including the media industry. In the half decade of 1986-1990, the Equal Employment Opportunity Commission received more than 5,500 filings of sexual harassment against radio, TV, and cable companies alone.[33] That is the tip of the iceberg; the vast majority of instances go unreported, even to supervisors, possibly because many instances involve them or other managers. A number of media personnel reported to *Electronic Media's* survey that they experienced sexual remarks (79%), suggestive looks (61%), deliberate touching (59%), pressure for dates (29%), letters and phone calls (15%), pressure for sexual favors (24%), and actual or attempted rape or assault (6%). But only 14% of them had ever filed a formal complaint. Two-thirds of them said their companies had procedures for filing complaints, but only one-third said they would feel comfortable following those procedures. The courts define illegal sexual harassment as conduct that makes the working environment intimidating, offensive, or hostile, including one person's explicit or implicit demand of another to submit to such conduct as a term or condition of employment or advancement. With men in the majority at top management, they must become more sensitive to such practices and their destructive impact on employees, both male and female; managers must responsibly strive to create a workplace free from damaging harassment.

NETWORKS

At the national network level, amid the surge of consciousness-raising civil rights movements in the 1960s and 1970s, CBS Inc. reported minority employment growth from 4.8% in 1965 to 14.3% by the end of 1971. The president of CBS Inc. cited a dramatic increase of minority employees as officials, managers, professionals, and technicians in all divisions of the corporation—including radio and TV networks and owned stations and news, as well as records, publishing, and others. By 1984, almost half (47%) the middle managers at CBS News were women. In 1986 and 1987, 13% of middle and upper management in all CBS divisions were women; at ABC 20% and at NBC 15% were women, at a time when 35% of the broadcast workforce were females (see chapter 4).[34] But in the 1980s and 1990s, severe budget cut-backs, with waves of terminations and early retirements—along with less emphasis on EEO concerns by the Reagan administration's deregulating FCC—reduced the momentum of equal employment opportunity efforts. For example, by 1991 CBS had done away with its Career Development and Education Training departments as well as career consultants who advised employees; but it did retain its Women's Advisory Council formed two decades earlier by corporate president Arthur R. Taylor.

Capital Cities/ABC actively promoted diversity among its network and owned-stations staffs with its Broadcast Management Training Program and the ABC Television Network Associates Program (recruiting minorities and women with senior management potential), plus a minority-oriented ABC News Correspondents Development Program, and founding and/or participating in Foundation for Minority Interests in Media, Inc. and the Minority Advertising Training Program. It summarized annual EEO reports for 1992 to 1994 (Table 5.2). In 1994 about 45% of some 19,000 employees were women. Among racial minorities, blacks made up 12.7% of ABC's corporate workforce, Hispanics 5.5%, Asians and Pacific Islanders 3%, while Native Americans and Alaskan Natives were 0.25%.

NEWS

From 1981 to 1993, some 1,100 stations dropped news operations; 17% of the nation's radio stations offered no local news programming in 1993; by 1996 one-third of radio stations offered none.[35] Table 5.3 tracks the growth, reduction, and reemergence of employment in radio and TV station news departments over selected years, including women and minorities. By 1993 white males made up 54% of the news workforce in TV and 63% in radio; the following year they dropped to 51% of TV news staffers.

■ For 1990 the RTNDA estimated the fulltime and parttime workforce in news at 22,800 in TV, with another 19,700 in radio. One of every three of those broadcast journalists was a woman (7,685 in TV, and 6,395 in radio), more than a three-fold increase since 1972. Minority persons constituted 18% (4,060) of newspersons in TV and 11% (2,130) in radio. Among the 42,500 working in radio and TV news in 1990, there were 3,225 blacks, 1,835 Hispanics, 840 Asians, and 290 Native American Indians. Minorities held largest percentages of news positions in the South and West (21% in TV, 16% radio), far smaller proportions in the East (16% and 4%) and Midwest (12% and 8%). Over fifteen years, percentages of blacks stayed at about the same level while Hispanics' almost doubled. Of all news directors in 1990, 10% in TV were minority persons, as were 9% of radio NDs; women held 16% of the 775 ND positions in TV and 25% of 6,250 NDs in radio.

In December 1994 the news workforce in television was 82.9% Caucasian; 10.1% African American; 4.2% Hispanic/Latino; 2.2% Asian American; and 0.6% Native American. In radio news 85.3% were Caucasian; 5.7% African American; 7.5% Hispanic/Latino; 1% Native American; and 0.6% Asian American. ■

Entering 1996, of 998 commercial TV stations, 704 (70%) programmed local news.[36] TV news operations averaged 30 fulltime and six parttime persons, for a total of about 21,000 fulltime and 4,200 parttime (table 5.3). Among PBS affiliates, 54 of 198 noncommercial stations also aired local news, employing

TABLE 5.1
WOMEN AS PERCENTAGE OF BROADCAST DEPARTMENT STAFFS, 1989

	RADIO	TV
Business	83%	77%
Data processing	76	83
Administration	55	64
Community service	43	65
Sales	42	50
News	34	32
Promotion	31	58
Production	23	28
Programming	20	46
Management	20	31
Technical	3	12

Source: Data reprinted in the Broadcast Education Association's "Leadership Challenge Committee Newsletter," September 1989, 6. *Journal of Broadcasting*: "A publication of the Broadcast Education Association, Washington, D.C."

TABLE 5.2
EEO EMPLOYMENT DATA, CAPCITIES/ABC, 1992–1994

YEAR	MALE	FEMALE	TOTALS	WHITE	BLACK	HISPANIC	ASIAN*	NATIVE AMERICAN**
1992	10,330	8,466	18,796	14,846	2,187	1,011	702	50
1993	10,204	8,159	18,363	14,383	2,304	1,038	586	52
1994	10,451	8,565	19,016	14,937	2,417	1,043	570	49

*Includes Asian and Pacific Islander
**Includes American Indian and Alaskan Native

Source: Corporate Affairs, Capital Cities/ABC, Inc., "The Capital Cities/ABC Commitment," May 1995, 11.

another 1,600 fulltime and 324 parttime newspersons. Meanwhile around two-thirds (63%) of the nation's 12,000 radio stations aired local news. With 1.6 fulltime and 2.6 parttime as average size news staffs, the 7,650 radio stations with news employed approximately 12,000 fulltime and 20,000 parttime—tripling previous numbers of parttime staff (if the figures are correct projections).

One cautionary note must be acknowledged—not to lessen support for hiring and advancing women in key media positions, but to anticipate serious practical consequences of opening all levels of employment to women. Felice Schwartz, president of a nonprofit research organization, cites the need for companies to be realistic in appraising the role of top-level female managers. She noted they are more likely than males to interrupt their careers or leave their job, citing a large industrial company's turnover for women in top management positions as 2.5 times that of males in similar posts.[37] She mentioned another large consumer company where half the women who took maternity leave returned to their jobs late or not at all. The point is that companies must become flexible, to accommodate specific patterns of female and family-related needs, such as day-care provision for children and alternative work schedules where possible. The labor pool of talented management-oriented persons is increasingly populated by women, with the steadily greater numbers (sometimes majorities) of female graduates from colleges and professional schools, including business and law programs offering advanced degrees.

TABLE 5.3
EMPLOYMENT AT RADIO AND TELEVISION STATIONS
SELECTED YEARS, 1987–1995

	TELEVISION					RADIO				
	Fulltime	Parttime	TOTAL	%F*	%Mi**	Fulltime	Parttime	TOTAL	%F	%Mi
1987	17,100	1,800	18,900		13	11,000	6,800	17,800		10
1990			22,800	34	18			19,700	32	11
1991	20,400	2,700	23,100			10,500	6,400	16,900		12
1992	21,000	2,800	23,800	35	19	8,500	7,100	15,800	31	12
1993	21,800	2,700	24,500	35	19	8,000	7,000	15,000	31	12
1994				36	17				26	15
1995	21,000	4,200	25,200			12,000	20,000	32,000		

*%F = % Females
**%Mi = % Minorities

Source: Data derived from RTNDA annual surveys (see notes 35 and 39), which did not always include all categories of totals and percentages (blanks above).

MINORITY STATION OWNERSHIP

Paralleling the movement by minority persons and women into the upper ranks of established broadcast companies was the increase in minorities holding half or more ownership in broadcast stations (described in chapter 4, and noted in chapter 10). Those stations generally offer formats and programming directed to African American or Hispanic audiences, so they tend to hire members of those minorities as managers and staff. Radio stations in the United States owned by blacks grew from 15 in 1972 to 182 (1990) then to 292 (1994); the first black-owned TV station started in 1972 and was joined by 16 more by 1990 and 31 by 1994. However, blacks in 1994 still owned fewer than 3% of all 12,155 licensed radio and TV stations, plus nine of 7,500 cable operations.[38] Companies owned by blacks and Hispanics share the burden of finding apt minority employees, then training them and advancing them to management positions. The goal is for all stations to vie against one another, openly competing for the most skilled and successful broadcast personnel—minority or otherwise.

On the other hand, some African American broadcasters cautioned early on that owning and operating an exclusively black or other minority-oriented station disenfranchised some of the total community who might otherwise be part of their audience—attracting advertising to the station. Instead, they counseled minority persons to consider joining white colleagues in ownership and management; many stations are 51% or more minority-owned, with the balance as non-minority ownership. In this view, "integrated stations" serve the entire service area, better attracting greater portions of mixed metropolitan audiences and thus more sales support.

HIRING, FIRING, AND INSPIRING EMPLOYEES

After three previous years of downsizing, the average TV station in 1989 expanded its fulltime news staff to 22, up from 19-20 the previous year; 120 independent stations' median news staffs grew from 2 fulltime personnel in 1988 to 6 the following year (3 parttimers were also employed in both years). The total news force at local TV stations increased by 1,900 (9%)

to 22,400 employees between 1988 and 1989, and to 22,800 in 1990. At the same time, although one of every ten major-market radio station discontinued news between 1988 and 1989, total fulltime and parttime news staffers remained the same—about 18,000 persons both years—and rose to an estimated 19,700 in 1990.[39]

As broadcasting turned into the 1990s, competition from cable and other media fragmented audiences and reduced the value of commercial time. At the same time, economic recession dried up significant portions of national and local advertising budgets. The mergers and acquisitions of the 1980s left media corporations with enormous debts to service (as a result of its merger, Time-Warner Communication in 1992 owed $3 million in *daily* interest on its multibillion dollar debt). The word in companies nationwide, including those in mass media, was to "downsize" operations and personnel to salvage budgets. This meant heavy and successive cut-backs in staffs at networks and stations, usually by attrition—not replacing retirees and enticing younger employees to take early retirement with cash buyouts.

■   Lay-offs and outright termination thinned broadcaster ranks: networks closed many news bureaus around the world, automated cameras (robotics) replaced studio crews, news teams were cut from a reporter with camera/audio/lighting crew to the reporter and just one technician. NBC's owned and operated (O&O) WMAQ in Chicago downsized 18% in 1990 and another 10% in 1991; within five years after 1986 the station's total staff was reduced from 440 people to 250 (it was also one of the first stations to use fully robotic cameras on the news set). One audio technician at NBC News in New York was offered severance worth $75,000 in 1991 but chose to remain with the company until scheduled retirement more than a decade later.   ■

Yet hope springs eternal, and openings for new employees continue to be created by normal personnel turnover—through retirement, promotions, or departures to higher positions in other companies. Financial and competitive pressures in 1990 resulted in 144 fewer radio stations and 29 fewer TV stations in the following year.[40] The U.S. Department of Labor estimated that between 1990 and 1991 the radio industry lost 1,800 jobs, but TV gained 300 employees while all forms of cable services grew by 5,300 jobs. The resurgence in station sales and network affiliation realignments in the mid-1990s, discussed earlier, helped news operations rebuild, with total full- and parttime newspersons at TV stations in 1995 numbering 22,600 and at radio stations 32,000 (two-thirds of them parttimers).

THE EMPLOYMENT INTERVIEW

It is important to have an opportunity to judge the poise, personality, and appearance of the applicant in addition to her submitted résumé, reference letters, and samples of previous work, including performing ability. There is no real substitute for the personal interview as a method of judging an applicant's potential. There is no substitute, either, for a skilled and experienced interviewer who can not only assess candidates objectively but can relate to them as individuals whatever their strengths or backgrounds—including ethnic and racial minority persons. The interviewer can judge an applicant's personality, manners, appearance, sincerity, emotional stability, intelligence, ability to articulate and to listen, as well as her opinions and attitudes about broadcasting. An effective interview session reveals the kind of questions asked by the candidate and the directness of her response to questions, and also her energy, enthusiasm and motivation.

Some broadcasters recommend, in addition to application forms and references, two separate interviews with applicants for employment—one by the personnel director and the other by the department head under whom the employee would work. Beyond these procedures for ascertaining the abilities and character of prospective employees, aptitude tests can be devised by the station staff to determine more precisely and with objectivity a person's qualifications for specific duties.

Even though no openings may exist at the moment or most interviewees cannot be hired for positions available, the fact that the station provides opportunity for interviews reflects favorably on it among those seeking to move to the station. While a time-consuming task, interviewing can also pay dividends by providing a list of prospective hires for future reference, and also of sources to check (college departments, local businesses, other media companies) when jobs do open up later.

When an immediate job is available, the interviewer should clearly explain specific duties, hours and working conditions, compensation, fringe benefits, and any other pertinent information. The actual hiring process, in most instances, should not be a part of the interview even though the interviewer may feel he has discovered the right person for the post. The interviewer may profit by a second thought. Usually, it is better policy to have the candidate wait a day or so before formally offering employment. It is wise policy to run a credit check on any applicants being considered for employment as cashiers, payroll clerks, mailroom personnel, or with responsibility for valuable equipment.

The one area where the director of personnel should have a less free hand is in selecting key upper-level personnel. In these cases, the personnel director may do some preliminary screening and referral of top prospects to the chief engineer, program manager, sales manager, or station manager for further interviewing. This saves administrators valuable time and better organizes the employment process. To fill top executive posts, a station usually conducts a search by invitation rather than through the application process. This may include consulting executive "head hunter" firms for apt candidates.

Job Indoctrination

In many instances a new employee reports for his first day of work to find no one assigned to introduce him to company policy or operations. Station practices and routines taken for granted by the employer and current staff can be complicated and strange to a new person. New duties can be downright frightening, especially when exercised among strangers and in unfamiliar surroundings.

A broadcast or cable company should prepare a brief printed or photocopied booklet with detailed information about the station and its operating policies that affect all employees.[*] This booklet can be given to new employees after they are hired and before they report for work. The Department of Broadcast Management of the NAB prepared a sample manual. It includes information about the broadcast medium, the station and its market, the management and department heads, plus specific practical points about paydays, checks, fringe benefits, work policies and practices, and leaves of absence.

Proper indoctrination of new employees also requires personal attention once they report for duty. On the first day, a new staff person should be given a tour

---

[*]This differs from a station manual that details principles and proper procedures required by federal agencies as well as corporate policy and local management's applications of those principles.

of the facilities and introduced to other employees. This is the time to discuss the station's administrative structure, its lines of authority, functions and philosophy, along with responsibilities of employees to one another. Reinsch and Ellis also recommend that a new employee spend at least a half-day in each department of the station to become familiar generally with how each operates and relates to other departments. Without being overly rigid, an organization chart of the station's executives and other personnel can graphically clarify relationships as well as titles and numbers of people employed at the station—those whom a new employee will be working with and reporting to.

Radio and television stations are not normally geared for extensive training periods. But many stations could profitably devote a parttime schedule over three to six months to thoroughly indoctrinate a few promising newcomers. Even the small station might arrange a few hours each week for new personnel to learn about station activities unrelated to their regular assignments.

Key personnel at every station must have special ability in dealing with people. Department heads need the ability to understand, direct, and guide employees for whom they have direct responsibility. Even a highly creative department head has limited value to a station if lacking skills in directing and motivating members of their staff—including women and minority persons.

Interpersonal Communicating

Initial misunderstandings can arise among personnel when minority persons are hired, often because it takes effort by all parties to communicate effectively beyond cross-cultural lines.

■   A three-year doctoral study of young black persons active in daily educational broadcast management responsibilities reported that most problems anticipated and stereotyped as "black" were actually common to people raised in a culture of poverty, whatever the race or ethnic background.[41] Such problems included poor orientation towards delayed goals and future time planning, lack of teamwork, suspicion of one another, preference for concrete and immediate experiences over abstract ones, dislike of reading and writing, and "fatalistic responses." More specifically identifiable as black-related matters, which initially hampered advancement of talented blacks but were eventually minimized, were: fear of failure in a white person's estimation, reservations in complying with sources of authority, loyalty to a "brother" even if he were wrong, disagreement about sources and directions of black awareness, and semantic misunderstandings of words that blacks considered derogatory or demeaning. Emotional and nonverbal communication by facial expression and tone of voice were often observed to be more significant to black persons than the cognitive content of the messages. Finally, the data implied that black young persons, especially males, initially learned more rapidly under the guidance of another black person, but that need lessened as they gained knowledge and confidence.   ■

Salaries and Wages

Major expenses for stations include offices, studios, transmitter, and complementary equipment. But once the capital has been invested and amortized over a number of years, the single greatest operating expense is salaries. This is particularly true of broadcasting because people are a station's greatest resource, in administrative decision-making roles as well as in news and other programming; and to the audience, on-air personnel *are* the station, especially in radio.

Station management can destroy its operation by either excess: underpaying staff and losing key members or at least undercutting their strong, enthusiastic support of the company; overpaying some staff members enormous amounts annually, thereby disenchanting less-rewarded colleagues while also dissipating potential profits. The obvious guideline is salaries equitable to each person's ability, creativity, length of service, and overall value to the total operation. While assessed partly on an individual basis, compensation can be grounded on representative averages. Clearly pertinent are the size and kind of station, the market size, the volume of advertising carried on the air, and the size of the station staff.

The National Association of Broadcasters has compiled through the years—now jointly with Broadcast/Cable Financial Management Association—average salary levels for ranges of market sizes and stations.[42] While figures vary from year to year, reflecting the national economy, they demonstrate relative pay scales for various positions and can therefore be compared internally (weighted to adjust to contemporary price and wage indexes).[43]

Tables 5.4 and 5.5 estimate salaries for selected positions in radio, television, and cable operations (compensation for managers and department heads was listed in table 4.4 in chapter 4.) While similar jobs in radio and cable received almost the same compensation, comparable positions at television stations brought about 50% more (in previous years, TV jobs had earned almost twice what radio paid). Sales personnel typically earned more than other staff members, often half again as much. On-air personalities in radio generally were paid less than half, often only two-fifths, of what their TV counterparts earned as meteorologists, news anchors, and sportscasters.

In 1991 about 17% of all employees in broadcasting and 16% in cable were supervisors. The industry's 83% who held nonsupervisory positions in radio-TV averaged almost $450 a week in 1991 or $23,350 a year; nonsupervisors in cable earned an average of $402 a week totaling $20,906 a year.[44]

"Gross weekly compensation" figures include the following: for nonsupervisory employees, the basic weekly wage plus premium pay for overtime; for announcers and producer/directors, the basic weekly wage, plus overtime pay and fees; for sales managers and salespersons, their average weekly income whether paid by straight salary, straight commission, salary plus commission, or any other method. To those salaries were added typically 15%–30% more in "fringe benefit" packages. The U.S. Chamber of Commerce estimated that American companies allocated an average of 37.6% of payroll to benefit programs: sick leave, holidays and vacations (10.8%); medical (9.3%); unemployment insurance and other benefits required by law (8.7%); retirement and pensions (5.1%); life insurance and miscellaneous other benefits (1.4%); plus paid break and lunch periods (2.3%).

WAGES AND HOURS: THE FAIR LABOR STANDARDS ACT

The U.S. Department of Labor sets guidelines for minimum wage, equal pay (for equal work by women and minorities), record keeping, maximum hours, overtime pay, and child labor.[45] Managers should be familiar with major considerations here; they must have competent bookkeepers and consult with a CPA or other auditors as well as with legal counsel on details of these federal requirements.[46]

TABLE 5.4
ANNUAL SALARIES, RADIO-TV-CABLE STAFF PERSONNEL
ESTIMATED NATIONAL MEDIANS, 1996*

Figures were computed by extrapolating Sherman's 1983 data and Head and Sterling's 1988 data, then projecting them proportioned to U.S. percentage raises for those five years plus the next three years, weighted to reflect media's previous 33% raises when the national private sector averaged 20%, and projected as 22% for media onto the next three years' national average of 15% increase (compounded); a final 15% was added as conservative inflation total 1991-1996 . Figures marked by † are ranges of median salaries, from smallest to largest markets, estimated by projecting 20% increases onto Stone's 1990 data (see note 43).

SELECTED POSITIONS	RADIO	TV	CABLE
Account executive	$39,600	$59,200	$36,800
Meteorologist			56,900
News anchor		†16–33,300	‡54,100
Controller		32,800	53,100
Sportscaster			44,200
Promotions director	33,500	43,900	
Assignment editor		†22–42,200	
Producer/Director		‡ 32,500	
Floor director			30,700
Technical director		29,400	
Service technician			26,000
Dispatchers			24,000
Music director	23,500		
Studio production technician		23,700	
News reporter	‡ 23,000	† 17–50,300	
On-air personality	22,200		
Customer service rep.			21,700
Account clerk			21,000
Traffic manager		21,566	34,300
Secretary	21,000	23,400	20,200

*Note: National "medians" lie midway between the lowest and highest levels of compensation. "National averages" would be somewhat lower than typical compensation for these positions in *mid-sized markets* (and often two-thirds or half of that in *large markets*), because of the number of positions at the great many stations in *small markets* where salaries are far below national averages. Estimates for key posts in large-market stations include $150,000 anchor, $99,000 sportscaster, and $60,500 news reporter. See table 4.4, chapter 4, for salaries of managers and department heads, plus note about salary levels as a function of station revenue.
‡Smallest-to-largest-market salary projections for 1996 (based on Stone's 1990 data): radio news reporter, $16,000–$24,100; TV news anchor, $26,000–$119,800; TV producer, $18,000–$42,500.

Minimum wages are set for all nonexempt employees in business enterprises with gross annual sales volume of $250,000 or more. In radio and television stations, newsgatherers, reporters and clerical employees, as well as office, maintenance and custodial employees are guaranteed minimum wages by federal law. But executive, administrative and professional employees as well as outside salespersons are exempt from minimum wage and overtime pay standards. Also exempt are announcers, news editors or chief engineers of broadcast stations whose major studio is located: (a) in a city or town of 100,000 population or less unless it is part of a standard metropolitan statistical area (as defined and designated by the Office of Management and Budget) having a total population in excess of 100,000; or (b) in a city or town of 25,000 population or less which is part of such an area but at least forty airline miles from the principal city in that area. "Stringers" and others paid only intermittently are considered freelance personnel who contract their services and thus are not employees as treated by this law.

TABLE 5.5
AVERAGE TOTAL COMPENSATION:
TELEVISION MANAGERS AND STAFF, 1996

National averages for total compensation packages include salary, plus benefits and bonuses, for TV station personnel. Estimates are projections of 18% (six years of annual 3% inflation) from 1990 data. As with all such data, totals range widely in each category depending on the market size and whether it is an affiliate or independent—always a function of each station's revenue.

DEPARTMENT HEADS		STAFF MEMBERS	
$158,000	General manager	$87,800	National sales manager
108,300	General sales manager	83,100	Local sales manager
90,250	Assistant GM/Station manager	76,400	News anchor
71,350	News director	56,800	Weathercaster
66,500	Marketing director	54,300	Sportscaster
58,000	Chief engineer	52,500	Account executive
55,000	Operations manager	34,200	News reporter
54,900	Business manager/Controller	33,300	Maintenance technician
54,000	Program director	32,000	Assignment editor
49,900	Research director	31,100	Producer/Director
45,700	Human resources director	30,900	News producer
43,000	Promotion/Publicity director	29,100	Technical director
41,200	Production manager	28,600	Film director
39,300	Company/Public Affairs director	27,700	Staff artist
37,400	Art director	26,700	News photographer
32,750	Traffic manager	26,600	Floor director
		26,400	Film/Tape editor
		26,100	Operator technician
		20,900	Production assistant
		20,700	Traffic/Computer operator

Source as basis for projected estimates: *1991 Television Employee Compensation and Fringe Benefits Report* by the National Association of Broadcasters and the Broadcast Cable Financial Management Association (Washington, D.C.: NAB, 1991); cited in *Broadcasting,* 25 November 1991, p. 50.

Wage differentials within job categories may be based on seniority, merit, quantity or quality of production, but not on gender, race, or age.[47] Unless specifically exempt, all employees receive at least one-and-a-half times their regular rate of pay for all hours over forty worked during a seven-day workweek. Employees' "regular rate" (on which overtime is based) includes all remuneration for employment except several specifically listed kinds such as discretionary bonuses, premium payments for overtime work, gifts, and talent fees to announcers if added to their fixed salaries (e.g., for special on-air services, spot commercial announcements). Specifically included by law are other payments such as production bonuses and commission. The federal act prohibits arbitrary discrimination in employing older workers, who are to be judged on ability rather than age regarding hiring, discharge, compensation, and other terms of employment.

■ Child labor provisions of the law set sixteen years as the minimum age for non-agricultural, nonhazardous occupations. But children fourteen to sixteen years of age are permitted to do clerical and other office work up to 3 hours on school days, 18 hours in a school week, 8 hours on a nonschool day, and 40 hours a week when school is not in session. Working hours cannot begin before 7:00 A.M. nor extend beyond 7:00 P.M. (except between June 1 and Labor Day when they may work up to 9:00 P.M.). The law permits minors from fourteen to sixteen years old to perform such broadcast duties as announcing, reading commercial advertisements, and cuing and introducing records.

(Federal laws about "child labor" do not apply to child employees as actors or performers in broadcast productions.) Employers should keep on file certification of any young worker's age, to document meeting minimum age requirements. ■

Federal regulations, Part 516, demand these kinds of specific information to be kept on file:

Full name of employee
Home address, including zip code
Date of birth, if under 19
Sex and occupation
Time of day and day of week when employee's workweek begins
Regular hourly rate of pay in which overtime premium is due; basis of wage payment (such as $5.00/hour; $40.00/day; $200.00/week; plus 5% commission)
Daily and weekly hours of work
Total daily or weekly straight time earnings
Total overtime compensation for the workweek
Total additions to or deductions from wages paid each pay period
Total wages paid each pay period
Date of payment and the pay period covered by payment

The Wage-Hour Division of the U.S. Department of Labor staffs regional and field offices that annually check a large percentage of broadcast stations, usually without advance notice. Inspections are either randomly determined, or are prompted by employee complaints. (Managers should not attempt to learn who may have filed complaints.) Inspectors must notify the manager or an authorized representative whenever they are present at the station. Obvious courtesies are mutually helpful: Provide the inspector with adequate work space if possible (otherwise the Wage-Hour Division can subpoena station records and remove them physically for proper examination); give access to documents pertinent to payrolls such as records of wages (but check with local counsel about access to documentation such as monthly financial statements); assist the inspector to meet with any employees, without participating in those confidential interviews. The investigator may copy whatever documents are needed, but the station is not expected to provide employees to do the copying. Nor may the investigator request any reworking of documentation; no single form is required for keeping records, as long as the station keeps complete records.

If the inspector's final report claims that money is owed to present or past employees, and the Regional Office supports this finding (and if the governmental Division has written authorization from the employees to press a court claim for payment), the manager can respond in one of four ways: pay all money claimed, refuse to pay, negotiate some settlement, or "wait and see" to negotiate at a later date.

HOURS AND WORKING CONDITIONS

The forty-hour week should be standard at any broadcast station or cable operation. Problematic for employees are split-shifts of two periods a day, or a continuous shift at hours that a person finds personally difficult to accommodate or even disruptive.

Working conditions should include office facilities that are comfortably arranged, clean, well lighted, warm in the winter months and comfortably cool in the summer. Workplace conditions must also be safe and healthy. The Occupational Safety and Health Administration (OSHA), created by federal law in 1970, issues guidelines to prevent hazards. Alert compliance with those provisions reduces a company's vulnerability to liability, including legal and insurance costs.[48] The NAB's Insurance Committee compiled a ninety-six-page booklet about minimizing accidents through risk management and loss control techniques for broadcast facilities; managers can apply it to develop standards and procedures for their operations, including fleet vehicles, crime prevention, and periodic safety audits for hazards.[49]

A few large stations have installed cafeterias or food services during periods of maximum working activity at out-of-pocket cost. This convenience can be a morale builder. Over a year's time, it can save thousands of hours of time otherwise lost for coffee breaks somewhere else in the building or across the street. Smaller stations install coin-operated machines or make arrangements for vending by caterers so food and beverages are available on the premises during working hours.

In most cities, free parking facilities are important. Many stations in large markets buy real estate adjacent to their studio buildings for this purpose (the land can also be useful for future building expansion). It should be well lighted and safe for parked automobiles and for employees at night.

Many television stations have relocated in outlying areas. Television operations always need more space and employees appreciate less obtrusive security precautions, relatively uncongested highway access to their place of work, and even landscaped grounds.

FRINGE BENEFITS AND INCENTIVE PLANS

No standard pattern exists throughout the broadcasting industry regarding fringe benefits and incentive plans for employees. Most workers now share in health and welfare plans at least partially supported by their employers. It is not unusual for companies to pay half or more of premiums for life insurance, hospitalization, and medical care. Almost all stations provide paid vacations and sick leave for jury duty, family funerals, and even short-term military service. Some companies also provide moving allowances, layoff benefits, severance pay, pensions and retirement plans, paid holidays, stock options in the company, deferred profit-sharing plans, and even company-financed college scholarships for dependents.

Many employees take fringe benefits for granted or are not aware of their extent. It serves the company's own interest to inform everyone on the payroll of substantial company contributions. Some stations give employees itemized costs of various benefits, along with their W-2 forms. In general, vacation periods and other fringe benefits tend to be more liberal in larger markets.

Biennial NAB surveys of broadcast employment do not include data for office workers. However, NAB recommends that station managers review the regular studies of office personnel conducted by: (1) telephone companies, (2) the Bureau of Labor Statistics and (3) the National Office Management Association. Their reports show the predominant pattern for office employees as a one-week paid vacation after six months of employment and two weeks after working a year. The general practice is to give office workers eight paid holidays each year.[50]

Employees drafted or enlisting for military service are entitled to regular pay up until the time they stop work. An employee who resigns and then is not accepted for military duty, or is released after having been accepted, is entitled to reemployment if she applies for reinstatement within thirty days. (Full seniority rights are retained, in accordance with provisions of Section 9 of the Universal Military Training and Service Act.) The personnel manager must keep abreast of governmental regulations and industrial developments. Matters of military leave, adjusted compensation for personnel in service, and the like, become vexatious problems only if solutions are improvised out of ignorance or neglect.

UNIONS AND LABOR RELATIONS

Radio and TV stations and cable systems in major markets as well as networks deal with labor unions active in electronic media. The programming department, especially announcers and other talent, comes under the jurisdiction of the American Federation of Television and Radio Artists (AFTRA) and the Screen Actors Guild (SAG) as well. The International Alliance of Theatrical Stage Employees (IATSE) and the American Federation of Musicians (AFM) are active in programming and production areas. In engineering and technical fields, the National Association of Broadcast Employees and Technicians (NABET), the International Brotherhood of Electrical Workers (IBEW), and Communication Workers of America (CWA) are the most active unions. *Broadcasting/Cable Yearbook 1995* listed forty-one unions and labor groups active in broadcasting and mass media.[51]

Smaller-market stations may be involved with only one union or none at all. About one-third of television stations in the United States deal with one or more unions, but only 8% of radio stations are unionized.[52] Broadcast employees join unions for many reasons, including inaction or unfair action by management, arbitrary or insensitive supervisors, poor working conditions, and unfair or inequitable wages. In recent years employees' concern for job security and protection has grown because of increasing changes in station ownership and formats. The NAB advises managers of nonunion stations to spend more time on their greatest assets, their employees, who may not be getting the attention, consideration, and rewards they ought. The 1980s saw increased efforts to organize nonunion stations, with station-wide labor units instead of unions for single departments or related job classifications. Another target was to unionize clerical workers who constitute the only large group of station employees yet unorganized in otherwise union shops.

Two other developments complicate union negotiations with station management. One is the continued growth of automation in program operations, where quality control (immediacy, flexibility, creativity) and especially personnel considerations (size of staff, jurisdiction over procedures, impersonal working conditions "with machines") can be greatly affected. Radio stations that program most of their daily schedules from imported satellite signals of program format syndicators have little need of on-air DJs or an in-station PD or music director. Television station and networks progressively reduce news team personnel as a single reporter on a field remote story can handle compact CCD TV cameras and audio while entire news studios can be "unmanned" by robotic cameras operated directly from the control room.

The other complicating matter—seemingly straightforward but complex to apply properly—is minority persons' and women's pay and opportunity equal to that of white males, as directed by federal legislation.[53] The Civil Rights Act legislates equal opportunity in matters such as hiring, firing, promotion, seniority, benefits, and conditions and terms of employment; it bans discrimination (based on sex, race, color, religion, or national origin) by private employers of twenty-five or more persons; in 1972 the law was extended to cover managers and professionals among others.

Commentators on unions and broadcasting offer guidelines for management attitudes and overall procedures in dealing with employees and their labor representatives.[54] They emphasize the essential relationship that should exist between persons who in their own right are individual human beings, some of them supervisory managers and others employees. In a climate of mutual collaboration in the business and creativity of a broadcast or cable operation, they should communicate freely with one another lest initially ignored minor difficulties grow into areas of grievance and eventually become issues of contention. In unionized stations, the contract should be honored exactly by both parties as a set of agreements mutually entered into, and not as a tool to take advantage of the other side. Nor is the time of negotiating a contract the only important period in these relationships; negotiations have grown out of the year-long context in the daily operation either of harmonious collaboration or of tension marked with altercations and misunderstandings.

CBS Inc. outlines, in its manual for employees in New York, procedures recommended for handling problems or grievances of nonunionized personnel (major excerpts follow).

■   (1) If an employee has a problem on the job, it should be discussed with the supervisor as the first step in the chain of command. If unable to give an immediate response, the supervisor will tell the employee when an answer can be expected, within a reasonable time period, but no more than five working days.

Although most problems will be resolved in discussions with the immediate supervisor, some may not. If an employee believes that a problem is still unresolved after the supervisor has given an answer, the employee should say so.

(2) The supervisor will then expect the employee to present the situation to the next level of management, normally the Department Head. The Department Head will listen to the problem and will also ask for information from the employee's supervisor. Whenever possible, the Department Head will respond within five working days. If the Department Head cannot give an answer within that time, the employee will be advised as to when an answer may be expected.

(3) If the problem is not resolved at this level, the issue is put in writing by both the employee and the supervisor and sent to the next higher level of management, usually a Vice President. At this stage, management must consult Personnel Policy and forward a copy of the issue to Personnel Policy.

(4) The appeal process may proceed up through the managerial organization to a member of Senior Management, such as a person with Officer status or the Division President.

At any time during this process, an employee or manager may consult with Personnel for advice and interpretation of Company policy. Personnel's involvement is that of a catalyst for communication between the employee and the supervisor, ensuring compliance with Company policies and providing consistency in the resolution of employee problems. Personnel's assistance is advisory. Problems must be resolved with line management.[55]   ■

The National Labor Relations Board outlined the rationale for supporting unionized workers: "Employees shall have the right to self organization, to form, join, or assist labor organizations, to bargain collectively through representatives of their own choosing, and to engage in other concerted activities for the purpose of collective bargaining or other mutual aid protection."[56] At the same time the NLRB noted the correlative rights of all parties involved in collective bargaining through union representation, and supported management's defense of its own policies and responsibilities:

> It is an unfair labor practice for either party to refuse to bargain collectively with the other. The obligation does not, however, compel either party to agree to a proposal by the other nor does it require either party to make a concession to the other.

The basic responsibility imposed by the Labor Management Relations Act on employer as well as labor organization is to bargain in "good faith" about rates of pay, hours, and terms and conditions of employment. This refers to negotiating a contract and also to resolving questions during the contract period. The goal is to avoid or eliminate unfair labor practices by either managers or staff employees.[57]

Signing the union contract does not mean management gives up fundamental authority for determining company policy and for exercising control and direction of the station.

Usually, the general manager and the station's attorney are directly responsible for dealing with unions. A merely moderately successful media company can find that one poor union contract makes the difference between profit and loss. A poorly written agreement can deny a company the flexibility it needs to develop its operation. Large companies, with multiple union contracts, frequently have a director of industrial relations who concentrates on the labor relations area. This individual must know every function of every union employee in the company. He must know thoroughly every agreement clause under existing contracts and must be aware of what can be accomplished in future negotiations to foster better relationships between union and company. Effective collective bargaining includes careful preparation, with precise knowledge of requirements by the National Labor Relations Board (NLRB) and the U.S. Department of Labor. It demands thorough knowledge and wide experience of techniques and procedures used at the bargaining table in the critical process of "give and take" toward mutually acceptable compromise. In addition to competent legal counsel for advice, the NAB recommends steps for planning labor negotiations: determine management's position on labor issues, whether economic or "noneconomic" (money demands, and demands involving either operational practices or matters of principle); study the current contract and how it affects operational procedures, then get supervisors' judgments about contract improvements; list proposals according to priority (those desired but optional, those very important, and those "strike-worthy"); review salaries of employees compared with comparable stations elsewhere and with all stations in local market; gather factual documentation (such as NAB's summaries of labor contracts from elsewhere); estimate probable demands by the union, and determine management's position on each; finally, draft management's proposal as the basic working document for bargaining (although some employers at this point wait for the union to present its demands first).[58]

The negotiator for management, or the negotiating team, must be selected for detailed knowledge of station policies, operations, labor laws, as well as for authority and ability to be firm yet flexible where appropriate. Management's goals when entering into negotiations should be realistic and reasonably attainable; and those initial proposals should be continually evaluated during the union's counterproposals so they can be modified when necessary. The NAB notes "the best negotiations are those in which there is no winner and no loser. If management did achieve that which it set out to secure, it is wise to be discreet."

Sometimes the broadcast industry has endured costly strikes when either labor or management or both did not approach negotiations maturely. Irresponsibility has been found on both sides. In union and labor relations it is essential to explain carefully the rationale for company policies, accompanied by thorough factual documentation. When people know the reasons for doing something in a particular manner, they can better understand the task as well as the issues involved. The bargaining table should be a meeting place of reasonable people who can reason together.

INTERNAL COMMUNICATIONS

Staff meetings provide a forum for internally communicating information, for training, and for shared decision making. These include sessions among managers and key supervisors, meetings of department heads with their staffs, and general meetings of the entire organization several times a year.

A weekly or monthly internal newsletter reporting the activities of the company and its people can be very valuable whatever the station size. An internal newsletter provides for an exchange of information among employees. It is an informal channel of communication between personnel and management. It can relay employees' suggestions for improving the station. Its columns can announce and adequately explain specific changes in working conditions or hours (thus avoiding inaccuracies and exaggerations of "grapevine" gossip). It can review news important to the media industry. The physical appearance and style of the newsletter are not as important as the content and the fact that it is available. Large operations might use attractive graphics layouts in "desktop" computer-generated publications. Small stations can achieve many of the benefits from circulating a photocopied newsletter.

Beyond recounting "chatty bits" of news about employees' travels, families, and hobbies, a newsletter should provide specific information that employees want: explanations of the company's general organization and operation; corporate and station policies directly affecting them; progress in achieving the company's stated objectives; corporate and station planning related to business environment, advances in technology, and political developments that might affect jobs or wages; how their own jobs relate to company strategy, including opportunity for advancement; reasons for any personnel reductions and how they will affect remaining employees.

A candid newsletter with accurate, pertinent information defuses unwarranted assumptions about all kinds of rumors looming as threats to employees and distracting them from the work at hand. People communicate far more freely to members of their own subgroups, and to those higher in status only when they are not supervisors; this spreads speculation based on partial data or misinformation. This occurs particularly when salaries, promotions, or other important personnel matters are at stake. Rumors erupt when employees are confused and unclear about what is happening and when they feel unable to affect matters touching their careers. Gossip and guessing are ways employees express anxiety about pending or fancied changes.

It is important to communicate often and fully with personnel through meetings, posted notices, circulated memoranda or a newsletter. But sometimes word-of-mouth carries information more swiftly and widely and with greater credibility than more formalized means. The obvious shortcoming of informal verbal communication—including the "grapevine"—is the possibility of distortion or misinformation. But management does have some control over the work environment in which the grapevine operates. Managers can even use feedback from this informal source to learn what information must yet be relayed to personnel, or where public clarification about misinformation is appropriate. No one can control the "grapevine" as a channel of communication, but it is possible to influence it properly. It should not be discounted as a significant factor in employees' understanding and support of management plans and goals.

Managers agree that staff workers should have information about things directly affecting them. Employees should also be told what they want to know rather than simply what management wants them to know, and as early and often as possible. This is particularly important in such matters as new ownership where jobs, assignments, formats, station schedules, and even some company policies may be subject to major changes.

Personnel losses, transfers, promotions, and replacements increased exponentially in the late 1980s and early 1990s. In addition to mergers, buyouts, acquisitions, and new ownership, a swiftly changing economic and regulatory environment spawned new media policies and procedures—most of them modifying people's relationships in a company.

Relationships between managers and employees should reflect and support mutual esteem and self-respect. A person's own self-respect is based on true ability, recognized achievement, and respect from others. Such collaboration, assisted by two-way communication among staff and management, breeds satisfied, productive employees as well as more efficient and creative work. (In contemporary argot, it "empowers" them at many levels.) This is essential to a well-managed broadcast station.

EMPLOYEE SUGGESTION SYSTEMS

Beyond contributing to productivity by supporting morale, the suggestion box or other procedure for employees to propose their ideas to management can seriously advance the pooling of skill and vision. Department managers should encourage creative suggestions by making everyone on the staff feel free to discuss ideas. Casual conversation can be partly structured by holding brainstorming sessions where all participants openly contribute whatever comes to mind about a project proposal or specific problem, so good ideas can be identified and evaluated.

A possible way to foster employees' sense of participation in the company, as well as to profit from their creative recommendations and opinions, is to organize an employee-management committee. This committee can be organized departmentally or with the entire staff; representatives can be selected by their colleagues to forward to management their constructive ideas, reactions, or inquiries concerning policies or points of factual information.

Some excellent program concepts and cost-control ideas at WGN radio and television in past years developed from suggestions made by station personnel. In an operation the size of WGN-AM-TV (when it had 600 employees in the Chicago offices), suggestions were generally confined to office procedures, engineering, or production techniques. But occasionally a splendid idea was submitted regarding general programming or community affairs. This opportunity created an environment where every person on the staff thought about ways to improve the station. Reinsch and Ellis emphasize this point:

Creative thinking develops only in a permissive atmosphere, where every employee knows about his job, respects his company, and understands what they both represent to the public. You cultivate creative thinking by real two-way communication, and by running a station where anybody on the staff feels free to walk through your open door, sit down, and give out his ideas.

More sound programming ideas have been hatched over a casual bull session than have come out of a dozen formal conferences.

If the atmosphere in a station is truly permissive, each employee knows he can drop in to talk whenever the inspiration strikes him and that his superior will listen to him. This requires listening carefully, attentively, sympathetically. It is vitally important to continue listening with courtesy and interest until the entire idea has been presented.

Remember, too, that, no matter how far-fetched or impractical a proposal may sound, it is not ridiculous. It has dignity and it demands a respectful hearing, because somebody on your staff thought enough of it, and enough of you, to express it to you.[59]

Most stations apparently operate on this principle. An early NAB study found over 85% of employees satisfied with the opportunity their stations gave to make suggestions and contribute ideas. Management's positive reaction to their interest supported further efforts to contribute to the company's success as well as to their own self-esteem (recall the "V" and "Z" theories of managing, chapter 2).

Activities of Personnel in the Community

The station manager must set the example for employee involvement in community affairs. She must show her interest in the well-being of the city, state, and general coverage area. She should encourage all employees to do likewise, whether by supporting a particular program of the mayor, city government, governor, church, school projects, or any other areas of community public service that the employee may prefer. Employees active in community affairs bring much credit to the station. Community involvement is vital for a business that depends so much on public support.

One benefit from employees' outside activities is their growth in confidence and self-respect. This development enhances their prospects for promotions in radio or television. More immediately, they become acquainted through first-hand experience with the kinds of people making up the station's audience; those interests, activities, and needs should be reflected in local news broadcast daily by the station and in other areas of local programming and scheduling. It may even reflect favorably upon the sales staff's efforts to win advertising support in that community.

At all times, while working in the offices and studios and when participating in or merely attending civic affairs, staff members must be instinctively aware they *are* the station to all those they speak with by telephone, meet personally, and collaborate with on community projects. Especially in smaller markets, the staff members' attire, attitude, vocabulary, and tone of voice represent the station itself. Their other personal characteristics of strength or weakness (courtesy, carelessness, unconcern, or attraction to alcohol) are often perceived as reflecting the station.[60] Management should make sure that all employees are aware of this unique, personal responsibility to themselves, their station, and their community.

Opportunities for Employee Growth

The station manager really doesn't develop anyone on the staff. Rather, he can provide an environment in which responsible and ambitious people can develop themselves. To this end, he should carefully plan a long-range program of continuing employee education. Staff members need opportunities to learn about licensee obligations under provisions of the Communications Act and other laws and regulations, program responsibilities, financial realities of budgeting and sales, and, in general, the basic concerns confronting management in a broadcast or cable company.

Few media employees know much about revenues and expenses, nor about problems of depreciation and obsolescence of broadcast equipment. Yet, they should be interested in this kind of information when given an opportunity to learn about it. Managers should explain their areas of primary concern to company employees and openly discuss with them mutual problems of professional responsibility. To help organize and assist with an employee development program, a station can employ the resources of a university or a consulting firm specializing in professional staff development.

Top executives in broadcasting have many opportunities to attend professional meetings, but they may overlook employees' need for a wider perspective beyond routine work assignments. An employee development program can pay dividends in employee appreciation and may expand their professional interests. Trade papers and magazines should be made available to all staff members to help them keep abreast of current factual data, trends, and significant issues in the field. Publications can be kept in a station library, with most recent copies in the reception area and employees' lounge. Beyond the studio building, further training can be made available through college courses, career seminars, participation in trade meetings and conventions, and attendance at important social and media events in the city. Those who attend might make informal reports on them to other staff members. Employees should be encouraged to visit stations in other markets.

Almost one-half of broadcast employees who told the NAB they did not want to make a career of radio or television intended to leave because they were dissatisfied with the field. They listed as reasons: "full capabilities not recognized, lack of advancement, and scheduling irritations." A decade later CBS Inc. found that more than two employees out of every five claimed they did not have the opportunity needed to develop their skills; only 37% judged their development opportunity as adequate.[61]

It is instructive to broadcasters to note that U.S. companies overall invest 2% of payroll on training; Japanese companies spend 6%.

Creativity in the Employee

In the early days of radio, it was exciting to work in broadcasting because it was all so new. Almost anything could be tried. It was easy to be different. Innovations in programming and production were encouraged. In those days, there was a real premium on creativity; imitated formats were rare. The early days of television brought a renewed emphasis on experimental programming and production; "live" programming was the standard for every local station as well as at the networks. Broadcast staffs were challenged by the opportunity to try something new and were generally given the freedom to do so. Outstanding advancements were made in programming, camera technique, lighting and staging, and performance. Early cable operations featured local access channels that, while relatively uncomplex, still offered a chance to try out simple program presentations. By the 1990s creative programming could be found more in production companies making commercials, syndicated radio services, and corporate video than in traditional cable and broadcast stations. However, the advance of cable services and direct broadcast satellite, with exclusively national information and entertainment, makes *local* programming produced in-house for and with a community the unique distinction terrestrial stations can offer audiences.

The opportunity to innovate has always been a prime motivating force for creative people. Standardized formats and routine assignments discourage men and women capable of new discoveries and new ideas. Stations that offer little or no opportunity for experimenting tend to attract a "plodding" type of employee satisfied by routine assignments. Highly creative people avoid this kind of broadcast operation. It is to a station's advantage to encourage development of new program formats and production methods, fresh ways to promote the station, innovative sales tactics and clever commercials, as well as effective procedures for computerized logging, billing, monitoring cash flow, and other necessary routine tasks. This atmosphere of innovation attracts and retains employees capable of producing new ideas to benefit themselves and the station. In an era of "downsizing" and cost efficiencies, paradoxically, a search for innovative yet practical ways of doing things acquires even more importance to keep a station staff aggressive and competitive as well as satisfied. Naisbitt and Aburdene stress that

> to lead others, managers who would create a vision sufficiently compelling to motivate associates to superior performance must draw on the intuitive mind. . . . [attracting] self-directed, talented people who seek a creative outlet for their personal goals within an organization which shares their values. . . . Ultimately, vision gets translated into sales and profit growth and return on investment, but the numbers come after the vision. In the old-style companies, the numbers are the vision.[62]

The process of creativity and the personal characteristics of creative people were studied under three grants from the Carnegie Corporation.[63] Several characteristics are particularly desirable in an employee of a radio or television station: easy adjustment to people and situations, an attraction to problems and their solutions, a high degree of sensitivity, and a wide range of interests. The creative person's preference for the new, the unknown, and the complex over the traditional, the known, and the simple should interest any station manager who wants her station to be different. The creative employee should be in demand in any station; here conformity is desired in behavior and social patterns but is not wanted in the field of ideas. Researchers noted that a person with only average intelligence can be creative if her environment is conducive to creativity. The three studies suggest that the creative employee needs an atmosphere of unfettered accomplishment in order to produce her best ideas. Any time an organization consciously or unconsciously inhibits a creative climate, it stands to lose important benefits—novel and challenging ideas, new and exciting program concepts, departures from established practices in sales and public relations, and (perhaps most important) the exhilarating atmosphere of the daring and different in almost any phase of the station's activities.

DOMINANCE AND CREATIVITY

The dominating person can inhibit staff creativity. Managers or supervisors who show concern for employees as people, both on and off the job, achieve better productivity than those who like to show their authority. The creative, productive person needs a considerable amount of freedom and a minimum of supervision. He appreciates constructive criticism when sensibly made and helpfully voiced. Supervisors who exercise authority to criticize, while having no ability to make constructive comments, can cause obstinate problems. The creative person is generally quite sensitive to criticism and cannot easily be objective about negative reaction. Perhaps an innovative person's morale and creative power can be most destroyed by management's indifference or even cynicism, much less ridicule, about an original and imaginative idea.

Creativity must be regulated by its own process of development. If it is forced by artificial deadlines or hectored by authority the results may be disappointing. Everyone knows how an idea may pop into the consciousness after struggling for days with a problem and then putting it aside. The solution may appear all at once during a drive home from work or while taking a walk or showering, or even while dozing. Once facts have been gathered and after considerable study and reflection, an inspiration often flashes into the mind when least expected. The truly creative idea cannot be regulated by deadlines in the same manner as the routine assignment. This raises the challenge of balancing creativity with the demands of standardization required by the realities of broadcast planning, scheduling, and programming. Management offers intelligent leadership by providing sound guidance and fulfilling its own responsibilities in overseeing policies and procedures, while at the same time "clearing the air for action and establishing a climate in which a creative spirit can live and stay healthy."

MOTIVATING CREATIVITY

What can management do to provide a favorable climate for creative people and to motivate them properly? Managers cannot force employees to be motivated, but they can create a context that stimulates employees to their best efforts. Reinsch and Ellis describe the process of motivation as "stimulating, listening, challenging, encouraging, inspiring, providing an atmosphere for self-expression and self-fulfillment," adding that substantive motivation for staff members to throw their creative energy into the broadcast enterprise depends on the station itself and its program service—"if it is sensitive to the needs and wants of the community, if it has a conscience and lets it speak."[64]

Countless broadcasters have made original contributions in writing, directing, selling, performing, and even market research and analysis usually because they received some encouragement from peers and, at the right moment, from superiors. Others have weathered misunderstanding, rejection and even opposition, but have developed their creativity in the broadcast, cable, and advertising fields—to their own fulfillment, but more importantly to the audience's satisfaction and to the advancement of their profession and their companies.

■ Edward R. Murrow and Fred Friendly fostered creativity by innovating in news and documentaries, largely encouraged and usually supported by CBS administration. Sylvester "Pat" Weaver's innovations when he was president of NBC lasted beyond his tenure at the network ("Today," "Tonight," "Monitor," the program spectaculars later dubbed "specials"). To counter TV's inroads, Gordon McClendon innovated specialized "Top 40" radio format that cycled pop hit records. Mary Wells and Stan Freberg took their creativity to their own newly-formed advertising agencies and succeeded beyond expectations. Joan Ganz Cooney developed for PBS the award-winning creative program series Sesame Street, as well as other children's programming. Roone Arledge built a dominant sports presence out of struggling ABC-TV, then developed a serious award-winning network news service as well. Grant Tinker oversaw development of award-winning, critically acclaimed programming series as head of M-T-M Productions, then revitalized the NBC television network by attracting and supporting creative, thoughtful producers who made it the #1 prime-time network for the first time in that company's history. Norm Pattiz built a radio syndication service, and later acquired NBC Radio and Mutual networks, by distributing custom formatted programming to stations across the nation—first by tape distribution and then by satellite. *Time* magazine acclaimed as its Man of the Year for 1991 Ted Turner, risk-taking entrepreneur of the first local independent station distributed nationally to cable systems by satellite; he also brought profitability and distinction to his all-news CNN/Cable News Network, adding other channels of news Headline Service, TNT/Turner Network Television featuring motion pictures, and others. Stanley S. Hubbard established CONUS (Continental United States) as the world's first satellite news-gathering cooperative, distributing satellite-relayed news to over 100 stations nationwide and 150 around the world; in 1992 he introduced a twenty-four-hour All News Channel in partnership with Viacom; he committed massive funding, in collaboration with Hughes Communications, to pioneer high-power direct broadcast satellite (DBS) transmission of national multi-channel program service by 1994.　■

Through the decades, where creativity was not supported by station, network, and advertising companies, some creative people—especially playwrights, directors, and actors from network drama, and even administrative executives—departed to other media such as motion pictures, legitimate theater, or cable. Not only the companies that lost them, but the entire broadcasting field suffered incalculable loss.

Important developments in broadcasting and cable continue to arrive on the scene, with DBS expanding and HDTV hovering in the wings. Whatever progress lies ahead will depend on the type of people who choose to work in the industry and on how they are treated. Creative people are the most important asset of any media system—radio or TV station, network, or cable system.

PROMOTION POLICIES

It is difficult to determine guidelines for promoting personnel. Regrettably, some stations and managers neglect to reward persons who make important contributions to the company. If they cannot promote deserving people when timely, managers should offer some other substantial recognition such as a cash bonus or reassignment to a more attractive post. By encouraging talented employees, managers can keep them with the company with the honest expectation of moving to an improved position. Proverbs 13:12 cautions that "Hope deferred maketh the heart sick." It also maketh the foot loose. Laggard promotions and grudging recognition can strip a station of its most promising people. In addition, much time and money is lost while replacements learn details of a job before they reach the level of first-hand experience and productivity of a departed employee.

Major stations, especially group-owned ones, often emphasize internal promotion not only to reward effective employees but also to expedite hiring competent personnel for job openings. CBS' procedure offers a model suitable for larger broadcast operations (excerpts follow).

■   *Departmental Posting*—If there are logical candidates available from the department in which the vacancy exists, the posting may be done first within that department. In these cases, the hiring executive should post the job opening notice for two days and should interview all qualified department members.

Central Posting—If the hiring executive wishes to interview other qualified candidates outside the department, a job opening notice is displayed conspicuously in prominent centralized locations for one to two weeks so that all employees have the opportunity to learn of the available position. Employees applying for the position are referred to and screened by Personnel Placement (in New York and Los Angeles), not the hiring department.[65]   ■

Only after those steps the CBS placement office recruits from outside sources. CBS employees are urged to notify Personnel of special skills, education, or other background not previously noted which might warrant consideration for transfer or promotion when openings occur in the company.[66]

In promotion as in hiring there is the knotty problem of adequately considering minorities and women who may have been systematically excluded from regular advancement in past years—if not individually, at least as members of a group traditionally undervalued or overlooked. Often this imbalance developed because of an organization's structure. Managers who themselves are willing to promote women and minorities may find it very difficult with the prevailing climate, real or imagined, within the management level of the company (where there might be authoritarian rather than egalitarian emphasis).

Although there are various "climates" within any organization, each company tends to have a general climate or culture characteristic of it. Loring and Wells suggest that the *exploitative* climate requires obedience, and all employees tend to be passive, dependent, and subordinate (McGregor's Theory X; see chapter 2). The *paternalistic* climate requires deference; patterns of advancement are traditional and sex-typed, although this benevolent authoritarian organization may occasionally advance women to positions in lower management. The *consultative* climate requires involvement; managers must perceive employees—including women and minority persons—less as stereotypes and more as individual persons (McGregor's Theory Y organization). Finally, the *egalitarian* climate invites participation, where women as well as men can contribute productively to the enterprise, with individual competence as the criterion rather than sex- or minority-role expectations (Theory Z and our own V Theory emphasis).

All persons intended for promotion towards managerial positions need appropriate training, at least by moving progressively through increasingly responsible positions where abilities can be demonstrated and where accountability for results increases. A special concern for top management is properly preparing minority persons for such advancement. Loring and Wells express this concern on behalf of women, but their caution applies equally to minorities and indeed all employees:

> To knowingly select a person who is underqualified without giving the essential support for probable success is a waste of time and effort on the part of the organization and the woman [or other minority] trainee. It will be better to take more preparatory time and adjust the timetable for accomplishing an increase in the number of women [and minorities] in management training. However, it will be necessary to convince most compliance officers that any

delay is not, in fact, lack of good faith. Objective evidence of the need to delay management training pending further preparatory training may be required. Surely some indication that such training is in process is essential. The pressures for promotion will undoubtedly increase as affirmative action programs are implemented.[67]

Reinsch and Ellis caution that employees, while keeping aware of opportunities for promotion, ought not overemphasize long-range advancement possibilities at the risk of distracting from what is most important—the need to do each day's job as well as possible. Robert D. Wood was sales manager of KNXT, the CBS-owned television station in Los Angeles, when he strongly advocated the philosophy of "Do the job you're in now the best way you know how, and you can't miss." He noted that one person in ten follows that policy and so stands out clearly above all other employees to managers seeking persons to promote. (Wood subsequently became general manager of KNXT, then president of the CBS Television Stations Division, and finally president of the CBS Television Network; he excelled at each position, and retained his personal concern for people; CBS executives could not overlook his obvious record of success at each step.)

RECRUITING AND DEVELOPING EXECUTIVES

The broadcast/cable industry needs well-trained management personnel—men and women who, early in life, can begin the study of station administration, mindful of their serious responsibilities and truly dedicated to broadcasting as a profession.

For many decades, experience as a successful salesperson was traditionally the path to managing a station. By the 1970s, as the arcane world of accounting and budgeting became central to operating an efficient, profitable broadcast property, a master's degree in business administration (M.B.A.) was looked upon as a key to effective managing. The following decade's mergers, leveraged buyouts, and station investing and trading called for astute financial skills in station, group, and network executives. By the late 1980s it was not uncommon for news directors to be tapped as TV station managers, reflecting the complexity of news staffs and budgeting along with the pivotal role of local news as a profit center for the entire operation.

Beyond seeking out productive new staff members, a broadcast company must give serious consideration to executive recruiting and training. The able and confident manager surrounds herself with the ablest talent she can find and afford. The fact is that people of real executive endowments and potential are in short supply in the age group from twenty-five to forty-five, the source of most executive potential. The past two decades of diversification and acquisitions in business and industry have created strong demands for aggressive yet flexible, competent, and productive executives.

It can be efficient and cost-effective for radio, television, and cable companies to develop their own executives. Top management must take the time to meet with and help develop young staff employees. They can institute an orderly and effective course of executive training to develop young men and women for key posts in the organization when future station leadership is needed.

Major stations and groups have conducted successful executive training in administration and programming and in developing news personnel. Some station and network middle managers (especially in news) have taken leaves for six months or a year for concentrated study of media issues in such programs as Columbia University's Graduate School of Journalism in New York City and the Gannett Foundation's center in Washington, D.C.

■   Broadcasters realize they must continually update themselves to keep abreast of the fast-paced media and information industries. Notable steps have been long in place. The National Association of Broadcasters annually offers special seminars and conferences at their Washington headquarters, around the country, and at their annual convention site, on topics of programming, sales, engineering, and regulation. The NAB conducts executive development seminars at Harvard and Notre Dame, designed to develop skills in analyzing and solving broadcast management problems. Other organizations such as The International Radio and Television Society, Radio Advertising Bureau, Broadcast and Cable Financial Management Association, Federal Communications Bar Association, and *Broadcasting & Cable* magazine regularly conduct customized sessions to help broadcasters keep informed about their industry and related fields.   ■

State associations of broadcasters, group-owned stations, network affiliate organizations, the wire services, music licensing organizations, and individual stations have arranged with colleges and universities to conduct program, engineering, sales and management seminars and workshops for radio and television employees. Instruction has been provided jointly by authorities in broadcasting and education.

Local managers need a strong conceptual grasp of issues, policies, patterns of possible actions, and the predictable consequences of given actions. This knowledge, coupled with sensitivity and skill in decision making, can be strengthened by special study and collaborative exploration with specialists. Such training sessions apply managerial concepts to innovative cultural and social-oriented programming, news and public affairs, minority topics, and contemporary issues in a changing society. This demands the highest quality of leadership. Effective leadership does not just happen; it must be sought out and groomed.

The manager seeking to enlighten and advance junior executives will find academic and professional opportunities for this purpose. Younger staff members will discover new insights readily transferable to advancing the station and their personal growth, which should more than repay the costs of enrolling in studies. Unfortunately, many broadcast stations provide opportunities for seminars and conferences only for the general manager and occasionally the sales manager. This is a shortsighted practice.

The well-run radio, television, or cable operation sees to it that young managerial talent is standing in the wings in each department. Each department head works with those persons to encourage development in various areas of responsibility, preparing them for possible future prime positions in the company.

Granted, the element of luck will be a factor. Many people with fine executive abilities have never had an opportunity to use outstanding talents because chance has not brought them to the right place at the right time. Still, the station must help. The station manager must be a student of human nature as well as a teacher of skills.

At the opposite end of the age continuum, many station managers find it desirable to keep retired employees on retainer as consultants, thus ensuring a continuity of wisdom and experience—but those veteran staffers' perspective and accumulated experience should be tapped long before they enter retirement. Too often management falls into stereotyped thinking that neglects profiting from elder employees who are sometimes relegated to routine tasks in corporate "backwaters." They are a source of creativity too frequently overlooked.

Separation Policies

The task of terminating an employee is never easy. Authority to take such action must be vested in the heads of departments, and their decisions need to be supported by the station manager.

Broadcast stations typically reflect continuing turnover of personnel. In radio and TV about 20% of management-level positions change annually; about 15% of all staff personnel leave their jobs or are fired each year (in smaller markets, almost 20% change jobs annually).[68] During the half-decade up to 1992, typical tenure of TV news directors was only two years; one study found one of every four on-air TV news personalities changed stations from one year to the next.[69] The precarious status of media "talent" was noted by former TV network star Tom Snyder; when dropped from a local news position (and $500,000 salary) by ABC's O&O station in New York, he cynically but realistically quipped, "I'm not disturbed about what happened to me; throughout my career, I have worked at the whim of management and I trust that will continue."[70]

Many employees leave of their own choice for presumably better job opportunities or for other personally initiated reasons. However, in an early survey, radio stations that terminated employees on their staffs discharged one out of every five of their personnel; small stations tended to discharge greater proportions of their staffs than did larger stations.[71] That national survey analyzed reasons for firing employees (table 5.6).

Respondents from radio and TV stations gave as reasons pertaining to "ability or training" for discharging one-third of all fired employees: lack of ability, incompetence, unsatisfactory work, unfit, not qualified, replaced by more competent employees, work not up to standard (including "poor sales records"), lack of education, and lack of technical know-how. The one out of five employees dismissed for deficiencies in "application to job" were appraised: inefficient, unreliable, irresponsible, unproductive, lack of interest, lazy, didn't improve, inattention to duty, or lack of effort. "Personal" reasons for one out of ten dismissals included the general references to simply "personal" and "conduct" as well as to drinking, personal indebtedness or not paying bills, dishonesty, emotional immaturity, maladjustments, conduct detrimental to station, and several single-instance reports of drug addiction, gambling, writing "hot checks," outside business interests, homosexuality, divulging sponsor information, moral irresponsibility, too many parties, bad credit, and "opium smoker." Characteristics catalogued under "Relationships with other People" (one out of twenty dismissals) included inability to get along with others, personality clashes, belligerent, insubordinate, noncooperative, incompatible with team, breach of sales policy, did not take direction, failed to follow policy, quarrelsome, trouble-maker. "Management factors," accounting for one-third of terminated employment, included factors over which employees seldom had any control, such as changes in operating procedure, economies in reducing size of staffs, job consolidation, combining AM and FM operations, automation, change in programming format and policies, and change of station ownership.

A quarter century later a sample of 277 TV stations representative of all parts of the United States identified kinds of employee behavior that led to reprimands and firings, as well as rewards.[72] The most prevalent behavior for reward or punishment was "attitude toward the job" (53% of rewards, 8% of reprimands, and 6% of firings). Reasons most common for sanctions related to "application of

TABLE 5.6
REASONS FOR TERMINATING RADIO-TV EMPLOYEES

REASONS FOR DISCHARGE	RADIO	TV
A. Ability or training	33.0%	41.6%
B. "Employee's Behavior"	36.6	36.8
Application to job	(12.8%)	(28.7%)
Personal characteristics	(17.6)	(6.1)
Relationships with others	(6.1)	(2.0)
C. "Management" factors	30.3	21.5
	99.9%	99.9%

Source: See note 71.

station policies" (30% of reprimands and 15% of firings) and "proficiency in working with others" (15% and 20%). Other major causes for terminating employees were failure in "punctuality" (20%), "proficiency in specialized skills/knowledge" (15%), and "reliability/dependability" (12%). Other reasons for firings included matters of "honesty" (5%), "interfering aspects of personal life" (4%), and "attitude of self-regard" (1%).

Other authors and broadcasters have commented on conditions that lead to eventual dismissal of an employee, such as failure to improve after repeated warnings, pressure on the supervisor from other supervisors, a culmination of a series of events, personality traits, and violating company policy and rules.[73] Although discharging an employee is unpleasant and reflects an expensive process of hiring and training anew, as well as possible severance payments that can be high, management must not overlook or be indecisive about an employee who fails to meet expectations. Close working relations in a media operation should not include involvement in the employee's personal life (unless embarrassing to the company), lest the warnings be extended beyond their proper limit. Reinsch and Ellis caution against two other mistakes: either ignoring a person's attitudes and actions because he is judged to be valuable to the station, or simply verbally criticizing and threatening, and even demoting, but never firing a person. One procedure undermines the confidence and morale of the person and merely forces him into a passive, defensive, nonproductive posture for the future. After carefully investigating the problem situation, "talk with him calmly, seriously, sympathetically" to learn what cause lies at the base of the employee's difficulty so as to correct whatever may be legitimate dissatisfaction or lack of proper participation in the company's work. At the same time, a manager cannot presume to be a special confidant with professional competence to deal with persons who have deeper problems of emotions, nerves, health in general, alcoholism, drugs, or the many personal situations that can affect anyone from time to time. The best advice seems to be to recommend that he seek adequate assistance from specialists.

Key to all this is regular, systematic appraisal of employees—especially new ones—by their immediate supervisors. Don Kirkley prepared for the NAB a helpful procedural manual of specific steps and forms to follow in tracking the work performance of both nonexempt and supervisory exempt staff members.[74] These routine reviews can be supplemented at other times by memos about instances of shortcomings in work, ability, or attitude—after such matters have been brought to the attention of the offending employee. This provides a documented record of lapses, together with notations about admonishments given, which may culminate in justifiable termination.

A survey of 100 vice-presidents of *Fortune 1000* companies in 1994 listed attitudes and behavior that caused most serious job difficulties: dishonesty and lying; irresponsibility, "goofing off," and attending to personal business on company time; arrogance, egotism, and excessive aggressiveness; absenteeism and lateness; not following instructions or ignoring company policies; a whining or complaining attitude; lack of commitment, concern, or dedication; laziness, lack of motivation and enthusiasm. Other traits that employers noted were lack of character, disrespect, making ill-informed decisions, and taking credit for work done by others.[75]

The entire process of monitoring, evaluating, and rewarding with raises and promotions or else penalizing, including firings, must carefully avoid discriminatory practices. Those range from subjective and biased assessments, to sexism (and related sexual harassment), racism, and ageism.[76]

Separation payments should be given as settlement whenever an individual is removed because their services or the position are no longer needed or for some other reason best known to the company. Such payments should not be made to individuals who are dismissed for cause, whether those individuals are union or nonunion.

Should an employee be relieved of her duties for other than a violation of good conduct or irresponsible performance, the station should try to find employment for her elsewhere. Sufficient advance planning will usually turn up another job opening; the good will that accrues from such gestures is of lasting benefit to the station. If displaced by automation or restructuring of positions, intelligent and dedicated staffers who have been with a sound broadcasting organization through much of their career can make a contribution with moderate retraining for another area at the same station.

The manager must also be prepared for possible reaction from clients if the severed employee is from the sales department, or from the public if the employee is a program personality. Honesty and brevity, with few or no details about the leave-taking, are appropriate for the client; and even less ought be communicated to audience members who might telephone the station, with no comment appropriate or necessary from the air personality's successor.

Finally, management must honestly respond to requests for references from a discharged person's prospective employer elsewhere. In fact, personnel who voluntarily leave a station are entitled to objective honesty in an appraisal by his former employer. It is appropriate to comment on their observable traits as persons in relation to other staff members and in performing their work assignments. Positive achievements as well as demonstrable, not merely suspected, shortcomings should be noted simply so that prospective employers can make reasonable advance appraisal of this potential addition to their staff.

Many problems cannot be predicted until a new employee has worked with staff colleagues, but perhaps many potential short-term employees might be identified more carefully and eliminated at the time of application and job interview, to limit as much as possible the inefficiency and loss of morale that accompany frequent turnover of less-than-apt personnel. John M. Kittross' editorial in the *Journal of Broadcasting* is as relevant in the 1990s as when he first wrote it:

> No industry can hope to operate successfully for long when it must continually hire and train employees who do not individually have a professional attitude, enthusiasm, and a deserved feeling of mutual loyalty and cooperation with their employer. An employee who feels he has no real job stability, whether

he is a union member or on a so-called long term contract or not, will have few compunctions about quitting his job, often with little notice and great dislocation to the station. A person who thinks he needs always to be alert to opportunities for a new job is rarely giving his best effort to his present employer. It is probable that the payola-plugola scandals developed in part from individual feeling that one had to look out for himself; that nobody else would. Sponsors have expressed mistrust of some salesmen who sell shoes today, station 'A' tomorrow, and station 'B' the following day. Extreme job mobility in an industry that has a slowing growth rate leads to ruthlessness, not the efficiency of a flexible operation. Although many employees do their best when 'running scared' or 'hungry,' some cannot keep the pace. Those who have proven 'track records' and enthusiasm need stability in their professional relationships for the sakes of their families if not for themselves, and that stability must be encouraged by management. Stability need not imply stagnation, and the manager does not lose his freedom of management by paying more attention to the quality of his greatest resource, his employees.

Because broadcasting cannot afford to recruit indiscriminately, and then delete 'rejects,' the solution seems to be not to let into the station in the first place the man who is incompetent, irresponsible, dishonest, or incapable of responding in the same way to trust and loyalty. Better yet, don't let him into the industry. The teachers of professional broadcasting already weed out misguided, unmotivated and incompetent potential broadcasters in numbers that might surprise the station manager. On the other hand, the normal entrance into the industry is through a small job in a small station. If management of these stations considered the eventual instead of the initial cost of people who are unable to 'cut the mustard,' these people should never get to the point of deserving to be fired. The possible consequent ill-will and loss of efficiency among both the departed employees and the remaining ones could be avoided.

It is not easy to give the employee a feeling of stability without also stunting his motivation or drive. However, indiscriminate hiring and firing can lead to tremendous waste of the ex-employee as an individual, of station morale, and of the station's 'image.'[77]

A Final Word: Employees and Broadcast Companies

The broadcast company cannot be disassociated from its close bond with its employees. If a station or network whittles away at its own creative and social integrity in the quality of its program service, it will at the same time erode the creative and ethical integrity of its personnel. When a broadcast company emphasizes economic strength to the detriment of its other capabilities and responsibilities to its staff and society as a medium of mass communication, it will inevitably force that staff, especially its management, to compromise their standards. The eventual result is the attrition of competent decision makers and creative artists who forsake the company and even the medium. Meanwhile, it is continually necessary to retrain successors to those departed staff members who as new executives must learn the corporate organizational structure and the people on the job in whom they can have confidence and to whom they can delegate work and responsibilities. When this is coupled with severed employment that calls for paying off the remaining portion of a contract, this is doubly expensive. This is waste and even economic suicide. It also erodes public confidence. It sometimes engenders more vigilance—even restrictive and punitive action—by courts and federal agencies, resulting in exhaustive legal proceedings, fees, and penalties.

The late Edgar Kobak—top-ranking executive of ABC, NBC, and Mutual, among other corporations during his long career in mass media—once noted that if employees as well as the company always do what they know or at least judge to be truly right, they will seldom lose business and will be able to sleep at night.

—⁓— CHAPTER 5 NOTES —⁓—

1. This total does not include stations employing fewer than five persons, nor does it include thousands of network employees. Federal Communications Commission, *59th Annual Report/Fiscal Year 1993* (Washington, D.C.: U.S. Government Printing Office, 1994), 26.

2. G. P. Green and E. H. Becker, eds., *Employment and Earnings* (Washington, D.C.: U.S. Dept. of Labor, Bureau of Labor Statistics (April/May 1991); cited by Kenneth Harwood, "Broadcasting and Cable as Employers," *Feedback* 33:1 (Winter 1992): 21–22. Federal estimates were based on samples, not actual counts.

3. Sydney W. Head and Christopher H. Sterling, *Broadcasting in America: A Survey of Electronic Media,* 6th ed. (Boston: Houghton Mifflin, 1990), 210.

4. Of 148,579 fulltime staffers at radio and TV stations, 58,725 (39.5%) were women and 26,990 (18.2%) were minority persons. Of 91,342 fulltime persons at local cable companies, 37,193 (40.7%) were women and 19,193 (21%) were minorities. FCC data were gathered from stations with five or more employees and cable companies employing six or more persons; network and cable corporate headquarters were not surveyed or included in these totals. Federal Communications Commission, *59th Annual Report/ Fiscal Year 1993,* 26.

5. These and following data are from the National Association of Broadcasters, *1984 Radio Financial Report* (Washington, D.C.: NAB, 1984); cited by Barry Sherman, *Telecommunications Management:* The Broadcast & Cable Industries (New York: McGraw-Hill, 1987), 94, 117. Later NAB reports for 1985, 1986, and in collaboration with the Broadcast Cable Financial Management Association in 1990 did not break out staff sizes by department.

6. "Monday Memo," *Broadcasting,* 6 January 1992, 119. See Michael Beer et al., *Human Resource Management: A General Manager's Perspective* (New York: The Free Press, 1985); and Michael Beer and Bert Spector, eds., *Readings in Human Resource Management* (New York: Macmillan/The Free Press, 1985).

7. John Naisbitt and Patricia Aburdene, *Re-inventing the Corporation: Transforming Your Job and Your Company for the New Information Society* (New York: Warner Books, 1985), 4, 11, 12; italics in the original.

8. These data are derived from Thomas Berg, "The Phenomenon of Employee Turnover: How Television Station General Managers and Department Heads Cope with Transition," [BEA] *Management and Sales Division Newsletter,* September 1990, 1–4.

9. Data from Glenn Starlin, ed., "Problems in Finding Qualified Employees: A Report from the APBE-NAB Employment Study," *Journal of Broadcasting* 7:1 (Winter 1962–1963): 63–67. The 167 TV managers responding to the survey represented all market sizes but 201 radio respondents skewed heavily to small markets.

10. "Electronic Media Career Preparation Study: Executive Summary," conducted by The Roper Organization, Inc. for International Radio and Television Society, Radio-Television News Directors Association, and National Association of Television Program Executives (New York, 1987), 6–7.

11. See Robert H. Coddington, *Modern Radio Broadcasting: Management and Operation in Small-to-Medium Markets* (Blue Ridge Summit, Pa.: Tab Books, 1969), 16–18,

150–158; also Jay Hoffer, *Managing Today's Radio Station* (Blue Ridge Summit, Pa.: Tab Books, 1968), 31–33.

12. Irving E. Fang and Frank W. Gerval, "A Survey of Salaries and Hiring Preferences in Television News," *Journal of Broadcasting* 15:4 (Fall 1971): 431.

13. William J. Oliver, "Documenting Training Need," *Feedback* 20:2 (November 1978): 5–8.

14. Data reported from a mail survey in December 1978-January 1979, reported in "TV Station Executive Salaries: 22nd Annual Report," *Television/Radio Age,* 29 January 1979.

15. Data from 1976 survey by Vernon Stone, "News Directors and Problems Surveyed," *RTNDA Communicator,* March 1977, 5–10.

16. Comments by Ted Koppel at the RTNDA (Radio and Television News Directors Association) international conference, Orlando, Florida; quoted in "Koppel Upsets Educators," *Static* [publication of Association for Education in Journalism and Mass Communication] 27:3 (October 1987): 1, 4.

17. Profile of Rose from CBS News press release dated 29 December 1983; data on Jack Smith and Joe Peyronnin in CBS News press release, 30 July 1987.

18. The Roper Organization, Inc., "Electronic Media Career Preparation Study: Executive Summary" (December 1987), 6–7. See also "Electronic Media Career Preparation Survey" (three-part report)," *RTNDA Communicator,* April 1988, 16–33.

19. Robert O. Blanchard, "Put the Roper Survey on the Shelf—We Have Our Own Agenda," *Feedback* 29:3 (Summer 1988): 3. Excerpts quoted in Head and Sterling, 218.

20. Gary Kaplan, "Management: Radio Should Go to College," *R&R* [Radio & Records], 10 May 1985, 20.

21. A comprehensive guide to FCC regulations and federal EEO laws governing media employees, including suggestions for avoiding and defending against discrimination claims, can be found in Stanley Brown and Jay Birnbaum, *A Broadcaster's EEO Handbook,* 2nd ed. (Washington, D.C.: National Association of Broadcasters, 1989).

22. Shu-Ling C. Everett, "Pondering 'Workforce 2000' in the Television Industry," *Feedback* 32:3 (Summer 1991): 10–13. The author cites T. Cox, "The Multicultural Organization," *Academy of Management Executives,* 5:2 (1991): 34–47; and C. Petrini, "How Do You Manage a Diverse Workforce?" *Training & Development Journal* (February 1989): 13–21.

23. Rosalind Loring and Theodora Wells, *Breakthrough: Women into Management* (New York: Van Nostrand Reinhold Company, 1972), 45.

24. Those laws and federal guidelines are discussed in detail, with particular reference to women's rights, in Rosalind Loring and Theodora Wells' excellently documented and reasoned analysis, *Breakthrough: Women into Management* (New York: Van Nostrand Reinhold Company, 1972). This is an excellent text for any manager wishing to understand better the evolving role of women in middle and executive management in the United States. See also a balanced, reasoned analysis supported by research and fully documented: Gary N. Powell, *Women and Men in Management* (Newbury Park, Calif.: Sage, 1988); his readable study offers concrete instances and court cases, fleshing out sound theoretical and legal analysis. Offering summaries of many studies of sexism in media, including extended quantitative data cited from them, is Matilda Butler and William Paisley, *Women and the Mass Media: Sourcebook for Research and Action* (New York: Human Sciences Press, 1980). Serving as a veritable directory of women in media with brief résumés of their careers is David H. Hosley and Gayle K. Yamada, *Hard News: Women in Broadcast Journalism* (New York: Greenwood Press, 1987). See also Marlene Sanders and Marcia Rock, *Waiting for Prime Time: The Women of Television News* (Champaign, Ill.: University of Illinois Press, 1994 edition). See chapter 11 in this text.

25. *The World Almanac and Book of Facts 1994* (Mahwah, N.J.: Funk and Wagnalls, 1993), 362–363.

26. "Vital Statistics: Changing Cities," *U.S. News & World Report,* 28 October 1991, 90.

27. Supporting groups can be contacted at: National Association of Broadcasters' Employment Clearing House, 1771 N St., N.W., Washington, D.C. 20036; Radio and Television News Director Association, 1717 K St. N.W., Washington, D.C. 20006; Broadcast Education Association, 1771 N Street, N.W., Washington, D.C. 20036; Minorities and Communication Division, Commission on the Status of Minorities, and Commission on the Status of Women, Association for Education in Journalism and Mass Communication, 1621 College St., University of South Carolina, College of Journalism, Columbia SC 29208; Association of Black Communicators, 295 Communications Bldg., University of Tennessee, Knoxville TN 37996; National Black Media Coalition, 38 New York Ave., N.E., Washington, D.C. 20002; National Association of Black Journalists, phone (703)648-1270, fax (703)476-6245; National Association of Hispanic Journalists, 1193 National Press Bldg., Washington, D.C. 20045; National Federation of Press Women, Inc., Box 99, Blue Springs MO 64013; Native American Journalists Association, Campus Box 287, Boulder CO 80309; Asian American Journalists Association, phone (415)346-2051 or fax (415)4671; American Women in Radio and Television, Inc., 1101 Connecticut Ave., N.W., Suite 700, Washington, D.C. 20036; Women in Communications, Inc., 2101 Wilson Blvd., Suite 417, Arlington VA 22201. Related internship and scholarship projects are organized and funded by group-owned stations such as Westinghouse's Group W and Cox Communications.

28. Mortimer B. Zuckerman, "Editorial: Black and White in America," *U.S. News & World Report,* 28 October 1991, 92.

29. Benjamin Hooks, the first black person appointed to the FCC, so stated in a half-hour interview on Channel 13, KCOP, Los Angeles, 6 May 1973.

30. "EEO Advice to Broadcasters," [NAB] *Radio/TV Highlights* 6:30 (4 August 1980): 3–4.

31. Data for 1988 were reported by the FCC, *55th Annual Report/Fiscal year 1989* (Washington, D.C.: U.S. Government Printing Office, 1989), 25, and also in its *Broadcast and Cable Employment Trend Reports* (Washington, D.C.: FCC, January 1989 [broadcast] and December 1988 [cable]); cited by Head and Sterling, *Broadcasting in America,* 6th ed., 212. Data for earlier periods were in "Parker Assails Hiring Practices at TV Stations," *Broadcasting,* 27 November 1972, 26, 31; citing the report by Ralph M. Jennings, "Television Station Employment Practices: The Status of Minorities and Women," United Church of Christ Office of Communication, November, 1972; plus 1995 Commission report cited in *Broadcasting & Cable,* 17 June 1996, 23.

Students studying broadcast management at the University of Southern California's School of Journalism and the University of Alabama's College of Communication reported employment data listed on FCC Form §395 in Public Inspection Files of various radio and TV stations on the West Coast from 1977 to 1982, and in the South from 1983 to 1991. Los Angeles TV stations in twenty-one cases employed 20%–28% minorities during that earlier period: 1977 (27.6%); 1979 (19.8%); 1980 (16.8%); and 1982 (28.6%); radio stations in thirty-five cases in those same years listed minorities on fulltime staffs as 20.7%, 19.8%, 21.9% and 18.1%. Those figures were double 1972 totals (12% average compiled for Los Angeles stations in 1972, and 10% nationally in FCC data computed by the Office of Communication, United Church of Christ).

In the South (14 small-to-medium sized cities in Alabama, 3 mid-sized markets in Florida, 2 in Tennessee, and 1 in Georgia) TV stations sampled in thirty-nine cases employed minorities in 1983-1986, 1988, and 1991 as 22.8%, 17.3%, 20.1%, 16.3%, 15.9%, and 26.9% of their fulltime workforce; sampled radio stations in sixty-four cases employed averages of 20.2%, 23.1%, 17.2%, 23.8%, 14.3%, and 9.1% through those successive years. While annual totals were not comparable because of the number of different stations sampled, averages often reached the low-to-mid-20% range and regularly exceeded 14%–15% (except the last year when only three stations in the 187th ranked market were reported).

In 1995 and 1996 seven radio stations in Alabama ranged from 11% minority staff, through 14%, 18%, to 30.5% and as high as 61.5% and 88% where formats were black-oriented; a small Mississippi station had no minority person on its eight-person staff. Mostly medium-to-large market TV stations surveyed included 9 stations in Alabama, 2 in Georgia, and 1 each in Louisiana and South Carolina; their staffs were comprised of 9.3% up to 31% minority persons (the median was 23.8%). Those same years students checked EEO Form §395 at a Milwaukee, Wisconsin, TV station staff with 11% minorities, and a Cincinnati, Ohio, station with 11.7%. Clearly the strong minority populations in the South prompted hiring far higher percentages of them than in other regions.

32. Data from the International Labor Organization reported by the Associated Press, "Gender Pay Gap Still Wide in U.S.," *Tuscaloosa News,* 21 January 1992, 1–D; "Most Women Overworked," *Tusacloosa News,* 7 September 1992, 1, 4–A; Michael Vermeulen, "What People Earn," *Parade,* 26 June 1994, 4–5.

33. Kate Maddox, "Sex Harassment in the Media," *Electronic Media,* 9 December 1991, 1, 40; see also Kate Maddox, "How Managers Deal With It," 40, 46.

34. Janet Stilson, "Net Femmes Facing Upper-Level Block?" *Variety,* 4 March 1987, 83, 110–111.

CBS had been the first network to hire a woman as an executive in news: in 1937 Helen Sioussat succeeded Edward R. Murrow as director of talks, a position she held for two decades. Early on, veteran administrators in network broadcast standards were women: Grace Johnson headed continuity acceptance for the ABC networks for many years, as did Dorothy Brown on the west coast from the 1940s (when the network was NBC Blue) until retirement in the 1970s. And Judith Waller ran NBC's flagship radio station WMAQ in Chicago in the 1940s. Pauline Frederick was ABC's correspondent to the United Nations in 1948, then moved to NBC in 1953 where she worked until 1974 when she retired from her four-decade broadcast career at age 65. For an exhaustive listing describing women and their work at stations and networks through the decades, see Hosley and Yamada, *Hard News.*

35. Data in this section are drawn in part from annual reports for the RTNDA published in the *RTNDA Communicator:* Vernon A. Stone, "Minority Share of Work Force Grows for Third Year," May 1991, 20–22; "News Work Force Holds Up in TV, Drops in Radio," May 1992, 18–19; "TV News Work Force Grows, Declines Continue in Radio," May 1993, 20–21; "Status Quo," August 1994, 16–18; "TV News Work Force Grows," May 1993, 20–21; "Fewer Budget Cuts for TV News," May 1994, 28–29; and from RTNDF/Ball State University surveys summarized in the *Communicator* by Bob Papper and Andrew Sharma, "Diversity Remains an Exclusive Goal," October 1995, 18–25; and by Bob Papper, Michael Gerhard, and Andrew Sharma, "More News, More Jobs," June 1996, 20–28. Stone's 1991 findings are summarized and factors behind slowing down of female/minority gains in broadcast employment are explored by Shu-Ling C. Everett, "Pondering 'Workforce 2000' in the Television Industry," *Feedback* 32:3 (Summer 1991): 10–13.

36. RTNDF/Ball State University's 1996 report of data gathered in December, 1995 (see previous note), listed 96% of affiliates of the three major networks as carrying local news, almost half of Fox affiliates, and less than one-third of independent stations. At ABC, 177 of 188 affiliates aired local news; CBS, 183 of 187; NBC, 189 of 196; Fox, 75 of 163; independents, 80 of 264. Among PBS-affiliated noncommercial stations, 54 of 198 produced local news.

37. Comments attributed to Felice Schwartz, president of Catalyst—a nonprofit group in New York which researches work-family issues—by Beth Brophy, "The Truth About Women Managers," *U.S. News & World Report,* 13 March 1989, 57.

38. Vice-President Al Gore's speech to National Association of Black Owned Broadcasters, cited in *Broadcasting & Cable,* 19 September 1994; Dwight Ellis, vice-president, Human Resources Development, National Association of Broadcasters, 31 January 1991; in personal correspondence cited by James Phillip Jeter, "Black Broadcast Station Owners: A Profile," *Feedback* 32:3 (Summer 1991): 14–15. The National Association of Black Owned Broadcasters in 1990 included 109 members who were licensees of 185 radio and TV stations.

39. "TV News Staffs Grow, Radio Uses More Part-Timers," *RTNDA Intercom* 7:5 (1 March 1990): 1; Vernon Stone, "Minority Share of Work Force Grows for Third Year," *RTNDA Communicator,* May 1991, 20–22.

40. In 1991 commercial AM stations had dropped 12 from the previous year to a total of 4,972; commercial FM dropped 114 to 4,258; noncommercial FM went down 18 to 1,420 stations; 23 commercial TV stations went off the air between 1991 and 1992, leaving 1,092, while noncommercial TV stations were reduced by 6 to a total of 348. Kenneth Harwood, "Broadcasting and Cable as Employers," *Feedback* 33:1 (Winter 1992): 22.

41. Elizabeth Shimer Czech, "Interaction Between Black and Corporate Culture in Broadcast Management" (Ph.D. diss., The Ohio State University, 1972). Her dissertation findings were published under the title "Minority Employment Guidelines for Broadcasters, Placement Offices, and Broadcast Educators, " printed by Kaiser Broadcasting and distributed by the Broadcast Education Association, 1771 N St., N.W., Washington, D.C.

42. Biennial reports through the years were variously titled; for example, "Wages, Hours and Employment/Television 19xx," "Radio 19xx/Employee Wage & Salary Report," and "1990 BCFM/NAB Television Employee Compensation and Fringe Benefit Report." Because published data are delayed more than a year in the process of gathering, compiling, and distributing reports, figures here are estimates for 1995 based on several patterns over previous years (tables 5.4 and 5.5).

43. Cf. Sherman, *Telecommunications Management,* 94, 118, and 140, for 1983 data from NAB's *Employee Compensation Report* plus *Cable Television Business,* 15 June 1983, 43; Head and Sterling, *Broadcasting in America* (1990 ed.), 214, for 1988 data in NAB's *Radio. . .* and *Television Employee Compensation & Fringe Benefits Report.* Rates of increased compensation were computed for positions between 1983 and 1988 figures provided by those sources; those totals were extrapolated further to reflect private industry's raises of 4.1% in 1989, 4.5% in 1990, and 4.1% in 1991 (making U.S. 1991 averages about 15% higher than 1988 figures when raise percentages were compounded); source: *U.S. News & World Report,* 24 December 1990, 66, reporting data and projections from Office of Personnel Management and Bureau of Labor Statistics. Projections were made from proportionate increases in respective positions from 1983 to 1988 in both radio and TV, which were respectively about 30% and 35%; to 1988 through1991—approximately 22% each. Computations were based on and resulted in:

U.S. AVERAGE SALARY: % INCREASES, 1983–1991

	1983–1988	*1988–1991*	*Composite, 8 Years*
Private-sector industries	20%	15%	35%
Radio stations	1%–45% [30%]	16%–32% [22%] (est.)	52%
Television stations	5%–45% [35%]	19%–32% [22%] (est.)	57%
Local cable systems	NA	NA	54%

Those estimated figures for 1991 were then projected further to 1996 by conservatively estimating 3% annual inflation rate in broadcasting (above the general 2%–3% national rate), or 15% noncompounded for the five years.

For Vernon Stone's data on news salaries for 1990, see "News Salaries Lag Cost of Living," *RTNDA Intercom* 8:3 (1 February 1991): 1, and "News Salaries Lose Out to Cost of Living," *Broadcasting,* 11 February 1991, 32. Also Randy Sukow, "Most TV Pay Lags Inflation, Study Finds," *Broadcasting,* 25 November 1991, 50.

44. Data cited by Harwood (1992, pp. 21-22) from report by Green and Becker (April/May 1991) for the U.S. Department of Labor's Bureau of Labor Statistics.

45. "The Communications Industry and News Media Under the Fair Labor Standards Act," Department of Labor, Employment Standards Administration, Wage and Hour Division; February 1972 (WH Publication #1355). A subscription to *Federal Labor Laws,* a looseleaf publication of regulations and amendments pertaining to wages and hours, may be ordered by advance payment to the Superintendent of Documents, U.S. Government Printing Office, Washington, D.C. 20402.

46. A clear and succinct summary of major points about compensation, including salary and benefits, can be found in Peter K. Pringle, Michael F. Starr, and William E. McCavitt, 2nd ed., *Electronic Media Management* (Boston: Focal Press, 1991), 73–76.

47. This became a serious practical matter for a station in December, 1971, when, for the first time, women employees at WRC-AM-FM-TV in Washington, D.C. asked the FCC's Complaints and Compliance Division to require owner NBC to reimburse them for back pay deprived them due to sex discrimination. A month earlier, the Washington office of the Equal Employment Opportunity Commission had presented findings implying such discrimination. The women also lodged a complaint against WRC and NBC with the Department of Labor. The twenty-seven-member WRC/NBC Women's Rights Committee recommended financial compensation, plus setting up timetables and goals for hiring women in all job categories and for developing training programs. See: "WRC Women Win at EEOC, Now Come Back to FCC," *Broadcasting,* 20 December 1971, 38. After two years, the U.S. Equal Employment Opportunity Commission returned the first finding ever against a broadcaster, supporting most of the charges of discrimination against women and blacks at NBC's WRC-AM-FM-TV. See chapter 11, "Managers and the Law,"; cf. "EEOC Upholds Charges Against WRC Stations," *Broadcasting,* 5 February 1973, 36; and "EEOC Also Upholds Blacks' Complaints Against NBC's Washington's Stations," *Broadcasting,* 19 February 1973, 32.

48. Physical steps to protect employees' safety and health must be accompanied by carefully filed documentation on those steps; see John S. Logan and Erwin G. Krasnow, *Record Retention Guide for Radio and Television Stations* (Washington, D.C.: National Association of Broadcasters, 1988).

49. NAB Insurance Committee, *A Broadcaster's Safety and Loss Control Guide* (Washington, D.C.: National Association of Broadcasters, 1990).

50. Employees of CBS Inc. in New York get 3 weeks of vacation after 5 years with the corporation, 4 weeks vacation annually after 15 years of employment, and 5 weeks after 20 years.

51. For an introductory study of labor unions and guilds in broadcasting, see the first book published on the subject: Allen E. Koenig, ed., *Broadcasting and Bargaining: Labor Relations in Radio and Television* (Madison, Wis., University of Wisconsin Press, 1970). This collection of essays and analyses surveys the historical development of unionism in radio and television, the role of federal mediation and arbitration, and selected problem areas (such as creative artists, minorities, and technical unions). Pringle et al., *Electronic Media Management,* 2nd ed. (1991), list categories of media workers who fall under various union jurisdictions; 77–79. See *Broadcasting/Cable Yearbook '95* (Washington, D.C.: Broadcasting Publications, Inc., 1995), I–20.

52. Ron Irion, NAB Director of Broadcast Management, in [NAB] *Highlights,* 30 November 1972, 3.

53. Federal laws include the Federal Equal Pay Act of 1963, an amendment to the Fair Labor Standards Act, and Title VII of the Federal Civil Rights Act of 1964, as well as President Lyndon Johnson's Executive Order 11478, and Revised Order 4 issued by Secretary of Labor J. D. Hodgson in December 1971. The first equal pay cases were decided by the courts in 1907; enforcement powers were given to the Equal Employment Opportunity Commission in March, 1972. See Loring and Wells, *Women into Management,* (1972), 20–23, and 26–28 for "sex bias in pay and position."

54. Hoffer, *Managing Today's Station,* 98–101, is cool to unions and suggests station policies that might preclude employees' choosing to organize. Reinsch and Ellis, *Radio Station Management,* 259–267, offer practical recommendations for collaborating positively with union representatives. Coddington, *Modern Radio Broadcasting,* 157–158, notes the bilateral responsibilities of management and employees toward each other, in consideration of each person's dignity, loyalty and professionalism, and adequate payment for work done. Specific negotiating procedures between AFTRA and broadcast employer groups are described by Allen E. Koenig, "AFTRA and Contract Negotiations," *Journal of Broadcasting* 7:1 (Winter 1962–1963): 11–22. AFTRA's internal processes on behalf of its 16,000 members nationally are described by Allen E. Koenig, "AFTRA Decision Making," *Journal of Broadcasting* 9:3 (Summer 1956): 231-248; he analyzes national convention procedures of AFTRA negotiators in determining their plans for demanding wage increases and better working conditions, and then how they decide whether to accept a particular three-year contract with national networks and advertising agencies.

55. CBS Inc., "The Company You Keep" (New York: CBS Inc., 1994), 2–3; this 43-page booklet summarizes personnel policies and practices affecting CBS employees.

56. National Labor Relations Board, *A Layman's Guide to Basic Law,* quoted by Department of Broadcast Management, *Negotiating the Broadcasting Labor Contract,* National Association of Broadcasters, Washington, D.C., 1969 edition, p. i, "Foreword."

57. See Howard M. Pardue and Human Resources Committee, *National Labor Relations Law Guidelines for Broadcasters,* (Des Plaines, Illinois: Broadcast Financial Management Association, 1987).

58. Early NAB publications about labor relations included *The Right of Free Speech, Collective Bargaining Units in Broadcasting, Important Clauses in Your Labor Contract,* and *The Eleventh Hour* (how to plan operations in the event of a strike). NAB's other services include reprints of union labor contract summaries, "Memo to Management" briefings about contracts, "Labor Relations Report" and "Arbitration Digest" which summarize current developments and sample cases. See NAB's twenty-six-page booklet, *Negotiating the Broadcasting Labor Contract* (Washington, D.C.: National Association of Broadcasters, 1969 edition), 2–3, and 11–24, for further procedures, such as when and where to meet, specific bargaining tactics, the problem of an impasse, use of mediators, drafting the contract, and follow-up details after the settlement. See also the NAB's *Getting What You Bargained For: A Broadcaster's Guide to Contracts and Leases,* 2nd ed. (Washington, D.C.: National Association of Broadcasters, 1989).

59. Reinsch and Ellis, *Radio Station Management,* 249–250; see also 251–252, 258–259.

60. See Coddington, *Modern Radio Broadcasting,* 16–18, and 157–158; Reinsch and Ellis, 49-50; also Hoffer, *Managing Today's Radio Station,* 28–30 about problems of "the boozer" and p. 39 about careful attention to the role played by a telephone answering service for audience members and clients who phone the station at non-office hours.

61. Glenn Starlin, ed., "The Broadcasting Employee," *Journal of Broadcasting* 7:3 (Summer 1963): 238–240; "CBS Looks Hard at Its Manpower Policies," *Broadcasting,* 20 December 1971, 38.

62. Naisbitt and Aburdene, *Re-inventing the Corporation,* 22, 27, and 21.

63. Carnegie Corporation, *Quarterly Report* (July, 1961), 2–5, describes a six-year study by the Institute of Personality Assessment and Research of the University of California about differences between highly creative people and those less creative. The study found that, while intelligence and special aptitudes are important factors in creativity up to a point, they are not perhaps the crucial elements. A person with even average intelligence can be creative if his environment is conducive to creativity. The Institute found highly creative persons unlikely to be interested in small details or in practical or concrete matters, but more concerned with what things mean and the implications of those meanings for new discoveries. All creative people in the study were consistent in choosing that which was complex over that which was simple. The creative person was usually a nonconformist in the field of ideas but not usually in the fields of behavior or social patterns. The creative person was generally independent, able to find satisfaction in those elements of challenge that other people might not even perceive.

A second study, undertaken by psychologist Morris I. Stein at the University of Chicago and at New York University, analyzed environments of creative people to determine working conditions most conducive to creativity. A third, by the Institute of Social Research at the University of Michigan, investigated conditions leading to high scientific creativity.

64. Reinsch and Ellis, 252.

65. "The Company You Keep" (New York: CBS Inc., 1994), 17.

66. A cheerless commentary on hazards of promotions is offered seriously but in a light-hearted style by Dr. Laurence J. Peter and Raymond Hull in *The Peter Principle:*

Why Things Always Go Wrong (New York: William Morrow, 1969; Bantam edition, 1970), 6–8, 19–21, and elsewhere. Competent employees are identified and promoted progressively until they prove incompetent in a given position and there they then remain. Thus the Peter Principle that "in a hierarchy every employee tends to rise to his level of incompetence" (p. 7). The inexorable corollary states that "in time, every post tends to be occupied by an employee who is incompetent to carry out its duties" (p. 8). Nevertheless, another irreverent best-seller highly encourages internal promotion for efficiency as well as humane consideration of employed human beings; see Robert Townsend, *Up the Organization: How to Stop the Corporation from Stifling People and Strangling Profits* (New York: Alfred A. Knopf, 1970); Fawcett World Library edition (Greenwich, Conn.: Fawcett Publications: 1970), 138–139.

67. Loring and Wells, *Women into Management,* 62.

68. Barry L. Sherman, *Telecommunications Management,* 94, 119, citing NAB sources.

69. "Major-market News Personalities Keep On the Move, Survey Shows," *Television/Radio Age,* 31 August 1987, 18.

70. Reported in *Variety,* 22 August 1984, 106.

71. Sherman P. Lawton, "Discharge of Broadcast Station Employees," *Journal of Broadcasting* 6:3 (Summer 1962): 191–196. Those dated figures were borne out in subsequent decades.

72. Gale R. Adkins, "Critical Qualities for Successful Employment in Commercial Television Stations," *Feedback* 27:5 (Summer 1986): 18–21.

73. *Advanced Management Journal* (October 1965), 69; quoted with approval and detailed comments by Hoffer, *Managing Today's Station,* 36–37. See Scott Snell and Kenneth N. Wexley, "Performance Diagnosis: Identifying the Causes of Poor Performance," *Personnel Administrator,* 30 (April 1985): 117.

74. Donald H. Kirkley, Jr., *Station Policy and Procedures: A Guide for Radio* (Washington, D.C.: National Association of Broadcasters, 1985), 11–17. Attorney L. Michael Zinser has outlined procedures for monitoring a staff member's questionable performance, plus practical steps and cautions about dismissing an employee; see Pringle et al. (1991), 69–71.

75. Accountemps, a temporary personnel service, conducted the nationwide survey; their list was published in "What Upsets Employers Most," *Job Choices in Business: 1994* (Job Placement Council, Inc.), 20.

76. See Kim Vance and Richard Lowe, "The Ten Commandments of Employee Evaluations," *Broadcast Cable Financial Journal* (January/February 1992): 26–28; also Howard M. Pardue, "Legal and Practical Implications of Employee Termination," *Broadcast Cable Financial Journal* (November/December 1990): 32–33.

77. John M. Kittross, ed., "Stability," *Journal of Broadcasting* 6:3 (Summer 1962): 189–190. This quarterly, later renamed Journal of Broadcasting & Electronic Media, is a publication of the Broadcast Education Association. See also Jay Hoffer, 31–32, for his constructive reminders to small-market managers about their valuable role as the "training ground for talent."

Managers and Media Audiences

Despite CPMs, we're not selling people to advertisers; we're in the business of attention rental.

DAN O'BRIEN, BROADCAST MANAGER, LAWYER

People who tune in to a radio or television station or cable system comprise its "circulation," much like the buyers of a newspaper or magazine.* Advertisers supply the station's operating revenue by purchasing airtime priced according to that circulation. Thus, the station's popularity, income, and consequent profit depend directly on audience receptivity to its program service.

However, *audiences* are a mixed, hard-to-define phenomenon. They are drawn from the larger mass public—or, more accurately, multiple *publics* in a community or market area. That regional population is not monolithic; it has widely divergent characteristics, and is heterogeneous as well as anonymous for the most part. There is not even a single homogeneous audience for a given station. Most stations (especially full-service TV, less so format radio and specific cable channels) have somewhat different groupings of people who listen or view at different times throughout the broadcast day, week, or month. At one time, for example, the audience may be composed mainly of teenagers. Another time, a station may attract a large proportion of highly educated people; at other times, most of the people in the audience may have limited formal schooling. Some audiences are made up mostly of home-bound spouses, or of employed single persons, or of farmers; other audiences are mostly high school students or the elderly. There are any number of demographic differences in the makeup of various audience groupings. Woven through all the audience compositions will be some people with widely different backgrounds from those dominating the local population or program listeners/viewers. Further the changing marketplace of radio, TV, cable, satellite, and videocassette media has been accompanied by shifts in lifestyles and tastes of audiences throughout the 1980s and 1990s.

What brings these people together to share common program experiences? What do these people, individually and collectively, really think about the programs they see and hear? What causes them to prefer one station's programming over another? How can their loyalty to any specific station or to cable channels be assured? How does their reaction to programs affect their response to commercials embedded in those shows? The station manager needs answers to these questions, so the sales department can sell airtime spots to advertisers whose products or services are geared

---

*In broadcast ratings measurements, *circulation* technically refers to a station's unduplicated audience or "reach" (contrasted with "frequency"), usually referred to as "cumes"—the estimated number of different people, as percentage of survey population, cumulatively tuned to a station in a quarter-hour period.

to various subgroups. (Media salespersons cannot really sell time itself or the programs, nor do they literally sell audiences, despite the "cost-per-thousand" formulation; as noted at the start of this chapter, broadcasters are actually in the business of "attention rental" of audiences attracted by programming.)

Most audience information available to media managers is *quantitative.* Surveys estimate the approximate size of their audience at any particular time. Several commercial research organizations supply information about numbers of homes tuned to a station.

While useful, data about the sheer size of audiences can lead to deceptive conclusions if used as the sole criterion for judging a station's popularity with the mass public and hence its value to advertisers. The local manager needs additional *qualitative* information, supplementing the findings of ratings services, to reveal the interests, opinions, and attitudes of listeners or viewers. Additional time and money must be invested to probe more deeply into audience make-up and attitudes by specialized research procedures.[1]

In any communication system, feedback in letters or telephone calls from the receiver to the sender is necessary for maximum efficiency. Feedback indicates beyond mere survey numbers how effective was the effort to reach people. It provides firm rationale for altering a radio format, TV schedule, or cable service line-up of channels. However, in mass media, communication is usually a one-way process that sends a regular supply of messages out to an anonymous public. Direct feedback from most of those people is rare; viewers and listeners regularly receive media programming without communicating any reactions back to those originating the messages.

The broadcaster may hear from a small segment of the audience when they dislike something. Seldom does she receive praise from people who appreciate what was aired. The media manager cannot easily meet with members of her unseen audience to learn their opinions and attitudes about her program service, although she attempts to do so through continuing contacts with community members and by ascertainment procedures. She cannot check her sales effectiveness with them directly as does the merchant with customers in a store.

Intense media competition—traditional newspapers and magazines and direct mail campaigns, plus cable and telemarketing—in the past decade have put a premium on far more detailed analysis of radio-TV audiences delivered to advertisers. Ad agencies demand more sophisticated and precise data beyond "nose counting" totals. Advertisers need to know demographics—age groups, income levels, and other differentiating characteristics. They also want information about "psychographics"—analysis of people's use of leisure time and their ranges of attitude and opinions. These have become common tools of effective marketing strategy in mass media.

EVOLUTION OF RADIO-TV AUDIENCE ANALYSIS

Over the years, as five, ten, or thirty radio stations plus four or more TV stations infiltrated single markets, competition for audiences and advertising dollars forced each station manager to find ways to demonstrate how large and receptive their audiences were. Each manager had to find out how many people preferred his station over competitors in order to sell advertising time at a profitable rate. Broadcasters began to purchase information about audiences from companies specializing in marketing surveys and analysis.

Thus developed the "ratings wars" that shaped how broadcasters crafted competitive program service and the way the industry does business. Decisions about content and scheduling were based less on whether a program was truly "good" than on its popularity as determined by numbers of people tuned in. Stations claimed dominance at various periods of the broadcast day, often on the basis of a decimal point or two. Although a single point rise or fall can represent thousands of people—millions in national surveys—two factors compromise absolute conclusions: (a) sampling procedures generally are accurate only to within plus or minus 3% (at 95% confidence level); (b) even a large rating does not necessarily mean the audience fully "likes" a program, but may merely reflect their choice by default because they found competing programs even less attractive.[*] The late Hugh Beville, life-long professional research specialist at NBC since 1930 (when the first radio ratings survey was conducted), underscored that "there can be no perfect rating service . . . [because] ratings users are provided with rough estimates by crude tools."[2]

DIMENSIONS OF THE BROADCAST/CABLE AUDIENCE

Both radio and television achieved almost universal popularity within a decade after each was introduced. They continued to expand in subsequent years. Community antenna TV (CATV) appeared early in the 1950s for limited, pragmatic use by those with poor reception of over-air broadcast signals. Decades later, satellite interconnection allowed scattered cable systems to collaborate nationally with program suppliers, causing that distribution service to grow exponentially in the 1980s.

In the 1970s FM radio first began to be profitable with sizable audiences. One-third of all audience listening in 1974 was spent with FM stations; FM grew to half of all listening in 1979 and to 74% in 1988. By 1992, half of all 11,000 commercial stations were FM; by 1995 FM attracted over three quarters of all radio listening. In most markets, FM stations draw far larger audiences than do competing AM stations. In the early 1990s half of all AM operators lost money; by the mid-1990s one-third of all stations—AM and FM—lost money, one-third broke even, and one-third were profitable.

By 1992 an average 5.8 working receivers were owned by every household (99.9%). Two-thirds of the 585 million radio sets in 1995 were in the home or used personally (transistorized sets such as the Sony Walkman), while one-third were outside the home in autos, offices, and places of business.[3]

By 1995, over 93 million homes (all but 2%) had TV sets, almost all with color; two-thirds of those homes had two or more sets (one-third had three or more sets); and four out of five homes also had videocassette units for tape-recording and playing back TV material.[4] Television's national coverage, extending to over

---

[*]Cf. Paul Kline's "Least Objectionable Program" (LOP) theory: people tune in television by habit, for the most part, so each weeknight will find about the same total viewers who vary their selections depending on what is on various channels. Viewers routinely turn on TV and, failing to find any attractive or satisfying fare, they dial around until settling for the "least objectionable program"—least uninteresting or least boring. Thus a high rating doesn't necessarily mean a program is highly desired, but merely less unattractive than other choices available at the time. "Objectionable" in Klein's phrasing does not refer to ethical, moral, or political values, but merely to lack of relative attractiveness among programs available at a given hour.

98% of all households, exceeds the 92% of American homes with bathtubs and showers and the 87% with telephones. The television rooftop antenna became a familiar characteristic of the American landscape over thirty years, until cable TV and satellite reception dishes began to replace it in the 1980s by offering a wide array of channels and stable picture signals. Since 1992 almost two-thirds of all U.S. homes received their television, including local broadcast channels, through wired cable systems rather than directly from over-air TV transmissions. In 1996 cable was available to 92% of U.S. homes that it passed; 64% subscribed with one-third of those households receiving forty or more channels. That year direct satellite TV served 5% of American homes.

With varied media program services and multiple sets, "family viewing" with a single receiver has splintered in many homes into individual viewing in separate rooms. A 1996 survey found over half (52%) of households with teenagers owned three or more TV sets, as did 42% of families with children under twelve years, but only 31% of homes without youngsters.[5] Almost all homes (95%) with children owned VCRs; half of those homes had multiple VCRs.

People have invested billions of dollars in purchasing and maintaining radio and television sets and antennas. In 1990 alone American consumers paid $6.6 billion for 23 million new television receivers, 94% of them color; that expenditure almost doubled to $11.5 billion in 1993 for 33 million new sets, and consumers continued to purchase about 25 million new TVs annually into 1996.[6] As purely voluntary expenditures, all the dollars people spend on radio-TV equipment, maintenance, and repair reflect a high degree of public acceptance, despite criticism of program service. The introduction of high definition television (HDTV) in the late 1990s will require progressive but complete replacement of all present TV receivers with units capable of receiving the wide-screen 1,025-line picture. That means further billions of consumer dollars—far outstripping the massive investments required of stations, networks, and cable systems to change over to the new HDTV standard.

Further, the average household (not individual viewer) gradually increased daily hours of television set operation from 5 hours and 6 minutes in 1960, to 5 hours and 56 minutes in 1970, then to 6 hours and 36 minutes in 1980, and 6 hours and 53 minutes in 1990 for all over-air and cable viewing; in the 1993–1994 season households averaged 7 hours and 15 minutes of viewing; in 1994–1995, households viewed 7 hours and 42 minutes (among adults, males averaged 4:17 daily, females 5:01, teens 3:14, children 3:26).[7] Thus the average American spends more time viewing television—over 4 hours daily—than on any other activity except working and sleeping.[*]

Projections of population samples provide estimates of "ratings": the number of households/people watching a station/network as a percentage of *all* households *that own TV sets* (almost 100%) in the area being measured—the total **potential** audience. Survey measurements also provide estimates of "shares": the

---

[*]Note the wide difference between average viewing *per household* (just over 7 hours a day) and viewing by *individual persons* (about 4 hours a day). That distinction has been overlooked on occasion by *Time* magazine; the *New York Times*; Morley Safer on CBS' "60 Minutes"; Norman Lear in formal testimony at Congressional hearings; a syndicated columnist for Cox newspapers; Pulitzer Prize-winning media critic Howard Rosenberg of the *Los Angeles Times* (speaking to college students); and other usually astute sources who should know better. For example, TV producer/executive and media analyst Fred W. Friendly once informally misstated on a Memphis TV talk program that television was "something people watch 6 hours a day, and kids 8 hours a day." Not so.

number of households/people watching a station/network as a percentage of those households *with TV sets turned on* at the time being measured—or the **actual** audience.[8]

Stations compete for the attention of all those actually watching TV sets at a given time; this results in a "share" of that audience for each station. The station that, at any particular moment, provides programming appealing to the largest segment of those with sets turned on has the largest share of that audience actually viewing TV.

Broadcasters consider the figure strong for a "day-part" such as early-evening fringe time when total homes viewing all stations in a market—that is, total sets-in-use or households-using-television (HUT)—reaches 40%–50% of all homes in the area; that represents one-half of the potential audience or total population in the area. A high figure for total households watching TV in prime time is 60%–65% of all homes in the area. The challenge for each station is to sustain its share of that local tune-in audience. The advent of cable, offering dozens of alternative program services to compete for viewer attention with several over-air TV stations in each market, made this a critical issue in the 1990s—just as first TV, then FM, had smothered AM radio's previous dominance of media audiences. Prime-time national audiences for the three major networks' 630 affiliated stations dropped from a total share of 91% of all evening viewers (HUT) in the 1979–1980 season down to 63% in the early 1990s, then plummeted further to 53% during the 1995–1996 season. That meant a loss of one-third of the country's households viewing in prime-time—mostly to cable (29% of viewing), plus independent stations including the Fox network (21%), and noncommercial Public Broadcasting outlets (3%).[9]*

GROWTH OF BROADCAST STATIONS AND CABLE SYSTEMS

To complete our look at the dimensions of broadcast-cable media in the United States, and the population that evolved into mass audiences for those media, the growth of radio and TV stations and cable systems serving the public must be noted.

In 1996, 4,906 AM radio stations were on the air.[10] FM gradually overtook the original AM service, first in audience and then in income; by 1996 over half of all stations were in the FM band—5,285 in 1996. They were augmented by another 1,810 noncommercial educational/public broadcast FM stations. Many FM properties are stand-alone operations; others are co-operated with AM

---

*Distinguished from computation of shares, drooping audience *ratings* were cushioned somewhat by the continued increase in U.S. population and homes, including proliferating single-person households. Lower ratings (percentages of all possible households) are based on increasingly larger total numbers of households. In 1980 the U.S. Bureau of the Census reported 80.8 million households; in 1990 that number had risen to over 93 million. Total population in the same period rose from 226.5 million in 1980 to approximately 250.5 million a decade later. Thus a "10" rating in 1980 meant 8 million homes or 22.6 million persons; it reflected slightly fewer households and about the same number of persons as a "9" rating in 1990 which represented 8.4 million households and 22.5 million people. The other factor is that the number of residents in households has decreased over the decades, from over 4 to 2.3—partly because of smaller families and the growing number of single-person households (almost one out of four). In 1995 ratings services estimated 1.6 viewers per home during evening prime-time, when 62.5% of all TV households in the country were using television (HUT). Total households in 1997 were 97 million.

interests, either duplicating AM programming or maintaining a completely different program service.[11]

By 1996 there were 1,181 commercial TV stations (622 of them UHF). Competing for local and national audiences with alternate programming were 363 noncommercial public broadcast TV stations (two-thirds of them UHF).

An emerging variation of TV service were local, low-power TV stations (LPTV). They numbered 561 VHF and 1,211 UHF operations in 1996. Those limited stations had small audiences; even cumulative audience impact was modest. That sector of the broadcast arena was awaiting its opportunity to grow into a measurable service, much like FM radio in its early years of struggling to survive.

Meanwhile, community antenna TV (CATV) had grown into a pervasive cable industry in its own right, not merely relaying over-air TV signals but developing its own local programs and leasing rights to material provided by national suppliers. Satellite interconnections made feasible national distribution of real-time program service by Home Box Office and then Turner Broadcasting's WTBS in Atlanta (the first to be dubbed a "superstation" reaching markets all across the United States via satellite and cable distribution). Dozens of other new program services were mounted for the many cable channels on proliferating local systems. By 1996 more than 62 million of the nation's 96 million households subscribed to cable. The country's 11,660 local systems "passed" or were physically available to 92 million homes, typically offering thirty or more channels of programming including relaying local and regional station signals.

Audiences, armed with remote switchers and VCRs to time-shift over-air or cable programs, were becoming program schedulers in their own homes. No longer were they locked in to the sequence of shows plotted out for them by several executives at three major networks or at their handful of local commercial stations. The battle for audience attention and loyalty had been taken to the formidable networks and to traditionally dominant local TV stations, just as early TV around 1950 had challenged and gradually wrestled away the nation's audiences (and advertisers) from classic radio networks and even some leading stations in major markets. By 1996 four out of five U.S. homes owned videocassette recorders.

RESEARCH IN BROADCASTING

Major Commercial Research Companies for Radio-TV-Cable Three commercial research companies entered the field of broadcasting in the decade of the 1940s: marketing research firm A. C. Nielsen Company; The Pulse, Inc.; and American Research Bureau (ARB). Nielsen and ARB (now called Arbitron) were joined in later years by Trendex, Videotex, Tele-Pulse, Sindlinger, Birch, and others that came and went over the decades. Acting in concert the radio networks formed RADAR (Radio's All Dimension Audience Report) to design surveys with highly dependable data; after 1972 they turned over surveying to Statistical Research, Inc. (SRI). RADAR remains the sole audience survey of national interconnected radio network audiences, although Arbitron compiles an annual national network analysis with its local radio ratings. Those firms led the way in broadcast measurement, with fifty-four companies engaged in research for broadcasting by 1973.

Media directories in 1995 listed 143 organizations providing research data to broadcasting and cable, including Arbitron, the Gallup Organization, Paul

Kagan Associates, Frank N. Magid Associates, Nielsen Media Research, Roper Starch Worldwide, Simmons Market Research Bureau, Scarborough Research, and Sindlinger.[12]

In 1992 Nielsen's local meter service in 29 markets represented over 50% of the country's homes; in 1996 it metered 33 markets (with Fox network pressing it to expand over several years to 70 metered markets). It supplemented those data with diary surveys. It provided local market ratings to 641 commercial stations (of which 318 also subscribed to Arbitron's service until it left the TV ratings business in 1993). To compete with impending "people meters" by AGB of Britain, in the mid-1980s Nielsen began to use such units for greater sampling detail drawn from people pushing buttons, recording how long each household member watched the home TV receiver.

In 1992 Arbitron provided local market audience data to 502 subscribing TV stations (318 of them also used Nielsen's service) and 1,950 radio stations in 263 markets.[13] The company did not compete with Nielsen in rating national network programming. As noted earlier, in 1993 it moved out of measuring local TV station audiences, leaving the entire TV field to Nielsen. It still measures 96 radio markets continuously, plus all other radio markets twice a year.

Dissatisfied with Nielsen's computation of TV ratings, the three major networks funded a three-year, $30-million project targeted for completion by 1998, to develop a more advanced system of determining ratings. A potential competitor for Nielsen's national rating service was Statistical Research, Inc. (SRI), encouraged by all four TV networks to offer an alternate to Nielsen; it would require $50 million–$100 million to develop. Responding to criticism of its data reliability, A. C. Nielsen planned to augment sometime after 1995 its local market diary sample by 15% (to achieve 7.5% greater reliability) or by 50% (for 25% more reliability) if subscribing stations would pay part of the cost of adding a million more diaries to its sample for a total of 2.3 million.[14]

Changes in Kinds of Research Information   In the early years, media researchers supplied subscribers with quantitative data including how many homes had receiving sets, the numbers of respondents listening to stations and programs, and the percentage who could identify sponsors of the programs.

By the early 1960s, audience information was more detailed and included some demographic characteristics. Demography as used here expands the academic concept from a statistical study of populations with regard to birth, death, health, and growth rates to include cultural, socioeconomic, and compositional data. In television terms, demography embraces a study of the audience as a potential viewing and buying group. Demographic disclosures proved increasingly valuable to advertisers needing more understanding of the audiences for their commercials. Trendex, Nielsen, Arbitron, and Pulse all began to publish additional data for their respective subscribers, including the number of viewers per home for each program; the proportions and various age ranges of men, women, and children in the audience; and income, occupation, and education stratifications. People-meters introduced prospects of highly detailed demographic data, relating to reach and frequency, available overnight fifty-two weeks a year (not just during "sweeps" periods), and providing comparable ratings figures for cable and VCR use.[15]

By the 1980s and 1990s, lifestyles and use of leisure time helped define clusters of audiences, and were predictors of their program preferences and buying habits. Computer-assisted analyses fed mountains of sophisticated data to marketers in direct mail, mass magazines, and broadcast/cable media. Although some systems of analyzing audience reaction to media were not widely accepted, others expanded into commercially viable services.

■   TVQ, a division of Marketing Evaluation, Inc., stresses qualitative data. It mails questionnaires to different families for each monthly national survey of attitudes and opinions about specific television programs.

Other sources of qualitative data about audiences, attractiveness of programs, and impact of viewing experience—in addition to Nielsen, Arbitron, and Scarborough —include Frank/Greenberg, Mediamark, Quantiplex, Simmons Market Research Bureau (SMRB), Television Audience Assessment, and R. D. Percy & Company's Vox Box. Analytical techniques for "geodemographics" are offered by Claritas Corporation's PRIZM and Reuben H. Donnelley Company; audience values and lifestyles are profiled by SRI International's VALS.

Specialists in analyzing audience reaction to local station programs and personalities include The Benchmark Company, Frank N. Magid Associates, McHugh and Hoffman, and Reymer and Gersin Associates.   ■

STUDIES OF AUDIENCE ATTITUDES AND BEHAVIOR

Scholars early on conducted comprehensive studies of listeners-viewers' motivations and attitudes. Glick and Levy found a certain degree of dissatisfaction among television viewers.[16] The excitement and enthusiasm in early years of set ownership had given way to greater discrimination in selecting programs. Shortly after TV's Golden Age, with its live drama anthology series as well as more than a dozen western series, FCC Chairman Newton Minow typified commercial TV schedules as a "vast wasteland."

Gary A. Steiner found that a viewer's college education was the most significant factor in amount of time devoted to television.[17] Those with no more than high school education were the most devoted fans, whereas those who had attended college were more selective viewers. They had a wider range of interests outside television and more sources of entertainment. Yet, they viewed few informational programs on either commercial or educational channels, despite saying they would like more of this type available.

A decade later Robert Bower replicated and updated Steiner's survey.[18] He found that better-educated viewers generally held the television medium in lower esteem, tended to prefer other media as dispensers of news, were more selective in choosing programs to watch, and were less likely to enjoy what they viewed. Yet better-educated viewers watched television about as much as other persons during evening and weekend hours; and they distributed their time among program types—such as comedy, movies, action, information, public affairs—in proportions similar to those with less education. Similarly, the better-educated viewer faced with a specific choice between a program of information or one of regular entertainment, tended to choose entertainment as much as other persons did. However, while confirming other surveys showing black Americans among the most positive supporters of television, Bower found that better-educated (including college graduates) and higher-income African Americans viewed TV more than other blacks; that is the reverse of the general public's pattern. Studies conducted in following decades generally confirmed those correlations between demographics and viewing habits.

Television homes in 1973 had TV sets turned on per day most (6 hours and 48 minutes) where the head of the household had one to three years of high school education; average household viewing was least (5 hours, 56 minutes) where the head of the house had one or more years of college. The average for *all viewers* sampled in 1974 was 3 hours and 2 minutes spent daily with television; *upper-income* people spent 2 hours and 47 minutes; and *college-educated* spent 2 hours 23 minutes a day with TV.[19] By 1995 Nielsen Media Research reported individual adults were spending significantly more time viewing TV daily, about four hours; cable subscribers watched more TV than those without cable, often up to five hours daily.

■ The Roper Organization (in the 1990s renamed Roper Starch Worldwide) conducted opinion studies for the broadcasting industry at approximately two-year intervals since 1959.[20] All its surveys found favorable attitudes toward television heavily outweighed unfavorable ones; the public's positive assessment grew through the decades from the first study up through 1991.[21] In 1963, for the first time most people reported receiving more of their news from television than from newspapers or any other mass medium. Television also led all other media in public believability of news reports, according to the Roper studies. Those surveys through 1991 reflect the public's steadily increasing reliance on TV over other media. (Whether this is altogether a good thing for society is another serious question; the point here is only to note how the public perceives and uses various media.) Between Steiner's 1960 survey and Bower's study in 1970, the percentage of respondents who perceived television as giving the *most complete* news jumped from 19% to 41%; by 1980 Roper reported 65% and ten years later 70% of the public claimed TV as their primary source of news (compared with 43% newspapers, 15% radio, and 4% magazines); by 1994 over 72% cited TV as their source of most news. While those figures reflected multiple answers permitted in the survey, fully 44% claimed TV as their *only* major source of news.[*] When asked which major medium they would most likely *believe* if they heard conflicting reports from those sources, more than 10% selected radio and magazines in 1959, but gradually dropped to 5%–6% in 1991; over three decades those relying more on newspapers dipped from 32% down to 20%, while those judging TV as the most believable medium rose from 30% to 51% between 1959 and 1994.

In the decade from 1982–1991, Roper surveys found the relatively high percentage of Americans who judged television favorably held the same or dipped only 1 point; viewers continued to find TV "entertaining" (59%), "informative" (52%), and "interesting" (50%). (This implies that almost half the population does *not* find TV informative or interesting, and about a third do not even find it entertaining!) In both 1982 and 1991 surveys, about 40% responded that the medium was "generally good." Favorable responses about network TV's "variety of programs" remained at around 30% in both studies. Largest losses in public favor over that decade concerned TV as "relaxing" (about 36%, down 6 points from 1982) and a "good companion" (down 9 points to about 22% in 1991). More cheering to broadcast managers was the major drop in all six

---

[*]During the Persian Gulf War in 1991, when CNN along with other networks provided continual live coverage, the public identified their primary news sources as TV (81%), then newspapers (35%), radio (14%), and magazines (4%) (multiple responses were permitted).

negative judgments about TV. Only about half as many people responded in 1991 that TV was "in bad taste" (17%, down from 29% in 1982), "simple-minded" (15%, down from 31%), and "dull" (14%, down from 31%). Other negative attributes were also cited far less in 1991: "programs all the same" (down to 12%, from 18% in 1982), "generally bad" (to 11% from 16%), and "annoying" (10%, down from 18%).

During the 1980s, the percentage of those saying *cable TV* was "generally good" rose from 26% (1983) to 34% (1989).[22] The fast growth of multi-channel satellite-programmed cable systems during that decade accounted for public assessment of *cable TV* as offering "lots of variety": from 34% of respondents in 1983, to 47% in 1988.

Over that same span of seven years, those judging noncommercial public TV favorably as "important" grew to over a quarter of the population: from 18% (1983) to 27% (1989); positive response grew similarly for public TV as "imaginative" from 19% up to 27%, and as "stimulating" from 23% in 1983 up to 27% of respondents in 1989. ▪

DDB Needham Worldwide summarized the Roper Organization's studies of how audiences assessed TV by eighteen criteria through the 1980s; excerpts are tabulated in table 6.1. The agency's demographic analysis of responses tracked patterns important to broadcast-cable managers:

Network TV was more likely to be considered **interesting** by those over 60 (11%), black (13%), living in the midwest and west (13% and 11%), and in B and D [less populated, rural] counties (11%). Cable, on the other hand, was considered **more interesting** by younger respondents, age 18-29 (11%), white collar workers (11%), and VCR homes (11%). Public TV was more likely to be thought **interesting** by 30–44 year-olds (10%), upper-income homes (18%), and those living in the west (27%). . . .

Perhaps more surprisingly, younger viewers appear to be more likely to feel that network television is **getting better** (37%), as do those living in the north east (16%) and south (16%). A belief that cable was improving was found to be stronger in affluent homes (25%), the western region (31%) and C counties (25%). . . .

Finally, network TV was more likely to be considered **generally good** by lower income homes (13%), and rural areas (13%). Meanwhile, cable was more likely to receive this rating by younger viewers (18%), blue collar workers (12%), people in the west (15%) and VCR households (15%). . . .

For the most part, both network and cable TV were likely to be seen as **getting worse** by the same demographic categories. These were 30–44 year-olds, those earning $35,000+ a year, executive/professionals, people in the mid-west, and those living in B counties. . . .

[Summary] Network TV finds greater favor among older, lower income and rural homes, whereas cable is championed more by younger, more affluent, and professional households. Public television continues to be regarded positively by middle-age, upscale, and western homes.[23]

A 1993 survey of 2,400 people across the country reported television as the most believable medium (34%) over newspapers (29%), magazines (23%), and radio (12%).[24]

Specifically noteworthy for managers of competitive networks/stations and cable operations was the Roper Organization's finding in 1991 that two-thirds of cable subscribers claimed they would probably cancel their cable service if among its program channels it no longer carried ABC, CBS, and NBC. Almost four out of five responded that their cable operators should charge only about half the current monthly fee if they no longer carried programs by the Big Three networks.

TABLE 6.1
AUDIENCE ASSESSMENTS OF TV: 1983, 1989
RESPONSES TO ROPER ORGANIZATION SURVEYS ABOUT
NETWORK, CABLE, NONCOMMERCIAL TV

	DECEMBER 1983			OCTOBER 1989		
	Network	Cable	Noncommercial	Network	Cable	Noncommercial
Generally good	29	26	25	32	34	35
Getting better	16	12	15	19	16	19
Getting worse	23	8	2	30	12	4
Interesting	36	35	41	38	47	51
Uninteresting	9	4	2	12	5	4
Dull	17	4	4	21	7	8
Lots of variety	24	34	13	28	47	19
All the same	15	6	2	20	8	2
Imaginative	7	9	14	10	16	27
Stimulating	8	9	19	9	13	27
Relaxing	26	15	8	30	24	19
Lots of Fun	8	7	3	11	15	7
Informative	25	20	40	29	29	51
Educational	17	21	61	18	25	70
Serious	4	3	13	7	6	21
Important	10	6	18	16	11	27
Too simple-minded	16	3	0	19	6	1
In bad taste	10	11	1	15	14	1

Source: Summary of Roper Organization's surveys in 1983, 1988, 1989; by DDB Needham Worldwide, March 1990.

Fewer than one in ten said they would not cancel their cable subscription if the only way to get network programs was to add a separate house antenna to receive local over-the-air affiliated stations.

PROFILE OF AUDIENCES

In general, what do broadcasters really know about the public? First, and of prime importance, they know it is composed of many diversified publics. Each of these smaller units has its own particular interests and tastes.

Broadcasters know that most audiences for radio and television are far more interested in light entertainment than they are in information or education. Whether a particular audience prefers comedy programs, baseball, music, or dramatic fare, they do want to be entertained. Even those favoring the fine arts find their own form of entertainment in that programming. A sure way to fail in any format is to neglect the element of showmanship that is an important ingredient of entertainment. Yet, while most people consider TV as the most entertaining mass medium, the public's attitude toward television reflects events and changes in the world during recent tumultuous decades.

The public's perceptions are important to the broadcaster. People have seen most forms of show business and all the established story plots and they have been eyewitnesses to the great news stories. Their expectation level for new and novel programs is insatiable. At any particular time, given too much repetition or imitation, they can turn their leisure attention to other forms of entertainment and recreation: traditional print and motion pictures, compact discs (CDs) and digital audiotapes (DAT), specialized interactive videotapes, and computer games.

CHANGING AUDIENCE CHARACTERISTICS

Broadcasters could more easily understand their public if it did not keep changing. These changes can be illustrated by examining society's basic categories: total population, age, education, income levels, and living contexts, plus minorities.[25]

POPULATION SHIFTS

The U.S. Bureau of the Census projected a 6.8% increase in population during the 1990s, from 250 million persons in 1990 to 268 million ten years later. Marketing experts project almost 100 million households by the year 2000, a 5% increase during the decade. Network and agency executives predicted that the three-network TV share of national viewing audiences would drop to 58%–62% by 1995, stabilizing by the year 2000 at around 55% (or including Fox network, at 70%).[26] But in fact the three major networks' prime-time shares—after a brief recovery from 60% back up to 63% of viewing audiences over the 1993–1994 season— plummeted over the next two years to 53% in the 1995–1996 season. Fox, cable, and others were drawing off increasing audiences. Another indicator was the erosion in news viewership; local TV news dropped in share of viewers from 72% (1994) to 65% (1996); the Big Three evening newscasts were down to 26% of viewers watching early evening television—contrasted with their 48% share back in the 1976–1977 season.[27]

As with ratings, dwindling shares are computed on an ever larger household base, meaning the number of viewers delivered to advertisers by traditional TV networks did not drop that drastically. Yet on the other hand, fewer persons per household partly balance off the growth in homes. By the year 2000, single-person households will increase to almost 29 million, compared with 24.5 million households of married couples with children under 18 years old.

AGE CATEGORIES

■   The Census Bureau reported those under the age of 25 dropped from almost half (46.1%) of the population in the United States in 1970 to about one-third (36.1%) by 1990. In the same two decades, those aged 25–44 increased from 23.3% to 32.5% of the population by 1990. People 45 and older increased slightly between 1970 (30.3%) and 1990 (31.3%). The past two decades of cycles in radio formats reflect those shifting concentrations to mid-age groups, with increasing numbers of stations featuring nostalgia, easy listening, and "golden" oldies as well as talk. The Census Bureau projected that by the year 2000 the mid-group will drop to 29.9%, while the oldest cluster (45 and over) will include more than one-third of all Americans (35.7%).   ■

Over the past six decades, the over-65 age group increased almost three times as fast as the total population. A century ago, the proportion of persons over age 65 was 1 out of every 35; now, it is more than 1 of every 10 (12.7%). (See tables 7.6 and 7.7 in chapter 7.) That consumer group's disposable income for leisure activities has also expanded—a growing target for major media advertisers' goods and services. Adults 50 and older, with mortgages nearly paid and early-family costs behind them, held almost half of all discretionary income among America's consumers. In 1992 the 35–54 age group had the largest proportion of discretionary spendable income, $138 billion.

EDUCATION

Elementary and secondary schools grew in enrollment as a result of the "baby boomer" population explosion following World War II. In 1940, some 10 million people in the United States had completed high school—almost 25% of those aged twenty-five or older; that percentage increased by 1993 to 80% of all Americans.

Of those twenty-five or older in 1940 only about one in twenty had completed four or more years of college, contrasted with one out of every ten in 1970, and over one out of five since 1988 (24% of all males and 17% of females). As noted in chapters 4 and 5, increasing numbers of women have earned college diplomas and advanced degrees; in some specialized fields they make up more than half of all graduate students. Media managers and advertisers know that larger audiences and greater viewing and brand loyalty are found among those with less advanced education; mass audiences and popular culture are roughly synonymous with less sophisticated or less highly educated sensibilities. By their nature, *mass* media are not *elite* media. The population's gradual rise in education level has important implications for what types of programming will attract tomorrow's audiences to radio, TV, and cable.

INCOME

Educators have long been aware of a direct correlation between education and income. An early study concentrating on people in the 25–65 range found that adults with four or more years of college earned an average annual income 50% higher than those who had completed only high school.[28]

The number of American people moving into higher income categories increased significantly over the last two decades. In 1995 one-fifth of the total U.S. population earned less than $17,000 (the lowest fifth of black population earned less than $8,000); the next fifth earned up to $30,000 (blacks, up to $16,000), the middle fifth had up to $45,000 income (blacks, $27,500), the fourth group earned up to $67,000 (blacks, $46,500), while the top 5% earned over $113,000 (the top 5% of blacks, over $83,500).[29]

■ Although the median family income rose from $10,000 in 1970 to over $30,000 in 1987, translated into constant 1987 dollar values, median incomes actually grew only slightly over those two decades (1970 income was equivalent to $28,880 in 1987 dollars). In the 1960s, President Lyndon Johnson mounted a government funded "War on Poverty"; the 22.2% (39.9 million) living below the poverty level in 1960 dropped almost in half by 1970 to 12.6% (25.4 million). Five years later 12.3% (25.9 million) of U.S. citizens, about 10% of all families, lived below the poverty level—almost one in ten white persons, one out of every three African Americans, and one of every four of Hispanic origin. Those percentages held about the same through the next two decades to the mid-1990s; but the total 15.1% of the growing U.S. population equaled 39 million people below the poverty level. People with lower incomes, whose education level also tends to be limited, are among the heaviest users of television.

Advertisers use media to seek potential buyers of their products and services. The cyclic recessionary periods in the early years of each decade in the 1970s, 1980s, and 1990s affected consumer buying potential, thus advertising that in turn provides income for radio, TV, and cable operations. Over two decades, Americans' disposable personal income (after taxes) rose five-fold from $1,096 billion in 1975 to $4,959 billion in 1994; allowing for a 150% inflationary rise during that period, 1975's disposable income would equal about $2.5 billion in 1994 dollars, making the actual increase in disposable income about double in constant dollars. Because of shifting trends in age, noted above, one market researcher predicted in 1990 that within ten years the annual discretionary income of the shrinking 18-34 age group would be $41 billion less, while the 35-54 segment would have grown by $281 billion.[30]   ■

These data have implications for planning varying kinds of radio formats to chase those drifting demographics. These data also influence changing genres of TV and cable programs which advertisers need to buy into to reach the largest segments of potential customers. These factors are critical for management's middle- and long-range planning, such as hiring radio personalities with extended contracts, acquiring record/tape/CD music libraries for one format or another, and leasing rights to syndicated TV program series for local schedules two to five years later.

LIVING CONTEXTS

The audience is a moving target in other ways.[31] Disappearing rural homesteads and burgeoning urban and suburban households, widespread education, women at all levels in the labor force, two-income families with "latch-key" children who return to an empty house after school until working parents get home—these are some characteristics of an increasingly mobile and independent society. Single adults, and unmarried or separated parents of children, account for the greater growth in number of households than in population during the 1980s. The average persons per household dropped from 3.33 in 1964 to only 2.63 by 1990 and remained at that point in 1993.

The 1990 Census found the household category that declined most since 1970 was traditional married couples with children where only one parent worked (down from 21% of all homes to only 9% two decades later). The fastest growing category since 1970 was single persons younger than 45 without children: up to 9% from only 3%. Childless singles 45 or older made up another 16% of households in 1990 (up from 14% in 1970). Thus one out of every four U.S. homes had a single occupant. Another 9% of households had a single parent with at least one child. Hardly the traditional mom, dad, and several kids gathered around the venerable TV set as a family! One in four children younger than 18 lived with a single parent. "Female-householder families" (no male spouse present) grew to 11 million. Those patterns directly affected listening and viewing habits.

MINORITY POPULATIONS

Audiences became more diverse in other ways as ethnic populations expanded in this multicultural nation. In 1995, African Americans continued to make up over 12% of the national population; but Hispanic (10%) and Asian (3.5%) immigration and births outpaced other segments in rate of growth, especially in urban areas such as the cities of New York (1.5 million Hispanics), Los Angeles (900,000), Chicago and San Antonio (500,000 each), and a quarter million each in El Paso, Houston, and Miami. (Of course, metro market areas extending beyond city limits have multiples of the above figures, such as 4.2 million Hispanics or 31% of the greater Los Angeles population.) The Miami/Ft. Lauderdale market was half Hispanic by the year 1995, as the Los Angeles metropolitan area will be by 2000. A quarter million Asians and Pacific Islanders lived in New York and in Los Angeles, and almost as many in San Francisco.

■   While the median age of white Americans was 33 years in the 1990 census, the mid-age of African Americans was 27.7, and of Hispanics 26; the average age of Asian

Americans was 30. Among the four groups Asian Americans in 1988 had the highest median household income with $31,578, compared to white ($28,661), Hispanic ($20,000), and black ($16,004) households.[32] By 1992 black and Hispanic consumers were spending more than $400 billion annually—an important market for all advertisers and thus for broadcast and cable programming.[33] Researchers have found purchasing characteristics among blacks and Hispanics to be similar: a high level of product awareness and brand loyalty, as well as purchasing response directly associated with broadcast commercials.[34]

A set of studies found African American households watched almost half again as much TV each week as the general population's average: 69 hours a week, contrasted with 47 hours by all U.S. households.[35] Categories watching more TV than any others in the United States were black women aged 35–64 (63% more) and black teens aged 12–17 (62% more). African Americans aged 12–17 watched 50% more late-night programming than did all other teenagers. Black viewers were bigger fans than non-black audiences of situation comedies (19.0 rating among blacks vs. 14.2 among all others) and action/adventure programs (17.9 vs. 13.1). These figures have implications for broadcast producers and program schedulers as well as for advertisers seeking to tap into the multimillion-dollar African American consumer market.   ■

The Hispanic/Latino population grew from 4.5% of the U.S. population in 1970 to 6.5% in 1980 and then to over 10% a decade later, totaling more than 27 million persons in 1995 (see table 6.2). By the year 2010, marketing researchers project Hispanics will replace blacks as the largest minority group in this country, and will make up one-quarter of the entire U.S. population by 2050.[36]

■   Because nine out of ten Hispanics speak Spanish, and almost 40% speak only Spanish, three TV networks and 14 TV stations in eight markets programmed in that language, while upwards of 300 radio stations and several networks were devoted full- or parttime to Spanish-language formats.[37] Two-thirds of all Hispanics watch Spanish-language TV programs; three-quarters aged 55 and over listen to Spanish-language radio. Hispanics spend 20% more time listening to radio than the rest of U.S. audiences. Hispanic population is concentrated in the top-30 markets, making it an identifiable target audience for local stations in those areas and for efficient regional advertising campaigns that can be highly selective.

Arbitron in the 1960s began sampling and weighing data for Hispanic survey areas. It introduced bilingual survey materials in 1974, revising the diary in 1981 to emphasize the Spanish language. It used Differential Survey Treatments (as for blacks) to strengthen response rates for audience surveys. By mid-1989 Arbitron was publishing annual or semi-annual Hispanic Local Market Reports on radio audiences in 16 markets. In 1995 Nielsen estimated that of the nation's 95.9 million households, 7.2 million were Hispanic.[38]   ■

Media managers have learned that network, station, and cable programming, as well as advertising campaigns, must reflect the multiethnic make-up of local communities and the national public—as sound marketing strategy in addition to social responsibility. Network programmers and ad agencies find practical implications for their strategies. Of the top-10 shows among U.S. households overall in the 1992–1993 season, not one was among the top-10 favorites among African American viewers. For example, "Murphy Brown" was 3rd among all U.S. homes but wasn't even among the top-50 among black households. Doug Alligood, vice-president of special markets for BBDO Worldwide, commented "If you buy just the top 10 shows on TV you will not be reaching blacks."[39] African Americans, at 12% of the nation's population, represented over $300 billion in purchasing power. Similarly, Hispanics' most viewed English-language programs included only two ranked among the top 10 nationally (they were ranked 6th and

U.S. POPULATION, PROPORTIONED BY RACE
1995 (ESTIMATED) TO 2050 (PROJECTED)

	1995		2050	
	MILLIONS	%	MILLIONS	%
Caucasian/White	193.5	73.7%	205.8	52.5%
African American/Black	33.0	12.6	61.5	15.7
Hispanic/Latino Origin	26.8	10.2	88.2	22.5
Asian & Pacific Islanders	9.2	3.5	40.4	10.3
American Indians, Eskimos, & Aleuts	2.4	0.9	4.3	1.1
Total U.S. Population*	262.6		392.0	

*Only the total population estimates and percentages are from source; subtotals by race are computed from rounded off percentages so are only approximations.

Source: Based on data from U.S. Bureau of Census, cited by *U.S. News & World Report,* 14 August 1995, 9; but in 1996 the same double source again projected percentages for the year 2050 at 24.5% Hispanic and 13.6% black, 8.2% Asian, and 52.8% white—a difference of 8 million more Hispanics (up to 96 million) and 8 million fewer African Americans (down to 53.3 million). Similar numbers, rounded off, from the Census Bureau and Department of Commerce were in *The World Almanac and Book of Facts 1996,* 392.

8th). The nation's 25 million Hispanics in 1994, with 9.8% of the population, represented purchasing power estimated by various sources at between $188 million and $206 billion—spending about half a billion dollars daily. Fox network's ethnic-oriented programs reached those consumers more effectively than other networks; the 7 million Hispanic households favored Fox shows as their first three plus fifth and ninth choices, while blacks' second, third, fourth, eighth, and tenth favorites were also on Fox that 1992–1993 season.

Austin McLean in 1991 profiled key U.S. demographic characteristics for the National Association of Broadcasters:

1. The population will continue to grow and is expected to increase by approximately 18 million persons by the year 2000.
2. The birth rate is expected to rise through the 1990s.
3. Consumption patterns are likely to be greatly affected by the aging of the baby boomers.
4. Regional population growth was recently greatest in the South and West regions, which saw 87% of the nation's growth occurring in these areas.
5. The percentage of minority and ethnic U.S. citizens grew at a faster rate than the overall population, with Asian Americans the fastest growing segment in the United States.
6. Two wage-earner couples are now the norm, with 58% of women in the labor force.
7. The influential baby boom segment of the population will continue to age, and the fastest growing segment of the population will be people 85 years and older.
8. The youth population will decline over the next 20 years, with as many as 38% of Americans under 18 belonging to a minority group.[40]

SIGNS OF CHANGING INTERESTS

The "average person" is not the only element that has changed through the decades since broadcasting began in 1920. Personal environment has changed, too, affecting outlook, interests, and tastes. Radio and television and then cable, along with other mass media, have helped shape the direction of those changes. Conversely, the social climate—more permissive, with fragmented families, both working fulltime in two-parent households, actively concerned about environment, women's and minorities' rights, the economy, and international issues—affects the media and how they shape entertainment, news, and commercials.

Leisure time has increased, with the public seeking forms of recreation, from sports and jogging to travel, computer games, and electronic "entertainment centers" in the home. Public interest has widened for music and the fine arts. How people use their growing leisure time and participate in cultural interests should concern every media manager.

Although some authorities are skeptical of the cultural trend as lacking depth and meaningfulness, others maintain that American society is in a period of change leading toward broader public appreciation of the fine arts. The station manager need not concern himself too greatly with the debate. He must appeal to the young, the elderly, and the working population. Culture as well as popular art are not for him to invent, but to understand, foster, and transmit.

THE MEDIA MIX

In 1996 one-third of U.S. homes had two TV sets, and 28% had three or more.[41] Almost two-thirds of all households had cable, offering more than 30 channels, and four out of five had VCRs (mostly for time-shifting aired programs for later play, less for renting prerecorded videocassettes). Almost half of the cable homes also subscribed to premium channels (or "pay cable") such as HBO, Showtime, or Disney.

■   Cable helped UHF stations achieve parity with VHF competitors in metropolitan areas; most U's were independents, with over 140 linked to the fledgling Fox network's evening service. Even before expanding its schedule to all seven nights weekly in Fall 1992, Fox was attracting significant audiences. In the May 1992 sweeps period Fox drew respectable ratings of 7.1, with a 13 share of national viewers (compared with NBC's 12/21, CBS's 11.9/20, ABC's 11.6/20); it placed first among networks in attracting men 18-34 and teens.[42] Two months later, for the first time in its six years, the Fox network moved out of fourth place (programming parts of six nights a week). For the week of July 13–19, 1992, ratings/shares for CBS were 8.5/16, ABC 7.7/15, Fox 7.3/14, and NBC 7.0/13. For the first week of September, Fox's big audience for the annual Emmy awards helped moved it into second place behind CBS. Fox led other networks in attracting adults 18–49, the demographic most highly sought by advertisers. The sweeps for Fall 1995 found the Big Three networks with a combined 54 share; and success in attracting 18–49-year-old viewers ranked the networks NBC, ABC, Fox, and CBS last. For the entire 1995–1996 season, the four major networks plus nascent Warners and Paramount drew a combined rating of 45% of all U.S. TV homes during primetime; their combined shares totaled 74% of all households watching TV, with the Big Three drawing a 53% share of evening viewers (table 6.3). One-quarter (26%) of viewing homes that season during prime time were watching cable, PBS, and independent stations. Still, when Warner Brothers' lowest-rated network had earned a 2.3 rating for one week at the end of 1995, it drew 600,000 more households in prime time than the top cable network, USA.   ■

Back in 1991 a program needed to draw a 17 rating to be among the top-10, but by 1995 the tenth-ranked show had only a 15 rating. Four years earlier 27 programs won at least a 14 rating, whereas in 1995 only half that many (15) scored that high with audiences.

TABLE 6.3
NETWORK PRIME-TIME RATINGS/SHARES, 1995-1996 SEASON

NETWORK	HOUSEHOLDS	VIEWERS 18–49
NBC	11.7/19	7.3
ABC	10.8/18	6.3
CBS	9.6/16	4.6
Fox	7.3/12	5.2
UPN	3.1/5	NA
WB	2.4/4	NA

Sources: Composite of similar Nielsen data for 35 weeks and for 30 weeks reported by *Variety*, 22–28 April 1996, 34; *Broadcasting & Cable;* and Broadcast Sales Training "TV Executive Summary."

■   Other national program services distributed by satellite to local cable systems were appealing to viewers around the country. During evenings in 1992, 1.2 million homes watched USA cable network; WTBS "Superstation" and TNT/Turner Network Television channels each attracted 1.1 million households, followed by ESPN (940,000) and Lifetime (627,000), while half a million viewers each watched TNN/The Nashville Network, CNN, Nick at Nite, and the Discovery Channel; A&E/Arts & Entertainment and MTV were each watched by over 300,000 homes.[43] Viewing levels over the full day were about half those prime-time levels for each cable program service.

One of the first quarterly Cable Nielsen Audience Demographics Reports in 1990 broke out income levels of prime-time cable viewers. Their incomes generally skewed higher than for audiences of the three traditional networks of "free TV" (table 6.4). Predictably, popular-oriented programmers attracted larger proportions of their viewers from those earning under $30,000, such as ABC (46.9%), CBS (52.7%), NBC (53.7%), TBS (57%), USA (56%), and especially The Nashville Network (63%) and Family Channel (64%). Cable services drawing largest proportions of viewers earning over $60,000 were A&E (27%) and Headline News (24%), followed by ESPN, CNN, the Weather Channel and VH-1 (each with 19%). Only about one-third of these cable services' audiences earned under $30,000, contrasted with half of the big three networks' viewers.[44] ABC's audiences were more upscale than CBS' or NBC's, with larger proportions from each higher income group.   ■

Studies confirmed that cable homes watched more TV than noncable homes. So while audiences for cable channels are partly drawn away from network/station programming, they increased viewing by adding cable programming to their diet of viewing broadcast stations. Cable home viewers still spent most time watching station/network programs. Broadcast stations fill only several channels available on most systems, but those station channels carrying the four major networks account for almost two-thirds of home viewing. In cable households almost all living room TV sets are wired to incoming cable; but two-thirds of the other sets in households (bedrooms, kitchens, dens) are not wired, receiving only over-air signals.[45] In homes with DBS service with around 100 or more channels via satellite, only 40% of viewing is of the four major networks.

PROBLEMS FOR COMMERCIAL RESEARCH

Despite developing additional forms for gathering audience data, various commercial research companies remained targets for successive waves of negative criticism. Broadcasters at first raised doubts about the value of quantitative research findings; then they questioned the validity of procedures and accuracy

—ɯ— TABLE 6.4 —ɯ—
PRIME-TIME VIEWERS' INCOME LEVELS, 1990

	PERCENTAGES OF EACH NETWORK'S AUDIENCE				
	Under $30,000	*$30-40,000*	*$40-50,000*	*$50-60,000*	*$60,000+*
ABC	46.9%	15.6%	13.0%	8.5%	15.8%
CBS	52.7	15.2	11.3	7.8	13.0
NBC	53.7	14.6	11.3	7.8	12.6
TNN/Nashville	63.0	17.0	8.0	4.0	8.0
A&E/Arts & Entertainment	32.0	19.0	13.0	9.0	27.0

Source: *Broadcasting*, 18 January 1990, 53.

of resulting data—mostly when ratings were low for their own operation. The president of Group W Television (Westinghouse) once lamented "Diaries were never terrific and the meter only transferred the problem from one place to another. But the people meter is absurd. . . . Nobody is going to press buttons for that length of time" as family members go in and out of the room while TV is on.[46] That was one reason offered for why network ratings dropped 3%–10%—and 27% for children's Saturday viewing—when Nielsen first introduced people meters.

The sometimes wide disparity of results from competing research companies bothered some broadcasters. The top-rated program or station in one research study might be considerably lower in the ratings of another organization. When different audiences were measured by different methods, it was not too surprising that results did not always agree. Robert Balon explained:

> Arbitron's method of keeping a diary over a seven day period bears little resemblance to Birch's, a telephone call where you're asked to go back and recall listening the previous 24 hours. They are inherently different research solutions. . . .
>
> The general rule of thumb is that Arbitron often reports much higher numbers for persons age 35+ while Birch generally reports larger numbers for persons from age 12–34. Not surprisingly, beautiful music formats, news/talk, AC, and country often do much better in Arbitron than in Birch while conversely, CHR, Top 40 formats, album rock and urban contemporary formats do significantly better in Birch than in Arbitron. This is again, a function of who fills out diaries and who completes telephone interviews. In national studies, the diary keeping response rate invariably goes up after the age of 35. Conversely, the cooperation rate for phone interviews drops markedly past age 35. This is a source of on-going frustration to broadcasters, but it is not particularly damning to either Arbitron or Birch. It is simply a reflection of how different the research methodologies are.[47]

Not only sales departments, but performers, writers, directors, producers, and others in programming became particularly sensitive to the variance in ratings. Often their careers as well as incomes depended on where they stand in comparative rankings.[48] Slavery to the decimal point has decided the fate of many a program, prompting renewal or termination of many contracts by advertisers; casualties are common every season in network schedules. Local radio personalities and formats often shift because of weakened ratings; so do local TV newspersons (chapter 5 noted the average term of TV station news directors as only two years).

The business of broadcasting looks inexorably to the bottom line: ratings determine advertising rates and that translates into income. A network's gain of a single rating point for prime-time programming throughout the year reflects increased national audience size that generates additional gross advertising revenues of about $130 million a year.[49]

■   Translating that into local TV station charges, for a commercial spot the cost per rating point in Hartford-New Haven in 1989 was about $94 (early fringe/news/access $74, prime time $120, late night/news $88); in Fort Wayne about $20 (fringe/news $14, prime time $26, late night/news $20). Each rise of a single point in local ratings added that much value to every 30-second time spot sold to advertisers. Multiplying all those spots in and between programs through the evening or day, times 52 weeks, equals "serious money" per rating point for a station. In the same markets, local cable TV's cost per rating point figured out to $390 in Hartford-New Haven for prime time (ESPN, CNN, USA channels) and $75 in Fort Wayne. That may sound like good money for cable, but the numbers mean poor efficiency for local advertisers on cable systems; cable spots cost them four times as much as local broadcast station spots when comparing audiences reached or costs per rating points.[50]   ■

The trend in radio research over the years moved toward cumulative studies. In addition to reporting audience shares during segmented parts of the broadcast day ("day-parts"), with dozens of competing stations splintering audiences in each market, average quarter-hour figures and "cumes" show how many people listen to radio, or to a specified station, over a longer period, usually a week's time. This information is comparable to newspaper or magazine circulation, where total numbers of sold copies of a particular issue are determined but not the readership of specific features. From a competitive standpoint with other media, "cumes" may be sufficient. Broadcasters need to use them effectively.

Writings on the nature, reliability, and influence of the rating system are voluminous.[51] Beville responded to major complaints about ratings, while acknowledging weaknesses and misuse of them. He noted questions of accuracy, bias (including client hyping of programming during ratings periods), rigging, and misuse by media managers. A reasonable, balanced evaluation—issued in 1966 by the Committee on Nationwide Television Audience Measurement (CONTAM), formed by the networks themselves—was simply that "ratings are an aid to decision making, not goals in themselves." Unless the computer takes over human affairs completely, management must still rely on courage and creativity as well as logic. Otherwise broadcasters will abdicate managing their own house.

Out of the welter of confusion over the ratings issue, a number of conclusions have emerged.

First, many station managers either have not properly understood research or have not had people on their staffs who could interpret research realistically for them. Such understanding has become more and more critical to every radio and television station.

Second, ratings are basically accurate in measuring what they claim to measure: estimates of audience size. In the absence of other means whereby audience response can be surveyed as rapidly and as comprehensively, ratings are essential to broadcasting. Stations and networks need to know how widely their programming is being received. The quandary of ratings stems not so much from their existence as from how they are interpreted and especially how they are applied in broadcast planning.

Some variation of the cumulative circulation tabulation seems sufficient for competing with other media already established on that kind of long-term average basis. Audience ratings do reflect comparative audience sizes for competing programs on rival stations, networks, and cable systems.

The necessity for realistic interpretations of ratings should be obvious. Gross numbers of people, even when stratified by demographic characteristics, reveal comparative sizes and little more. Reading other meanings into ratings can produce false conclusions that may undercut responsible decision making by management. "Psychographics" and other qualitative data can refine how managers interpret percentages and fractions of estimated audience numbers.

Further, ratings surveys extrapolate from sampling; the resulting data are approximate *estimates* projected from those samples. Again, Balon cautions:

> For example, if your market currently has an in-tab report of 2,000 completed diaries, and the standard error factor is .5 this means that if you have a 4 share it could actually be as low as a 3.5 or as high as a 4.5 and still fall within the 68% confidence level for that kind of statistic. The problem with Radio is that the word 'estimates' is usually forgotten when dealing with the final product of printed numbers.
>
> . . . And so radio broadcasters will continually be subjected to the dreaded 'wobbles' of the rating process that occur from too small samples within individual cell analysis. The 2,000 sample survey may indeed be fine to measure gross trends in the report but when you start talking about hour-by-hour analysis in men 18–24 or the performance in a particular daypart, those samples become drastically reduced and hence the potential for error becomes much greater.[52]

A single rating has little meaning without a broader context. Media planners must obtain a sequence of ratings figures for a given program over time to identify a true trend. Nor can they attribute a rating solely to the content and form of programs measured. Strong or weak audience numbers are also a function of total audience available at the time period (season, day of week, daytime/nighttime), competitive programs at the same hour, lead-in and lead-out programs, plus previous promotion of the show, and also of its network, station, or radio format.

Third, other factors are at least as important as ratings in determining success in reaching people in the coverage area. Such factors as station image, creativity in programming, flexibility and relevance of station performance, the availability of the station to the community, and costs of station services to advertisers relative to results delivered should all be related properly to any ratings used. Audience composition is far more important than total audience size. The race to be more popular—in sheer numbers—than the competition throughout the broadcast day has perhaps caused more distortion or misuse of media research than any other single factor. Agencies and their advertisers should be alerted to the difference between cost-per-point selling of 25–54 demographics when promoting commodities through TV, contrasted with radio's more segmented audiences bridging that same age span. Unistar Radio Networks Vice-President Kirk Stirland reminded that those in their mid-20s often have parents in their early 50s—with far differing lifestyles, interests, and financial standing. One group tends to prefer album oriented rock (AOR) while the other tunes more to contemporary hit radio (CHR) or talk/news formats; both are quite different target subgroups within that one broad demographic. An agency executive reported clients were starting to look more to 35–64 than 25–54 spreads in 1991, as the nation's 76,000,000 "baby boomers" aged and joined the older group with few fixed expenses and more discretionary dollars to spend.[53]

Fourth, the industry must exercise continual surveillance over research conducted among its audiences. That oversight must be prompted by complete objectivity and not by the profit motive. Scientific research must take precedence over the numbers manipulation game, and methodologies must be validated for their accuracy. Media leaders continually challenge Nielsen and Arbitron or other companies' procedures and findings, especially when discrepancies occur between competitive reports or within methodologies (diaries, meters, and especially people meters). For example, in the 1990s, radio clients complained that Arbitron's audience sample was too small because of the high rate of refusal (as high as 50%) by potential diary keepers. William S. Rubens, who headed NBC Research from 1972 to 1988, recalled that

> people meters were an outgrowth of politics, technology, economics and competition, not research. One can only caution the industry that it ought to add research to the mix. The passive people meter should be subjected to rigorous methodological review.[54]

Because people meters—both active and especially newer passive "imaging"-sensitive—measure *viewing* rather than merely *tuning,* they can offer more precise data about actual audiences for media.

Fifth, the industry cannot bargain for cheap research; meaningful research is expensive but worth supporting. Often experimentation is necessary in order to discover verifiable facts; such experimental projects demand patience. The major TV networks each contributed $200,000 and other media clients $10,000–25,000 each to attempt testing the AGB people meter in the early 1980s (motivated partly by dissatisfaction with Nielsen's national diary and partly by ad agency complaints about charges for Nielsen's service).

■   The NAB funded a scholarly study of the impact of people meters by tracking prime-time network ratings from 1971 to mid-1990.[55] It reported no difference between pre- and post-meter success rates of new programs, no changes on how people watched network TV, no increase in programs' rating instability, no effect on the percentage of time-slots that gained or lost ratings from year to year. On the other hand, the research found that people meter data reported demographics changes of many more young males, slight increase of women 35 to 48, but declines in young women, those over 50, and children (as much as 20% or more loss in youngsters, partly because of nonuse or misuse of meter when alone). The report underscored a permanent loss of 2 ratings points throughout prime-time, with CBS and NBC most affected (-2.5 and -2.3 ratings points respectively) but with no loss by ABC; and genres of programs were affected differently by people meter measurement of audiences. Losses to networks were estimated at over $300 million a year, plus diminished value of commercials sold by affiliates during prime-time. Conversely, cable channels and independent stations gained significantly in audiences measured after people meters were introduced. That is why, after a year of the new meters, 58% of independent station general managers (GMs) judged people meters better than diaries while only 17% at affiliates agreed; 45% of affiliate GMs thought people meters worse than diaries, 31% were "not sure," and 6% found them about the same.   ■

Prompted by Congressional hearings in 1963, ratings services and national media strove to refine methodologies and reporting techniques. Major media companies collaborated to form the Broadcast Ratings Council after the hearings, later retitled the Electronic Media Rating Council. They have worked with committees on National Television Audience Measurement (CONTAM) and Local Television and Radio Audience Measurement (COLTRAM), and the Advertising Research Foundation. Coalitions of local broadcasters also monitored ratings services in some major cities.

Audience measurement methods have been in transition as the radio-TV-cable service continued to evolve in recent decades. Interviews, diaries, and meters have been joined by people meters—including "passive people meters" with image-recognition sensors, and Arbitron's short-lived ScanAmerica people meter that also measured product purchases linked to commercials viewed or "buyergraphics" (abandoned in late 1992 for lack of financial support by advertisers). Auditorium testing, field studies, and focus groups have been added to the marketing research mix. Advertisers and agencies continually seek more definitive information about networks, stations, cable systems and about the people they reach. More than measuring households, research has trended toward concentrating on individuals. Multiset homes in both radio and television involve varied selections of programs, plus shifting during programs—especially at commercial breaks—with constantly active remote tuners. (Passive meters are expected to record second-by-second, including moments when commercial messages are on the screen.)

Out-of-home viewing remains a massive unmeasured audience in hotels, college campuses, bars, restaurants, hospitals, military bases, country clubs, vacation homes, airports, places of work, day-care centers, and with transportable battery-powered sets. In an early test market study, Nielsen estimated that hotel and motel viewers added from 4% in low-viewing day-parts to as high as 16%–24% in prime hours.[56] Commercial ratings companies, academic researchers, networks, and ad agencies have explored how to track these uncounted viewers.[57] Studies during the 1980s estimated additional out-of-home viewers included up to 10% more adults and 20%–30% more in specific target audiences. Each week 4 million viewers watched one or more network TV programs outside of home and not in someone else's home. College students and women watching daytime TV were estimated at half a million per quarter-hour, and 5.7 million on a weekly cume basis. In 1989 R. H. Bruskin surveyed working women; one out of four had a TV set available where they worked, and 20% of them did watch at least one serial program a week. That corroborated the previous year's estimate that 2.8 million working women watched one or more daytime serials a week while at work; they were uncounted in ratings measurements. An estimated 5 million additional viewers of ABC's "Monday Night Football" were out-of-home and not counted in traditional ratings surveys. ABC's surveys projected that hotel and motel guests alone added 3.5% viewership to early morning network news and almost 10% to early morning network entertainment programs.

About 2.5% of the U.S. population lives in "group quarters" such as college dormitories and military housing; they are not surveyed by the ratings companies. Late in 1989 one study estimated almost half of all 8.5 million college students live away from home; 96% of all students watch some TV each school week. NBC's 1989 survey of college students found almost half (49%) of all 18–24 year-olds watch network late-night entertainment programs during the week.[58] Those late shows plus daytime serials together reached nearly three out of four college students each week. NBC projected that adding unmeasured out-of-homes audiences would raise current ratings for "Saturday Night Live" by 10% (again, not by 10 rating points), "The Tonight Show" by 15%, and "Late Night with David Letterman" by 30%.

College students earned an average $3,800 a year in 1989; of collegians' total $21 billion income, about $9 billion was considered "discretionary income" to spend during the school year—directed in large part to products advertised by broadcast media.

■   A three-year study by University of Michigan and NAB's COLTAM produced a specialized diary intended not for the household or its TV sets but for the individual viewer wherever that person watched TV; it was to be tested in the early 1990s. In 1992 Nielsen estimated 2 million males 18 and older watched weekly NFL football outside the home, which would raise ABC's ratings for those broadcasts an additional 7% (not 7 gross points) per game. Among those viewers, 33% were in restaurants, bars, and clubs; 32% on college campuses; 12% in hotels and motels; 12% at work; and 11% in various other locations. The Big Three networks and Nielsen in 1992 researched ways to verify out-of-home audiences, estimated at almost 28 million previously uncounted viewers per week—about two thirds of them watching some network programs.[59] Almost one-quarter of all their viewing was done out-of-home and goes unmeasured: about 27% on college campuses and the same percentage at their workplace; 17% at hotels and motels; 10% at restaurants and bars; almost 3% at second/vacation homes; and 14% at other places such as hospitals, airports, boats, and recreational vehicles.[60] If able to be confirmed, these figures would affect CPM rates. Networks and local stations want to charge more for bigger audiences delivered than ratings currently reflect, whereas agencies in the past paid only for actually measured audiences, with out-of-home viewing a bonus because it is unable to be verified.   ■

The increase of UHF and FM stations, and especially of cable channels, has expanded competition resulting in greater audience selectivity. Fragmentation of the total mass audience by more stations, networks, cable and DBS satellite services, multiple sets and VCRs, makes obsolete a system that equated success with merely the largest audiences. Increasingly, listeners-viewers' opinions and attitudes are as important as the simple fact that they choose a specific program or station at a given time. Advertisers need information about predefined groupings of people who are their targeted potential consumers amid what Balon refers to as this "demassification of the American marketing process." Astute research is essential to identify a station's niche among a market's audiences, and to measure the effectiveness of programming strategy.

Not to be overlooked are what these services cost. Nielsen offered 1987–1990 contracts for people-meter ratings services to each network for just under $5 million a year, almost a 50% increase from the prevailing $3.5 million for traditional audience surveys. ABC and CBS responded by not renewing contracts, partly because of dissatisfaction with results of early people-meter reports (often 7% lower than ratings by conventional meters and diaries).[61] However, by 1992 networks were each paying about $5 million annually for Nielsen data, advertising agencies somewhat less. Charges for nationally syndicated services are based on market size and, in radio, on a station's cumulative audience. Birch in the late 1980s charged radio stations from $3,500 a year in the smallest market to $84,000 in the largest, with all stations within each market paying the same rate.[62]

LOCAL STATION RESEARCH

A station needs to undertake local audience research to supplement syndicated research services it buys. Over time some stations add or drop contracts for national services.

■   In the twelve months to mid-1991, 51 TV stations canceled one or other service (32 left Arbitron, 19 Nielsen). At that time, the CBS Television Stations division dropped Arbitron, whose services included $2 million worth of local-market ratings for CBS' five owned TV stations. During the next twelve months to mid-1992, another 15 stations dropped Arbitron while Nielsen gained 37 exclusive clients. By late 1992 almost two-thirds (62%) of 825 TV stations subscribed to only one national ratings service: 323 to Nielsen and 184 to Arbitron, while 318 stations used both services.[63] As noted earlier, in 1993 Arbitron abandoned researching audiences of local market TV stations and cable systems. ∎

National audience surveys do not provide information about particular station audiences. Qualitative information about local audiences must still be discovered through local effort. The station manager needs to learn the real image of his station in the community. To provide the best possible service to the local public and to local advertisers, he must learn the people's true attitudes and reactions.

Management finds local station research an important aid toward achieving more efficient operations: providing basic marketing data, determining competing stations' and other media's audiences and revenues, developing sales, evaluating programming, planning for expanded operation, and measuring patterns of progress or erosion. Local audience research may reveal shortcomings in station service while also confirming station achievements. The public's true public attitude toward the station is of far greater value to a good administrator than distorted impressions based on guesswork or on comments from a closed circle of friends and colleagues. Local audience research, then, should be conducted primarily for the assistance it offers to station administration. Dr. Sydney Roslow of The Pulse, Inc. cautioned against misusing research by letting it replace judgment, when it is rather a tool "to complement other tools used by managers in making decisions."

The NAB has always emphasized research as an information-gathering tool for management and decision making, but never as a substitute for judgment:

1. It can resolve differences of opinion among equally competent members of management as to what the facts really are.
2. It can help management assign a weight or an order of importance to a set of known factors.
3. It can disclose relationships among what were hitherto thought to be unrelated facts.
4. On occasion it can uncover things that no one had previously thought of before.[64]

What does local research cost? A professional in media marketing estimated in 1991 that a radio station in the 20th to 25th markets might budget from $97,000 to $113,000 a year for in-house research, contrasted with $121,000 to $154,000 for research by firms hired from outside. Those totals include salaries and overhead, computer support, syndicated research (such as Arbitron or former Scarborough/Birch), and specialized research including focus groups to study station image and programming, music, and sales. A television station in markets ranked 50–65 might budget about $155,000 for annual in-house research or $253,000 for studies by outside firms.[65]

Who should be in charge of local research depends on its purpose. That may include confirming the effectiveness of past operational patterns and providing guidelines for the future. Since any circulation of data outside the station compromises the diagnostic purpose and also may alert competitors to possible future moves by the station, research should be conducted by a station staff member. Then, whether findings are favorable or unfavorable, the information can be used only internally by station administrators.

Some larger stations designate a director of research or marketing to coordinate various research efforts. These persons are not always given the opportunity to initiate and supervise significant independent local station research, but minimal rearrangement of job functions would permit them to feed important information to management. It is imperative that local research studies be completely divorced from daily pragmatic activities of sales and promotion departments, to avoid conflicts of interest when developing research to test the effectiveness of sales, programming, and promotion.

Many stations unable to add a fulltime research director can employ someone competent for regular staff duties who is also qualified to conduct qualitative research. Young graduates from colleges and universities, especially those with master's degrees, have often received research training while majoring in radio and television and in business courses.

In addition to contracting with an outside research firm, a station finding it impossible to support staff-conducted research might initiate studies with universities or state broadcasters' associations. A radio or TV station is often located near a college whose research facilities and resources have not been adequately utilized by broadcasters. Educational researchers are particularly suited for the kinds of audience studies that local stations need. They generally have less interest in quantitative "ratings" type of research than in projects proposing to study audience opinions and attitudes. These "farmed-out" or cooperative efforts may lose the advantage of confidentiality of station-conducted studies because the results will be known by those conducting the research. However, the information can be very useful to management and such studies certainly are better than none at all.

How often should local research be conducted? Over-researching a local audience could be as wasteful for a station as no research at all. A thorough, continuing study of the local audience once each year should provide a realistic venture for any station. The returns to management should far outweigh the investment.

The possibility of cooperative studies with leading merchants in the community should not be overlooked. Some stations have had great success with return-postcard mailings in monthly statements of banks, public utilities, or savings and loan associations. Information can be gathered about station or program popularity. Typically, the station pays for the mailings and the financial institutions contribute personnel time for tabulating results. Local businesspeople accept the findings of such surveys because of the position of trust that major business institutions hold in the community. The element of respect also bolsters public response to the mailings.

STATION RESPONSIBILITY

If broadcasters are both leaders and servants of the public, then a question of proportion must be raised. When, how often, and under what conditions should the leadership function be exercised? How much of the broadcast schedule should be what the public wants and how much should be devoted to what the public needs? Presumably the professional broadcaster knows her audience's general wants and specific needs. (This remains true, despite the FCC's deregulation in the 1980s of formerly mandated procedures such as community ascertainment by broadcast licensees and public access channels on cable franchises.) A media manager is often aware of some of the public's needs and wants that they themselves may not realize. The problem is further complicated by changing patterns of the public's desires. If programmers could ever efficiently identify those shifting, even fickle desires, there would be only hit series on TV; no new programs would ever have to be canceled.

Two problems arise from attempts to give the audience what it wants. First, it is doubtful whether most people know what they want at any particular time. Tuning usually involves two kinds of choices: whether to turn the set on or leave it off, and which station or cable channel to select from among the many available. The latter choice may not offer what the consumer wants at that moment. So she makes her selection from whatever is being broadcast, more by default than positive choice. In this sense, the listener-viewer adjusts to the media rather than vice versa. Paul Klein, former research executive with NBC, coined this exercise of options as the "least objectionable program" theory (see footnote early in this chapter), which critic Les Brown agreed was a weakness of the program service. People tend to turn on TV and then flip around the channels until something catches their eye (not necessarily their interest or even their fancy). Thus a program scoring the highest rating or share in its time-spot may not be popularly desired; it may merely be the least disliked from among the generally unattractive programs available at that hour.

A related problem is posed when ratings are used as the basis for giving the audience what it seems to want. This reasoning is flawed because ratings show what was selected yesterday or last week or last month. Ratings represent measurement of *past* choices. There is no assurance that the listener wants more of the same today. This holds for radio, although for longer time-frames sometimes measured in months. People are fickle; they become jaded as they seek ever-new diversion.

It is no easy task to determine the best methods for satisfying most of the people most of the time. Certainly, ideal combinations have not yet been discovered.

The FCC reasserted in the 1990s (after Mark Fowler's deregulatory chairship disavowed the fiduciary role of licensees regarded as a "public trust") what the Radio Act of 1927 and the Communications Act of 1934 mandated: broadcasters are explicitly licensed to serve "the public interest, convenience and necessity." This is something like discussing the weather; we'd all like it improved but how to improve it escapes us.

A balance must be sought to reconcile proper responsibility toward the public good with appropriate stewardship toward commercial obligations. Perhaps the problem should not be expressed in terms of *giving* but rather *selling* the station's programs to the public since commercial media do operate in a sales atmosphere. If so, then the test of effectiveness, as in sales endeavors, should be the mutual satisfaction of both parties. A network or station cannot merely make quality programming available; it must vigorously promote that programming to attract a sizable audience.

Negative Social Effects

A dichotomy exists in the area of alleged effects of programs on people. Some critics claim that some broadcast materials make radio or television addicts of some people, or that they delay or prevent normal maturation or cause people to seek unrealistic goals or to become generally apathetic. Other critics maintain the very same material provides healthy relaxation aiding people in their normal maturation and serving as an escape valve to defuse aggressive impulses.

Broadcasters must admit these powerful media do have effects on people. Such an admission is simply realistic. Certainly, sales departments assert this fact whenever they try to sell air time to ad agencies' clients.

The effects of media, particularly in depicting violence—whether fictional (drama, comedy) or nonfictional (news, documentaries)—have been studied for decades. Research reports by scientific, academic, and governmental sources have provided supporting data for several sides of the debate about media's influence and effects on the audience. Early appraisals claimed that media mostly reinforce preexisting attitudes and behavior—with major formation of ideas and actions coming from family, peer group, school, and church. In later years that judgment was modified to note that a small percentage (but a large number nationally) of children or not truly mature adults whose behavior or thought patterns are imperfectly developed may be triggered to emulate what they see and hear in media presentations. Scientific support has grown for the probability of media's negative social effects on a marginal but large group of the nation's children. The Surgeon General's commissioned studies and final five-volume report in 1972 interpreted evidence available that televised violence and depicted aggression increase aggression in the real world. Yet, even the senior research coordinator and science adviser to the Scientific Advisory Committee on Television and Social Behavior noted that those implications were subject to further research for corroboration, modification, or even reversal.[66] For example, Dr. Floyd Cornelison, Jr., head of the Department of Psychiatry and Human Behavior at Jefferson Medical College in Philadelphia suggested in 1973 that children be allowed to watch all the aggressive TV programming they want: "Kids who grow up aware that life is full of violence and horror are apt to be normal because they're prepared to deal with reality." People with unstable characteristics who watch considerable television quite possibly could get destructive ideas and miss the constructive elements completely. True, they may get the same ideas in other places but perhaps not as graphically or as attractively demonstrated. Programming that seems quite innocent to a normal person can be interpreted in an altogether different manner by an abnormal person, unwittingly reinforcing antisocial and self-destructive predispositions.

The authors of a widely used textbook on broadcast and cable media summarized the state of the question in the early 1990s by asking:

> [W]hat can we conclude about the effects of violent TV on antisocial behavior? Lab experiments demonstrate that under certain conditions, TV can have powerful effects on aggressive behavior. Field experiments provide additional, although less consistent evidence that TV can exert an impact in the real world. Surveys show a consistent but somewhat weak pattern of association between violence viewing and aggression. Longitudinal studies also show a persistent but weak relationship between the two and suggest a pattern whereby watching TV causes subsequent aggression. Unfortunately, few findings in areas such as this are unambiguous. Nonetheless, a judgment is in order. Keep in mind that some might disagree, but the consensus among social scientists seems to be:
>
> 1. Television violence is *a* cause of subsequent aggressive tendencies in viewers; it is not *the* cause since many other factors besides TV determine whether people behave aggressively.
> 2. The precise impact of TV violence will be affected by many other factors, including age, sex, family interaction, and the way violence is presented on screen.
> 3. In relative terms, the effect of TV violence on aggression tends to be small. . . .

The majority of researchers concede that there is some kind of causal link between viewing video violence and aggression, but several argue that the link is too weak to be meaningful.[67]

When a president of the United States or a civil rights leader is slain by an assassin's bullet, when eight student nurses are murdered, when forty-five people are hit by the bullets of an unbalanced person on a major university campus, or a dozen are shot in a fast food restaurant, the question rises "What and who will be next?" It is also time to ask whether the pleasurable rewards some people find in violence depicted in programs are worth the consequences to others. The question remains partly unresolved about how much influence broadcasting has, if any, on these and other criminal acts. To the extent that mass media may be contributory factors, media managers are challenged to respond seriously.

Also questionable is how the opposite extreme of naive, Pollyanna types of programming affect people. Some people prefer "good news" programs that do not unsettle them, instead of a display of battlefield corpses or swollen bellies of youngsters starving in foreign lands. Broadcasters have a serious challenge to discover a balance in media presentations that neither narcotize and insulate the public nor place audience members in danger.

The relation of mass media to crime and violence is only one aspect of its social role. A body of literature reports studies and similar findings for related social topics such as media stereotyping, racism, sexism, language, materialism (including hedonism and sensuality), and other areas of human values depicted appropriately or distorted by broadcast/cable entertainment, news programming, and commercials. They are more subtle and potentially more insidious matters than observable violence in programs. They cannot be dismissed by serious managers whose audiences, especially impressionable youngsters, faithfully tune in.

For much of the public, the opinions of Candice Bergen or Tom Cruise or a rock music star can be more important than those of a college professor or social researcher because media personalities are more physically attractive, are seen more often, and more closely represent the social aspirations of many admirers. Whether those desires are in the people's best interest is moot. They do exist and the public reacts accordingly. Broadcasters may not determine social standards, but celebrities and the roles they play do influence the public through the status that mass media confer on them. Most of the public project too much credibility in radio and television sources. Audiences tend to believe whatever famous personalities, network anchorpersons, and local newscasters say.[68] In most cases, this reliance is well placed; but it is an excessive burden of responsibility for a broadcaster. Nevertheless, since that tendency does exist among those unaccustomed to look much beyond the headlines of a single newspaper or of second-hand conversations (gossip and rumor), the trust of those people can never be ignored. Exploiting that trust is seriously irresponsible.

Because broadcasting and cable are so pervasive, reaching into homes where parental guidance of children is weak or nonexistent, they must exercise greater responsibility than other media. Yet the question arises whether commercial media can be expected to act *in loco parentis,* either by self-regulation or external governmental oversight, assuming responsibilities of absentee or uncaring adults who fail to exercise their role in guiding children. Can a market-driven sales medium truly be the nanny for a nation? Should a governmental agency staffed by civil servants impose restrictions on media content and scheduling to protect impressionable youth? No less a body than the Supreme Court of the United States has given a qualified "yes" to the need to require over-air broadcasters to modify some program patterns to preclude contaminating the young. They based their judgment in the "Seven Filthy Words" decision on the pervasiveness of broadcast programming unable to be anticipated by unsuspecting audiences. (See chapter 11, "Managers and the Law.")

Broadcasters can help shape the values people live by. Through radio and television people can come to feel a sense of participation in current history. They can learn life's deeper satisfaction and, as a result, become active in helping to reduce the many injustices of modern society. They can come to believe in the value of compassion for fellow humans.

Radio, TV, and cable media can stimulate the thinking of their audiences. Raising significant questions, beyond merely "setting the agenda" of topicality, is a major accomplishment of electronic media—even more than attempting to provide deceptively neat answers. Both can be achieved to some extent within entertainment programs as well as information or education formats.

Each local station and cable system needs to take initiative in establishing rapport with its audience. Involvement in community affairs can be followed by on-the-air reports of those activities. The station should regularly report to its audience of local citizens what various staff personnel are doing for community improvement and how the station attempts to serve them better.

If the broadcast or cable manager expects public support on important issues affecting the industry, she must begin to encourage a more vocal response from audiences about relatively minor matters. She cannot expect support when she needs it if the station cultivates a one-way communication system. A lack of audience response reflects poorly on station management. When the public does contact the station, management cannot afford to be overly defensive about opinions and criticisms that do not happen to coincide with those of management. It is through the conflict of ideas that any open, robust, democratic institution is able to grow and even achieve greatness.

Media managers regularly face a common anomaly. Apathetic audiences are of little practical value to the broadcaster when determining program schedules. Yet those same audience members can be highly critical of shows once they air. The situation is somewhat comparable to the person who does not vote on election day, but months later stands on the street corner to condemn the administration. After all the research is analyzed and staff and department heads make recommendations, managers must weigh key factors and finally make decisions that affect the station, personnel, advertisers, and audiences.

Manufacturers have discovered that products and promotion campaigns are successful when they are "people-centered." Marketing research is regularly conducted to determine whether manufacturers' perception of the "public" is accurate to ensure continued favorable response from consumers. Similarly, the station or cable manager needs to maintain personal involvement with the people in his area. He won't necessarily find representatives of the majority at country clubs or among his business associates. Lest he become isolated from the mass audience his station serves, he must keep in continual touch with as many of them as feasible. He might even ride a bus and talk with passengers. He can converse with garage attendants, laborers on construction projects, taxicab or truck drivers, radio-repair men, store clerks, and the myriad of people who live outside his ordinary world. An enlightening and profitable experience is to participate for a day or two in one of his station's personal interview surveys by phone or in the homes of his audience.

Letters and telephone calls are a valuable source of continual feedback from the community. Alert managers have receptionists and operators log all comments phoned in, usually by irate listener-viewers but occasionally by those offering thanks or praise for programming. Even mere inquiries about station personalities, local and network news or entertainment programs, can provide some sense of shifting patterns of audience response to the station's service.

Managers can also learn from audience members who take the trouble to write a letter and affix postage to tell a station or network what they think of its product. While not representative of the entire audience, this self-selective subgroup of letter writers can often bring to management's attention matters that busy, distracted staff overlook or more apathetic members of the audience ignore. Again, a running tabulation of the pattern of letters helps to identify problem areas or weak spots in scheduling, content, or format of programming. It is usually not difficult to distinguish chronic cynics or lonely isolated viewers from thoughtful, sensitive audience members with valid observations about media performance. Networks for decades monitored all mail with bi-weekly or monthly summaries about favorable and critical audience mail; those summaries were distributed to top administrators and division heads to help them get a "feel" for the pattern of audience qualitative reactions beyond mere quantitative ratings and demographics. A prime-time hit program on a national TV network typically attracted only a dozen or two letters a week from the national audience; but over many months clear patterns of reactions could be discerned for various programs or kinds of content (language, situations, topics, scheduling, and so forth).[*] Even while it reigned among the top-10 on CBS-TV, controversial sit-com "All in the Family" often drew only ten to twenty letters a week, many of them inquiring about re-runs or requests for photos, some praising an episode, and fewer than half criticizing the program. Despite the far-flung national audience, networks attempted to answer all letters in some way, at least by a form letter but often with detailed responses. Stations and cable systems have even more to gain than networks in goodwill and audience support from letter writers who live within reach of their signal in that community.

OTHER STATION-AUDIENCE RELATIONSHIPS

A station manager cannot stop at acquiring knowledge about her audience. She needs to use all the methods at her command to maintain cordial relationships between the station and the public. Her attitude toward the audience will be reflected in the attitudes of her employees and how her station serves the community.

Stations with the best records of successful audience relationships usually are led by managers with true concern about the audience, and who work to maintain a feeling of mutual understanding and respect between station personnel and the people they serve.

---

[*]Those numbers were for letters addressed to the networks; other correspondence addressed to programs by name, producers, or stars were forwarded to the appropriate persons or production companies. Totals included only those for entertainment and "special" programming; news correspondence was handled separately. Of course, a particularly challenging issue or treatment of a topic in comedy or drama might prompt hundreds of letters (from an audience of multimillions!); but those occasions were rare.

As a practical matter, ensuring proper handling of telephone calls and mail from the audience can reflect to the advantage of the station. The cordial telephone voice and manner can win people and influence them favorably. A lack of courtesy and consideration can destroy goodwill that has taken years to establish. The "personal touch" is essential for building support and loyalty among the public. As already noted, it is useful to tabulate each critical call, noting the program or practice the caller complains about. This provides one form of informal, nonscientific, but partly representative feedback from the otherwise anonymous audience the station depends on for its success.

Unanswered mail is inexcusable. The use of form letters should be avoided except when mass mailings are necessary for contests or promotional efforts. Even letters that come to the station from apparently "marginal" personalities deserve the courtesy of a reply, although they may be difficult to answer. Again, letters complaining about programs should be tabulated to supplement more systematic research data.

Management can achieve tangible benefits from organizing a working advisory council made up of responsible men and women from the coverage area of the station—those active in civic affairs and organizations throughout the company's service area. Members must be fully aware their role is advisory only and that station officials make all final decisions. In a smaller community, such a council can have an immediate and highly personal impact. It can aid broadcast operations whose program resources are limited by directing favorable attention to the station; and council members can report back to various organizations they represent how the broadcaster is contributing to the community.

When broadcasters, audiences, and critics pool their resources, splintered self-interests can evolve into mutual cooperation. By action, reaction, and mutual interaction among senders and receivers in the total communication process, broadcast/cable managers and a community can benefit one another.

—⁓— CHAPTER 6 NOTES —⁓—

1. For an analytical summary of major forms of audience research, see Gerald C. Hartshorn, ed., *Audience Research Sourcebook* (Washington, D.C.: National Association of Broadcasters, 1991). Chapters I and II succinctly explain concepts and kinds of primary and secondary research, including various methodologies for studying station image and on-air talent, music, formats, and sales. Another useful review is Valerie Crane and Susan Tyler Eastman, "Promotion and Marketing Research," in Susan Tyler Eastman and Robert A. Klein, *Promotion & Marketing for Broadcasting & Cable,* 2nd ed. (Prospect Heights, Ill.: Waveland Press, 1991), 78–98.

2. Hugh Malcolm Beville, Jr., *Audience Ratings: Radio, Television, Cable* (Hillsdale, N.J.: Lawrence Erlbaum Associates, 1985), xiv.

3. Sources: RAB data cited in *Broadcasting & Cable Market Place 1992* (New Providence, N.J.: R.R. Bowker/Reed Reference Publishing, 1992), E–16, xxiii; and *Broadcasting & Cable Yearbook 1995* (New Providence, N.J.: Broadcasting & R.R. Bowker Publications, 1995).

4. These and other data, unless noted otherwise, are from annual compilations in the successively named *Broadcasting/Cable Yearbook 1989* (so named again in the mid-1990s), *The Broadcasting Yearbook 1991, Broadcasting & Cable Market Place 1992,*

Broadcasting & Cable Yearbook 1995. Also from Funk and Wagnalls annual *The World Almanac and Book of Facts, 1994* and *1996.*

5. Data from 16th Annual Television Ownership Survey by Statistical Research, Inc., underwritten by the Committee on Nationwide Television Audience Membership (NAB and three major networks); reported by Lynette Rice, "Multiple Sets Rule in U.S.," *Broadcasting & Cable,* 19 August 1996, 35.

6. *Broadcasting & Cable Market Place 1992,* E–109; *Broadcasting & Cable Yearbook 1995,* C–224; *Variety,* 29 April–5 May, 1996, 36.

7. Source: Television Bureau of Advertising and A. C. Nielsen Media Research, cited in annual editions of various titled predecessor yearbooks and *Broadcasting & Cable Yearbook '95,* C-226; and *Broadcasting,* 2 March 1992, 60. Household usage for a full week in the 1993–1994 season was broken down by day-parts: 11 hours and 13 minutes (6 A.M.–1 P.M.); 7:45 (1 P.M.–4:30 P.M.); 9:53 (4:30 P.M.–7:30 P.M.); 2:04 (7:30 P.M.–8 P.M.); 13:08 (8 P.M.–11 P.M.); 4:52 (11 P.M.–1 A.M.); and 3:53 (1 A.M.–6 A.M.). Figures for 1994-1995 from Nielsen Media Research, cited by Broadcast Sales Training, "Executive Summary," 28 August 1995, 2.

8. For a detailed descriptive analysis of "ratings" and "shares," see Sydney W. Head, Christopher H. Sterling, and Lemuel B. Schofield, *Broadcasting in America: A Survey of Electronic Media,* 7th ed. (Boston: Houghton Mifflin, 1994), 403–405, and Chapter 11, "Ratings."

9. *Los Angeles Times,* 7 April 1992, D–1; *Broadcasting & Cable,* 25 March 1996, 10.

10. *Broadcasting & Cable,* 1 January 1996, 65; 19 August 1996, 89, carried the same totals, citing as sources FCC, Nielsen, and Paul Kagan Associates.

11. Code of Federal Regulations, Title 47, #73.242, limited commonly owned AM and FM stations in cities over 100,000 to no more than 50% duplicated program service. That ruling was deleted as part of the FCC's deregulation in the late 1980s and early 1990s.

12. *Broadcasting & Cable Yearbook 1995,* H–88–92.

13. "Room for Two? AccuRatings Hopes So," *Broadcasting,* 7 September 1992, 54.

14. *Broadcasting & Cable,* 1 April 1996, 16.

15. See Steve Behrens, "People Meters' Upside," *Channels,* May 1987, 19.

16. Ira O. Glick and Sidney J. Levy, *Living with Television* (Chicago: Aldine Publishing Company, 1962).

17. Gary A. Steiner, *The People Look at Television* (New York: Alfred A. Knopf, 1963).

18. Robert T. Bower, *Television and the Public* (New York: Holt, Rinehart and Winston, 1973), 179.

19. The Roper Organization, Inc., "Trends in Public Attitudes Toward Television and Other Mass Media: 1959–1974" (New York: Television Information Office, 1975), 6.

20. Originally named Elmo Roper and Associates, the public opinion research company eventually became part of Starch INRA Hooper marketing firm, renamed Roper Starch Worldwide. It conducted the studies sometimes annually but usually every two years, under contract to the commercial industry's public relations agency, the Television Information Office; in the 1990s it contracted with the NAB and the Network Television Association formed by ABC, CBS, and NBC. Some scholars have questioned the wording and thrust of questions as inherently fostering responses favorable to commercial broadcasting, but the similarity of questions and methodology over the thirty-three-year span permit internal analysis of those data to identify trends. See M. Mark Miller, Michael W. Singletary, and Shu-Ling Chen, "The Roper Question and Television Vs. Newspapers as Sources of News," *Journalism Quarterly* 65:1 (Spring 1988), 12–19; also Charles C. Self, "Perceived Task of News Report As a Predictor of Media Choice," *Journalism Quarterly* 65:1 (Spring 1988), 119–125.

21. Following data are partly derived from The Roper Organization/NAB/National Television Association, *America's Watching: Public Attitudes Towards Television 1991* (Washington, D.C.: NAB, 1991); and from *Broadcasting & Cable Yearbook 1995,* xxi.

22. Data compiled from surveys by Roper (later renamed Roper Starch Worldwide) in 1983, 1988, and 1989 by DDB Needham Worldwide advertising agency, March 1990.

23. Roper data summarized by DDB Needham's Media Research department in *DDB Needham Bulletin:* "What People Think of Television: The 1989 Roper Report on Attitudes Towards Television," March 1990, 3, 8–9.

24. Broadcast Sales Training "Executive Summary," 17 January 1994, 1; citing Communication Briefing, January 1994, 4. That rating of "most believable" at 34% in 1993 was far below the 51% figure (cited earlier) in *Broadcasting & Cable Yearbook 1995,* xxi.

25. A useful summary of some of the data below, applied to broadcast contexts, can be found in Austin J. McLean, ed., *RadioOutlook II: New Forces Shaping the Industry* (NAB: Washington, D.C., 1991), Part II: "Demographic Changes," 21–38.

26. Census estimates from the *World Almanac 1989,* 536; media estimates for 1995 cited in "TV in 1995: Looking Good," *Broadcasting,* 2 January 1989, 39–40; network shares for 2000 estimated by Pier Mapes, NBC president, reported in *Broadcasting,* 8 June 1992, 18.

27. Associated Press article, *Tuscaloosa News,* 13 May 1996, 8–B.

28. Radio Advertising Bureau, "The Listening Habits of Better-Educated Adults," August 1963, 4, citing a study that year by R. H. Bruskin Associates.

29. Robert Famighetti, ed., *The World Almanac and Book of Facts 1996* (Mahwah, N.J.: Funk & Wagnalls, 1996), 394. Following data from U.S. Department of Commerce, Bureau of the Census, "USA Statistics in Brief 1989: A Statistical Abstract Supplement" (Washington, D.C.: Government Printing Office, 1989).

30. *Broadcasting,* 24 September 1990, 66; *The World Almanac and Book of Facts 1996,* 127.

31. Following and related data are from Mark S. Hoffman, ed., *The World Almanac and Book of Facts 1991* (New York: Pharos Books/Scripps Howard, 1990), 550–555; NAB *Radio Reports,* 25 November 1991, 1.

32. "Mktg/Tech News: Audience Outlook—U.S. Population Changes," *NAB News,* 5 November 1990, 33–34.

33. The Hispanic consumer market was estimated at $207 billion; NAB, *Radio Week,* 2 March 1992, 6. The American black population's $218 billion of purchasing power in 1989 was equal to the ninth largest free market in the world, greater than the gross national product (GNP) of Australia, Mexico, Spain, or other countries; NAB, *Radio Week,* 12 June 1989, 5.

34. Robert E. Balon, *Radio in the '90s* (Washington, D.C.: National Association of Broadcasters, 1990), 23.

35. A.C. Nielsen analyzed November 1989 viewing; *Variety,* 14 January 1991; "Nielsen Makes Demos of Black Viewers Count," *Variety,* 11 November 1991, 25.

36. NAB "Info-Pak," December 1988, 7, citing The Lempert Co. data in *Television/Radio Age,* 25 July 1988; Census Bureau projections cited by *U.S. News & World Report,* 14 August 1995, 9—as in table 6.2.

37. These and following data from *Television/Radio Age,* 18 August 1986; 23 November 1987; July 1988; 24 July 1989; *Broadcasting,* 10 December 1990; 20 May 1991.

38. Cited in Broadcast Sales Training, "TV Executive Summary," 4 September 1995, 1.

39. Quoted by Bruce Horovitz, "A Case for Different Strokes," *Los Angeles Times,* 7 April 1992, D–1. See *Broadcasting & Cable,* 5 April 1993, 434; also John Carman, "FCC Ruling May Have Explosive Impact," *San Francisco Chronicle,* syndicated column published in the *Tuscaloosa News,* 11 April 1993, 18–H; and Sharon D. Moshavi, "Fox Shows Rank High in Hispanic Survey," *Broadcasting & Cable,* 22 March 1993, 41.

40. McLean, ed., *RadioOutlook II,* 37–38.

41. Data reported by Nielsen Media Research as of May 1992: 99% of all homes had color TV, 37% had two sets, 28% three or more; 61% had cable, 75% own VCRs; *Tuscaloosa News,* 2 August 1992, 7–H.

42. *Broadcasting,* 25 May 1992, 3; 27 July 1992, 76.

43. Data based on Nielsen Media Research for second quarter period (April/June) of 1992; *Broadcasting,* 13 July 1992, 24.

44. *Broadcasting,* 18 January 1990, 53.

45. Philip D. Jurek and Norman S. Hecht, *Television in Transition: The Values of Broadcasting Television in a Changing Marketplace* (Washington, D.C.: National Association of Broadcasters, 1989), 34–46; *Broadcasting & Cable,* 22 January 1996, 125.

46. Tom Goodgame, president, Group W Television Stations, quoted by Robert Sobel, "The Tip O'Neill of Broadcasting," *Television/Radio Age,* 28 September 1987, 152.

47. Balon, *Radio in the '90s,* 27–47, discusses inconsistencies among radio ratings services; he explains how broadcasters can resolve some apparent discrepancies by not confusing differing methodologies.

48. For example, as early as 1972 WPIX-TV in New York City offered a news anchor, in addition to a base salary of $75,000, additional salary per rating point (over 2 points) on the following scale: a rating of 3.0 would bring total compensation to $94,200; 4.0 would bring $113,400; 5.0, $132,000; 6.0, $151,800; 7.0, $171,000; plus almost $20,000 per each additional rating point after successive years with that independent station's local news operation. *Variety,* 9 August 1972, 38.

49. At the national network level, each full rating point represented 921,000 homes in mid-1992 (that is, 1% of 92,100,000 U.S. homes, according to Nielsen Media Research). If a program's cost-per-thousand is $10, an average "12" rating representing 11 million homes would bring in $110,000 per 30-second spot; six spots per half-hour program generate $660,000. Assuming a 12 rating throughout an evening's three hours of prime-time, a network would earn a total of $3.96 million. For seven evenings, a week's sales of commercial time in those programs would bring in $27.7 million; 52 weeks with that rating in all prime-time programs would earn the network $1.44 billion. Further, an increase in a single percentage point (from 12 to 13 rating) brings an additional income of 9.2% at each step computed above, or an annual *increase* of $132,000,000 (more with higher CPMs for key demographics). That *additional* $132 million a year is for one rating point difference throughout each evening's programming! So one network's annual ratings of only a point or two higher (or lower) than the others—such as CBS 13.4, NBC 13.0, and ABC 12.1—translates into many tens of millions of dollars advantage for the leader over the third-place competitor that year. While other day-parts—early evening national news, daytime, weekends, late-night—are not as lucrative, they cumulatively gain or lose tens of millions of dollars depending on audience ratings that determine how much they can charge for commercial time. William Rubens, head of NBC Research for almost two decades, estimated one prime-time rating point in 1989 was worth $140 million a year to the network (*Broadcast Financial Journal,* March/April 1990, 33); Mick Schafbuch, chairman of CBS Television Network Affiliates Association, informally estimated that CBS made about $120 million from one prime-time rating point (*Broadcasting,* 29 June 1992, 37).

By dropping sets of zeros, similar comparative figures can be computed for ratings and related revenues of local stations competing in each of the nation's 200+ markets.

50. Jursek and Hecht, *Television in Transition,* 57.

51. Among most comprehensive studies of media audience research are Belville, Jr., *Audience Ratings;* David F. Poltrack, *Television Marketing: Network/Local/Cable* (New York: McGraw-Hill, 1983); and James F. Fletcher et al., *Handbook of Radio and TV Broadcasting: Research Procedures in Audience, Program and Revenues* (New York: Van Nostrand Reinhold, 1981). An early exhaustive federal investigation into the validity and reliability of ratings research was "Evaluation of Statistical Methods Used in Obtaining Broadcast Ratings," a Report of the Committee on Interstate and Foreign Commerce (Washington: U.S. Government Printing Office, 1961).

The National Association of Broadcasters (Washington, D.C.) regularly commissioned reports, analyses, and descriptive booklets: NAB, *Audience Research Sourcebook* (1991)—updating three 1983 publications, *Why Do Research?;* James Webster, *Audience Research,* and Judith Saxton, *Audience Research Workbook;* Robert E. Balon, *Radio in the '90s: Audience, Promotion and Marketing Strategies* (1990); Robert E. Balon, *Rules of the Radio Ratings Game* (1988); Jhan Hiber, *Winning Radio Research—Turning Research into Ratings and Revenues* (1987); Jhan Hiber, *Hibernetics: A Guide to Radio Ratings and Research* (1984). The NAB published an "intermediate-level, self-instructional" text with exercises using actual Arbitron and Birch reports including computer programs to calculate statistical ratings data, CPM efficiencies, and so forth: *Profiting from Radio Ratings* (1989).

Of course, media research companies publish booklets explaining their services and methodologies as well.

52. Balon, *Rules of the Radio Ratings Game,* 30. Further, the response rate to Birch surveys is about 60%, and to Arbitron about 40%; Balon, 39.

53. Dennis McGuire, vice-president/associate director, local broadcasting, N.W. Ayer, New York. Both he and Kirk Stirland were cited in *Broadcasting,* 5 August 1991, 31–32.

54. William S. Reubens, "A Personal History of TV Ratings, 1929 to 1989 and Beyond," *Broadcast Financial Journal,* March-April 1990, 33–38 (quotation, p. 38); originally published in *Feedback,* Fall 1989.

55. William J. Adams, "Ratings—Before & After the People Meter," *NAB News,* 5 November 1990, 15–18.

56. A.C. Nielsen Company conducted the pilot study for WDBO-TV and WFTV(TV), Orlando, Florida, August 9–16, 1979; *Broadcasting,* 24 December 1979, 35.

57. For example, Dick Montesano, vice-president, network research, ABC Television Network, in *Television/Radio Age,* 12 December 1988, 67, and 17 July 1989, 68; Timothy C. Nichols, vice-president, media research director, Chiat/Day/Mojo, New York, in *Television/Radio Age,* 7 August 1989, 47; Survey Design and Analysis, Ann Arbor, Michigan, for NBC, in "Of Special Interest" (NBC Corporate and Business Relations publication), December 1989, 3; University of Michigan study for NAB's COLTAM, and Nielsen study with Capcities/ABC and NBC, in *Broadcasting,* 3 December 1990, 76; Nielsen study for ABC, in *Broadcasting,* 4 May 1992, 62.

58. Nicholas P. Schiavone, vice-president, Marketing and Sales Research, NBC, "NBC College Study 1989" (presentation speech, 18 October 1989).

59. "Missing Persons" (editorial), *Broadcasting,* 13 April 1992, 82.

60. Data from the Network Television Association and Nielsen Media Research reported in *Broadcasting & Cable,* 15 March 1993, 54.

61. *Broadcasting,* 15 July 1987, 34–35.

62. TV network ratings costs, *Broadcasting,* 18 May 1992, 8; Birch's radio charges, *Television/Radio Age,* 24 November 1986, 52.

63. These and other data in this paragraph are from *Broadcasting,* 27 May 1991, 80; 3 June 1991, 39; 27 July 1992, reporting stations subscribing to ratings services.

64. National Association of Broadcasters, *A Broadcast Research Primer* (Washington, D.C.: NAB, 1966/1971), 6.

65. Susan Korbel, Korbel Marketing, San Antonio, Texas; in Hartshorn, ed., *Audience Research Sourcebook,* 29–30. The book lists fifty-one firms specializing in media research; media directories identify other marketing and research companies who hire out their services to local stations.

66. Dr. George A. Comstock, cited in *Broadcasting,* 28 August 1972, 31–32. For example, early in the debate social psychologists Dr. Stanley Milgram and Dr. R. Lance Shotland reported a unique field study of viewers of a national network program with different endings in different cities; they noted that while violence depicted on television may seem repugnant, it does not necessarily lead to antisocial behavior among the viewers: "We have not been able to find evidence for this; for if television is on trial, the judgment of this investigation must be the Scottish verdict: Not proven." But those investigators did note that their noncorrelation between TV violence and real-life aggression in single-program exposure did not preclude possible cumulative, long-term effects of viewing many programs over a long period of time. *Television and Antisocial Behavior* (New

York; Academic Press, 1974). The substance of their findings had been presented verbally in 1972 to a section of the New York Academy of Sciences, and reported in *Broadcasting,* 22 May 1972, 47.

67. Joseph Dominick, Barry L. Sherman, and Gary Copeland, *Broadcasting/Cable and Beyond: An Introduction to Modern Electronic Media* (New York: McGraw-Hill, 1990), 364–365. They cite twenty-two major publications for substantive reporting of research findings. See also Jennings Bryant and Dolf Zillman, *Perspectives on Media Effects* (Hillsdale, N.J.: Lawrence Erlbaum Associates, 1986).

68. Walter Cronkite, revered by most Americans as "the most trusted man in America" while he anchored the nightly CBS news 1962-1981, ironically fed the public's misconception of newspersons as omniscient. He ended each evening's newscast with "And *that's* the way it *is,* [date]; this is Walter Cronkite, CBS News, New York." It would have been more apt to echo his standard caution in speeches about the public's need to synthesize information from a variety of sources, including print and first-hand experience. A variant closing might have been "And that's the way it *looks* . . .," which metaphorically offers a range of interpretations: the way it looks to professional journalists with finite resources in a limited time period, or to allegedly liberal East Coast network personnel, or on your TV screen at home. Cronkite's successor Dan Rather backed off from such assertiveness by closing the CBS nightly newscast with "That's part of our world tonight."

When posed with this suggestion in a letter and again in person, Mr. Cronkite graciously demurred. Subsequently, his superior Richard Salant, CBS News president, responded in a personal letter to author James Brown: "As to your comments about Walter's closing signature, I can only say 'Amen.' I have been arguing the point with him for years but have been unable to persuade him and am reluctant to order him." But *USA Today,* 23 May 1996, reviewed a new CBS program "Cronkite Remembers" in which the 79-year-old newsman acknowledged that "Salant hated it [his closing line], calling the very idea arrogant. Cronkite now thinks he was probably right."

7

Managing Local Programming

The same fundamental philosophy of programming pertains
to television as to radio: provide genuine service to the public
audience.

HAL NEAL, ABC OWNED STATIONS

The programming department has evolved over the decades. It began with casual flexibility in the innovative era of radio in the 1920s, emerging into the structured, creative "golden age" of radio networking and full-service stations in the 1930s and 1940s. Then in post-television decades following the 1950s it narrowed to tightly patterned "formula" programming with the proliferation of competing radio stations. Finally, radio programming split up into distinctive, highly specialized services by the 1960s, especially as FM became dominant with its quality stereo sound. Meanwhile television programming inherited from radio nationally serviced schedules along with scattered efforts at local shows (usually news, plus some public affairs). Then television met head-on newly competitive program forms offered by cable systems, including pay-TV. Except for early years of local access channels, cable's program planning has mostly involved negotiating for satellite services from suppliers such as ESPN, CNN, Discovery, and HBO, along with relaying local and distant over-air station programming.

The program manager's responsibilities changed successively in radio and then in television, as the industry developed over the years. By the 1990s some broadcast stations began phasing out the position of program manager or director (PD). General managers (GMs) had taken over budgeting and acquiring and even scheduling high-priced syndication series leased by TV stations; news directors oversaw budgets, personnel, and equipment in that area of local programs.[*] Promotion managers also began to assume some responsibilities involving programming.

This chapter initially explores areas common to both radio and television local programming, then notes characteristics distinctive first to radio and then to TV and cable systems. The next chapter studies national aspects of radio and TV networks and syndication, and national cable program services.

---

[*]We refer to the head of a station's management team as the general manager. Of course, where a company operates multiple stations (AM-FM and TV) in the same market, each operation typically has a station manager, both reporting to the general manager who oversees the entire multilicensed property.

BROADCAST STATION PROGRAMMING

THE PROGRAM DEPARTMENT

This department used to be the "showcase" of a radio or television station. It produced the only *product* the station had to offer, whether entertainment or information. It usually employed more people than any other department. More than any other part of the operation, it created the station identity or personality. That continues to be true for radio. But in television the local news with its on-air anchors and reporters establishes a TV operation as distinctive from competing stations, even more than do its syndicated and network program series.

The program department has two broad areas of activity: (1) planning the overall program philosophy and schedule; (2) producing local programs and leasing syndicated programming to fit that schedule. The general manager is concerned about both activities, but is principally involved with the first. (As noted earlier, GMs at smaller stations may also act as "film/video buyer" of syndicated packages—in effect partly serving as program director.)

The GM is the architect of the station's program framework. She provides the philosophy (microintentionality) out of which grows program policy (directives). She knows the expectations (macro-intentionality) of owners/stockholders and of the public and accepts responsibility for making the station's program structure successful.

The manager's programming blueprint needs to be clear and realistic, and flexible enough to be adaptable to future change. It is finalized only after carefully appraising a station's objectives and resources.

While the program director (PD) is assigned to implement that blueprint, the general manager must continually evaluate the station's programming. She spot-checks concepts and performance. She notes places where improvements can be made. She confers with the program manager to suggest changes or corrections.

The ability to recognize where change is needed and to offer constructive advice is developed by experience in programming. General managers with earlier broadcast experience confined to sales or news or engineering must make special effort to nurture programming. In radio especially, leadership is necessary in this department where much of the station's success depends on the inspiration and energy of sensitive, creative people. Selecting an effective program director is an important consideration.

THE PROGRAM DIRECTOR

In radio's early days, when the head of station programming generally had the title of program director or program manager, this was a key position.[*] The assignment was challenging because so many programs were planned, rehearsed and broadcast live from local studios. Even with heavy network schedules, stations had to fill many hours locally. PDs considered it a challenge to match the quality of network production. They created an aura of excitement in the radio stations.

With the advent of "formula" radio, which was radio's first response to the competition from TV, there was little need for live production. In a majority of

---

[*]The terms "program director" and "program manager" are used interchangeably. In some TV stations, the title is "operations manager" or "director of operations" with added responsibilities over production facilties, studio crews, and other personnel, as well as program development and scheduling.

stations, the PD served as a kind of chief office clerk. He arranged announcers' schedules, counted the number of times a particular record was played in a day, tried to maintain a measure of authority over the disc jockeys, and looked for direction from the station manager. He became a follower, overseeing mountains of paper. Today, with radio stations striving for a distinctive identity, a premium is again placed on station creativity. A PD's paramount duty is to oversee programming staff while monitoring budget. Concrete tips for monitoring air personalities include:

> Set short- and long-range goals with staff
> Conduct staff meetings about policies, practices
> Listen to their ideas and problems
> Give advance notice about shifts in plans, schedules
> Keep available and flexible for all staff members
> Show serious concern for them as persons
> Monitor on-air mix of music and DJ "chatter"
> Constructively critique air talent
> Schedule DJs with at least one weekend free each month
> Provide lounge, snacking amenities, and reduced prices[1]

A similar scenario might be written for TV stations. In the dynamic chaos of early television, creativity was the hallmark of a program director who could be inventive with meager materials. But expanding segments of the broadcast day supplied by national networks (66% of an affiliate's schedule) and syndicated materials (most of the rest except local news) narrowed the opportunity for local program innovation. Except for mounting news and rare public affairs programs, program management centers on scheduling ready-made product, not building personalities and program forms in local studios. Proliferating independent stations (especially on UHF channels), as well as early cable TV, brought the medium full circle to its exploratory early days of imaginative production with small staff, limited space and facilities, and minuscule budgets.

The program director is responsible, ultimately, for the sound of the radio station and the look-and-sound of the television station as broadcast through programs to the public. His goal is to produce the best program service possible, given the staff, facilities, budget, and time available in the schedule. His work and the sales department are mutually dependent because programming that attracts audiences provides a marketable product for the sales staff. While having a sense of showmanship PDs must also be competent businesspeople, combining creativity with practical judgment. The programmer conscientiously assesses data about trends in music and talk (radio) and about the audience's shifting tastes for comedy, drama, and sports (TV) for acquiring syndication packages. Retaining both independence and creative sensitivity, his decisions achieve the goal of attracting local audiences—kinds of audiences sought by ad agencies for their advertiser clients. Sydney Head identified "common strategic themes" for programmers as: broad appeal to subsets of mass audiences, compatibility of programs and scheduling with audience lifestyles, helping listeners-viewers form habits of tuning in, controlling audience flow through the broadcast day/evening amid competing media offerings, and conserving program resources through efficient production and distribution.[2]

The good program manager needs a sense of show business. The world of entertainment in all its forms should be her key interest. She also needs to be concerned about communicating information. In addition she must know intimately the public making up her community—tracking patterns of audience preferences for music, talk, news, and entertainment generally. Continual research of audiences, music playlists, competition, and practices in other markets is a key responsibility of the PD.[3] Company product—in radio and TV, the schedule of programs—is only as strong as the person supervising its growth and development. Projecting the future of station programming depends on the vision and wisdom of the manager of this department. Given a PD with these and other qualifications—such as ability to budget program acquisitions, staff, and equipment—combined with top management's support, a station can continually develop and improve its programming. This leads to growing audiences and enhanced revenue from advertising.

As noted earlier, the role of television PD in the past decade has cycled from almost complete command over program content, to the point where general and sales managers often develop competitive programming strategies and negotiate for syndication packages (such as at annual NATPE conventions).[4] As stations have come to depend more and more on promotion and marketing to compete successfully, the role of program director has diminished. PDs must integrate their planning and decision making at every step with sales and promotion departments to achieve success for the station in its market.[5] While entertainment and news programming together consume most station dollars, the TV program department often requires far more of the station's budget than the news department despite the latter's need for many more personnel.

BALANCED PROGRAMMING

"Program balance" means different things to different programmers, just as it does to various audiences. If there were a standardized definition universally accepted, then programming would tend to be similar on all stations. Nor does "balance" apply to radio and cable as it does to local and network television.

Obviously it makes little sense for a station with a completely urban audience to devote part of its schedule to farm programs (not the same as country music). On the other hand, farm programming serves listeners of clear-channel and some regional stations and local stations in rural areas. Urban contemporary music would find meager audiences in mountain and farmland communities. Spanish-language programs attract listeners in the Southwest, southern Florida, and New York, but would be less apt for communities of central-European heritage in Great Lakes and New England cities.

Program balance lies in a variety of offerings. A radio station format that provides a day-long diet of one minute of news headlines and the rest of the hour devoted to a single type of music cannot be said to offer much variety. At least the rotation "clock" will offer varied kinds of music within a format (such as current hit selection, then upbeat tempo instrumental, followed by more mellow vocal, and a "golden oldie" from the same genre). "Day parting" matches music and tempo to segments of the day and night; one pattern is to air hits and midtempo soft pop tunes in morning hours, ballads in the afternoon, and hard rock late in the day and into the night.

In radio's early years, the several stations in every market were each mandated by their FCC license commitment to serve diverse subgroups of the total audience, each scheduling different kinds of programming throughout their broadcast day. They were "full service" stations, programming schedules much like TV stations do today. The Federal Communications Commission over the years required percentage tabulations of various categories of programs, as an index of diversified programming to varied interest-groups in a station's market. However, as the number of AM, and then FM, transmitters proliferated in even middle-sized and smaller markets, the FCC eventually permitted individual stations to provide specialized program service: all rock or country music, all news, even all classified ads. Taken together, varied program formats of a market's many local stations complemented one another, more or less providing every market with a total spectrum of broadcast service. Contemporary radio schedules in a market are similar to the mix of specialized channels on local cable systems, each offering different kinds of programming for audiences to choose.

And yet, except for the early Top-40 tightly produced formats, no single radio station attracts the same individual persons all day long. A flowing sequence of different people, with differing schedules of work, play, relaxation, and travel, listen to a station purposely or by chance throughout the broadcast day. Thus every station must be alert to "day-parts" with demographics of a continuously changing population available through the cycle of hours each day (see radio section, later in this chapter). This calls for some range or variety even within fairly homogeneous format content. In 1973 the trade press reported that "the idea of programming a consistent sound all day has, for some programmers, lost many of its former applications"; instead they tried to appeal to different, though not widely diverse, audiences at different times of the day.[6] By the 1990s that perspective had revolved almost 180 degrees; even subcategories within larger forms of music constituted the sole fare of many radio stations, such as *progressive rock* or *album oriented rock* within the larger area of rock 'n' roll music. Still, crossover artists and varied music forms—such as acoustic, amplified electrical, country melodies backed by strings—provided a range of musical treatment even within a single program niche.

By 1996 even many traditional full-service stations, highly successful for decades, had swung to news/talk formats, including local and syndicated news/talk/personality programs. They included powerful AM outlets WJR Detroit, WBAP Dallas, WCCO Minneapolis, WSYR Syracuse, and KOMO Seattle. But WGN (AM) Chicago continued to offer full-service programming—since 1924— and always stood among the top three stations in that megamarket, often ranking first ahead of scores of local competitors.

The question of balanced programming in television is partly like the early days of radio: A few mainstream, over-air broadcast channels serve each market (except for several massive metropolitan centers), so every channel attempts to provide a cross-section of kinds of program content throughout the broadcast day. This attracts successive segments of the total local audience and also justifies its license to serve *all* the public's "interest, convenience, and necessity" in that signal area. Yet, there is the anomaly that most TV stations are affiliated with national networks, so offer their local markets mostly national network service in major day-parts of early morning, daytime, prime-time, and late-night; even 140 independents carry the part-time Fox network. (*New York Times* TV-radio critic Les Brown noted before cable became widespread that because national programs are virtually interchangeable among the networks—and even imitatively indistinguishable from one another within the few categories of TV program types— there is really only *one* national program service through three conduits, plus the alternative noncommercial Public Broadcasting Service.)

So any real balance of television programming at the local station level must be achieved by the programming department's efforts to produce and sell local news, plus rare local public affairs, children's, or even entertainment offerings, and in leasing various syndicated TV series. The major opportunity to achieve some kind of balance in television lies in scheduling programs rather than in creating variety of content and form.

Thus any discussion of program balance will emphasize the variety and diversity possible in programming *formats* for radio but mostly in *scheduling* for television. Cable systems offer both approaches to balance: diversity of program types scheduled within some channel services (superstations, Arts & Entertainment) and diversity among various specialized channels (such as ESPN sports, CNN news, Discovery travel, MTV contemporary music).

Philosophies of programming for radio remained similar over past decades, with only a few relatively short-term modifications or emphases as various "fad" forms appeared on the scene and then melted into the broader spectrum of fairly standard programming styles. While finding a format hole in a market's radio services, and specializing in that niche to target segmented audiences, there remains room for adjusting within program styles and between differing genres. Appraisals spanning three decades could be made with equal aptness in the mid-1990s:

> In general, as one analyzes programming, it is easily perceived that no one philosophy emerges as dominant. Radio is in a state of program flux, with operators trying all kinds and types of programming. Radio program planning is always in motion and today it must move at a far greater speed than in the past. Research is beginning to play a major role and should serve as a stimulating guide in the future.[7] [1956]

> The senior broadcast medium [radio] is now so alive, so flexible, in so constant a state of change, that the broadcaster who would stay ahead of the game has got to be running full tilt each day of his competitive life. . . . The truth is that, in 1971, it is radio which is outpacing its larger kid brother, television, in terms of innovation, excitement and—perhaps—communication itself. Radio is once again on the leading edge of broadcasting.[8] [1971]

> Another clue to excellence in radio station management can be found in the way a station targets its programming. Rarely has the station whose ratings excel year after year, built its audience on one narrow target demographic. Invariably their eggs are in two or more baskets. By widening the spread and targeting a bit more broadly, there is far less risk that a major bailout by one segment of the listening audience will leave the station in the lurch. . . .
> Building a coalition audience based on a number of groups, each of whom perceives the programming as their own, also allows for stability as people age, but are able to retain their loyalty to the station. They remain, as new listeners feed in at the low end.[9] [1985]

Even a tightly formatted "formula" station's music and news programming cannot remain routinely predictable and ultimately dull. It must continually incorporate new creative ideas to attract audience listenership, goodwill, and support.

TECHNICAL FACTORS

To serve the public adequately and to be profitable, broadcasters must know the total market in which they operate. As mentioned earlier, the number of competing stations and their respective programming affects the kinds of audiences able to be attracted to a complementary or alternate (or even "counter") program service.

In both radio and television operations matters such as frequency and channel allocation, innovations in broadcast technology, and the question of music licensing all determine feasibility of various kinds of service to a community.

Compact transistorized battery-powered AM radios in the 1950s joined auto radios in expanding that medium's market against inroads by nonportable television. In the 1970s efforts were mounted to enable FM broadcasters to compete more equitably against AM stations for potential audiences by legislating "all-channel" radio receivers. Meanwhile audiences initially were attracted to high-quality FM programming with few commercial interruptions (because advertisers were uninterested in the small audiences). Growing audiences for FM gradually drew manufacturers to include FM along with the AM band on radio sets. This resulted in more intense competition among all stations in a market. It led to a wider range of program formats, including highly specialized services by stations that attracted highly segmented radio audiences. The total public's many kinds of interests and needs were served by "niche" programming with FM's superior sound reproduction, eventually surpassing AM station listenship and advertising.

By the 1990s FM slightly outnumbered AM stations, but it far surpassed AM in audiences: three quarters of all listening was to FM. AM moved more to news and talk which did not need stereo enhancement. Hundreds of weak AM stations with dwindling audiences left the air over several years—200 in 1992 alone—as FM dominated the scene with superior stereo reception (see chapter 10 for financial data on both services).

TELEVISION CHANNEL RECEPTION

Similar issues affected television. "All-channel" VHF-UHF receivers and then detente tuners mandated by law made it feasible to establish hundreds of new UHF stations in intermixed markets, competing with strong VHF stations. While legislation did make it physically possible for enormous numbers of people to tune in UHF signals, VHF channels sustained their dominance because they transmitted more stable, far-reaching signals; and they had established familiarity in their communities with local news as well as network programs.

Because it was difficult for early UHF stations to compete with established and prosperous VHF stations in bargaining for feature and syndicated films, the UHFs attracted audiences by showing old films and off-beat episodic series, plus children's programs and local sports. The FCC's 1970 Prime Time Access Rule and other federal regulations modified broadcast processes and business operations, opening affiliates' lucrative early-evening hours to non-network program sources. Production companies introduced more first-run syndicated programs such as game and talk shows and episodic drama. Those, plus increasing inventories of off-network reruns entering syndication, helped independent UHFs fill out their schedules.

As cable systems grew to pass more than 90% of U.S. homes (two-thirds of which subscribed), UHF stations almost achieved parity with VHF because viewers could tune both groups of stations among the same cluster of channel numbers on their receivers and with equal picture quality. Then Fox Broadcasting started up its parttime network; by 1992 it linked

140 previously independent stations five nights a week and on Saturday mornings, and by 1996 it had 169 primary affiliates (including some VHFs formerly with CBS and NBC). Programming schedules of independent stations more and more resembled those of stations affiliated with the traditional Big Three networks. For two-thirds of American households, all forms of over-air TV stations were equally part of the panoply of channels available through their cable-connected family TV sets.

CABLE CHANNELS

In the early years of commercial television, homes in most markets could receive only one or two broadcast signals. By 1973 almost two-thirds (60%) of all U.S. homes could receive seven or more commercial and noncommercial over-air TV signals; and 3% of the population could receive as many as thirteen or more stations. As cable developed, subscribers soon had access to dozens of TV stations imported from distant cities, including national "superstations," in addition to scores of special channels programmed nationally by satellite to cable systems. The proliferation of program sources reaching the typical home's TV screen forced programmers to cater to specialized subgroups of market audiences to compete for attention and support. This generated widely diversified program services to meet the public's many kinds of interests and needs.

ALTERNATE SYSTEMS OF PROGRAM SERVICE

Few argued about the benefits of early CATV/cable to the public and the industry as long as it concentrated on bringing more choices, without duplicating programs, to markets with no service or with one or two stations. But in multistation markets, broadcasters found a different story. There, additional channel activation resulted in fragmenting existing mass audiences. Smaller audience shares meant less sales revenue.

By the mid-1970s, the initially sanguine prognosis for cable TV had become less optimistic as audiences found them offering few specialized services other than clearer reception of already existing channels. Program sources were not all that available to cable operators and costs for cable installation and service kept mounting. The prospects for cable then seemed to lie in truly diverse video services such as professional information retrieval, two-way purchasing, and voting. At that time it was similar to wired-music services that never supplanted commercial radio broadcasting. But cable boomed after satellite interconnects fed national program services mounted by major entertainment corporations like Time Incorporated's HBO and major media companies such as ABC's co-owned ESPN and Turner Broadcasting's CNN (see chapter 8). With cable available to 90% of U.S. households early in the 1990s, even smaller markets received twenty or more cable program services in addition to half that many over-air stations; major markets offered almost 100 channels. In 1993 Tele-Communications, Inc., the nation's largest cable corporation, planned digital compression of cable signals, making possible up to 500 channels on each of its systems.

In 1984 the FCC deregulated the cable industry to help it expand and stabilize. By 1992 cable had swept across most of the country while traditional networks' audiences and revenues shrank. Cable is a programming issue as far as the general public is concerned because resolving controversy about subscription fees and governmental oversight will determine the kinds and cost of offerings available on home television receivers. That was demonstrated by the hotly contested two-year battle about reregulating cable in mid-1992, when Congress overrode President George Bush's veto of the cable bill (the only veto blocked in his four-year term). Any eventual settlement will need to reaffirm or redefine the American right of free enterprise applied to all parties involved: networks and stations, producers of programs, and cable systems. The issue involves questions of prior rights and ethical business practices.

Pay television, unsuccessfully experimented with decades earlier in many North American locations, reappeared as one form of cable television service—providing first-run motion pictures and sports for added monthly fees or specific per-play charges to the viewer. There seemed to be no discernible adverse effect on commercial television viewing. Pay or subscription television's attractions appealed to selective interests; the amount of expenditure per home varied according to the quantity, diversity, and quality of programming offered. Nevertheless, broadcasters feared that pay-TV—first basic cable program systems, and eventually pay-per-view services—would gradually siphon off program features from free television, especially sports and major motion pictures. As with early AM radio, over-air TV could find itself forsaken by audiences and thus advertising revenues.

COMMUNITY-CENTERED PROGRAMMING

Traditionally, radio and TV programmers found some of their greatest satisfaction in developing community-centered programming, featuring people living in the area. Local program coverage can build a position of solid respect in the community. As much care should go into occasional public service programs as goes into news. Well conceived and produced, not merely routine, those program efforts should be regarded as station investments in the area's welfare and future growth.

The station should be known as dedicated to civic improvement and promoting public safety, racial and religious understanding. It ought to take a strong stand supporting constructive projects as well as regularly exposing community problems. Mere notices, brief announcements, colorless paragraphs sandwiched in the day's schedule or newscasts are not enough. Whenever a station becomes a true mirror for the community, the manager who guided it to that level can take pride in this achievement. The station will enjoy widespread civic acceptance and approval.

Local programming—primarily news/public affairs (or informational)—is the cornerstone of effective, responsible, and successful station service to communities they are licensed to serve. Not only the FCC, but also audiences and local advertisers, tend to respond to the presence of a healthy community-oriented public service effort in announcements and local programs. Sammy Barker, former owner and manager of an AM (later AM-FM) station, urged small-market managers to "change programming to a concept of service, community events, local news and special event broadcasts, with less emphasis on music-driven concepts"—targeting programming "to listeners' real needs" to achieve "clear and decisive [results] in both revenues and loyal listeners."[10]

The FCC in the deregulatory 1980s began to look less at service in renewing licenses, except when applying criteria in comparative hearings or in increasingly rare instances when consumer or other special-interest groups filed petitions to deny renewals. "Substantial" or even "superior" service had been the criterion of the Commission in past decisions and of the appellate court (the U.S. Court of Appeals in Washington, D.C.), Congress, especially through its Senate Subcommittee on Communications, expressed concern about erosion of locally originated community programming among television stations. "Localism," which implies community-oriented programming, was once a matter considered seriously by the Federal Communications Commission but has become less important in the licensing process, although the FCC still mandates listing of community problems along with programs responding to those surveyed problems. Before leaving his post as FCC chairman, Dean Burch had advocated adopting gross percentages of broadcast time in several broad programming categories as criteria for determining what constitutes "substantially acceptable" service to a community. Those categories included local programming and informational programming (news and public affairs—not the traditional program type categories such as agricultural, religious, instructional, and so forth).[11]

Local coverage is a factor in media evaluation not only by the federal government but by a public concerned about local news and by advertisers who need local target marketing. Neighborhood newspapers, city magazines, radio stations reflecting metropolitan suburbs, and local TV news services of one to two full hours each early evening provide citizens with information about local events such as neighborhood crime, deteriorating public services in their area (transportation, schools, garbage disposal, utilities), housing, narcotics, unemployment and strikes, racial problems, and ecology. Alert programmers in some markets have bypassed competitors by promoting local activities such as voter registration, free rock concerts, civic events including minority projects, basketball workshops, and drug addiction centers or phonelines.

■   WVOX-AM/FM in New Rochelle, New York, capitalized on candid coverage and comment about community affairs. The station used "stringers" not only in the immediate area but throughout the state. WVOX President William O'Shaughnessy involved the station in community service activities and in strong editorial positions on controversial issues. He made it an outlet for expressing widely divergent viewpoints on local and national issues—with invited guests, discussion programs, and wide-open, telephone call-in shows. WVOX generated community response; that attracted advertiser response, making the operation very profitable. WVOX was praised by a national trade magazine as "the national prototype of the suburban station that succeeds by integrating with local 'gut' issues."[12]

Perhaps the strongest record of local TV service has been set by WCVB-TV in Boston. In less competitive years before cable's inroads, that station programmed seven half-hours in prime-time access periods with subjects of special interest to Greater Boston citizens. Including news at 6:00 A.M., noon, 6:00 P.M.(one hour) and 11:00 P.M., it programmed a total of fifty-one hours of local production weekly by 1974, increased to sixty-two hours in 1981, including a daily 90-minute live program, "Good Morning," with live remotes of features and entertainment as well as news. Its Public Opinion Research Unit and management's regular meetings with community leaders and organization representatives assisted the station in responding to community interests and concerns. That local program commitment led not only to annual profits but to the value of the station itself: It sold for a record $220 million to Metromedia in 1981, with GM Robert Bennett becoming president of the new parent corporation.[13] Bennett noted that for the decade up to its sale, station owner Boston Broadcasters, Inc.'s board of directors was committed to doing good programs even if it resulted in less profitability; they were not intent on absolutely maximizing profits at the expense of quality programming. He judged that other stations at least in the top 50 markets could follow WCVB's lead, which would not be as feasible in mid-size or smaller markets. That station also collaborated with four others in different cities to produce series shared by one another—locally produced programming interchanged regionally, for cost savings. Corporate groups such as Capital Cities Communications (subsequently Capital Cities/ABC) and Westinghouse Broadcasting/Group W (later Westinghouse/CBS) regularly exchanged special programming among their sister stations. In 1993 WCVB-TV continued to develop 30- and 60-minute local dramatic productions, often with national actors. Station manager/PD Paul La Camera said these were profitable ventures, partly because they were syndicated nationally for barter coupled with partial sponsorship by Heinz. ■

Speaking to TV station programmers at the 1993 NATPE convention about competing with a "500-channel universe" spawned by digital compression technology and fiber optics, former NBC Entertainment president and Paramount Studios head Brandon Tartikoff stressed local production in addition to news and public affairs:

> As the channels grow around you, your local production needs to grow as well. Just as a network needs to look like a network to the itinerant grazer, so must a local station seem local. Your signature shows and productions may need to multiply by the same factor as channel capacity does.[14]

Procedures for determining community needs are effective not only for meeting governmental guidelines but also for determining continuing shifts in the public's needs and interests. Public affairs program planning can be built on these surveys if properly conducted to truly reflect local citizen's concerns. Half-way measures, inadequate sampling, improper procedures, or careless handling or interpreting of data contribute nothing to the station's programming efforts nor to its responsible reporting to the FCC about its stewardship in a community; more serious than facetious is the caution to avoid superficial "half-ascertainment," superficiality or manipulating.

Inherent value lies in conscientious public service and community-related programming that can attract and even entertain audiences. A broadcaster should not grant free time merely because a topic or cause is "worthy." Airtime is valuable. It must be used to draw and hold audiences if it is to sustain that value—whether measured in social impact of information and motivating messages to the public, or in terms of dollars generated through advertising. Quality in production and content is necessary to justify airtime in a medium of mass communication. Poor quality disserves the content and the "cause," the participants, the programmer, and the audience.

Behind all radio or TV programming, and pushing it to expand audiences and revenues, is the concept of making a profit by attracting and serving the community—programming that not only interests the public but also serves that public's needs. Managers of the most successful stations through the decades have emphasized public service as the hallmark of a professional operation with balanced programming. The NAB's national search for outstanding radio stations reported:

> Excellence in service to the community . . . is probably radio's foremost social responsibility and it may well be what radio does best. Excellent

stations know their communities well and they become an integral part of their communities.[15]

In his speech accepting the NAB's Distinguished Service Award, veteran broadcaster Martin Umansky, chairman of KAKE-TV, Wichita-Hutchinson, underscored the point:

> Public interest, convenience and necessity [is] what our industry is about. That license requirement never was a burden to good, serious broadcasters who do not live in a vacuum. They ascertain...they see the problems and address them. They know that serving the public well is good broadcasting and that's good business. . . . It's the key to success.[16]

To compete with multichannel cable services—including channels devoted to twenty-four-hour national and international news and to nonstop contemporary music—the key to distinctive programming not offered by cable is for radio and TV stations to emphasize local news and other programs closely identifying them with the community. Stations owned by the three major networks have done just that by expanding news programs to strengthen local ratings.

NEWS

Certainly the expansion of national network and local news programming in prime evening hours—coupled with attractive visual effects and mobile technology for immediacy, vividness, and impact—contributed to the ascendancy of TV news over the decades since 1970.

What was radio doing all this time? For the most part, it continued to depend on the wire services for its news and it concentrated on one- to five-minute summaries of news highlights or headlines. The radio industry showed little disposition to seek out local news on its own. At many radio stations with a program policy of music and news, newscasts of one or two minutes make the policy a misnomer; music gets almost total attention with only token recognition of news. In 1992 only half the AM stations surveyed nationally aired newscasts longer than three minutes; FM stations typically had shorter news segments. Many of the nation's radio stations employed one person for news, or even split their duties between news and other program chores. As deregulation in the 1980s removed the threat of license challenge for lack of news service, many radio stations abandoned news altogether. (In 1981 the FCC removed minimum requirements for nonentertainment programming by radio stations, and did the same for TV stations in 1984.) A 1991 survey reported no news operations at half of independent TV stations (50.3%), 1.2% of network affiliates, 3.4% of commercial radio outlets, and 20.3% of noncommercial radio stations.[17] Yet few stations surveyed responded that deregulation prompted dropping news; instead they cited financial factors as a main reason for dropping it. Stations also said that they drew audiences with programming that differed from other stations that did carry news in their markets.

QUALITY VS. COSTS

Audience attention and credence generated by news service are important for a station as a sales medium. A responsible, competent news service can benefit a station in audience numbers and revenue. With the cycle of turmoil in local, national, and world conditions, the American public wants news. Sponsorship of TV news programming has increased through recent decades. Many small- and medium-market radio and TV stations must struggle to meet costs of news operations. But in larger markets, radio news operations produce revenue, while news tends to break even or earn revenue at two-thirds of the nation's TV stations.[18]

During the late 1960s into the 1970s news began to drop its negative mantle of being an automatic dollar write-off by management. Serious, competitive, comprehensive news service began to pay its own way and more, where management supported an adequately professional staff and news-gathering facilities.[19] Depending on dual accounting procedures, and whether radio and TV staffs overlap in combined operations, spots in news commercial time more than pay for the service; additional revenue is indirectly generated by the prestige and audiences that news attracts to a station's total program schedule, including its commercials. Some broadcast managers claim that the most important program commodity they have is news; although it can be the most expensive programming, one station manager put it simply "a good news operation makes a good station while a bad news operation makes a bad station."[20]

As mergers and buyouts proliferated in the deregulated 1980s, new owners often looked to broadcast media as investment opportunities rather than as communications systems serving society. Veteran news director (ten years) and then general manager (eight years) Ron Handberg of respected WCCO-TV, Minneapolis, took early retirement in 1989 when that station was offered for sale, noting that "the revenue strings are tightening" with "impact on what we can do in news and public affairs programming; it's discouraging to me to see this happening."[21] News divisions at the major networks were caught in heavy cut-backs annually, termed "downsizing" by new corporate executives. General Manager Bob Morse of NBC's owned-and-operated WMAQ-TV in Chicago, formerly news director and then GM at WHAS-TV in Louisville, noted that within five years his station staff had been cut from 440 to 250 people; fully robotic cameras replaced studio crew members for local news production on October 1, 1989. He commented on how cost-cutting steps affects managers:

It adds to the burden of any manager. It's not just WMAQ-TV, or the broadcasting industry. It's happening in the auto, banking, and airline industry. Virtually everywhere you look. The question of how a manager keeps up morale in a period of downsizing is the huge challenge today. No matter what industry you're in. . . . This isn't something that's unique to NBC.[22]

Echoing assessments cited earlier in this chapter, he voiced his major concern about shifts in broadcast ownership around the United States:

It's the erosion in the concept of the public trust. We're not only bottom-line companies. We're granted that license and we're granted certain constitutional protections and consequently we do have to recognize the public trust. I would do anything I could to help new ownership of major broadcast properties understand that.

Morse's comments were published in *RTNDA Communicator;* the following month's edition carried the notice that he, too, had been terminated by GE/NBC top management.

Ideally, management should be concerned first about the quality of the news service, and then about the audience and costs-vs.-revenue of that service. But the reality of broadcasting as a business demands attention to expenses and profits. One way to sustain or even increase news profitability in the face of a soft economy with tight ad dollars is to cut back expenses—usually personnel.

■   During the soft economy in 1990–1991, half of TV news operations and one-third of those in radio reduced budgets. About 20% of TV stations cut back staffs as did 25% of radio stations—half of them by attrition rather than layoffs. About one in ten TV stations cut back reporting and overtime; about 5% dropped news wires or other services, only 2% reduced news airtime.[23] However, during the same period one out of four TV news operations added staff. By 1993 the total workforce in TV news had grown in medium and small markets, while radio staffing remained stable nationally; but still one of five TV stations and one of four radio stations reduced their news staffs. In 1995 news staffs grew at two out of three TV stations while remaining the same at most of the rest; and 14% of radio stations increased news staff, with 78% remaining stable with no cut-backs.

Responding to a survey in 1995, three out of four TV news directors reported they made money with their news operations, 7% broke even, 4% lost money, and 17% were not sure. Among radio respondents, 18% of news directors claimed a profit, almost one-third (31%) broke even, 5% lost money, and almost half (46%) said they didn't know.[24]   ■

The most obvious characteristic distinguishing competing television stations from one another in a community is their local newscasts and personalities. That sets them apart from cable channels with no news or only national news. Radio has come to depend on music and the overall "sound" of specialized services or format, plus snippets of network or syndicated news; but the quality of its local news service could play a more dominant role if it wants to stand out among competing signals to attract audience and *local* advertisers—who typically provide 80% of radio revenue.

To build strong news capability, management must attract and support a capable news director, experienced in journalism and broadcasting. The news director, with support of top management, must be able to gather a staff—whether it be one other fulltime person in the small radio station, or 100 in a major-market TV station—who are competent and dedicated to the principles of good journalism.

■   In 1992 the Radio Television News Directors Association reported that responses from 506 TV stations and 315 radio stations reflected news carried by four out of five TV operations (most affiliates but only half the independents) and almost nine out of ten radio stations. About 250 radio outlets ceased offering news in 1990–1991. The average size of TV news staffs in 1991 was twenty-two fulltime and three parttime (almost double the size of two decades earlier); statistical average for radio was one fulltime and one parttime staff (one-sixth of the stations, typically AM or AM-FM combos, had six or more on their news staffs). Tightened budgets for news operations were reflected in the slowing turnover rate among newspersons in major markets as the 1990s approached. Shifts in TV news personnel dropped from 13% of staffs in 1987 to 10% in 1988, and less in the top five markets. Remaining constant through those years, one of every five news staffers moved annually in mid-sized markets, and one out of four in smaller markets. RTNDA surveys reported the average tenure of news directors at TV stations was two years (KING-TV in Seattle had eight news directors in a nine-year period).   ■

To support competence and integrity in news reporting, management must support budgets for adequate staffing and entrust the news function to professional personnel free from interference or pressure.[25] Managers should hire personnel with strong academic and professional credentials; they should encourage them to up-date and grow intellectually by reading journals, attending colleagues' meetings, and advanced formal studies if possible. Increased specialization has become important in larger markets; beyond an undergraduate degree in journalism or broadcast news, some advocate a graduate degree in another field such as political science or economics. A well-trained, conscientious staff is needed to work effectively with the news director who is responsible for overseeing the station's news operation.

After the bubble of the lucrative 1970s to mid-1980s burst or at least shriveled, corporate economics and intense competition forced trimmed news staffs to spend their time covering breaking stories, with little opportunity to develop special projects or features or to follow-up on previous stories. Compounding the daily pressure is the fact that most market ADIs include a range of ethnic and racial communities in urban centers, affluent suburbs, and rural farmland communities. Local news department have limited personnel and equipment to try to cover all municipalities in their viewing area. It is counterproductive for "cost efficiency" to reduce news staffs to too few reporters, producers, and editors to do investigative pieces that can distinguish a station from its competitors, helping to attract loyal audiences and higher advertising rates. In 1993 KIRO in Seattle sought cost efficiencies with enhanced news service by combining its AM, FM, and TV news operations into a single staff of 160.

■   General manager Hoyle Broome at WBMG, CBS affiliate in Birmingham, Alabama, supported better research for substantive stories by rotating each reporter's week with only four days of duty at the station; each reporter spent their "free" day gathering background information and thoroughly studying sources (library, experts, officials) to develop solid, well-founded, and informed reports. This carried over to the quality of each reporter's other stories during the week; they also became proficient at efficiently producing *more* stories throughout the week.[26]   ■

FORMATS AND STYLE

Managers, including the news director, often engage a research firm to appraise the local news operation to help make it more effective. A consultant provides data for managers and news staff to assess, not to replace their independent judgment. Outside consultants might recommend "packaging" news program formats that work in other markets but may not be the most appropriate form for the local market, station, and news team. Some consulting firms for television news re-created the patterned, packaged formats that had dominated "Top-40" radio. They sampled local markets scientifically, then used those data to creatively restyle news presentations. They imitated staccato radio news, shortening TV stories to 90 seconds, and increasing the number of news items from the previous 10 to a new standard of 18 or 20 per half hour program. They emphasized personalities and magazine-style segments within the news broadcast, linking partners in news presentations by informal conversation among on-air presenters (née newspersons). Media philosopher Marshall McLuhan acknowledged what he called the "friendly teamness" (*sic*) format; he claimed it was truer to the nature of the medium of television—its immediacy and spontaneity constituting much of the ongoing process that is television—rather than the formalized package associated with more static print media. McLuhan posited that "when the news team seeks to become the news source by means of direct dialogue rather than by remote report of the event, they are being true to the immediacy of the TV medium in which comment outranks the event itself."[27]

Although large audiences were usually attracted to these faster-paced, less formalized news presentations ("happy talk" formats), professional newspersons were often displeased with this emphasis on performance over journalistic reporting. Others noted that only transitions between news stories were lightened, not the news content itself. Also, with increased audiences, more people were being exposed to news content—albeit briefer and more visual, deemphasizing abstract and complex issues in the news. But fierce competition for audiences among radio, TV, and cable spawned mixes of information and entertainment ("infotainment") in prime-time network programs and local station schedules. Some panel discussions became less public affairs than verbal shouting matches, to vent emotions rather than illuminate issues; news magazines and "reality-based" programs included amalgams of staged re-creations and whodunit intrigue in dramatic form. The Radio-Television News Directors Association's journal posed the question in 1992: "Has public discourse become the verbal equivalent of mud wrestling?"[28]

Management can augment the local relevance of programming by preparing editorials about issues involving the community of license. By the early 1970s almost two-thirds of the nation's television stations editorialized, but that number dropped in following decades. A survey in the mid-1980s found 1,400 radio and TV stations aired editorials, 244 daily and 414 once a week.[29] By 1990 full- or parttime editorial directors were rare; many stations instead offered investigative news coverage with commentaries reflecting the newsperson's view rather than station management's. Great care must be taken with the serious responsibility of commenting about events and issues of concern to a community. Management itself might best make statements of opinion and interpretation, especially when advocating community action on an issue. The practice of having news directors or anchors deliver editorials is questionable because viewers should be able to distinguish clearly the editorial comment from journalistic reporting and editing. Editorial directors or writers report to management and present viewpoints reflecting management's judgment. In recent years former fulltime editorial directors have had to take on other duties such as public affairs programming and consumer services. Nicholas DeLuca, public affairs director of KCBS-AM in San Francisco, was shifted from full- to half-time researching and writing editorials; although editorials were moved out of "too commercially valuable" drivetime, he asserted that "we'd be the last to give up editorials because we are a network-owned station and we have always been players in the public life of this city."[30]

PROFESSIONALISM

All news departments, but especially those at key stations in the largest markets and at networks, must continually strive for objectivity and professionalism in their reporting. Management must support news directors' efforts to avoid sensationalism in content and presentation of news stories. "Gang journalism" contributes to stereotypical interviews of available celebrities and politicians; rather than seeking out centrally significant persons, street reporters can become "walking mike stands" merely thrusting microphones before prominent persons at airports, outside government buildings, or at non-news-generating "press conferences." Nor should broadcast media artificially create personalities, becoming almost common carriers for sophisticated news makers and news manipulators—whether they represent status-quo establishment or anti-establishment adversaries. Media should not influence the course of events in any substantive way. Newspersons must be observers and reporters, not participants in the event either emotionally or physically—even though, as surrogates of the audience, newspersons conducting interviews are also human beings interacting with other human beings. Presence of news crews should be muted, keeping a "low-profile" of neutrality. If at all possible, station policy should oppose newspersons' doing on-air commercials or voice-overs. Despite advertiser and sales department (and even audience) interest, the news function should not become mingled with the sales function of broadcasting. The integrity of news coverage, and the believability and impartiality of radio and television newspeople, must be protected.

News staffers should understand the traditions of journalism and of journalistic integrity. A good newsperson may rebel occasionally against some traditional rigidities of broadcasting, including formats, time segments, and commercial scheduling; he or she may at times react against some established station practices or impositions by federal regulation. On the other hand, when they believe in the importance of news seven days a week, they bring energy and enthusiasm that can be contagious throughout the station.

■   **Syndicated News Services.**   Most independent stations rely on the radio wire services of Associated Press (AP) and United Press International (UPI) for national and international news. Most small-market stations purchase the services of one or the other. Large-market stations use both and often supplement them with Reuters, CNN, and network and syndicated services as well as with their own remote correspondents. Network affiliates are aided by "hot lines" through which they are alerted whenever an important story is breaking. Syndicated services maintain daily voice news and actuality reports to radio stations, supplying spot news, commentary and features. Fulltime staffs of domestic and overseas correspondents are supplemented by "stringers" on assignments around the world. Syndicated news companies provide similar services for independent television stations. The satellite has made instantaneous regional and worldwide TV news service economically feasible. In recent years coalitions of independent stations and network affiliates have interchanged news on an ad hoc basis through regional satellite feeds. Affiliates receive daily closed-circuit feeds of hard news, features, and sports stories from their respective networks. Hubbard Broadcasting's Conus Communications since 1984 has provided satellite news service to hundreds of TV stations around the country. It developed regional news cooperatives among stations in 1991—providing 500–700 stories a week through 16 national and regional daily news feeds—plus a twenty-four-hour All News Channel with Viacom Satellite News for cable systems and cable networks as well as Hubbard's own USSB DBS satellite service.   ■

LOCAL COVERAGE

Too much dependence on wire services and syndicated suppliers limits coverage of local news. The radio or TV station that sends none of its news staff into the community on a regular beat misses a good chance to increase circulation. Direct lines to police and fire departments and city hall, and "stringers," have their place in complete news coverage, but they are not substitutes for regular first-hand reports by a member of the news staff. The practice of inviting listeners to telephone news tips to the station may be useful, but it can create problems of inaccuracy or at least of redundance or even triviality. Local station news should emphasize just that: localness. People are interested in information affecting their daily lives in the community—road conditions, closed schools, utility shut-downs, and conditions and service in local hospitals.

Mobile units equipped with cameras, tape recorders, cellular phones, beeper devices, and satellite up-link units have given news reportage a new dimension. In larger metropolitan markets, cruisers and helicopters provide continuing coverage of traffic conditions, informing commuters of highly congested areas and street conditions during bad weather. To add immediacy to television reporting, stations employ extremely light-weight portable TV cameras (ENG or electronic news gathering equipment) for live and videotape relay, comparable to compact transistorized microphones and audio recorders. This equipment makes possible "instant" reports during the broadcast day as important events develop. Microwave and satellite transmission make such sites accessible to live TV, now as flexible as mobile radio has been. Reshaped news formats, regional news cooperatives, and collaboration between TV stations and local cable systems have reconfigured news service in many markets and at the national level.

Major radio and TV stations go beyond essential news programs to produce documentaries dealing with community problems. At least less ambitious "mini-docs" can contribute substantially as feature pieces for regular newscasts. They can be some of the most exciting production experiences on the air. Radio documentaries are relatively inexpensive compared to costs in live or videotaped television.

Covering community events and producing in-depth features, however, is no guarantee that anything worthwhile will be reported. Stations need newspersons who can discover news stories overlooked by others not trained as journalists. A special kind of writing and editing for the ear and eye is needed. Often, the journalist with no training in broadcasting may be too well trained in how to write for print media. He or she may also be ineffective as an air personality. What is needed is a news personality with training and experience in both journalism and broadcasting. Many universities provide that dual training, supplemented by internships in professional print, radio, and television operations. Beginning newspersons often start at newspapers and local radio stations and continue their careers in those media; others use that experience as a foundation for moving over to the television side.

A. Radio Station Programming

Beyond characteristics common to both radio and television programming, some aspects of program content and scheduling pertain specifically to the sound medium.

The program director's role in radio changed radically since the early days of broadcasting. It continued to evolve in the 1990s, even to the point of extinction at some stations. Some history about how radio programming developed helps to understand changes in the PD's role.

The 1950s brought radical changes in radio programming as the number of radio stations grew and television service expanded. Many network stars and lavish productions moved from radio to the new TV medium, along with advertisers and national audiences. Throughout the country, radio stations began to shift away from network affiliation. Within a few years two-thirds of all radio stations in the United States operated independently of any network, faced with more hours in their schedule than they knew how to fill. Most were not budgeted or staffed for live production, unable to produce much original programming even if they had wanted to.

A bandwagon effect began among non-network stations. As local outlets hit upon a successful format, other stations hastily duplicated that format.

Into management positions in radio during this period of change came eager young men without background in the great traditions of radio and with little interest in the industry's professional status. They were blessed with considerable energy, egotism, promotional ability, and a consuming interest in generating profits.

In their search for successful formats, some scanned lists of the best selling phonograph recordings and then featured those records on their stations. It was not long before the "Top-40" recorded songs of the day were played repeatedly.* Record manufacturers recognized a potential gold mine complete with free advertising for their product. Station after station began to adopt the formula. Eventually most radio stations sounded alike.

Much of radio's programming became cheap and blatant. Music was largely rock 'n' roll, disc jockeys were verbose and even arrogant, news was surrounded by the noise of teletypes, trains, or rockets, and commercials were hard-sell. What appeal the format had was largely confined to teenagers and to adults with the interests of teenagers. The formula seemed to be: do it cheap and without imagination for people whose desires are simple and undemanding.

Most broadcasters by the early 1960s were aware the radio industry was operating far below its potential. Some managers changed their station formats. A few managers had never accepted formula programming.

Payola

The radio industry was hit hard by the "payola" investigations. Much of management's authority and initiative was forfeited to the "play for pay" people. Some managers claimed to have learned for the first time that their stations were involved in unethical practices by their disc jockeys. In effect, they had neglected to supervise their programming. In 1958 and 1959, record distributors had their recordings artificially promoted by radio DJs in twenty-three cities for secret payments totaling $263,244.[31] By carelessness or lack

---

*Innovative radio broadcaster Gordon McLendon is credited with starting the first "Top 40" station in the late 1950s; Todd Storz pioneered the format with his group of stations. McLendon also innovated the all-news radio format.

of oversight, managers abdicated proper control of their stations when staff on-air personalities accepted hidden payments to promote recorded songs. The public was misled, assuming frequently aired recordings were chosen for their quality or popularity. Owner-licensees and managers were cheated because DJs or music directors pocketed bribes while using station facilities and prestige to artificially promote record companies' stars and songs.

In May of 1963, the FCC issued final rules against payola. Stations were allowed to accept free records from manufacturers or distributors only as needed for regular programming. Radio personnel could not promise to play any of the free records on the air. Any form of payment or service to a disc jockey or to management in return for a free air "plug" was prohibited. AM, FM, or TV stations that accepted "money, services or other valuable consideration" in return for an agreement to broadcast any materials must announce on the air what program materials were sponsored and must inform the listeners "by whom or on whose behalf such consideration was supplied." Licensees and managers also must "exercise reasonable diligence" in discovering all cases where considerations are made. Both giver and receiver became equally liable for criminal prosecution under a law enacted by Congress following the investigations. Regrettably, the problem continues to recur in cycles every decade.

Station managers must watch relentlessly for any intrusion of payola by disc jockeys. Management must remind staffers, especially newly hired air talent, with explicit rules and instructions. There is no excuse for keeping on a station staff any employee unable to resist this sort of temptation. Payola is "sellola." The things sold are station quality and integrity. Ultimately it is management's responsibility to know about and to correct any abuses or unethical activity by station staff members.

REAPPRAISAL AND REASSESSMENT

Radio has unique resources. Some broadcasters, caught up in the hectic day-to-day struggle to dominate a market (or merely to survive competitively), do not develop radio's inherent strengths that still make it a distinctive medium.

No other medium of mass communication can capitalize on imagination as radio can. Radio can isolate and mix sounds that generate strong emotional impact. Copy can be written that makes a person hungry or happy or excited. While some pictures may be worth a thousand words, in many instances words may be worth a thousand pictures. Speech, persuasion, and eloquence—coupled with adroitly blended music and sound—are still prime movers of human response.

The illusion of intimacy in informal person-to-person communication is another of radio's advantages. The listener can select a favorite kind of program and listen to it alone, at home, work, or outside. The ability of radio to reach the listener in a personally selective environment should not be discounted.

Important factors unique to radio include *economy* of transmission and personal reception. Radio is more cost-effective than any other mass medium. Radio's *portability* cannot be matched by television. The radio listener can take her chosen programs wherever she goes: walking, jogging, shopping, driving, at the beach, office, or ballgame, or waiting for a bus or subway. Radio is the most portable source of instantaneous news and entertainment, especially at times of natural disasters when electrical power is cut off. Battery powered radios regularly prove their unique value in hurricanes and electrical storms, after floods or tornadoes. Radio is *flexible*. It can adapt faster than any other medium to changes needed in program content, whether breaking news or last-minute commercial or political strategies. It has the freedom to experiment with new program formats and to create new modes not common to other media. On radio a voice and an idea can initiate a promising experiment that need not be an elaborate or costly innovation.

All these advantages inherent in radio can be exploited effectively only with original thinking and hard work.

Music and news formats are admirably suited to the radio medium. They are the basic program elements that radio communicates best. For example, National Public Radio has offered brisk, authoritative, creative, and memorable hard news as well as features—blending news remotes, studio announcements, pretaped interviews, feature pieces, and aptly selected musical bridges—in its daily hours of *Morning Edition* and *All Things Considered,* plus *Weekend Edition.* The task ahead for radio managers is to search for new ideas and for techniques to best present them. Stations that lead the way in programming can expect strong audience loyalty and thus advertiser support. To achieve this, radio needs leadership by top managers who support strong program directors.

STATION FORMATS

The first step in creating a station image is to determine the overall format. Most radio stations are independent of network affiliation or have contracted short-term syndicated music/talk/news services, so are flexible for changing formats. With ten times as many radio as TV stations and most of them searching for a distinctive sound-image, various formats have emerged and continue to evolve.[32] As musical tastes and patterns of popular music change and "cross over," shifting radio formats reflect those merging or dividing forms. For example, the Recording Industry of America reported rock 'n' roll dropped to one-third of all music sales in 1992, down from half of sales four years earlier; country music rose from 9% of all sales in 1990 to 17% in 1992; rap and soul (or urban contemporary) was also 17%, up from 12% four years before, while other percentages of recorded music sales were easy listening 11%, classical and jazz, each 4%, and gospel music 2%.

Radio programmers keep attuned to shifting tastes and music forms and adjust their play menus and even entire formats to capture the continually shifting audience market. In 1990 alone 1,050 stations changed their formats—almost one-tenth of the entire radio industry; they adjusted within genres, as from hard rock to classic rock, or across genres from adult contemporary music to all talk.

A station's manager and program director must work closely to identify a "format hole" that is underserved or not yet tapped in a market, if they are to draw listeners. Of course, they can also compete head-to-head with dominant formats, if one or two stations attract big numbers that might be split off among them all. Or the "new station on the block" can fail badly, with audiences loyal to the originally formatted stations freezing out any newcomer.

Changing tastes and continually emerging new trends make it difficult to offer a definitive listing of program formats aired by U.S. radio stations at any given time, much less attempting to compare formats over time. Further, crossover artists and evolving musical forms merge previously distinct genres. For example, folk/country moved from acoustical instruments to electronic amplification and synthesizers, as mainstream pop singers added up-beat orchestrations of country melodies to their repertoire; R&B/rhythm & blues divided into rock, soul, and urban; "oldies" or "nostalgia" depend on who is listening and what songs were hits in their earlier lifetime: big bands? the Beetles and rock? pop songs of five years ago? For youngsters, "oldies" are what was popular last year!

■   In 1996 *Radio & Records* listed fifteen major formats, ranked in order of national audience share, as surveyed by Arbitron in fall 1995: country, adult contemporary (AC), and news/talk led the line-up.[33] (See figure 7.1 for national shares of audience listening.) That line-up differed widely from the Spring 1992 survey rankings when AC was first, followed by country and album oriented rock (AOR)/new rock (NR). Predictably preferences differed from region to region (figure 7.2). In 1995 Country was the most listened to format in the Midwest and South, while AC was first in the East and news/talk in the West.[34]   ■

Throughout radio's format permutations over recent decades, several formats clearly dominate over others, either in number of stations or in share of national audience, or both. Late in 1991, *Radio Business Report* listed formats most widely aired by 9,281 commercial AM and FM stations (formats of 1,719 noncommercial stations were not tallied).[35] (See table 7.1.)

Similarly, *Broadcasting & Cable Yearbook 1995* listed radio formats by numbers of stations airing them. Many stations had mixed or multiple formats, such as news plus public affairs, classical, and jazz, so the total came to 15,495 format listings for the listing's 4,706 AM and 6,753 FM stations—1,791 of them noncommercial (table 7.2).

The 1991 to 1994 listings had some similarities as well as changes in rankings. The country format clearly dominated the number of stations with over 2,500. Adult contemporary had almost as many stations, followed by the oldies format. Combination news/talk remained among the top five formats; by 1994 exclusively all-news and all-talk formats were added bringing the total to 2,019 stations airing those three genres. The first list in table 7.1 excluded noncommercial stations, so two religious formats showed only 795 (447 + 348) stations, while table 7.2 counted 472 noncommercial stations among a total 1,455 stations (1,227 religion + 228 gospel). CHR/Top 40 dropped from the top five. Adult standards (408) related to MOR (550) in the rankings. Rock and rock/AOR came midway through the second five formats in both listings.

Typical of the problem with categorizing types of formats, the broad "Spanish" listing usually refers to all Spanish-language programming, which actually includes many of the other-named musical and information formats. One report cited "at least 10 totally distinct formats programmed in Spanish . . . from news/talk to contemporary international hits."[36] The generic designation "Spanish" can mislead audiences and potential advertisers, complicating life for program directors and sales staffs.

Formats such as country are featured on a great many stations with low power or in rural areas; the mere numbers of those stations, many of which reach modest-sized audiences, can obscure the enormous audiences of formats carried by powerful metropolitan stations. This is evidenced by comparing format rankings in tables 7.3 and 7.4, and also in figure 7.2 where among regionally ranked formats in 1992 A/C dominated the East and West parts of the country and was just ahead of Country in the Midwest, while Country was Number One in the South.

■   The Arbitron Company and *Billboard* reported shares of audiences for major formats in 1991. Among the nation's total audience aged 12 years or older, country attracted 11.5% of all listening, while adult contemporary drew almost twice as many (18.6%); news/talk and Top 40/CHR scored slightly higher with audiences than did hanks to its enduring strength in the South as well as in key markets in the North, Country continued as radio's top format in the 100 largest markets during fall 1995. AC was the second-most popular, but thanks to select defections to other formats in recent years (most notably NAC and Oldies), its national audience has fallen 28% over the last five years. The other big winner was News/Talk, which rose despite '95 being a non-election year (**O.J. Simpson**, no doubt, helped take up the slack). Even CHR showed a modest rebound, but its national listening profile fell by a third since 1990.

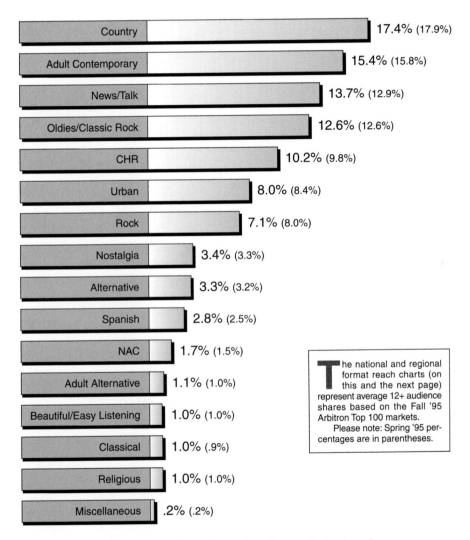

The national and regional format reach charts (on this and the next page) represent average 12+ audience shares based on the Fall '95 Arbitron Top 100 markets.
Please note: Spring '95 percentages are in parentheses.

FIGURE 7.1   Radio formats by national audience listening shares
Data © 1996 Arbitron Ratings Company; quoted with permission of Arbitron.
Graphic copyright © 1996 Radio&Records, Inc. Reprinted by permission.

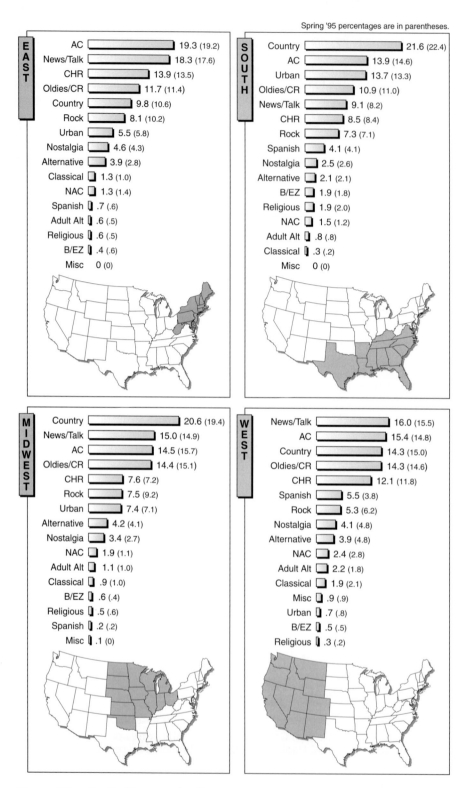

Spring '95 percentages are in parentheses.

EAST

Format	%
AC	19.3 (19.2)
News/Talk	18.3 (17.6)
CHR	13.9 (13.5)
Oldies/CR	11.7 (11.4)
Country	9.8 (10.6)
Rock	8.1 (10.2)
Urban	5.5 (5.8)
Nostalgia	4.6 (4.3)
Alternative	3.9 (2.8)
Classical	1.3 (1.0)
NAC	1.3 (1.4)
Spanish	.7 (.6)
Adult Alt	.6 (.5)
Religious	.6 (.5)
B/EZ	.4 (.6)
Misc	0 (0)

SOUTH

Format	%
Country	21.6 (22.4)
AC	13.9 (14.6)
Urban	13.7 (13.3)
Oldies/CR	10.9 (11.0)
News/Talk	9.1 (8.2)
CHR	8.5 (8.4)
Rock	7.3 (7.1)
Spanish	4.1 (4.1)
Nostalgia	2.5 (2.6)
Alternative	2.1 (2.1)
B/EZ	1.9 (1.8)
Religious	1.9 (2.0)
NAC	1.5 (1.2)
Adult Alt	.8 (.8)
Classical	.3 (.2)
Misc	0 (0)

MIDWEST

Format	%
Country	20.6 (19.4)
News/Talk	15.0 (14.9)
AC	14.5 (15.7)
Oldies/CR	14.4 (15.1)
CHR	7.6 (7.2)
Rock	7.5 (9.2)
Urban	7.4 (7.1)
Alternative	4.2 (4.1)
Nostalgia	3.4 (2.7)
NAC	1.9 (1.1)
Adult Alt	1.1 (1.0)
Classical	.9 (1.0)
B/EZ	.6 (.4)
Religious	.5 (.6)
Spanish	.2 (.2)
Misc	.1 (0)

WEST

Format	%
News/Talk	16.0 (15.5)
AC	15.4 (14.8)
Country	14.3 (15.0)
Oldies/CR	14.3 (14.6)
CHR	12.1 (11.8)
Spanish	5.5 (3.8)
Rock	5.3 (6.2)
Nostalgia	4.1 (4.8)
Alternative	3.9 (4.8)
NAC	2.4 (2.8)
Adult Alt	2.2 (1.8)
Classical	1.9 (2.1)
Misc	.9 (.9)
Urban	.7 (.8)
B/EZ	.5 (.5)
Religious	.3 (.2)

FIGURE 7.2 Radio Formats by Regional Audience Listening Shares
Data © 1996 Arbitron Ratings Company. Graphic copyright © 1996 Radio&Records, Inc. Reprinted by permission.

— TABLE 7.1 —

RADIO FORMATS BY NUMBER OF STATIONS, 1991

2,473	Country	323	Spanish
1,784	Adult contemporary	313	Soft adult contemporary
710	Oldies	248	Urban
664	Contemporary hit radio	200	Easy listening
543	News, talk	178	Classic rock
447	Religion (talk, services)	129	Variety/Other
408	Adult standards	66	Urban adult contemporary
348	Religion (music)	52	Jazz
346	Rock	49	Classical

Source: *Radio Business Report,* 8:46 (Report 303), 18 November 1991, 14.

— TABLE 7.2 —

RADIO FORMATS BY NUMBER OF STATIONS, 1994

2,767	Country	406	Classic rock
2,077	Adult contemporary	387	Jazz
1,227	Religious	384	(All-) talk
1,009	Oldies	334	Progressive
975	News/Talk	301	Urban contemporary
660	(All-) News	245	Educational
624	Rock/Album oriented rock	240	Beautiful music
604	Contemporary hit/Top 40	161	Black
506	Variety/Diverse	160	Big band
465	Classical	97	Nostalgia
414	MOR—middle of the road	95	Agriculture and farm
412	Spanish	65	Other
408	Gospel		

Other formats carried by fewer than 75 stations: foreign language/ethnic (74 total, plus 7 Portuguese programming, 4 each American Indian, Polish, and French; 3 Korean, 2 Japanese; 1 each Italian, Jewish, Filipino, Vietnamese, Serbian, and Arabic); public affairs (60); New Age (42); blues (39); bluegrass (19); children (17); folk (15); polka (7); reggae (4); drama/literature (2); 1 each disco and comedy.

Source: *Broadcasting & Cable Yearbook 1995.*

country (table 7.3).[37] Broken down further into subgroups, almost half (46%) of teen audiences aged 12–17 listened to Top 40/CHR, about one out of five to urban, and one in ten to AOR; only 4.4% of U.S. teenagers listened to country formats. Listeners aged 18-34 also favored AOR (20%), AC (17%), and Top 40/CHR (16%), with 9.4% favoring country programming. ■

While the number of stations with country format in 1991 totaled 2,473 (table 7.1) and adult contemporary (AC) numbered 1,784 stations, AC attracted almost double the audience (18.6% share) for country (11.5%). That year only 543 news/talk formatted stations attracted 12.8% of the nation's radio listening. Strong audience interest in talk radio prompted hundreds of stations to get aboard. By 1994 talk radio had taken off with phone-in shows and nationally distributed programs hosted by conservatives such as Rush Limbaugh, G. Gordon Liddy, and Lt. Col. Oliver North, plus outspoken and even raunchy talk hosts following the lead of Don Imus and Howard Stern. Within three years the 543 talk-format stations had quadrupled to 2,015 stations; 971 of them offered a blend of news/talk, 660 all news, and 384 all talk.

TABLE 7.3
NATIONAL SHARES OF RADIO AUDIENCES FOR MAJOR FORMATS, SPRING 1991

18.6	Adult contemporary	3.8	Spanish
12.8	News/Talk	3.5	Classic rock
11.6	Top 40 [CHR]	2.5	Easy listening
11.5	Country	2.1	Adult alternative
9.9	Album oriented rock	1.9	Religious
8.7	Urban	1.8	Classical
6.7	Oldies	0.9	Modern rock
		0.1	Remaining formats

Source: NAB citing Arbitron and *Billboard* (Spring 1991).

TABLE 7.4
RADIO FORMATS BY AUDIENCE SHARES, IN TOP-25 AND IN 90 MARKETS, FALL 1995

TOP-25 METRO MARKETS		90 ARBITRON MARKETS	
News/Talk	17.2	Country	8.8
Urban	11.1	A/C	5.9
A/C	9.4	News/Talk	5.6
Country	7.6	AOR	4.6
Spanish	6.4	CHR	4.5

Source: Interep Radio Store and Katz Radio Group, cited in *Broadcasting & Cable,* 5 February 1996, 41.

Differing listening preferences in top metro markets and in broader areas of the country in the Fall of 1995 can be seen in table 7.4. In the nation's top-25 markets news/talk far and away topped other formats, drawing 17.2% of audience listening. Urban music attracted an 11.1% share of listening, and adult contemporary just under 10%; country attracted 7.6%, just ahead of Spanish with 6.4% Those figures contrast with the rankings in more than 90 markets surveyed by Arbitron: country was king with 8.8% of all listening, adult contemporary and news/talk each drew just under 6%, album oriented rock (AOR) and contemporary hit radio (CHR) each about 4.5%.

Those data reflect formats shifting over time both in number of stations carrying them and in popularity among the nation's listeners. Further, the regions of the country and urban/rural signal areas are factors in the success of some formats over others. Finally, demographics of age, education, and income-levels all add to the mix from which program directors and their managers must choose to determine a target audience or market niche where their station can be competitive.

In Winter of 1992, audiences in ninety-five metropolitan areas (measured as Average Quarter Hour shares in the Winter 1992 Arbitron ratings report) favored top formats by demographic groups (table 7.5). The taste of older populations grows towards AC and country plus oldies (including classic or golden rock) and middle of the road (MOR) including nostalgia. News/talk also is heavier than with younger audiences.

It is relatively easy to track the "bubble" of baby boomers (born after World War II ended, 1946 to the mid-1960s) as that massive population segment

~~~ TABLE 7.5 ~~~
Top Format Preferences by Audience Demographics, Winter 1992, 95 Metro Markets, Average Quarter Hour Shares,

	ALL PERSONS 12+	ADULTS 18–34	ADULTS 25–54	ADULTS 35–64
AC/Soft Rock	15.0%	15.8%	18.0%	17.5%
Country	10.6	9.5	11.3	12.7
CHR	9.3	13.5	7.0	Oldies 8.0
AOR	9.0	17.7	9.9	MOR* 5.8
News-News/Talk	7.1	Urban 8.3	Oldies 7.6	News/Talk 8.4

*(includes Nostalgia)

Source: Interep Radio Store, using Arbitron data; cited in "Top Five Formats by Demographic Group," *Broadcasting,* 1 June 1992, 31.

aged. Table 7.6 charts percentage increase in population each decade: While percentage change in population of all ages progressively lessened (18.7% to 13.5%, for example), a single age-group each decade surpassed that by two or three times, while all other age groups only matched or fell behind that rate of change. The baby boomer crowd clearly progresses from decade to decade, accounting for 35.3% change in population of those under 16 years of age (the 1950s); 48.9% of 16–24-year-olds in the l960s; 48.6% of those 25–34 in the 1970s; 46.3% of 35–44-year-olds in the 1980s; and accounting for a 46.1% increase in those 45–54 years old in the decade of the 1990s. You can plot the rise and fall from dominance of different formats that cater to age niches. As teenagers of the Top-40 1960s decade grow into young adulthood, less frantic pace and more melodic music attracts them; and as they mature into middle age, their tastes evolve, leaving behind some kinds of recreational listening and seeking out more relaxed and informative radio. More discretionary income, grown families, paid-off mortgages and increased leisure time in later years make yet other forms of radio listening attractive to older groups.

Meanwhile, the pool of young teenagers available as audiences in the 1980s grew only slightly while older teenagers and young twenty-year-olds dropped significantly during that decade. In the 1990s the first group was expected to increase only 3.3% and the latter group was not to grow at all, whereas 45–54-year-olds were to increase by almost half. Clearly, formats catering to that older age range had prospects of a fast growing audience. The accelerated growth of news/talk radio forms reflect those shifting demographics.

The shift from Top-40 to adult contemporary as the dominant format over recent decades is reflected in the sheer size of population subgroups, coupled with advertisers' preferences for target audiences in their late-20s to late-40s. Table 7.7 reports that almost half the U.S. population in the 1960s and 1970s was under 25 years of age, while only one-fourth were aged 25–44. In the 1980s and into the 1990s the under-25 age group dropped to one-third of the nation's total (although the absolute numbers remained large). Baby boomers now 25–44 years old expanded that group from 23.3% in 1970 to 32.5% in 1990. By the end of the 1990s the U.S. Census Bureau projected 25–44 year olds would be just under 30% of the total, with those 45 and older making up more than a third of the U.S. population. Radio formats will not be far behind, if they wish to capture audiences "where the bodies are."

─ TABLE 7.6 ─
Population Percentage Changes, by Age Group, 1950–2000

AGE RANGE	1950–1960	1960–1970	1970–1980	1980–1990	1990–2000
All Ages	*18.7%*	*13.5%*	*11.1%*	*9.6%*	*7.3%*
Under 16	**35.3**	5.3	–10.8	4.6	3.3
16–24	9.8	**48.9**	19.2	–16.6	–0.2
25–34	–4.6	10.5	**48.6**	15.7	–16.3
35–44	11.9	–4.4	11.7	**46.3**	15.6
45–54	17.9	13.3	–2.4	11.6	**46.1**
55–64	16.6	19.6	16.5	–3.3	12.9
65+Over	34.5	20.6	27.8	23.3	10.2

Source: Robert E. Balon, *Radio in the '90s: Audience, Promotion and Marketing Strategies* (Washington, D.C.: National Association of Broadcasters, 1991), 22; he cites the U.S. Bureau of the Census as source for these data.

─ TABLE 7.7 ─
U.S. Population by Age Groups, 1960–2000

	ALL	AGE GROUP AS PERCENT OF TOTAL		
Year	*(Millions)*	*Under 25*	*25–44*	*45+*
1960	180.7	44.6%	26.1%	29.2%
1970	205.1	46.1	23.6	30.3
1980	227.8	41.3	27.9	30.9
1990	249.7	36.1	32.5	31.3
2000	268.0	34.3	29.9	35.7

Source: Based on tabulation compiled by Balon from U.S. Bureau of the Census; Robert E. Balon, *Radio in the '90s: Audience, Promotion and Marketing Strategies* (Washington, D.C.: NAB, 1991), 21.

Dominant Radio Formats

Top-40 and CHR/Contemporary Hit Radio

Audience ratings for various radio formats shift over time, reflecting changing age clusters among the population, as noted in the previous section. The post-World War II baby boomers provided a massive teenage audience in the 1960s and 1970s, who had growing discretionary dollars. Contemporary "sound" depends on anticipating and responding to swiftly shifting likes and fads of teenagers and young adults, reflected in the sale of records, audiotape cassettes, and CDs. Advertisers find teenage consumers a significant market to reach.

■   Out of this grew the teen-oriented Top-40 or "countdown" format that is fast-paced, loud, with tight production; in turn it spawned less frantic CHR/contemporary hit radio. The Top-40 sound is the same throughout the broadcast day. The top 20 or 30 popular recordings are repeated more frequently than the last dozen songs on the list. News is minimal, consisting mainly of headlines, time, and temperature. Many on-air promotional techniques include contests.   ■

Once dominant when teenagers were one-fifth of the U.S. population, by 1992 this form had dropped to a distant fourth place among station formats and audience listening (27 million teenagers made up only 11% of the population). But teen audiences remained attractive to radio advertisers with their annual spending of some $80 billion dollars in 1990. Top 40 and CHR still plays a role in promoting hit single records, but MTV on cable has taken its place as the dominant influence in popularizing contemporary hits and promoting sale of CD recordings.

AC or Adult Contemporary

As previous tables demonstrated, the population bulge progressed decade by decade through successive age blocks, so that in the 1990s the baby boomers were 40- and 50-year-olds. That partly accounts for AC being the most listened to format in the 1990s, along with the baby boomer age range being highly sought by advertisers. This age group liked softer contemporary music with less novelty and abrasive sound.

■   Depending on a station's definition, AC included "soft rock," "mellow rock," and even "oldies" or "easy listening." As noted earlier, those oldies or golden records were often the songs of their youth: early rock classics that once made up Top-10 lists several decades earlier. Strong melody lines with familiar musicians and vocalists such as Barbra Streisand and Neil Diamond or Kenny Rogers are introduced by subdued deejay/announcers who remain more in the background; the songs drive the format. Music "sweeps" run 10 to 12 minutes between commercial clusters. But some AC stations emphasized personalities in morning drive time, including teams with zany comedy and stunts, sometimes referred to as "the zoo." This format offers news segments during morning and late afternoon drive times. Because this format draws the largest audiences almost everywhere in the nation, AC stations in the same market push contests and promotions as they compete intensely for listeners. Variations include more rhythmic "hot AC" or "mix" (Phil Collins, Sting, Billy Idol) and rock AC (including country rockers, Jackson Browne).   ■

AOR/Album Oriented Rock

Just as similar sounding names for music formats arise, merge, then break away over time, so the descriptions for those categories vary. For example, *Broadcasting* magazine's editors reflected in 1991 on the ambiguity, diversity, or richness of the AOR format:

> AOR: In Search of a Definition
> Album-oriented rock encompasses much more by definition than it does in practice. Based on the free-form, progressive radio that got its start in the late-1960s, the format generally culls its playlist from current product and high-appeal classic cuts. Within this broad spectrum are such splinter formats as Classic Hits [Detroit], Z-Rock [Cincinnati], plus adult rock, hard rock, soft rock, new wave, new music, eclectic rock, rock hits and the new Pirate radio in Los Angeles.[38]

■   This format emerged in the early 1970s by breaking away from Top-40's individual songs interrupted by DJ chatter and commercials. Instead it aired complete sides of longplay albums, featuring music with heavier metal sound like Guns 'n Roses and Motley Crue. It offered an alternative to mainstream radio. But it also turned away from the harsh Jimi Hendrix to Bruce Springsteen and David Bowie (whose songs eventually migrated to pop hits and CHR stations). The format covers a quarter century of contemporary musicians—Led Zeppelin, the Grateful Dead, R.E.M.—playing longer and often obscure pieces, with emphasis on consistency of sound in a given program segment (unlike wide ranging current hits) and special features.   ■

Country

By the late 1970s, rural/folk country-and-western music had evolved into an up-tempo "modern country" sound with elaborate production and sophisticated orchestrations. It partly merged with the sounds and musicians from rock music and big bands and even orchestras with strings.

When stations with low popularity in their markets switched to the early country format, they typically increased their ratings and billings. Success bred imitation and competition. Small stations programming country formats with low power and high frequencies were eventually joined by major stations that dominated local airwaves.

The 81 fulltime country music stations in 1961 increased to over 650 by 1970. The drift of traditional country music towards modern "pop," rock, and MOR sound began to concern musicians by 1975. But it didn't bother country radio; by 1993 it was the format programmed by most stations (over 2,500). Country became the most popular format among the U.S. audience with 13.2% share in 1991, moving farther ahead of other formats with 16.1% of all listening in 1992.[39]

Country music listeners originally tended to be among rural populations or those with little spending power. But in later years the range of artists and styles of music within the "modern country" genre appealed to ever-widening audiences, including those in the largest metropolitan markets.

MOR/Middle of the Road

This music format plays popular melodies, with conservative pace and light music. It features popular instrumental and Broadway show music, favoring older tunes. Newscasts may be more complete. It airs promotions in a restrained manner, and seldom stages contests. Often it is a full-service station, especially on AM, offering news, talk, interview, and feature programming.

While rock, country, or classical formats offer a clearly identifiable image, managers with a middle-of-the-road music policy have a more difficult task impressing listeners and advertisers with their individual images. The MOR station has built-in variety right on the shelves of its music library. Imagination plus time devoted to planning can produce real rewards.

■ The big bands of Glenn Miller, Benny Goodman, Count Basie, Duke Ellington and others still have a following today as do vocalists like Frank Sinatra and Barbra Streisand; they are good sources of program material, along with music by Simon and Garfunkle, and the young reviver of the big band sound, Harry Connick, Jr. Modern jazz appealing to the "middle" group is available from such artists as Oscar Peterson, Andre Previn, Dave Brubeck, Pete Fountain, and Branford and Wynton Marsalis. Broadway original-cast albums and original motion-picture sound tracks are often requested by MOR audiences. Albums of routines by stand-up comics can be made into a series of popular features to balance musical offerings. ■

Oldies or Nostalgia

These music formats are relative terms. Depending on the targeted age group, play lists may include big bands of the 1930s and 1940s, plus revival forms such as Harry Connick, Jr. in the 1990s; or, they may refer to the Presley or Beatles rock classics. Air personalities can play a strong role, and music syndicators often provide the songs (Bill Drake and Chuck Blore developed the syndicated format in the 1960s).

Good Music, Beautiful Music, EZ/Easy Listening

This format programs instrumental standards and string arrangements of show tunes or familiar classics such as the Beatles, often as "background" music. Music sets are long, with few breaks for commercials. There is little talk, including news; consequently, the role of the disc jockey and "personality" is diminished. Half the stations with this format air programming packaged by syndicated music services. They commonly use automated music systems to program their local schedule.

Classical

In direct contrast to the Top-40 station, this format has a calm, deliberate pace. It often departs from lengthy programs of symphonies and classics by the masters, to present discussions of fine arts and contemporary affairs. It usually features news reports and commentaries. Though small, its upscale audiences attract selective advertisers. Most noncommercial public radio stations feature classical formats with strong news service; many others offer eclectic mixes of classical music balanced by jazz, folk, alternative/experimental, bluegrass, and information features such as interview and panel discussions as well as concert and theater reviews. The absence of commercials on public stations enhances the long-form classical listening experience.

All-Talk

The economic necessity for stations to specialize with distinctive image and audience appeal helped generate the all-talk format. Through the decades most commercial stations had broadcast occasional talk programs in addition to news in their full-service schedule.

■   In 1960 ABC-owned KABC in Los Angeles was the first to program talk exclusively. By 1994, more than 13% of U.S. radio stations carried a talk format; those 2,219 stations included 1,545 AM and 574 FM outlets. Of those, 484 stations featured all-talk. FM's wide-channel quality with stereo makes it the favorite for music listeners, so AM radio embraced formats of talk and news where broad-frequency signal quality is not essential.   ■

"Talk" formats refer to all kinds and styles of content: hard news, telephone conversations, play-by-play sports, editorials, serious interviews with prominent persons, and informal interviews with passers-by or phone-ins about current local and national issues or trivia and gossip. Many stations program talk during the daytime and shift to music through the night; others program the opposite pattern.

■   Talk hosts range from informed, urbane (Michael Jackson on KABC in Los Angeles), to neutral and inquisitive (Larry King), to outrageous and provocative (Howard Stern). Conversation programs with guests or panels, and telephone call-in portions, fill the broadcast day with local audience involvement. News is presented in longer segments, augmented by specialized informational programs dealing with sports, business, health and medicine, women's affairs, and the like. Many carry nationally syndicated talk features including news features of Paul Harvey, conservative and liberal hosts, and personal advice counselors.

Phone-in talk programs became increasingly popular in the 1990s. By 1993, over a thousand stations featured call-in portions in their schedules. The populist movement invigorated in 1992 by presidential candidates Ross Perot and Bill Clinton promoted direct feedback from listeners to local stations, and to national radio talk shows conducted by hosts such as Larry King (fifteen years late-night on Mutual Broadcasting System, adding CNN cable later) and conservative commentator Rush Limbaugh. ■

Studies have shown that listeners to talk programs, once involved, give closer attention to what comes over their receivers than do those listening to music. Talk radio that attracts and holds audience attention attracts increased advertising. But a station can also lose revenue because of irritating or extreme statements by listeners, guests, or air talent. In 1973, FCC commissioners and members of Congress objected seriously to uninhibited conversations on the air about intimate details of sexual activity (so-called "topless radio" programs). Broadcast deregulation and liberalized standards generally in society allowed more leeway by the 1990s; but indecency and libel remained a problem for spontaneous interchanges over live radio (see chapter 11 on regulation). The only protection against profanity, obscenity, and libelous comments is a sharp, intelligent, and genuinely witty announcer who can think extremely fast. Instant editing by tape-delay procedure gives a few seconds to delete inappropriate words or comments before the tape hits the air.

All-News

A variant of the all-talk station, the all-news station appeals even more to the information-seeking public. Hard news is intermittently updated and cycled around information segments and features, along with weather and traffic reports. Segments follow as tight a "clock" schedule through each hour as do all-music formats. Instead of featuring a succession of talk hosts and syndicated chat programs, the all-news station is staffed by working reporters for local coverage, supplemented by syndicated news services (wire services, audio news satellite feeds, networks) for regional and national stories. Expenses mount with remote vehicles and helicopters beaming back on-the-spot traffic coverage during morning and late afternoon drivetimes. This format can attract continual tune-in and-out by large portions of a metro population. Audiences come to rely on it for updates on traffic patterns, weather conditions including school closings, and coverage of local and regional disasters such as floods, explosions, railroad accidents, and civil disturbances. The all-news station can establish a respected position in a market by aggressively and competently "being there" with around-the-clock coverage.

In 1974 some 68 stations reported news as their total programming schedule or a major part of it. By 1994 that number had grown to 660 all-news stations, plus another 975 with intermixed news/talk format.

As stations proliferated even in small markets, competition for the listener's ear and the local advertiser's dollar forced program directors to become adept at marketing and audience research techniques to win a share of the audience. Long gone are the days of full-service programming in small markets where few radio stations easily shared the community's attention. Some radio professionals have noted that competition is just as intense in small as in large markets; only the number of zeros are fewer—in audience figures, budgets, salaries, time sales, and charges for syndicated services. One might say the big-metro media "rat race" is matched by a proportionately challenging "mouse race" in smaller markets. Keith and Krause (1989) charted the multiplying formats introduced since the 1950s when full-service radio segued into specialized formats (table 7.8). To that listing can be added the 1990s' emerging formats, including rap, hip-hop, rock AC, mix, and adult urban contemporary.

TABLE 7.8
EMERGING RADIO FORMATS, PRE-1950 TO 1990

PRE-1950	1950s	1960s	1970s	1980s
Full service	MOR	News	A/C	Arena rock
Classical	Top-40	Talk	AOR	Hot adult
Country	Beautiful music	News/Talk	Easy listening	CHR
Hit parade		Progressive	Contemporary country	Urban contemporary
Religious		Acid/Psychedelic	Urban country	New age
Black/R&B		Jazz	Mellow rock	Eclectic/Rock
Hispanic		All request	Disco	All sports
		Oldies	Nostalgia	All motivation
		Diversified	British rock	All comedy
			New wave	All Beatles
			Public radio	Classic hits
				Classic rock
				Male adult contemporary

Formats emerging in the **1990s** include rap, hip-hop, rock AC, mix, and adult urban contemporary.

Source: Slightly adapted from Keith and Krause, *The Radio Station,* 2nd ed. (Boston: Focal Press, 1989), 48.

SPECIALIZED PROGRAM SERVICE AND FORMATS

Specialized appeal stations program to specific tastes and interests. Examples are foreign-language, ethnic, jazz, and sports stations. In an effort to attract audiences competitively, many stations reach out to subgroups within the total market, which have distinctive interests and needs, as well as considerable purchasing power.

■   Programming formats have been designed for ethnic groups such as African Americans or Hispanics, Native Americans (Indians), and central European or Asian nationalities. In 1990, 59 non-Hispanic stations aired foreign/ethnic programming full-time, and another 553 at least one hour per week: Polish (182 stations); German (102); Italian (94); French (75); Irish (56); and Greek (44). That same year 261 stations offered Spanish programming fulltime, plus 361 stations with at least one hour of special Spanish program weekly (see below). The 1990 Census reported that almost 16% of those living in 284 U.S. metropolitan areas spoke a foreign language at home; 41.6% of residents in Miami did so, 38.3% in Los Angeles, 28.4% in New York, 26.8% in San Francisco, and 25% in San Diego. ■

Urban Contemporary

This format partly grew out of black radio which originally emphasized rhythm and blues and popular "soul" music and personalities.[*] WDAI(AM) in Memphis was perhaps the first station to direct its program schedule to African American audiences. Black radio gradually reflected wider ranging diversity along with more professional style, to serve the diverse tastes of black Americans. By the last decade, cross-over music and black musicians widely popular also among nonblacks merged with the short-lived disco beat plus break dancing and rap rhythms to generate the urban contemporary sound with up-tempo, strong beat, and progressive black programming.

---

[*]After the 1960s, the designation Negro changed to Black and then in the 1990s to African American (in the 1970s Afro-American was preferred by some). Terms similarly evolved, based partly on country of origin, from Spanish to Hispanic (South and Central America) to Latino (Caribbean and Central America) and Chicano (Mexico).

Some stations stress black music other than soul, such as gospel or jazz. Other parts of the black experience are reflected in programming of African music and talk. An early purpose of radio by and for blacks was to link the African American community by programming music and talk of black awareness, and also to provide employment in the radio business. However, as general-audience stations included music and discussion of interest to members of minorities, the distinction between specifically "black" stations and others became blurred. This was so partly because many white teenagers supported the upbeat tempo of black station programming. In 1990 black, R&B, and soul were favorite formats of 86% of black teens and almost 56% of Hispanic teens (but hard rock/AOR formats were preferred by 48.5% of white teenagers, 48% of Hispanic teens, and 10% of black teens).

In 1974 only 17 stations programmed fulltime black-oriented content. By 1988 black and urban contemporary formats ranked among the top stations in some major metropolitan markets. *Broadcasting Yearbook 1995* listed 161 stations whose primary format was black or rhythm & blues/soul, and 301 with urban contemporary format; hundreds of other stations offered partial black programming from 1 to 20 hours a week. In the early 1990s those stations attracted about 10% of all listening across the country.[*]

■   The commercial viability of urban stations focusing on the African American population is clear. By 1974 nearly 75% of all blacks lived in cities, with almost two-thirds of all blacks living in the 50 largest metropolitan areas. In the mid-1990s more than 25% of the population in New York, Chicago, Philadelphia, Washington, and Los Angeles were black. The 1990 Census reported a total of 30 million African Americans—more than 12% of the U.S. population. National advertisers must reach this massive part of major-market populations if their products are to succeed competitively (by 1991 black America's purchasing power approached $250 billion). Blacks listened to radio an average of 3 hours and 46 minutes a day, far more than whites (2:37) and other groups (2:23).   ■

For economic reasons of attracting large audiences and advertising, as well as for total community service, many black-format stations reflect the sound and commercial outlook of any professional general-market station—seeking a broad audience both black and white. In 1991 one study reported the cumulative audience for urban format as 68.7% black.[40] More than half (58%) of urban format listeners were in the key 25–54 demographic, and almost half (45%) had household incomes higher than $30,000 annually with 19% above $40,000.

Several black-oriented radio network services provide feature programming and news services emphasizing information and issues particularly relevant to the African American population (see next chapter on national radio).

Hispanic

As described in the previous chapter on audiences, the nation's fastest growing population group are persons of Mexican, South American, or Caribbean heritage—Spanish-speaking Chicanos, Hispanics, and Latinos. In 1947 KCOR-AM, San Antonio, became the first Spanish-language station. In 1974, 270 stations programmed part or much of their schedules to this audience, but only four stations

---

[*]Only 16 stations were owned by minority persons or companies in 1974, up from five in 1968. By 1988 the number of black-owned radio stations had risen to 150 (plus 11 TV stations) and to 292 in 1994 (plus 31 TV stations). Management positions for blacks in those stations slowly increased, with most program directors being African Americans.

devoted their entire schedules to Hispanic programming. By 1995 there were 412 stations exclusively devoted to Hispanic programming! Multiple stations air Spanish-language formats in metro markets such as Miami (9); San Antonio (7); Los Angeles (6); Houston (6); and New York (5). They air varying formats to attract demographic subgroups among vast Hispanic populations. In 1992 for the first time a Spanish-language station became the highest rated, most-listened-to station in the Los Angeles megamarket (4 million Hispanics live in the Los Angeles metro area of almost 9 million persons).

Most Hispanic-oriented stations are AM. They range from full-service to contemporary formats; news is an important component for Hispanic audiences. Balon notes that two major patterns of musical programming dominate: traditional Spanish music including Mexican ranchera (country), plus conjunto, norteno, salsa, and mariachi; and more current music featuring "musica romantica" that is similar to soft adult contemporary. Other forms include bolero, ballads, Top 40/CHR (contemporary international hits), Tex Mex, Caribbean, urban, news/information, and Christian Spanish.

In 1984 Spanish-language radio attracted $35 million in national advertising and another $65 million from local advertisers. In 1988 time sales approached $20 million by Spanish stations in New York where almost 3 million Hispanics are estimated to live. By 1991 almost $210 million advertising dollars were spent with Spanish radio.

■　In the late-1980s, nine out of ten Hispanics spoke Spanish—almost half spoke only Spanish, and 66% preferred to speak in Spanish—providing a unique mass audience. By 1990 the 23.7 million Hispanics constituted 9% of the U.S. population and continued to increase at 6% per year; about one-third lived in California. Their purchasing power was estimated at $200 billion in 1992. Hispanic households had 3.48 persons (contrasted with 2.58 in non-Hispanic homes); 92% of all Hispanics lived in urban areas, but 23.4% were at the poverty level—more than twice the percentage of non-Hispanics.

Groups of stations collaborated on Hispanic programming through some 15 regional networks with a dozen or so stations each. (See next chapter about traditional suppliers such as CBS Hispanic Radio Network programming to 35 U.S. affiliates plus 150 from Mexico to South America.)　■

American Indian

Programming for Native Americans was featured by fewer than a dozen stations with that format; another 72 stations featured programs for the Indian population from 1 to 5 hours a week, up to as many as 20 hours weekly. Stations were located in the West, Southwest, and Alaska.

Farm

The Communications Act that became law in 1934, provided for clear-channel frequencies to insure that rural areas would receive some program service, at least by ionosphere-reflected AM skywaves at night from distant cities. In the six decades since then, radio stations have proliferated throughout the land, providing local service to communities even in some of the most remote areas. By 1994, 95 stations carried agricultural/farm formats (50 stations fewer than five years earlier). More than a thousand others reported programming for farming communities' special interests and needs from several hours weekly to 15 or 20 hours a week.

The business of farming depends on radio for livestock and grain market information during all periods of the day. Livestock producers rated radio above other mass media and even above telephone as the most useful means for deciding when and where to sell and the price to pay for grain or other feed. Grain producers put only personal contacts with grain elevator operators ahead of radio as most useful for planning transactions.

Regional groups of stations in the Dakotas, Oklahoma, Wisconsin, Pennsylvania, New York, and other areas collaborate to share information and other programming pertinent to farmers.

Religious

Radio has provided religious programming since its earliest days. The federal regulatory agency once institutionalized this part of a station's service by listing "religion" as one category among sixteen in the early "composite-week" summaries compiled for license renewal applications. Because people have sustained their interest in religion and in religious broadcasts, not only has devotional and inspirational content been part of most stations' schedules through the decades, but by 1974 some 120 stations devoted significant portions of their schedules to religion throughout the week; five of those stations programmed religion almost exclusively. As commercial stations proliferated in the next two decades and narrowed their formats to niche entertainment with little time for religion (no longer required by federal guidelines), 1,227 stations sprang up by 1994 to program exclusively "religious" format—including sacred, inspirational, and Christian subcategories; another 408 aired Gospel programming (usually directed to black audiences). Many of these stations have low power, in the upper part of the AM frequencies, with signals limited to local areas.

In addition to traditional sermons, hymns, church services, and inspirational talks, religious stations feature commentaries about world events, call-in shows (local and national) for faith sharing and prayer requests, and Christian rock music with contemporary beat and rhythms coupled with religious and inspirational lyrics. Many religious programmers request donations to support their religious work and "broadcast ministry"; others offer religious goods and booklets to listeners. Religious stations often generate income by selling 30- and 60-minute blocks of airtime to local churches and clergy and to nationally syndicated religious programmers.

Underground or Progressive

Demonstrating radio's diversity has been the rise of "underground" or alternative radio, also calling themselves new wave, free form, or hard core radio. These stations grew from 25 in 1974 to 127 by 1990. Many of them were located on college campuses, operated by students; some were affiliates of public radio networks. They programmed loosely as progressive-rock or as "free form" with few traditional structures in format or content. The formlessness, while encouraging creativity by performers and announcers, at times resulted in public criticism and FCC warnings—including the caution that free-wheeling radio "gives the announcer such control over the records to be played that it is inconsistent with the strict controls that the licensee must exercise to avoid questionable practices." That echoed the FCC's public notice of March 5, 1971, about licensee's responsibility in playing drug-oriented music. Some progressive-rock and contemporary format stations permit the on-air disc jockey latitude in selecting the sequence of recordings, but all recordings intended for air play are approved by station officials, including program directors.

Noncommercial or Public

Early educational radio eventually advanced (after a decade of wrangling in the 1930s) when the FCC reserved FM frequencies for noncommercial organizations. Instruction and religion initially filled those dedicated channels. After the government formed the Corporation for Public Broadcasting in 1967 and established National Public Radio in 1970, noncommercial stations moved beyond education toward broader programming: classical music, alternative forms such as jazz and bluegrass, comedy, and especially public information and discussion—enhanced by NPR network with professional worldwide news service.

In 1996, 1,810 noncommercial FM stations were on the air, plus a small number of nonprofit operators of AM stations on nonreserved frequencies. Noncommercial licensees include colleges, civic groups, organizations, school boards, and municipalities. They ran stations primarily as a public service for their communities. They offered special-interest programming that commercial radio could not afford to air when competing for mass audiences and advertising. In addition to classical, jazz, blue grass, and cultural talk programming, they offered broadly popular shows such as live variety broadcasts (*Prairie Home Companion*), humorous but authentic auto repair call-in programming (*Car Talk*), and daily 90-120 minute award-winning worldwide news (*Morning Edition, All Things Considered* in the late afternoon, and *Weekend Edition*).

Niche Programming

The preceding chapter sketched the impact of population trends on broadcast audiences. Program directors must monitor those trends as numbers shift among age groupings, along with their leisure time and changing income levels. Stations need to develop tightly segmented programs to catch the attention and listening loyalty of large subgroups in each community. To compete aggressively against other stations in the market, PDs must target precise audiences with "niche programming" formats, determining a format hole that is underserved or not yet even tapped. This is similar to cable television's offering specialty channels that compete with one another among dozens of channels on a cable system, each vying for audience attention by attracting different subgroups among cable subscribers. For example, they may subdivide urban contemporary format into adult urban, current urban, rap urban, and jazz urban. By 1992, the country format's national ratings prompted stations to narrowcast by splitting into such forms as young country (KRSR-FM, Dallas) and easy country (KKBQ-FM, Houston), the former directed to younger listeners and the latter to the 35-plus crowd. Instead of battling head-to-head for audiences of other stations, a station's strategy may be to devise more narrowly defined format variations to counter-program against established formats. Balon referred to this as "niche or micro marketing." He speculated that by 1999, *national* radio services will again be surpassed by intensely *local-oriented* radio stations. This means customizing a station's program appeal to precise groups within larger demographic clusters, and developing unconventional ways to reach that narrowly targeted audience.

> To be an effective niche marketer means isolating the environment in which your listener uses your station and then creating exciting ways for that listener

to remember listening. It means you have to truly know your listeners. You have to be able to understand their language, what they want, their attitudes, opinions, desires. . . . While agencies may request 25–54-year-old buys, there isn't a station in America that genuinely delivers a 25–54-year-old audience. Stations deliver audiences that are 23–29, 30–36, 37–44 (core numbers). . . . The way to be strong 25–54 (as long as that outmoded buying mechanism exists) is to dominate a programming niche or demographic niche *within* that category.[41]

Tactics for reaching potential listeners require the program director and promotion director to collaborate on both on- and off-air campaigns to reach listeners where they live, work, recreate and in the style and terms they use daily—at home, the ballpark, the jogging trail, in cars, at points of purchase such as shopping malls. It also involves community involvement where the station is identified with serious local issues and activities as well as entertainment events.

In some intensely competitive instances, station management may consider repositioning the station by developing a fresh format, new on-air talent, even different call letters, logo, and identifying slogan. In 1992 "local marketing agreements" (LMAs) grew, permitting one licensee to rent or lease another station in that market—usually a weak one with financial pressures. Station owners did this partly to eliminate a direct competitor, partly to program both stations with a trimmed-down administrative staff and other economies of scale while selling both schedules to advertisers. The FCC estimated that 6% of all radio and TV stations were involved in LMAs. That same year the FCC established new rules whereby one owner may become licensee of two AMs and two FMs in markets with more than 15 stations, or licensee of three stations (no more than two of either AM or FM) in smaller markets. These rules permitting "duopoly" began to change the configuration of competition among local radio stations in the 1990s, even prior to further liberalized federal standards with the Telecommunications Act of 1996, which permitted single ownership of up to eight stations in a major market with over 45 stations.

DAY-PART SCHEDULING STRATEGIES

Whatever her station's format, a program director must schedule personalities, news inserts, features, and network/syndicated services around predictable patterns of audience flow through the broadcast day. Radio's largest audience everywhere is during morning drive time, followed by late afternoon drive; daytime offers modest ratings, and evening and late-night segments draw fewest listeners. Strategy involves programming against directly competing formats for those four levels of audience available at different times of day. Younger listeners tend to tune in during evening hours more than do others who lean towards TV viewing after the work day. Some stations modify their basic format to play to the somewhat different audiences available at night. Even strong music stations insert weather, time, and traffic reports during A.M. and P.M. drive times, so listeners don't tune away for that information elsewhere on the dial. Of course day-part listening patterns reflect out-of-home routines noted in a previous chapter: A large proportion of radio listening is in cars (morning and late afternoon especially), with much listening at work or elsewhere away from home during the day. Of 585 million working receivers in 1994, 373 million were in-home and personal receivers, while 212 million were out-of-home.

Advances in sound recording and reproduction through sophisticated studio facilities, including Dolby stereo sound, mixing equipment, and compact disks plus satellite distribution played directly into the waiting hands of wide-channel, high fidelity FM radio. At the same time it further eroded AM as an attractive sound medium except for news/talk, including sports coverage.

An important factor in staffing, salaries, operations, and programming was the introduction of computer automation. Initial high capital investment made feasible a lean day-to-day operation, especially where music was the primary program staple. Tightly formatted radio provided a natural matrix for introducing automation into programming. Transistorized electronics with computer-programmed schedules and on-air operation began to remove the announcer-disc jockey as the station's "living link" with listeners. Initial capital investment solved long-term expensive staffing and scheduling problems in small- and medium-sized stations—especially stations licensed to a single owner, who could format all his stations by computer playlists, program logs, and switching cues by computer disks. However, the demands of competitive radio "on the move" with localized, believable spontaneous sound of the human voice undercut excessively ambitious efforts to replace staffs with electronic robots.

Packaged syndicated music services (see the next chapter) did offer turnkey operation for those who wanted it. Complete schedules of music, features, and individualized station IDs were distributed around the nation by satellite, along with national spot ads inserted. Local ads could be automated by cart machines to kick in on coded signal from the program supplier in Dallas or Los Angeles. Local news could also be inserted if a station wanted. Skeleton staffs consisting mostly of salespersons joined the billings clerk and manager at the station site.

Processing of marketing statistics, time availabilities, fiscal records, and programming logistics of bookkeeping were apt for the computer (the "living local sound" of radio's on-air performance was another thing). Early on, the precision and cost efficiency of automation did help FM radio achieve stability against its big-audience and better-budgeted AM competitors, especially when early FM featured more conservative formats and "background music" for sparse audiences. When the FCC's nonduplication rule had required twin AM/FM stations to program separate schedules, computerized satellite program delivery systems freed management from staffing a completely separate operation. One staff ran the primary station; automation operated the second. After the FCC rescinded that prohibition against duplicating programming in 1986, sister stations forsook imported services and simulcast much or all of their program schedule.

Still, to gain sophisticated audiences in competitive markets, the spontaneity of "live" announcing keyed in to local people, places, and current events can make the difference between just one more radio station and a station of distinction—with audience and advertisers.

FM vs. AM

In 1972 only 25% of the radio audience's listening was spent with FM stations. By 1989 listening patterns had exactly reversed: FM's share had grown to 77% of all tune-in, with AM relegated to one-fourth of listeners, usually older ones, who preferred the news and talk formats that did not suffer from lack of quality sound. In 1990 adults aged 18–34 spent only one-tenth of their listening with AM, while those aged 35–64 listened to AM stations one-third of the time. The broad middle portion of 25–54-year-olds spent 21% of their listening with AM.[42] Managers and PDs of AM stations must develop strategies to win uncommitted portions of their market audience. One study reported that, while less than one-fifth of listeners loyally tuned in AM programming (mostly for news and talk), another one-third never intended to listen to AM radio; but the remaining one-half in the market listened to FM "largely because they've been given no reason to tune to AM."[43] That largest segment offered potential audience for assertive, well programmed and well promoted AM stations.

THE BOTTOM LINE

As audience patterns reversed over 20 years, so did radio's financial fortunes (chapter 10 details the financial structure of broadcasting). Successful traditional AM found itself bypassed by FM, so that in the 1980s, stand-alone AM stations averaged annual losses, AM-FM combos earned marginal profits, and FM stations earned substantial profits. Then the national economy and subsequent recession in the early 1990s undercut the industry as a whole. Average pre-tax profits reported in 1990 were $78,507 for FM-only stations; $53,501 for AM-FM combinations; and $103,953 for fulltime AM stations; but hordes of daytime-only AM stations averaged pre-tax losses of $9,582.[44] In 1992, 58% of all radio stations lost money; 200 radio stations went out of business. As noted earlier, many other stations were all but taken over in LMA transactions.

Most managers try to genuinely serve their community through program service and nonbroadcast activities. However, every manager must make a commercial station "pay"—not merely break even, but be as profitable as possible, especially if it is part of a group that is publicly owned by stockholders. The NAB/BCFM fiscal study cited in the previous paragraph broke out profits and losses of radio stations in 1989 by format. Complete data are in chapter 10, but excerpts here indicate how radio fortunes shift by formats—always depending on market size, signal strength, local competition, and effectiveness of management.

■   Profits at country stations in 1989 ranged from $22,000 for fulltime AM outlets, to $70,000 for FM stand-alones, to $155,000 for AM/FM combos; daytime-only AM country stations averaged losses of over $5,000. Highest profits were earned generally by twin AM/FM stations, but beautiful music stand-alone FMs reported $177,000 profits while urban format FMs averaged $319,000. Even higher profits came to news/talk stations, averaging $622,000! The most profitable of all formats was CHR AM/FM combos with $846,000. Which formats averaged losses that year? Almost tied at the bottom were AM/FM combos airing either a mix of nostalgia and country, losing $238,000, or AOR on both, losing $222,000. Smaller average losses were reported by AM/FM combos linking nostalgia and AC—$106,000; AC and CHR AM/FMs—$56,000; news/talk and AC AM/FMs—$47,000; and nostalgia on fulltime AM—$24,000. The two formats reported by daytime-only AM stations both reported average losses: religious—$20,000, and country—$5,000. No average losses were reported for any formats by FM-only stations.[45]   ■

As competition heated up with new stations entering already crowded markets, radio managers presided over continual changes in station formats. During 1990 alone 1,050 stations—almost one-tenth of all commercial stations—switched from one format to another.

■   In Fall 1992, audiences in the nation's top two markets put "soft AC" formats in second position (WLTW-FM with a 5.2 share in New York and KOST-FM with 5.1 in Los Angeles). But Arbitron reported the most popular stations as New York City's "urban" WRKS-FM with a 6.5 share and recently reformatted "Mexican/ranchera" KLAX-FM, Los Angeles, with 5.3—the first time a Spanish-language format was #1 in that sprawling megamarket. Early in 1993 New York received a third all-news station, emphasizing business news, competing against similar formats operated by Westinghouse/Group W and CBS Radio (the latter were joined in 1996 when Westinghouse bought CBS Inc.). Even in niche categories, competition continually stiffened.

NewCity's WZZK-FM perennially dominated the Birmingham, Alabama, market, far outdistancing the #2 and #3 stations. PD and operations manager Jim Tice strongly emphasized continuous research of the entire market audience, competitors' listenership, and their own audience, coupled with tracking playlists and published rankings of current artists and songs. Music was generally researched on a weekly basis, "oldies" twice a year. To hold its competitive leadership, the station allocated an unusually high percentage (20%) of its annual budget to promotion, in addition to 30% to sales.[46]   ■

Station managers must keep abreast of multiple factors affecting the radio business, many of them only indirectly related to programming. In Fall 1991 the National Association of Broadcasters listed issues of common concern to managers as topics for sessions at their annual radio convention:

Skills and personal qualities for management and ownership
Station group managers' assessment of competitive radio markets
Marketing through alternative means of advertising and promotion
Strategies for "a winning positioning and image campaign"
Augmenting sales volume
Improving facilities
Planning technical projects with engineers
Cooperative ventures between stations
FCC enforcement and fines
Legislative update about pending Congressional actions (DAB, talent
     royalties, spectrum fees, taxes, political advertising)
Digital audio broadcasting (DAB), including its impact on revenue
     budgeting and forecasting
Radio financing with investment bankers
The psychology of successful negotiations
Guidelines for hiring, firing, and other employee issues

The lone session devoted to "Programming, Ownership and Financial Strategies for Survival and Success" analyzed time brokerage agreements in joint sales and combination rates, plus potential changes in ownership rules by the FCC.

That's show business? It definitely is radio business, even if not radio as some used to know and love it. It is the business of radio management in the 1990s. (Later chapters look at sales, finances, law and regulation, and technology.) The editors of *Broadcasting* magazine asserted that radio "has grown up" in the two decades from 1970 to 1990, relying more on research, reflecting the business climate, as well as embodying traditional resilience and enjoyment. Dan Halyburton, GM of KLIF(AM)-KPLX(FM) in Dallas, described "the consummate manager [as one] who has great people skills, creative juice, understands the bottom line of things, can appreciate the marketplace, and can come into work every day and have some fun."[47]

B. Television Station Programming

Apart from a network, a television station has two principal sources of programs: film/videotape syndication and in-house local production, mostly news. Stations use syndicated product for most of their "locally-originated" (or non-network) program schedule. For affiliates that means most of the one-third of their schedule not programmed from the network; for independents that means almost all of their schedule. With widespread deregulation in the 1980s, serving the public interest evolved over the years into serving up what interests the public. FCC Chairman Mark Fowler campaigned for marketplace forces (of economics more than of ideas) to replace the original trustee concept behind broadcasters' planning of program schedules. Once the dominant motive, economics became almost the sole criterion: the product that attracts the largest audience for the most reasonable charge to advertisers.

Historical Trends

Between 1963 and 1993, network programming grew from filling over half (57%) to two-thirds (66%) of affiliates' schedules—about 16 hours every twenty-four-hour day. In addition to prime-time and daytime, network service expanded in early morning, late-night and overnight. During the decade 1963–1973, syndicated videotape programs increased five-fold in schedules (up from 2% to 9% of weekly programming).[48] The pace accelerated after the FCC's 1971 prime-time access rule removed an hour of early fringe time from networks (except for news); that opened a lucrative big-audience spot for non-network programs. Although the ruling was intended to promote local programs, stations found audiences preferred better produced shows by national distributors. Syndicated game shows and controversial talk programs joined off-network reruns to fill those early-evening slots.

Local Programming

FCC data for 1973 offer managers of the 1990s some sense of the Commission's historic (and potentially reemerging) criteria for preferred licensee service—contrasted with shifting expectations of the deregulatory 1980s. The Commission reported that VHF network affiliates in top-50 markets annually aired from 22% to 30% of their schedules as nonentertainment programming (news, public affairs, and so on). That is a far cry from current schedules of stations. Even in the 1970s, critic Les Brown harshly appraised the status of local television programming. His comments could have been written in recent months:

> There is no such thing any more as a staff writer, except in the news department; the resident directors are usually involved with cuing up commercials within the local movies; the production staff busies itself with commercials for local automobile dealers or department stores; and the director of programming is little more than a film buyer. Having toured numerous stations throughout the country in markets of all sizes, I have been impressed only with the size of their sales and clerical staffs.[49]

As noted earlier, when the FCC implemented its prime-time access rule (PTAR) in 1971, stations at first attempted to program more locally originated shows; but most dipped into the reservoir of off-network reruns or the growing assortment of syndicated programs that usually imitated previous network successes, especially quiz/game programs and music and variety formats such as *Hee Haw* and *Lawrence Welk*. Predictably, economics dictated program strategies.

SYNDICATED FILM AND VIDEOTAPE

All stations—from healthy network-owned VHF operations in the largest markets to tightly budgeted small-staff UHF affiliates in the smallest markets, and independent stations in all markets—build program schedules by leasing rights to film/video product.

■   An affiliate receives from its network about 16 hours of programming a day—more than 5,800 hours a year. It must provide the remaining 8 hours of its daily/nightly schedule. An independent is responsible for programming its entire schedule, a total of 8,736 hours throughout the year if on the air 24 hours. (Stations interconnected with the Fox network fall between those totals, depending on Fox's schedule, which by the mid-1990s expanded to parts of all seven evenings plus some afternoon and weekend service.) Staffs at affiliates prepare daily news and a smattering of other local programs. Most independents air little or no news; but the number of indies airing news and the length of their newscasts increased as the Fox network began to establish news service.[50] The rest of those thousands of hours every year are filled with off-network and first-run syndication. Local stations lease filmed or videotaped syndicated series and theatrical feature movies from motion picture companies, production houses, and national syndication distributors (discussed in the next chapter).*   ■

In the late 1980s, TV general managers of affiliates estimated expenditure for syndication (including barter) approached $1 million per station; indy GMs estimated over $3 million a year. Syndication spending by all stations (almost 1,000 then) totaled more than $1.3 billion, about 16% of that for packages involving barter.

■   In the late 1970s *M*A*S*H* episodes leased for $3,500 in major markets; in 1977 WTNH in New Haven, Connecticut, paid $1,738 per episode for exclusive rights to 119 episodes to be aired between 1979 and 1985, the total license fee coming to $205,832. But by 1983 half-hour episodes of *Laverne & Shirley* leased for $32,000. Five years later program rental fees peaked, with the record-setting *Cosby Show* leading the way: in 1988 New York indy WWOR paid $350,000 per episode or almost $45 million for the 182-week run of the package; the same off-network series went for $180,000 an episode to both WCAU in Philadelphia and KPIX in San Francisco. That same year Fox's WNYW, New York, paid $165,000 for each of 130 episodes of *Golden Girls,* an investment totaling more than $21 million.   ■

Annual conventions of the National Association of Television Program Executives (NATPE) bring together station programmers and more than 250 syndicators to explore available properties and negotiate costs. Syndication company

---

*As of 1993, networks were still prohibited from owning more than 40% of programs on their schedules (up from 5% for the previous twenty years) and from having any control over syndication of programs they carried. The FCC fully rescinded the "fin-syn" (financial interest/syndication) rule two years later, again allowing networks to be another supplier from which stations can lease syndicated programming.

salespersons also visit local stations to promote their catalogs of programs. In 1996 in the nation's top market, WPIX-TV's director of programming Julie Nunnari estimated that every day she spent over 45 minutes each with at least three syndication sales reps—more than a dozen each week.[51] She reviewed movie packages with 20 to 25 titles (perhaps half of them only B or C films); WPIX typically bought two to three new packages each year from among 10 or 12 major packages. Their computer system indexed WPIX film library's 2,500 titles and scheduled them for optimum rotation efficiency (including frequency of repeat, day-part, demographics). The stand-alone computer was independent of sales and traffic, monitoring cost control and amortization of film library holdings, and providing confirmation reports for syndicators-owners. Some program buys, especially co-ventures on new productions by outside producers, were coordinated with parent-company Tribune Broadcasting in Chicago.

As early-evening fringe time was wrestled away from the traditional network prime-time line-up by the FCC's Prime-Time Access Rule (PTAR) in 1971, and as UHF stations grew in number, West Coast production companies developed new properties exclusively for local TV stations. These "first-run" programs included efficiently produced talk/chat programs that interacted with audiences (Merv Griffin, Phil Donohue, Oprah Winfrey) and eventually moved into expensively mounted series such as action-adventure and science fiction (Star Trek: The New Generation). Replacing expensive "bicycling" of cumbersome film and videotape reels from station to station, satellite distribution offered instantaneous nationwide distribution of a high-quality signal at an efficient cost. At the same time, exploding costs for producing network shows made it imperative for series creators to recoup expenses and become profitable at the "back-end" by renting them to stations market by market for five- or six-year leases. The growing availability of programming stimulated the entry of more new UHF independent stations, which in turn expanded the market for first-run as well as off-network product. It became feasible to program non-network stations with a wide variety of programs that attracted audiences.

A syndicated package of off-network programs typically includes 100 or more episodes; a station leases rights to run each episode up to five times over a period of five or six years. When episodes have reached maximum replay, or the time has expired, all rights revert to the producer-syndicator owner. The property can then be offered for renewed licensing or to competing stations who bid higher in that market. Some stations experimented with "checkerboarding" different series through five weekdays at the same hour, with limited success. The most effective strategy has been to "strip" episodes of one series five days a week at the same time so audiences build loyal viewing habits.

Syndicated programs should not be selected for a station on the basis of how they appeal to station personnel. The program manager, film buyer or whoever is designated to pass final judgment must evaluate all purchases based on probable local audience response. Creative judgment and instinctive estimates must be undergirded by thorough research into the programs' ratings/demographics track-record during the original network run and in syndication to other markets.

However, a syndicated package with high ratings from its network run and in later syndication may not translate into high ratings in every market. It depends on the make-up and "culture" of each market, on the day-part and time-period when it is scheduled, and what airs against it on competing stations. Some extremely successful syndicated series may not always justify high enough rates for advertising spots to pay back the original cost, much less show a profit. Some stations have passed by highly desired syndication packages when competitors' bidding got so high they could not recoup their investment; by foregoing expected high ratings they earned more in the same time-slot with less successful but also less expensive programming. On the other hand, it rarely makes sense to buy a film package simply because it is the cheapest available.

The same value judgments apply to approving local live-production efforts as to selecting film/videotape programs to lease. To the extent that all local originations—whether locally produced (live/tape) or leased through syndication—blend into a planned design, the station image becomes clearly recognizable. A hodgepodge of program offerings, some of which fit into no overall pattern, can blur that image. A logo or catchy slogan can help identify a station and attract viewers, but the programs themselves are what keep them tuned in and coming back.

Action-adventure, comedy, travel, game shows, variety, musicals, documentary—the range is as broad as show business. Somewhere among all the programs available is the most effective combination for each television station. Within each category is a wide range of quality. No syndicated or network re-run material was created for the specific public of any local television station but rather for a broad cross-section of the U.S. population. Most syndicated programs can become popular locally; but many others do little to build a station's audience. Of course, no programming can succeed without strategic scheduling and apt promotion.

Off-network re-runs generally find favor with many audiences, particularly series that had extended runs of four or more seasons on a network. Less successful network programs formerly scheduled against popular shows on competing networks can still offer good prospects for local reruns, because original audiences were relatively small. Actually, any off-network series can still be fresh, because less than 20% of potential viewers see the average original telecast of a series episode.[*] Any re-run that the other 80% see for the first time is a new show as far as they are concerned.

Barter Syndication

Escalating costs of top-line syndicated series bring great financial risk to investing in a long-term contract for a property that may not do well enough to pay for itself when it finally hits a station's schedule. (Locally originated programs can be canceled overnight if unsuccessful.) Softening the cash outlay for stations by barter (or advertiser-supported programming) is a practice once looked down upon but now accepted even among network owned and operated (O&Os) and major groups.

■   In barter transactions, syndicated packages are distributed nationally at no cost in return for each station's air time and most of the commercial minutes in each program. The distributor pays nothing for national exposure and earns income from national spots inserted into those programs. The local station spends nothing on program material and makes money from the several spot availabilities left for it in those programs.

---

[*] A national rating of 20 would be high for a popular network series in the 1990s; the more typical prime-time rating of 15-or-so (percent of all TV households) often translates into less than a 25 share of all households watching TV at that hour. That means 85% of all U.S. homes did not watch the program, and even three-quarters of all those with TV turned on missed it.

Variations include combining cash payment plus some barter commercial time in program segments. Many barter packages are new, untried, or only moderately popular programs. They are an inexpensive way for a station to fill its program schedule. But if they don't draw enough audience to make spots worth much, and offer weak lead-ins to programs following in the schedule, they may not justify the up-front cost savings. ■

A station's rep firm can provide helpful data about the track record of syndicated packages carried in other markets. Station reps can break out computerized information such as audience ratings and package costs. A rep might even help a station's GM or program director negotiate contracts with major syndicators. Syndication companies expedite their own business by getting current data out through reps to their station client lists; that is more efficient than trying to contact every program director at 1,100 U.S. stations.

In addition to Hollywood-based syndicators and first-run distributors such as Viacom and Multimedia Entertainment, various group-owned stations distribute their own productions, among them WGN Chicago, KING in Seattle, and Group W especially with its successful trend-setting *PM Magazine*. The high cost of local production can be amortized by sharing shows with stations in the group, by program exchange with other stations, or by outright syndication leasing. An ambitious station like WCVB in Boston has produced distinctive high-concept programs suitable for many other markets (see later section on station-produced programs).

Feature Film

Most feature films for television are entertainment originally made for theater exhibition. Made-for-TV and made-for-cable feature-length "movies of the week" are also in distribution. They are attractive purchases because they fill large blocks of time and are popular with audiences. A properly promoted feature film can often bring a higher rating than competing programs on other stations, whether network or local. Theatrical feature films are show pieces, many of them brilliantly conceived, splendidly cast, and elaborately produced. They have star quality and built-in audience awareness from previous nationwide screenings, especially if they earned critical acclaim and awards.

The value of feature films as a program product is amply demonstrated by their success on network schedules (see next chapter). But this program genre (two hours or more per film, including commercials) has given way over the years to proliferating syndication packages with their more flexible, easily scheduled 30-minute episodes. Cable channels including "super stations" compete for TV audiences with large libraries of classic films; video rental shops also compete in three-quarters of American homes with VCRs on which they view current and classic motion pictures. Independent stations make heavy use of feature films to fill out their long weekly schedules.

■ Stations typically purchase the rights for five to seven showings over four years for each film leased in a feature package. They can attract good-sized audiences for each showing by scheduling six to eight months between showings and by presenting each feature at different days and different times of day whenever it is aired. By the late 1980s, each title in a good package cost about $30,000 to Boston market stations. The same films cost twice as much for Houston stations, triple in Chicago, and ten times as much for stations in the massive Los Angeles market (over $8 million for 22 titles in a Paramount package).

Leaner, more efficient station operations in the 1990s prompted less warehousing of movies. Down from the 2,500 film titles in a station's library, as noted earlier for WPIX in New York, the typical film inventory shrank in recent years closer to 1,500 titles, with independent stations owning several hundred fewer than that. ▪

STATION-PRODUCED PROGRAMS

If expert knowledge is needed for selecting syndication packages, it is just as necessary for planning the station's locally produced programs. By and large, all networks stake equal claims on audience appeal. Cable channels add to the competitive mix with their potpourri of nationally distributed program services. A local station often achieves dominance in its viewing market more through its local schedule than by relying on network affiliation. If a station turns exclusively to syndicated programs and feature movies, market dominance can be hard to achieve; it is not distinguishable from its competitors. Syndication/film are available to every station, so the difference in appeal between any two stations' offerings is minimal. The only chance to acquire a clearly identifiable local image, distinguishing a station from competitors and look-alike national programming, lies in local production including news.

Before the networks provided extensive program schedules, when film was scarce and generally inferior, local stations had to produce most of their own programs. Local program departments showed much ingenuity, even with only one or two cameras and no remote facilities. Today, with more knowledgeable personnel and a vast array of highly sophisticated equipment available, the same degree of resourcefulness should result in outstanding local programming. Yet, far too many stations do not take advantage of this opportunity.

Producing local live or videotaped programs is not easy. It involves more personnel, more time and more expense than are needed to plug into the network or to videotape machines loaded with syndicated tape/film. Sometimes the return for all the extra investment seems hardly worth the effort, particularly if audiences are small, generating only modest ratings and revenue. Yet, a TV station manager who hires a program director with true dedication to the medium can never be satisfied unless she engages in the kind of activity that motivated her to enter this form of communication in the first place. The thrills of creating a program, discovering new talent, prompting favorable comments by audiences and critics, are gratifications that feed one's sense of fulfillment and personal contribution. A manager should never be satisfied with merely supplying facilities to carry the creations of other sources. If she is truly imbued with a sense of show-business, her staff will necessarily be involved with production because it is a part of her way of life; and the product her station displays must have professional quality or she will not be satisfied.

The authors of this book find it difficult to accept the short-sightedness of some TV station owners. Owners with millions of dollars invested in each enterprise have entrusted their capital to management that, in many instances, either does not understand creative programming and production or is not interested in it. Some managers advise owners that it is impossible to appeal to local audiences with station-produced programs other than news. In our opinion these managers use this argument to cover their deficiencies or lack of vision in the program area. (We wrote this in the 1976 edition of *Broadcast Management* before the onslaught of cable channels and VCRs peaked in U.S. homes.) But in the early 1990s that very competition for fragmented audiences make the point even more pertinent.

Jim Coppersmith, GM of ABC affiliate WCVB in Boston, in 1988 endorsed the call for "localism, localism, localism; we believe that the best thing you can do is to be unique, and that means local news, local sports and local entertainment."[52] During the decade of the 1970s and 1980s, WCVB produced 40 to 50 hours every week of local programming (more than 25% over what typical major market stations air). That required big production staff and updated equipment to sustain audience-grabbing production values, coupled with heavy spending on promotion and sales. Parent conglomerate Hearst Corporation helped initiate the ambitious effort—not as feasible for stand-alone or small-market operations. Most of their local shows made money, including news magazine *Chronicle,* syndicated also to Arts & Entertainment cable network (partly owned by Hearst). The executive producer of *Chronicle,* Charlie Kravets, stressed:

> It's not just a public affairs gesture. It's the second-largest operation at the station, after the news department. The theory was to do a quality local program . . . I don't want to denigrate the syndicators, but you cannot buy the public relations in the community resulting from doing your own thing.[53]

But there is the opposite view that heavy local news—much less other local non-news programs—requires large, expensive staffs and technical facilities that mid- or small-market independent stations cannot afford. KSAS in Wichita avoided local programming and also high-cost top syndicated properties, going instead for cost-efficient, modestly rated syndication series that delivered enough audience and spot time sales to turn a profit.

At some stations, whatever local programming does get on the air is hardly ever distinguished or even distinguishable. One example to be deplored is the so-called public affairs program that relies on a static set consisting of a background curtain and a chair in which a guest from the community, untrained in television presentation, stiffly talks in monotone about a community project or controversial issue. No one but charitable family members and friends stay tuned in. The spot loses audience that might have carried over from the preceding program. Simple but creative use of the standard TV studio, plus apt film clips, video excerpts or slides, can easily enhance and clarify points being made—at least holding the inherited audience through to the next program. Free time granted to spokespersons which is vapid and inconsequential reflects indifference, lack of commitment, or simply incompetence on the part of station personnel, including management.

Capability of stations to create local programs varies widely. Major metropolitan stations obviously have far more resources available than do small-market stations. But differences between the two should be only in quantity of output, not necessarily quality of what they put on their air. In local production, resourcefulness is a far more important commodity than a huge budget. Sometimes, mere attention and creative experimenting is what is lacking. Even sign-off segments or other routine material offer opportunity for a dedicated, innovative production team to create aesthetically pleasing and meaningful TV pictures and sound, without excessive expenditures. Everything that a station originates should reflect the same care and professionalism that usually go into promotion spots and commercials.

It is advisable to develop no more than one or two local program formats at a time and concentrate on improving them before attempting additional efforts. One outstanding daily local show with general audience appeal in addition to one or two well-produced daily news programs would be preferable to a number of mediocre programs.

Many stations have not realized the potential for audience popularity in locally produced daytime programs. Karen Miller, vice-president of programming for CBS-owned TV stations, commented in 1993:

> I think there is a lot more local programming being produced than most programmers or station executives realize. There are certain station groups who have been very ambitious about it and those that are able to connect with a major distributor [for subsequent syndication]. But I'm particularly interested in some of the things that are happening for once-a-week-type programs. You know, they see an opportunity, they have a good idea and they are able to take that good idea, maximize it locally, and then they have found an opportunity outside [for subsequent regional distribution or national syndication].[54]

There is sufficient interest in good locally produced programming to warrant experimenting with new ideas and new techniques. By observing other stations' local programs with creative originality, broadcasters can develop ideas for originating new and different programs for their own schedules.

Surveys, opinion polls, local responses to local questions can be basic ingredients of good, pertinent program material. Telephone questions and conversations can interest a community. As a variation, studio guests can speak on selected problems during the first ten minutes of each program, then the viewing audience may be invited to ask questions. The format caught on during presidential campaigns in 1992, first on local and national radio then on national cable. Candidates even used the techniques for telecommunicated town meetings, responding to callers in a market or from around the nation, often with a studio audience.

Preempting Network Programs

Approaching the 1990s, many affiliates were replacing some network feeds with their own programs, preempting even prime-time offerings to air syndicated or locally produced shows.

■   After researching its Denver market, NBC affiliate KCNC programmed locally, including high school sports and prime-time preemption for championship games. General sales manager Rick Wardell confirmed that "our game plan was, and still is, to stay out of the high-cost syndication business, with all its audience fractionalization, and concentrate on local news, local productions and local sports."[55]

Among independents, KCAL and KCOP in Los Angeles scheduled a full evening of news, while WWOR New York offered a morning block of live local programs.   ■

The networks responded to the threat of losing national carriage of some of their programs by reversing their early 1990s' reduction in compensation to affiliates. Not only did they increase payments, but to ward off Fox Broadcasting's snatching key VHF affiliates, the networks entered into lucrative long-term contracts with many affiliates in the mid-1990s. The traditional three-year affiliation agreements were extended to as long as ten years once the FCC lifted limits. As licensees were permitted to own dozens of TV stations (as long as they reached no more than 35% of the nation's households) the networks' parent corporations bought or invested in stations to assure program distribution through loyal affiliation in selected markets. To further confirm their identity among audiences, against the onslaught of hundreds of cable and DBS channels, networks campaigned with "branding" by having local stations forego call letters in IDs and promotions, for the network letters and the local channel number. So WVTM (channel 13) in Birmingham became instead NBC 13; WBMG (channel 42) became CBS 42. That pattern was introduced around the country, tightening the bond between network and affiliate for a consistent recognizable logo in audiences' perception.

Local News and Public Affairs Programs

More stations place news programs in the top-rated local category than all other types combined. A station should budget as much as it can afford for news because costs can usually be covered through sponsorship. RTNDA noted that in 1966 network and local news began to be profitable. By 1990 over 70% of TV station news operations reported profitability. (See earlier portion of this chapter.)

In 1992 stations affiliated with networks spent an average of about $2.5 million annually on their local news operations (NBC affiliates spent an average of $2.7 million, ABC and CBS affiliates $2.4 million, and Fox affiliates $700,000). Fox Broadcasting introduced national news to its network and its "NewsEdge" news service to its affiliates, numbering more than 180 in 1996; partly due to affiliation shifts in 1995–1996, Fox affiliates that produced local news grew from 53 in 1995 to 67 stations the next year.

■   A strengthened economy coupled with competing news operations in Phoenix and the loss of ABC to independent TV status prompted Phil Alvidrez, news director at KTVK, to expand news from three hours up to five in 1994, with a further half hour in 1995; he expanded the news staff by 25 to 120 people. Sinclair Broadcasting assembled the nation's largest TV station group by 1996 and proceeded to start news operations at those independents and Fox affiliates. President David Smith reasoned that start-up expenses would be worth it because one-third of a market's TV revenue comes from commercial spots in news; "so if you want to play for that piece of the pie you have to be in the news business."[56] (Of the Pittsburgh market's $200 million annual TV revenue, about $70 million came from news; similarly $50 million of the Columbus market's $150 million TV revenue came from commercial sale of spots in newscasts.)   ■

At times of civic unrest or disaster, such as Midwest tornadoes, Florida hurricanes, the Oklahoma City bomb destruction of a federal building, or a crashed passenger jetliner, local stations mobilize to keep citizens informed, often at enormous out-of-pocket costs. A classic instance came in January 1994 when an early morning 6.6-level earthquake struck the metropolitan Los Angeles area, causing several dozen deaths and $30 billion in property damage, while shutting down large parts of the sprawling city for weeks. During the first two days each of L.A.'s seven TV stations aired some 30 hours of nonstop coverage. Their total revenue loss from canceling commercial time was estimated at $3 million, plus another $5 million to $7 million for production expenses and crew overtime costs.

Light-weight portable electronic news gathering (ENG) equipment in 1974 brought TV coverage closer to highly mobile radio's access to news events. Noncommercial public TV stations led the entire industry in 1977 as the first to use satellites to exchange news with other PBS stations. In the 1980s five public stations—WGBH Boston, WTVS Detroit, WNET New York, KCET Los Angeles, and WETA Washington, D.C.—formed a consortium for daily satellite exchange of news and features to use in their respective local programs. By 1983 some 375 stations (and over 6,000 local cable systems) owned earth-receiving stations, including about 80% of top-50 market stations; within a year 90% in top-50 markets and 70% of all U.S. TV stations had 600 satellite receivers. Hubbard Broadcasting's KSTP in Minneapolis initiated satellite news gathering (SNG) locally, then expanded to serve over 160 other stations with its Conus (Continental United States) satellite news gathering network. Other syndicated TV news services include TVDirect by the Associated Press, Viacom, Spacenet, Group W's NewsFeed, CNN, and Visnews International. These alternatives for national and regional news were matched by the networks which formed services to link affiliates with satellite capability: Absat (ABC), Newsnet (CBS), and Skycom (NBC). Networks fed off-air news material to their affiliates over a dozen hours a day via satellite, through renamed ABC NewsOne, CBS NewsPath, NBC News Channel (based in Charlotte, North Carolina), and Fox NewsEdge, plus CNN Newsource to 412 stations.

Mobile units—from compact cars to four-wheel-drive vehicles to large trucks (all emblazoned with station logos)—are a must in any area where significant news occurs daily; they provide remote pick-ups for local inserts, promotions, and out-of-station originations. By 1991 approximately 300 stations owned up-link satellite SNG trucks or SNV (satellite news vans) equipped with complete production controlroom and editing facilities, valued at over $250,000 (to as high as $5,000,000 at KCOP in Los Angeles). Helicopters and fixed-wing planes have proved advantageous in major markets for reporting traffic conditions and special news events, despite high expenses for aircraft, pilot, hangar fees, and insurance.

Adequate equipment is good to have, but adequate people are critically important. The programmer who relies primarily on technical equipment and exotic electronics overlooks or misuses a station's most enterprising and inventive resource: its staff and crews. Essential ingredients of quality local programming, according to Paul Dicker of WOWK, Huntington, West Virginia, are "good people, no budget and a desire," with "very talented, young and hungry producers out there" contributing to successful production.[57]

■   At KLJB in Davenport, Iowa, a staff of only thirty-four produced a farm report, weekly comedy show, hour-long documentaries and other local programs with one three-camera studio and a single editing suite. GM Gary Brandt said "we do all of this for profit; we can go home each night with a satisfied smile on our faces."[58]
WCCO-TV, Minneapolis-St. Paul, earned the RTNDA's Edward R. Murrow Award for "overall excellence." Five daily half-hours of news including documentaries, editorials, investigations, news series, and community projects were produced by a staff of 120. News director Reid Johnson explained: "If there's excellence, most of it starts with a station ownership that goes all out in its support of our news and public affairs operation. . . . [T]hat's not cheap . . . but [it] goes beyond raw dollars, as necessary as they are, and extends to conflicts with advertisers and to support in the face of lawyers . . . who threaten us with legal action if we don't back off."[59]   ■

NBC's O&O TV station division president Al Jerome reiterated the imperative he saw for local stations for the 1990s:

I'm looking for the unique selling proposition to viewers that we have and that others do not. It's our judgment that our niche in the competitive battle is local programming. . . .

The station that neglects public affairs, community service and editorials because of deregulation or because the station has changed hands at high prices is looking for trouble. Once again, it is one of the ways that we can relate to the community, a way that many of the new delivery systems cannot. The medium that will survive is the one that has credibility in the market, and you cannot have that without public affairs.[60]

■   Many stations have discovered that public affairs offer the greatest opportunity to be different in creating local programming. Their aim is to inform, to broaden understanding, and to stimulate community thinking and involvement. Carolyn Wall, GM of Fox's independent WNYW in New York, chose to get away from expensive syndication by expanding local programming. She scheduled brisk information-oriented "Good Day New York" with what she described as distinctive "content, style, format, spontaneity and localism," to compete with network morning shows she assessed as predictable and bland.[61] The show had a staff of 45, including from three to five remote crews; it took two years before it paid its own way. On the opposite coast, Disney-owned independent KCAL in 1990 moved to all news during three hours of prime-time, increasing staff from 38 to 145 to compete against six other VHF stations; it successfully maintained that schedule into 1993. Over three years of three-hour prime-time news in Los Angeles, KCAL's ratings had risen 29%, 50%, and 40% for each of its three 60-minute blocks. But, reflecting market differences, within a month of KCAL's expanding its news, ABC affiliate WQOW in Eau Claire, Wisconsin, let go 30 full- and parttime employees as it dropped all local news for "economic reasons." Throughout the industry the see-saw continued into the mid-1990s, depending on local market conditions, with many stations expanding news programming while others compressed or eliminated news.

A different intermedia synergism was developed by Cox Enterprise's WSB-TV in Atlanta. It shared the audio of its evening news hour with 24 northern Georgia stations that used 17 of the minutes for selling local spots or inserting local news; the stations also fed regional news back to WSB-TV. In 1993 WNBC-TV (NBC O&O) in New York planned co-ventures with WNET-TV (PBS) on news, public affairs, and other programming including election coverage.   ■

Integrity in News/Public Affairs

Negative criticism and public reaction to one style of gritty news coverage, coupled with widespread concern about "family values" in TV programming generally, prompted many stations in the mid-1990s to avoid "body counts" and "if it bleeds it leads" graphic reporting of incidents of violence. Some dubbed that the ambulance-chasing style of video journalism. Instead they sought to provide perspective on reasons behind homicides, drug dealing, and failures of local government or business. Investigative reporting takes newspersons' time and patient hard work but can result in feature pieces that truly inform and illuminate audiences. Consumer news pieces inform viewers about business, health, medicine, and education as well as crime. News directors realize that violent events, especially where there is a pattern, reflect a community, so they must be covered. But they monitor *how* violent events are presented on their air, keeping off the screen images of graphic violence—blood spattered on building walls, bodies mangled in car accidents.

In television, reality exists side by side with the illusions of show business. Whenever a station presents a public affairs program, it must make clear to the public that they are witnessing truth rather than the fiction seen at other times. (This distinction became muddied in the 1990s by some network "reality-based" programs using news coverage techniques and re-creations of crimes and other news events, and by several major blunders by NBC News in 1992–1993.) The station is responsible for making the distinction clear by eliminating elements of make-believe or artifice whenever it presents the factual. Every television station should have one individual, even on parttime assignment, designated to handle all public affairs programming. The function should not be assigned to a group of individuals nor should the duties be passed from one person to another from time to time. Public affairs programming is too vital to be given casual treatment.

Unfortunately, many stations have taken the easy way by giving lip service and time allotment to a public affairs concept that failed to interest or alert the public to a genuine community problem. A good public affairs program need not be devoid of interest and even excitement. As noted earlier, these programs should be produced with all the skill, verve, and appeal of commercial programming. If well done they can appeal to many local advertisers. Qualities necessary for sponsorship are: importance of subject, imaginative presentation, good production-direction-writing, wide viewer appeal, timeliness, and news worthiness; in other words—*good television.*

■   Intense competition in the 1990s spawned innovative ways to make news more profitable for local stations. For example, WFLD Chicago and WWL New Orleans collaborated with local cable systems, supplying them with news and weather segments, entire news features, and even repeating the station news programs on cable. NBC's O&O WRC in Washington bought time on an independent UHF to air a 30-minute newscast. Such "news cooperatives" found local TV outlets sharing news staff coverage with cable systems, as WBZ in Boston did during elections in 1990. The *Chicago Tribune* with WGN formed ChicagoLand Television News, providing twenty-four-hour news on a regional cable channel to 600,000 households. Similar ventures were mounted in New York City, Long Island, New England, Washington, and Orange County in southern California.

Independent WCVX in Vineyard Haven, Massachusetts, adopted the counter-strategy of reducing its news staff by two-thirds, replacing its local nightly newscast with a CNN package, and adding five-minute ad-supported updates every hour to emphasize its local commitment. In nearby WHLL, Worcester, Massachusetts, GM Michael Volpe followed a similar plan—with 2-minute hourly news inserts throughout the day—explaining that a regular evening newscast would require fifteen more people and up to $2 million a year. Some news consultants partly agreed with the tactic, but feared in the long run it could damage a station's identity and news reliability. The counter-pattern also appeared when McGraw-Hill's CBS affiliate KMGH in Denver in 1988 abandoned its only non-news local program, a daily magazine format. It released nine production staffers, for an eventual annual savings of $400,000.[62] The #3 station in that market, it then moved the programming department from under marketing to the charge of the news director.   ■

Another pattern began earlier expanded in the early 1990s as network audiences continued to drop: more stations preempted weak network programs for locally produced public affairs "specials" and sports (and also for strong first-run syndication). As early as 1988 one-third of all program directors responded they were preempting more network shows than before, 90% of them in prime-time and the rest in daytime slots.[*]

---

[*]Stations preempting prime-time shows in the 1990s also increased their negotiating leverage with networks that were threatening to eliminate compensation because of dwindling profit margins. Country-wide station clearance for national programming and commercials is the foundation of network operations.

Children's Programs

Programs for children must be done *well*—not necessarily expensively—or not at all! Children are loyal viewers but merciless observers. They are also sensitive and suggestible young human beings. Programming concepts, formats, vocabulary, actions, film segments, cartoons, personalities, and commercials must respect and even protect youthful viewers whose perceptions and values are being reinforced and even formed by television.

Locally produced series range widely in costs for production, from a few hundred dollars to thousands per program; commercial time within kid shows has varied from none to 16 minutes.

■ Recouping production expenses became more difficult as government rulings limited commercial time permissible when children dominate viewing audiences on weekends. After 1974, the former NAB code limited "non-program" material in children's shows to 9.5 minutes an hour on weekends and 12 minutes on weekdays. After the NAB dissolved its code (when the Justice Department ruled against its ad standards as restraint of trade), public criticism eventually resulted in the Children's Television Act by Congress in 1990. That law set a limit of 10.5 minutes of commercials in programs aimed at youth audiences during any hour on weekends, 12 minutes per hour on weekdays. Stations were also told to offer some of their programming, whether local or syndicated, suited to inform and educate children. Three years after the law was passed, stations tended to justify their regular cartoons (e.g., *The Flintstones*) and syndicated re-runs as having "educational" values. Critics and federal agencies balked, threatening to appraise license renewals in the context of serious, substantive programming response to the requirement to serve children's cultural and educational interests and needs. ■

Early in 1993 the FCC levied $10,000-$20,000 forfeitures (fines) on each of three stations, and reprimanded three others for exceeding commercial limits in kids' shows, including "program-length" infomercials. The Commission delayed renewal of seven stations pending further inquiry into their children's programming. In 1996 President Bill Clinton, supported by long-standing Congressional pressure on the media, and the FCC hammered out with broadcasters a mandate to provide at least three hours weekly of programming directed to children which emphasized substantive educational and instructional values.

Programming to Minority Audiences

As minority populations expanded in urban areas, stations began to look at local, shared co-op, and syndicated programming directed to massive minority populations. Most notable examples were Hispanic audiences in Los Angeles (projected to be 50% of metro L.A. by 1999) and Miami, and African Americans in Washington and New Orleans. While Hispanics watch almost 32% more television than do other populations, two-thirds of them—including young adult Hispanics—watch Spanish-language TV stations.[63] Philosophies differ. Some stations in markets with heavy Hispanic populations assume that Spanish-language stations serve them already (nor do they wish to program away from their core Anglo viewers). Others reach out to growing Hispanic audiences by simulcasting on second audio channels Spanish-language versions of their programs.

■ Tribune Broadcasting's KTLA in Los Angeles by 1990 employed a fulltime translator for 22.5 hours of simulcasting weekly, including the evening newscast, California Angels baseball games, and syndicated series like *Loveboat* and *Columbo.* KVIA in El Paso, Texas, did the same on Sunday mornings with *A-Team* and *Knight Rider,* via its second audio program (SAP) channel. New York stations programmed

Spanish-oriented public affairs programs on weekends: *Visiones* by WNBC, *Tiempo* by WABC, and *Horisons* by WWOR. KHOU in Houston permits Spanish-language KXLN to rebroadcast video excerpts of its newscast.[64]

Tribune Broadcasting in 1992 established a separate division to attract advertisers to its syndicated programming directed to African American and Hispanic communities. Their syndication included several series such as *Soul Train, Apollo Comedy Hour,* and *2nd Generation* plus a half dozen specials, one distributed to 100 stations, another to 140 markets. The Tribune's six O&O stations reached over a quarter of all African American households and 40% of Hispanic households in the U.S.[65] ■

TELEVISION PROGRAM SCHEDULING

Central to programming strategy is inventory control: carefully planned buying and scheduling of program properties. This involves what a station already owns, how long each licensing contract yet has to run, what it needs to strengthen its schedule competitively, how to identify and negotiate feasible prices for new product, and then when and where to run new acquisitions for maximum audience and sales impact.

As in radio, scheduling television in various day-parts is tied closely to the availability of kinds of audience, and to attracting those audiences from one program to the next ("audience flow"). Careful analysis of demographic market data and of competing stations' ratings and program types must guide the strategy of television programmers. Basic options include (a) noting the most successful competitive program at a given hour and attempting to "woo away" that already-available audience by going "head to head" with a similar show, or (b) counter-programming by offering a type of presentation no competing station schedules at that hour in the hope of attracting the nonviewing audience to their home receiver and your channel.

Even apart from the early evening prime-time access ruling (which was voided by deregulation in the mid-1990s) responsible station management should still make some effort to reach large evening audiences with programs pertinent to those living in their signal area.

COSTS OF LOCAL/SYNDICATED PROGRAMMING

The economics of local programming can be frightening when a PD considers the need for live studio facilities, engineering, production, and all other required elements. Some stations' program expenditures (total live, taped, and film programming including syndication packages) are less than 25% of their total operating costs. This incredibly low figure usually reflects a lack-luster programming effort or a weak competitive situation, or both. The average television station should expend at least 35% of its total operating budget for programming, apart from news. Stations not affiliated with a network may expect costs for programming to go as high as 55%–60% of overall expenses. In 1990 the NAB reported programming and production averaged 34.6% of U.S. TV stations' annual expenses, plus another 13.9% for news—together almost half the average station's entire budget (see chapter 10 on finances). The Tribune Broadcasting group of stations, the largest buyer of syndicated programs, annually spent over $200 million for that programming in the mid-1990s.[66]

The most effective means of program cost control in a well operated television station is through long-range planning, especially with contracts for film/video licensing. Any station involving itself in one crash program after another in an attempt to find winners from the available syndicated half-hours and features will probably disappoint an unhappy audience and an unhappy auditor.

The overall cost of film/video product depends on the type of station and its competition for syndicated shows in that market. Obviously, a market with seven VHF stations such as Los Angeles sparks far greater competition for available packages and series than a three-VHF commercial market such as Cincinnati or Atlanta. Pricing depends on supply and demand—on the product's successful track record, potential, and how it fits into each station's scheduling needs and budgets.

■   Although there is no standard asking price for syndicated film, the average cost for each half hour episode in a package includes rights to five runs over a five- or six-year licensing period. A station in a small market might be able to lease the same package for about one-fifth of what the more competitive medium-market pays. Of course, stations in the nation's largest markets pay premium prices for syndication and motion picture properties (see samples quoted earlier). In 1993 a station in the 75th market might pay approximately $3,500 per half-hour episode of a popular off-network entertainment series like *Cheers,* with rights to five runs over six years; the total cost for 200 programs in a syndicated package comes to over $700,000 (to be paid within 36 to 48 months).[*] Thus the cost can be amortized over each of the five runs of a 30-minute episode at $700 each, or more usually, successively at $1300, $1000, $700, $400, and $100 (with advertising rates weighted to reflect the decreased audience interest as those runs are repeated over six years and gradually lose their revenue-producing value). Those 200 episodes are scheduled as "strip" programming—the same hour every day Monday through Friday—to build audience support through the year.   ■

The smaller the market, the greater the chance that live production could exceed the cost of leasing syndication. In a medium-sized market, both the costs and the drawing power of live production and film might be somewhat comparable. Here, the rental fee for a syndicated half-hour might equal the investment in a local production of a half-hour studio program. If that local origination is part of a regular news staff's output, in addition to putting together daily newscasts, then equipment and facilities overhead and salaries are already covered (and excerpts of some material prepared for the special program might be suitable for regular newscasts as well). The costs for additional equipment, part- or overtime personnel or nonstaff (e.g., musicians) and out-of-pocket expenses for sets, videotape, and graphic materials would be far less than for purchasing syndication rights.

■   A Los Angeles independent station once preferred to pay a small franchise fee for the concept and title *Bowling for Dollars,* with local participants taped on a lane built at one end of the studio by Brunswick *gratis* (a trade-out, with its logo prominent) rather than invest big bucks long-term in an uncertain syndication package that did not enjoy the ease of cancellation or replacement if subsequent competitive factors dictated. Station management found the local strip series popular and flexible for their week-to-week needs with little investment. It wasn't drama or documentary, but at least it was local—and very cost effective. The converse, of course, is that local production may limp in audience attractiveness while most properties "in the can" have proved performance in previous network and syndication exposure.   ■

---

[*]Typically 10% is paid immediately when rights are contracted, with the rest paid over three years starting when the episodes are first aired. This permits longer range planning; a station can "warehouse" some packages without paying for them until they are scheduled and begin to produce sales revenue.

Expenses of local production in proportion to film costs can be kept under better control by stations in major markets. Even here, a decision between local live shows and film programming is difficult because of the risk involved in popular acceptance of the local production when the popularity of film is to some degree predictable.

Purchase of a large off-network or other syndicated package commits the station to a multimillion dollar investment over more than half a decade—during which time audience and advertiser interests may change, or competitive program scheduling on other stations might upset the availability of audience. On the other hand, a locally produced low-cost series can be mounted for a trial thirteen weeks to determine audience acceptance. It can then be continued or canceled, depending on its initial success; there is no long-term investment in shelves of expensive film properties.

■   The *Bowling for Dollars* program concept mentioned above was franchised by its East coast originators for $50 a program. The production was mounted in Los Angeles in 1974 for $300 a program by taping a week's five shows successively on one evening a week—several hours of work a week for a single production crew already on payroll. Another audience-interview format cost that Los Angeles station about $600 a day to produce—compared with costs from $3,000 to $5,000 a day for an off-network entertainment series programmed by the same station. In both those cited instances, the ratings quickly rose to 5, then to 6, 8, and 9 (enormously successful for early evening "fringe time"), to 11 and finally to 12—far outdistancing the high-investment syndicated re-run packages on competing stations.   ■

Planning for one to three years down the road, a PD with the GM selects syndicated programs available which will attract local audiences.

■   For the 1995-1996 season, beyond already circulating syndicated programs, major *new* first-run program series included 13 talk shows (all but two an hour long), 8 reality-based dramas (all half-hour episodes except for two), 2 game shows (half hours), 19 first-run weekly programs (all but three were half hours), 11 action-adventure or drama series (all hour-long except one), and 27 programs directed to children and teens (all half-hours except two 120-minute programs). All newly introduced series, they had no track record for a PD to check. Less chancy were familiar off-network series becoming available for the first time: eleven half-hour series and six hour-long series. Available for straight barter were 67 of the first-run series and five of the off-network series. The rest could be leased by stations for straight cash (eight first-run packages, two off-network) or cash plus 1 to 9 commercial spots—typically 5—filled by the distributor (eight first-run and eight off-network series). One off-network series was the hugely popular *Seinfeld* sitcom; earlier episodes brought near-record license fees (similar to the *Cosby* sitcom) and local stations found them attracting strong audiences with prime demographics—justifying the initial outlay. *Home Improvement* cost stations high fees and also performed very well for them. Available off-network for 1996–1997 were *Mad About You, Friends,* and *Frasier* which won big national audiences and critical acclaim.   ■

Station managers had to project astutely what audiences those shows would win in their local markets before deciding to invest enormous sums for syndication rights. Costs of rights rose with the initial success of network series, and went higher when several stations in a market bid for the hottest properties. The entry of cable networks into leasing off-network hits increased competitive pressure and further boosted leasing rights. (Each episode of the sitcom *Mad About You* was expected to generate about $2 million in cash license fees plus barter revenue; 30-second national spots distributed with the program went for $100,000 each in 1996–1997.)

A further risk, of course, is that a series that proves unsuccessful in a local schedule destroys that time period for the station. The audience shifts viewing to competitive channels. Subsequent efforts at rescheduling the time period demand enormous promotion for the replacement series, to win back the audience. Meanwhile there is the audience-flow problem; programs following that time period suffer from a small inherited "lead-in" audience. The risks are great in any programming strategy, whether scheduling locally produced or syndicated properties.

Theatrical feature film production, as noted earlier, has not kept pace with the use of those features by television and cable. The limited supply for strong demand caused prices to skyrocket. Costs for leasing rights to film features increased heavily over past decades in large markets, more moderately in medium-sized and small markets.

"Paid" Programming

A genre growing in the 1990s was complete programs promoting products and services, which were placed in entire time segments (usually between 1 A.M. and 6 A.M.) bought by the program supplier. Some referred to them as "infomercials," such as those demonstrating house products or touting real estate or get-rich-quick investment plans.

■ A limited survey reported 300 station responses in 1991: nine out of ten carried paid programs, but half also said they were "bad" for the industry, one-third were neutral, while 6% judged them beneficial (at least as a revenue source). John Rohrbeck, GM of NBC-owned KNBC in Los Angeles, said stations in that market were offered $10,000 for early morning half-hours on weekdays, and up to $40,000 per program hour Saturday or Sunday early morning or daytime. But he rejected them:

> I am not a high-brow zealot, but it is a simple fact that [paid] programs are limited in their audience appeal and can have overall negative impact on our programming flow—lead-in and lead-out programming. I would like to think that most broadcasters maximize their audience and not exchange [it] for a few extra quick bucks.[67] ■

Managers whose stations do carry paid programs must be wary of misleading or noncredible claims in "infomercials." The president of CBS O&O stations said their general managers were required to view any paid programming they might accept. An independent syndicator urged stations to air graphics with disclaimers (identifying a paid program-length commercial) every five minutes during such programs. Stations regularly air such notices prior to and at the end of what the FCC termed "program length commercials."

There is a distinct advantage in involving the general manager, program manager, sales manager, and head of financial affairs in decisions about all syndication series and motion picture packages bought for the station. Each of these administrators has different concerns about what is selected. Interacting with their ideas at this point in a group-decision atmosphere can pay dividends to the station.

C. Cable Program Service

What began in 1948 in eastern Pennsylvania as a wired hilltop antenna to extend TV station signals down into valley homes with poor or no over-air reception, expanded into community antenna (CATV) operations and eventually into local cable systems. As described in chapter 1, national interconnection of cable operators via satellite—initiated by Time, Inc./HBO in 1975 and Turner's WTBS superstation in 1976—blew the lid off cable expansion, which rapidly grew to 11,000 local operations, most owned by regional and national multiple system operators (MSOs). This part of the chapter focuses on local cable operations; national cable program services and "network" channels are covered in chapter 8.

HISTORICAL BACKGROUND

As cable operators obtained other program material, they gradually increased the range of viewing choices. But the few thousand isolated cable operators depended on retransmission of TV station signals (brought in by master antenna and microwave relay) or a narrow range of cheaply available programs (circulated on film/videotape reels via express freight) on otherwise unused channels. The object was to earn enough in monthly subscriptions to pay for the heavy initial capital outlay of elaborate receiving antennas at the cable headend, distribution equipment, and miles of coaxial cable (including recurring right-of-way charges by ground or pole).

The 700 local community antenna TV operations in 1961 grew during the next decade to 2,750 local systems linking 6 million homes.

With national program creators and distributors eager to get their wares onto the new wired medium, instantaneous high-quality national feeds by satellite in the late-1970s broke open the supply/demand impasse. The local cable operator could choose among dozens of program suppliers, many establishing distinctive program services for specialized interests. Unlike full-service national broadcast networks, cable programmers built individualized program streams such as sports (ESPN/Entertainment & Sports Network), news (CNN/Cable News Network), ethnic (BET/Black Entertainment Television), rock music (MTV/Music Television), and premium-fee services free of advertising such as uninterrupted movies and specials (HBO/Home Box Office, Showtime). Competing program sources multiplied to more than 100 by 1992, including seven so-called superstations—local TV station schedules relayed by satellite to hundreds of local cable systems and tens of millions of potential viewers. Local operators could pick and choose, negotiating cost-per-subscriber carriage fees: 15 cents per month per subscriber for one service, pitted against a more widely popular program channel at 25 cents per month; a new, untried, or narrow specialized service might charge a cable system only several cents a month per subscriber. As the variety of program services expanded, consumers found large-screen TV sets with remote units gave them more flexibility and ease of selecting. As home videotape recorders proliferated exponentially, consumers could "timeshift" by taping shows for later replay—watching what and when they wanted, even skipping past commercials.

Cable changed the face of the TV screen in sharpness of picture and range of choices. Remote dialing devices helped cable program watching become almost interactive as viewers explored choices available first on 12 channels, then 36 (available on 80% of all systems by 1990), and eventually 54 to 108 addressable channels in some communities.* By the late-1990s, 500-channel systems became possible through digital compression and fiber-optic's enhanced properties. Time Warner's cable system in Queens used coaxial cable and fiber optics to expand from 75 to 150 channels (including 40 pay-per-view); digital compression offers the possibility of 600 to 1,200 channels. The arrival of direct-to-home broadcast satellite (DBS) and high definition television (HDTV) promised to contribute to or compete with cable systems and over-air broadcast stations with enhanced picture quality. Meanwhile wireless cable or MMDS—multichannel multipoint distribution systems—strove to complement or compete with local cable companies; they transmitted low-wattage microwave signals of 25 or more program channels to subscribers' special antennas connected to household downconverters and descramblers.

■ **Local Franchises/Systems.** Local governments typically grant one franchisee per area the right to establish a cable system. Major markets may divide the city into sectors to be cabled separately by two or more systems. Although loosened regulation permitted "overbuilding" of an established system by a new cable operator, huge costs of laying cable make competing systems unfeasible. Only 100 cable systems have direct competition for subscribers; the rest of the 11,000 systems enjoy a virtual monopoly in their service areas (which is why the FCC in 1993 ruled rollbacks for mounting basic cable fees). But this began to change with government rulings permitting entry of telephone companies into providing information services by wire, especially with sophisticated fiber optic capability. ■

Local cable staffs consist primarily of business and technical personnel, with few if any creative staff for programming or news. Medium and large cable systems may put a person in charge of programming, including local origination, public access, on-air promotion and printed program guides, traffic, and commercial production; smaller systems combine these duties with sales and marketing.[68]

Tele-Communications, Inc., (TCI) the largest MSO, centrally programs all its systems. The next largest, ATC (American Television and Communications Corporation, now merged into Time Warner Communications), lets each local system decide its own programming. Major programming responsibilities involve selecting over-air signals to retransmit and contracting with satellite program services for each available cable channel, plus originating any local programs. Highly specialized or esoteric program services do not produce the audiences or revenues generated by more widely popular programming—the same bottom-line factors as in over-air broadcasting. However, ad hoc local/regional sports feeds can produce income beyond basic subscription rates. In 1992 professional sports teams in twelve major cities were on pay-per-view (PPV) cable; cable systems scheduled from 11 to 50 games a season at $4.95 to $13.95 per game (depending on the sport and the team).

Because of continual "churn" or turnover in cable subscribers that provide a system's lifeblood of monthly subscriber revenues, marketing and research play important roles in selecting what each channel will carry so the system's program service is attractive to as many as possible. That is why managers and directors

---

*Addressable systems permit the headend to send one-way signals of customized program channels to converter units of individual subscribers; otherwise, signals of premium and pay-per-view programs must be intercepted by mechanical "traps" outside nonsubscribing households. Two-way customized signals, mutually addressable, permit interactive data transmission between individual subscribers and the headend, making possible various kinds of electronic data services.

of sales or marketing plan programming at many independent systems; corporate administrators of MSOs often dictate most of what their various systems carry. Economies of scale make central negotiating efficient; for example, Cablevision, among the ten largest MSOs, contracted to carry the Disney Channel on all its systems for a minimum of $75 million over ten years (1985-1995).

Local systems offer up to three dozen or more channels, most of them "basic" for a monthly fee, others "premium" or "pay" (without commercial advertising) for an additional fee each. The typical headend operation uses satellite dishes to receive most of the program services, and master antenna or microwave to receive local and regional TV signals. Some originate programming in their modest building (some as small as 200 square feet), often automated weather or news "crawls" or program listings.

Local origination channels differ from public access channels in that cable operators exercise editorial control over local content. Regulations prohibit such control over content of community access, except (as of 1996) for content that cable managers judge indecent. Cable systems originate local sports or information, usually from suburbs and neighborhoods or surrounding municipalities that receive little coverage from metro TV stations. Cable operations reaching only a portion of a large market can serve subscribers with news specifically pertinent to them, unlike full-service TV stations covering the total metro area. But splinter audiences of local/public access channels attract little advertiser support and challenge the feasibility of program production for severely limited viewership.

A strategic public relations value of originating local programming involving community members is to appease local government bodies or subscribers critical of poor maintenance service and increasing monthly fees (see later section on customer assessment). Aggressive cable systems have programmed local sports, arts and entertainment, comedy, and newscasts (sometimes in conjunction with over-air TV stations, as noted earlier). In 1986 Rainbow Programming Enterprises on Long Island created the first twenty-four-hour local cable news service; supported by advertising, its start-up costs of $6 million included a staff of 100 to cover exclusively Long Island news. Some systems share programs within an MSO, as Cox Cable does among 62 of its local cable operations. As noted earlier in this chapter, some metro cable systems collaborate with local TV stations to re-run TV newscasts or for special feeds of unaired news material; KHQ in Spokane, Washington, provided Cox Cable with five-minute news segments inserted every hour into CNN Headline News service. Between 1991 and 1996 regional twenty-four-hour all-news cable channels started up in fifteen areas of the country, reaching 12 million subscribers. While varying in their service, they often teamed up with local newspapers or broadcast stations' TV news departments.

CABLE CHANNELS

A local cable system offers three tiers of service: *basic* cable—perhaps two dozen or more advertising-supported channels for a monthly fee; *premium* channels—additional monthly fee for each channel of unedited, commercial-free movies and specials (scrambled for nonsubscribers on the system); and *pay* channels—pay-per-view specialized programs and ad hoc networks for regional sports, and others. (The following chapter discusses national program services for cable.) Roughly half of all subscribers to basic cable add one or more premium channels. Only about 10% of subscribers make use of special pay programming. Almost 60% of an urban cable system's revenue comes from basic subscriber fees, more than one-third from premium/pay channel fees. By far the greatest expense (after capital invested in laying cable) is for programming, almost one-third of the annual budget; the other two-thirds is divided among technical operations, sales and marketing, and administrative and general expenses.

As with over-air broadcast strategy, cable operators seek a balance in kinds of programs scheduled on their channels. This helps attract various segments of the total community. Most subscribers find material they like on some channels. Another motive is to sustain a positive image for the cable system. Susan Tyler Eastman, researcher-author of media program strategies, has written:

> One way to balance a system is to bury controversial programming and create an overall image of responsible community service. Consumer promotional materials, for example, can stress the family-oriented content of many channels, even though the system contains adult-oriented programming on other channels. Cultural channels are often marketed more for their balancing effect than for any audience lift they create. News, public affairs and community access channels have a positive image effect. A further strategy, adopted by the industry as a whole, has been to locate adult programming only on pay tiers, which makes good economic as well as political sense because so many people are willing to pay extra for adult fare.[69]

■   In Chicago, TCI and Continental both carried to the 650,000 homes on their two systems a 24-hour news channel programmed in part by staffs of the *Chicago Tribune* with WGN. Cablevision Systems Corporation took substantive steps in negotiating its franchise with the city of Boston. In addition to offering cable to low-income residents at half the regular subscriber price, it earmarked $250,000 for job training (over five years) and hiring of minorities and another $750,000 to set up a studio for access programs; it was to sustain its 300+ employees at 60% city residents, and 30% each of women and minority persons. In return the franchise fee to the city was reduced from 3% to 1.5% of revenue during the four years of half-price subscriptions.[70]   ■

Another challenge to cable programmers is assigning program services on the channel dial. One tactic is to position VHF stations on channel numbers corresponding to their over-air channel so audiences can benefit from on-air promotion and print ads featuring logos and channel numbers. Local UHF stations may be slotted the same way or else put on lower cable channels to cluster them on equal footing among VHF competitors. Some cable systems attempt to group related program services near one another on channels. Because local listings vary widely, except for some major MSOs that coordinate channel assignments on their systems, national promotion must be by program service and not by channel placement. (Traditional networks always faced that dilemma, each with over 200 VHF and UHF affiliates on variously numbered local channels; but with only three network services for decades, audiences easily located them on their TV dial, unlike the many dozens of channels scattered across cable dials.)

Customer Assessment

After contracting franchises, constructing headends, and installing cable trunklines to pass a market's homes, cable operators stress marketing and promotion. In 1988 many cable systems penetrated only 30% of households in their market; but in the early 1990s most had grown beyond the 40%–50% level needed to be profitable, with over 62% as the national average by 1996. Cable managers had to continually promote attractive program channels to gain new subscribers, to replace those who dropped the service, and to expand their revenue base. A study found only one of four cable subscribers canceled because dissatisfied with programming; most dropped cable because of poor service from repair or salespersons or simply because of the cost. Another study reported half of subscribers judged cable an "excellent" or "good" value—the diversity and quality of programs for the monthly charge—with only 13% rating cable a "poor" value.[71]

Unlike the three traditional networks plus Fox and PBS which attempt to reach broad cross-sections of the mass public, cable channels provide twenty-four-hour access to various kinds of programming preferred by different viewers, such as sports (ESPN), music videos (MTV), culture (A&E), information (C-SPAN), news (CNN), nostalgia (Nickelodeon), exploration and nature (Discovery), country music and variety (TNN), and movies (TNT, HBO, Showtime). Cable's specialized channels follow the format concept of a market's multiple individualized radio stations. Many people subscribe in order to receive specific channels of niche programming unavailable from over-air stations or video rentals. With abundant channel resources, programming directed to children has found a place in cable schedules. Popular have been Nickelodeon on basic cable, and the premium Disney channel, plus old movies and cartoons on other program services.

All three local forms of electronic mass media—especially television and cable, less so radio—depend on networks and syndicators for part or most of their program service. The following chapter looks at those national suppliers of entertainment and information, with whom station and cable managers do business.

CHAPTER 7 NOTES

1. Adapted from Michael Langevin, "Using High-Performance Air Talent Motivation," *[NAB] RadioWeek,* 10 June 1991, 8.

2. Sydney W. Head, "A Framework for Programming Strategies," Chapter 1 in Susan Tyler Eastman, Sydney W. Head, and Lewis Klein, *Broadcast/Cable Programming: Strategies and Practices,* 3rd ed. (Belmont, Calif.: Wadsworth, 1989), 4–43.

3. Noted by Jim Tice, program director and operations manager, WZZK-AM-FM, Birmingham, Alabama, December 12, 1991, speaking to visiting college students majoring in broadcasting.

4. "Program Directors: Getting Back in Decision-Making Loop," *Broadcasting,* 30 January 1989, 39–40. The National Association of Television Program Executives (NATPE) began in 1963 when five PDs met in New York; by the 1990s NATPE conventions include as many station managers and salespeople as program directors.

5. Comments in this paragraph made by management of WVTM (NBC affiliate) in Birmingham, Alabama, in interviews and group presentations November 29, 1989, and December 12, 1991.

6. "It's Back to the Tried and True for Top-40 Radio," *Broadcasting,* 29 January 1973, 41–50.

7. Richard M. Mall, "The Place of Programming Philosophy in Competitive Radio Today," *Journal of Broadcasting* 1 (Winter 1956–1957): 21–32.

8. Donald West, "On the Leading Edge of Broadcasting—Special Report: Radio '71," *Broadcasting,* 21 June 1971, 41–80.

9. Rick Sklar, "Programming for Excellence in Radio," in Research and Planning Department, *Radio in Search of Excellence: Lessons from America's Best-Run Radio Stations* (Washington, D.C.: National Association of Broadcasters, 1985), 64–65.

10. Sammy Parker, "Monday Memo," *Broadcasting,* 4 March 1991, 10.

11. Dean Burch, address before the International Radio and Television Society News-makers Luncheon, New York City, September 14, 1973. (Copy from FCC, #06608). Prominent communications lawyer and former president of the Federal Communications Bar Association, Marcus Cohn, argued strongly against any such restrictions as ultimately inhibiting (see chapter 11 on regulation). Acknowledging complications in such rulemaking, FCC Chairman Burch recommended that within those categories percentage *ranges* should be established based on market size and station strength (in revenue, audience, and so on); he would apply such criteria to television only, and only to VHF stations whether network affiliates or independents.

12. "Gutsy 'VOX' of New York's Westchester County Gets a Proper Home," *Broadcast Management/Engineering,* December 1973.

13. "Growing Public Interest in Community Affairs Sparks Local TV Shows," "The Gallagher Report," 5 February 1973, p. 2; "Boston's WCVB-TV Sets 51 Hours of Local Production a Week Including New Morning Strip, Six Access Series," *Variety,* 22 August 1973, p. 28; "The Happy Hookup," *Broadcasting,* 24 June 1974, 32; "WCVB-TV: The Trend in Post-Production?" *Broadcast Management/Engineering,* October 1981, 35–40; "The 'Can-Do' Style of Robert Bennett," *Broadcasting,* 11 April 1983, 134–143.

14. "Grabbing the Grazers in a Crowded Field," *Broadcasting,* 1 February 1993, 14.

15. James H. Duncan, Jr., "American Radio: Beyond Excellence Toward Perfection," in NAB, *Radio in Search of Excellence* (1985), 148.

16. Speech reprinted as "Viewpoints," *Broadcasting,* 13 April 1987, 57.

17. Michael L. McKean and Vernon A. Stone, "Why Stations Don't Do News," *RTNDA Communicator,* June 1991, 22–24.

18. "News Makes Money in TV, Pays Its Way in Radio," *RTNDA Communicator,* April 1991, 5, 32.

19. So noted by news directors and broadcast managers at a four-day seminar conducted by the International Radio and Television Society in Tarrytown, New York, February 11-15, 1974. The following comments were made at the same conference by experienced broadcasters from large and small markets.

20. George Lyons, station manager, WZZM-TV, Grand Rapids, Mich. He spoke at a seminar conducted by the International Radio and Television Society in Tarrytown, New York, February 11-15, 1974.

21. Quoted in "WCCO G.M. Ron Handberg Calls It Quits; Veteran Newsman Deplores Belt-tightening," *Variety,* 12–18 July 1989, 43.

22. Lou Prato, "Bob Morse Interview: Management in the 1990s, *RTNDA Communicator,* April 1991, 16.

23. Vernon A. Stone, "Downsizing Myths: What the Numbers *Really* Show," *RTNDA Communicator,* April 1992, 32–34. Other data in this paragraph are from Vernon A. Stone's analyses of annual surveys in RTNDA's *Communicator:* "Cutbacks Strike Again," April 1993, 32–33; "TV News Work Force Grows, Declines Continue in Radio," May 1993, 26–27; "TV News Work Force Grows," April 1994, 20–21; "Few Budget Cuts for TV News," May 1994, 28–29; plus Bob Papper, Michael Gerhard, and Andrew Sharma, "More News, More Jobs," *RTNDA Communicator,* June 1996, 20–28.

24. Bob Papper, Andrew Sharma, and Michael Gerhard, "News Department: Producing Profits," *RTNDA Communicator,* April 1996, 18–26.

25. See Roger LeGrand (vice-president, general manager, WITI-TV, Milwaukee), "Management—News, Public Affairs, Editorializing," *Broadcast Management* (Washington, D.C.: Association for Professional Broadcasting Education, 1973), 111–118.

The debate continues whether cross-ownership among several media in the same market inhibits journalistic autonomy by undue concentration of economic and, by implication, news "power." Studies have noted that multimedia ownership tends to have a better record of locally originated programming, including news, than do separately-owned stations. See: James A. Anderson, "The Alliance of Broadcast Stations and Newspapers: The Problem of Information Control," *Journal of Broadcasting,* 16 (Winter 1971–1972):

51–64; Paul W. Cherington, Leon V. Hirsch, and Robert Brandwein, eds., *Television Station Ownership: A Case Study of Federal Regulation* (New York: Hastings House, 1971). Both the report and the book were prepared for and/or by broadcasters as background presentations to the FCC, with conclusions favoring the broadcast industry's status of multiownership by group owners and by multimedia, cross-channel owners. See: "Old Arguments, New Impetus for Action on Crossownership," *Broadcasting,* 29 July 1974, 16–17. In 1975 the FCC concluded that sixteen single-newspaper small markets with broadcast properties owned by those newspapers had to divest. Major markets with multiple print and broadcast companies were judged not to constitute a threat of monopoly control of media information or economics, so cross-ownership was permitted to continue; but no new acquisitions of cross-ownerships between newspapers and radio or television stations would be approved. "FCC At Last Defines Policy on Broadcast and Newspaper Ownership," *Broadcasting,* 3 February 1975, 23–26. That policy unraveled as the 1990s began, with newspaper-station cross-ownership once again permitted.

26. Obviously, some monitoring helped ensure serious use of the one-fifth released time from studio duty; initial results were sometimes uneven, but overall quality and quantity of news stories by the entire news staff increased over months.

27. So quoted in a full-page advertisement consisting entirely of McLuhan's analysis, by ABC Owned Television Stations, in *Variety,* 25 August 1971, 35. On July 21, 1975, the daily production log (not the official program log) for a major TV station in Detroit listed at five points during the half hour line-up of news items the symptomatic directive "bs"; that cryptic cue matched multiperson camera shots of the on-air news team in spontaneous conversation.

28. Emerson Stone, "Marrying Entertainment with News," *RTNDA Communicator,* September 1992, 32; cited in *RTNDA Communicator,* January 1993, index.

29. Ernie Kreiling, "Hollywood Television Report," 31 December 1973, 1; "Editorializing Falling Out of Favor?" *Broadcasting,* 6 February 1989, 45.

30. "More Work for Stations' Spokespeople as Editorialists Take on New Chores," *Variety,* 12-18 July 1989, 43.

31. That figure was cited by the House Legislative Oversight Subcommittee; quoted by Steve Millard, "Special Report: Radio at 50—An Endless Search for Infinite Variety," *Broadcasting,* 16 October 1972, 34.

32. Beyond standard textbooks, useful descriptive summaries of various formats may be found in: Edd Routt, James B. McGrath, and Fredric A. Weiss, *The Radio Format Conundrum* (New York: Hastings House, 1978); Ed Shane, *Programming Dynamics: Radio's Management Guide* (Overland Park, Kansas: Globecom, 1984); Michael C. Keith, *Radio Programming: Consultancy and Formatics* (Boston: Focal Press, 1987); Susan Tyler Eastman, Sydney W. Head, and Lewis Klein, *Broadcast/Cable Programming: Strategies and Practices,* 3rd ed. (Belmont, Calif.: Wadsworth, 1989); Lewis B. O'Donnell, Carl Hausman, and Philip Benoit, *Radio Station Operations: Management and Employee Perspectives* (Belmont, Calif.: Wadsworth, 1989), Chapter 3, "Formats and Programming"; David T. MacFarland, *Contemporary Radio Programming Strategies* (Hillsdale, N.J.: Lawrence Erlbaum Associates, 1990); Robert E. Balon, *Radio in the '90s: Audience, Promotion and Marketing Strategies* (Washington, D.C.: National Association of Broadcasters, 1991); and Michael C. Keith and Joseph M. Krause, *The Radio Station,* 3rd ed. (Boston: Focal Press, 1993); Raymond L. Carroll and Donald M. Davis, *Electronic Media Programming: Strategies and Decision Making* (New York: McGraw-Hill, 1993).

33. Radio & Records, *Ratings Report & Directory,* Special Supplement, vol. 1, 1996; national rankings, p. 6; regional rankings, p. 8.

34. During the two decades between 1971 and 1992, books on mass media analyzed characteristics of major formats and their relative popularity, either by shares of total audience or by number of stations airing them. Routt, McGrath, and Weiss (1978) commented on 22 musical formats plus 7 talk/religious/educational/variety, etc.; Keith (1987, reproducing *Radio&Records* data) listed 17 genres; Keith and Krause (1989) listed 12; O'Donnell, Hausman, and Benoit (1989) noted 8 major and 7 more specialized formats; Eastman,

Head, and Klein (1989) cluster them under 6 headings; McFarland (1990) categorized 6 broad areas; Pringle, Starr, and McCavitt (1991) discuss 12 music formats, 3 more under information, plus 3 under specialty formats.

35. "Format Track #8," *Radio Business Report* (Report 303), 18 November 1991, 14.

36. "Following the Formats," *Broadcasting,* 24 September 1990, 66.

37. The Arbitron Company and *Billboard* (Spring 1991), quoted by National Association of Broadcasters, *Broadcast Marketing & Technology News* (Washington, D.C.: NAB), October 1991, 1.

38. "AOR: In Search of a Definition," *Broadcasting,* 14 April 1991, 72.

39. "Country Extends Its Lead in the Format Field," *Broadcasting,* 24 August 1992, 20; data were cited by Katz Radio Group Analysis of Spring 1992 survey data by Arbitron for MSAs (metropolitan statistical areas), persons 12+, total week.

40. "Radio: Interep Study Pushed Value of Urban Format," *Broadcasting,* 26 August 1991, 29.

41. Balon (1990), 65–66; emphasis added. In Chapter 4, "Marketing in the '90s," 59–82, he offers concrete proposals for marketing tactics to achieve identity, parity, and eventually dominance in a radio market.

42. Data from Radar Volume 1, 1972-1989, Statistical Research, Inc. and by Arbitron, Spring 1990; cited by Austin J. McLean, *RadiOutlook II: New Forces Shaping the Industry* (Washington, D.C.: National Association of Broadcasters, 1991), 64–65.

43. "While Improving Technology is Important, in AM the Programming's the Thing," *Broadcasting,* 9 October 1989, 66.

44. Data from National Association of Broadcasters and Broadcast Cable Financial Management Association, *1990 Radio Financial Report* (Washington, D.C.: NAB, 1990), viii. Because fewer than one in three radio stations in the United States responded to this survey, Price Waterhouse used statistical procedures of weighting and imputation to arrive at representative figures for nonreporting stations as well.

45. *Ibid.,* data excerpted from tables, pp. 85–115.

46. Briefing sessions by department heads at WZZK-AM-FM, Birmingham, Alabama, November 29, 1989, and December 12, 1991.

47. "Radio '89: A Tale of Five Cities," *Broadcasting,* 11 September 1989, 75–88 (cited pp. 75–76).

48. *Broadcasting Yearbook 1974* (Washington, D.C.: Broadcasting Publications, Inc., 1974), 70.

49. Les Brown, *Televi$ion: The Business Behind the Box* (New York: Harcourt, Brace, Jovanovich, Inc., 1971), 178.

50. Fox offered over eight hours of daily feeds in 1993, with news footage supplied by 34 Fox owned or affiliated stations, plus 160 non-Fox stations around the country, and free-lance stringers in other cities. "Fox Expands News Sources," *Broadcasting,* 5 October 1992, 27.

51. "Julie Nunnari: Getting a Fair Bite of the Big Apple," *Television/Radio Age* (n.d., 1988), 32–36.

52. Quoted by J. Max Robins and John Flinn, "Local Television Under Fire," *Channels: The Business of Communications—Field Guide 1989,* December 1988, 10–22, 14.

53. "TV Production on a Local Level," *Broadcasting,* 25 March 1985, 50. WCVB's *Chronicle* series cost $1.5 million a year to produce in the mid-1980s; it had a staff of twenty-five, three camera crews, nine field producers, and four fulltime on-air personalities. Still, it made money, and was second in its time period. Compare that with KMBC-TV in Kansas City, Missouri, which budgeted under $70,000 annually for several specials and series of locally produced programs.

54. Mike Freeman, "CBS's Karen Miller On the Marketplace," *Broadcasting,* 18 January 1993, 61.

55. Quoted by Merrill Brown and John Flinn, "9 Rules for the '90s," *Channels: The Business of Communications—Field Guide 1990,* December 1989, 14–20, 18.

56. Quoted by Steve McClellan, "Sinclair Makes News Moves," *Broadcasting & Cable,* 19 August 1996, 33–34. Phoenix data from Joe Flint, "Local News Gets New Lease on Life," *Variety,* 10–16 October 1994, 29, 33.

57. "Local Programming with a Flair," *Broadcasting,* 21 January 1985, 58.

58. "Broadcasters Demonstrate Impact of Local Programming: Even Low-Budget Productions Can be Socially Responsible and Profitable," *Broadcasting,* 30 January 1989, 37.

59. "In the Picture," *Television/Radio Age,* 17 August 1987, 79.

60. Quoted in "NBC O&O's Make Major Push in Local Community Programming," *Variety,* 4 June 1986, 106.

61. "WNYW Outfoxes the Big Boys with Strong Dose of Localism, Original Programming," *Variety,* 20–26 September 1989, 119. "Profit Begins at Home," *Broadcasting,* 22 April 1991, 31.

62. Kathy Clayton, "Denver Affiliate Lays Off 9 Staffers," *Electronic Media,* 6 February 1989, 2.

63. National Association of Broadcasters, "Research & Planning: Broadcast Marketing & Technology News," 5 (December 1988): 9.

64. Station examples cited from "'Mainstream' Television is Still Changing Course," *Broadcasting,* 10 December 1990, 74–76.

65. "Tribune Targets Minority Audience," *Broadcasting,* 9 March 1992, 26.

66. *Broadcasting & Cable,* 22 January 1996, 18.

67. "NATPE Debates Paid Programming," *Broadcasting,* 21 January 1991, 39.

68. For an organization chart see Thomas F. Baldwin and D. Stevens McVoy, *Cable Communication,* 2nd ed. (Englewood Cliffs, N.J.: Prentice Hall, 1988), 229. This book offers a thorough, readable description of cable operating structures, procedures, and services, including business and regulatory aspects.

69. Eastman, Head, and Klein, *Broadcast/Cable Programming,* 273.

70. "Cablevision Gives Low-income Discount as Part of New Boston Franchise Agreement," *Broadcasting,* 26 September 1988, 73.

71. The R. H. Bruskin study was summarized in "Identifying Cable's Demographics," *Broadcasting,* 3 October 1988, 63.

Managers and National Programming

*Broadcasting in the United States has always been more of
a delivery system for, rather than a producer of, mass
entertainment.*

DON LE DUC, BROADCAST EDUCATOR, LAWYER[1]

*It depends upon how [broadcasters] view their business. If they
see their business as emitting nonionizing radiation from the tops
of red and white towers, then they may be in trouble [from telco
entry with fiber-optics]. . . . But if they see their business as
producing, selecting, and packaging television programming
that attracts audiences to be delivered to advertisers, then they
may benefit from a universal broadband network that reaches
nearly all homes.*

ROBERT M. PEPPER, FCC[2]

This book focuses primarily on managing local radio and TV stations and cable systems. But because managers of those operations depend on national suppliers of programs and specialized program services, this chapter looks at the structures and processes of networks, national syndicators, and cable program services. Effective management is crucial for all those national services. However, they involve corporate managing which deserves a complete book in its own right, so we look at them only briefly. Our purpose is to sketch national systems that influence planning and budgeting at the local level.

Historical Trends   As described earlier, original, full-service radio networks gave way to highly localized formatted stations in the 1950s and 1960s, only to reemerge in 1968 with multiple specialized services to radio stations across the nation airing talk/information, top-40, rock, or country music. Syndicated services developed to serve local stations with programming tailored to their specific sound. Some syndicators merged with former networks, such as Watermark acquired by ABC, and Westwood One which bought Mutual and NBC radio networks.

After television began intercity networking in 1948, programming on "the tube" was really televised radio—a visual extension of radio. Music, comedy, and variety formats of popular radio shows were transposed to the new medium with little change; comics read from their scripts as they

Much of today's radio programming originates in local stations, supplemented by nationally syndicated format services. In contrast, most TV programming (except for local news) originates with West Coast production studios and is distributed to stations by networks, syndication companies, and satellite-fed cable program services. For the most part, networks dominated American television almost from the beginning—plus aggressive group-owners such as Westinghouse/Group W and WGN Continental Broadcasting (later renamed Tribune Broadcasting) which also developed distinctive programming. Despite competition from cable systems, satellite services, PBS, Fox and independent stations—which divide among their several dozen channels fewer than half of all viewers—the three major TV networks in 1996 still commanded more than half (53%) of the nation's home audiences in prime-time.

As recounted in chapter 1, the broadcast industry in the 1990s was swept up in swiftly shifting corporate structures and reconfigured program development and distribution systems. *Broadcasting & Cable* magazine editorialized in mid-1993 that "the entire telecommunications landscape—the multimedia landscape, they're starting to call it—is going to go through more upheaval in the next five years than in the past 25."[3]

A distinctive change in national program distribution arrived as three direct broadcast satellite (DBS) companies entered the competition in 1994: Hubbard's U.S. Satellite Broadcasting, Hughes DirecTV, and cable-MSO consortium's PrimeStar. These program services to DBS subscribers competed directly for audiences and advertising with networks and local cable, to a lesser extent with terrestrial over-air broadcast stations (DBS did not siphon off stations' local advertising revenue as cable systems did). On the other hand, local cable retransmitted regional TV station signals including their network programming, expanding their reach.

A. National Radio Programming

often in the titles (e.g., *Chesterfield Supper Club*). ABC's radio network survived and then thrived during TV's ascendancy by splitting on January 1, 1968, into first four, and later seven, separate network program feeds to distribute various kinds of program services suited to local stations with specialized formats of music or talk; American Contemporary (227 stations), Entertainment (259), FM (180), and Information (368) networks were augmented by ABC's Direction, Rock, and Talkradio networks. Other networks subsequently followed that lead. To its basic CBS Radio Network (440 stations), CBS added CBS Radio Programs and RadioRadio, later merged into Spectrum (600), and CBS Hispanic Radio. NBC expanded with The Source (for young adults). In the late 1980s syndicator Westwood One bought MBS and NBC radio networks.*

Into the early 1980s radio networks continued traditional compensation payments to stations carrying their national programs supported by national advertising. Eventually this would change—in part to the barter system and to payment by stations to syndicator-networks for program services.

RADIO SYNDICATORS

In the 1960s, companies formed to syndicate program formats and station ID (identification) musical jingles. Their program services were tightly formatted with strict guidelines for performers. More music, less talk between songs, and fewer commercials—at higher rates, and in clusters—became the hallmark of stations that subscribed to Drake-Chenault's trendsetting syndicated format service. The lower-keyed pace of on-air disc jockeys and the narrow-range play-lists of current hit songs eventually gave way in the 1970s to renewed emphasis on personalities who talked to and with their audiences. Whereas the 1960s stressed production, the 1970s again emphasized personalities. Even nationally syndicated radio packages featured personalities such as Wolfman Jack, Dick Clark, and Casey Kasem, plus "live-on-tape" rock concerts, and also comedy series. Talk/interview programs were hot going into the 1990s, especially Larry King, Rush Limbaugh, and Howard Stern. Expanding beyond the previous decades' "niche emphasis" of narrow demographic target listeners, these national stars attracted broad national audiences. Local stations paid syndicators for their programming, unlike network service.

Total radio revenues in 1992 approached $7 billion–$8 billion, about 75% of it in local advertising. Of that only about $400 million was in national network (ABC, CBS, Westwood One, and Unistar) and syndicated radio spots.

In 1996 radio revenues totaled about $10 billion, almost $500 million of that with networks. That year ABC's multiple radio networks served 2,900 affiliated stations, Westwood One over 1,000 stations, and CBS radio networks 485 stations.

---

*NBC Radio Networks employed 250 people and constituted 30% of the network marketplace at the time (ABC was 40%), with losses estimated at $10 million a year. The sale included a licensing agreement whereby Westwood One/Mutual used the name "NBC News" and the TV audio of NBC News, and controlled the news it programmed. See *Variety,* 22 July 1987, 49.

Satellite Distribution

Expensive long-line interconnections formerly leased from the telephone company were replaced by high quality signals at lower costs via satellite transponders serving the nation's stations. In 1981, 13 radio networks and syndicators transmitted programs via satellite; by the end of that decade 75 distributed programming over eight different satellites.

■ By 1993 compression technologies let more signals use a single transponder, resulting in 30%-50% drop in costs to networks and syndicators for satellite time; local radio managers found satellite receivers dropped in half from $10,000 in 1980 to about $5,000 in 1993. Some 4,000 radio stations were linked to Satcom C-5 in 1988, increasing to 6,000 stations by 1993. ■

Nationally syndicated radio services and the reemergence of radio networks via satellite provided programming for the whole range of music and talk formats by local stations around the country.

Most services offer various packages such as major day-part programming segments including features of talk, interviews, and news; music-only; or complete "turnkey" automated program schedules. This provides an efficient way to program a station with minimal expense, employing almost no staff. It also offers professional quality sound for remote small-market outlets, including "jingle" packages for station IDs and on-air promotion. Major syndicators invest in extensive research to track ever-changing audience tastes in music and performers which small stations often cannot afford to do effectively on their own. National program services carefully select music and rotation patterns to match trends and niche subgroups targeted by client stations. On the other hand, centrally distributed services can limit a station's local orientation. The unique marketing advantage of a radio station is its distinctive sound and close link with community people and events. That takes knowledgeable and involved local staffs—on-air personalities as well as salespersons and management.

After the FCC adopted its nonduplication rule for AM/FM combos in 1964, many AM signals passively carried automated program services imported from Dallas or Los Angeles while its dominant sister-FM station originated its own schedule. Since that rule was eliminated in 1986, many AM-FM combos simulcasted the same locally originated program service—often offering advertisers AM at discounted rates or as a bonus for purchasing spots on the FM facility. (This completely reversed previous decades when mighty AM ruled the radio roost and FM struggled for scattered listeners with little commercial time sold even at minimal rates.) By 1992 half the nation's radio stations subscribed to national tape or satellite program services, partly to cut costs (especially with barter arrangements) and partly because of station consolidation in the 1990s through local marketing agreements (LMAs) and expanded multiple ownership, including duopolies.

National Program Services

The difference between a national network and a nonnetwork national program service or syndicator became blurred as companies crossed over from one to the other. A practical distinction might be that networks pay compensation to affiliated stations for their local time; radio networks pay as much as $1 million to a highly rated station in a top market, but to others a fraction of that, sometimes even none. Nonnetworks typically sell their program services at negotiated prices, most often including barter agreements whereby they give stations some commercial time in the programs in exchange for their local airtime and audiences.[4]

At the top of the scale in 1995 were clever "bad-boy" Howard Stern, whose outlandish and often vulgar comments drew massive audiences and thus the highest cash charges to stations carrying him, and archly satirical and humor-laced conservative Rush Limbaugh whose popularity commanded fees that in some markets reached $100,000 plus part of each program's advertising time inventory. A hybrid is the half-century-plus Metropolitan Opera weekly broadcasts, first on networks and later on ad hoc syndicated networking. Started in 1931, the series of music plus commentary has been sponsored solely by Texaco since December 7, 1940; it continues today via satellite linking over 300 stations to the Texaco-Metropolitan Opera Radio Network with estimated 6 million to 10 million listeners weekly—made possible by more than $100 million from Texaco over those fifty-plus years.

A broad definition of "network" used by the FCC is "a national organization distributing programs for a substantial part of each broadcast day to . . . stations in all parts of the United States, generally via interconnection facilities."[5] The Radio Network Association limits members to those distributing programming to all forty-eight contiguous States: ABC, CBS, Westwood One, and Unistar. Statistical Research, Inc. conducts the only radio network audience surveys; its criteria include continuity of program service, written contracts with affiliates, ability to distribute to all affiliates instantly, and a system of clearance for determining what programs are carried by which affiliates. SRI's biannual RADAR surveys include only the multiple networks of ABC, CBS, Westwood One, Unistar, and American Urban Radio Networks.

■ More than fifty major specialized radio networks offered program services by satellite and tape, including news, talk, music, entertainment, religion, and education. Radio & Records, Inc. took seventeen pages of its *R&R Ratings Report & Directory for 1993* (volume 1) to list addresses and key personnel for 310 "Program Suppliers and Networks" offering programming to radio stations.

Radio news services in 1996 were provided by ABC, CBS, and Westwood One network/syndication feeds; American Urban Radio Networks; and Accu-Weather, Inc., Associated Press Broadcast Service, Bloomberg Information Radio, Metro Networks, Monitor Radio, Shadow Broadcast Services, UPI Radio Network; and by radio networks formed by Salem, USA, *Wall Street Journal,* and Weather Channel. The Radio Television News Directors Association (RTNDA) listed pages of news services available. ■

Station managers and program directors (PDs) contact these and other format-oriented satellite radio syndicators to negotiate charges for carrying program services to fill out their local schedules. Some services above distribute radio music formats by audiotape and compact discs—offering more flexibility and local control than satellite; they include major distributors BPI/Broadcast Programming, Inc. (which acquired Drake-Chenault in 1991) and Bonneville Broadcasting System.

Another 100 regional radio networks, many of them statewide, also link stations for programs about agriculture, recreation, news, and other special interests.

Networks Serving Minorities

In 1972 the Mutual Broadcasting System inaugurated the Mutual Black Network with sports, news, and feature services of interest primarily to the African

American population. MBS took over operation four years later until Sheridan Broadcasting absorbed it in the early 1980s; at that time its 89 affiliates reached potentially 17 million black listeners (70% of the black population), increasing to 137 AM and FM affiliate stations by 1991. In mid-1973, the National Black Network began serving black communities with news and sports coverage; 42 radio stations, many in the nation's largest markets, initially subscribed to the network, growing to 95 AM and FM affiliates by 1990. The Black Audio Network served four out of five black-oriented stations, 90 of them in major markets. In 1992 Sheridan and National Black Network merged into American Urban Radio Networks.

Hispanics are served in part by Spanish-language programming from Westwood One (concerts and interviews for its *Radio Español*), and UPI's *Nuestras Noticias* with 17 daily seven-minute newscasts on the hour to stations reaching 85% of U.S. Hispanics. CBS Hispanic Radio Network began in 1990; its entertainment and sports play-by-play served 35 affiliates by 1993; under the name of CBS Americas, by 1996 it reached 60 U.S. stations plus others in Mexico and Latin America but with so few affiliates it was dissolved that year.

PUBLIC NONCOMMERCIAL RADIO

National Public Radio, a noncommercial service carried by 460 public stations in 1993, was the first fulltime network to distribute programs exclusively by satellite (1980), followed by MBS a year later. Other networks began switching over to satellites as their contracts for terrestrial lines came up for renewal. In 1988 NPR moved away from centralized networking to more of a syndication service, "unbundling" its schedule and offering shows individually or in clusters so local stations could budget the ones most attractive to their audiences.[6] A coalition of five noncommercial station managers developed American Public Radio in 1982 as an alternate service to NPR—partly in reaction to NPR's high fees and its non-carriage of programs sought by some station managers. APR was renamed Public Radio International in 1994; it offered over 450 affiliates a schedule of mostly music plus news/talk, comedy, and variety programs developed by the stations and other sources. Other distributors of specialized noncommercial programs include the Public Radio Cooperative, Beethoven Satellite Network, Concert Music Consortium, the Longhorn Network, and WFMT Fine Arts Network. While public radio stations in 1991 originated local programming for over half (54%) of their schedules, they devoted one-third of their airtime to NPR (21%) and APR (13%) programs; WFMT provided 7%, Seaway 2%, with another 3% supplied by various individual stations.

■   NPR features widely admired, award-winning national news in uninterrupted news-magazine blocks of 90-120 minutes: *Morning Edition* weekday mornings, *Weekend Edition* on Saturday and Sunday mornings, and *All Things Considered* every late afternoon. Public station managers must specifically subscribe to each of them, in addition to a flat fee for the basic schedule of cultural programming such as recent concerts on *Performance Today,* humorous car-repair phone-in show *Car Talk,* daily interview *Fresh Air,* and others. NPR centrally produces most of its program service, similar to traditional commercial networks. Its counterpart PRI (formerly APR) instead distributes programs produced by local stations or independent sources. PRI charges a basic fee plus extra rates for each selected program such as Garrison Keillor's widely popular comedy-variety series *Prairie Home Companion* which after a decade became *Good Evening* and then *The American Radio Company of the Air.*   ■

Prevented by law from carrying direct advertising, NPR and PRI programming is funded partly by federal support through the Corporation for Public Broadcasting, partly by state or municipal or institutional subsidy, and especially by corporate grants and listeners' donations to local public stations. That funding covers expenses including payment of membership fees to receive network programming (public radio's fiscal operation is detailed in chapter 10).

NATIONAL RADIO AUDIENCES

Statistical Research, Inc. rated full-service commercial radio networks and syndicated programming networks with common day-parts in its "RADAR 46" survey for Fall 1992. The top 15 networks based on numbers of listeners 12 years and older per average quarter hour are ranked in table 8.1. Network names vary from earlier and subsequent lists, reflecting the continual churn in radio network configurations as they strive to customize specific program services for local stations' niche audiences across the country.

■   In 1996 RADAR 53 reported smaller total audiences for many networks, partly because they were drawn off to other newly formed ones. For example, ABC Prime attracted 3.8 million listeners 12 and older, down from 5.3 million in 1992; CBS Spectrum dropped from 2.4 million in 1992 to 1.7 million in 1996. New networks and their average quarter hour listeners in 1996 included Westwood One's Variety (2,124,000) and Country (907,000); CNN+ and Adult Contemporary and Young Adult (over 1 million each); and ABC Advantage (438,000). With their multiple networks, ABC's share of the national radio audience in 1996 was 46.5%; Westwood One 34.9%; CBS 14%; and American Urban Radio Networks 4.6%. Almost 22 million listeners aged twelve and older tuned in radio/programming networks during an average quarter hour in 1996; SRI estimated the weekly cumulative audience at 135 million—about two-thirds of all teenagers and adults.   ■

Audiences for individual program personalities far exceeded these full-schedule music and news services. Most popular were news-features veteran Paul Harvey, chat-host Larry King, and conservative commentator Rush Limbaugh. Limbaugh's audience was 300,000 (12+ AQH) when he began national syndication in 1988; by 1993 his 571-station audience numbered almost 4 million—surpassing Harvey who led all of radio for decades, with an estimated 6 million listeners a few years earlier.

After its first decade on the air, National Public Radio's *Morning Edition* in 1988 was drawing a daily audience of almost 5 million listeners, about twice that of the noncommercial network's older afternoon news magazine. NPR's entire schedule was estimated to reach cumulatively over 20% of the potential national radio audience.

Audiences mean subscriber-donors to public broadcasting, and can justify support by corporation grants as well as federal and state tax dollars. For commercial broadcasters, of course, audiences directly affect the value of time sold. Over a three-year period commercial radio network revenues grew by almost 17%, from an estimated $370.5 million in 1987 up to $432.5 million in 1990 (despite a sagging fourth quarter that year with a weakened economy coupled with war in the Persian Gulf). The total grew to about $500 million annually in the mid-1990s.

━ TABLE 8.1 ━

TOP 15 RADIO AND PROGRAMMING NETWORKS, 1992
BASED ON NUMBERS OF LISTENERS 12+
(AVERAGE QUARTER HOUR)

NETWORK	LISTENERS 12+ (AQH)
1. ABC Prime	5,340,000
2. Westwood Mutual	3,030,000
3. ABC Platinum	2,980,000
4. ABC Genesis	2,550,000
5. CBS Spectrum	2,420,000
6. Unistar Ultimate	2,320,000
7. Unistar Super	2,230,000
8. CBS Radio Network	1,790,000
9. Westwood/The Source	1,560,000
10. Westwood/NBC	1,410,000
11. W.O.N.E.	1,340,000
12. Unistar Power	1,310,000
13. ABC Excel	1,190,000
14. ABC Galaxy	1,080,000
15. American Urban	1,060,000

Source: RADAR 46 data reported in *Broadcasting,* 8 March 1993, 33.

B. NATIONAL TELEVISION PROGRAMMING

NETWORKS

CBS/Broadcast Group President Gene Jankowski sketched the history of TV networking in three phases. "Television One: Coming of Age" was the period of rapid growth from 1950 to 1976 with two dominant networks, leading to FCC regulations in the 1970s severely restricting them (prime-time access, financial interest and syndication, cross-ownership). "Television Two: The Territorial Imperative" in the latter 1970s found ABC fully competitive, adding with inflationary times to radically escalating costs in all phases of intense network rivalry in affiliations, talent, production, sales, and marketing. "Television Three: Action and Reaction" in the 1980s brought new media technologies (satellites, cable that also expanded UHF signals and syndication) and loosening of government restrictions (on maximum station ownership, etc.), which formed new structures and relationships in television broadcasting. Jankowski offered a graphic analogy:

A good symbol for this new order might be the 'double helix'—the famous model of the DNA molecule. It looks like a spiral staircase without the steps. We still have discrete industry segments—stations, networks, production companies, agencies, clients and so on at the core, but the functional elements of our business no longer relate themselves to these segments in the traditional manner.

Instead, those elements wrap themselves around the core, interfacing with it in different ways at different places. For example, VCRs affect commercial broadcasters one way, cable operators another and production companies yet

> another. So do various regulatory and legislative developments. And all of them play into the financial world differently. In these circumstances, competitive postures are not rigid or consistent. Instead, combination and recombination become the style.[7]

But the core remains the same for creating and delivering media product to mass audiences: there must be a reliable way to develop, produce, distribute, and fund media programming. In the 1990s traditional TV networks were still reliable for broadest mass impact.

> Network competition intensified as a fourth network began to emerge after Rupert Murdoch's News Corporation in 1986 bought six Metromedia stations and the 20th Century-Fox studios.
>
> ■   Fox's broadcasting division in 1987 linked its owned stations with 108 independents around the country, to program two prime-time hours on Sunday and then Saturday nights. It gradually expanded to Mondays in 1989, Thursday and Friday nights the following year, Wednesdays in 1992, and finally filled out the weekly schedule with Tuesday nights in June 1993. Fox Broadcasting's staff of 400 employees contrasted with five times that many at each of the other networks. Its line-up of 140 stations reached 95% of U.S. homes in 1993; in markets where it had no affiliates it added another 546,000 homes over 212 cable systems of Tele-Communications, Inc. In 1996 its affiliates numbered 211, many of them strong VHS stations that switched from the major networks. By stressing youth and young-adult oriented programming, plus risking innovations in concepts and content—sophisticated animated *The Simpsons,* risqué language and situations in *Married...with Children* and *In Living Color*—Fox drew strong audiences, including black and Hispanic minorities, and on some shows and on occasional evenings scored higher ratings than the established full-service networks.   ■
>
> The first TV network to serve the Spanish-speaking population was SIN/Spanish International Television, formed in 1962 by a consortium of thirteen UHF stations, including six owned by SIN. It was succeeded by U.S. Spanish-language TV network Univision, which by 1987 programmed twenty-four hours a day by satellite to 463 affiliates. In January of that year competitor Telemundo started up with five owned stations and several affiliates. Both relied heavily on *novelas* (Hispanic serial dramas or "soap operas," mostly imported), plus talk, game shows, sports, and news to attract national audiences among 19 million U.S. Hispanics, two-thirds of whom preferred the Spanish language or spoke it solely.

■   By 1990 Univision reached 85% of U.S. Hispanic households and Telemundo 70% (Hispanic cable network Galavision reached 25%). National advertisers directed to these networks far less than 1% ($150 million) of their $23 billion annual network/spot TV ad budgets that year. One-third of all Hispanic media ad dollars in 1990 went to Univision and its owned stations, whose network revenues had increased 25% from the previous year.[8] In 1995 Univision attracted three-quarters of Hispanic household viewing while Telemundo's share of that audience was 24%. The Census Bureau estimated the Hispanic population, including undocumented aliens, at almost 26 million by 1995, equal to one-tenth the U.S. population and expected to pass blacks as the largest U.S. minority group by 2010. Their median family income was three-quarters that of average U.S. families. Hispanic consumer spending in 1995 exceeded $250 billion. That year ad spending on Hispanic TV totaled some $475 million—over half of that to network/national, the rest to local. (Another $300 million went to buy commercial time on national and local Hispanic radio.) ∎

Late in 1993 Time Warner Communications' Warner Brothers announced a $2 billion plan to establish a fifth network, possibly in association with Paramount; it depended on collaborating with major broadcast groups that owned independent stations.

∎   In January of 1995 Warner Brothers, linked with Tribune Broadcasting, premiered its network with two hours of programming two nights a week; by 1996 it had 82 affiliates (reaching 83% of U.S. households), plus interconnected cable outlets in some markets. One week later Viacom-Paramount, with Chris Craft/United, launched its network's two-nights weekly service over 102 stations, which increased to 152 by 1996 (93% of the country). In Fall of 1995 the WB network expanded programming from two hours to thirteen, including Sunday evening and children's programs on Saturday and weekday mornings. In the 1996 fall season, WB aired seven hours of programs during Monday, Wednesday, and Sunday prime-time, while UPN scheduled eight hours over Monday, Tuesday, and Wednesday evenings; Fox offered 15 hours over all seven evenings, against the other major networks' regular 22 hours. For the week of September 16–22, 1996, WB's prime-time audience ratings averaged 2.6% of the nation's 97 million households; UPN attracted between 2.4% and 4.3% (*Star Trek: Voyager*), averaging 3.3%. Fox's ratings ranged from 4.6% to over 8.1% by two shows (*Melrose Place* and *Beverly Hills 90210*), with an average of 6.3% of the nation's homes tuned in. That same week NBC's prime-time programs attracted 11.8% of the nation's households, CBS 10.8%, and ABC 9.9%.[9] ∎

Noncommercial PBS also put a dent in the networks' prime-time audiences, attracting 2.2% of the nation's homes to its overall schedule during the 1990s. Special programs earned national ratings of 6.0 ("The Kennedys"), 5.1, and 3.7. Award-winning *The Civil War,* an eleven-hour mini-series, attracted 9% of U.S. households for a 13 share of TV viewers, an estimated 14 million people each of five nights.

Structures

CBS/Broadcast Group President Jankowski noted that networks (under the FCC's financial interest/syndication restrictions) as distributors did not own the means of producing, exhibiting, or financing the process: "The networks will stay in business for as long as the owners of programs, the owners of the affiliated stations, and the advertisers benefit from the service they provide."[10] Unlike other media organizations, networks interconnect those three elements "continuously and completely" to reach national audiences. Looking from the outside, author-critic Ken Auletta put it in lay terms:

> [A] TV network is a renter pretending to be a landlord. A network rents its programs from Hollywood studios or sports team. And then it rents space on local stations to circulate this product. . . . [T]he networks own neither the manufacturing process, nor the means of distribution.[11]

(Part of that equation changed as FCC deregulation in 1985 and again in 1991 and 1993 gradually readmitted networks to producing and owning programs on their own schedules.)

Networks are organized similar to stations with four basic areas—administration, programming, sales, and engineering. As national broadcast operations and servicing clients grew more complex, network management became more specialized.

■ By 1976 CBS had a corporate president, 5 more presidents of broadcast divisions, and 3 presidents in the record division, plus 218 vice-presidents; ABC had 10 presidents in the broadcast area, several presidents in nonbroadcast areas, and 90 broadcast vice-presidents; NBC had 5 presidents and 100 vice-presidents.

Approaching the 1980s CBS (following ABC's lead, as NBC did later) split its original TV network into separate areas, each with its own management team, each headed by vice-presidents eventually designated separately from the TV network division. In 1986, parent CBS, Inc. retained a corporate staff of 1,400—including 100 executives in business development, 129 in personnel, 159 attorneys in the legal department, 42 in corporate affairs, 15 in industrial relations, and 545 in finance (including management information services and computers). That was after reducing its broadcast division personnel from 8,600 to 7,600, but prior to selling off its publishing group of 4,000 employees and CBS Records with 10,500 jobs (many blue collar). By 1992 CBS had "downsized" the entire staff of all its remaining divisions to 6,650 employees.

In 1993 ABC formed a new corporate division, reflecting the synergism among various media and new technologies, CapCities/ABC Multimedia Group. It was made up of the ABC TV Network Group, CapCities/ABC Broadcast Group (owned stations), and CapCities/ABC Publishing Group. The new division was headed by a former ABC general counsel who was executive vice-president of ABC News. Yet further reorganization came with Disney's purchase of the company in 1995. ■

CapCities/ABC Chair Thomas Murphy acknowledged the tightly competitive media arena when he candidly told the Association of National Advertisers that networks "are not as good as we once were, but we're still the best buy in town."[12]

Affiliated Stations

Each of the 600-plus network-affiliated local stations give up two-thirds of their schedules to carry network program service. That assures them of programming including news events and sports, plus compensation (15% to 30% of their rate-cards, based on audiences reached) with some local commercial time within and between big-audience programs. It also assures that popular shows with major stars will be promoted in national publications and on the air. FCC regulations ensure that affiliates can accept or reject any network's program, can carry a different network's program if offered, can set their own rates for local commercial time in or around network programs, and since 1989 are no longer bound to two-year periods for network contracts. Typically affiliates clear time for 97% or more of their network's program feeds, except in the early 1990s when stations for several years pre-empted a number of weaker daytime and late-night and occasional prime-time series for more lucrative syndicated shows or local productions.

Licensees of local stations—owners and their surrogates the managers—retain legal responsibility for everything transmitted by their broadcast signal, whether locally originated or distributed to them by networks or syndicators. So each network must keep alert to how over 200 affiliate managers and their communities react to their programs (see later section in this chapter on program critics and network departments of standards and practices). Local managers rely on the network to exercise prudence regarding programs and commercials, lest local viewers and advertisers be "turned off" and thus their station channel be literally turned off.

One index of mutual roles in the affiliate-network relationship has been the networks' long-term desire to capitalize on news operation efficiencies while offering fuller evening coverage of world events. Despite network executives' stated intentions and even specific announcements about expanding to full-hour evening national newscasts, affiliates adamantly refused to cede the lucrative half-hour of prime-time access, first in 1976 (ABC), then in 1981 (80% of NBC affiliates rejected the proposal) and 1982 (61% of CBS affiliates opposed it, 89% in the top-25 markets) and again in 1987 by CBS. Syndicated game and talk shows as well as local news drew large audiences in their markets and good local spot rates.[13] The networks' power was clearly limited to what affiliates would agree to, so the hour news concept failed repeatedly. Similarly, when networks warned that waning profitability called for drastic reduction or abandonment of compensation, so many affiliates threatened to pre-empt freely any weak network programs that the networks backed off, although they were able eventually to reduce some payments.

Approaching 1993, among the nation's markets measured for TV ratings, ABC had 227 affiliates, CBS 214, NBC 209, and Fox 156. Those numbers were in continual flux, of course, as affiliates and networks negotiated changes and as Fox Broadcasting first added independents to its line-up, and then in effect raided major networks' traditional affiliates in the mid-1990s.

The cyclic horserace among competing networks prompted some station managers to switch affiliation. Local managers sought the top network's strong prime-time programming line-ups to bolster local ratings and sales ratecards. Prior to 1976 there had been few such changes. But ABC's rise to first place over CBS' decades-long dominance prompted eight affiliates to shift to its line-up. A total of 21 affiliation changes occurred in 1975–1980, another 37 in 1981–1990, and 69 stations in 34 markets switched affiliations in 1990–1995.[14] Most of the last group involved Fox network's involvement with New World Communications whose 12 stations dropped previous networks in favor of Fox. In all, 21 affiliates of the Big Three moved to Fox; the three major networks were downgraded to UHF outlets in 19 markets (ABC nine, CBS and NBC five each), severely affecting their national ratings. In many markets across the land, a domino effect resulted in successive shifting among network affiliations. Coupled with snagging the NFL football package from CBS (who had carried it for a third of a century), Fox strengthened its affiliates' audience base in the nation's markets. By 1996 Fox had assembled 180 affiliates, plus some secondary ones (on channels already affiliated with their primary network) and via local cable system channels for a total of 211 market outlets.

Although some individual affiliates can dominate a market regardless of their network's success or weakness by vigorous local news and strategic scheduling of strong syndicated programming, most rely on network service to attract and sustain large local audiences. For example, NBC led the networks in the latter 1980s, sold a record $3 billion of commercial time in one year, and was first in twelve consecutive quarterly ratings "sweeps"; by 1988, 137 of its affiliates led their markets, 28 were second, and only 18 were third.

Affiliates became restless with their networks' dwindling audience shares of total viewing through the late 1980s. Many pre-empted selected low-rated prime-time shows in favor of audience-attracting syndicated programs that also gave them local commercial time to sell. Further, a recessionary economy grew worse in 1990 and 1991. For the first time in thirty years, total advertising spending declined from one year to the next (although cable ad revenues increased 18%, by almost $500 million, to $3 billion in 1991).

Network Compensation

In 1991 the networks continued cutting back payments to affiliates in compensation for the major portion of their daily and nightly schedules given over to network programming. ABC cut its total compensation by 10% down to $100 million; depending on market size and standing, average station reduction was 5%–10% (WCVB in Boston was cut 12%–18%). CBS reduced compensation by $24 million, distributing $130 million annually to its 200+ affiliates. NBC cut back by $13 million to $132 million; its affiliates were foreseen doubling preemption of programs with that reduction. One financial tactic was to end compensation for weekend sports, worth $6 million annually; instead, that network offered affiliates an extra sixty seconds of commercial time for local ads. In subsequent years networks continued to renegotiate compensation, at times even returning some previous reduction, to induce affiliates to carry entire network schedules. In the mid-1990s Fox's offensive strategy, plus the entry of UPN and WB networks seeking affiliates, brought renewed vigor to local stations' role. Networks vied for their loyalty, even contracting for ten-year affiliations (instead of the customary two years) and enhanced compensation to forestall further raids by competing networks.

Beyond providing current programs for affiliates' schedules, networks also serve them by providing off-network feeds of national and international news stories for use by local TV news departments. This is partly to forestall affiliates' linking with other news suppliers (as many as one out of four NBC affiliates are associated with CNN). Networks also serve as interchanges for regional affiliates to share breaking news among one another through the network center; NBC's original Skycom news service relocated in 1992 from New York City to Charlotte, North Carolina, as NBC News Channel twenty-four-hour service. Networks also assist their stations with research data, promotion, marketing, and other resource material.

TV NETWORK ENTERTAINMENT PROGRAMS

Competitive Programming Strategy

The three full-service television networks provide nearly 17,000 hours of programming a year, almost 6,000 hours each—the equivalent of ten years of motion picture films produced in Hollywood, or twenty-five years of theatrical performances on Broadway. They are the backbone of local affiliate schedules. Most of those hours are filled with entertainment genres that consistently attract the largest audiences. How to determine which kinds of programs appeal to mass audiences is a vexing challenge to network management. Their decisions affect the success of every local station manager. For example, a news consultant with Frank Magid Associates reported in 1993 that late evening local newscasts (11:00 eastern time) retain about half the audience from the last hour of prime-time when a network airs an entertainment program, but almost 85% of viewers from a newsmagazine or "reality" show lead-in.[15] (Whether the resulting carry-over audience is larger depends, of course, on the original number drawn to each of the entertainment or news/"reality" programs in that previous hour.)

Management strategies for researching, developing, scheduling, and overseeing production of the wide range of leisure oriented programming are complex and unending. Details would more than fill another book—as the present long chapter attests.[16] Former CBS programming chief Michael Dann claimed that "what is fundamental and has never changed is that where a show is placed is infinitely more important than the content of the show."[17] Just as their PD counterparts at local stations, network programming vice-presidents employ scheduling strategies to attract the greatest possible numbers of viewers while maneuvering them away from competing shows. Tactics include counter-programming (different genres, as pitting comedy against drama); or head-to-head blunting (same genres on at the same time); and plotting audience flow with strong lead-in shows before new or weak ones, or hammocking an untried program between a strong lead-in and a popular program following it.

Ratings alone do not always determine the fate of scheduled programs. The demographics of older/younger, male/female, affluent, educated, minority, and other characteristics important to advertisers are part of the equation. Even programs of substance or with widespread national audience support also succumb to the realities and strategies of network television's competitive programming. Classic instances were *The Firestone Hour* and *Lawrence Welk* musical programs sponsored by satisfied advertisers and viewed by large loyal audiences; but ABC canceled the shows because they attracted not only smaller audiences than other kinds of programs, but also the "wrong kinds" of audience—too old, rather than many of the product-purchasing 18–49-year-olds whose viewership generated higher value for commercial spots.

Another factor in recent years is that the sheer cost of a program may outweigh its mass audience because commercial time cannot be sold at high enough levels to make up the expenses. Even top-rated series that earn critical acclaim reach a point of loss. For the 1992-1993 season, having dropped from first back to third place, NBC in desperation renewed highly successful *Cheers* for an eleventh and final season. But it did so reluctantly because rights to license it from Paramount were more than four times the usual half-hour sitcom rate; star Ted Danson alone received over $400,000 per weekly episode, with each episode costing Paramount $2 million (the studio reportedly first sought $4 million per episode for licensing rights), while other sitcoms typically cost $500,000.

The erosion of national audiences for networks means fewer viewers for programs. In 1991 a program had to earn at least a 17.0 rating percentage of U.S. households to be among the top 10 prime-time shows; in November 1996 the tenth-ranked program had only a 14.2 rating. In 1991 some 27 programs had ratings higher than 14.0, but only 11 scored that high in October 1996.

The Program Planning Process
Programming involves three stages: conception (ideas, proposals, scripts, pilots, or sample episode); execution (production of pilots and subsequent episodes); and scheduling (strategies for maximum audience within a network's line-up and competing with other network shows). In the executive ranks, a typical network entertainment or programming division headed by a president had as many as eleven vice-presidents, one each for prime-time programs, comedy development, current comedy programs, drama development, current drama programs, mini-series, television movies, production, specials, daytime, and children's programs. Richard Lindheim, first a researcher with NBC and then many years in program development at MCA/Universal before moving to an executive position at Paramount television, commented that going into the 1990s the program development process had not really changed much over previous decades.[18]

Despite reconfigured corporate structures, new kinds of high-tech equipment and more distribution outlets through first-run syndication, cable, and Fox, the search for new programs and the tactics for bringing creative ideas to fruition had not been displaced. Inter-network rivalries and negotiating with program suppliers as well as client advertisers still require management to go through the complicated process of research, insights, plans, budgets, strategies, negotiations, and luck as in the past. CBS Television Network President Robert Wood recounted the decision-making process that weighed factors in CBS' major shift from older-skewing programs popular more in rural counties to social-oriented sitcoms for younger urbanites, introduced by *All in the Family* in 1971. Little has changed since then.[19]

> What finally controlled the decision was a combination of many factors—not only the appeal of the programs among various audience segments but the mix and balance of our entire nighttime schedule—that is, program sequence, audience flow, competitive scheduling, innovation and experimentation, as well as program cost, advertiser interest and station acceptance.[20]

These multiple interacting concerns of management hover in the background when we later discuss entertainment programming, sports, and news. They then become more explicit in looking at costs and outside suppliers of programs. The final part of this chapter looks at other national programmers: syndicators and cable channel services.

Program Development

A network executive once remarked that if the elements making programs successful were predictable, every program chief would be a genius and rich. Notably, the majority of pilots contracted from thousands of proposals and hundreds of scripts annually end up failing when put on the air. The three major networks each year consider up to 6,000 submissions of ideas, outlines, or scripts; perhaps 500 are developed further, with about 150 produced as pilots or sample episodes (usually costing twice what a single episode would cost). Depending on scheduling needs, each network may commit to 10 new series—with short-term contracts for 13 episodes, or as few as 6, until initial audience reaction is confirmed by ratings. Almost three out of four new programs fail, often canceled within a few months.

New programs introduced by the four networks for 1993–1994 are categorized in table 8.2. They replaced series that earned only mediocre ratings, despite critical praise for several (*I'll Fly Away, Homefront, Civil Wars*), plus the highly rated *Cheers* terminated by the producers and cast rather than by the network.

Total new projects for series, specials, and made-for-TV movies for the 1993–1994 season—including those on the schedule for Fall 1993—were 63 (55%) to be produced by major Hollywood studios, and 32 (29%) by independent production companies. The latter were down from the previous year's 41% because networks were increasing in-house production with 19 projects or 17% of the 1993–1994 total. Including newsmagazines and evening sports along with entertainment programming under relaxed FCC rules, in 1993–1994 the networks produced almost one-third of all their prime-time: CBS 39% of its schedule, ABC 35%, NBC 25%, and Fox 20%. For the 1994–1995 season—a year before the fin/syn restrictions fully expired—the three networks produced in-house 14 hours of their own prime-time programming, in addition to their news divisions, each producing three

TABLE 8.2
NEW NETWORK PRIME-TIME PROGRAMS, 1993–1994 SEASON

PROGRAM TYPE	ABC	CBS	NBC	FOX
Comedy	7	5	7	4
Drama	4	3	2	2
News magazine	-	1	1	1
Variety	-	-	-	1
Mystery anthology	-	-	1	-
TOTAL PROGRAMS	11	9	11	8
TOTAL HOURS	*8*	*6.5*	*8.5*	*6*

Source: *Broadcasting & Cable,* 17 May 1993; 24 May 1993; 31 May 1993.

news magazine shows; they also invested in some productions by outside suppliers. Fox's in-house production division, Twentieth Television, produced four hours for its own network.

The lives of most TV shows are short; turnover is constant. For the 1990–1991 season the networks introduced a record 34 new programs to their prime-time schedules; five of them were ranked among the top 10 programs by mid-season in January, but many of the rest fell by the wayside. The 1995–1996 season saw 42 new shows introduced by the networks. For 1996–1997 they introduced another 40 new programs. Including shifting airdates and times, first-place NBC changed 40% of its schedule, CBS 45%, ABC 48%, and Fox fully 50% of its line-up. To come up with their changes, ABC invested in developing 28 pilot projects (20 comedies, eight dramas) and Fox 33 projects (18 sitcoms, 12 dramas, and 3 reality shows). From those pilots ABC selected five new comedies and five dramas for their fall schedule, Fox ordered three comedies and two dramas. NBC and CBS equally leaned on the sitcom form for their revised schedules, as all networks did the previous season when two-thirds of all new shows were sitcoms.

On the Big Three networks, during the decade from 1984 to 1994, sitcoms remained the top genre by far at 40% of prime-time shows—almost twice the percentage of drama and far outpacing action-adventure which had been a TV staple of earlier decades (and which continued to be very popular in foreign distribution into the late 1990s). Hour-long dramas dropped over that ten-year period from 37 to 23 shows while hour-long reality-based programs increased from 4 to 11. But reality shows migrated to cable and to syndication in such numbers, they blunted their attractiveness as network fare and lost their hold by the mid-1990s.

Annual creative costs for developing and licensing all entertainment programs in prime-time totaled $1.25 billion for the three networks in 1975; $3.5 billion for their 15,000 hours of programming in 1980; in 1988 CBS alone spent $2.25 billion.[21] Costs dropped in following years as license fees were held in check or even reduced and entertainment shows gave way to low-budget, reality-based programs and newsmagazines, which grew from five in 1988 to twenty-four in 1992. In 1993 the four networks spent about $3.6 billion on programming, and expected to increase funding by almost $250 million annually to a total of $4.6 billion by 1997.[22]

Network Programming Patterns
Distinctive tendencies characterize each network's programming choices over the decades. They evolved out of their executive managers' strategies to compete for the national audience's attention. One analyst typified ABC as programming "sexy

action-adventure shows and trendy sitcoms aimed at teenagers," CBS scheduling "well-wrought sitcoms of broad appeal" to families and older viewers, and NBC aiming at young marrieds and singles with "innovative programming: anthology dramas . . . 90-minute series" and being "a strong 10 o'clock network" with hour dramas like *Police Story, Quincy, Hill Street Blues, L.A. Law.*[23] The decade since that was written generally bears out the descriptions, with the Fox network added as relentlessly committed to youngsters' programs and teenager dramas that walk along the margin of "smart and sassy" material plus shows with strong appeal to African American and Hispanic audiences. Fox network executives were daring, taking risks to offer something distinctively different from traditional network fare. Author and trade magazine editor Alex Ben Block claimed

> their intent was to come out with outrageous, on-the-cutting-edge shows that go to the very limits of what taste is all about. They had to wave all those red flags just so viewers would find them. . . . If you go to the movies and read books, what Fox is doing is still relatively conservative. It may be outrageous for television but not for society.[24]

The top-10 programs of the total U.S. audience in Fall 1992 differed totally from the favorite 10 shows among African Americans, and found only two on Hispanics' top list. The Fox network had no shows among the ten most watched by the total national audience, but it had five of the top-10 on both minorities' listings, with *Martin* and *Married...with Children* appearing on each (table 8.3). Different genres shift in dominance over any decade, as detailed below.

The Costs of Network Programming

The programs of the three networks in prime nighttime hours alone cost multi-millions of dollars every evening. For the advertiser, the cost of the program's production and personnel is only part of the total expense; the value of airtime and charges for network facilities must be added. Time and facilities charges vary with the time of day, the number of stations on which the program is carried, length of contract, and other variables. In the early 1970s a potential sponsor had to decide whether to buy *NBC Reports* for around $150,000 for the entire hour, or one of the top hour-long variety programs at a cost of $225,000 to $250,000 to sponsor the whole show, or co-sponsor ABC's NFL *Monday Night Football* at a figure of some $325,000 for half the game. The 1990s would find those same amounts spent on merely a single 30-second spot in such programs (the 1994 Super Bowl's 56 commercial spots were 80% sold out six months ahead, at $900,000 per 30 seconds; spots sold for over $1 million each in 1996).

If ratings of different programs were comparable, the less expensive series might at first appear to be the better purchase. However, other variables to be considered include demographic characteristics of audiences reached, the type of program in relation to the product advertised, competition on other networks at the same hour, and a program's prestige value for the advertiser.

Ideal program spot buys are those with the lowest budget that deliver the most people with demographic characteristics sought by the advertiser. Determining this ideal combination is one reason why contracting for, scheduling, and buying/selling commercial programming is such challenging work.

Network programming costs rose steadily through the decades into the 1980s when they leveled off somewhat. At least seven factors beyond the nation's general inflationary trend raised network program costs: TV's success which until

TABLE 8.3
Top-Ranked TV Programs, Fall, 1992,
By Total U.S., African American, and Hispanic Households

	TOTAL U.S.		AFRICAN AMERICAN		HISPANIC
1.	A *Roseanne*	N *Fresh Prince of Bel Air*	F *Simpsons*
2.	C *60 Minutes*	F *Roc*	F *Martin*
3.	C *Murphy Brown*	F *In Living Color*	F *Beverly Hills 90210*
4.	A *Coach*	F *Martin*	A *ABC Sunday Night Movie*
5.	A *Home Improvement*	N *Blossom*	F *In Living Color*
6.	C *Murder, She Wrote*	N *A Different World*	A **Roseanne*
7.	A *Jackie Thomas*	N *Out All Night*	N *Fresh Prince of Bel Air*
8.	A *Monday Night Football*	F *In Living Color*(re-runs)	A **Monday Night Football*
9.	C *Love and War*	A *Hangin' w. Mr. Cooper*	F *Married. . .with Children*
10.	A *Full House*	F *Married. . .with Children*	N *Blossom*

A=ABC C=CBS N=NBC F=Fox   * = Same as Total U.S. Preferences

Source: A.C. Nielsen data and BBDO study reported by New York Times News Service, 11 April 1993, and *Broadcasting & Cable,* 22 March 1993, 41.

the 1990s enabled it to pay more as its suppliers demanded more; the demand for increased production values; unions' rising wage scales and benefit packages; geometrically expanding fees to lead actors and production-direction staff; changing economics of programming, especially in the syndication and cable market; widespread acceptance of the inevitability of continuing price increases; and sheer show business.

■   Current TV program costs are worlds apart from almost a half century earlier. Ed Sullivan's first variety hour in 1948 cost only $1,350 to produce, including technical crew, Sullivan's salary, payment to every act, and a fourteen-piece orchestra. In 1977 hour-long action-adventure programs moved beyond $400,000 per episode, while 30-minute sitcoms for the first time exceeded $200,000 each. In the 1985–86 season producers licensed half-hours to networks for an average $275,000, and in 1991–1992 for $450,000 (those totals fell below studio production costs, each 30-minute episode resulting in producers' deficits of $50,000 in 1985 and $250,000 in 1991).[25] Payment to major sport leagues and franchised teams for rights to cover their games multiplied over the last two decades, until networks lost so many hundreds of millions on their baseball contracts in 1991–1992 that CBS chose not to renew and ESPN paid $400,000 to be released from the remainder of its deficit-ridden contract.

The growing costs of television programming early on had resulted in a shift away from full program sponsorship to participation advertising whereby various commercial accounts appear within a single program. In 1960, single sponsorship was first surpassed by multiple advertisers in the majority of network programs; single sponsors dropped from a third of all programs in 1970 to one-fourth the next half-decade and far below that subsequently. More advertisers sharing costs made it feasible to schedule longer feature films, 90- and 120-minute program forms, and extended mini-series of four to eight hours or more over successive evenings. (ABC produced its own twenty-hour mini-series *War and Remembrance* in 1986; contrary to early projected costs of $40 million, *Television/Radio Age* reported it ended up costing $105 million.) Program charges, prohibitive for a sole sponsor, were divided among many participating advertisers.

The continually spiraling costs for leasing rights to television programming resulted in fewer new programs per season in a series, with re-runs occupying fully a third of the former fall-winter-spring broadcast season. Network radio

series had always been scheduled for thirty-nine-week seasons; television followed suit with three sets of thirteen weeks, with summer schedules given over
to re-runs and experimental programming. Since the 1970s, networks ordered
only 22 to 24 new programs from a series producer, leaving more than six months
of the year for re-runs and pre-emptions. In a two-week period during summer
1982, 126 prime-time hours included 76.5 hours of re-runs of regular-season programs, 24.5 hours of movie re-runs, and 25 hours of original programs including
five hours of baseball and 12.5 hours of pilots and episodes that did not make the
regular season.   ■

Network cost accountants explain that the first showing of a national program does not generate enough revenue to pay all the costs—including production license fees, compensation payments to local stations for their airtime,
telephone company line charges or satellite transponder rental, advertising agency
costs, plus the network's percentage for overhead. Re-runs became necessary for
a network to profit from a program. Creative people (writers, production staff)
and unions (technical crews) in Hollywood complained to the government about
fewer newly produced programs. The White House Office of Telecommunications
Policy criticized the networks for oligopolistic competitive practices that contributed
to soaring production costs leading to the efficiency move to half a year of
repeated programs. Broadcast spokespersons countered that economics demanded
such policies; they added that when the average network program is first broadcast, 131 million people or 86% of the total potential audience do not see it and
therefore have another opportunity when it is rebroadcast later in the same season. Ratings indicated in 1996 that re-runs of dramatic programs drew between
66% and 85% of numbers that viewed new episodes; top sitcom re-runs attracted
90% the size of original audiences.

As noted earlier, networks license rights to air an original episode and one
re-run for perhaps 80% of what a production company spends to mount that episode
(see tables 8.4 and 8.5). The program supplier risks those weekly losses "up front"
with the hope that a network will extend its initial limited commitment to the entire
season, and then renew for succeeding seasons, thereby promoting the program
series to a national audience whom local stations can later attract by off-network
re-runs in syndication. It is in that "back-end" domestic and also foreign distribution that production companies recoup their earlier losses many times over. However, there is the constant risk of a short-term cancellation, which is the common
fate of new shows, when the program does not attract large enough audiences.
The network pays a penalty fee (sometimes up to half the original license fee)
for any episodes originally ordered but canceled when a series is dropped in mid-
season. A program supplier must have enough hits to subsidize failed projects.

■   In 1992, average cost of filming or taping a half-hour program was similar, but
networks paid somewhat more for taped programs (depending on specific programs'
ratings popularity). Deficits for each episode averaged $391,500 for a filmed show
and $273,000 for a taped show; hour-long filmed programs "earned" average deficits
approaching one-third of a million dollars per episode. Over a season's twenty-two-
episode contract, an hour-program supplier stood to lose over six and a half million
dollars. Meanwhile the network whose ratings were low for that series lost the potential
value of national advertising within the program and also found its schedule weakened
for that prime-time evening as well, lessening audience lead-in to shows following the
weak one. Further, the network had earlier invested in the development process, having
paid the program supplier as high as $1 million for preliminary scripting and produc-
tion of a half-hour pilot. Companies producing shows for networks ran up total deficits

NETWORK LICENSE FEES, PRIME-TIME PROGRAMS, 1981–1992

YEAR	120-MIN. MOVIE	30-MIN. SITCOM	60-MIN. DRAMA	120-MIN. SPORTS
1981-82	$ 1,800,000	$ 300,000	$ 600,000	$ 1,800,000
1985-86	2,300,000	375,000	750,000	2,100,000
1991-92	2,600,000	550,000	990,000	2,500,000

Source: Estimates based on listings in *Variety,* 30 September 1981; 25 September 1985; 26 August 1991.

—⁓⁓— **TABLE 8.5** —⁓⁓—

AVERAGE COSTS AND DEFICITS, PRIME-TIME EPISODIC SERIES, 1991–1992 PRODUCTION SEASON

	30-MIN. FILMED	30-MIN. TAPED	60-MIN. FILMED
Production cost	$ 950,000	$ 959,000	$ 1,432,000
License fee, network	558,500	686,000	1,125,000
Deficit	(391,500)	(273,000)	(307,000)

Source: *Broadcasting,* 8 June 1992, 4–5.

of $464 million for all programming in the 1991–1992 season, according to the Alliance of Motion Picture and Television Producers; that was up 31% from the previous season. Over the previous half-decade production costs had continued to escalate annually while license fees dropped somewhat. ∎

On the other hand, when a series is successful, the network schedule attracts huge audiences, so advertisers pay premium rates; a strong show also builds audiences for others following in the evening's schedule. The production company stands to earn multiples of its original licensing fees from potentially endless recycling of popular shows through subsequent syndication. Syndication fees from local stations can total from one to three million dollars *per episode* in a half-hour sitcom package for a five-year re-run contract (see later section in this chapter).

Production/distribution companies also lease their off-network series to national cable program services. For example, packages of hour dramas brought $200,000 per episode of *L.A. Law* and $475,000 for *Murder She Wrote,* while sitcom *Wings* earned over half a million dollars for each half-hour episode in the package. Further such "back-end" revenues (after original airing by networks) come from distribution to foreign countries and sometimes from domestic videocassette sales.

Program Attrition: Casualty Rates

Program mortality rates are always high. All forms of show business are "chancy." CBS/Broadcast Group President Gene Jankowski alluded to the one in ten shows on Broadway that becomes a hit, perhaps one in twenty motion pictures. For books and records the ratio might be only one in a thousand; Jankowski underscored that no other medium had even remotely comparable demands for material. Casualties of new programs within or after one season on the air averaged twenty or more each year since 1960; over the past two decades about three-fourths of new network shows failed.

In the 1990s most new series were dropped within five to ten weeks when ratings reflected inadequate audiences. This was due to insufficient audience interest coupled with too many similar shows, as well as other viewing options available on cable channels, PBS, and independent stations. Also, some new programs were simply not strong enough in concept or presentation to last very long.

Programs with extended licensing contracts from the networks sometimes become so successful that lead cast members and other principals demand higher salaries (the multiple young cast of highly successful sitcom *Friends* in 1996 sought to have their $40,000 weekly salaries raised to $100,000). Renegotiated license fees have at times become so high that networks refused to renew them despite strong audience ratings.

■   From 1970 to 1984, the three networks regularly canceled two out of three new series during their inaugural seasons. From 1984 to 1987, CBS canceled 84% of its new series between September and May of their first season, ABC 78%, and NBC 71%.[26] Eighteen of 37 new programs failed in the 1992–1993 season; a dozen ad agencies predicted 22 of 39 new ones would fail in 1993–1994 (they were generally correct).

Going into the 1980s, very successful programs might last ten years; by the end of that decade, five years was more the norm for success, although several series exceeded that. The phenomenally popular *The Cosby Show* ran from 1984 to 1992, *Cheers* remained a top-rated show when it ended its eleven-year run in 1993—it was the first comedy to hold the #1 position in its ninth season (matching the record set by the western *Bonanza*). Late-night was the domain of Johnny Carson when he hosted *The Tonight Show* for three decades from 1962 to 1992. ■

Those who criticize networks for scheduling mostly light entertainment must consider the intense competition in the medium. Television programming has evolved which can attract audiences needed for survival; executives select shows that appear to have the least chance of failing. Eventually, there was little to distinguish one network from another. All networks appeal to the public in similar terms for approximately equal shares of the viewing audience. The parity of 30 audience shares for each network in the 1970s eroded to lower than 20 share points each in the 1990s; cable, Fox network, independent stations, public TV, direct broadcast satellite program services, and videocassettes drew off one-third of networks' earlier audience shares.

Time is sold and charges based on statistical rating points. In 1996 a single rating point equaled 970,000 households or potentially some 2 million persons in those homes. It was estimated that over a full season in the late 1980s, one rating point represented about $60 million in network pretax profits (if that point was drawn from competing networks' audiences). By the mid-1990s that estimate rose to over $140 million per season's rating point.

■   Having moved up from third to first place in 1991–1992, CBS again won the next season with an overall prime-time rating of 13.3 and a national share of 22% of audiences. ABC's 12.4 rating/20 share put it second, ahead of NBC with 11.0/18; Fox's incomplete schedule earned it a 7.7/13 share. In 1993 each rating point represented 931,000 TV households. The networks' individual ratings and shares continued to drop annually. When NBC regained the ascendancy, in the 1995–1996 season it scored a rating of 11.7 and a national audience share of 19; ABC reached 10.6/17; CBS 9.6/16; and Fox 7.3/12.3; plus UPN 3.1/5; and WB 2.4/4. By May of 1996 each rating point represented 959,000 U.S. households. That year the big three's overall rating totaled 31.9% or 30,592,100 U.S. homes—a serious drop from the 36.7 rating or 34,167,700 homes posted just three years earlier. Adding Fox, the four major networks together had reached a season average rating of 44.4 (41,336,400 homes) in 1992–1993, but only 39.2 (37,592,800 homes) in 1995–1996.   ■

The instant "hit" occurs occasionally in television entertainment. When it happens, even program executives are at least mildly surprised. The hit series is partly a stroke of luck that every producer hopes to achieve. In seeking to develop a successful series, it is common to appropriate a format that has already proved popular and to add a few new twists. Comedian Fred Allen observed that "imitation is the sincerest form of television." Even so, the new "spin-off" venture based on a current program often fails.

Occasionally, a special individual program will be so well received that gears will start in motion for more of the same type. A Christmas dramatic special "The Homecoming" attracted critical and popular acclaim, and quickly returned as a weekly series, *The Waltons,* which eventually became one of the top-five programs nationally for several seasons. When reality-based re-creations found favor with audiences early in the 1990s, all three networks hastened to mount regular prime-time offerings based on police files and emergency medical incidents.

Program Genres

Despite former CBS programming chief Mike Dann's assertion that scheduling determines audiences, the content of programs is what ultimately "brings people into the tent" as viewers tune in TV and cable channels for their favorite national programs. Those programs fall into genres that cycle through successive years of network programming.

Through most of the 1980s, serious drama accounted for between 26% (1983) and 20% (1990) of prime-time programming, while action-adventure shows rose and fell through the decade from 27% in 1984 to only 8% in 1990. Those two categories plus sitcoms made up more than three-fourths (77%) of network prime-time during most of the decade. Theatrical and "made-for-TV" movies dipped slightly to about 5% of the schedule by 1990 after being a point or two higher in previous years.

Serious hour-long dramas dropped over the ten-year period 1984–1994 from 37 to 23 shows. Action/adventure dramatic programs dropped from 32 in 1983 to only 19 a decade later. Networks aired six nightly movies a week throughout that ten-year period. The success of *Cosby* and *Cheers* in the 1980s helped revive the failing sitcom form. In 1983 the networks scheduled 29 sitcoms; by 1993 they aired 42 of them. For the 1996–1997 season, the six broadcast networks scheduled 64 sitcoms, airing 16 of them within a two-hour period on Wednesday evenings. Newsmagazines produced by networks' news divisions held steady at about 5% of network prime-time in the 1980s, then proliferated in the 1990s. "Reality-based" programs grew from 1% in the early 1980s to the third-highest category in 1990 (11%); their number increased from 4 in 1984 to 11 in 1994 before dropping back.[27] Partly because that genre migrated to cable and to syndication in such numbers, their attractiveness waned as network fare and they lost their hold there in the mid-1990s.

Drama

The quality of many weekly dramatic series on television in the 1990s prompted Charles McGrath, editor of *The New York Times Book Review* to laud them in his twelve-page article in *The New York Times Magazine* (October 22, 1995): "TV is actually enjoying a sort of golden age—it has become a medium you can consistently rely on not just for distraction but for enlightenment." The magazine featured the article on its cover with prominent type: "Want literature? Stay tuned. More than movies, plays and even in some way novels, television drama is making art out of real life." Multiple-plot programs drawing these accolades included issue-oriented and character-driven medical dramas *E.R.* and *Chicago Hope,* police/crime dramas *N.Y.P.D. Blue, Law and Order, Homicide: Life in the Streets,* and the earlier *Hill Street Blues* and *L.A. Law,* plus *Picket Fences* and *My So-Called Life.*

Beyond costs and control, a key factor favoring original TV films over theatrical movies was the larger audiences they drew. Made-for-TV movies and theatrical motion pictures, especially ones of action-adventure, also attract 18–49-year-old audiences most desired by advertisers.

Whether caused by a growing necessity to schedule dramas less expensive than theatrical feature films, by the audience's growing sophistication, by maturing of the industry, or by a demand for excellence, network programming grew steadily more "adult." Provocative themes and language that would have met mass resistance in television's early days began to be aired. The general audience's acceptance of these changes accelerated with the entry of cable services introducing uncut theatrical films replete with "R"-rated raw language and graphic scenes of violence and partial nudity.

Mini-Series
The "serialized dramatization" derives from public television's BBC-produced 26-part *Forsyte Saga* in 1970 and *Masterpiece Theater* presentations since 1971 whose various dramas over the past quarter century have run from five to thirteen hours. *New York Times* critic John J. O'Connor called this genre "perhaps television's most original and exclusive programming vehicle."[28] Commercial networks carried these extended dramas, especially after ABC's enormous national impact in 1977 with the twelve-hour *Roots* about the heritage of American slaves and black society. (Uncertain about its potentially ponderous prospects, ABC executives decided to put the entire series on nightly through an entire week of prime-time, prior to the regular season, so it would not interfere with the fall program schedule. To their astonished delight the series built an enormous audience up to and including its final evening's 51.1 rating and 71 share of the viewing audience; it set a record average rating/share of 44.9/66 for the week's episodes.)

Serious themes in nonseries drama also appeared in the schedules of the Arts & Entertainment cable network, pay-cable services HBO and Showtime, as well as PBS. Enormous audiences and critical praise supported producer Ken Burns' riveting multipart mini-series on the Civil War, baseball as a reflection of American society and culture, and the West—all drawing on contemporary photographs, diaries, reports, and other sources blended into a seamless moving chronicle of the individual people and their times.

Ratings reflected American audiences' enduring interest in entertainment presentations, and their significantly less support for public affairs "specials" about news, special events, and politics. The economics as well as the very nature of a mass medium reaching a vast audience makes reasonable—albeit not desirable by critics or more selective viewers—the kind of popular programming and scheduling associated with national networks.

Situation Comedy

Sitcom programs early on were slapstick (*I Love Lucy,* Jackie Gleason's *The Honeymooners*) or moralistic (*Father Knows Best, Make Room for Daddy, Andy Griffith Show*)—and still continue to reach audiences via cable's endless re-runs. Sitcom series increased steadily in number and popularity over the years, particularly with the "character" and "topical" comedy of the 1970s, such as *All in The Family, M*A*S*H, Sanford and Son, Maude, Jeffersons,* and *Good Times.* A temporary sag caused some to pronounce the form dead; whereas ten years earlier 8 of the top 10 shows were sitcoms, not a single one was among the top 10 programs of the 1983–1984 season. *Cheers* had drawn only modest audiences its first two seasons, 1981–1983, but NBC kept it until audiences began to grow, pushing it to #3 by 1986 and among the top 5 or so for almost a decade, including #1 in 1990–1991; it won 26 Emmy awards, with 111 nominations over the decade.

■ In 1983 NBC Chairman Grant Tinker defended renewing the 25th ranked program: "In the past a lot of good, solid, long running and successful shows took a long time to find their audiences. With *Cheers,* we've made an absolute judgment, a creative judgment, that television these days doesn't get any better. And we are confident that the audience will come along eventually and make it a good business judgment." His strategy of patience more than paid off. When the series ended after its eleventh season in 1993 still rated near the top, its final 90-minute broadcast averaged a 45.5 rating with two out of every three U.S. homes watching TV tuned in to the program; more than two-dozen 30-second commercial spots each cost $650,000, a record price for an entertainment program. Ten years earlier another eleven-year sitcom series *M*A*S*H* ended with a two-and-a-half hour program that drew 125 million viewers (60.3 rating, 77.0% share), the largest program audience in TV history, with spots selling for over $400,000 each. ■

Beginning in fall 1984 and on into the 1990s *The Cosby Show* and *Cheers* became perennial blockbusters; their success was reflected in the 11 new comedies added to 17 returning ones in 1986–1987. In 1988–1989, 11 of the top 15 rated series were sitcoms, spawning a succession of the genre into the mid-1990s, some featuring young adults and African Americans. Six of the top-10 regular series in the 1991–1992 season were sitcoms that drew ratings from 20.2 to 17.0, although the #1 position was held by the quarter-century-old newsmagazine *60 Minutes* with a 21.9 rating for the season. (Ratings for 129 series on four networks ranged from 21.9 down to 4.1, with the median 10.5; below 5.0 were the bottom seven shows, all on the limited Fox network.) On the Big Three networks, during the decade 1984 to 1994, sitcoms remained the top genre by far at 40% of prime-time shows—almost twice the percentage of drama and far outpacing action-adventure which had been a TV staple of earlier decades. In the mid-1990s sitcoms continued to dominate the top-10 prime-time series each season.

Quiz and Game Shows

These audience participation formats borrowed largely from radio were hugely popular nighttime attractions in television's early years. They were canceled after quiz scandal revelations of 1959–1960, but reemerged with great strength in network daytime schedules. A decade later they also became staples of early-evening "prime-time access" periods and other day-parts, through syndication to local stations.

Musical-Variety

Until the mid-1970s, this format attracted large, regular audiences. In this category were such widely different kinds of programs as *Lawrence Welk, Sing Along with Mitch, The Dean Martin Show, The Bell Telephone Hour, The Smothers Brothers Comedy Hour,* and specials featuring Perry Como, Bing Crosby, Harry Belafonte, Liza Minnelli, Johnny Cash, and Tennessee Ernie Ford. By 1974, the musical-variety program had become an "endangered species," scheduled by networks as one-time-only specials instead of as weekly series. Rock concerts, star-studded award shows, and other special music events succeeded the regular musical-variety genre. Several attempts to revive the form failed in the late 1980s, including one featuring popular country singer Dolly Parton. Audiences showed little interest in weekly programs of musical numbers interspersed with guest performers and comedy sketches. Perennial legend Bob Hope, approaching a half century of high-ratings TV performances, continued to present two or three music-variety hours every year, until his final retrospective special (at the age of 93) on November 23, 1996.

Some national cable program channels succeeded with nightly music shows, especially The Nashville Network, MTV, VH-1, and BET (Black Entertainment Network) channels.

Daytime and Late-Night Programming

Daytime network TV offers dramatic serials ("soap operas" introduced to TV in the 1950s when they had almost run their course on radio), game shows, various talk formats, and network re-runs, plus morning news-interview programs and occasional late-afternoon offerings for school-age viewers. CBS' *Guiding Light* was in its 49th year in 1996, with "As the World Turns" in its 46th year. Following prime-time and syndication patterns in the early 1990s, networks developed talk-and-topic shows and celebrity interviews.

■   All networks' daytime program schedules earned between $1 billion and $1.25 billion from year to year during the 1980s, but dropped from their traditional 60% profit margins. From 1990 to 1991 revenue from daytime sales rose about 7%–9% at ABC and CBS, to $370 million and $345 million respectively, but dropped 18% at NBC to $200 million.   ■

Late-night programming of interview-comedy-variety, news-talk, special entertainment (contemporary music, drama), and feature films pushed network service back into the overnight hours, while daytime news, interview, and featurette programming thrust network service forward into the early morning hours. NBC's long-time exclusive service with *Today, The Tonight Show,* plus *Tomorrow* and *Midnight Special* programs (subsequently *Late Night with David Letterman* and *Later with Bob Costas*), was eventually joined by CBS and then ABC who competed for early-morning and late-night audiences. By 1993 all three networks met CNN's twenty-four-hour competition by providing their affiliates with continuous news service throughout the night.

The estimated national viewing audience after local news (11:30 P.M. Eastern time) was 24.5 million in 1988 and rose to 26.8 million by late 1992. Key to drawing late-night audiences for decades had been informal talk/interview-comedy-music programs. From Steve Allen and Jack Paar in the 1950s, through Johnny Carson from 1962 to 1992, and since then Jay Leno, *The Tonight Show* drew large audiences. Meanwhile ABC had successfully counterprogrammed its highly praised in-depth news analysis with Ted Koppel on *Nightline,* and CBS scheduled original crime dramas until it aired David Letterman's *Late Show.*

Stakes in the late evening hours are high. *The Tonight Show* in the 1980s brought in about $100 million annually; in 1992 the show was estimated to cost $18 million a year to produce while grossing up to $200 million a year. Competing programs on ABC and CBS were widely pre-empted or delayed by local affiliated stations for more lucrative syndicated programming at that hour. Until Arsenio Hall dropped out of the late-night rivalry, his program was estimated to earn $1.2 million a week for syndicator Paramount, totaling $62.4 million a year—more than Paramount had gained from its highly expensive perennial prime-time leader Cheers.

Newsmagazines and "Reality" Programs

The early 1990s abounded with nonfiction forms of programming. Newsmagazines followed the lead of venerable *60 Minutes,* which introduced the form in 1968, and also the satellite-syndicated *Entertainment Tonight's* lighter view of show business and celebrities. Real-life video and dramatic re-creations included compilations of amateur camcorder videos of humorous incidents or tragic events; gritty recounting and reenactments of police, firefighters, rescue teams, heroes, and just plain folks in jeopardy. By 1993 the success of 14 "reality" shows already on the air prompted networks to plan additions for the fall season: Fox developed five more series, ABC and CBS two each, while NBC stood pat with the four already on its schedule. These were in addition to similar series distributed through first-run syndication. In the 1993-1994 season 15% of prime-time was reality-based programming, although their luster was beginning to fade. Beyond audience and advertiser support for networks' in-house newsmagazine programs, news divisions could produce them for two-thirds or even half the cost of licensing hour entertainment programs from outside production companies (see "Network News and Public Affairs" below).

Network Sports Programs

While a form of entertainment, sports were split off from the network programming division because of their specialized negotiations and scheduling and their impact on expenditures and time sales. Costs for sports coverage more than kept pace with mounting inflation and spiraling fees by leagues and clubs that benefited from vast audiences viewing the games. As early as 1970 when a total of 13 million people paid to attend pro football, 20 million regularly viewed those same games on TV. Stadium seats in several dozen major cities could hold only so many ticket-holders, but TV brought games to expanding audiences in the nation's millions of rural and urban households.

Production costs rose swiftly as more sophisticated coverage demanded elaborate technical facilities and engineers. Every week ABC used more than a dozen color cameras, and clusters of videotape recorders plus slow-motion machines, including a second unit crew of director and cameras for "isolated coverage" (repeat inserts of excerpted action). Satellite links permitted live inserts of other games in process around the league and the world, interspersed with updates on sports scores from New York studios. New York Times News Service writer Joe Lapointe noted ESPN's technical contribution to college basketball as "the development of the 'whiparound' format of cutting back and forth among various games taking place simultaneously; at its best, the technique blends seamless editing with tremendous pace."[29] Sports have intrinsic appeal with their live action and element of the unexpected. The isolated camera, instant replay, stop-action freeze-frame, slow-motion and directional microphones plus prerecorded mini-features and interviews add to the interest in TV coverage, along with informed (although at times garrulous and intrusive) play-by-play analysis and commentaries about players.

This section on network sports notes *patterns* and *trends*. Fact-based management makes decisions based on analyzing such data, not merely on creative insights or "gut" feelings. Reviewing statistics is a key step for managers in assessing future program strategies.

The National Cable Television Association countered claims that cable was siphoning off sports from free over-the-air TV. In 1989 it reported that sports programs on networks had risen the previous year to a record 1,753 hours, including Olympic coverage, and that local TV stations carried 1,647 Major League Baseball games in 1989 (up 111 games from four years earlier). In the preceding two years the networks had increased college football coverage of regular season games by 52% (up from 27 to 41 games in 1989) and of NCAA regular games and post-season bowls by 21% over four years (up from 86 to 104 in 1989–1990). In 1990–1991 CBS alone televised 700 hours of sports events. In addition to competing with NBC and ABC sports coverage, it faced national ESPN and superstations and 32 regional all-sports cable ad hoc networks, plus local and regional sportscasts by over-air television stations.

In 1993 the FCC made public its similar findings from data covering a decade: no migration to cable by NFL or college basketball, and only modest movement to cable by Major League Baseball, the National Basketball Association and National Hockey League. Cable often covered games passed over by the major networks; ESPN signed a four-year contract with Major League Baseball for $400 million to broadcast 175 games annually starting in 1990, supplementing CBS's 12-game regular season coverage (but after early heavy losses ESPN sought to buy out of that contract for a $400,000 penalty fee). On the other hand, after eleven years of carrying NCAA Division I basketball tournaments, ESPN lost the rights to CBS in 1991.

Football

In 1936 the CBS radio network *was paid* $500 by the Orange Bowl Committee to broadcast the game; in 1972, NBC purchased the rights to telecast the Orange Bowl for $700,000 and in 1988 ABC paid $10 million for the Rose Bowl (and the same amount annually during each of the nine years in the contract). Since 1952, the National Collegiate Athletic Association (representing amateurs, be it noted)—just as the professional NFL—sought competitive bids every two years until Federal courts removed game negotiations from NCAA control in 1984. Until the U.S. Supreme Court's decision in 1984 concurring with lower courts, the NCAA had controlled how many times a year teams could appear on national television and determined colleges covered and rights to be distributed among colleges; after 1984 broadcasters could negotiate directly with individual colleges and conferences, the Big Ten/Pac Ten and the new CFA/College Football Association. The result: from only two college games per weekend, broadcast and cable soon carried twelve or more on Saturdays.

CBS had carried NCAA games through the 1962 and 1963 seasons; NBC bid $13 million for the next two seasons. Then, without announcing any open bidding, the contract for 1966 and 1967 was awarded to ABC for $15.5 million. By 1988 CBS alone paid $80 million for rights to college sports which made up one-third of that network's sports schedule; all broadcasters' TV fees totaled $150 million. Between 1988 and 1993 TV networks, cable, syndicators, and local stations expected to pay more than three quarters of a billion dollars for rights to air collegiate sports.

■ Those early years set the foundation for subsequent decades of competitive bidding for TV rights. From 1980 to 1990, total attendance at NFL games grew 25%, while NFL players' average salaries increased 343%! Box office ticket prices and spiraling TV rights reflected and contributed to those figures. (So did merchandise endorsements, with the NFL grossing $1 billion from 1990 sales). NFL rights for the 1990 season went to ABC, CBS, NBC, ESPN, and Turner Broadcasting System for $910 million, almost double the previous year's contract.

For a four-season contract through 1994, the three major networks paid $2.74 billion for an average of $664 million per season, including preseason and playoff games and three Super Bowls; adding ESPN and Turner's TNT, the total four-year NFL rights package cost $3.65 billion. For those four years CapCities/ABC spent a total of $1.35 billion for football broadcast rights for ABC-TV ($900 million) and 80%-owned ESPN ($450 million). At renewal time, Fox outbid CBS by 50% for the National Football Conference two-year package, paying $1.58 billion for broadcast rights to the 1996 and 1997 seasons. The remaining rights to NFL football went to ABC for $920 million, NBC $868 million, ESPN $524 million, and TNT $496 million—for a two-year grand total of around $4.4 billion to the football league. ■

NBC paid approximately $3 million to cover the Super Bowl on January 12, 1975, charging advertisers $214,000 for a commercial minute— the highest rate-card price for any once-only attraction up to that time. In the 1996 Super Bowl XXX, 30 seconds of commercial airtime cost advertisers over $1,000,000. Those charges reflected audiences delivered—the largest audience of any program in television history, 138.5 million viewers (a 46.1 national rating, 72 share of all U.S. homes watching TV those hours).[*] For 1997's Super Bowl XXXI the Fox network sold a half-minute of airtime for $1.2 million; all but five of the 58 slots were sold to advertisers five months before the January game.

Networks sometimes paid more for rights than time sales would bring in, partly for prestige, and also to promote their other programs, while also enhancing overall ratings. With heavy payments for rights surpassing revenues from airtime spots, in the 1993 football season alone the networks lost about $250 million: CBS

---

[*]The top-10 shows of all time were Super Bowls except for the final episode of the eleven-year sitcom *M*A*S*H* in 1983; that two-and-a-half hour special show delivered a 60.3 rating and a 77 share (121.6 million viewers in that earlier, smaller TV universe).

lost $100 million, NBC $75 million, ABC $50 million, and ESPN and TNT smaller losses on their split Sunday night package.

These paragraphs of numbers and statistics demonstrate the challenge to network programmers and sales staffs. They fiercely compete for viewer attention to their schedules and for advertiser investment in spot availabilities in these expensive long-term sports packages.

Baseball

Local and regional radio and TV stations carry the bulk of baseball coverage of the expanded two leagues' 26 teams; and networks negotiate multiyear and shared network packages.

■ Media rights after 1989 rose significantly when NBC-ABC were outbid for the 1990–1993 rights by CBS ($1.06 billion) and ESPN ($400 million). CBS wildly outbid its rivals because it stood third in rankings and needed the prestige and promotional advantages of summer and fall baseball to bolster its position among national audiences. It was the first time that network carried baseball in a quarter century, and the first time NBC had no baseball at all since the late 1940s. ■

Both CBS and ESPN lost heavily in the first year of their contracts, and more in subsequent seasons. In 1991, 30-second spots sold for under $25,000, only half of what commercials brought the previous year. (In Fall 1991, CBS reported sports losses to that point were already $604 million. That was only the second year of its four-year contracts with Major League Baseball for $1.06 billion and also with the NFL as well as the second year of its seven-year $1 billion contract for exclusive coverage of the NCAA basketball tournament.) Industry estimates were that CBS and ESPN only earned sales revenues amounting to about half their baseball rights fees; thus over the four-year contract CBS lost nearly $500 million and ESPN $200 million.

That bitter fiscal failure marked an abrupt turning point in network negotiations for professional baseball rights. MLB knew it could no longer draw lucrative contracts from networks that lost half a billion dollars on them. Rather than exercising its $235 million option to televise major league baseball in 1994 and 1995, ESPN instead opted to pay a $13 million buyout to MLB to be free of further nine-figure losses. ABC and NBC offered a combined strategy that eliminated all rights fees, replacing them instead with a joint venture of MLB and the networks to share airtime revenue, with MLB receiving over 80% of sales income; however, that arrangement was canceled in 1995. ESPN was able to reduce its previous contract by renewing at the same $400 million level, but for six years rather than only four, from 1994 to 1999. Major League Baseball administrators reversed an earlier comment made in the context of lowering rights fees to networks, instead claiming they would refrain from contracting for pay-per-view telecasts at least until after this contract expired prior to the 2000 season. In 1996 national network fees for broadcast rights totaled $244 million, while local/regional TV, cable, and radio rights fees brought in $239 million to MLB clubs around the country.[30]

Basketball

Basketball coverage grew gradually at first, then geometrically expanded in the 1980s. By the 1987–1988 season the three major networks plus ESPN, USA Network, FNN/Score, and superstation WTBS televised more than 450 professional and collegiate games (ESPN aired 182 of them). Regional cable sports networks carried another 60 games, and local TV stations aired countless local contests.

■   When NBA contracts expired, NBC won rights for the next four years, 1990–1994, by tripling the bid to $600 million. Turner Network Television (TNT) and TBS on cable renewed its contract for another four years to 1994 for $275 million, almost double its previous rate for each two years; it subsequently paid 25% more for 10 fewer games annually when it paid $350 million for NBA rights from 1994 through the 1997–1998 season. Once again, NBC won rights to another four-years, 1994–1998, for $750 million, plus sharingwith the NBA advertising revenues above a certain level ($1.06 billion). Meanwhile CBS in 1994 committed $1.7 billion to extend its rights to NBA post-season coverage for seven more years until 2002.   ■

This revenue-sharing factor in mid-1993 was the first between professional sports and a network. CBS a few weeks later made a similar offer but lost to a combined NBC-ABC bid for the next round of Major League Baseball rights through 1999, as described earlier. The tactic was prompted by losses suffered by networks paying enormous fees for sports contracts which could not earn enough revenue from time sales in those games to cover their costs. Modest audience viewing levels coupled with advertisers facing a recession economy kept commercial sales soft in the early 1990s. Professional sports leagues realized they had to accept reduced fees, or less coverage by the networks. The contract meant that networks could at least be sure to meet their expenses plus some profit, with any extra profits shared with the pro leagues.

The Olympics

Every four years since 1960 the networks have competed for the rights to cover the Olympic Games. CBS aired the first Olympic broadcasts that year: Winter Games in Squaw Valley, California, for $50,000 and Summer Games in Rome for $394,000.

■   CBS won North American TV and radio rights to the Winter Games in Lillehammer, Norway, in 1994 with its bid of $300 million, and to the Nagano, Japan 1998 Winter Olympics for $375 million. NBC won rights to televise 168 hours over seventeen days of the 1996 Summer Games in Atlanta with a bid of $456 million, plus splitting with the International Olympic Committee all time-sales revenue over $615 million. (ABC had unsuccessfully bid $450 million and CBS $415 million, with no revenue sharing.) With strong ratings for those Atlanta Games, NBC estimated a record $70 million profit (the six-nights prime-time 22.8 rating surpassed the 18 guaranteed to advertisers, avoiding "make-goods"). Ad revenues from forty-eight sponsors came to $685 million, against the rights fees of $456 million and another $159 paid for production costs and revenue-sharing with the Atlanta Olympic committee. NBC paid $715 million for Sydney's Summer Games in 2000 and another $555 million for Salt Lake City's Winter Games in 2002. In what trade magazine *Broadcasting & Cable* called the richest deal in sports history, NBC also paid $2.3 billion for rights to the Summer Games in 2004 and 2008 and Winter Games in 2006.   ■

Other Sports

Network commercial television covers a wide range of sports. During the mid-1990s Fox paid $150 million and ESPN $100 million for a half decade of National Hockey League games through the 1998–1999 season. What networks do not broadcast, cable channel services cover, in addition to regional and local TV. Especially on weekends the networks offer broadcasts of major golf tournaments, tennis championships, auto racing, and thoroughbred races including the "triple-crown" of the Kentucky Derby, Preakness and Belmont Stakes. The sports fan has a permanent seat in the television grandstand, especially with the addition of cable's ESPN2, and DirectTV's comprehensive multi-channel pay-TV coverage of professional and collegiate sports, along with other cable sports coverage. Premium and pay channels, including HBO and Showtime, offer championship boxing; USA Network airs tennis and golf. And more than 20 regional sports cable operations around the country also held rights to the NBA, NHL, and Major League Baseball in the late 1990s.

High rights fees led to coalitions among broadcast and cable operators. Some corporations offered those opportunities internally: Disney/ABC/ESPN, CBS/Group W, Fox/Liberty, and Time Warner/Turner. CBS and NBC also looked to cable partners in sports coverage.

TV NETWORK NEWS AND PUBLIC AFFAIRS

Television News

Only gradually did television news become a distinctive service of national networks. It was not always that way, starting with inauspicious 15-minute evening newscasts pioneered by CBS with Douglas Edwards in 1948. Producer Fred Friendly and others echoed media philosopher Marshall McLuhan when they called network news the central nervous system of the nation. CBS producer Don Hewitt who created *60 Minutes* bit the hand that fed him when he reflected on shifting priorities of mergers, buyouts, and cut-backs due to debt servicing in the latter-1980s: "We [news people] don't belong with the entertainment conglomerates. We belong in the journalism business. The networks are nothing but an outmoded delivery system [for news]."[31] But David Burke, executive vice-president of ABC News and later briefly president of CBS News, disagreed with his perspective: "We don't feel we are part of an entertainment conglomerate. We believe we are the leaders of the network" whose "extraordinary access to the people" makes it "a basic national delivery system to deliver their [news] message." James Lynagh, president of Multimedia Broadcasting and chairman of NBC's affiliate board of governors, noted at that time:

> Individual affiliates may complain from time to time about news pre-emptions or various other things, but the network news operation is a wonderful resource that this nation has. . . . When there's an Iran-Contra hearing, late night specials, documentaries, late night news programming, prime-time and early evening, the *Today* show, the weekend programs—all combined, they are a resource that we [local stations] could not duplicate.[32]

Broadcast managers are torn by the twin responsibilities of informing their public with accuracy, truth, and swiftness while not being a financial drain on the total operation—and even generating income beyond newsgathering costs. Historical perspective lets us understand better how this all came about.

History   As the three networks' entertainment offerings became more and more similar, each network increasingly depended for its distinctive image on its sports packages and on its news and public affairs programming. In retrospect, the season of 1963–1964 was the breakthrough year for TV network news and public affairs. It was the year the Roper poll first disclosed that television was the public's major source for news; it was the year NBC and CBS doubled the length of their evening newscasts to 30 minutes; it was the year the three television networks committed themselves to spend $70 million for the production of news and documentary programs for the season, prompting *Broadcasting* magazine to editorialize that the rise in information programming was "the healthiest trend to develop in broadcasting's recent history" and would "do more than most others for broadcasting and for the country." Further, 1963–1964 was the year the three networks averaged about six times as much "hard" news coverage as they had in 1950 and about 100 more hours of documentaries than they had carried as recently as 1959 (with over half of the documentaries sponsored).

It was also the year of covering devastating events involving the assassination of a president of the United States.

The Weekend of November 22nd, 1963

That awful weekend brought the heaviest possible challenge to electronic journalism. The cancellation of almost all regular programming and advertising, the medium's thorough participation at every step, and the dignity of its coverage of all events marked television's coming of age as a major news medium that earned the public's respect. NBC's instant coverage began at 1:53 P.M. on Friday and continued for 70 hours and 27 minutes until 1:16 A.M. Tuesday morning, replacing all entertainment programs and commercial advertising. ABC and CBS had similar records.

The television audience on that weekend approached universality for the first time in history: 96.1% of all homes with TV sets, according to Nielsen, watched the weekend coverage an average 31.6 hours per home. The weekend's largest audience was assembled during the requiem funeral mass, "attended" through television by more than 97 million persons. Almost as many watched the burial service at Arlington National Cemetery. At no time, day or night, during the entire weekend was the audience less than 14% of all homes, representing at least 7 million people.

The public's reliance on television in those hours of crisis is reflected in Nielsen's report of the expanding audience throughout the afternoon and evening of the assassination. At 1:30 P.M. eastern standard time, on Friday, November 22, television sets were on in approximately 23.4% of American TV homes. Fifteen minutes later, after the first reports of the assassination, more than a third of all TV homes in the country were tuned in. By 2:00 P.M. the audience had increased to 42.2% of homes. Two and a half hours later, sets were on in 75% of all homes. By 11:00 P.M. the people in 92.6% of American homes had watched an average of almost six hours of television coverage. But these figures are almost irrelevant to the tragic event that surpassed statistics, surveys, and cost accounting.

The tragedy in Dallas, the moving rituals of mourning, and the outpouring of the nation's love and grief bonded through simultaneous viewing, were unparalleled in American experience. That much of television can be trivial and tawdry—as much of quite ordinary life can be trivial and tawdry—is beyond question. But the tragedy evoked a common dignity and a common greatness. In shock and trial and bereavement, no medium of communication served so selflessly and so well.

Regular News Coverage

By coincidence, not many weeks before that November weekend, CBS and then the other networks had expanded their quarter-hour evening newscasts to a half hour. During the previous fifteen years limited technical facilities and budgets had supported the shorter news form. For the next four decades network evening news remained at 30 minutes (actually 22 minutes, minus credits and commercials), despite satellite and efficient ENG capability, epochal events of civil rights marches, intensely controversial Vietnam war, recessions, near impeachment and resignation of the president with "Watergate" revelations, widespread famine, the rise of drugs and violent youth gangs, "Iran-Contra" scandals, and the collapse of the Soviet empire, along with recurring natural disasters and intermittent classic international sporting events and scientific accidents. (Earlier we recounted how news managers and network executives repeatedly proposed hour-long newscasts but local affiliates each time rejected them.)

■　　In the late 1950s NBC's *Huntley/Brinkley Report* from New York and Washington attracted the largest national audiences. After Walter Cronkite became evening news anchor in 1962, *CBS Evening News* sustained first place for fourteen of the next nineteen years; successor Dan Rather saw that lead diminish in 1985–1986. NBC briefly led the next season. In 1988–1989 *ABC World News Tonight* with Peter Jennings led in audience favor, and continued to hold its lead through 1996, while CBS and NBC vied for second position. Mirroring the audience's gradual drift to cable and other media, average evening news ratings for all networks—which had grown from 13.5 in 1976 to 15.1 in 1980—dropped by 1984 to 12.4% of all U.S. households with TV (CBS was 14.4, ABC and NBC tied at 11.4).* By 1988 that overall average dropped a further 20% to 9.4 (ABC 9.9, CBS 9.6, NBC 8.6). Put another way, during that decade the three networks' combined ratings, which totaled 38.2 (76% share of all households viewing TV at that hour) in the 1979–1980 season, dropped to 31.6 (59% share) in 1988–1989, and to 30.4 in 1992–1993 (the network news' shares continued into 1995 at about 60% of viewers for that half hour of early evenings).　■

A factor, of course, in network news is the lead-in audiences supplied to each by their respective 200+ affiliated stations. In markets where aggressive managers schedule respected and widely viewed local news as well as top syndicated programs in access time, their networks' news programs generally lead competing network news in that area. All those 200 markets added up together constitute a network's national rating.

---

*Through those same years, however, the country's population continued to grow (more people, more homes, more TV sets, somewhat more hours of daily viewing), so decreasing ratings were percentages of a rising universe. CBS executives pointed out that when the three-network evening news programs together reached 29.6 million homes in 1981, five years later they reached 29.7 million homes despite diminished ratings/shares.

Top management sets the standard and the style for broadcast operations, including news. In 1977 Roone Arledge began to oversee the ABC news division in addition to being head of ABC sports. He soon moved its news into contention with CBS and NBC by innovation including creative use of new technology, graphics, and satellite-interconnected live interviews (*Nightline*). He successfully held that executive post for over a decade. Lawyer Richard Salant headed CBS News for two of its most respected decades, then in the retrenching years of mergers and cut-backs in the 1980s, a succession of news presidents struggled to retain or repair CBS' news position. Similarly at NBC some five news presidents rotated through a decade—about the same two-year average as local stations—followed by a sixth and seventh in the early 1990s.

In their effort to attract audiences and to compete with services like CNN and CNN Headline News on cable, the networks' evening news of the 1990s moved away from reporting the day's headline stories exclusively. They introduced "soft" feature pieces and more in-depth analysis of continuing stories, from three to five minutes each out of the 22-minute program.[33] The "special reports on social issues" received as much as one-third to almost one-half of the 22-minute news block. Topics included investigations of serious matters such as homelessness, unemployment, health care, education, the military, family values, and religion. But they also included lighter items about show business and celebrities. (One study found that all networks in 1990 devoted 68 minutes a month to arts and entertainment, up from 38 minutes the two previous years.)[34] CBS News anchor Dan Rather publicly regretted this trend and TV tabloid newsmagazines in a speech to the Radio and Television News Directors Association convention in 1993. He urged news colleagues to "fight the fear that leads to 'showbizification'" and lamented that the broadcast news medium has often "been squandered and cheapened" resulting in lowered credibility and professional reputation.

Meanwhile, PBS continued its highly respected nightly *MacNeil/Lehrer News-Hour;* with Robert MacNeil's retirement in 1995 it continued as *The News House with Jim Lehrer.* It offered perspective on each day's news events and on continuing stories by lengthy interviews often with the principals involved as well as expert opinion. In 1989 alone, for its previous year's news and documentaries PBS won 18 Emmy Awards from the National Academy of Television Arts and Sciences, while CBS won 16, ABC 8, and NBC 5. (Lehrer's competence and reputation for even-handedness prompted rival political organizers to place him as sole moderator of the two Clinton-Dole presidential debates and the Gore-Kemp vice-presidential debate in October 1996.)

Early Morning News Service

Competition gradually grew for the early morning national audience long after NBC's two-hour *Today* began in 1952. Although the *CBS Morning News* was respected for its focus on serious journalism, it eventually lost favor with the masses who preferred the side-bar feature pieces and celebrity interviews on NBC and later on ABC's entertainment-oriented *Good Morning America* (anchored not by a newsperson but by actor David Hartman). With titles, anchors, and casts shifting through the decades, NBC and ABC took turns leading in ratings with CBS always far behind. Caught in a succession of CBS News presidents, that news division lost the morning news/talk program after three decades when it was turned over to the network's entertainment programming division.

By mid-1993 ABC's *GMA* with a 4.0 rating moved ahead of NBC *Today*'s 3.8; *CBS This Morning* still trailed with 2.6% of U.S. households. To bolster affiliates' position in their own markets, CBS turned back all or portions of the first hour to stations preferring local news—with three national news inserts—while keeping the second network hour intact; 70% of its affiliated stations cleared their schedules to carry that cooperative program.

Links with Cable

In 1995 all three networks planned 24-hour domestic news services for cable channels. But in May 1996, ABC backed off of that plan. (Thirteen years earlier it had joined with Group W to offer Satellite News Channel, but lost $35 million the first year; it sold the operation after sixteen months to competitor Ted Turner, who had created CNN Headline News as a direct response to their entry into the field.) NBC joined Microsoft to build its MSNBC cable news channel in 1995. Rupert Murdoch inaugurated the Fox News Channel in October 1996, with $150 million start-up funding (and a projected loss of $400 million over the first five years). The service was fed to 16 million home subscribers to TCI, Continental, and Comcast MSOs—to whom Murdoch paid an unprecedented $10 per subscriber as a bonus for carrying the service. But it lost 12 million more households when Time Warner/Turner changed previous plans and instead carried MSNBC—prompting a $2 billion lawsuit by Fox's parent News Corp. In mid-1996 MSNBC reached 24 million homes, and veteran CNN 67 million (nearly all the country's cabled homes).

Sunday News and Public Affairs

The Sunday morning standard-bearers of network news continued through the medium's history. NBC's venerable *Meet the Press* panel of newspersons interviewed its first national figure in 1947; it initially aired on weekday late evenings, moving to Sundays in the mid-1950s. CBS's *Face the Nation* joined it in 1954. ABC's *Issues and Answers* began in 1960 and was succeeded in 1981 by *This Week with David Brinkley* whom that network had lured away from NBC. All based in Washington, the programs through cross-examination and discussion explored currently breaking stories usually with those in top government positions. They regularly made Monday morning newspaper headlines, partly because most other offices of business and government were shuttered on the weekend. In 1990 *Brinkley* attracted 3.6 million on Sundays, *Face the Nation* 3 million, and *Meet the Press* 2.6 million. Also on Sundays, CBS presented *Sunday Morning* and *The Wall Street Journal Report,* NBC *Sunday Today,* and ABC *Business World.*

Produced by network news divisions for a quarter century, CBS' Sunday morning religious programs left the air in 1978, when only 44 of 210 affiliates were clearing local time to carry them (many instead aired paid religious syndication). Rotating topics and representatives of major faith groups, *Lamp Unto My Feet* began in 1948, *Look Up and Live* in 1954. The two half-hour programs were succeeded by *For Our Times* which presented religious themes for ten more years.

Innovative and experimental *Camera Three,* a cultural affairs series carried since 1956, also ended its run in 1978. CBS replaced it and the religion programs with a 90-minute newspaper of the air, *Sunday Morning* with Charles Kuralt.

Special Events Coverage

Perhaps as much as any other single influence, television prompted public awareness of the civil rights movement and efforts by minorities to be recognized and respected. Although networks and their news divisions in the early 1970s met with considerable criticism (especially from the Executive Branch of the federal government), broadcasters were widely praised for their continued coverage of the series of revelations and hearings about the "Watergate" scandals and for their deft presentation of the abrupt transition of power from the Nixon to Ford administrations in 1974.

Network television news divisions have perhaps offered their finest service to the national public when world events swiftly triggered concrete manifestations of important issues. Networks not only reported them in regular news programs but also created "instant specials" consisting of partly news, partly documentary, and partly special events real-time coverage. Some occurred with advance warning, others broke on the scene unexpectedly: civil disturbances and riots in Newark (New Jersey), Watts (California), Detroit, and Washington, D.C. in the 1960s; bloody tumult in the streets of Chicago during the Democratic National Convention in 1968; a nation torn by Vietnam and mass demonstrations into the 1970s; turmoil in the White House, leading to resignation; terrorists holding hostages for over a year, releasing them at the moment of President Reagan's inauguration ceremony in 1981; bombing of U.S. Marine installations in Beirut; guerilla warfare in central America; Iran-Contra hearings with Lt. Col. Oliver North in 1987; the fall of the Berlin wall in 1989 and violent wrenching of nascent freedom in Tiananmen Square, Beijing; South African struggles against apartheid; famine in Ethiopia and Somalia; bombing atrocities in Northern Ireland; the Persian Gulf War with "Operation Desert Storm's" coalition of world nations allied against Iraq in 1991; and natural disasters around the world from Mexico City's earthquake in 1985, another in San Francisco at the start of a World Series game in 1989, to hurricanes thrashing Florida in 1992 and destructive record floods in the Midwest in 1993. There were also more benign special events such as America's Bicentennial celebration, visits of popes, and space shots with live TV of astronauts in their spacecraft or floating alongside it orbiting around the planet.

Swift compilation, editing, and mounting of historical footage on persons and former events, scripts, guest specialists, graphics, studio sets, crews, and newspersons—as well as AT&T long lines, satellite feeds, and other remote technical equipment, plus program schedule changes—have been marshaled by network staffs on short or no notice, often to critical acclaim.

■   Networks provided extended coverage of "Desert Storm," the short, intensive combat in Iraq in 1991. CNN anchored early coverage of the initial engagement because only it had correspondents and equipment in Bagdad. That night and the next day NBC's Tom Brokaw in New York spoke in detail with CNN's Bernard Shaw in Bagdad, live-by-satellite, almost as if he were NBC's correspondent. Brokaw acknowledged "the very enterprising Bernard

Shaw representing the very enterprising CNN." CNN's ratings were limited to cable audiences; ABC outdrew its competitors for the national broadcast audience.

All networks turned over their complete schedules to on-going news coverage, as they had the weekend of President Kennedy's assassination. No network went longer than sixteen hours without commercials; they had started to insert spots of advertisers who were willing to have their products showcased in the midst of war coverage. Several radio networks cut all commercials during the first twenty-four hours of coverage. *Broadcasting* magazine estimated the networks could lose $5 million to $9 million a day without commercials or with a reduced number. (That year each network earned roughly $8 million a day from commercials.)

Major TV and radio networks sent some 325 reporters, producers, camera-persons, and technical personnel to the Middle East; almost half (150) were from CNN. It was estimated that over the seven months of Persian Gulf coverage up to the war, the four networks total $145 million in news expenses. Continued coverage after the brief war cost each network $1.5 million a week (beyond each network's annual newsbudget of $260 million to $350 million). The three broadcast networks lost some $125 million in canceled commercials; CNN lost $18 million over six weeks from the war's outbreak. NBC put its total loss over the two months of Persian Gulf war coverage at $44.7 million, including costs of news coverage plus commercial revenue lost from pre-empted entertainment programming. CNN estimated total coverage of the war cost it $12 million to $15 million. ■

In January 1994, networks relayed nationally live local station coverage of Los Angeles' major earthquake. Because of its twenty-four-hour flexible schedule of news, CNN first reported the disaster from that city at 7:39 A.M. Eastern time, eight minutes after the earthquake struck. National attention pushed up the networks' combined ratings for that evening's news by 38%, and the following morning's news shows by 48%.

Documentaries

Early documentary series included NBC's twenty-six-part *Victory at Sea* in 1952–1953, and Edward R. Murrow's series *See It Now* that began on CBS in 1952 and three years later became *CBS Reports*. NBC established a documentary unit to produce its *Project XX* series; CBS did the same with *Twentieth Century*. In those days, a documentary competing with an entertainment program almost always showed a dismal rating. Sponsors anxious to reach large audiences considered the documentary a poor investment. But sponsors changed their opinions before the public did. By the end of 1960, there was increased advertiser support for all informational programming. On the heels of the quiz show scandals, with threats of government intervention into program practices, part of the networks' response was to mount well crafted analyses of social issues, regularly scheduled in prime-time.[35]

Two important network decisions support the documentary's growth as an important format. The first decision was to increase documentary budgets to permit more time, personnel, and facilities to be devoted to their preparation. As a result, a type of program once dull and one-dimensional in its scope became exciting and broad in treatment. Documentaries began to appear regularly in prime-time evening schedules. While their ratings did not compare with popular entertainment, the size of their audiences showed as much relative growth as other kinds of programs.

The second network decision was to deal with controversial subjects previously judged too risky to present on national television. As a result, network documentary units broke barriers of public sensitivity on various topics by bringing analyses into the open. Edward R. Murrow and Fred W. Friendly had pioneered this investigative reporting on substantive issues from the mid-1950s on, in the *See It Now* series on CBS-TV. *CBS Reports* and *NBC White Paper* led the way. Subjects examined by network documentary units included: integration, poverty, urban development, campus morals, welfare inequities, unemployment, menopause, the population explosion, birth control, bookies, problems of adoption, traffic in drugs, juvenile delinquency, abortion, cigarette smoking and cancer, and divorce. International issues treated included the Cuban crisis, the Berlin Wall, the conflict between India and Red China, the Panama revolt, revolutions in South America and the Middle East, the Kremlin, the Vietnam and Laos situations, the common market, and emerging nations of Africa. Not all documentaries dealt with serious subjects but even the lighter ones informed, broadened understanding, and stimulated thought. They explored topics like the decline of royalty, the American woman, the festival frenzy, the circus, the decade of the twenties, and American humor.

Some investigative studies amounted to exposés and generated stiff reactions from their subjects, often highly placed in government and industry. Major lawsuits were brought against networks and producers at times. In most instances the general accuracy of documentaries was supported, but the turmoil and legal expenses took their toll on network management.

In 1970 the three networks broadcast 79 documentaries in prime-time, in 1980 only one-third that many, and by 1986 just over a dozen. CBS aired 20 to 24 hour-long prime-time documentaries each year in the 1970s; that number was reduced to about 15 hours in the early 1980s. The entertainment division jealously preserved prime-time for maximum audience-generating shows, allotting some 15 or 16 hours for documentaries plus several hours for unplanned news specials. Even then the rare hour allotted often ended up pitting the documentary against another network's blockbuster program, further diminishing audience potential.

Documentaries grew in the 1950s and 1960s when network evening news was only 15 minutes long, and local stations aired only a half hour nightly. The FCC's demanding licensing requirements fostered news and public affairs programs by stations. The traditional form all but disappeared as networks went to half-hour evening news plus morning news hours and weekly newsmagazines (*60 Minutes* in 1968, *20/20* in 1979) and stations expanded local news to two or more hours daily.

Pragmatic concerns of costs vs. audience size in intensely competitive prime-time, plus general "downsizing" of networks and their news operations, caused the documentary form to fade. It was gradually replaced by the more flexible, faster paced format, the newsmagazine.

ABC aired its *Closeup* documentaries; CBS offered an occasional *CBS Reports*. All networks produced special-event programs as national and world affairs dictated.

PBS took up the cause of documentaries about controversial social issues, nature studies, human exploration, and arts and culture generally. Its *Frontline* documentary series continued in the mid-1990s. Oscillating between PBS and CBS for over two decades, Bill Moyers regularly offered thoughtful investigations, at times almost poetic in their sensitive portrayals. *Bill Moyers Journal* offered 26 to 39 weeks of probing with intellectual curiosity, as did his *A Walk Through the 20th Century with Bill Moyers;* between 1971 and 1990 Moyers developed over 600 hours of programming for commercial and noncommercial networks. Filmmaker Ken Burns contributed impressively to the genre, especially with his widely honored documentary mini-series *The Civil War* followed by multipart *Baseball* and *The West*—each series airing on successive evenings for 12 to 18.5 hours.

Newsmagazines

This form emerged into its own with CBS' *60 Minutes* which in 1968 began to alternate with *CBS Reports* during the last hour of prime-time on Tuesday nights. It ranked 101 that season.

> ■ The *60 Minutes* hour was divided into three major investigative pieces, with a light feature at the end. The pace and subject matter eventually attracted large audiences interested in nonfiction, but only after it was moved first in 1971 to late Sunday afternoon where it was often pre-empted by runover football games, and finally in 1975 to an hour later at 6:00 ET as counterprogramming to *Disney's Wonderful World.* That eighth season it ranked 52nd. Two years later it was fourth and has remained through 1996 among the top five or ten programs every year, in some years ending in first or second place—after over a quarter century on the air! In 1985 the series brought in $70 million, almost one-fourth of CBS annual profits. In 1987 each hour cost less than $800,000 to produce, while its thirteen 30-second commercials brought in $3 million for each broadcast. Producer-creator Don Hewitt estimated in 1991 that during its 23 years, *60 Minutes* had netted more than $1 billion. ■

The newsmagazine genre settled in somewhere between traditional documentaries and the newer investigative and exposé tabloid syndicated half-hours *Hard Copy* and *Inside Edition.* In the late 1980s and into the 1990s news staffs at CBS produced *60 Minutes, 48 Hours, Street Stories,* and *Eye to Eye with Connie Chung* (the first two succeeded); ABC countered with *20/20, Prime Time Live, Day One,* and *Turning Point* (its first two also remained). After several years of failed attempts including *American Almanac, Yesterday, Today, and Tomorrow,* and *Exposé,* NBC's news division successively introduced *Dateline NBC, I Witness Video,* and *Now;* only *Dateline* succeeded—so much so that the network expanded the hour magazine to three nights weekly in the mid-1990s. The Fox network joined them temporarily with *Front Page.*

The newsmagazine format filled nine hours of network prime-time in the 1993–1995 seasons. Most were produced for less than $500,000 a program, half what networks paid for an hour entertainment program that often failed. *60 Minutes* costs more because its half dozen veteran correspondents each averaged almost $1 million annually and creator-producer Don Hewitt earned $2.5 million—all of whose compensation was covered by revenues from just three episodes.

Support by Advertisers and Affiliates

A key factor in the success story of network news and documentary units has been their insistence on independence from sponsor control. No other network program group has been so divorced from advertisers' advice or restraint in selecting and treating controversial subjects. In a few cases, advertisers have threatened cancellation. In most instances, the networks have backed their documentary and news producers by announcing the programs would be aired with or without sponsorship. They rarely lost advertisers; because a dozen different companies bought the half-minute spots, none was wholly identified with the series. Research shows that news-documentary programs typically reach far more professional and white-collar people than entertainment programs, far more with one or more years of college, and more at upper-income levels—optimum target audiences for selective advertisers.

The problem of clearing time on affiliates' local schedules has been difficult for documentaries, less so for audience-attracting newsmagazines. In 1965 almost half the CBS affiliates refused to carry *CBS Reports.* The following year substantial numbers of network-affiliated stations did not carry news programs on the war in Vietnam. Specials on February 12 of that year were broadcast by only 42% of CBS stations and 30% of NBC stations. Fewer than half of CBS stations carried a Vietnam special on February 18 while 65% of NBC stations carried a similar special the same evening. Although sponsored evening newscasts can count on affiliate acceptance, individual weekly public affairs programs were often rejected or else carried on a delayed basis by a large number of affiliates.

It was more profitable, and avoided controversy, for a station manager to schedule a syndicated film in place of a network documentary program. Widespread refusal to carry their networks' public affairs programs was a major factor in lower national ratings for those programs, leading to their diminished frequency.

News Budgets

Every network manager with responsibilities for news-related programming works within budgeted parameters. These are determined partly by corporate financial status, partly by other broadcast divisions competing for allocated dollars, and partly by prospects of audiences and sales for resulting news products. In a word, will the investment pay off, or at least pay its own way? (See following chapters on sales and finance.) In earlier years networks assumed their news divisions were deficit operations, providing balance to the total program schedule and assuring public esteem for the network as well as adequate public service for its over-200 affiliated stations federally licensed to serve the "public interest, convenience, and necessity." With deregulation in the 1980s the urgency of that statutory "public trustee" concept faded considerably. Network executives more closely scrutinized operating costs of their news divisions, making them accountable as other divisions toward corporate profitability.

■ In 1981 both ABC and CBS expected to operate just above break-even, while NBC News continued to lose money. CBS' news division provided 17.4% of the network's programming, including 237 hours of documentaries and public affairs presentations. In 1983 each network was earning about $25 million annually in pretax profits from their evening news programming. In 1985 CBS News budgeted $188,700,000 for hard news (which it exceeded by $10 million over the twelve months); $35,600,000 for public affairs including *60 Minutes* and documentaries (which came in at $2,400,000 under budget); $7,100,000 for special events; $5,100,000 for political coverage; $10,900,000 for "by products" (worldwide distribution of videocassettes, and so forth); and $30,900,000 for executive staff and overhead.

From 1985 to 1986 CBS increased its documentary programming from 7 to 9.5 hours; its competitors programmed the same number both seasons, ABC 6 and NBC 5. In mid-1986 the three evening news programs were virtually tied with NBC's rating 10.1, CBS' 10.9, and ABC's 9.5. Through the next decade ABC consistently led the others in evening news ratings, with CBS in third place.

By way of comparison, young CNN had begun in 1980 with 300 staff members and a budget of some $30 million; 1985 was its first profitable year with $12.5 million before taxes; by 1987 it employed 1,500 with eighteen bureaus and generated $60 million in profit on $205 million in revenue. It prided itself on producing six times as much news as any other network, for about one-third the cost. Over on PBS, the *MacNeil/Lehrer NewsHour* had cost $1 million a year in its 30-minute format, $6 million by 1982 as an hour news program, and $22 million during 1987 for the nightly hours. In 1988 it cost about $1,350 a minute to produce the broadcast, contrasted with $17,500 a minute average for commercial network evening newscasts. Those kinds of comparisons were not lost on incoming executives at the Big Three networks. ■

Constricted Budgets/Staffs

For a decade starting in the mid-1980s, mergers and corporate management changes brought cuts in staff and expenditures. In 1986 CBS terminated almost 10% of its news staff, usually by layoffs and early retirement. NBC's personnel cuts that year were expected to save $6 million annually in the news division alone. That year total expenses for news operations at the three networks plus CNN were $900 million, but revenues totaled only $830 million. Revenues remained flat, with no increase the following year. Corporate executives mandated managers in all network news divisions to bring expenses more in line with income because of the softening network economy.

In 1987 CBS News still employed 1,225 with 23 bureaus around the world; ABC news and its eighteen bureaus employed 1,200; NBC had thirty bureaus and 1,250 on its news staff. NBC in 1988 introduced robotic cameras and automation to its evening news and some other operations; eliminating camera crews and stage hands was expected to save $2 million a year in salaries. CBS reduced its news budget by $30 million to $270 million.

NBC News President Michael Gartner estimated that in 1989 it cost roughly $800,000 a day to run his division with a $272 million annual budget. That same year ABC reported a $100 million profit in its news operations, which dropped by half to $50 million in 1990. Unscheduled crisis coverage of the Persian Gulf war cost the network about $1 million each night of six nights of lengthy broadcasts (the preceding build-up at the end of 1990 had cost the network another $1 million a week). NBC's news president used rough approximations in an interview when he stated that the 7,000 stories reported during the year's newscasts came out to about $63,000 each.[36]

Bureaus in some U.S. and foreign cities were replaced with a lone fulltime correspondent. After several years of cutting back personnel, overseas bureaus, and kinds and extent of events covered, network news budgets in 1991–1992 were estimated to stand at $385 million at ABC, $285 million at CBS, and $260 million at NBC. NBC's Gartner did not expect his news division to break even until the next year, 1993, helped by the prime-time newsmagazines. News staff had been reduced from 1,500 six years earlier to about 1,000 in 1992.

By 1994 all network news divisions were profitable: ABC News earned $100 million after $400 million in expenses; the other two networks spent $350 million–$375 million on news, leaving $80 million in profit at CBS News and $20 million–$30 million at NBC News.

Stations' Appraisals of Network News

How do news directors at affiliated stations appraise their networks' news service? A representative range of 75 news directors (NDs) at local TV stations around the country, affiliated about equally with the several networks, responded to annual surveys by *Broadcasting & Cable* magazine in the mid-1990s. When asked which of four network news organizations they considered the strongest overall, one out of three consistently named CNN while fewer than one in ten said CBS News (table 8.6). But over that three-year period preference for ABC News dropped in 1996 from over 40% to 29%; most benefiting from that shift was NBC News which tripled its favorable response rate from the previous year.[37] Comparing evening news broadcasts, as news directors' preferences for ABC dropped, both competing networks gained substantially, NBC more than CBS. Similarly over the survey years, Peter Jennings dropped annually in local NDs' estimation while Dan Rather and Tom Brokaw advanced considerably in their eyes, but Jennings still remained favored by two-thirds of all news directors. The 1996 survey was conducted just prior to the new prime-time season, so ABC's precipitous drop in entertainment program ratings that fall could not have influenced professional newspersons' evaluations of network news services. After eight weeks that fall, NBC continued its previous year's top position in prime-time (10.9 rating/18 share), CBS climbed into second (10.0/17), and ABC slipped to third (9.8/16). Their evening news programs held steady in the ninth week of the 1996–1997 season (ending November 17) at ABC 9.2, NBC 9.0, and CBS 7.7.

At the height of "downsizing" broadcast corporations in the late-1980s, news divisions at times appeared in disarray and discontented newspersons publicly criticized some of the trends. Former news leaders such as anchor Walter Cronkite and producer-executive Burton Benjamin lamented the shift in priorities that left news no longer supported by the entertainment division as a network's public service but instead expected to pay its own way and even contribute to corporate profitability. That threatened to replace substantive journalism with popular fast-paced stories to attract mass audiences and high advertising rates. In a word, some felt quality and integrity were being replaced by profitability. Critic Les Brown deplored what he called corporate businesspeople's "tampering with the nervous system" of the nation.

Network/Affiliate News Collaboration

The news divisions set about redesigning their operations. They relied more on their owned and affiliated stations to gather spot news in their regions and relay them by satellite to the network. This saved overlapping crews from local stations and multiple networks converging on the same events with access to similar scenes and interviews. (Between 1987 and 1993 about 60 NBC affiliates added satellite uplink remote trucks, while 50 ABC and 50 CBS affiliates did the same; each network split the $300,000 cost per truck with their affiliates.) These cooperative services were dubbed NewsOne by ABC, NewsNet (later NewsPath) by CBS, and Skycom (later News Channel) by NBC. NBC established its twenty-four-hour affiliate up/down-link service and overnight *NBC Nightside* in Charlotte, North Carolina—a $20 million venture in the spirit of "redesigning" national news processing. Noted earlier was pooled coverage among networks to avoid wasteful

LOCAL TV NEWS DIRECTORS' ASSESSMENT OF
NETWORK NEWS DIVISIONS, 1994–1996

National news organization considered strongest overall

	ABC	CBS	NBC	CNN
1994	45.3%	8.0%	13.3%	32.0%
1995	42.7	6.7	9.3	34.9
1996	28.4	9.5	29.7	31.1

Network evening news broadcast considered best

	ABC	CBS	NBC
1994*			
1995	56.0	9.3	24.0
1996	43.2	18.9	37.8

Considered best national news anchor

	ABC P. Jennings	CBS D. Rather	NBC T. Brokaw
1994	70.7%	8.0%	16.0%
1995	62.7	9.3	14.7
1996	59.5	14.9	23.0

*Not asked in survey

Source: Surveys by *Broadcasting & Cable*, 10 October 1994; 7 October 1996.

duplication at political conventions, presidential addresses, and the like. At times they replaced fulltime reporters and technicians with free-lance professionals stationed around the world, and they shared newsfeeds with news agencies in Europe and elsewhere. (This introduced American viewers to clipped British voices narrating newsclips of scenes from overseas, replacing the familiar tones of American network correspondents.)

With the success of proliferating newsmagazines in prime-time—nine in Fall of 1993, at least one every night—some even pondered the feasibility of moving the network evening news itself up into prime-time.

PROGRAMMING FOR AND ABOUT CHILDREN AND MINORITIES

Children-Oriented Programming

This genre has been under fire from critics and educators for decades. The cycle recurs about every five or six years. Criticism mounts about TV's effects on impressionable youth—especially violence, but also language and suggestive sexual situations, as well as merely vapid noncontent. The FCC makes noises about licensed stations' responsibilities, and networks become defensive about their youth-attracting ad supported programs. Then Congressional committees hold hearings, industry representatives deflect direct government action by promising some reforms, and after sometimes concerted and sometimes sporadic demonstration of better program fare, things eventually slide back to the status quo ante. Then the cycle begins anew.

First Senator Estes Kefauver of Tennessee, then Senator Thomas Dodd of Massachusetts and later Senator John Pastore of Rhode Island through the 1960s and 1970s regularly prodded regulatory agency commissioners and network executives to respond to growing evidence of some correlations and even causal connections between TV viewing and youngsters' aggressive behavior. Further concern has regularly been leveled at the quantity and kinds of advertising found in programs geared for young people.

In 1973 public pressure prompted the National Association of Broadcasters to issue detailed principles about content and persuasive as well as technical tactics used to advertise products in kids' shows. (All Code advertising restraints and later the NAB Code itself were eventually abolished after Justice Department action.)

In 1974 the FCC adopted guidelines for increasing the number and educational value of children's TV programs. In 1979 a Commission staff reported "there has been no broadcaster compliance in the area of educational and instructional programming" because "only eight percent of all programming by network affiliates and 11 percent of programming by independent stations is devoted to children."[38] The staff suggested that the FCC require seven and a half hours per week of educational or instructional programming scheduled between 8:00 A.M. and 8:00 P.M., five hours for preschoolers and the rest for those of school age. The Commission staff noted that PBS should also put greater emphasis on such programming; they added that some three new cable services offered children's programming, especially Nickelodeon's 13–14 hours every day of the week.

By 1979 networks claimed they had effectively responded to complaints about violence and also racial and sexual stereotypes in youth programming. They then carried informational spots, afternoon programs for youngsters, and Saturday programs; 1979 also marked *Captain Kangaroo*'s twenty-fifth year as a daily program on CBS, airing for a full hour since October 3, 1955.[*] That was also the eleventh year for Action for Children's Television (ACT), a private citizens' interest group to better TV programming for children, which continued its highly public critical monitoring and often effective lobbying into the 1990s.

The issue still simmered in 1983, with deregulatory-oriented FCC Chair Mark Fowler objecting to governmental dictates about program content or minimums, including children's programs. Economic factors in the competitive marketplace determined what broadcasters could feasibly schedule, so he judged that public broadcasting might be funded to pick up the slack in TV service to children: "Public broadcasting has an impressive track record in children's programming, and it should be given further opportunity to fill gaps in the commercial sphere."[39] Meanwhile Congressman Timothy Wirth introduced H.R. 4097 to require commercial TV stations (and, indirectly, with their networks' support) to schedule an hour every weekday for children.

---

[*]In Fall of 1981 the program was reduced to 30 minutes so CBS could expand its *Morning* news show; a few months later it was moved to 6:30 A.M. so the network could schedule two hours of *Morning* to compete with NBC's *Today* and ABC's *Good Morning America;* some major affiliates no longer carried *Captain Kangaroo* which was eventually canceled in 1984—television's longest-running children's program.

The Children's Television Act of 1990 put into law widespread entreaties to limit advertising in children's programs (10.5 minutes per hour on weekends, 12 minutes on weekdays) and to provide content serving children's educational and informational needs. But no minimum standard was set. The FCC issued on April 12, 1991, a report and order, plus a small-print double column "Memorandum Opinion and Order" on August 1, 1991, to clarify its previous points about "programming that furthers the positive development of the child in any respect, including the child's cognitive/intellectual or emotional/social needs."[40]

In 1990 competition for youth audiences on Saturday mornings heated up when Fox introduced a block of programs from 8:00 A.M. to 11:00 A.M. In 1992 three-quarters of all children's TV programming was on cable networks; not counting syndication, between 80% and 90% was on cable (depending on whether Turner's projected Cartoon Network is counted in; see table 8.7).

Of 36 children's programs aired by the four networks in 1993 on Saturday mornings, CBS scheduled 11, ABC and Fox 10 each, and NBC 5; no network dominated in popularity—the shows of each were split between the upper half and lower half in ratings (4.5/15 down to 1.4/5). Most were animated cartoons. Network schedules for the 1993–1994 season included more programs reflecting mandates set in the Children's Television Act of 1990 and reinforced by the FCC's guidelines issued in 1991 and by public hearings of Senate committees. Public interest groups once again called for government to mandate at least one hour a day of primarily educational/instructional (not merely entertainment, especially cartoons) for each TV licensee. Affiliates looked to their networks to assist them in meeting such potential demands. ABC was doing its part; its educational programs *ABC Afterschool Specials* and *ABC Weekend Specials* were respectively in their twenty-second and seventeenth years in 1993, to which it added new educational shows *Cro* and *CityKids*. Also meeting FCC criteria for pro-social children's fare were programs developed in 1992 by NBC, *Name Your Adventure,* and CBS, *CBS Storybreak,* which added *Beakman's World* that it acquired from syndicators. Fox added *Where On Earth is Carmen Sandiego?* to its four-year old *Bobby's World.*

Social and cultural concerns can compromise economic ones (or vice versa, depending on one's perspective). The obvious immediate goal of commercial broadcasters is to attract audiences sought by advertisers. One estimate is that children under 12, who view between 24 and 28 hours a week, directly or indirectly have impact on $100 billion of goods and services sold annually.[41] The nation's 42 million children were said to control $32 billion of disposable income annually— $750 each in 1994. To reach that young audience national advertisers bought about $550 million of national television time, plus $250 million in spot TV. Of those dollars, $200 million went to kids' TV on cable.

In 1993 one-third of children's viewing was with network programs, over 60% with cable such as Turner's several channels, Nickelodeon, USA Networks, The Family Channel, the Learning Channel, plus PBS—with long-time classics *Sesame Street, Mr. Rogers' Neighborhood, The Electric Company,* and many others, plus Children's Television Workshop's newest learning-oriented series *Ghostwriter* for pre-teens added that year.

Congressional mandates were implemented by the FCC in the mid-1990s. The agency levied stiff penalties on stations exceeding limits set by Congress for commercials in children's programs. Those limits were established not by a trade association like the NAB nor even by the regulatory agency but by a law passed by Congress. That action was unprecedented in broadcast affairs. Further,

CHILDREN'S TV PROGRAMMING: NATIONAL, 1991-1992*

(HOURS PER WEEK)

BROADCAST	FALL 1991	FALL 1992	CABLE	FALL 1991	FALL 1992
Fox	10.5	19	Turner TBS	17.5	17.5
NBC	4.5	2	" TNT	32	32
ABC	5	5	" Cartoon Network	-	168
CBS	5	5	USA Network	26	26
Disney@	10	12.5	Nickelodeon	126	126
	35	43.5		201.5	369.5

*Not including syndication
@Buena Vista Television's ad hoc network of "Disney Afternoon."

Source: *Broadcasting*, 2 March 1992, 40; 19 April 1993.

all stations were to offer three hours of regularly scheduled shows weekly which had "as a significant purpose" (not tangential or indirect or minor) educating and informing children. New and returning programs on networks—plus Nickelodeon cable channel and PBS—include entertaining personalities and animation presenting science, pro-social and educational sitcoms, and other content presumably contributing to children's cognitive growth.

Minority Interests

These concerns were addressed more and more in national television programming after the civil rights initiatives of the 1960s and 1970s. Critic-historian Les Brown nicely summarized key results of the public reform movement:

> The networks, in response to efforts by broadcast reform groups, reduced violence-oriented programs on Saturday mornings, created children's programs with pro-social messages, cut back the number of commercials carried in children's programming, altered their hiring practices to include more members of minority groups, gave ethnic identities to protagonists in entertainment programs and accelerated the promotion of women and blacks to managerial posts.[42]

African American Viewing

Another factor was that while African American households were 11% of the U.S. total, they accounted for 17% of all viewing. With a "viewing index" 25% higher than for nonblack households, they had impact on programs' ratings, as discussed in chapter 6. In 1989–1990 black households viewed TV 72 hours weekly, half again more than nonblacks; those homes had 49% more youths aged 2–17 than other U.S. homes, while blacks aged 12–17 viewed 50% more latenight TV than nonblacks their age.[43]

In the 1986–1987 season, of 10 network programs most watched by African Americans, only 4 were top favorites among nonblack households (*The Cosby Show, Amen, Golden Girls,* and *The A-Team*). Nine of those 10 were on NBC, as were 16 of the top 20 network shows favored by blacks (three shows were on CBS, one on ABC).[44] But by the 1991–1992 and 1992–1993 seasons (see table 8.3) the top

10 prime-time programs viewed by blacks wholly differed from Nielsen's national listings; many were on Fox and on cable channels targeted to minority audiences.[45]

Emerging networks WB and UPN, with affiliates located primarily in larger urban markets, offered program alternatives to the established networks. Those two networks aired 11 comedies with black casts—more than half their networks' limited program schedule; all 6 of UPN's comedies and 5 of WB's 11 comedies featured African Americans.

Hispanic Viewing

Fox network's programs targeted not only the young but minority viewers as well. By 1993, of the top 15 English-language programs viewed by U.S. Hispanic families (typically larger and younger than others) Fox aired 4 of the top 5, including the first 3, plus 9th and 15th; ABC had 7 shows, NBC 2, and CBS none among the top 15. Fox's *The Simpsons* drew 20% of the nation's 6.25 million Hispanic households; that came to almost 5 million Hispanic viewers, almost twice as many as for shows on other networks.[46]

Perhaps the most successful Hispanic-oriented prime-time program on the major networks was NBC's *Chico and the Man* starring Freddie Prinze, from 1974 to 1978; the final season failed when it tried to write new characters around the absent role of the young original lead who committed suicide in 1977.

■ Heaviest viewing by Spanish-speaking viewers was to Univision and Telemundo networks (plus Galavision cable service).[47] The favorite format was *novelas,* dramatic serials somewhat like *Dallas* and soap operas. Some sixteen of them were scheduled 1988–1990 in large blocks on afternoons and also prime-time every night; most were imported—production of this form in the United States expanded in the early 1990s. By 1990 about 30% of Univision's twenty-four-hour schedule of programs were produced in the United States, as were almost half of Telemundo's sixty-five-hours weekly. Both networks leased half of their imported programming from huge Televisa in Mexico; those 7,000 hours included 2,000 hours each of *novelas* and of news, plus comedy, sports, and other programs. Game shows, plus music-variety and celebrity interviews, made up much of Saturday Hispanic network TV. ■

Of $550 million spent on media advertising directed to Hispanic audiences in 1988, 22.4% or $123 million was spent on national TV. Two years later the percentage was about the same, 22.7% or $142.8 million of the total $628.2 million. Ad dollars in 1990 also went to: 20.8% local TV, 24.3% local radio; 8.7% national radio; 9.4% print (down from 22.2% two years earlier); and 13.5% to promotion, outdoor, and transit. By 1992, U.S. advertisers devoted about 1.4% of their total television budgets to Hispanic TV which served almost 10% of the nation's population. That was up by half from the almost 1% several years earlier.

PUBLIC BROADCASTING SERVICE

Wholly different from commercial television networks in structure, goals, target audiences, funding, and programming procedures, the Public Broadcasting Service was created by the Public Broadcasting Act of 1967 under the Corporation for Public Broadcasting (CPB). It was formed *not* as a centralized *system* but rather as a coordinating *service* for the nation's several hundred noncommercial TV stations. Most of CPB's TV funding went directly to stations that then proposed productions to be distributed through

PBS if enough affiliated stations agreed to carry them. PBS did arrange for some programming, especially imports from the BBC like *The Jewel In the Crown* and *Brideshead Revisited.* After a period of confusion to potential audiences when individual local stations carried, delayed, or pre-empted PBS distributed programs as they wished, in 1979 all but 22 of the 156 affiliates agreed to schedule a core of at least two hours a night, four nights a week.

True to the federal establishing mandate, PBS programs were "produced" (or arranged for, proposed, and endorsed) by local public stations: notably WGBH Boston, KCET Los Angeles, WTTW Chicago, plus others around the country. More than three-quarters of PBS programs were co-produced with those individual stations, independent production companies, or foreign broadcasters.

Those costs were borne not by commercial advertising but by grants from various corporations, foundations, governmental agencies, and dues from member public TV stations plus contributions by viewers. But federal government funding remained constant each year while production costs almost doubled.

PBS coordinated funding, stations' votes on selecting shows to schedule, and program distribution. That very complexity hindered top PBS management; former PBS President Lawrence Grossman lamented public broadcasting as "so diffuse, duplicative, bureaucratic, confusing, frustrating and senseless, that it is a miracle [it] has survived at all."[48] At the end of the 1980s PBS established a firm prime-time schedule and instead of waiting for local station proposals began actively seeking projects that fit PBS needs. For fiscal 1992 PBS had a budget of $105 million ($113 million in 1994), including $78 million of member-station assessments, for program decisions by the recently created central position of executive vice-president of national programming and promotion services. The first to hold that position, Jennifer Lawson (a female African American), succeeded in wooing back audiences that had begun to drift from PBS to cable. She sought multicultural programs and arranged for unprecedented joint election coverage with NBC in 1990, airing anchor Tom Brokaw throughout prime-time when he was free of NBC's evening news or its own election coverage inserts.

Public television in general was described by John Wiseman as a "billion-dollar-a-year enterprise [that] employs almost 10,000 people at more than 300 stations nationwide but seldom gets more than three per cent of the audience."[49] Although 100 million people in the United States view public TV at some time, only one in ten viewers contribute to their local noncommercial stations. Stations forward some of the funds to PBS for their program service, keeping the rest of the donated funds to develop their own programs and to pay overhead and salaries. (The Twentieth Century Fund reported that 80% of public TV's $1.8 billion went to 350 local stations' operating costs including programs they underwrite, only 20% to produce national programs.)

PBS was established to provide an alternative to mass audience driven commercial TV. But public TV's management shifted perspective over the decades. They needed to justify huge financial outlays for programming (including tax money from the U.S. Treasury via CPB's $286 million for 1994) by attracting large audiences. Further, stung by criticism of elitism they added popular entertainment fare to their traditional schedule of classics and thought-provoking explorations of nature, the arts, the human condition, and major news trends.

As cable program services proliferated, entire channels were devoted to material previously televised only by PBS: Arts & Entertainment, Discovery, the Learning Channel, Nickelodeon, C-Span, C-Span II, and even CNN. In the 1990s some members of Congress challenged the role of public TV as a grantee of federal funds (which made up 14% of public broadcasting's funds); their criticism was prompted in part by PBS programs offensive to conservative views. But the *New York Times* in major feature articles in the 1990s repeatedly lauded PBS for its sustained record from season to season of quality documentaries and dramas.[50] Managers scrambled to sustain public TV's mandate and image, while at the same time accommodating for increasing competition from new forms of video distribution with specialized formats and content. PBS directed its emphasis to history, education, and public affairs programs.

In the 1990s, noncommercial PBS and its 350 stations attracted 2.2% of U.S. homes to its programming. Special programs earned national ratings of 6.0 (*The Kennedys*); 5.1 (*The Dinner Party*); 3.9 and 3.7 (two episodes of *The American Experience*). The award-winning twelve-hour mini-series *The Civil War* drew double-digit ratings (11.2 rating/16 share in top four Nielsen metered markets the opening night); all five evenings the mini-series attracted 9% of U.S. households for a 13 share of TV viewers nationally, an estimated 14 million people each night. It was the highest rated series in PBS history. Twenty years earlier in 1971 PBS had introduced *Masterpiece Theater* which the *New York Times* claimed "institutionalized the miniseries, or 'serialized dramatization'—running from 5 to 13 hours—perhaps television's most original and exclusive programming vehicle."[51]

CRITICS AND PROGRAMMING

There are two kinds of critics. One is the professional or scholarly critic of media whose criteria are founded on social and economic data and aesthetic principles. The other kind criticizes media solely on subjective, sometimes irrelevant and even invalid grounds. It is in the confused area of public "taste" that broadcasting's critics find their most fertile ground. Regularly, critics denigrate the programming of radio and television, usually through competing print media.

What troubles media managers is the apparent influence of some critics who are either poorly qualified commentators about broadcasting, or are cynical stylists bent on condemning the all-assimilating "bastard medium" for having sold its birthright for a pot of message (to paraphrase scripture). Too often, critics indicted first radio and then TV for not having enough of the kinds of programs critics themselves prefer; any wider appeal of such programs to a majority of the public would be purely coincidental. Les Brown, successively chronicler and critic for *Variety* and the *New York Times,* aptly noted the differing perspectives of critics and broadcasters:

> To the critic, television is about programs. To the broadcast practitioner, it is mainly about sales. This explains why most critics have nothing important to say to the industry and why, among all the critics in show business and the arts, the television reviewer is probably the least effective.[52]

The fact is that many innovative, creative broadcasters receive very little critical support; nor did the mass public support those programs. Ironically, when alternate systems such as PBS do air thoughtful, penetrating programs of quality they are accused of being elitist. Prime-time specials favored by national audiences are typically sports and comedy shows, rarely serious drama, with documentaries and other public affairs programs ranked near the bottom of hundreds listed annually.

The commercial message is a particular target for published critics. Yet print media carry advertising messages just as do electronic media but in greater volume (broadcasting airs commercials in 15%–25% of airtime; newspapers typically devote 65% and more of their space to advertising. Although broadcast ads have greater salience because they interrupt program continuity, the ever-present remote or "mute" switch makes changing channels or silencing commercials as easy as flipping magazine pages past print ads.)

Broadcasters have been instrumental in broadening public understanding of culture and of society itself, including issues of poverty, hunger, race, gender, and international affairs. Most major broadcasters have contributed positive achievements. But a few deviant station operators and some poor judgments by networks and cable systems, coupled with biased criticism, distort that positive record. Broadcast and cable managers must act against industry failings to retain popular support.

Of all social institutions, broadcasting and cable must be as current as the people they serve. To the extent that people are becoming harder to satisfy and easier to bore, there is profound risk in operating a station by yesterday's standards. No one will detect out-of-date methods faster than the public, who can always shift to other more advanced, responsive media systems available through new technologies. American society—with significant shifts in age, income, formal education, and developing interests in contemporary culture—will hardly settle for the same values of earlier years. It is vital to the future success of any radio-TV station or network or cable system to discern the direction that people are moving.

Given the opportunity to provide either popular entertainment or self-conscious cultural offerings, most network and station managers would not hesitate to give audiences what the majority prefer. The manager, charged with responsibility to generate profit for owners and investors, has small choice but to attract the majority audience. On occasion, when popular programs are canceled in order to carry special informational or cultural programs, reactions from the audience do not encourage repeated efforts. Networks and stations have been flooded with complaints, for example, when regular comedy series or soap operas were pre-empted to carry congressional hearings, political conventions, space exploration ventures, and even war coverage. Large numbers of people react negatively when programs are interrupted to carry news bulletins and "specials" of urgent timeliness.

A growing issue in the 1980s that partly burned itself out in the 1990s was proliferating tabloid and talk shows which for the sake of ratings outdid one another in dredging up individuals with bizarre and intimate details of personal crises. Some programs reveled in shouting matches and, on occasion, physical blows. History has demonstrated that degrading spectacles always draw big crowds, from burning at the stake to townsquare stocks and lynching. Some syndicated TV programs and radio talk/phone-in shows pandered to those baser instincts. But the surfeit of shows, numbering in the dozens over the decade, wore out its novelty and its hosts as well as much of the audience. Taste and better sense began to prevail in the mid-1990s as the worst of those unenlightening confrontational meleés dropped from the airwaves. Those tasteless talk programs that remain continue to draw criticism.

Whether commercial broadcasters have an obligation to try to upgrade audience taste remains a debated issue. Much can be said for knowing the audience well enough to be able to aim consistently a little higher than existing taste and thus raise standards by degrees. Retired network anchor Walter Cronkite in 1992 reminded broadcasters to "lead culture, rather than pander to the moment."[53] But it would be financial suicide for most programmers to move too fast, too soon in that direction. There is equal danger in playing down to an audience; condescension has a way of alienating people. Franklin Dunham (early NBC executive and then chief of the Radio-TV division of the U.S. Office of Education) put it to media classes decades ago: "Give the people what they want *most* of the time, and a little of what they *would* want if they only knew about it."

It makes sense for broadcast managers to make a sincere effort to learn genuine interests of specific audiences, through various forms of research and analysis. It is folly to make decisions exclusively on the basis of critical (including academic) diatribes, merely personal preferences, or aesthetic guesswork.

C. STUDIOS/PRODUCTION COMPANIES AND NATIONAL PROGRAM SYNDICATION

PROGRAM SUPPLIERS

Robert Harris, president of the MCA Television Group, noted the relevance of program production sources to the TV/cable industry as a whole:

> If you want to see what television is going to be in the next decade, watch the program suppliers because the delivery system [networks and cable] is secondary to the product. Every delivery system is going to be product-driven. And it's still going to be a business of hits.[54]

TV divisions of motion picture studios and independent production companies provide networks with most of their prime-time programs, both series and one-time-only shows. Among principal suppliers in 1996–1997 were 16 production companies that supplied 39 program projects for networks (table 8.8). Some of their individual producers developed projects in association with one of the major studios that mounted 49 projects. Unlike earlier decades when studios retained large groups of writers, directors, and producers under long-term contracts, by the 1980s all except Universal hired outside creative talent for each project. An unnamed executive once described a studio as "nothing more than a central place where a lot of independents come to work." Production and news divisions within the networks also supplied their own programming divisions with shows.

■ Networks' own production divisions supplied 21 hours (17 returning programs, 5 new)—roughly split between their news divisions and entertainment production arms. Four of the five major studios had their own networks—WB, UPN, Fox, and Disney (ABC). The FCC's removal of restrictions on airing in-house entertainment productions resulted in networks' creating more prime-time hours of their own, including reality-based shows and newsmagazines. Of each network's weekly 22 prime-time hours, CBS produced 9 hours for its own schedule, ABC and NBC 6 hours each in the 1996–1997 season. ■

Many production companies successful in placing their properties on network schedules also arrange to distribute them later market-by-market to stations as off-network re-runs. They also develop new programming expressly for first-run syndication. (This was discussed in chapter 7 from the perspective of local station operations; also see the next section.)

TABLE 8.8
Series Program Suppliers for Networks, 1996–1997
Ranked According to Hours of Prime-Time Programs Scheduled for Fall 1996

Hrs.	# SHOWS N	R	Production Source	Hrs.	# SHOWS N	R	Production Source
10.0	7	10	Warner Brothers	2.0	0	2	CBS News
7.0	5	5	Columbia-TriStar	2.0	0	1	ABC Sports
7.0	3	5	CBS Productions	1.5	2	1	Brillstein-Grey
6.5	3	7	Paramount-Viacom	1.5	0	3	Castle Rock
6.5	4	4	Universal	1.5	2	0	MTM
5.5	3	3	Twentieth [Fox] TV	1.0	0	2	Barbour-Langley
4.5	4	4	Disney-Touchstone	1.0	1	1	HBO Independent Productions
4.0	1	3	Aaron Spelling Productions	1.0	0	1	ABC Productions
3.5	3	4	Carsey-Werner	1.0	0	1	Cosgrove-Meuer Productions
3.0	2	2	NBC Studios	1.0	0	1	New World
3.0	0	3	ABC News	1.0	0	1	Rysher
3.0	0	3	NBC News	1.0	0	1	Vin DiBona Productions
2.5	1	2	Steve Bochco Productions	0.5	0	1	Big Ticket
2.0	2	2	Witt-Thomas	0.5	0	1	Worldwide Pants
2.0	2	1	Dreamworks SKG				

(N = New shows; R = Returning shows)

Source: Tabulation by *Variety,* 27 May–2 June 1996, 33.

This enabled networks to control better the spiraling costs of production; filling total prime-time schedules cost each network more than $2 billion annually. When the FCC removed financial interest and syndication ("fin-syn") rules in 1992–1993—subject to dissolving the networks' consent decree with the Justice Department in 1976—networks were also eventually permitted to put their projects directly into syndication instead of having to sell rights to them to outside syndication firms. These regulatory and economic developments began to impede traditional suppliers of national programs—movie studios and independent production companies—from selling as many properties for network airing. By 1984, over 58% of MGM/ UA's $707 million revenues had come from television—including network and syndication. Even the motion picture studio most successful in theaters, Paramount, earned over half its revenues from television. That is why Disney bought ABC and why Fox, Paramount, and Warner Brothers mounted their own networks: to ensure an outlet for their studio TV product.

Program Suppliers as Persons

Broadcasting and creative TV production is a "people" business. Norman Lear, Grant Tinker, and Lee Rich were among the most successful of U.S. program suppliers to network schedules and subsequent national syndication. Others include Aaron Spelling and Steven J. Cannell who both named their production companies after themselves; they specialized in action-adventure fare, each selling multiple series year after year to the networks. Stephen Bochco generated cutting-edge police dramas with *Hill Street Blues, L.A. Law, NYPD Blue,* and *Murder One,* among other series. Their companies' prolific productivity of successful programs set the standard for various network genres. Their fortunes shifted amid the complex of economic forces and audience preferences which saw independent producers build production companies, then merge or be bought out by larger entities in the entertainment business.

The production company has certain advantages in licensing programs to a network rather than to individual stations through syndication. It recovers production costs sooner, lowers distribution costs, needs far fewer film or videotape prints, and networks usually absorb the costs of promoting the program. Even with these advantages, creating a pilot film for a prospective series is a heavy risk. The cost of a pilot, due largely to contract restrictions on talent and overtime payments as well as first-time costs for building sets and mounting an elaborate and compelling trial episode or pilot, runs considerably higher than (sometimes double or more) the cost of each subsequent program once a series is sold. Investment in a half-hour pilot averages close to a million dollars. An hour pilot, usually involving shooting on location, can cost well over $2 million. Each year dozens of pilots are not accepted for network schedules (although the network underwrites much of their cost, and some are utilized on a one-time-only basis as replacements for canceled shows or for summer fillers). Many independent producers understandably hesitate to invest so much capital on a highly speculative risk, so they enter into co-production arrangements with other companies and major film studios. It is also clear why so many producers in this competition tend to play it safe by developing programs similar to those already successful.

On the positive side, once a pilot is accepted, there are prospects of selling rights to more programs in the series, usually 22 to 24 in a season from September to April; there is the further possibility of network renewals for succeeding seasons, eventually leading to off-network syndication to U.S. television stations and subsequent distribution to the foreign market. That is where massive profits potentially lie, after initial deficit financing is recovered from the loss-structured network run. As described earlier, networks pay program license fees equal to perhaps 80% of production costs (occasionally as low as 65%–70%), to protect themselves somewhat when three out of four series fail with small audiences that draw limited ad revenue. For example, Paramount's sitcom series *Dear John* with Judd Hirsch cost over $650,000 an episode to produce in the late 1980s, but its network license fee was less than $450,000; comedy series in their first season often run deficits of $2 million. To recover its investment at the "back end," a production company's network series must be renewed for four or more seasons to make it a viable package to syndicate and recover investment. Ideally, 100 or more episodes make up a saleable package that local stations can strip daily for several months without replaying any episodes.

In the early and mid-1980s about 80% of syndication was off-network, often leased with barter, among the hundreds of independent stations. They were the one or two out of every ten network shows whose successful runs with national audiences set them up for later syndication. On the other hand, original first-run programming did not have that previous exposure and track record; some estimated an 80% to 85% failure rate for first-run projects.

Station managers could choose from an ever widening range of syndication programs available, categorized in table 8.9. For example, in the mid-1990s program directors could lease rights, if no other station in their market already had contracted for syndicated exclusivity, to 95 off-network half-hours averaging 126 episodes. *M*A*S*H* is such an example, with 255 episodes.

The challenge to station management, of course, is two-fold. First, they must select from the national programming pool those properties best matching their schedule needs, strategies, and community preferences. Secondly, they must

SYNDICATED PROGRAMMING AVAILABLE 1993–1996

NUMBER OF EPISODES

Type	No. of Series	Range	Total	Typical Avg.	No.Runs
CURRENTLY Available					
Off-network half-hour	95	20–274	11,996	126	5–8*
Off-network hour	63	13–402	7,020	111	4–6
First-run half-hour strip	28	65–260	4,584	164	1–6
First-run hour strip	13	65–240	2,385	183	1**
First-run half-hour weekly	39	13–80	1,475	38	1,2,6,8
First-run hour weekly	35	6–96	1,045	30	1,2,6
Children's animation	53	8–209	717	14	2–8, Unlimited
Children's animated segments	6	130–308	1,344	224	Unlimited
Children's live action	3	39–130	273	91	1,2,4
Mini-series	13	4–15	72**	6	2–6
Available for FUTURE Airing (after 1993, 1994, or 1995)					
Off-network half-hours	12	65–136	1,294	108	1,2,5–10
Off-network hours	3	88–108	119	40	5,6,10
First-run half-hour strips	10	65–260	1,420	142	1–4
First-run hour strips	4	190–200	780	195	1–2
First-run half-hour weekly	8	13–52	247	31	2
First-run hour weekly	6	22–52	191	32	1–2
Children's animation	13	13–73	504	39	2,4,6,8

*Some series permit only 2 runs, others 5–8, one 12; most common is 6 runs.

**MCA listed "Various" number of episodes for *Best Sellers I & II.*

Source: Based on data from Petry Television in *Variety,* 25 January 1993, 110–116.

bargain astutely for license fees that can be more than covered by commercial time sales when they finally broadcast the programs, often years after they signed contracts.

Television production companies with off-network and first-run series available for syndication meet local TV program directors and station managers at the annual convention of the National Association of Television Program Executives (NATPE). There syndicators negotiate from market to market, and competitively among stations within markets, for the best licensing fees possible. These "back-end" sales typically marked the move into profitability for off-network programs that had originally appeared on networks at a 20% loss to the producers. (See further details in chapter 7.)

Barter

In this arrangement, one or more national advertisers or a production company pays all production costs, and the programs are distributed at no charge or for a modest fee to stations in exchange for commercial spots placed in the program; the stations sell the remaining ad positions for their own income.

Advertising agencies deal with barter distributors much as they do with networks. Any syndicated show that reaches at least 70% of the country's population through local stations can be a useful vehicle for some national advertisers. For example, Bristol-Myers bought $5 million worth of advertising time on *Family Feud* for the 1988–1989 season; its ads were inserted in all the shows by distributor LBS Communications for syndication to local stations on a barter basis.

By the late 1980s local stations bartered an average of one-tenth of the commercial units in their schedules. While saving outlays of cash for programming, stations nevertheless lost partial control over their key asset—inventory of time to sell to advertisers. Over a decade advertiser-supported TV series available for barter tripled from 50 series in 1983 to 162 series in 1995; their airtime hours multiplied from 80 to 275 hours. Revenues from barter syndication more than tripled over that same period from $400 million to $1.8 billion—almost half of all syndication revenues that year. Including worldwide distribution, total syndication was expected to approach $10 billion that year.

Costs, Revenues, Income

Expenses for producing programs, including deficit financing of productions licensed to networks, were described earlier in this chapter. Examples here show how off-network and first-run series fared financially in syndication.

■   In 1986 WOR-TV (later renamed WWOR-TV) in New York bid $43,680,000 for 182-week rights to 125 half-hour episodes of *The Cosby Show*. That came to $349,440 per show, plus one minute of barter time reserved for the syndicator in each program. Their bid was $10 million more than the second highest bidder in the market; and it helped boost station fees around the country by as much as 25%. Syndicator Viacom required stations to pay a 10% downpayment of the total license fee; with the series not scheduled to begin airing until Fall 1988, that meant WOR-TV paid $4.3 million two years before it earned its first income from ads. In its 5.5 minutes of commercial time, the station needed to get about $6,000 per 30-second spot—in each of five runs of all episodes over the life of the license—just to break even; to earn that figure per spot, its household rating in New York would have to be 15 but industry predictions were for 11 to 14 for a loss. Within seven months of negotiating distribution of *The Cosby Show* Viacom attracted $300 million from stations in 80 markets. The heralded series first aired on October 3, 1988, on 15 independent stations and 159 affiliates covering 97% of the country. Those 174 contracts brought in $550 million over the three years, plus more than $100 million Viacom earned from selling one barter minute in each episode over the license period (for up to $80,000 per 30-second nationwide spot). Most stations scheduled it as a lead-in to early local news; but many found sitcom audiences not fully compatible with news so they had to shift the costly series to other parts of their schedules. When subsequent audiences and station time sales did not match *Cosby*'s lofty fees, Viacom found heavy resistance for its second round of bidding; many markets passed on the offer, others paid far lower rates. The syndicator cut fees by 25%–75% for renewals in the second cycle that began in Fall 1993; it expected to earn about $100 million on series renewals and extensions for new episodes, less than one-fifth of the first round's revenue.   ■

Revocation of the Prime Time Access Rule made syndication even more lucrative; now for the first time key stations in the top 50 markets including network O&Os could air them in access time. In the mid-1990s popular sitcom *Mad About You* expected to return $2 million per episode in total cash license fees and barter revenue from around the country.

Indicative of the investment made by local stations to get quality off-network fare were per-episode license fees paid by individual stations in major markets. In 1995 sitcoms licensed to New York stations included *Frasier* which commanded $125,000 per week plus one minute of barter (five episodes, stripped Monday through Friday) over the 169-week license term for a total of $21 million. That was almost equal to *Seinfeld*'s $128,000 per week licensing fee in that city; *Mad About You* went for around $60,000 for each week's five-episode run.

Syndicators at the same time contracted with national cable program services, especially when they were linked with those studio syndicators. Warner Brothers' *Fresh Prince of Bel-Air* and *Friends* were licensed to Turner's TBS cable network (later merged into Time Warner); Paramount's *Wings* went to cable, as did Fox's *NYPD Blue.* For its seven major-market stations Tribune Broadcasting (with a 12.5% equity stake in WB network) paid more than $250,000 per episode to Warners for 96 episodes over 3.5 years; each episode had 1.5 minutes of barter time, the other 7 minutes of ad time for the local stations.

In 1990 syndicators' barter ad sales in first-run programs totaled $1 billion in pre-season up-front buys, plus another $250 million in later scatter sales of inventory. That doubled the ad revenue from syndicated programming four years earlier. Ad agencies could buy for their clients a 30-second spot in *Arsenio Hall* for $35,500, in *Oprah Winfrey* for $40,000, and in *Star Trek: The Next Generation* for $135,000. But beginning that same year national advertisers began to cut back on budgets for syndication time sales, shifting some dollars to cable and also back to networks whose soft sales lowered rates. Of course, the supply-and-demand factor affected prices; syndication inventory in 1992–1993 increased 9% to 214 series available to stations, up from 196 the previous year, softening prices for rights.

Syndicated Talk Shows

In 1996 the longest running series in TV syndication history, pioneering audience talk-show *Donohue,* ended with its twenty-ninth year after some 6,000 shows. *Hee Haw*'s country music and comedy aired continuously for twenty-seven years, the last twenty-five in syndication. It was first distributed to stations in 1971 after completing two seasons on CBS; its $6.5 million in gross revenues for 1990–1991 were half again more than annual production costs of $4.25 million. Previous long-running talk/interview syndication programs were *The Mike Douglas Show,* 1963–1982, and *The Merv Griffin Show,* 1965–1969 (after one season on NBC) and 1972–1986 (after three years on CBS, 1969–1972). Talk programs continued in popularity into 1995, with 20 syndicated talk series in daytime and late-night. Respected host *Oprah* far outpaced the others in the mid-1990s, regularly attracting 10% of all U.S. households. Most others drew national ratings of 2.0 to 4.0; the five lowest had ratings of around 1.0. Long-lived *Regis & Kathy Lee* earned 4.1, *Donohue* 3.6, and *Geraldo* 3.2; newcomer conservative Rush Limbaugh drew a 3.5 national rating. As such productions proliferated, ad revenue from syndicated talk shows bounded from $76 million in 1987–1988 to $275 million in 1992–1993.

The intensely competitive form spawned imitations that tried to outdo one another on intimate topics such as sexual dysfunction, addiction, abuse, and other hitherto private matters. Many of the shows drew criticism as "trash TV," prompting some advertisers to defect. In 1995 Proctor & Gamble and Sears, Roebuck & Company withdrew millions of ad dollars from the genre, particularly *Sally Jesse Raphael* and *Jerry Springer* which some claimed exploited guests. A majority of local station managers in surveys agreed that much content in the newer entries was "sleazy" and in bad taste. The peak of twenty-three talk shows in 1996 fragmented the increasingly alienated audiences, who shifted viewing to reality-based shows and especially to very popular sitcoms just released off-network.

Reality-Based Syndication

In 1993–1994, reality-based syndicated programs multiplied, cutting into one another's audiences. Two "cops-and-robbers" reality series competed with four tabloid and newsmagazines.

■   Among the latter, Paramount and Cox Enterprises' *Entertainment Tonight* was in its thirteenth year, the first daily strip syndication series to air on the same day and date around the country—breaking from film and tape delay to satellite distribution. The series cost from $500,000 to $600,000 for weekly production; that $31 million annual expense was far surpassed by yearly gross revenues of $110 million to $160 million. Weekly production costs for rival newsmagazines were $250,000 to $400,000 each, and for reality-based police and rescue syndicated programs, $150,000 to $250,000 a week.   ■

In 1995 major studios, independent production companies, and station groups syndicated 10 daily strips and 13 weekly series of "reality" journals and action programs, to which they added a new cluster of six strips and nine weekly series.

Syndicated Children's Programming

Prospects for distributing children's programming rose in the mid-1990s. The FCC was pressed by Congress to enforce its Children's Television Act of 1990 mandating TV stations to provide youngsters with three hours weekly of educational and informational programming where entertainment was secondary. At least 23 big and small syndicators offered 28 series they claimed met federal criteria.

In recent years syndicators who until then leased programming exclusively to over-air broadcasters began to sell rights to cable companies and services such as USA Network and the Family Channel. One reason is that up to two-thirds of some program license rights are based on high residuals (or royalties) in union contracts of actors, writers, and directors for those original properties. But contracts with cable operations call for about one-tenth the level of broadcasting's residual payments to program creators. When negotiating original contracts with unions and agents, major program suppliers like MCA-Universal, Time Warner, Paramount, and others with heavy holdings in cable program channels and systems apparently ensured favorable residual terms for cable (MCA and Paramount jointly own basic-cable USA Network) and not for broadcast networks or over-air stations.

To compete with the quality of network programs, movies, and original productions by HBO and Showtime, producers began to invest large budgets in first-run syndicated action-adventure series. This also distinguished them from cost-effective talk and game shows in first-run syndication. Paramount in 1993 introduced new weekly hours of *The Untouchables* and *Star Trek: Deep Space Nine;* it budgeted both at $1.5 million per episode. They were intended to compete in prime-time not only on independent stations but on affiliates that pre-empted weak hours in their network's schedule. Local stations, instead of only the minute of commercial time provided in a network hour plus modest compensation, received 4.5 minutes for local/regional and national spot ads in *The Untouchables* (syndicator Paramount retained 7.5 minutes that it sold to national advertisers).

Foreign Ventures and Distribution

As costs of mounting programs with audience-attracting quality escalated far beyond what networks would pay in licensing fees, major studios and independents sought co-ventures with foreign partners who would distribute products in their own countries. They co-produced or shared financing on some series, movies, and specials with French, Italian, German, British, and Australian companies. These deals are often initiated at the annual MIPCOM international TV market in Cannes, France, where production companies and broadcasters of many countries— including U.S. networks and movie studios—meet to lease programming rights to one another for foreign distribution.

In 1988 U.S. distributors earned over $318 million from private commercial telecasters in Japan and $800 million from Western Europe; total revenue world-wide was $1.3 billion. By 1993 that total tripled to $3.5 billion as American TV program sales generated $715 million from Japanese broadcasters and $2.7 billion from Western Europeans despite their import quotas. That "back-end" overseas distribution income was added to the expected $4.8 billion to program suppliers from domestic syndication and cable markets. Together they more than made up for the deficit incurred from the predicted $2.7 billion in license fees for prime-time programs produced for network schedules.

D. NATIONAL CABLE PROGRAM SERVICES

During the past decade cable program services have attracted increased audiences, drawing them away from network viewing as well as from among nonviewers. Basic cable subscribers in 1975 were fewer than 10 million, 13% of U.S. homes. By 1996 subscribers had expanded to more than 56 million of the nine out of ten U.S. homes passed by cable. Almost 8,000 systems with 11,340 operating head-ends delivered TV service to those 56 million homes. Three out of four cabled households subscribed to pay TV channels.

■   The proliferation of specialized "niche" program services cater to youngsters (Nickelodeon), teens (MTV, VH-1), women (Lifetime), men (ESPN), older viewers (Discovery, The Travel Channel), those interested in religion (EWTN, VISN), news junkies (CNN, C-SPAN 1 and 2), families (The Disney Channel, Family Network), minorities (BET/Black Entertainment Television, Univision), and movie lovers (TNT/Turner Network Television, American Movie Classics, The Movie Channel, Showtime, and HBO). Channel services were solely dedicated to sports, fine arts, finance, cartoons, or other specialty programming.   ■

By 1992 cable systems across the country paid more than $3.5 billion dollars annually to national program networks supplying their multiple channels. Those dollars came from monthly subscription fees. An additional $2.2 billion in advertising was generated by ad-supported national cable program services (local cable systems also earned $761 million in local and national spot revenue). Of the cable industry's total advertising revenue of $3 billion in 1991, cable networks earned 70%, local systems got 25% from national/local spot advertising, and regional sports brought 5%.

Cable has also had an impact on the international social and political scene. In 1993 MTV reached more than 200 million homes in eighty countries. Turner Broadcasting's CNN/Cable News Network was received by almost all cable-connected homes, plus in over 150 countries. Its impact was noted earlier, when

NBC and other broadcasters used its feed during early stages of the Persian Gulf War in 1991. It became a quasi-participant as well as reporting observer of world events with its worldwide correspondents and twenty-four-hour coverage. In 1992 Secretary of Defense Richard Cheney told a national academic gathering about watching CNN with his military staff as the Gulf War began, and also when Boris Yeltsin stood atop a tank in Red Square to urge the crowds and TV audiences worldwide to support Mikhail Gobachev against the coup plotters. Cheney acknowledged that "as an administration official, it's gotten increasingly difficult to sort out what we know from Intelligence [CIA/Defense Department] and what we know from CNN. . . . It clearly is a factor now in the way we do business."[55]

Cable "Network Channels"

More than seventy national cable program distributors—also referred to as cable networks or cable program services—provide programming exclusively to local cable systems via satellite. A dozen such as HBO and Showtime offer premium services, requiring local subscribers to pay specifically for each one received. Except for pay-per-view, the rest deliver basic programming included by local cable systems in overall monthly subscriber fees. Programs are sometimes clustered into several tiers, available for progressively higher fees to home subscribers.

In 1992 the National Cable Television Association reported 69 cable program services and 7 superstations, plus 11 audio cable services, and 10 cable text or other services. National cable networks were listed by type of service, according to form of license fees charged by networks:[56]

> BASIC (38 channels)—Per-subscriber fee charged to cable operator.
> BASIC/NO CHARGE (13 channels)—No per-subscriber fee to cable operator.
> PREMIUM (8 channels)—Monthly charge to operator and subscriber.
> PAY-PER-VIEW (6 channels)—Charge per event or per movie.
> BASIC or PREMIUM (2 channels)
> PREMIUM or PAY-PER-VIEW (2 channels)

Basic Cable Services

Basic cable networks are ranked in table 8.10 by number of subscribers in 1995. Actual viewing, of course, was far different from households reached by a cable service; for example in 1996 when the MTV channel reached 63 million households, its average viewing audience was about 295,000 households (worldwide MTV reached 266 million households).[57]

Premium and Pay-Per-View Cable Services

Premium channels offer commercial-free service of major sports events, early-run unedited motion pictures, original dramas, and comedy specials. For each premium service, in addition to their cable system's basic charges, local subscribers pay an extra monthly fee or else a pay-per-view charge for each premium program such as championship boxing matches or live rock concerts.

Eight national pay-TV services and five pay-per-view services fed their programming by satellite to local cable systems in 1992. Most widely distributed and viewed was HBO/Home Box Office, started on November 8, 1972 by Time Inc. (later Time Warner Communications) to 365 subscribers in Wilkes-Barre,

TABLE 8.10
MAJOR BASIC CABLE PROGRAM NETWORKS, 1995

SUBSCRIBERS (000,000)	CABLE SERVICE	SUBSCRIBERS (000,000)	CABLE SERVICE
66.0	ESPN	42.8	BET
66.0	CNN	37.5	Prevue
65.8	TBS	35.8	Comedy Central
65.2	USA	32.6	E!
65.2	The Discovery Channel	32.6	Faith and Values
64.5	TNT	39.5	Learning Channel
63.6	Nickelodeon	38.1	WGN
63.0	Nashville Network	29.7	CMT
62.4	The Family Channel	23.7	Sci-Fi
62.0	Lifetime	23.5	HBO
61.7	A&E	22.9	ESPN2
61.4	MTV	22.2	Court TV
59.3	The Weather Channel	22.0	fX
58.0	CNN Headline News	17.5	Travel Channel
55.0	CNBC	12.5	TV Food Network
52.3	VH-1	12.1	Showtime

Source: Nielsen Media Research, National Cable Television Association, *Broadcasting & Cable*, 8 May 1995; 10 July 1995.

Pennsylvania. In 1995 HBO had 23.5 million subscribers, several times more than Showtime or Cinemax and almost four times as many as The Disney Channel.

From 1983 to 1989 the average total time spent watching television in homes with premium/pay cable channels was over 8 hours a day per home, compared with the general U.S. average of slightly more than 7 hours a day per household.[58]

Hispanic Channels

In 1979 Galavision started as pay then changed over to basic cable for Spanish households; after eight years its eight-hour daily schedule had 170,000 subscribers. Over-air broadcast networks Univision and Telmundo were made available to cable systems. In 1990, 526 systems carried Univision to 9.3 million households (2.1 million of them Hispanic), with another 801 systems getting the signal directly from satellite for no license fee. New Hispanic cable programmers came on the scene: In 1990 Viva Television Network featured music and children's shows for 5¢ per subscriber license fee to systems. In 1991 Cable Television Nacional achieved 34% penetration in Hispanic households, with 5¢–7¢ to cable operators; it expected losses in the first three years to total $8 million–$10 million. Unlike most Hispanic TV it did not stress novellas and news but classic films, music videos, children's animated, public affairs, and sports, with 5%–8% of the programs bilingual. Late in 1993 HBO introduced HBO en Español for 25 million Hispanics—a Spanish-language version of HBO, not a separate program service. That same month MTV Latino came on the scene; and GEMS Television directed to Hispanic women expanded to twenty-four-hour service (to 1.5 million homes in eighteen countries including the United States). In 1996 Westinghouse/CBS purchased Spanish-language news channel TeleNoticias which Telemundo had launched two years earlier; it reached 20 million cable, MMDS, and broadcast homes in the United States, Caribbean, Latin America, and Spain.

New, Developing Cable Networks

By 1996, many of 163 proposed cable program networks were in operation, but 26 in development had been withdrawn by entrepreneurs; 44 others planned to launch by then were postponed past mid-year. *Cablevision* in April 1996 listed how many of those 163 new and proposed cable networks featured various program services.

22 news/information	8 science/technology
20 entertainment	7 cultural and performing arts
16 hobbies/how-to	7 home shopping
15 lifestyle/special audience	5 health/fitness
14 sports	5 music video
12 foreign language	4 pay-per-view
9 interactive	3 talk
9 public affairs/self-help	7 other

MTV (reaching 62 million U.S. homes) and VH-1 were joined by eight other music channels by 1996. Meanwhile pay TV's Showtime added five commercial-free channels, and Encore expanded to eight channels.

Turner Broadcasting, USA Networks, MTV Networks, Fox, and other major programmers were developing new program channel services. One cable executive estimated start-up costs for a new cable network would total about $30 million, apart from charges for acquiring any significant programs. Others estimated CBS might spend almost $100 million to start its twenty-four-hour cable news and public affairs channel—as Fox invested in its twenty-four-hour cable news service. Cable distributors in 1993 agreed to set aside a channel for CBS to program in lieu of paying for rights to carry its O&O stations and network programming on their systems; earlier that year they negotiated similar trade-offs with NBC for "America's Talking" (later MSNBC) cable service and with ABC for ESPN-2 sports oriented to younger audiences aged 18–34. In 1996 NBC and Microsoft each budgeted over $200 million to set up and operate the MSNBC news channel over five years (it reached 22 million cable homes, compared to CNN's 66 million, and the Big Three over-air networks' 96 million U.S. homes).

Major Cable Companies

Discussed in chapter 10 are multiple system operators (MSOs) who controlled local systems that received national cable services. Obviously they had an impact on cable program suppliers who relied on them for carriage and subscription fees as well as an audience base for selling national advertising in those programs. In 1996 giant Tele-Communications, Inc. (TCI) owned local systems serving over 13 million subscribers, about 20% of U.S. cable homes. Time Warner Cable served 6.6 million households and Time Warner Entertainment-Advance/Newhouse reached another 4.5 million. Together those MSOs provided cable service to one out of every three cable households in the country. The other 65% were divided among a few large MSOs—Continental Cablevision (4.1 million), Cox Communications (3.3 million), Comcast (3 million), and Cablevision Systems Corporation (2.7 million)—and hundreds of other MSOs and individual operators.[59]

By far the largest, TCI's growth since 1985 reflected cable's rapid expansion during that period, as satellite-distributed programming channels eroded traditional over-the-air networks' hold on audience viewing. Further, TCI and Liberty Media Corporation have common ownership, with investments in over ninety cable programming services including CNN, Discovery, Encore, BET, and Home Shopping Network. Their programming assets in 1996 totaled $7.5 billion, including 100% ownership of 12 national and regional cable/broadcast companies, and from 5.5% to 90% of 34 other media program suppliers.

Cross-over activity increased between cable and broadcast corporations. Among major players providing national program services to cable systems were broadcast networks and station groups, motion picture studios, syndicators, and cable MSOs.

■ Capital Cities/ABC owned 80% of ESPN and one-third of Lifetime and Arts & Entertainment; Turner Broadcasting owned CNN and Headline News, TNT, and superstation WTBS; NBC owned CNBC and 33% of Arts & Entertainment, plus America's Talking/MSNBC channel. Others included Liberty Broadcasting Network's cable Family Network, Univisa's (Hispanic TV network, Los Angeles) Galavision, the Tribune Company's superstations WGN (Chicago) and WPIX (New York), and Gaylord Broadcasting's KTVT (Dallas). Among entertainment corporations in Hollywood, Paramount Pictures and MCA Inc. owned USA Network; MCA Inc. owned superstation WWOR in New Jersey-New York; Walt Disney Company operated The Disney Channel; Time Warner Communications ran HBO and Cinemax; TV syndicator Viacom owned Showtime and The Movie Channel, and owned or had holdings in MTV, Nickelodeon, Lifetime, and VH-1. Cox Cable, United Cable, and TCI along with Newhouse Broadcasting owned The Discovery Channel; TCI and HBO owned much of BET; Cablevision Systems owned American Movie Classics and Bravo channels. ■

Top management kept their eye on all sectors of the larger TV field, fusing motion pictures with broadcast stations and national cable program services. The synergism among media companies was personified by the migration of broadcast executives over to cable in the 1980s. Cable's fledgling national program services could utilize former network executives to manage these similar business ventures that were engaged in planning program content and schedules to attract national audiences.

AUDIENCES

Between 1987 and 1991, superstation WTBS and ESPN vied back and forth for top rating position among cable program networks, according to fourth quarter ratings based on number of homes reached by each network rather than total U.S. TV households (see table 8.11). The top five basic cable services, in terms of viewing audiences, often rotated rankings with one another from week to week in national ratings.

It is important to note that ratings were based on potential viewers in areas reached by each cable network (table 8.10), not on the total universe of TV households in the country, nor even on the total homes penetrated by cable (two-thirds of U.S. households) because cable services were not carried by all existing systems. For the entire first quarter of 1993, USA was first in *prime-time* viewership with 1.4 million households (2.3 rating), WTBS followed closely with 1.2 million (2.1 rating); then came ESPN with 1 million (1.7), TNT with 900,000 (1.5), and CNN with 722,000 households (1.2). But ranking for *full-day* viewing found WTBS first with 841,000 homes (1.4 rating), USA next with 724,000 (1.2), then TNT with 572,000 (1.0), and ESPN and CNN virtually tied with 490,000 and 480,000 homes (0.8).[60]

The aggregate of cable viewers significantly reduced broadcast networks' audiences by the mid-1990s, with all cable programming attracting almost 35%

TABLE 8.11
PRIME-TIME RATINGS, TOP BASIC CABLE PROGRAM SERVICES
(BASED ON TOTAL HOUSEHOLDS REACHED BY EACH CABLE NETWORK)

NETWORK	1987₄	1988₄	1989₄	1990₄	1991₄	1992₂	1993₂	1994ᵧ	1995ᵧ	1996₂
(W)TBS	*2.6*	*2.6*	*2.6*	2.3	2.5	2.0	2.2	1.9	2.0	*2.1*
ESPN	*2.6*	*2.5*	*2.7*	2.5	2.4	1.6	1.5	1.7	1.7	1.5
USA	*1.7*	2.1	2.2	2.2	2.2	*2.1*	*2.3*	*2.3*	*2.3*	*1.9*
CNN	1.3	1.3	1.2	1.2	1.1	1.0	0.9	1.0	1.2	0.7
CBN/Family	0.8	1.2	0.8	0.8	0.8	0.9	0.9	0.9	0.9	1.1
Nashville	1.3	1.1	0.9	1.2	1.4	1.1	1.0	1.0	1.0	0.9
Lifetime	0.7	1.0	0.8	1.1	1.2	1.1	1.0	1.1	1.5	1.4
TNT	-	-	1.4	1.9	1.7	*2.0*	1.9	1.8	*2.0*	*2.1*
A&E	0.6*	0.8*	0.6	0.6	0.7	0.7	0.9	1.1	1.0	1.3
Discovery	-	-	0.8	1.0	1.0	0.9	0.9	1.0	1.1	1.1
Nick at Nite	0.4*	0.6*	0.9	0.8	0.9	1.0	1.1	1.2	1.6	1.7

#1 ranked cable network ratings boldface italic, #2 in boldface
2nd or 4th quarter reported, noted by subscript after year; "Y" = year (all 4 quarters)
* = Full-day rating, including prime-time

Source: Network estimates, Turner Broadcasting System data based on Nielsen Media Research, cited in *Broadcasting,* 9 Jan 1989; 8 Oct 1990; 14 Jan 1991; 13 Jan 1992; 13 July 1992; *Broadcasting & Cable,* 19 July 1993; 1 January 1996; 1 July 1996.

of the national audience and the three major networks about 53%. Nevertheless, in 1995 the biggest audience for advertisers on the top-ranked USA Network provided a rating of only 2.3% of the two-thirds of U.S. homes hooked to cable systems carrying that network, while any one of the three broadcast networks drew national ratings of 9% or more of all U.S. homes with TV sets; even the lowest rated WB broadcast network still drew 600,000 households more than the top cable network in prime-time. National advertisers could reach significantly larger mass audiences through commercials placed on one traditional over-air network rather than on all top five cable networks combined. Respective advertising rates were justified to advertisers depending on whether they sought broad mass audiences or far smaller but sharply defined target audiences.

Cable penetration seemed to plateau around 1992–1995 with 62%–65% of U.S. homes subscribing. As growth in potential audience began to stabilize, cable management faced the challenge of retaining subscribers and increasing the time they spent viewing cable by developing strong program attractions to compete with broadcast networks.

In 1991 almost two-thirds of all cable homes subscribed only to basic TV channels, with another 29% adding premium service, and 7% pay-per-view (table 8.12). The next year, almost half of all cable subscribers added premium and/or pay service. Then more cable homes added pay and/or pay-per-view in 1993. This strengthened the position of national premium and pay services, boosting their annual revenues from millions of homes paying monthly and per-event fees.

TV audiences across the country increased the time they viewed national cable services, at the expense of over-air broadcasting. From the 1983–1984 season to 1990–1991, viewing of basic cable channels almost tripled from 9% to 24% of *total* viewing (not just prime-time) by U.S. TV households; pay cable viewing hovered between 5% and 7% of total U.S. viewing. During

TABLE 8.12

CABLE SERVICES SUBSCRIBED TO, 1991–1993,
AS PERCENTAGE OF ALL CABLE HOUSEHOLDS*

	BASIC ONLY	+PREMIUM	+PAY-PER-VIEW	= TOTAL CABLE HOMES*
1991	64%	29%	7%	100%
1992	51%	39	10	100
1993	47%	41	12	100

*Approximately two-thirds of all U.S. TV households (62%, or 58 million homes)

Source: *Variety,* 22 February 1993, 76.

that same period, day/night viewing shares for broadcast network affiliates dropped from 60% to 53%, while independent TV stations remained about the same at around 20% as did public TV stations with 3% (see tabulation in table 8.13). Networks and their stations, of course, drew larger shares in prime-time hours—some 62% of all national viewing. Counting only the two-thirds of TV households with cable (table 8.13), basic cable networks attracted one-third of all viewing, broadcast network affiliates far less than half (46%). But pay-TV channels dropped from 11% of all viewing in cable homes in 1983–1984 to 9% in 1990–1991.

The decade's consistent growth in cable viewing encouraged MSOs and local system operators to repeatedly raise monthly fees. The average monthly bill for basic cable alone rose from $8.46 in 1982 to $18.85 in 1992. Rising fees caused widespread public complaint, triggering Congressional action for the FCC to rollback most rates in 1993.

A continuing challenge to premium cable management is to limit churn among subscribers. In the early 1990s HBO annually lost about 5% or 850,000 of its 17.4 million subscribers; to replace and add to them it aggressively promoted its program service—current movies, original dramas and comedy, and other award-winning special presentations. It matched and in some categories surpassed broadcast networks as well as rival cable services in Emmy Award nominations. In 1993 for the first time a cable program service won more Emmys than any network; HBO won seventeen, four of them for *Stalin,* while two other HBO shows won three awards each.

Revenues and Programming Costs

Audience ratings translate into income from advertising and subscribers' fees. Basic cable programmers depend on the twin revenue stream of license fees from local cable systems and national advertising (see following chapters on sales and finance). In 1993 they earned one-half to two-thirds of their revenues from advertisers, the rest from their contracted share of the systems' per-subscriber fees.

Between 1979 and 1985 Showtime increased its investment in original programming from $6 million to $35 million—about one-third of HBO's budget for original shows. By 1990 some premium networks scheduled original series, specials, and made-for-cable-TV movies as almost half of their service, the other half being theatrical motion pictures and syndicated or off-network material. Both premium and basic services such as ESPN substantially increased their bidding for rights to cover major sports, drawing those events and audiences away from broadcast television as described earlier in this chapter.

PERCENTAGES OF TOTAL TELEVISION HOUSEHOLDS*

	1983 –84	1984 –85	1985 –86	1986 –87	1987 –88	1988 –89	1989 –90	1990 –91
Basic Cable Networks	9	11	11	13	15	17	21	24
Pay Cable Services	5	6	5	4	7	7	6	6
Broadcast Network Affiliates	69	66	66	64	61	58	55	53
Independent TV Stations	19	18	18	20	20	20	20	21
Public TV Stations	3	3	3	4	4	3	3	3

PERCENTAGES OF ALL CABLE HOUSEHOLDS*

	1983 –84	1984 –85	1985 –86	1986 –87	1987 –88	1988 –89	1989 –90	1990 –91
Basic Cable Networks	17	19	19	23	25	28	32	35
Pay Cable Services	11	11	10	10	11	11	10	9
Broadcast Network Affiliates	58	56	56	53	52	49	46	46
Independent TV Stations	17	17	17	17	17	16	16	17
Public TV Stations	3	3	3	3	3	3	3	2

*Each column may total more than 100% because of multi-set use and rounding off of numbers.

Source: "Cable Television Developments: Directory of Top 50 MSOs/Directory of Cable Networks," National Cable Television Association, (May 1992), 5–A.

■ The heavyweight championship fight between Evander Holyfield and James (Buster) Douglas attracted a million cable households—7.3% of the 17 million homes equipped for pay-per-view. At $36.50 per home the single event brought in $40 million. The following year cable systems charged $35–$40 for the Holyfield-George Foreman championship bout; about 9% of pay-cable homes tuned in, paying a record $60 million.

Highest gross revenues of six nonsport rock music events up to 1994 ranged from $3.1 million to $5.5 million each; average ticket prices for a viewing household ranged between $19.95 and $24.95.

Revenues from pay-per-view TV events, movies, and adult services increased steadily from $149 million in 1988 to an estimated total of $648 million in 1996. ■

Starting a Cable Channel Program Service

Managing a prospective national cable service begins with identifying a program service that local systems will choose to carry because it attracts subscribers. This calls for distinctive programming filling an underserved niche—much as with developing a format in local radio or a specialized network/syndicated program service in national radio. The next step is to market the service to MSOs and local systems—through individual selling, at national conferences, and through promotional mail-outs and advertising. The proposed channel's unique programming, highlights of special offerings, and fee schedules for licensing are important considerations. To interest local system "buyers" a new program service often offers start-up support payments to cable operators, often 25¢ per subscriber, to help promote the channel to current and prospective subscribers in their area. Key to negotiating with local systems are license fees charged for carrying the service. Those can range from nothing per subscriber (EWTN), to 11¢ (CNBC, A&E), 15¢–20¢ (TNN/Nashville, MTV, USA), or even more (CNN 23¢ and ESPN 50¢ in 1992), depending on the attractiveness of the program service—the programming's value to local subscribers, and cost to cable services that must set fees at a level to ensure adequate revenues. Premium channels, of course, bring in no revenue from advertising and depend entirely on fees, typically splitting monthly fees of $8–10 or more, with local operators keeping over half the monthly fees paid by their subscribers. The twin keys to negotiating between national program service management and local system managers, then, are the programming itself and the fees local systems must pay to get that service.

On the other hand, total national cable revenues from advertising passed $4 billion in 1996 and were expected to rise over the next four years to $7 billion in 2000.

PROGRAMMING

Competition

Cable program services compete with traditional networks and with one another for national audiences. Their program content and how they schedule it is chosen to attract audiences away from other alternatives in a given day-part and hour. Against one another, cable networks strive for distinctive service distinguishing them from competing services. That makes them attractive to local cable systems for initial carriage, and for viewers with thirty to fifty program options at any given time. Basic cable networks specialize in different kinds of programming; and they compete within each category for niche target audiences.

Programming Strategies

■   A useful though atypical case study of programming strategy is the Arts & Entertainment network, as profiled by John Dempsey of *Variety* in an extended interview with Nick Davatzes, A&E president and CEO.[61] Formed out of the failed CBS cultural arts channel and Rockefeller Center's Entertainment channel, A&E built its audience by an amalgam of programming, countering the fare of rival cable networks. Offering no science, technology, or sports, it gradually expanded a wider audience base than its predecessors. Originally cultural with informational and educational elements, A&E managers developed a mix of old documentaries, BBC drama and mystery mini-series, World War II themes, action movies, slapstick comedy, and older motion pictures. It avoided high costs of top-line theatrical films or original projects— co-producing more programs with the BBC than any other video company in the world. Its 1992 budget of $50 million was far below that of competing channels. It reversed its original mix from two-thirds British, one-third American to opposite proportions. To develop viewers' habits it favored series with predictable scheduling rather than anthologies. Its frequent historical military documentaries of the late 1930s and 1940s prompted criticism as the "World War II network" (or, less benignly, the "Nazi Network" with its weekly Stuka bombers, storm troopers, Hitler-Goering conferences, and the like); but only 14% of its schedule carried such programs, compared with more than twice that percentage on the Discovery channel. Charged with looking much like public broadcasting, it actually devoted 66% of its programming to entertainment (vs. 15% on PBS). In the early 1990s A&E ratings increased by one-third in prime-time, and 25% in total day/night, each successive year. Its monthly license fees to local operators matched those gains. A&E charged systems 13¢ per subscriber in 1991, 15¢ in 1992, 17¢ in 1993, and 19¢ in 1994 (with volume discounts to massive MSOs carrying the service). Some 7,600 cable systems carried A&E to their 53 million subscribers (up from 30 million in 1988). At 19¢ each in 1994, that meant over $10 million a month from fees or $120 million a year—about 40% of the year's revenues. The other

60% came from advertising (actual ad revenues in 1990 were $48.6 million, the next year $69.3 million). Its audience profile became 55% men (the reverse of several years earlier), 35–45 years old, with four years of college and an average income of $40,000–50,000. ■

Programming Costs

Cable gradually invested heavily to create original programs as well as leasing packages of other shows to fill twenty-four-hour channels.

■ In the 1992–1993 season, basic cable networks spent $1.8 billion on original programming, and pay-TV services $1.2 billion. Major program commitments in 1993–1994 included over 500 hours each of first-run series and specials by both A&E and The Discovery Channel, TNT's original movies (including a $15 million four-hour special on the Civil War battle of Gettysburg), USA Network's original animation and sitcoms, and The Disney Channel's animation, movies, concerts, documentaries, and specials, among other cable program services. HBO devoted 30% of its schedule to original programming, events, movies, series, and documentaries. Showtime put $120 million into first-run production including long-form anthology and 18–20 original theatricals.

From 1993 through 1995 cable spent $7.3 billion to produce original programming. Sampling the field in 1995–1996: A&E's schedule was 85% original programs; Discovery and Lifetime spent $160 million on its own shows including 800 original prime-time programs; Showtime produced forty original movies in 1995 with budgets of $2.5 to $5 million each, devoting almost 40% of its schedule to original programming; USA Network put $40 million into newly created product; HBO and TNT continued their prolific output; and of course ESPN and other cable program channels produced thousands of hours of live sports coverage.[62] ■

Cable's expenses for original programs in the 1995–1996 season totaled $3 billion. That year, when movies produced for broadcast networks had budgets of $2.5 million–$3 million, HBO and Showtime and a few others commissioned motion picture dramas for cable costing $7 million–$9 million—often more daring, socially relevant, and intellectually ambitious as well as more expensive than their network counterparts. (Meanwhile theatrical motion pictures driven by movie stars and laden with spectacular special effects were running up budgets of $20 million–$30 million—some huge successes, but many of them failures critically and at the box office as well.)

Original programs for cable programs were not only watched by larger mass audiences, but won awards for outstanding quality programs on lists formerly the preserve of major network fare. In 1993 HBO received 17 Emmy awards for programming, NBC 16, CBS 14, and ABC 12, with three awards to other cable programmers.

Costs of programming were met partly by co-ventures with broadcast networks; USA and CBS shared costs of *Silk Stalkings,* aired first on CBS and then later the same week on USA; Nickelodeon scheduled *Hi, Honey I'm Home* on Saturday and Sunday after co-funder ABC aired it on Friday. Cable also began to carry shows first aired on over-air networks. The first top-10 program to move over to cable from network TV (and while still airing there!) was *Murder, She Wrote;* USA Network paid $30 million for rights to telecast 111 episodes over six years. Into the mid-1990s those shows remained among the top 20 weekly shows in all of cable. While hardly innovative programming, the cable service provided a program that the mass public obviously wanted, scheduling it at times convenient for audiences to view it.

Cable News

In 1988 critic Les Brown publicly lamented that steady viewers of pay channels such as HBO, Showtime, plus MTV, VH-1, Nick-At-Nite, and others would not have a "window on the world" keeping them posted about any truly major events from day to day. Only established news operations on a few other cable channels, and the traditional over-air networks and local stations, provided full service to keep viewers informed about urgent weather information, stock market swings, urban disturbances, and overseas wars or other calamities, not to speak of major communications by government leaders. Brown urged HBO officials to consider incorporating some intermittent news service, possibly brief feeds from CNN, at least when important news incidents occurred that viewers should be informed of. He reflected sociology studies when offering "my theory that people most often tune into television or radio from an unconscious desire to be plugged into the outside world. . . . In the back of the viewer's mind, those [network-affiliated broadcast stations] connect him to a vast news-gathering apparatus" about any catastrophe, emergency, or crisis about which one should know swiftly and reliably. Instead, he told HBO cable channel's officers that "HBO's failing . . . is that it's as unresponsive to events as the VCR."

Through the 1990s, cable continued to offer a number of news-oriented channels, described earlier, such as CNN and Headline News, CNBC, Discovery Channel's "World Monitor" (by *Christian Science Monitor*), C-SPAN I and II, MSNBC, Fox's news channel, and The Weather Channel's reports and special features, plus inserts in some other services. Some cable operators collaborated with local TV stations' news operations (noted in chapter 7).

Both CBS and NBC had previously explored the possibility of establishing a cable news channel. NBC once attempted to buy out Turner's CNN news operation; instead Turner bought the nascent satellite cable news operation mounted by ABC and Westinghouse. NBC tried to start a channel news service in 1985 but dropped it the following year; in 1988 it bought a cable program service from which it developed CNBC/Consumer News and Business Channel, into which it folded FNN/Financial News Network. In 1996 it introduced the MSNBC news/talk channel in collaboration with Microsoft computer software manufacturer.

DIRECT BROADCAST SATELLITE PROGRAM SERVICES

No longer only figuratively on the horizon, but literally there orbiting 22,300 miles above the equator DirecTV, USSB, PrimeStar, EchoStar and other newly launched satellite program services added to the mix of program distributors and suppliers competing with cable and terrestrial television. HBO and Showtime came with sharp digital signals, more attractive than not always dependable local cable service. DirecTV also offered scores of dedicated channels with its own phalanx of pay-per-view uncut recent theatrical motion pictures and dozens of sports channels along with the established line-up of national program channels available on cable.

By the end of 1996 about 6.25 million subscribers owned small dish, high-powered satellite receivers. After nine years (in 2005) 33.6 million U.S. homes are expected to have the service. A subsidiary of General Motors and Hughes, DirecTV's 1.6 million subscribers in 1996 brought it just up to the level of the top 10 cable MSOs. USSB estimated a DBS provider needed 3 million subscribers to break even. Subscriber-owners multiplied when competing suppliers dropped initial costs of receiving equipment from $800 to $200 during 1996.

As noted in this chapter, the traditional major networks created much of television as we know it today and they still continue to dominate the programming of American commercial television. Production companies and cable channel services have built on that base. The public is most familiar with network programs because these are what most people see most of the time, even with prime-time divided 55% to network viewing and almost 45% to all other cable channels combined plus public broadcasting and independent stations. But throughout the day and night syndicated and locally originated programs along with station breaks for spot commercials between network programs all contribute to broadcast revenues. How these revenues are generated by local and national sale of time are discussed in the next chapter.

CHAPTER 8 NOTES

1. Don Le Duc, *Beyond Broadcasting: Patterns in Policy and Law* (New York: Longman, 1987), 89.

2. Robert M. Pepper, FCC Office of Plans and Policy, in report to the Commission; cited in *Television/Radio Age,* 12 December 1988, 78.

3. "Editorials," *Broadcast & Cable,* 5 April 1993, 66.

4. For details of distribution contractual fees and processes, and descriptions of program services, see Raymond L. Carroll and Donald M. Davis, *Electronic Media Programming: Strategies and Decision Making* (New York: McGraw-Hill, 1993), 254-271: Chapter 10, "Radio Program Suppliers."

5. Quoted by Sydney W. Head and Christopher H. Sterling, *Broadcasting in America,* 6th ed. (Boston: Houghton Mifflin, 1990), 190.

6. For details of the struggle between NPR and its member stations, which that year received funding directly from the Corporation for Public Broadcasting, see Wallace A. Smith, "Public Radio Programming," Chapter 19 in Susan Tyler Eastman, Sydney W. Head, and Lewis Klein, *Broadcast/Cable Programming: Strategies & Practices,* 3rd ed. (Belmont, Calif.: Wadsworth, 1989), 503–523; also Stephen L. Salyer, "Monopoly to Marketplace—Competition Comes to Public Radio," in The Freedom Forum Media Studies Center's *Media Studies Journal: Radio the Forgotten Medium* (Summer 1993): 176–183.

7. Gene F. Jankowski, "Remarks Before the International Radio & Television Society Newsmaker Luncheon" (New York, 15 January 1986), pp. 10–11 of speech text.

8. Of $628 million ad dollars spent with Hispanic media in 1990, $55.1 went to national radio, $131.1 million to local TV, $156 million to local radio, $58 million to print, and the rest to other forms of advertising; *Broadcasting,* 10 December 1990, 67–84.

9. Nielsen Media Research and CBS Research provided data reported in *Broadcasting & Cable,* 30 September 1996, 40.

10. Gene Jankowski, "The Network Connection" (speech presented before the Hollywood Radio and Television Society, Beverly Hills, Calif., 24 May 1988), p. 4 of speech text. Some legal restrictions against networks as producers or distributors were loosened by the FCC in 1991 and 1993 in doing away with the FCC's financial interest/syndication ("fin/syn") regulations.

11. Ken Auletta, "Where Renters Pretend to be Landlords: Visiting 'Planet TV,'" *RTNDA Communicator,* April 1992, 18–19. Auletta authored an exhaustively researched and well written study of the network "culture" and their restructuring in the mid-1980s, *Three Blind Mice: How the Networks Lost Their Way* (New York: Random House, 1991).

12. Quoted by Elizabeth Jensen, "Networks Fight Back," *New York Daily News,* [April 1990 (n.d. on clipping)], 88.

13. When the hour network news proposal was considered in 1981, their 150 affiliates in the top-50 markets, often with first-run syndication in access time, stood to lose up to 15% of their revenues by ceding that total of 29,000 hours. The top-50 stations then grossed about $550 million in national and local spot ads during one year of access programs. For example, the three affiliates in Dallas would lose about $8 million a year by giving the access half-hour over to expanded evening news by their networks.

14. See Julie A. Zier, "Fog of War Engulfs Affiliation Battles," *Broadcasting & Cable,* 5 December 1994, 50-56; Terry Likes, "Affiliate Switching Now and Then," *RTNDA Communicator,* March 1996, 19-21.

15. Cited by J. Max Robins, "Rush Hour for Newsmags," *Variety,* 11 January 1993, 35, 40.

16. Analyses of planning and production by networks and studios can be found in Sally Bedell, *Up the Tube: Prime-Time TV in the Silverman Years* (New York: Viking Press, 1981); Richard Levinson and William Link, *Stay Tuned: An Inside Look at the Making of Prime-Time Television* (New York: St. Martin's Press, 1981); Mark Christensen and Cameron Stauth, *The Sweeps: Behind the Scenes in Network TV* (New York: William Morrow, 1984); and Richard A. Blum and Richard D. Lindheim, *Prime Time: Network Television Programming* (Boston: Focal Press, 1987). Still worth reading is the earlier classic by Les Brown, *Televi$ion: The Business Behind the Box* (New York: Harcourt, Brace, Jovanovich, 1971).

17. Quoted by the *New York Times,* 26 February 1989, H-42.

18. See Blum and Lindheim, *Network Television Programming;* for descriptive analyses of executives listed in programming departments, see Chapter 3, "The Buyers: Network Television Management," 25–42.

19. Sociologist Todd Gitlin described in a dozen pages in personalized, concrete detail this sequence of planning by Wood, "when a single network made a brilliantly correct guess about public moods and how to package them for TV consumption. That was trend-spotting in its moment of commercial glory, and much of Hollywood still remembers it fondly as the moment when television threatened to grow up. . . . [Bob Wood's] regime was singular in television history, and his successes changed the tone and texture of television comedy"; *Inside Prime Time* (New York: Pantheon, 1983/1985), 205, 219.

20. Robert D. Wood, "The Decision-Making Process in Television" (address to Graduate School of Business, University of Southern California, 23 November 1970), 7–9 (booklet published by CBS Television Network).

21. Jankowski, "The Network Connection," p. 2 of text.

22. An industry analysis by investment banking firm Veronis, Suhler & Associates was cited by Joe Flint, "Network to Spend More for Programs, Says Study," *Broadcasting & Cable,* 26 July 1993, 91.

23. Michael Pollan, "Can 'Hill Street Blues' Rescue NBC?" *Channels* (March-April 1983): 30–34.

24. Alex Ben Block, *Outfoxed: ... The Inside Story of America's Fourth Network* (New York: St. Martin's, 1990); cited by Mary-Ann Bendel and Paulette Walker in *USA Today,* 25 June 1990, 13A.

25. "The Rise and Rise of Program Prices," *Broadcasting,* 23 September 1991, 44–45. Cf. *Broadcast Financial Journal,* September/October 1987, 23, which cites the average cost of 30-minute prime-time programs as $73,000 in 1965–1966, $115,000 in 1975–1976, and $365,000 in 1985–1986.

26. Robert F. Lewine, Susan Tyler Eastman, and William J. Adams, "Prime-Time Network Television Programming," in Eastman, Head, and Klein, *Broadcast/Cable Programming: Strategies & Practices,* 3rd ed., 134–172; annual statistics for new programs canceled by each network, Table 5-5, p. 155.

27. See Alan B. Albarron et al., "Trends in Network Prime-Time Programming, 1983–1990: The Emergence of the Fox Network," *Feedback* (Fall 1991): 2–5.

28. John J. O'Connor, "A Very Model of a Modern Major Mini-Series," New York Times News Service, in the *Tuscaloosa News,* 13 January 1991, 8H.

29. Joe Lapointe, New York Times News Service syndicated article "ESPN Says Goodbye to NCAA Tournament," in *The Tuscaloosa News,* 18 March, 1990, 4–C.

30. See the cover story "Baseball 1996" of several articles in *Broadcasting & Cable,* 25 March 1996, 24–32.

31. Quoted in "Network Newsmen Look at Future of TV News," *Broadcasting,* 7 March 1988, 89–90.

32. "Clutter, Local Time the Major NBC Affiliate Concerns," *Television/Radio Age,* 25 May 1989, 48, 71–72.

33. An apt analogy was described by Danny Duncan Collum, "CBS News Anchors Adrift in Shallow Waters," *National Catholic Reporter,* 16 July 1993, 19: "Serious broadcast journalists, which most of the folks at the networks still are, today face the same dilemma that late 19th century painters faced with the advent of photography. Suddenly artists found it was no longer enough to objectively reproduce the surface of reality as it appears to the eye. A photograph could always do that better than even the most skilled and fanatical brush person. The painters' answer to this dilemma was, of course, Impressionism. Don't paint what things look like, paint what they feel like. And the visual arts were off on a century-long joyride of exploration and risk.

"In broadcast news today, if you want strict objectivity and comprehensiveness—the old journalistic standards—then you'll never beat the round-the-clock video vérité of C-Span and CNN. C-Span especially represents the 'objective journalism' ethos taken to its ultimate height. They just turn on the camera, and leave it running."

34. The Tyndall Report newsletter, which monitors evening newscasts, was cited by Walter Goodman, "Nightly News Looks Beyond the Headlines," *The New York Times,* 7 July 1991, Sec. 2, p. 23. During the week of June 17–21, 1991, of the total 22 minutes, national and international news items received 10.93 minutes (ABC); 13.02 (CBS); and 13.09 (NBC); while "in-depth reports on social issues" received 9.99 minutes (ABC); 6.73 (CBS); and 7.81 (NBC). Entertainment and human interest "soft" stories received 0.77 minute (ABC); 1.98 (CBS); and 0.89 (NBC).

35. For descriptive summaries of documentaries in the context of networks and advertisers, see Brown, *Televi$ion,* and Eric Barnouw, *The Image Empire: A History of Broadcasting in the United States Since 1953* (New York: Oxford University Press, 1970). See also the analysis of the network documentarian's work in Fred W. Friendly, *Due to Circumstances Beyond Our Control. . .* (New York: Random House, 1967); also Kendrick, *Prime Time: The Life of Edward R. Murrow.* Friendly wrote a detailed analysis of one documentary that brought grief to CBS: "Television: The Unselling of *The Selling of the Pentagon,*" *Harper's Magazine,* June 1971, 30–37. The efforts of a creative, investigative documentarian are recounted by David G. Yellin, *Special: Fred Freed and the Television Documentary* (New York: Macmillan, 1972).

36. But that would total $441 million whereas the annual news budget was actually $250 million, so either the number of stories was far fewer or the pro-rata average came closer to $36,000. Perhaps the *Broadcasting* article inverted the figures, which would be for them a most unusual typographical error, "The Futurist in Charge at NBC News," *Broadcasting,* 29 February 1988, 44–54 (questionable figure cited on p. 48). Equally rare were two articles in successive issues, one signed "MF" reporting a speech by NBC News head Gartner who expected his division "to end up $15 million to $20 million in the black for the season," whereas four weeks later an article signed by "RB" stated "the division should close $15 million to $20 million in the red . . . as NBC News President Michael Gartner recently told RTNDA Convention attendees": "Gartner Backs Network and Affiliate Pools," *Broadcasting,* 1 October 1990, 47–48, and "News Divisions Get In-House Edits," 29 October 1990, 68–69. This note is not to cavil at a discrepancy, but rather to note the phenomenon of an extremely rare lapse by that respected publication, whose full and accurate reports set the standard and have proved invaluable in preparing many portions of the present book.

37. Special RTNDA report by Harry A. Jessell in *Broadcasting & Cable,* 7 October 1996, 42–43; cf. issue of 10 October 1994, 50–54.

38. Quoted in "FCC Opens Up Possibility of Children's TV Requirements," *Broadcasting,* 5 November 1979, 24–27.

39. Mark Fowler spoke at Arizona State University, quoted in "Fowler No Fan of Federal Pre-emption on Children's TV," *Broadcasting,* 14 February 1993, 85.

40. FCC, re MM Docket No. 90–570, Policies and Rules Concerning Children's Television Programming, 6 FCC Rcd 2111; "Memorandum Opinion and Order," adopted August 1, 1991; Released August 26, 1991 (FCC 91–248), 7.

41. Cited by Steve McClellan, "It's Not Just for Saturday Mornings Anymore," *Broadcasting & Cable,* 26 July 1993, 38. Bert Gould, senior vice-president of Fox Children Network, cited the Yankelovich Youth Monitor as source for the following estimate; *Broadcasting & Cable,* 13 June 1994, 19, 24.

42. Les Brown, ed., *Les Brown's Encyclopedia of Television,* 3rd ed. (Detroit: Gale Research Inc., 1991), 73.

43. A.C. Nielsen data reported by John Dempsey, "Nielsen Reports on Black Viewing," *Variety,* 14 January 1991.

44. BBDO Special Markets analysis of Nielsen data, cited in "Black Viewers Turn to NBC," *Channels,* November 1987, 64.

45. "Monday Memo: Blacks are Gravitating to Programs with Black Themes that Relate to Them," *Broadcasting & Cable,* 26 April 1993, 75.

46. "Fox Shows Rank High in Hispanic Survey," *Broadcasting,* 22 March 1933, 41. Data were for November 1992 when A. C. Nielsen first surveyed Hispanic viewing of English-language broadcasts.

47. Data for much of this section were drawn from special supplements in *Television/Radio Age,* November 1987, July 1988, July 1989, and 3 April 1989; also from special reports in *Broadcasting,* 3 April 1989, 10 December 1990, 26 October 1992.

48. John Wiseman, "Public TV In Crisis: How To Make It Better," *TV Guide,* 8 August 1987, 26–40; quoted excerpt, 34.

49. *Ibid.*

50. For example, see the *New York Times'* major pieces by Walter Goodman, "Public TV Juggles a Hot Potato," 3 September 1989, 1H; Constance L. Hayes, "Overcoming Obstacles to a Civil-Rights Chronicle," 14 January 1990, 31H; Jan Benzel, "Short-Form Films Bolster Long-Range Objectives," 25 March 1990, 33H; Michael Shapiro, "Reliving the War Not Fought in the Living Room," 11 November 1990, 35H; Walter Goodman, "The Documentary Hits Its Stride on PBS," 25 November 1990, 31H; John J. O'Connor, "Celebrating 20 Years of Elitism," 4 October 1992, 31H.

51. John J. O'Connor, New York Times News Service, "A Very Model of a Modern Major Mini-series," in the *Tuscaloosa News,* 13 January 1991, 8H.

52. Brown, *Televi$ion,* 58.

53. Cronkite spoke at the conference of American Women in Radio Television (AWRT) in Phoenix, March 29, 1992; cited in *Broadcasting,* 8 June 1992, 19.

54. "At Large: The Play's Still the Thing," *Broadcasting,* 14 December 1987, 73–78; quoted excerpt, 77.

55. Secretary Cheney responded to a question from the floor after his speech to Pi Sigma Alpha, national political science honor society, in Washington, D.C.; speech rebroadcast by C-SPAN, Fall 1992.

56. Research and Policy Analysis Department, National Cable Television Association, *Cable Television Developments: Directory of Top 50 MSOs, Directory of Cable Networks* (May 1992), iC, 16A–17A.

57. *U.S. News & World Report,* 29 July 1996, 10, cited MTV and Nielsen Media Research as sources for these figures.

58. "Pay Homes' Viewing Habits," *Channels,* 13 August 1990, 52; citing data by Paul Kagan Associates and Nielsen.

59. Data from *Cablevision,* 15 April 1996, 50.

60. "First-Quarter Ratings: 1993 vs. 1992," *Broadcasting & Cable,* 5 April 1993, 30. Similarly switched rankings for prime-time (USA, TBS) and all day/night (TBS, USA) were reported in mid-1990, *Channels,* 13 August 1990, 52.

61. John Dempsey, "Highbrow Cabler Comes of Age," *Variety,* 15 April 1992, 25, 32.

62. "Special Report: Original Cable Programming," *Broadcasting & Cable,* 19 February 1996, 32–57; cf. previous report, 20 February 1995, 22–48.

Managing Marketing and Sales

*That has been one of the favorite subjects of radio's critics,
ever since the medium's beginnings: that the entire mysterious
process, the curious appeal of that sound, is placed in the service
of commerce. American radio is a marketing tool; love it, live
with it, analyze it, the whole bloody sound is there to move goods
on the shelf. Broadcasters do not 'admit' this charge; that would
be like admitting that Tuesday follows Monday. . . . 'And anyone
in this business who isn't sales oriented is in trouble.'*

STEVE MILLARD, BROADCASTING MAGAZINE[1]

*The broadcast networks act as brokers and consolidators for
local affiliated television stations in the business of selling access
to audiences.*

BRUCE OWEN AND STEVEN WILDMAN[2]

Since 1922, when WEAF in New York broadcast the first advertising message as
a source of revenue for operating the radiotelephony service, commercials have
paid the way for radio and then for television in this country. Since that time, the
broadcast industry has been supported not by income from selling receivers (as
originally conceived by the Radio Corporation of America) nor by subscription
fees from listeners (as cable and pay television) nor by a government tax on purchase or annual use of receivers (as in many countries today). Instead, revenue
has come from businesses wanting to reach potential consumers for their products and services; they paid a "toll" to have their messages broadcast to widespread
audiences. The public and the federal government, as well as broadcasters and the
business community, have endorsed this system of advertising-supported broadcasting. Despite some criticism of this orientation toward the commercial marketplace, the general public over the decades has continued to accept the basic
structure of American commercial radio and television.[3]

Other chapters of this book discuss management's concerns for the programming and engineering departments. But neither could operate in a commercial
station or network were it not for the sales department. Time sales produce the
revenue that pays for all the station's and network's activities and, through good
management, makes a profit for the broadcast company's owners/investors. By
contrast, commercial time sales accounted for about 5% of cable systems' revenues, and 10% of national cable program services' income; the other 90%–95%
was generated by monthly fees paid by cable subscribers. Public noncommercial

broadcasting's counterpart to sales is fund-raising "development" which seeks private corporate funding, government grants, and contributions from viewers-listeners.

WHEN SALES BECAME MARKETING

In the intensely competitive arena of over-air and cable distribution of program services in the 1980s and 1990s, stations and networks could no longer simply offer time to buyers who came to them wanting to advertise their goods and services, as in earlier decades. CBS Broadcast Group President Howard Stringer acknowledged that "for years, we were price negotiators, and forgot how to be salesmen."[4] Aggressive print media, outdoor advertising, direct mail and other forms of promotion vied for advertisers' budgets. Beyond that, major companies' advertising dropped from half their marketing budget in the mid-1970s to about one-third by 1990; although the absolute totals expanded annually, a decreasing portion went to traditional forms of media advertising.

Revenues from all forms of national advertising in 1996 were expected to total $102 billion. Of that massive ad pie $45 billion were spread over print, broadcast, and cable. ($35 billion went to direct mail, $1.5 billion to Yellow Pages, and $21 billion to other national media.) National advertisers' dollars spent with four major TV networks were projected at $13 billion, plus $10 billion on spot TV, $5.3 billion on cable and syndicated TV, $2.7 billion to radio, $9.6 billion to magazines, and $4.3 billion to newspapers. Media sales staffs vied against one another to expand their share of advertisers' expenditures.

The media marketing mix became increasingly complex as local radio and TV stations proliferated, along with cable services; bartered syndication and co-op advertising cut further into national budgets for network program spots. Just as AM radio discovered when TV and FM came on the scene, each medium today must identify its specific claim on ad dollars, offering distinctive ways to reach cross-sections of consumers in the mass audience. The buzzword became *niche* programming: filling "format holes" in a radio market, and offering highly desirable demographic and psychographic profiles among target audiences reached by a radio or TV program or cable channel.

Further, relaxed regulations permitting duopoly—actually, operating up to eight stations in a single market—and local marketing agreements called for innovative managing of sales and programming in radio, such as merging separate sales departments into a single multistation team. Ways of selling differed for national radio network feeds of block personality/talk programming, and for barter TV syndication and unwired networks through station rep groups and brokers. In 1991 CBS Television Stations and Meredith Corporation formed a cross-media ad sales alliance, as did CBS Television network and the New York Times Magazine Group. Networks aligned themselves with major chain distributors like K-Mart for collaborative marketing campaigns promoting new program seasons. CBS began to group its salespeople in teams assigned to different ad agencies, authorizing them to negotiate all day-parts to meet various clients' specialized marketing needs.

As noted in chapter 4, networks created new executive positions to oversee their entire marketing effort, including sales and promotion. Media companies often merged sales departments into wider-embracing marketing divisions. They backed up all sales planning with astute data gathering and analysis, heavily fortified by computerized databanks. Gaining and then protecting "market share" among vigorously competitive media became the touchstone for sales operations and the standard for success. TV sales managers followed the path already set by aggressive radio counterparts in creative promotion and selling, market positioning, and targeting audience niches. A number of stations, including Gannett Broadcasting's WUSA-TV in Washington, D.C., created a new position of vice-president of marketing, added to a vice-president of sales or general sales manager.

Selling focuses more narrowly on program product and spot announcement times available, while *marketing* looks to segmented audience preferences and to advertisers' needs in the total media mix. Management in the first instance is oriented to sales volume, but in the latter more precisely to profitability—cost/income efficiency. Sales planning looks more to present product and conditions, whereas marketing strategy assesses long-term factors and addresses product segmentation, market differentiation, target and core audiences, socioeconomic marketing environments, and positioning strategy.[5]

Ries and Trout stressed positioning as the tactic of placing one's product or service (here: a station generally and program commercial spots on its schedule) in the consumer's *mind* as strongest among competitors and most advantageous to the consumer because of a feature or benefit distinctive to that product. They describe the 1950s as the **product era,** which fostered a unique selling proposition based on a product's features. That period gave way to the **image era** that promoted favorable contexts and characteristics creatively associated with a product. The next era focused on **positioning**—looking not to the product but to the consumer's perceptions. Ries and Trout claim that "positioning is what you do to the mind of the prospect; that is, you position the product in the mind of the prospect." They underscore that marketing is not a battle between products or services but a battle of perceptions among targeted consumers. Promotion is essential to successful positioning. Promotion takes management's commitment of money and extended periods of time to be effective.

> It takes money to build a share of mind. It takes money to establish a position. It takes money to hold a position once you've established it. . . . Positioning is a concept that is cumulative. Something that takes advantage of advertising's long-range nature.[6]

The goal is to create a category in which your station or program or spot avails can be first, in the mind of the prospect—the agency and its advertiser-client.

The terms *selling* and *marketing* reflect not only shifts over time but also different perspectives at any one time about advertising as the economic base for American broadcast/cable media. For example, publications in the early 1970s stressed the selling function, but gradually shifted to broader concepts of marketing and positioning.[7]

CBS research executive David Poltrack compared the broadcast advertising marketplace to the commodity market with its volatile price changes. Audience impressions, which radio-TV-cable ads create, are complicated and dynamic, measured indirectly by ratings estimates and predictions interpreted subjectively by media planners and sales reps. Meanwhile a radio or TV ad campaign, whose time-buys are stalked by competing stations and networks, is "but one component of a multimedia advertising campaign, and that campaign but one element of a total marketing program; the planner must be familiar with the synergistic quality of various media when used in combination with television" and radio or cable.[8]

BUDGETS

The general sales manager works closely with the general/station manager and other department heads in planning annual budgets.

Sales budgets include personnel compensation and material and operations costs along with supplies and portions of twelve-month utilities. Staff are paid in large part by commissions based on the billings each handles, to which are budgeted any base salaries, plus benefits (such as health and unemployment insurance, social security).

In 1996 when TV general managers' salaries averaged $160,000 and program directors $54,000, general sales managers earned over $110,000, and promotion managers $43,000; that same year in radio general managers (GMs) earned an average $73,500, program directors (PDs) $39,000, and promotion managers $33,500.[9] TV promotion managers' salaries ranged from $20,000 in some small markets to more than $65,000 in top markets.

How do expenses for running sales departments compare with other departments in local stations? During the half-decade from 1985 to 1990 (table 9.1), about one-fifth of the average radio station's annual budget went to the sales department, while less than one-tenth of TV stations' annual expenses were for sales. Put another way, in 1990 radio's expenditures for sales did not quite equal the amount spent on programming (not including news), but was less than half of general and administrative costs. TV's sales expenses came to less than one-quarter of programming costs (excluding news), and slightly more than a quarter of general/administrative expenses. In 1990 the typical radio station budgeted 19% to sales; supporting the sales function as well as programming, the 9% for advertising and promotion would bring the total to more than one-fourth (28%) of the station's annual expenses. Those same years a TV station typically budgeted 8.5% for sales plus another 4.5% for advertising and promotion, totaling 13% of the year's expenditures.

Promotion budgets vary among media, markets, and the size of stations within those markets (sales promotion is detailed later in this chapter). In 1976 half of all radio stations had annual promotion budgets less than $10,000, as did one out of five TV stations. By 1988 promotion departments at half the radio stations and almost two-thirds of TV stations had budgets over $250,000. Among AM and FM stations, 17% budgeted $1 million–$2 million for promotion, as did 13% of TV stations; another 5% of TV stations in large markets earmarked over $2 million annually. When Gannett-owned KIIS radio in Los Angeles had one of the largest billings of any U.S. station, in 1985 it spent $3 million on promotion.

■ GM Mike Kettenring of WSMV, NBC-TV's Nashville affiliate, in the mid-1980s reserved 10% of the station's best spots in each day-part for promotion—worth $3 million of airtime annually. Station computers were programmed to schedule them as nonpre-emptible confirmed spots. The station moved production of all promotional spots to an in-house creative services team, formed with money re-allocated from billboard and print ads. Within three years, sales revenues rose from 32% to 36% of the TV market, the highest in the area. ■

Promotion staff around the country vary in size from one to two people at stations in markets of all sizes, to more than ten persons in top-10 markets, with the average between three and five in promotion. The promotion manager uses budgeted funds and staff to oversee audience promotion, sales promotion, research, community relations, station image through graphics and award presentations, press

TABLE 9.1
DEPARTMENTAL EXPENSES, RADIO AND TV STATIONS
AS PERCENTAGE OF STATION BUDGET
REPRESENTATIVE AVERAGES, EARLY 1990S

EXPENSES	RADIO AM, AM/FM, FM	TELEVISION Affiliates	TELEVISION Independents
General and Administrative	41%	35%	24%
Programming and Production	24	25	51
News	3	20	5
Sales	19	8	8.5
Advertising and Promotion	9	5	4.5
Engineering	4	7	7

Sources: Various, including trade magazines and surveys by NAB and BCFM.

relations, news, and general administrative functions, plus direct marketing activities at cable systems. KING-TV in Seattle clustered research, creative, public relations, and sales promotion into a single marketing department of fourteen persons with a "seven-figure" budget.

Network TV affiliates put 75% of promotion budgets into promoting their local news, and 20% to promoting early fringe and prime access syndicated programs—usually in newspapers and *TV Guide*. Independent stations direct three-fourths of their promotion to movies and local sports coverage, 15% to syndicated series, and 10% to Fox network shows.[10]

Noncommercial radio stations typically devote 4% to promotion; unusually high was public radio KUSC-FM in Los Angeles, with $130,000 of its $1.3 annual budget directed to promotion in the late 1980s. Public TV stations averaged 5% in the 1980s, increasing that to 6%–7% into the 1990s, mostly for local newspaper ads.

Cable systems vary. Some use one-third of their promotion for basic cable network programming and two-thirds to promoting pay channels, including one-sixth for pay-per-view. But others stress promoting basic cable with up to three-fourths of the budget, leaving the remainder for pay channel promotion. Some cable specialists recommend 20%–25% of avails should go to promoting the cable system, including customer service and repairs. Local systems in 1992 allocated an average of 4% of revenues to marketing.

Neither ABC nor CBS accepted promotional advertising from basic cable networks, while NBC carried ads only for ad-supported cable networks but not for premium or pay channels. Basic cable's heaviest expenditure on regional and network airtime in 1991 was Family Channel's $2 million; premium HBO bought $30.6 million of spots on broadcast networks. CNBC in 1992 budgeted $20 million for local broadcast spots promoting its business/talk cable channel.

On the other hand, basic cable channels carried plenty of promotional spots by the major TV networks. In 1991 ABC spent $2.4 million, NBC and CBS half a million each, for spots promoting their shows placed on CNN, ESPN, Family Network, MTV, WTBS, and USA. In 1992 ABC again spent $2.5 million, NBC increased to almost $750,000 plus an additional $1.6 million to promote its Olympic Games cable "Triplecast," CBS halved its cable spending to $280,000, and Fox put $127,000 into promoting its programs on the six cable program networks. Those cable channels also carried $700,000 of promotional spots bought by syndicators of *The Oprah Winfrey Show*.[11]

This chapter focuses mostly on sales departments in local radio and television stations, with only limited data about national networks and cable operations. Some sales-related material for the latter, including annual revenue figures, will be found in the next chapter on media economics and finance. Data directly pertaining to network program series, syndication, sports, and news operations were in chapter 8 on managing national programming.

A. THE SALES STAFF

The sales department is the station's principal contact with people in the world of business. A strong radio or television station in a community offers a needed service for merchants and institutional advertisers. Those businesses depend on mass communications media to inform as many people as possible about their products and services. Few business enterprises question the value of advertising. Their concern is what media to use and which specific offerings of each medium most effectively reach potential customers.

A station's owners and managers must initially develop technical facilities and programming that reach apt audiences, offering excellent returns for the business community. Management must astutely select a sales staff. A good station can become better with good salespeople. Competent sales reps can become even more productive by working under good sales managers.

Sales as a Career

The number of salespersons employed by a radio station is, on average, fewer than by a television station; but the income from selling local advertising in either radio or TV can be attractively high. The sales department offers higher individual incomes than any other department, except for top on-air performers. Further, sales experience in broadcasting is the most common route to station management.

A career in sales offers definite advantages: (1) in any industry, comparable grades of professional status are usually better paid in sales than in any other department; (2) work in sales brings constantly changing challenges; (3) opportunities to learn human nature are boundless; (4) the good salesperson has little worry about a job since there are always more positions open in sales than in most lines of endeavor; and (5) one who travels about meeting other people learns of opportunities that never come to the attention of those working in an office with the same people day after day.

Despite both immediate and long-range opportunities in sales, it is more difficult to locate good salespeople than to find most other kinds of broadcast personnel. There is often a surplus of applicants for most other station positions, but people who want to sell are usually scarce.

Young persons with college experience will discover that there are many openings and opportunities for advancement in radio and TV for the right people. But nothing can be worse than frustrations experienced by those not naturally inclined toward a sales career. No one should ever go into sales work motivated exclusively by the monetary rewards. Unless he or she truly likes to sell, they will be doomed to failure before they start.

WOMEN AND MINORITIES

By 1981 women were more than a third (36%) of salespeople in radio, and in 1991 they were almost half (48%). In TV during that decade the percentage of female sales reps more than doubled from 21% to 48%, while the percentage who were national sales managers rocketed from 0.4% in 1978 to one in four (26%) by 1991.[12] That year half the sales staffs at TV stations in the top-100 markets were women; in all other markets about one-third of staffs were women, with the national average of 39% female salespersons in TV (see table 9.2).

In 1991 in the top-100 markets minority persons made up 14% of radio sales staffs and 13% in TV sales, but for all markets only 7% (radio) and 6% (TV) overall.

Table 9.2 identifies where women and minorities stood in sales management in 1991. In both radio and TV, women held more than one out of every four management positions in sales—including general sales manager (GSM) and national and local sales manager (SM); in TV women were only 7% of GSMs but almost one-third (31%) of local SMs. Proportions were higher in top-100 markets than in others. Minority persons in sales management did not reflect their larger presence among sales staff members in broadcasting: Except for 4% of radio national SMs and 4% of TV local SMs, minorities served as 2% of all radio GSMs, and as 1% each of radio local SMs and TV national SMs; less than 1% were TV general sales managers.

QUALIFICATIONS AND FUNCTIONS OF THE SALES MANAGER

Major duties and qualifications were discussed in chapter 4. Details are added here.

The National Association of Broadcasters listed primary responsibilities for radio sales managers, which apply to TV as well:

1. Carry out sales policy of the owner/general manager.
2. Staff, direct, and control the sales department.
3. Develop the station's local sales and service.
4. Establish and review rates and rate structure.
5. Coordinate national and regional accounts.
6. Provide sales promotion packages for the sales staff.
7. Maintain personal sales efforts.
8. Supervise collection efforts assigned to sales staff.
9. Coordinate sales activities with appropriate departments.
10. Promote community involvement and visibility, [e.g.] service club groups, fund-raising.[13]

The key to efficient selling is keeping on top of inventory, knowing what spots are still available, how the market is currently buying, and assessing the current value of yet-unsold time. This helps determine which time-charge patterns on the ratecard grid the sales staff should apply as daily and weekly supply-demand factors constantly shift.

The sales manager must be willing to join his sales reps on the street, if appropriate, and to work with them in conceiving and making presentations to agencies. But radio consultant Don Beveridge cautions that sales managers who visit clients with a salesperson tend to do most of the talking and usurp the selling job. The roles of department head and staff member are different; a sales manager's role

SALES STAFFS AND MANAGERS: WOMEN AND MINORITIES
U.S. RADIO AND TV STATIONS, 1991

	MARKETS 1-99		MARKETS 100+		ALL MARKETS	
	Radio	*TV*	*Radio*	*TV*	*Radio*	*TV*
TOTAL STAFF						
% Female	57%	50%	40%	30%	48%	39%
% Minority	14	13	*	*	7	6
GENERAL SALES MANAGER						
% Female	27	10	25	4	26	7
% Minority	3	*	*	*	2	*
NATIONAL SALES MANAGER						
% Female	46	28	*	23	24	27
% Minority	8	2	*	*	4	1
LOCAL SALES MANAGER						
% Female	33	33	20	25	29	31
% Minority	2	6	*	*	1	4

* Less than 1%

Source: Warner and Spencer for TVB and RAB, reported in *Feedback* (Summer 1992), 9 (see note 12).

is coach, not player. In Schulberg's phrasing, "salespeople should deal with their customers' needs; sales managers should deal with their salespeoples' needs."[14] Further, sales staffers need their manager available for advice, motivation, and monitoring rather than as a competitor. Yet in 1992 three out of four radio sales managers carried a list of clients they serviced, while about half of TV SMs did so.

Along with customary interpersonal skills, the sales manager must have both poise and aggressiveness to gain respect from staff members who are successful salespeople themselves, and from other department heads.

SELECTING A STATION SALES STAFF

Two-thirds of all sales staffers are less than thirty-five years old; through recent decades, average age has remained in the mid-thirties for both radio and TV salespeople. Turnover of personnel was predictably higher for radio and in smaller markets; nationally, one out of three in radio sales and one of every five in TV leave their positions each year.

Researchers Charles F. Haner and Givens L. Thornton sought early on to discover behavioral traits distinguishing successful salespersons from unsuccessful ones by studying four medium-sized companies.[15] Their findings have relevance to broadcast-cable media reps. Of several hundred behavior traits analyzed, the most important was their willingness to study carefully and in detail the customer's needs. Next was their ability to set goals and quotas for themselves which they were willing to follow. Other factors, in order of importance, were: keeping customers informed, persistence, analysis of sales efforts, knowledge of the competition, keeping the company informed, and coverage of the territory. Factors least useful in distinguishing good salespersons from poor ones were: social activities, drinking habits, personal traits, home life, and group activities. Thus job-oriented behavior traits were best predictors of successful selling, while personal and social traits were less important.

The successful salesperson in broadcasting is the one who, after convincing clients to use the station's facilities, follows through and makes sure that sales results please the clients. Long-term relationships between sales staffers and clients are crucial to the success of a station's sales department.

THE SALES MANAGER AND THE SALES STAFF

Bases for Assigning Salespersons' Accounts

Several factors to be considered before assigning a list of local accounts to a sales staffer include: knowing retailer's stores and services, understanding particular problems of specific kinds of businesses, and previous experience in dealing with certain types of retailers. Most important, though, is the probability of compatible relationships between the salesperson and each account. The sales manager (SM) needs to use careful judgment in matching sales reps with clients according to their temperaments, interests, and probable general rapport. A major factor in determining how long an account executive remains assigned to an account is his or her compatibility with that client.

Assignments of choice station accounts are usually based on seniority or successful sales experience. A younger, less experienced member of the station sales staff must build an initial list into lucrative accounts by providing excellent service to clients after they are sold. Accounts, of course, should not be transferred among staff members once a good relationship has been established, unless there is no alternative. It takes a good sales rep, with plenty of patience and thoroughly trained in media, to establish conversational rapport with those assigned advertising and time-buying positions in retail establishments and agencies.

Intermedia Competition

Some store buyers and advertising managers seem to favor print media almost exclusively; unfortunately for broadcast account executives, retail management listens to those buyers and advertising managers. Radio-TV-cable salespeople assigned to such contacts should be experienced or trained in retailing as well as in broadcast media. They must be able to talk intelligently with those engaged in every conceivable category of retail sales. Sales reps must be able to suggest media solutions for problems of inventory, cash-flow and pricing, retail promotion, and moving merchandise. Otherwise they are merely order-takers or peddlers of spots and programs instead of serving the client as a form of consultant and problem-solver with a "client-centered attitude."[16]

Store managers, as well as buyers and advertising managers, feel comfortable with the printed ad. They can look at it, admire it and show it to their co-op people. It is tangible; broadcasting is not. The electronic media salespeople will make little headway by attacking print media. Radio and television are so strong with such a success record in advertising for all categories of retailers that the "positive sell" is the only route to take. Newspapers will always be basic buys for some retailers because of the nature of their business, for example, those that depend on detailed price-lists and coupon promotions for department and variety store goods and supermarket product sales. But those same retailers can gain additional business through supplementary use of radio and television.

The broadcast salesperson's role is to make the client feel comfortable with electronic advertising and to show them how they can earn profits from using it. The challenge is to open their minds to the selling strengths of radio, television, or cable.[17]

Staff Sales Conferences and Reports

The sales manager holds daily sales conferences with the sales staff. (In large stations, the local sales manager may meet with staff account executives apart from the general sales manager and national sales manager; in smaller stations, the sales manager might also handle national sales through liaison with the station rep firm in addition to overseeing local sales.) Early each morning she should meet with them to check their contacts of the preceding day. Together they work out strategies for that day to move time availabilities in the commercial schedule. Martin Antonelli summarizes items the SM covers with salespeople in daily sessions:

> Pending business, establishing shares for specific buys, what salespeople are saying to clients, how clients respond. Are the appropriate areas being pitched on specific buys? Are specials being used? Are packages being used? What does the package look like? What are the client requirements in terms of CPP [cost per point], target demo, number of spots, reach, frequency, traffic building, etc.? What kind of order does it look like the station will get? Determine the problems and propose possible solutions. Time thus spent with salespeople can prove invaluable in determining strengths and weaknesses of the sales staff as well as providing direction and leadership for the team.[18]

Each Monday morning, the SM gets individual reports from each staffer, recapping their previous week's activities and results. From those reports, the SM is able to spot troublesome cases where she can point out approaches or other tips to make a sale. She may even personally assist a salesperson with the client. Sometimes, where on-air talent is involved in a program with spots available, the sales manager may have that talent participate in a follow-up presentation to help a salesperson close the sale.

At least four times a year, the general sales manager (GSM) should hold half-day or day-long meetings, perhaps on weekends, to engage in long-range planning. The program director with others from that department and the outside station rep attend these sessions along with the sales staff. Outside consultants may also be included. Topics for major meetings—also for weekly, even daily general sessions—include: sales policies, the ratecard, special events packages, client list (present, new, and potential), sales aids, oral sales presentations by staff members, internal station procedures that affect the sales operation, programming developments, a review of commercial copy style, and sales service and client cooperation. Sales managers often use these sessions for continuing training, to keep salespeople alert to new developments in marketing and research methodology, computer tools, sales techniques and strategies, economics and finance, and motivational exercises. Sessions can include sales-training videotapes, guest speakers, and data handouts from sales organizations such as the Radio Advertising Bureau (RAB), Cable Advertising Bureau (CAB), Television Bureau of Advertising (TVB), Broadcast Promotion and Marketing Executives (BPME), Broadcast and Cable Financial Management Association (BCFM), and the National Association of Broadcasters (NAB). Sometimes speakers are invited from major ad agencies, a local business, the Chamber of Commerce, or local advertising clubs. Many stations invest in sending selected salespeople from time to time to national conferences or to specialized seminars ranging from Dale Carnegie presentations to behavioral institutes and statistical analysis training sessions.

The sales manager keeps records of each salesperson's performance with assigned clients and agencies, and prepares regular progress reports for the station manager. These are not to be merely paperwork exercises; they supplement and support brainwork.

Sales Quotas and Source Materials

The sales manager is responsible for setting sales quotas for the department and for each staff member. By realistically evaluating each salesperson's list of clients and agencies, she can establish proper quotas. These expectations set goals that promote everyone's energetic striving, but they cannot be absolutely rigid requirements.

Many advertising budgets of large retailers, particularly department stores and supermarkets, emphasize print media almost exclusively. These retail stores run page after page of newspaper ads every day, while they settle for a few spot commercials in radio and television. How can the local station increase its share of that retail advertising? It takes determination, enterprise, and imagination by the sales staff.

A radio or TV station's story about programming schedule, audience ratings and demographics, cost efficiencies, ratecard, and time availabilities are supplied to the station rep. These tools provide persuasive selling data that attracts time-buys by agencies for their national and regional advertisers. Just as competing airlines vie hour by hour for passengers to fill their aircraft seats on daily flights, a station's continually changing avails listing must be in the hands of reps by telephone and electronic interconnection by computer and facsimile. Those same data, constantly updated, are used as a selling tool with local advertisers.

The station's master sales plan consists of basic information about the station's circulation strength, its image in the community and within the industry, and its success stories. The sales kit includes descriptions of technical facilities and special equipment which enable the station to perform its services efficiently, including mounting local commercials. It relates in detail the program and production repertory of the station, including whatever live talent is available and, in the case of television, the syndication and film library. It mentions whatever merchandising services are provided. Added to this general sales material, the sales plan includes special applications designed specifically for each individual advertiser.

In local sales, data are used differently than for national sales. Locally, contacts are more direct and personal. Very few negotiations are closed over the telephone. Local selling involves salespersons calling on those whom they have contacted many times before. Tailor-made presentations and specially constructed commercials or programs must be designed to meet special needs of specific merchants.

The local radio or TV salesperson used to rely simply on the medium's inherent effectiveness to make good on his promises to the retailer. Today's salesperson knows there is no magic in a broadcast campaign unless it is targeted to a particular client's needs and objectives. The modern salesperson becomes as interested in helping the advertiser to grow and prosper as he is in gaining commission from his sales efforts. The modern broadcast salesperson is in a form of marketing, working as much for clients as for the station. If truly informed about retailing, he should be able to offer suggestions that will favorably surprise the merchant and make the salesperson's counsel so valuable that the businessperson will come to depend upon him for ideas.[19] The retail-oriented salesperson will know that the on-air advertising he can supply will be successful only if the store offers up-to-date services, facilities, and customer conveniences. The salesperson needs to be able to offer tips to the retailer on how to accomplish that. So the station's local account executives, its sales staff, must have some expertise in helping the retailer to grow. While negotiating time sales is a major responsibility, it is only one service the effective salesperson provides as a station prepares and schedules effective advertising on radio or television for clients.

Combination Talent/Salespersons

On-air talent such as announcers, disk jockeys, and weather persons can also be effective in small markets as sales "combos." In fact, they are employed in some of the nation's larger cities but don't usually bear the combination title. Even a neophyte DJ in a small town is somewhat of a celebrity within the coverage area. He or she becomes a familiar name in homes throughout the trade region. Small-town merchants, especially those new to using radio and TV, are frequently motivated more rapidly by sales visits from on-air personalities than from other members of the selling staff.

Many small stations can offer retailers only modest assistance in continuity and production. On-air talent doubling in sales get to know a client very well, often writing and producing that client's commercials as well as servicing the account. Announcers or DJs who sell should visit their clients' stores regularly to become thoroughly acquainted with the merchandise and service. Then they can be true sales representatives of the stores when presenting commercials over the air. The announcer/salesperson can be useful, too, in store promotions.

Complaints of Account Executives

Salespeople have a legitimate complaint when availability schedules are not ready, when their clients' copy is revised without their knowledge, or when rating reports are not delivered on time. Such procedures can cost them sales commissions and reputation.

An efficient broadcasting organization demands complete cooperation between sales and continuity, traffic, billing, and administrative units. The latter offices must never contact a client or agency, local or national, without first clearing with the salesperson on the account and with the sales manager. It is disturbing for an account executive to discover that other people at the station have established relationships with one of her clients without her knowledge. Such action causes confusion, ill will, and eventually results in inept handling of the account.

Salespeople also complain legitimately if a new corporate sales policy is implemented without their prior knowledge. Management should make policy changes affecting the sales department only after consulting each person in sales.

Another cause for complaint is when staffers in other departments show little interest in the work of the sales staff. The station manager must make sure that every employee understands how sales are made and how accounts are retained or lost.

Compensation Plans for Salespersons

Sales staffers, unlike sales managers, are rarely paid a regular salary. Around 80% of stations offer straight commission; 7% of radio stations and 13% of TV stations pay a base salary plus incentive commission.[20] Market size does not seem to be a factor in the form of compensating salespeople. Most commissions are disbursed after billings; a few are paid after a logged account is aired, while some are not paid until after collections. Almost all TV stations, but only two-thirds of radio stations, reimburse salespeople's expenses.

The Infinity radio group paid straight 6% commission with no limit to earnings. Many salespersons earn over $100,000; their top salesperson made $300,000 in 1992. Infinity President Mel Karmazin reasons that "if a salesperson is going to get rich on a 6% commission, Infinity is going to get very rich on the other 94%."[21] The goal of sales managers in each market is to outsell their rating—if their station draws 5% share of the local audience, the SM works to gain a 10% share of market media revenue.

A method of monthly payment is to draw one-twelfth of a salesperson's anticipated commission on billings placed for the year, adjusting eventually for sales that fall short or exceed that target by year's end. One station following that pattern pays out about 80% of the monthly figure, holding back the remaining 20% from all staffers until the end of the year; then the pooled funds are distributed among the sales staff on a "merit" basis according to criteria relating to account administration—billings, paperwork, discrepancies, meetings attended, special sales, among others.

Incentives help generate new business. WDIV (TV) in Detroit offered larger commissions on sales to new clients. KCCI-TV, Des Moines, Iowa, gave an extra 4% commission for bringing in any new account not on the air during the previous year. A similar bonus plan at WICS-TV in Springfield-Decatur-Champaign, Illinois, helped spur sales staffers to attract ninety-two new accounts with billings totaling over $250,000.

Warner and Spencer caution that broadcasting's dominant straight-commission plan may contribute more to administrative than strategic goals. It is uncomplicated and does keep sales costs directly proportioned to actual revenues. But that emphasizes immediate volume and total revenues rather than salespeople's overall performance for long-term sales gains, which includes servicing accounts and cultivating new prospects while keeping abreast of marketing and media developments in a complex competitive selling environment. A base salary supports a staff member's total sales effort, with additional commission and bonuses reflecting productive selling. That permits flexibility and contributes to managing sales efforts by setting performance/compensation standards to achieve company objectives.

Senior account executives typically service established accounts from major advertisers and agencies while newer salespeople, less experienced and with fewer high-level contacts, work with less lucrative accounts. That results industry-wide in some 15% of sales staffs, usually veterans, generating as much as 85% of a radio station's sales; TV's pattern is somewhat similar. In the late 1980s WBBM-TV in Chicago had twelve local account executives who brought in 55%–60% of the station's billings (the rest was national sales); the general sales manager estimated the top five or six of them produced two-thirds of those billings.[22]

What do salespeople earn in local broadcast media? Warner and Spencer's survey for TVB and RAB in 1991 found compensation around the top-100 markets ranged from senior TV salespeople making $80,000, through middle incomes of $51,000, to $30,000 at the bottom. Their radio counterparts in those larger markets earned $70,000, $39,000, and $26,000 respectively. In markets sized 100–200+, top TV salespersons earned $52,000, with middle-range compensation at $24,000, and the lower end at $19,000—obviously in the smallest markets. Radio sales incomes in those markets topped at $40,000 annually, through $24,000 midway, and $15,000 at the bottom.

B. Time Selling

The sales manager oversees the sales staff's efforts and evaluates their performance in selling the station's commercial time. By private conversation with individuals and by informal discussion with staff members, general sales department meetings, memoranda, and by circulating printed materials (reports, statistical data, sales success stories, and tips on selling techniques), the SM can motivate and support the staff's efforts to sell successfully.

Duties of Sales Staff

Warner describes the central role of salespersons as (a) creating a "differential competitive advantage" in the client's mind—or positioning, and (b) managing that account for long-term effectiveness by fostering a close relationship with the client.[23] He lists a salesperson's specific functions: develop new accounts and manage current ones; obtain and process orders; maximize revenue by matching apt packages with clients; servicing clients by monitoring proper scheduling and billing and by relaying current station policies and market information about competing media; forecast projected sales revenue and recommend tactics for enhancing packages and prices; and cooperate with fellow salespersons in achieving department sales goals and also participate in promotions, social functions, and trade organizations.

Warner nicely lays out five major steps in media selling and offers estimated percentages for each step over an average month, based on a weekly 25–30 hours devoted to actual selling, not including another 10–25 hours on administrative paperwork, sales meetings, and travel.[24] Note that the typical sales work-week far exceeds 40 hours, even up to 55.

Prospecting for future calls on potential clients *(10% of actual selling time, averaging about 3 hours a week).*

Qualifying which prospects best fit the station's audiences and programming for effective selling, plus assessing the person who is that company's contact *(15% or 4.5 hours).*

Researching and targeting the client's business and field, needs, and potential specific media packages budgeted to meet those needs *(15% or 4.5 hours).*

Presenting written and verbal proposals, anticipating the client's concerns and doubts, and planning strategies to negotiate effectively to close the sale *(40% or 12 hours).*

Servicing the account to inform and satisfy the client, building a long-term relationship for sustained sales business *(20% or 6 hours).*

White and Satterthwaite assert that inadequate research is a major factor in salespeople's failing to complete sales. They pithily summarize this stage of sales planning: "*product* knowledge plus *prospect* knowledge plus *psychology* knowledge plus *perseverance* equals sales *productivity.*"[25] Warner emphasizes organizing one's office and briefcase and plan of attack, along with client accounts, while carefully managing time and money—including "to do" agenda lists and methodically evaluating results.[26]

Preparing for the Sale

Before any presentation to local retailers or to agency media buyers for national/regional advertisers, a salesperson studies that firm's experiences in all forms of advertising and its business in general. She carefully learns about the retailer's field so she can discuss intelligently the prospective advertiser's interests and problems.

Such study will disclose, for example, that the problems of a shoe store owner are completely different from those of a jewelry store owner, say, in the month of April. The sales rep will discover which are peak business months in the year for other enterprises such as grocery stores, hardware stores, drugstores, and dry cleaners. She learns seasonal fluctuations that differ among various retail categories or national advertisers in order to make the right sales calls at the right times.

The sales rep must talk to the point in her presentation, knowing a specific client's retailing methods, including patterns of store traffic, uses of merchandise displays, the personalities of store clerks, the store's conveniences and services, and the kinds and quality of merchandise sold.

The basis of all effective selling to local retailers is complete information about the business of each prospect or client. Studies of retail establishments includes research into other local firms and those in other markets in the same category. For example, in the case of men's clothing stores, the salesperson needs to know the problems of a particular store but also needs to be familiar with the men's clothing industry. A salesperson calling on a men's-wear establishment or its agency should be well acquainted with current ads in *GQ/Gentleman's Quarterly, Esquire, Playboy,* and leading sports publications. Additionally, she should study trade publications in the men's-wear field. Sources of information for every retail field are available in most large libraries.

The more a media account executive knows about each retailer and their particular business category, the more successfully she can help move their merchandise. This must be the first objective of any good radio, TV, or cable salesperson: not merely selling station time but moving clients' goods and services through those commercials.

Salespeople have not understood the advanced planning patterns of store buyers, merchandisers, and advertising personnel. In many instances stations have spent heavily on elaborate presentations to department stores six weeks to two months before the Christmas shopping season. But advertising budgets for those particular stores probably were established in the first quarter of the year, or even in the last quarter of the previous year. Advertising plans in most department stores and many retail establishments are finalized six months or more in advance of their actual campaigns. This is especially true for seasonal types of advertising where certain lines of merchandise have consumer demand for a brief time. A media salesperson making calls after the advertising budget has been allocated for the year simply wastes her time and weakens her credibility with potential clients. By anticipating store decisions on seasonal merchandise, some sales reps have formed long-term business relationships with retailers. The success of initial short-term contracts resulted in year-round broadcast advertising by those stores.

The Sales Presentation

Few sales are made in one call. It is not unusual for a salesperson to average five to six calls on a retailer, whether large or small, before completing an initial sale. Much of what follows about the sales presentation equally applies to a succession of visits by the sales rep.[27]

If the retailer has an agency, even one that is small or inexperienced in using broadcast media, the broadcast salesperson should consult that agency before making a presentation. Smaller agencies may resist buying radio or TV time for some clients. Then it is appropriate for the sales rep to take the case directly to the client, after explaining his intent to the agency. Regardless of the agency's attitude toward broadcasting, whether cooperative or reluctant, the station should have direct contact with the client. The sales rep contacts the same person that would have been consulted had no agency been involved—the advertising manager, the head buyer, or the owner-manager of a retail establishment.

In calls on retail firms, and even to some agencies, the salesperson must first sell the merits of broadcast media in general. He must also convince the retailer or its agency that expanding their advertising policy to include radio or TV will produce increased business. After selling his medium and concept, the broadcast account exec must then demonstrate how the station's programming, geared as it is to the community, can help the retailer reach present and potential customers.

The sales presentation illustrates the station's coverage, circulation, and audience composition. The account executive explains which time and program availabilities can do most for the client. The retailer needs every bit of service, advice, and guidance the sales rep can provide. It helps to furnish the retailer with information on media buys and campaigns used by similar business establishments in other cities. The Radio Advertising Bureau (RAB), Television Bureau of Advertising (TvB), and Cable Advertising Bureau (CAB) have inventories of such success stories. Industry trade magazines such as *Broadcasting & Cable* and *Advertising Age* regularly feature cases. Evidence of local successes, especially by competitors, has motivated many a nonuser to embark on broadcast advertising or to allocate part of their advertising budget to a previously untried radio or TV outlet.

Note that little has been said about using ratings in the sales presentation.[28] Unless a station is completely dominant in its market—at least in some demographic category—its ratings are not as important in making a sale as some broadcasters think. Time-buyers in agencies have access to sophisticated quantitative data and do rely on those audience statistics for much of their decision making, but most local retailers have only modest interest in a station's rating service. Their audience index is the cash register in the days following the start of an advertising schedule. But clients must be reminded that media ads have cumulative effect rather than overnight results; over time a campaign of well produced and skillfully scheduled commercials can generate significant public awareness of a company's service which will carry over into purchases. A strategic schedule of persuasive broadcast ads further reinforces initial impressions and sustains audience response.

Effective sales presentations include at least five characteristics. Presentation material should be organized clearly and logically. The presentation must be attractive and neat, supporting the "story." Creativity must be applied to ideas, but without superficial gimmicks. The presentation must be brief, taking only the time actually necessary for key points and significant supporting data; and it must be direct, concluding with the salesperson asking for the order.

FOLLOW-UP ACTION AFTER THE SALE

Some sales reps resent the task of servicing a sale after it is made. They would rather close a sale, forget about it, and get started on another prospect. Such an attitude runs counter to the purposes of selling. As noted earlier, the object of the sale in the first place is to perform a real service for the client. If the advertiser's air campaign is not successful, it will take more effort to get him back on the air than it did to sell him originally. While true that in servicing accounts the salesperson does not place immediate dollars in his pocket, in the long run he stands to gain by it. The salesperson interested only in immediate financial gain is not really interested in what advertising can do for the buyer. His selling methods can offset the very sales claims the station makes. His quick personal gain can be a longtime station loss.

Traffic and Continuity

Two indispensable units in client service by any station are traffic and continuity. In television, almost half of all traffic managers reported to the general sales manager in 1991, whereas only one in four radio traffic managers did so, mostly in top-100 market stations. The numbers reverse in both media for traffic managers reporting to general managers: 23% in TV and 44% in radio. More than one out of three radio-TV traffic managers report to someone other than GMs or GSMs, such as programming or the business office, especially in smaller markets.

Traffic is the very heart of the operating organization and is vitally important to airtime inventory control. Its primary responsibility is to prepare the daily log listing all program and commercial placements minute by minute throughout each day. The people in traffic check for commercial separation of similar products, usually keeping two brand names for the same kind of product or service at least ten minutes apart in the schedule (better, quality-oriented stations allow fifteen minutes). At most stations, traffic also prepares weekly assignments for announcers and DJs and, in some cases, production assistants and even engineers. Traffic ensures that commercial spots—carts, tapes, copy—supplied by agencies or produced in-house are properly listed and physically in place for on-air production. Subsequent billing depends on accurate records kept of every commercial spot, including those failing to air or pre-empted and requiring make-goods for the client. The traffic department must coordinate closely with sales, programming, and engineering departments to ensure continuous flow of the air schedule.

Continuity, in addition to writing copy for the station, often is assigned the responsibility for rewriting agency or client copy for better air presentation. Those who work in continuity also check copy for compliance with station policy standards.

Before making any call on a client, a salesperson needs to check with the traffic unit to learn what airtimes are available for spots in current and future schedules. This assures that what is sold will be aired on specific days and times. After making a sale, the salesperson informs the traffic unit immediately so blocks of time sold are removed from the availabilities list.

For every sale, the account exec supplies the traffic department with complete contract details, filled in on printed forms. Traffic then makes out requisition orders to be circulated to all departments affected by the order. Once a sale is made and reported, much work must yet be done before the campaign goes on the air: writing format scripts, formulating preliminary production plans, arranging for music beds, artwork, slides, video, and film.

A client or agency must provide copy in adequate time for processing prior to airtime. Stations typically set a 48-hour deadline for all audio and video copy before air use. For certain retail establishments, this lead-time has to be shortened to compete with other media. Some daily general-circulation newspapers and shoppers' guides have as little as an 18-hour copy deadline in advance of publication.

But after using a broadcast facility for a period of time, most retailers find it not too difficult to comply with a 48-hour deadline on all copy.

Servicing the Sale

After closing a sale and prior to the schedule's start on the station, the salesperson needs to maintain contact with the client, as noted earlier. Normally, the nature of the contacts is left to the discretion of the salesperson. The sales manager should encourage him to use budgeted funds for this purpose. Luncheon dates or tickets to sports events or other entertainment favored by the client are among many ways to let the new advertiser know the station appreciates their business. Telephone calls from the salesperson, the sales manager, and, with large accounts, even from the station manager help maintain a working relationship during this period. Of course, regular contacts should continue throughout the contract's duration.

Once an ad campaign begins on the station, the local client may begin to look for almost immediate results. There is no single conclusive yardstick for measuring the effectiveness of broadcast or cable advertising. It is a salesperson's duty to use every method available to evaluate the campaign's success; that appraisal should begin as soon as the commercial "flight" starts and should continue throughout the campaign. Techniques include call-out survey interviews, focus groups, over-the-air and cable-control testing of commercials, pre/post attitude and awareness studies, market audits, and in-store interviewing to compare effectiveness of print ads vs. broadcast commercials for retail advertisers.[29]

The client will expect to measure results of their commercials by noting increased sales, as noted earlier. Even though other measures of success are equally important to a business in the long run, the media salesperson must also be alert to actual sales results. If results fail to develop as planned, the sales department may need to make adjustments by placing programs or spots in different time periods, revising commercial copy, or replacing air personnel involved in the commercials.

In any case, sales reps must maintain positive, supportive relationships with clients. Advertisers should feel broadcasters are an ally helping to build company business, consumer strength, and image in the area.

The station's accounting unit handles all billings and collections. Management should avoid assigning salespeople to collect overdue accounts. A salesperson is employed to sell; selling time should be so valuable that it is wasteful and inefficient to have them engage in bill collecting. After a spot schedule has aired, payment was traditionally expected in thirty days; as the economy slackens or competition increases, past due accounts are usually kept sixty or ninety days, then turned over to a collection agency. Most stations have overdue accounts from less than 20% of their total business. Late or nonpayment sometimes results from clients' changing agencies, from advertisers or agencies going out of business or pressed by cash flow problems, or more commonly by budget and schedule changes from what was contracted for (usually verbally and often with no signature on the contract), resulting in discrepancies that client and station must resolve.

C. Local Sales

Sales are either local or national (including regional). *Local*—also called retail— accounts with local merchants and businesses are usually placed directly by those companies but sometimes through local agencies. *National*—or general—accounts with companies doing business nationally or regionally are placed with advertising agencies and usually through station representative companies. Total revenue in broadcasting is derived from local, national spot, and network advertising. A portion of the last is forwarded to stations in the form of network compensation, equal to about one-third of TV station ratecards in return for airtime turned over to network programming (over 60% of a station's schedule). (Sale of network time is treated only indirectly in this section of the book.)

Local sales are the largest source of income for radio stations. By 1995, local sales brought in four out of five dollars (81%) of a typical radio station's revenue. TV station revenue in earlier decades derived most income (typically 60% or more) from national/regional spot; but a softer national economy and increasing number of stations and cable operators vying for that business led in the early 1990s to an equal split between local sales and national/regional sales, then shifting more towards local revenue (see table 9.3). For decades network-affiliated TV stations earned far more ad revenue from national/regional sales than from local, while indies relied heavily on local revenue. The standard in the 1980s became around 5% of station revenue from network compensation (comp), and the rest roughly split between national/regional and local. By 1995 national had dropped to 46% ($9.1 billion) and local grew to 51% ($10 billion), with network comp adding 4% of revenues. That was due to leaner years at traditional networks with diminishing shares of national audiences while advertisers also diverted dollars to competing cable programming networks, and also to a recession economy affecting national advertisers' tightened budgets.

In the 1970s and 1980s both local and national sales by TV stations almost doubled every half decade until 1990 when the rate decelerated (table 9.4). Network compensation to stations bounded through the 1980s, then skidded as the next decade opened.

In those same decades, local radio sales increased along with multiplying AM and FM stations: from $272 million in 1950 to $428 million (1960), $881 million (1970), over $2 billion (1980), to $6.6 billion in 1990—when national spot was adding another $1.6 billion to the nation's radio stations (that year network radio had under $400 million in sales revenue). In 1995 local sales by the country's 11,000 radio stations totaled over $9 billion.

SPOT ANNOUNCEMENTS

Commercial spots are often more effective over a period of time on radio than on television—assuming, of course, that the "right" stations are bought. This is because radio is still very much a habit medium. Listeners tune in their favorite radio station with its program "sound" in the morning and often return at various times of day, whether at home, in the car, or at the office or beach. Frequency of impressions on the same listeners is complemented by the station's reach to cumulative audiences. In contrast, TV viewers are interested in specific programs and tune to whatever station has the particular show they desire at a certain hour. Cable television is a hybrid: Like radio, niche program channels attract habitual viewers throughout the day, and also like traditional TV its basic, premium, and pay channels offer specific shows that viewers seek out.

Ad agencies assist clients in developing a mixed schedule of buys for maximum impact. Station sales staffs and their national reps must provide current data

— TABLE 9.3 —
Sources of Time Sales, as Percentage of Revenues
TV and Radio Stations, 1985–1995
($ in billions)

Revenue Sources	TELEVISION				RADIO			
	1985	*1990*	*1995*	*1995 Revenue*	*1985*	*1990*	*1995*	*1995 Revenue*
Local Sales	49.2%	50.4%	50.7%	$10	76.1%	77.3%	81.5%	$9.1
National/Regional Sales	46.7	46.4	46.2	9.1	2.4	21.2	17.0	1.9
Network Compensation	4.1	3.3	3.1	0.6	1.5	1.5	1.5	0.16

Averages for all TV stations (affiliates, independents) and for all radio stations (AM, AM/FM, FM)

Sources: Surveys of U.S. TV and radio stations for annual *Radio Financial Report* and *Television Financial Report* by National Association of Broadcasters (NAB) and Broadcast Cable Financial Management Association (BCFM), 1986, 1991; and NAB, *Broadcast Marketing & Technology News* (Washington, D.C.: 1991), 3; McCann-Erickson in *Broadcasting & Cable,* 16 September 1996, 93; Radio Advertising Bureau, in *Broadcasting & Cable,* 11 March 1996, 39. TV and radio network compensation in 1995 estimated by authors.

offering strategic patterns of radio-TV-cable ads to serve those clients' needs. Spots can be very effective movers of some merchandise and services; they can be good reminders and can build brand awareness. On the other hand, for some kinds of high-priced items, spots alone may prove ineffective without the strength of at least co-sponsored programming—among several major advertisers jointly, as in sports coverage.

Rates and Rate Increases

A review of radio and TV properties across the country will find few with over-priced ratecards. Most are underpriced for the services they deliver.

CPP and GRP

Rather than CPMs used in national selling, local stations sell airtime by contracting packages based on cost per point (CPP) and gross rating points (GRP). These offer reliable planning and buying tools for television, less so for radio stations. With only a handful of TV stations in a market, each reaching most of the people much of the time, GRPs are useful for comparative measuring. But GRPs are less reliable in buying radio time because narrowly formatted stations serve limited sectors of the total audience. For example, 100 radio GRPs spread over several stations might not get good reach with a broad demographic base; and 100 GRPs on one station would deliver high frequency but very low reach (regular listeners segmented from the total population). Thus, a radio salesperson contracting with an agency for 400 GRPs based on 100 local audience rating points a week over four weeks might actually deliver a 50 reach with 8 frequency, or 80 reach with 5 frequency, or 20 reach with 20 frequency—depending on format and local competition as well as time-of-day commercial scheduling.

Similarly, cost per point makes sense for TV station sales: A buy at $50 a point means 100 TV points will cost $5,000; and because based on the broad Arbitron ADI or Nielsen DMA definitions, it relates closely to total area cost per thousand (CPM). But radio stations in a market differ widely in how much of the metro or total area their respective signals cover. Weaker stations serve audiences in the metro area, strong ones have up to one-quarter or more of their total audience beyond the metro area. Radio sales should emphasize total geography rather than only metro

TABLE 9.4
TV STATION TIME SALES REVENUES, 1965–1995
(IN MILLIONS)

	LOCAL	NAT'L SPOT	NETWORK COMPENSATION
1965	$ 303	$ 786	$ 230
1970	563	1,092	240
1975	1,080	1,441	258
1980	2,484	2,920	369
1985	4,665	5,077	446
1990	6,500	7,000	450
1995	10,000	9,100	600

Sources: FCC, *Television/Radio Age; Broadcasting,* 11 January 1988, 50; TVB, *Broadcasting,* 10 February 1992, 67; McCann-Erickson, *Broadcasting & Cable,* 16 September 1996, 93.

area, to account for total audiences. Another tactic is to recommend that an agency average two or more stations to achieve their campaign goal. For example, for an agency's-targeted CPP of $50, select a schedule combining one station selling at $40 a point with one selling at $60 a point. Still, radio stations differ from one another in format, image, sound and personalities, role in the community, news service and community standing. Stations should sell their distinctive strengths rather than mere numbers. The sales account executive must be able to demonstrate the qualitative worth of their station—its programming and commercial environment—to determine the value or cost per point of a particular station.

Ratecards

While in small markets there is some limited justification for a double ratecard with both local and national rates, most major-market radio stations and almost all TV stations sell from a single ratecard, making no distinction between national and local sales. Warner offers an excellent analysis of ratecard structures and time-selling strategies.[30] He traces how discounts for volume or frequency and pre-emptible listings along with separate national and local rates evolved into a single card with times arranged into a grid to reflect shifting pressures of supply-and-demand. In effect, the grid card offers an established set of three to five or so different rate-cards to be variously invoked as market conditions warrant. A useful sales management tool, the grid outlines different levels of rates for the same clusters of time periods; the salesperson tells a client which rate-level applies to given spots at the time when a sale is about to be made and the actual value of those avails has been determined by current demand—rather than by a previously published rigid listing.

Some expedient station operators not only have a double rate structure but depart from it to satisfy certain clients, distributors, jobbers, and brokers. A vicious development in the history of commercial broadcasting that greatly harmed the industry was "double billing." This practice bills a local co-op retailer according to higher national ratecard listings (to pass on to his wholesaler or national advertiser) while actually charging that retailer lower local ratecard prices, so he can keep the difference in the national co-op payment. This harms stations because national accounts have become prejudiced against broadcast media. The FCC formally barred this fraudulent billing procedure, levying sanctions of possible license revocation and forfeitures of $10,000.[31]

Barter and Tradeout

Another station practice that reps deplore is tradeout of goods and services, technically called *barter*. The Station Representatives Association once judged that "bartering and/or brokerage of radio and television time is the most destructive practice that broadcast licensees have to contend with in their relationships with legitimate advertisers and their agencies who are willing to pay published ratecards for their facilities." Although these were once highly secretive transactions, the FCC estimated that as early as 1972 barter and tradeout accounted for $54 million in television and $38 million in radio. Not confined to lower-income stations, it is a familiar practice in most markets. The classic tradeout transaction involves trading station time for goods and services instead of money. The usual commodities are merchandise for station contests, studio and office equipment, automobiles, travel, and hotel or motel accommodations.

Syndicators barter radio and TV programs not for merchandise but for station airtime. They distribute programs with some commercial minutes given to the station for local sale; the syndicator retains two or more of the other commercial minutes to sell to national advertisers. Thus, in a sense, "free" program product is traded for "free" airtime; both syndicator and station derive income from selling their respective commercial minutes. Norman Pattiz, chairman and CEO of syndicator Westwood One (and the Mutual Broadcasting System and NBC Radio network), praised the form:

> Barter syndication became one of those rare exchanges in which everybody won. The syndicator was supported by the advertiser; the advertiser was able to generate tremendous national exposure to its targeted audience at a fraction of the cost of buying television time, and the local radio station received a program it could not have produced itself, to increase its ratings and sell local advertising time.[32]

In earlier decades, many TV stations simply did not accept even this form of barter. But the practice became part of mainstream broadcasting, expanding 16% yearly between 1986 and 1991, then 6%–8% annually to 1993 when total barter in nationally syndicated programming was valued at $1.5 billion. Over $1 billion of that came from up-front buying (the rest was scatter ad placement), nearly equaling one-third of the $3.3 billion up-front sales revenues combined by ABC, CBS, NBC, and Fox networks that season.[33] Also called advertiser-supported programming, barter accounted for one-third of the total $1.275 billion syndication market in 1992; the rest came from cash license fees. The Advertiser Syndicated Television Association projected barter would double by 1995 to $2.5 billion—almost half of the $5.5 billion syndication market.

> ■ In the 1983–1984 season, 60 series produced for first-run syndication brought in $400 million in barter; by 1987–1988, 116 series generated almost $1 billion. In 1992 syndicators distributed 214 weekly hours of programming; in 1993 that rose to 231 hours a week in 139 daily and weekly barter-supported series. As networks' average CPMs dropped to around $10 in the early 1990s, top syndicated series like off-network *Cosby* commanded up to $9 CPM from national advertisers. Off-network syndicated "Roseanne" in 1992 drew $20 CPM for women 18–49, with each spot selling for almost $100,000 as it approached a 9 national rating with about 90% clearance in U.S. markets, most of that in prime-time access.[34] King World Productions' Camelot Entertainment offered 30-second spots in *Wheel of Fortune* for $50,000, in *Oprah* for $35,000, and in *Jeopardy* for $25,000. ■

> Total national ad revenues by barter in syndicated TV programs quadrupled over a decade, from $550 million in 1985 to $2 billion in 1996.

A comparative index of barter syndication's strength can be found in children's programming. Between 1987 and 1991 network ad sales in kids' TV dropped from $141 million to $132 million, while barter ads doubled from $100 million to $200 million. (This reflected 2–11-year-olds' viewing levels, half of it to cable, a third to syndicated programming, and only 14% to network programs.)[35] In the 1993–1994 season, upfront buying in kids' TV totaled over $550 million in national sales, almost $250 million in spot, about $140 million in cable, and $180 million in barter syndication sales.

Powering barter's growth was the proliferation of AM, FM, and UHF TV stations that fragmented local audiences, resulting in revenues often too slim to recoup expenses for daily program costs. Even major stations typically sold out only 80% of availabilities. Both programming and commercial airtime inventory were helped by the influx of no- or low-cost shows without many time spots that needed to be sold.

On the other hand, bartered programs and especially tradeouts of goods and services often devalue station airtime, bringing in shows and merchandise not equal to the actual value of commercial time if sold at proper ratecard levels. The ratio of merchandise value to airtime value can go as high as one-for-two, meaning one dollar's worth of merchandise for every two dollars' worth of airtime. An unenergetic or inept sales staff along with competitive market conditions conspire against a radio or TV property's earning its rightful revenue from advertising. Lapsing into straight-trade/no-cash deals, some stations unwittingly undercut themselves in the local sales market. The NAB estimated tradeouts and barter by radio in 1987 totaled between 3% (markets with population under 25,000), 4%–5% (populations over 1 million), to almost 7% (250,000 to 500,000 population) of total station time sales. Several industry organizations estimated that barter in 1989 cost TV stations more than $350 million in saleable inventory. That year independent stations gave up about 15% of their sales inventory to barter, and affiliates about 5%. Furthermore, the Internal Revenue Service requires stations to report bartered transactions as income: their value must be reported in gross income.[36] Failure to report on IRS form 1099-B and related filings can bring penalties of up to $100,000 yearly, with additional penalties possible.

So long as the station and agency/advertiser arrange a straight tradeout deal, the practice does not offend station representative companies as much as a deal made with a third party. Reps insist that such a transaction is the same as the old practice of "brokerage" of station time.

Whether called barter or brokerage, the real risk to the system is that broadcasting can sell itself short. Radio and TV are legitimate business enterprises that can effectively get results for advertisers. They should not maintain an inequitable system of charges. Under some forms of barter, all advertisers are not treated equally. One advertiser pays the full card rate at the station; another gets the time in exchange for a marked-up retail value of his merchandise or service. Further, who wants to pay full published rates after discovering alternative forms of compensating a radio or TV outlet?

Stable Ratecard Structures and Practices

It would be unrealistic, within the confines of this book, to set up guidelines for rate structures for individual radio and television properties in the country. With or without frequency and volume discounts or price levels for fixed and pre-emptible positions, or a grid of schedules to be activated based on available inventory (that is, supply) and demand, all lists of airtime charges are tentative and can be negotiated under certain conditions. But the standard is to set a realistic value on airtime and then consistently offer it at those levels to all agencies and advertisers.

In some markets with four or more TV stations and more than fifteen radio properties, weaker operations treat the ratecard as a "rubber yardstick." However, a stable ratecard openly sold to all advertisers on the same basis—with any discount sales for volume and frequency of commercials announced publicly and adhered to—reflects a stable station that knows the honest and true value of its airtime.

Rate increases, when justified, seldom pose a problem to a good property. As far in advance as possible before implementing it, the station's salespeople and rep firm should explain to clients the reasons for and the merits of a proposed rate increase. An early explanation precludes problems. Whenever a rate increase is to take place, three to six months' protection should be given to current advertisers. The fundamental justification, of course, is not that the station needs more revenue but that the station's airtime audiences are worth it. Properly handled, higher rates can actually serve as a sales incentive; they create an aura of success and instill advertisers' confidence in the station.

SALES DEVELOPMENT AND PROMOTION

In earlier decades when demand often exceeded commercial time available, broadcasters basically *managed* sales. But with intense market competition these days, as noted at the start of this chapter, broadcasters must also *develop* sales. In past years a market with three TV affiliates scheduled 265,000 30-second spots annually; an independent station added another 166,000 half-minute units available to advertisers. But in the 1990s that same market of three affils and an indy also has a 30-channel cable system which expands total commercial time available to 1,100,000 30-second units.[37] The business of broadcasting requires continual sales development for a station to remain competitive. Although broadcasting's advertising dollars nationally have been substantial, their 13% share of total local ad expenditures has not matched the true impact offered by radio-TV ads. In co-op advertising alone, of 1989's $10 billion co-op spending, radio and TV got only 18% compared with 65% by newspapers. The Radio Advertising Bureau and the Television Bureau of Advertising have achieved industry-wide progress in sales development nationally; but more progress must be made on the local level. Some broadcasters spend so much time attempting to conquer one crisis after another they don't take adequate time to plan how to broaden the base of advertiser participation in their station's schedule.

Since the late 1980s direct-response advertising has become respectable, with cable channels and even traditional TV networks accepting ads that solicit phone-in purchasing. What Mark Higgins, vice-president and director of sales at WTVT in Tampa, called "direct marketing television" expanded from cable to local stations with "infomercials"—30- and 60-minute informative selling.[38] These program-length commercials, or "marketing programming"[39] became feasible with deregulation and the NAB's 1984 removal of limits on advertising time, thus permitting entire blocks of time to be sold for ads. The genre accounted for almost $400 million in time sales in 1993, about two-thirds of that to TV stations and the rest to basic cable channels. Nine out of ten over-air stations carry infomercials, mostly in late-night schedules, but 15% of stations also run them in prime-time, one out of four in daytime. In New York City, infomercials generate some $50 million of the annual $1 billion in billings by that market's independents and network owned and operated stations (O&Os). The number of different infomercial programs rose from 80 in 1988 to 175 in 1993, selling $1 billion worth of products annually by 1994.

Promotion is a strong ally of the sales department, and sometimes housed with it. The promotion director may report to the general sales manager or station manager. No matter how good a station or cable system's programming, it cannot sell commercial time unless audiences know where and when it is scheduled and unless advertisers are impressed with its impact. Bob Wussler, then executive vice-president of Turner Broadcasting System and former president of CBS-TV, described successful television as "50% promotion and 50% everything else." Noting that almost one-third of a network affiliate's revenue is in news, he continued:

> It's up to the marketing and promotion department to work overtime to provide their news product with a perceived distinction, and to generate the best possible numbers during the sweeps. . . . But we possess a fragile and unspoken mandate: the public trust. We must never do anything to put that trust in jeopardy. We must not as an industry promise more than we can deliver. You can always build back ratings or reposition a station for a specific program, but you can never fully restore someone's faith in you once it has been called into question. The media cannot afford to squander credibility. As promotion and marketing executives, you must stand guard to protect that credibility.[40]

In addition to airing spots and buying ads, promoting a station's identity and expanding its audience involves news, local programming, and community service along with its entire network and syndication schedule.

The Broadcast Promotion Association reported that radio and TV GMs surveyed in 1983 wanted their promotion directors to become marketing executives.* By the 1990s, station managers looked to marketing/promotion to contribute directly to generating revenue, by helping meet audience projections and sales goals. To maximize sell-out of available inventory, GMs put promotion directors on the executive team with sales managers and program or operations directors—offering research support and active relationships with other media, as part of managing marketing strategy.

KNBC-TV in Los Angeles integrated in-house sales and promotion by cross-promoting with Universal Studios' Hollywood Tour; the $100,000 project was estimated to have generated $900,000 in "incremental advertising revenues." BPME President Gary Taylor explained:

> If I were a program director, I would learn as much as I could about promotion and marketing. The days of programming specialist are over. I know this may upset some of my friends at NATPE, but in today's market reality, the general manager is making the purchasing and programming decisions at the station level. And, increasingly, the GM has turned to his sales, marketing and promotion managers to bring more viewers to the TV set and develop incremental revenues.[41]

---

*BPA in 1985 changed its name to BPME/Broadcast Promotion and Marketing Executives, and in 1993 to Promax/Promotion and Marketing Executives in the Electronic Media. In 1992 it began collaborating with the Cable Television Administration and Marketing Society (CTAM).

By mid-1993 a Promax survey reported 42% of all promotion managers were involved with major programming duties. These included working with national syndicators who can assist with their own promotion budgets for co-op advertising to help build audience awareness of station programs. Syndicators often contract with a station to provide a set number of gross ratings points (average rating times number of spots—such as 300 GRPs) for airtime given to program promos, as a factor in syndicators' allocating promo co-op funds to that station. Such arrangements must be monitored closely to avoid depleting avails needed for other promotion and, of course, for placing local and spot commercials. The sales department can sell airtime at higher rates when programs before and after station-break spots (adjacencies) are well promoted by the station.

As for cable, local systems need effective promotion to minimize churn among current subscribers as well as to extend penetration into ever more homes in their markets. They are aided by national cable program networks that spend an estimated 5% of revenue for promotion.

■ Premium cabler HBO spent some 10%–12% of gross revenue on promotion in 1990; over a six-month period it improved its rate of installation by 24% and reduced disconnect rates by 7%. National cable program services spent $627 million in 1995 on advertising, marketing, and promotion. The previous year, apart from promotions on their own air, ESPN alone spent $24.8 million on promotion; Discovery spent $19.2 million, TBS $16.7 million, USA $12.9 million, A&E $10.9 million, and TNT $10.7 million. ■

Promotion Staffs and Budgets

Every radio and television property, whatever its size, should have at least one person under the sales manager who concentrates on developing sales through promotion. He or she should be energetic, aggressive, and able to chart a course of potential business for the station. Certainly, if any property is to grow in revenue and profit as it should, it must extend its range of advertisers. A sales promotion department needs those trained under experienced promotion leaders in the industry. (*R&R [Radio & Records]: Ratings Report Directory* in 1993 listed 231 marketing and promotion companies.)

Promotion has been described as a sales tool for the broadcast sales department. Sales promotion works on what the station can do for the client, especially retail advertisers. Thus it is distinguished from audience promotion, whose purpose is to build audiences for programming through print and press relations and advertising. The station demonstrates to clients that, once they sign a contract for airtime, the station and its staff begin to go to work for them. The station offers to promote their time-buys and programs associated with their product or service, through mailings to retailers, conducting conferences, promoting tie-in purchases and buying by telephone, and various other forms of merchandising. Sales promotion is a station's "advertising of a client's advertising."

A good sales promotion department works with clients and their agencies through salespersons assigned to each account throughout the course of a contract. It supplies regular reports on what is being done on behalf of a program series or a spot announcement campaign.

What can a director of sales promotion and sales development be expected to accomplish?[42] She can gather materials for and produce fact sheets on the market and the station and its programming, write sales presentations, develop slide and film\video materials, take charge of all station exhibits and make trade show and convention arrangements, negotiate trade deals for contests, prepare promos, design and produce marketing aids, write articles for trade publications, analyze and summarize ratings reports, handle public appearances by talent, prepare advertising for the trade press, develop client brochures, keep station mailing lists up to date, prepare newsletters, assemble advertiser success stories, take charge of all station and personnel publicity, build audience promotion, and prepare a salesperson's handbook. If this isn't enough for one or two people, then she can be responsible for keeping the atmosphere charged with new ideas for increased sales activity.

A station's sales promotion activity will be stronger if its director is professionally trained in the field of research. Every year broadcasters are called upon by clients and agencies for more and more research data. Here is a field that merits consideration by some young men and women eager to enter the field of broadcasting. The work of good research people is necessary for broadcasting's growth and future development.

In the late 1980s, half the TV stations in large and medium markets had three to five persons on promotion staffs; some TV stations in top-50 markets had more than ten. Its growing role was reflected in 1989 when promotion and advertising expenses accounted for almost 5% of radio budgets and one-tenth of TV budgets, in addition to sales expenses. The previous year, promotion budgets at almost half of all radio stations and one-third of all TV stations ranged between $500,000 and $2 million or more annually.

Promotion managers at TV stations typically earn one-third of general manager's salaries, and about three-quarters of that paid to program directors; at radio stations they earn between half and three-quarters of GM's salaries, and one-quarter more than PDs.

Merchandising is an often misunderstood, misinterpreted, and ill-employed term in advertising. It has many meanings but in general refers to a station's assisting an advertiser's selling efforts. The station offers this cooperation as a bonus to attract national and regional as well as local advertisers. This service should be a "value-added" bonus to clients buying commercial time on the station; it is not a form of rate-cutting or hidden rebate of dollars.

Agencies find many broadcasters inept in handling merchandising, vending mistaken concepts of merchandising rather than the real strengths of their broadcast facilities. Some stations only briefly mention in client presentations their program impact, ratings, facilities, and station personalities, and then devote an enormous space or time on what they can do in merchandising, including in-store merchandising.

Generally, merchandising costs are at a proper level when considered as a supplementary service to the station's real source of impact: its broadcast facilities and programming. The great error in merchandising occurs when stations spend huge sums of money to put specific merchandising services ahead of what they have to offer in programming, which is what attracts audiences and serves the community.

Merchandising, when well directed within a station, can be an outstanding service for a client. Some clients, especially in the food and drug field, not only desire this service but demand it. But others have in-depth elaborate merchandising plans of their own and do not want the broadcast entity involved.

The owner running a local business correctly counts on help from advertising media in publicity and promotion. A station should offer assistance to the merchant, without selling those side benefits as more valuable than its facilities and airtime. Sometimes a store is willing to include the costs of such services in its advertising budget. Most stations try to help by having their sales promotion departments suggest selling aids such as window displays, point-of-purchase advertising, and on-air promos—all of which call attention to the sponsored programs and station as well as assist the retailer.

The concept of in-store merchandising and similar forms of promotion involves station-client cooperation. It helps ensure the client's success with their on-air advertising schedule; and it keeps the station's call letters before the public as well as the client. Among promotion tactics for stations to cooperate with clients are: in-store participation, air tag lines for local retailers of distributors' goods, newspaper ads, product sampling, direct mail pieces, playlist advertising, distributors' sales meetings, participation by station personnel, retailers' sales prizes, air salutes of "best salespeople" of client's goods, public transportation ad cards, billboards, station newsletters and magazines, disc jockey appearances, station contests, technical facilities (such as spotlights, sound systems, "live" remote units) at client openings, and in-store checkups in collaboration with Media Survey, Inc.[43]

Many types of cooperation with advertisers may be called merchandising at the point of purchase. It might be in the form of promotion cards on behalf of an advertiser, such as a distributor for a beer or a wine, with signs appearing in restaurants and other public places; or pump islands with banners, flags, and pennants at gas stations; or special shopping bags at a department store or specialty shop for a particular week, with the printed material on each bag calling attention to the store's sponsorship of radio or TV programming. Merchandising can also include direct-mail promotion to dealers carrying a product.

Only the largest stations doing considerable national and regional business can afford a fulltime merchandising director. More often the promotion manager or, in smaller operations, the sales manager oversees the station's merchandising efforts on behalf of clients, especially for national sales business.

Merchandising can bring returns if done in good taste and if it represents, in the final analysis, a contributory rather than a major service of the station. The most valuable support a station can give clients remains its basic business of programming that attracts audiences who respond to effectively produced commercial messages.

D. NATIONAL SALES

Major radio stations and most TV stations generate a large percentage of their revenue from advertising airtime purchased by companies not in the local area. To negotiate this national and regional business, most stations—except major ones with traveling sales forces or branch offices in metropolitan centers—rely on two groups of middlemen. They are advertising agencies and station representatives who transact business on behalf of their respective clients: nonlocal advertisers and local stations.

After many mergers consolidated clusters of smaller firms into megacompanies, *Broadcasting & Cable Market Place 1992* listed 132 rep firms working on behalf of U.S. radio and TV stations and cable systems (plus another 19 Canadian rep firms); they placed non-network national spot billings of $7 billion with TV stations, $840 million in non-network radio, and $160 million in national spot buys on cable. *The Broadcasting & Cable Yearbook 1995* listed 123 U.S. and 19 Canadian rep firms that negotiated with 169 ad agencies handling major broadcast accounts; they were joined by another 19 independent media buying and planning services.

In the first half of the 1980s, sales revenues for local and national/regional spots doubled, network sales increased at half that pace, while cable ad revenues multiplied more than ten-fold with expanded cable programming channels and local systems (table 9.5). From 1990 to 1995 cable more than doubled its ad revenues, while local TV sales grew by almost one-third; national spot and network time sales each grew by one-quarter. In 1995 total TV ad revenues for the first time almost matched newspapers ($35 billion).

■   Expenditures by national advertisers on all broadcast and cable media totaled over $28 billion in 1995; including local advertisers, that total came to $48 billion. Ad agencies placed $12.4 billion of those national advertisers' dollars with four national networks (plus $255 million on UPN and WB networks), another $9.1 billion of national spot through reps for local stations, plus over $2 billion worth of ads in syndicated television programming; TV stations' own sales staffs brought in $10 billion from local advertisers. Cable TV drew $5.3 billion in national ad dollars, while radio's local ads plus national spot and radio networks attracted $11.4 billion. In fifteen years, from 1980 to 1995, while radio and TV ad revenues tripled, fast-expanding cable multiplied its revenues from advertising almost a hundred-fold (in addition to the bulk of its revenues coming from subscription fees).   ■

Advertising Agencies and Broadcasting

Ad agencies might be New York-based giants like Young & Rubicam with 8,000 people in 152 offices worldwide, or local with as few as several persons fulltime. Charles Warner outlines three functional areas in agencies: (a) *account management*—executives and supervisors who collaborate with client advertisers; (b) *creative*—writers and art directors who develop ideas, campaigns, and produce commercials; and (c) *media*—planners, buyers, and media directors who work with station reps and network/cable/station sales staffs.

After an advertiser has determined its annual corporate budget for advertising, its ad agency's media department weighs the relative effectiveness of using various media. Then it selects specific properties within each medium. To some advertisers, the agency recommends allocating most of the budget to print media, plus direct mail. To others, the agency advises placing the largest share in broadcasting. Typically the overall budget is strategically distributed among several media. The advertiser's own salespeople, of course, influence the decision by expressing media preferences based on goals and tactics applied to the current ad campaign.

After determining the amount to be spent in broadcasting, the agency allocates portions of the budget to purchase TV, cable, and radio time. It identifies what kinds of vehicles (types of programs or participations) are desired and available at what rates. It determines how much should go for network or cable services and how much to spot advertising in selected markets and on individual stations.

Agency time buyers are ready with information about available time, talent, and facilities at networks and stations. They know from track records which availabilities are best suited to the needs of each of their clients. They continually study audience composition—demographics and psychographics—of TV and cable network programs, and stations in various day-parts. They know the reach

TABLE 9.5
Total Ad Revenues: Television, Cable, Radio, 1980–1995
(in millions)

	1980	1985	1990	1995
Local TV	$ 2,967	$ 5,714	$ 6,500	$ 10,000
National/Regional Spot TV	3,269*	6,004*	7,000*	9,100
TV Networks	5,130	8,285	10,020±	12,410+
Total Television	$ 11,366	$ 20,003	$ 23,520	$ 31,510+
Total Cable TV	58	767	1,953±	5,342
Total Radio	3,600	6,000	8,800	11,450

*Includes syndication barter. Syndicated TV ad revenues in 1995 totaled an additional $1.9 billion.
±Estimates/projected by industry
+Includes ABC, CBS, NBC, Fox. Add estimated sales of UPN $250 million, WB $65 million.

Sources: TVB and CAB; data reported in *Channels,* January 1988, 82; *Advertising Age,* 2 May 1994, 4; McCann-Erickson in *Broadcasting & Cable,* 16 September 1996, 93; Radio Advertising Bureau, in *Broadcasting & Cable,* 11 March 1996, 39.

of stations, their programming and community image. They determine costs, estimating cost per thousand (CPM), gross rating points (GRPs), or cost per point (CPP). Finally, they negotiate to buy network and station time that offers greatest return by capturing the attention of potential buyers of the advertiser's product or service. Selections must be as productive as possible, utilizing mountains of data from computer banks coupled with analyses by sales reps and sales staffs. The goal is to reach efficiently a maximum target audience (see chapter 6 on media audiences). Today's successful time-buyer is a well trained professional with a wide range of knowledge and skills.

The 1970s were a decade of transition in ad agencies as well as in many sectors of broadcasting and society generally. Added to media buying by numbers—by demographic statistics—was consideration of less tangible but often measurable characteristics of audience attitudes and reactions to programming and commercials, including their number, placement, content, and style of presentation. Media researchers began to give attention to what they termed "psychographics"—defining prospective consumers by attitudes, opinions, beliefs, preferences, and lifestyles. (In print media special-interest magazines increasingly replaced general-audience publications, and they flourished.) Advertising media entered a period of major readjustment, with innovative marketing essential to justify continually higher rates. Advertisers paid a premium for larger numbers of prime prospects, greater impact, and more effective psychographics. The goal was greater cost efficiency in the face of escalating ad expenditures.

A weakness developed when computer-generated numbers replaced imaginative, personalized interpretive buying of time. In effect, time-buyers often became "bean counters" or number crunchers. Long-time New York sales executive Robert Innes described the process in media departments of major agencies:

> The [media] planners, using various sources of information (sales reports, individual market information, etc.), figure out what spot television weight they want in a given market for a product or products. This information is sent in written form to the person who will buy the schedules for given markets. The buyer frequently does not know the reason for a given rating point goal in a market. The buyer then tries to achieve the goal set. For example, in Chicago his goal may be to get 95 rating points in homes or rating points in demographics per week for a limit of [$30,000] per week. He naturally tries to achieve his

goal for fewer dollars. The buyers, in effect, are negotiators, which some agencies actually call them.

The danger in this system is that the buyers will blindly try to achieve their rating goals without really knowing the aim and rationale of the client and planners. Frequently we run into situations where, say, for the given [$30,000] the buyer may be able to achieve 125 points per week instead of his goal of 95. This confuses them. As they say, they don't need 125.

The result of this structure is, naturally, that the planners must be pitched by the salespeople to okay any deviation from their "master plan." For exceptional media proposals—sponsorship of baseball, special events like parades, football games, etc.—the planners and the clients must be covered . . . with the agency's knowledge. This system instituted by large agencies frequently makes the buyers or negotiators merely spot buyers and not in the true sense media buyers. In effect, they turn into non-knowing human computers who eventually will probably be replaced by electronic computers for purely spot buys.[44]

The truly bottom line is that selling time through agencies tends to be ratings-oriented, while retail selling directly to advertisers is results-oriented. Quantified numbers offer security—measurable and tangible and neat. But ratings are only one tool in planning and evaluating the success of an ad campaign through mass media.

The changing role of agency buyers made new approaches necessary for station and network salespeople calling on them to influence their buying decisions. For small-market stations, this function is usually performed by their station rep organizations. Whenever direct presentations are made by a station, the station manager or sales manager usually makes them. Many larger market stations assign specific sales staffers to call on the list of agencies, to ensure that calls are made on a consistent and organized basis.

The Sales Manager's Relationship with Agencies

The sales manager either calls on agencies herself or assigns account executives to make calls. An efficient sales manager soon learns the particular abilities of each staff member and what each can accomplish at various agencies. She sees to it that contacts are maintained with all-sized agencies at national, regional and local levels. Some agencies place a high value on quality in advertising, often favoring an institutional approach. Others, generally local, have little interest in an elusive "image" created by their ads; they confine their focus to fast returns noted in the client's cash register. Knowing these variances at agencies the sales manager must consider the make-up of her sales staff as well as of the leading timebuyers at the agencies. She needs to match her salespeople with specific agencies to ensure effective results in their assignments.

After every agency call by an account executive, the sales manager should receive a report from that salesperson early the following morning. This verbal report or briefly written statement is necessary so the department head is kept posted on progress or on problems anticipated.

By maintaining month-to-month tabulation of all accounts with various agencies, the sales manager can determine gains or losses and strengths or weaknesses not only of each salesperson but of the station's relationship to agencies and clients.

The sales manager also maintains relationships with those agencies called upon by her staff. She should plan periodic trips at least every three months to major advertising centers in New York and Chicago plus, when possible, Detroit,

Atlanta, Dallas, Los Angeles, and San Francisco. Many stations are located in or next to states with regional advertising centers that generate considerable sales traffic. A personal visit by the sales manager every thirty days to all agencies in that market is extremely important. Some stations near those centers contact agencies as frequently as once a week. This practice usually brings rich dividends in added time sales.

It is perfectly in order for a salesperson to call on a national advertiser directly, but under no circumstance without first advising that advertiser's agency. There are times when a station's call on an advertiser can greatly help the agency, particularly when the sales presentation involves an approach far different from that agency's normal approach. The call may open up the relationship between the advertiser and his agency personnel.

Importance of Station Image

The image a station conveys to advertising agencies is important for placing spot business. With so many stations trying to make impressions on them, agencies find it difficult to identify outstanding features of particular broadcast properties. Time-buyers often simply contract for commercial airtime on leading stations in a market; in radio that often means buying only the top three outlets to reach a substantial share of the audience without digging into ratecards and availabilities at dozens of lesser stations that together do not deliver as many listeners as the top several. Sales managers and their reps need to communicate station assets not found in ratings, directly by visiting agencies and sending information sheets and promotion material, and also by placing eye-catching ads in trade magazines. Some agencies check on radio and TV operations in each market by collecting descriptions and verifications of qualities and services usually not reported. Stations should be quick to cooperate in gathering such data. A station wanting increased spot advertising needs to give agencies proof of its standing in the community, demonstrating characteristics that make it different, desirable, or unique in a market, such as less commercial clutter or higher standards for acceptable advertising, personal appearances by popular on-air talent, merchandising effectiveness, and mail response on promotions.

Agency Problems and Progress

Advertising agencies face constant challenges beyond that of competition. Rising costs for advertisers, switching of agency accounts, increased operating expenses, and greater emphasis on proof of performance are some of the problems agencies have encountered in recent years. For example, in 1994 IBM terminated its array of accounts placed among forty ad agencies, consolidating all $400 million worth of billings with Ogily & Mather. Despite these challenges and maintaining a traditional standard commission of 15%, advertising agencies continue to grow and their billings continue to set new records each year.

■   By 1992 the largest U.S.-based ad agencies by total billings in all media were Young & Rubicam with $8 billion, Ogilvy & Mather Worldwide with $5.3 billion, Saatchi & Saatchi and Foote Cone & Belding with $5 billion each, BBDO with $4.7 billion, and J. Walter Thompson which placed $4.4 billion of clients' budgets in media advertising.[45] (Typically, top agencies place from 60% to 80% of billings with broadcasting and cable, the rest with print and other media.)   ■

In the past decade some major agencies merged and expanded into international ad agencies, such as BBDO Worldwide, DDB Needham Worldwide (Doyle Dayne Bernback plus Needham Harper Steers), DMB&B (D'Arcy Masius Benton & Bowles, Inc.), and Saatchi & Saatchi DFS Inc. (Dancer Fitzgerald Sample). Other familiar major players expanded into multi-named Bozell, Jacobs, Kenyon & Eckhardt; Campbell [Ewald]-Mithun-Esty, Inc.; while some multiplied their overseas offices such as FCB International (Foote Cone & Belding) in thirty-five countries on all continents, and Interpublic's McCann-Erikson Worldwide with affiliated companies in sixty-four countries.

Computers

Agencies have improved service to their clients by using computers that reduce significantly the mountain of paperwork involved in buying spot advertising. Analyses, media decisions, and billings that formerly took weeks or months of human labor can now be accomplished in minutes. Differences between media and relative strengths of stations—in audience reach, demographics, and cost efficiencies—can be determined accurately without human guesswork. Once the computer could turn audience data into sales predictability figures, time-buying became a science or at least an activity capable of mathematical expression.

Broadcast and agency executives supported efficient processing of routine work through standardized procedures, permitting sales staffs more time for qualitative judgments and human negotiating needed for their primary job of selling. Among services supplied by computers were preparing campaign budgets, plotting out many hundreds of different media plans, checking for discrepancies between order and delivery of commercial spots, speeding up billing and collections, and media evaluations. By 1987 more than 400 stations were also linked to their rep firms' central computers; another 300 stations subscribed to ratings data air-expressed on floppy discs which they used in their local personal computers. By 1993 buyers and sellers of radio-TV airtime—agencies, reps, stations—were all becoming linked electronically for instantaneous interchange of schedule changes, avails, and rates. That year several cable networks began to transmit contracts electronically; they employed software with electronic contracts meeting industry standards.

Computerized systems reduce manual paperwork by providing electronic affidavits of spot buys and confirmation of airing. Computers permit creative flexibility in placing time-sensitive commercials. A Petry study in 1992 reported that almost half of all national TV spot ads placed through station reps were confirmed only one to five business days before they were to be aired by local stations.

In the volatile area of political ad spending, computer software provides "media-buy matrixes" outlining markets, stations, and schedule position where ad budgets should be directed to reach key targeted voters. Satellite facilities can up- and down-link to stations in the region or across the country political commercials just minutes after being updated in production studios.

In 1992 cable TV ad revenues totaled about $3 billion, two-thirds by cable networks and $1 billion in spot and local ads on the nation's 11,000 head-end systems. But almost all the $10 billion in spot TV ads that year went to over-air broadcasting; only $150 million were spent on cable for national/regional spots (compared with the $750 million generated by local ads on cable). That is partly because placing national spots in the sprawling, fragmented local cable universe has been unfeasible for ad agencies and reps until the industry fully harnesses the swift flexibility offered by sophisticated computer-assisted time-buys. (11,000 systems with an average 36 channels times 24 hours daily comes to almost 10 million program hours within which commercial spots can be inserted across the country every single day!) Multiple-system operators (MSOs) help coalesce some cable time-buying, again using computers for traffic and billing systems and for inserting commercials into operations schedules. Computers and digital compression technology together can provide a national interchange and reservation system, so agencies can place billing orders for specific head-ends' schedules and can transmit commercial spots that local cable operators store for later retrieval at given times into specific channels.

In the mid-1990s the industry developed a sophisticated electronic interconnect system for instant negotiating, ordering, confirming, modifying, trafficking, invoicing, and payment between and among ad agencies, station reps, and stations sales departments.

■   The computerized transactions through this electronic data interface (EDI) coupled efficiency with accuracy. It allowed swift interchange of data and decisions, freeing reps and agencies to focus on qualitative analysis rather than on painstaking, time-consuming details of negotiating by successive telephone and fax hardcopy in successive requests/replies/responses. Rep account executives would be less tied down in agency spot buying units, spending more time selling qualitative factors of specific markets, stations, and marketing packages to senior decision makers at agencies and even directly to advertisers and back to stations. Petry and other rep companies collaborated in a joint venture between Donovan Data Services and Jefferson Data Services in a proprietary application called DARE for Direct Agency Rep Exchange. Late in 1994 J. Walter Thompson and Saatchi & Saatchie ad agencies first installed the Unix equipment in their New York offices to test ad contracting in six Petry markets: Washington, Miami, Boston, Pittsburgh, Atlanta, and Orlando. The system was to be fully operational in 1995, and available to all other reps and agencies. The system would facilitate complex cable buys as well as for radio and TV. Meanwhile Group W and others developed their AdValue Network version partly because some 70% of invoices between agencies and stations had some discrepancy in them; more than half the revenue lost from errors would be eliminated by electronic tracking of spot sales transactions. Katz Television also developed a similar computerized system that it called Katz Initiative. (Initially, only AdValue Network linked agencies directly with stations.)   ■

SPOT BUSINESS AND THE STATION REPRESENTATIVE

The system of station advertising sales representatives, commonly called station reps, began in 1932. That year Edward Petry started the first of the modern station-rep firms. Petry organized his company on sound business principles and ethical practices, thus establishing high standards for the firms that were to follow. Refusing to do business with time brokers, his company instead dealt directly with station managers. Petry insisted that the one station in any market which the company selected to represent must be concerned with quality operation and fair business practices. Nationally, this made a select list of radio stations with great appeal as advertising media.

Within three years 27 additional station rep companies started up. The number of national and regional station reps in broadcasting steadily increased until the early 1990s when the total started dropping—to the total in 1995 of 123 rep firms in the United States plus 19 in Canada. As with ad agencies, that smaller total than previous decades was largely due to corporate buyouts and mergers of multiple companies into single giants like John Blair & Company's Blair Television, and also Katz Communications, Inc. (called Katz Media Corporation after 1994) with its separate operating divisions for Hispanic Radio Sales, Radio, Television, American Television, Continental Television, and Independent Television.[46] Katz Radio alone in 1994 included five companies that represented 1,700 radio stations, generating for them 18% of their ad revenues. In 1994 Katz accounted for almost half of radio's $1.6 billion national spot market. Katz also repped 325 TV stations.

The role of the rep firm is to provide national and regional agency time-buyers with up-to-the-minute information about rates and time availabilities for their client stations. (See figure 9.1 for a schematic diagram of the relationships among station reps, national ad agencies, and local stations.) Clusters of several stations up to fifty (or, rarely, even more than a hundred small stations) are served by rep offices staffed proportionately with from five or ten people up to 250 persons at the largest organizations (Blair, Katz, Petry Television, and TeleRep).

■ Katz is the oldest rep firm and the only one active in both radio and TV. In 1988 it employed 1,400 employees in twenty-two sales offices to represent 193 TV stations (30 in the top-50 markets) and 1,440 radio stations. It placed $1.5 billion of advertising, two-thirds of it with TV stations, thus earning almost $100 million revenue from commissions.

By 1992 most of the $5 billion in the television rep business was handled by eleven firms. Their billings and revenues depended not so much on the number of stations they repped but by the markets and quality of those TV stations. With far fewer stations, 68 in 1991, TeleRep serviced all but one of the top 50 markets and outpaced Katz in total TV billings (see table 9.6). In 1993 Katz repped 200 TV stations for $1.25 billion in billings (that translated into $160 million revenue on commissions, with a $33 million cash flow within the company); that same year TeleRep brought $1.1 billion of business to its 68 TV stations. ■

Commission rates to station reps for local broadcast and cable vary from company to company and from station to station. In earlier years, owners of several station properties paid only 5% commission to their reps while stations since then have paid from 8% up to 15% to be represented in national and regional time-buying centers. Although there is no set standard, some rep firms do maintain rate policies. Usually, radio stations are charged more than TV stations for representation; the usual rate for radio is 15% of all billings that the rep firm generates for the station. Most large TV stations pay 8% to 9% to their reps; medium-sized TV stations pay from 10% to 12%; and TV operations in smaller markets pay 15% to be represented. Cable operators typically pay up to 20% commission on billings to rep firms that sell their local cable ad time to agencies representing national and regional accounts.

The rep firm is sometimes strictly a sales organization; but often it also provides other support services such as research, marketing, estimating, traffic, merchandising, sales service, promotion, press relations, and even programming (strategies and syndication buys—no rep firms are involved with a station's production or engineering).

The Sales Manager and the Rep

The station sales manager should work closely with his station representative in every market where that company has an office.[47] The sales manager arranges for the rep to receive regularly all updated sales tools essential to selling the station's

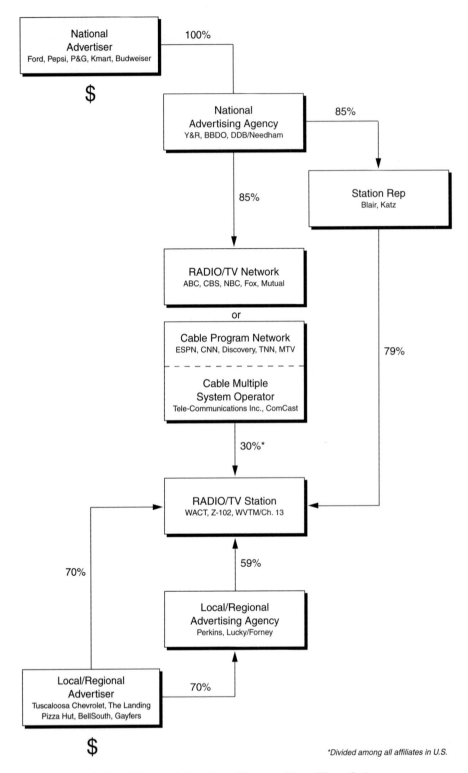

FIGURE 9.1   Dollar Flow = Advertiser/Agency/Rep Time Sales

TV STATION REP FIRMS, RANKED BY BILLINGS, 1992
(MILLIONS OF DOLLARS)

REP FIRM	BILLINGS ($000,000)	# STATIONS	REP FIRM	BILLINGS ($000,000)	# STATIONS
TeleRep	$ 950	68	Seltel	$ 290	111
Katz	900	196	MMT	285	30
Blair	760	140	Group W	200	5
Petry	675	106	CBS	150	5
HRP	475	45	Adam Young	55	26
ABC	315	8			

Source: *Variety,* 13 January 1993, 90.

airtime availabilities. Materials include program schedules with all changes noted, station brochures and routine sales presentations, and all special campaigns prepared for specific clients. The sales rep can be no more effective in selling the station than the data supplied by the sales department permits. Unfortunately, stations do not automatically receive spot business merely by signing with a rep firm.

The sales manager regularly must stimulate—even agitate—the rep organization, particularly when they lack aggressiveness. Rep organizations manifest widely varying levels of initiative. Most rep firms work around the clock to obtain business for stations on their lists; station sales managers energetically keep these sales companies supplied with data and give them maximum assistance. But some station rep organizations hardly extend themselves beyond an occasional "strike," otherwise waiting for agencies to call them about availabilities.

The successful sales manager insists on a weekly call report from each office of the station's representative. A sales manager failing to oversee the rep is managing station sales poorly, derelict in his obligation to top management. With these weekly reports, the GSM is able to keep current on national sales activity, compare the rep's monthly and annual performance, and assess the station's relationships with agencies and their clients.

Station reps also service television stations by recommending which syndicated packages to lease. Pairing the right program purchases with appropriate advertisers can bring added national spot business for a station. In these decisions, reps can play an important role because of their knowledge of client desires and their accumulated data on kinds of off-network and first-run packages available. Most maintain track records of ratings in various markets against competing station schedules, and of asking prices contracted for elsewhere.

Reliance of Stations on Sales Reps

It is disconcerting to note that some GMs and sales managers rely on their station representatives for management advice having nothing to do with sales. In some cases the reps have, in effect, become station consultants without fees for extra services. Today, it is possible for a person to "manage" a radio or TV station practically in name only. When problems arise, he has only to turn to his station rep. There he looks for solutions to problems he is unable or unwilling to confront.

Just as with consultants contracted for news, programming, or research, giving such authority and responsibility to sales representatives virtually results in an informal transfer of considerable policy control from the station manager to rep firm personnel. This sort of referral has led to poor decisions damaging to individual stations and harming the entire industry. Further, it can contradict federal regulations about licensee control and management supervision of a station. This is not to criticize a rep firm's involvement in occasional discussions pertaining to broad station policy; nor does it advocate that the station rep should be absent from all station planning about sales promotion and sales development. But willful or witless transfer of a manager's basic responsibilities to an absentee station rep is inexcusable. It is one reason for lackluster performance by certain stations.

Inappropriate and improper involvement of rep firms in stations' affairs usually originates not from the rep's desire, but from the ineptness of poorly trained, ill-prepared station management lacking competence to manage a broadcast property properly. So they use the rep as a crutch.

Self-Representation

At the opposite pole from stations overrelying on station reps, are others that have terminated relations with national sales rep companies.

Starting in 1959, a few stations began to represent themselves exclusively. Some corporations owning groups of stations, such as the major networks' O&Os and Westinghouse's Group W, continue to sell their own local avails in the national market through their own in-house rep operation.

SPOT SALES

How does a buyer in an agency differentiate between stations in those markets where there are fifteen to twenty or more? Does the small market station have a chance at influencing the agency buyer? Media buyers must negotiate for spot radio with hundreds of salespeople from a dozen rep firms representing over a thousand radio and 250 TV stations in the top-50 markets alone. Now cable is added to the mix.

With so many stations, it is hard to distinguish among them. It is essential that sales staffs, with promotion and research support, assemble attractive presentations of material essential for reps and agency time-buyers with authority to spend national advertisers' budgets. Meaningful data include complete demographic and psychographic data on audience composition in various day-parts, appraisals of the station's program service by audiences and other advertisers, estimates of kinds of products and services best suited to audiences reached, and comparative data about other media in the same market.

Exaggerated statements about station effectiveness impress very few people who count. Such distortions, especially half-truths or manipulated statistics, hurt the station's long-range revenue possibilities. There is no substitute for fact. The crux of the matter is a realistic and honest appraisal of exactly what the individual station can do for specific advertisers. Once the station representatives and agencies get this information regularly and know they can rely on it, a station will be in a good position to receive the spot accounts it deserves.

Factual data are a necessary supplement to, but no replacement for, effective personalized selling. Every station needs to campaign vigorously and steadily with the agencies and reps to achieve its goals.

E. NETWORK SALES

Although this book concentrates on management at local broadcast stations and cable operations, some explicit treatment is appropriate here about sales at broadcast networks and national cable program services.

ADVERTISING ON NATIONAL NETWORKS

By the early 1990s cable networks were siphoning major advertisers' dollars from traditional networks. Procter & Gamble's advertising agencies budgeted almost 20% of that company's national TV budget to cable; other major advertisers often allocated 10% of network spending to cable (see table 9.7). The TV network divisions of ABC and CBS each reported revenues of around $2.73 billion for all of 1993, while NBC reported $2.42 billion.[48]

Advertisers' budgets predictably increased annually as the nation's economy remained firm. Three corporations were the top investors in commercial time both in broadcasting and in cable networks in 1994 and cable TV in 1995. Procter & Gamble put four times as much into broadcast commercials ($680 million) as into commercial spots on 17 cable networks ($166 million); General Motors six times more in broadcasting ($509 million vs. $84 million); and Philip Morris seven times as much ($428 million vs. $61 million). Two of them also led in ad budgets for barter syndication in 1993.

Cable Sales of Ad Time

In 1987 total cable advertising revenue passed the $1 billion level, and in 1995 rose to $5.3 billion. Agencies budgeted cable time-buys partly because network audiences were eroding, partly because audience demographics could be better targeted by specific cable program networks, and fairly large audiences could be reached by combining clusters of cable channels at relatively low cost.

■   In 1992 six top cable networks attracted $1.35 billion in advertising revenue, up 10% from the previous year; ranked by revenue in millions of dollars they were ESPN ($302), USA ($281), WTBS ($270), MTV ($188), CNN ($174), and Family ($129). The following year the twenty-one largest basic cable networks had ad revenues of $2.43 billion. (By comparison, the three major TV networks together generated a similar amount of $2.2 billion in three months, April–June in 1992.)   ■

Taken all together, national cable network plus national spot and local cable advertising in 1992 totaled $2.1 billion—still modest compared with cable's $17.5 billion revenues from subscriber fees. That more than doubled in three years; basic cable's ad revenues in 1995 were $5.3 billion—$3.6 billion from national cable network advertising, and $1.4 billion in local and spot advertising (table 9.8).

Upfront and Scatter Sales

Sales divisions begin selling each new season (fall-winter-spring) during the previous spring. Sales offices must estimate the audiences for each program series and thus their value, to set charges for commercial airtime inventory. Those estimates can be accurate or off the mark depending on the actual success of weekly programs once the competitive season is under way. Strategy calls for getting the best payments possible in advance, in "upfront" selling that concludes in June, while still leaving enough inventory open for advertisers' later "scatter" buying,

━ TABLE 9.7 ━
Top Advertisers

TOP ADVERTISERS IN NETWORK TELEVISION, 1991 ($ In Millions)			
General Motors	$ 527.8	Kellogg Company	$ 236.6
Procter & Gamble	515.6	Ford Motor Company	217.5
Philip Morris	390.2	Pepsico	204.6
Johnson & Johnson	239.8	Sears Roebuck	187.9

Source: The Arbitron Company data cited in *Electronic Media,* 13 April 1992, 48.

TOP ADVERTISERS, 17 NATIONAL CABLE NETWORKS, 1992 ($ In Millions)			
Procter & Gamble	$ 121.6	Anheuser-Busch	$ 42.7
General Motors	59.6	AT&T	42.7
Philip Morris	53.4	American Home Products	32.4
General Mills	45.7	Kellogg Company	29.4
Hasbro	43.0	Mars	27.9

Source: Monitor-Plus data cited in *Cablevision,* 10 May 1993, 7A.

TOP ADVERTISERS, BARTER SYNDICATION, 1993 ($ In Millions)	
Procter & Gamble	$ 132.8
Philip Morris	118.2
Unilever	55.7
Kellogg Company	42.2
Warner-Lambert	35.0

Source: *Electronic Media,* 13 June 1994, S–30; cited by "Broadcast Sales Training Executive Summary," 4 July 1994, 1.

TOP ADVERTISERS, BROADCASTING AND CABLE NETWORKS, 1994 ($ In Millions)	*Broadcasting*	*17 Cable Networks*
Procter & Gamble	$ 680.2	$ 166.5
General Motors	509.4	84.0
Philip Morris	428.9	61.0

Source: Competitive Media Reporting, cited in *Broadcasting & Cable,* 27 March 1996, 53.

TOP ADVERTISERS IN CABLE TV, 1995 ($ In Millions)			
$ 213.5	Procter & Gamble	$ 48.2	General Mills Corporation
94.7	General Motors Corporation	47.9	MCI Communications Corporation
77.4	Philip Morris Companies	45.5	Johnson & Johnson
62.4	Kellogg Company	45.2	Grand Metropolitan PLC
59.5	AT&T	44.7	Pepsico, Inc.
50.8	McDonald's Corporation	44.4	Sprint Corporation

Source: Cable Advertising Bureau data in *Broadcasting & Cable,* 25 March 1996, 65.

with the hope of getting higher payment for them—because programs sometimes surprise with better audiences than predicted in spring and because the scarcity of avails drives up the value of remaining spot time. On the other hand, if programs fail to perform as expected, advertisers get rebates for the shortfall in audience ratings from what they paid for upfront—usually in the form of "make-good"

TABLE 9.8

SOURCES OF ADVERTISING SALES, AS PERCENTAGE OF REVENUES, 1995, BASIC CABLE NETWORKS AND SYSTEMS, TV & RADIO NETWORKS AND STATIONS

Ad Revenue Sources	CABLE (BASIC)		TELEVISION		RADIO	
	($ billion)	%	($ billion)	%	($ billion)	%
Local Sales	$ 1.442	27.0%	$ 10.000	31.7%	$ 9.100	79.5%
National/Regional Spot	*.215	*4.0	9.100	28.9	1.900	16.6
Network Sales	3.685	67.0	12.410	39.4	.450	3.9
TOTALS	$ 5.342		$ 31.510		$ 11.450	

*Cable national/regional spot included in local totals above; added are spot sales by regional cable sports.

Source: Radio Advertising Bureau, in *Broadcasting & Cable,* 11 March 1996, 39; McCann-Erickson in *Broadcasting & Cable,* 16 September 1996, 93.

free spots inserted later in the schedule; and scatter prices must be lowered, offered at discount or even "fire-sale prices."*

■   CBS' coverage of the Winter Olympic Games in January 1994 pulled 40% larger audiences than expected. That was fine for competing in the sheer ratings race, but the heavy upfront and even some early scatter buys received a bonanza of bonus audiences for which they paid nothing extra. Only last-minute time purchases paid a premium because of few avails, as massive audiences began to build during the week's coverage. Upfront sales are tied to program track records; during CBS' years of struggling with hard news to compete with "infotainment" morning magazine shows, its 1986–1987 upfront sales for *Morning News* came to only $4 million, contrasted with NBC's $32 million for *Today* and ABC's $24 million for *Good Morning America.*

The 1990 total prime-time upfront sales by the four networks totaled $4.35 billion—NBC $1.5 billion, ABC $1.4 billion, CBS $1 billion, and Fox $550 million. The following year that total dropped to about $3.4 billion (excluding Olympics spending). The reversal of respective networks' program fortunes, literally, was reflected in 1992's upfront sales of about $3.7 billion, with CBS on top with $1.4 billion, ABC $1 billion, and NBC reduced to only $800 million, with Fox holding steady at about $500 million in upfront prime-time sales. The strengthened economy in following years brought total network prime-time upfront buying successively to $3.7 billion in 1993, $4.4 billion (1994), $5.6 billion (1995), and $5.7 billion in 1996. In barter syndication, upfront sales came to $1.6 billion in spring for the 1996–1997 season.

In 1990 upfront buying of spot in syndicated barter programming reached a record $1 billion by July 4, leaving about $250 million in inventory to be sold in scatter sales later in the year after the fall season was under way. In barter ad sales, first-run syndicated programmers typically sell 80% of their next season's inventory by midsummer. That year advertising revenue from syndicated programming totaled about $1.25 billion; 30-second spots in "Arsenio Hall" sold for $30,000, in "Oprah" for $40,000, and in "Star Trek: The Next Generation" for $135,000.   ■

The value of one rating point in network prime-time in 1990 was about $8,000 per 30-second commercial, or $16,000 a minute. A hit half-hour show (with seven 30-second commercials) with a 17 rating that season thus was worth $32,000 more

---

*In the mid-1960s, with the development of audience demographics, ABC and then other networks began guaranteeing CPM households, and eventually sold prime-time with guaranteed costs for cost per thousand women 18–49. See William S. Rubens, "A Personal History of TV Ratings, 1929 to 1989 and Beyond," *Broadcast Financial Journal,* March-April 1990, 33–38; and BEA, *Feedback,* Fall 1989.

per half-minute commercial and $224,000 more each weekly episode than an average series with a 13 rating. NBC research veteran William Reubens estimated each prime-time rating point for the 1990 season was worth $140,000,000 to a network. Half a decade later with the stakes even higher, no wonder that networks competed intensely with program selection, scheduling, and promotion strategies to build audiences. As noted earlier, the networks finished the 1996–1997 season: NBC 10.5 rating, CBS 9.6, ABC 9.3, and Fox 7.8, plus UPN 3.2 and WB 2.6. Cable and DBS continued to draw away some of their audiences.

Commercial time in prime-time shows on the networks was based directly on audiences delivered (or, in upfront buying, expected to be delivered). In 1994 a 30-second spot in top sitcom *Seinfeld* sold for $390,000, *Home Improvement* $350,000, *Roseanne* $310,000, *Monday Night Football* $285,000, and *60 Minutes* $225,000. In 1996 *Seinfeld* had risen to $500,000 a half-minute; a spot in *Mad About You* sold for $300,000. The highest price paid for a 30-second spot in a theatrical movie replayed over a network was $650,000 for Steven Spielberg's "Jurassic Park"; the 1993 film was broadcast for the first time by NBC over the entire three evening hours during the May 1995 ratings "sweeps" period (it had paid $50 million for rights to multiple showings).

The Big Three networks sold almost $8 billion of commercial airtime in 1994, more than $5 billion of that in prime-time shows and sports (table 9.9).

Specialized Audiences, Programs, Ads

Sales must be directed, of course, to target audiences sought by advertisers. Generally that means 18–49-year-olds as a key demographic. Of course, different kinds of programs attract varied audiences to whom advertisers seek to sell their products and services.

■   **Children.** During the decade 1990–2000 children ("kids") were expected to increase from 45 million to 49 million—20% of the total U.S. population. In the early 1990s they annually spent $8.8 billion of their own money and, according to research estimates, influenced almost $507 billion of parents' spending—the largest being food at $82 billion. Times sales in programs directed at youth or at least seen by youngsters are crucial to many advertisers.

Sports. For the 1994 Super Bowl game in 1994, advertisers paid $69.7 million for 31 minutes of commercial time. For Super Bowl XXX in 1996 NBC sold all 58 commercial spots at a record $1 million or more each. CBS charged $280,000 per 30-second spot in the 1992 Olympic Games coverage; to buyers it had guaranteed an average prime-time rating of 17, but actually delivered 18.6—a bonus for advertisers. Olympics in Atlanta in 1996 went for $380,000 for each prime-time 30-second spot. NBC had paid $456 million for broadcast rights; its revenue from time sales to 48 advertisers totaled $675 million (out of which it also paid production costs and some shared revenue with the Atlanta Olympic Organizing Committee). Predictions were $351,000 per spot in the 1998 Winter Games and $608,000 each in the Summer Games in 2008 (the final one covered by NBC in its $2.3 billion contract for exclusive broadcast rights to all games from 1998 to 2008).

Political. In 1992 presidential political advertising amounted to $130 million— $74 million network and $56 million spot; spot spending by state and local candidates added $226 million. The 1994 campaigns brought in $400 million to broadcast and cable sales departments. Cable in 1996 expected to sell more than $30 million in spot political ads; digital ad insertion equipment made it possible for cable operators to update and change commercials more rapidly than in the past. Television looked to a record $500 million in political spending that year, 80% to spot, and the remaining $100 million to network ad buys.

NETWORK REVENUE BY DAY-PART, 1994: ABC, CBS, NBC

(IN MILLIONS)

Prime-time	$ 3,581.9	Late-night	$ 360.8
Sports	1,784.3	A.M.	275.0
Daytime	950.7	Children's	119.6
News	828.3		

Source: Ernst & Young for Broadcast Cable Financial Management Association; in *Broadcasting & Cable*, 6 March 1995, 53.

Hispanic. TV spending to reach the fast growing Hispanic population grew to $452.2 million in 1994: $240.8 million in advertising on networks and national spot, and $240.8 million in local TV. Advertisers reached out to them with ad budgets for all media totaling almost $1 billion in 1994. The goal was evident: Consumer spending by those Hispanic audiences was expected to top $260 billion in 1995. To reach that segment of the U.S. population Procter & Gamble spent $37.6 million in targeted advertising, AT&T spent $19.2 million. ▪

F. SALES AND ADVERTISING PRACTICES

COMMERCIAL CLUTTER

Sale of airtime is affected by trends in broadcast practices. By 1970, 30-second spot announcements began to replace the minute-long commercial as the virtual standard for the television industry. In the next decade, 15-second spots became the norm; in 1986 about 20% of network commercials were 15-seconds, increasing to 30% in 1987. More short time-units per minute proliferated announcements about different products and brands. Thus a two-minute break in daytime programming offered the viewer not two but four to eight commercials. Adding to criticism by viewers generally and published critics, some advertisers and agencies voiced concern about how effective their commercials were with audiences whose attention was diffused by multiple messages.

It is notable that in 1984 when NAB code limits on commercials were eliminated, AM stations aired an average of 14 commercials filling 11 commercial minutes an hour, and FM stations scheduled an average 11.4 spots during 9 minutes and 54 seconds an hour—all far below NAB Code standards (used as unofficial guidelines by the FCC). The next year with restrictions gone, commercialism dropped a bit: AM averaged 12.7 commercials in 10:06 out of every 60 minutes; FM's number of commercials per hour rose slightly to 11.9 during 9:17 in each 60-minute period. Westinghouse Broadcasting Company's chairman and CEO Daniel L. Ritchie lamented in 1987 that during the two decades since 1965 the number of TV network commercials in an average week almost tripled from 1,800 to 5,400, while in just the previous two years local stations increased the number of commercials by 20%.

By 1991 network weekly commercials had risen to over 6,300, according to Broadcaster Advertiser Reports. In 1992 two advertising associations computed that nonprogram time (commercials, credits, promotions) in network programs had grown from under 10 minutes per prime-time hour in 1983 to over 13.5 minutes in 1991. In 1992 Don West, editor of *Broadcasting* magazine, wrote:

> Would programming be enhanced by having fewer than 36 commercials in a two-hour movie? Would the public interest be served by having more than 22 minutes of news in the evening news? Would [a radio station] sound better with fewer than 300 spots a day?
>
> We don't have the answers to those esthetic questions, but we do know there'd be a quantum change in the Fifth Estate. Enough, perhaps, to turn the industry around. For the cruel fact is that radio and TV are awash in a sea of inventory that threatens to swamp them both.[49]

■ In 1992 commercials in prime-time were clocked at an average of 7 minutes and 19 seconds per hour on each major network, while overall nonprogram material per hour averaged 12 minutes and 35 seconds (Fox had 14 minutes and 38 seconds). Nonprogram material during prime-time on cable averaged 13:36 per hour, on independent stations 13:53 per hour, and in syndication product 16:58.[50] The following year all three networks aired over 13 minutes of nonprogram material every 60 minutes in prime-time (ABC 13:24, CBS 13:53, NBC 13:43); in daytime programs CBS had the heaviest load with almost one-third of each hour taken up by nonprogramming material (19:26).[51] In 1994 ABC added 11 half-minute commercial spots to its weekly prime-time schedule, although it still remained behind the other two networks in commercialization. Those added spots were expected to produce an additional $1.3 million revenue each week. A study of network, syndicated, and cable schedules in 1994 by Competitive Media Reporting found daytime the most cluttered day-part, with almost one-third of every hour taken by commercials—averaging 18 minutes and 56 seconds an hour in May and 18:52 in November. ■

When advertisers such as Procter & Gamble first pushed for 15-second spots (which they found about 75% as effective as half-minutes in terms of audience recall), television networks generally charged half the price of a 30-second spot; they later increased the charge to 55%–60% of a 30. But about one in five radio stations introduced unit pricing in the mid-1980s, charging the same for each commercial whatever its length. This helped slow down the swing to shorter spots; the station achieved a cleaner sound with fewer commercials without losing revenue.

Overcommercializing may generate increased revenue for a period of time, but it can undermine advertising's effectiveness with the public; it might even occasion governmental intrusion into business aspects of broadcasting. Ironically, the NAB's commercial standards occasioned advertisers' charges of anti-competition, supported by the Justice Department, for raising the cost of airtime because artificially limiting the industry's total avails inventory. The NAB was ordered to jettison those guidelines and subsequently dismantled its entire Code of broadcast standards and its office for monitoring compliance.

But as a "commercial break" came to mean many minutes of successive ads, wary and wily viewers armed with remote controls flipped away to browse among other channels, including the growing dozens of cable services. A CBS study in 1993 claimed only 5% did so when watching networks, while 10% of cable audiences grazed during program breaks. Audiences developed a canny sense of timing, prompting them instinctively to return to their original program just as commercial clusters were ending, thwarting advertisers' goals. In the 1990s some prime-time program formats and schedules tried to counter between-program grazing by shifting commercials away from the beginning and end of shows; the closing scene of one show ended with brief if any credits and jumped directly into the first scene of the following program—without even a formal station break at the half-hour mark—as NBC's successive sitcoms on Thursday evenings in 1993–1994, *Mad About You, Wings, Seinfeld,* and *Frasier.* Further, in 1993 the FCC considered reviewing the elimination (in 1984) of commercial time limitations.

CHILDREN'S PROGRAMMING AND COMMERCIALS

The FCC seemed to threaten involvement in programming in 1974 by its initial proposals against advertising support of children's programming. Consumer groups such as Action for Children's Television (ACT) lobbied throughout the 1970s to eliminate all commercials from programs scheduled for children. The NAB precluded direct federal action by amending its Code provisions restricting advertising time (as well as content) of children's programming. Typically, the Commission did not terminate its proceeding but instead exhorted broadcasters to continue to upgrade programming practices for youngsters through self-regulatory measures; and the Commission cautioned that rule-making might yet be in order if broadcasters were not up to the challenge. That earlier threat by the Commission seemed a distant anachronism in the deregulatory 1980s; but it came back to roost when Congress itself looked into the issue and passed the Children's Television Act of 1990, mandating limits on commercials in children's programming: 12 minutes an hour during weekdays and 10.5 minutes an hour on weekends. The FCC later backed that up with stiff forfeitures (fines) for violations by some stations including Turner's superstation WTBS (see chapter 11).

DECEPTIVE ADVERTISING

Related to children's programming is the matter of deceptive advertising. The Federal Trade Commission kept increasingly closer watch on false or misleading commercial presentations during the 1970s, then was less vigorous in the 1980s, becoming more active once again in the 1990s. ACT criticized what it called deceptive ads in programs directed at impressionable youngsters. Although the original "fairness doctrine" for handling controversial issues of public importance was in force until the mid-1980s, the FCC refused to apply it to products other than cigarettes (while they were still advertised in broadcast media until January 2, 1971) in so-called counter-advertising announcements; the courts supported that stand. But the Commission reminded broadcasters that its public notice of November 7, 1961, had already mandated the duty to protect the public from false advertising by taking all reasonable measures to eliminate such from their air.[52]

After the NAB dismantled its code review board for commercials, national advertising intended for broadcast continued to be screened by network "standards and practices" staffs. (Each network checks approximately 36,000 commercials a year, including various stages of some commercials, and revised versions of perhaps 10%–20% initially not accepted; it ultimately rejects about 1% of them.) Obviously local and regional advertising messages still had to be scrutinized by individual radio and television stations. Alert local stations' efforts in continuity clearance will pay off in the station's reputation for reliable service to the community and in minimum time and expense spent responding to criticism and even lawsuits.

A decade or two ago, continued progress and prosperity in media sales spawned the danger of complacency. Deregulation in the 1980s ushered in wildly expanding sale prices for TV and radio stations and networks, as monied corporations expanded their holdings by paying multiples of up to 15 times or more annual cash-flow for local properties (see next chapter). But economic recession in the early 1990s smothered that affluence as multiples shrank to 7-10 times cash flow. That shrinkage came on the heels of inroads from new competing media such as cable. Nor did competition on national and local levels from other media like print and direct mail stand still. In the national spot market as well as locally, television and radio are competitive and must often sell against each other and cable as well as against other advertising media.

A rising curve of business in the broadcasting industry offers obvious advantages, but there is also a danger. Increased revenue is accompanied by a rising spiral of costs—for equipment, production, program leasing, salaries, and the cost of doing business generally. A regularly ascending curve of revenue from year to year may mask compounding expenses. Extra effort must be expended to increase revenue to offset added costs, just to maintain a station's income level. But owners and stockholders do not want merely level income; they seek continually increasing profitability. Either costs must be held in check and even lowered (most difficult) or else revenue must expand faster than expenses do. This calls for ever more energetic and creative selling by local sales staffs, by station representatives, by agencies and networks, and by industry sales promotion organizations such as Radio Advertising Bureau, Television Bureau of Advertising, and Cabletelevision Advertising Bureau. The chief foe of progress is not really *competition,* but rather *complacency!*

The next chapter of this book discusses managing income and expenses in order to produce those profits. Revenue figures from sales show only one aspect of the financial picture. But it is central, for without a healthy income all else in commercial broadcasting and cable can be written off as futile.

⤙⤚ CHAPTER 9 NOTES ⤙⤚

1. Steve Millard, "Radio at 50: An Endless Search for Infinite Variety," *Broadcasting,* 16 October 1972, 31.

2. Bruce Owen and Steven Wildman, *Video Economics* (Hillsdale, N.J.: Lawrence Erlbaum Associates, 1992), 153.

3. For example, the Roper studies regularly reported that most citizens not merely tolerated the commercial system, but preferred it to alternative forms of broadcast service supported by direct payment per program, by subscription, or by taxes. In later decades, of course, the public in large numbers did come to accept subscription fees and pay-per-view for ad-free programs offered by cable and direct broadcast satellite services. Cf. Robert McChesney's *Telecommunications, Mass Media, and Democracy: The Battle for Control of U.S. Broadcasting, 1928–1935* (Cary, North Carolina: Oxford University Press, 1993) which analyzes the struggle in radio's formative years by nonprofit religious and educational groups to wrest part of the spectrum away from commercial interests; their limited voices were drowned in the sea of free-enterprise, profit-oriented business that carried the day and set public policy with only oblique glances at alternative structures for the still-forming American system of radio. Eventually in the 1940s and 1950s first radio frequencies and then television channels were set aside for exclusively noncommercial use, initially to serve education and ultimately after the 1960s for information, culture, and the arts generally.

4. Howard Stringer, "Living with Competition" (speech presented at the Royal Television Society Symposium, Birmingham, England, November 26, 1988), 4 of speech manuscript.

5. See William J. Stanton, Michael J. Etzel, and Bruce J. Walker, *Fundamentals of Marketing,* 9th ed. (New York: McGraw-Hill, 1991); these and related concepts are analyzed by Carroll and Davis, *Electronic Media Programming,* 28–47.

6. Al Ries and Jack Trout, *Positioning: The Battle for Your Mind* (New York: McGraw-Hill, 1981; Warner Books edition, 1986), 196–197; the preceding quoted excerpt is on p. 2. See their related book *The 22 Immutable Laws of Marketing: Violate Them at Your Own Risk!* (New York: HarperCollins, 1993).

7. Charles Warner, *Broadcasting and Cable Selling* (Belmont, Calif.: Wadsworth, 1986); Barton C. White and N. Doyle Satterthwaite, *But First These Messages. . . The Selling of Broadcast Advertising* (Boston: Allyn & Bacon, 1989); Josh Gordon, *Competitive Selling: A Fundamental Approach* (Stamford, Conn.: Cowles Business Media, Inc., 1991); Bob Schulberg, *Radio Advertising: The Authoritative Handbook* (Lincolnwood, Ill.: NTC Business Books, 1989); Al Ries and Jack Trout, *Positioning: The Battle for Your Mind* and *The 22 Immutable Laws of Marketing;* David Poltrack, *Television Marketing: Network/Local/Cable* (New York: McGraw-Hill, 1983); Susan Tyler Eastman and Robert A. Klein, *Promotion & Marketing for Broadcasting & Cable,* 2nd ed. (Prospect Heights, Ill.: Waveland Press, 1991).

8. David F. Poltrack, *Television Marketing: Network/Local/Cable* (New York: McGraw-Hill, 1983), 3–4.

9. See chapters 4 and 5, tables 4.4 and 5.5, for composite estimates drawn from various sources.

10. Data in these paragraphs are from Eastman and Klein, *Promotion & Marketing,* 188–189, 213–215, 252–255, 301–303.

11. "1992 Cable Network Ad Dollars," *Broadcasting,* 15 February 1993, 53.

12. Data reported by Television Bureau of Advertising, cited in *Broadcasting,* 20 July 1987, 70; also a survey by Charles Warner for the TVB and RAB was summarized in *Broadcasting,* 11 November 1991, 71, and reported fully in a Broadcast Education Association publication by Charles Warner and James Spencer, "Television Station Staff Profiles, Compensation, Practices, and Motivation," *Feedback,* 33:3, (Summer 1992): 8–13, 20–21.

13. Donald H. Kirkley, Jr., *Station Policy and Procedures: A Guide for Radio* (Washington, D.C.: NAB, 1985), 63–64.

14. Don Beveridge quoted by Bob Schulberg, *Radio Advertising: The Authoritative Handbook* (Lincolnwood, Ill.: NTC Business Books, 1989 [1990 printing]), 173.

15. Charles F. Haner and Givens L. Thornton, "How Successful Salesmen Behave: A Counseling Guide," *Personnel,* May/June, 1959, 22–30.

16. See White and Satterthwaite, *But First These Messages. . .* 48–54, and "Appendix I: RAB Consultant Call Interview," 352–364.

17. Warner, *Broadcasting and Cable Selling,* 187–218, compares strengths and weaknesses of each medium and describes how each is sold; see also White and Satterthwaite, Chapter 9, "Selling Against (and With) Other Media," 224–268.

18. Antonelli, president of Antonelli Media Training Center, in "One Seller's Opinion," *Television/Radio Age,* 20 March 1989, 51.

19. See Warner, Chapter 9, "Retail/Development Selling," 254–277; in the latter pages he outlines co-operative advertising between local retailers and national manufacturers and merchandisers.

20. See Warner and Spencer, "Television Sales Staff Profiles, Compensation, Practices, and Motivation," *Feedback,* 33:3 (Summer 1992):12–13. Among non-media companies, many pay straight salary to entry-level salespeople, then shift to mixed compensation after they complete training and are actively making sales. Half of all companies surveyed in 1993 compensated salespeople with straight commission, one-third with salary and commission, 7% commission plus bonus, 4% salary and commission plus bonus, 3% straight salary, and 2% salary and bonus. Data reported in *Personal Selling Power,*

July/August 1993, 34; cited in "Executive Summary, Broadcast Sales Training" newsletter, 9 August 1993, 4.

21. "Infinity Unbowed: Top-Dollar Acts Power Infinity Stations," *Broadcasting,* 30 November 1992, 4, 8, 35.

22. See White and Satterthwaite, 335–336, quoting John F. Lee, general sales manager, WBBM-TV, Chicago.

23. Warner, 11–12.

24. Ibid., Chapters 4–5, pp. 83–155. Those estimates can be compared with the Dartnell Corporation's survey of salespeople generally; they spend just over half their time selling—30% face to face and 21% by telephone—plus 20% traveling and waiting, 17% on administrative tasks, and 12% making service calls on already-sold clients; cited by The Marketing Communications Group, "Executive Summary: Broadcast Sales Training," 6:28, 12 July 1993, 4.

25. White and Satterthwaite, 60, in Chapter 3, "Research and Prospecting," 60–73; italics in the original.

26. For concrete details and tips for preparing sales presentations, see Warner's Chapter 14, "Organizing Individual Sales Effort"; see also White and Satterthwaite's Chapter 4, "Organizing the Sales Effort," outlining proper handling of account lists, planning sales calls weekly, keeping records, organizing daily and weekly agendas and goals.

27. Very practical behavioral tips for meeting and interacting with clients or agency representatives are offered by White and Satterthwaite, 149–174, and 186–197.

28. See earlier chapter 6, "Managing Media Audiences," for discussion of ratings services.

29. Poltrack, *Television Marketing,* describes these methodologies for documenting advertising effectiveness for marketing managers and retail advertisers in Chapter 12, "Measuring Television Advertising Effectiveness."

30. Warner, 220–229.

31. Frederick W. Ford and Lee G. Lovett, "Interpreting the FCC Rules & Regulations: Fraudulent Billing Practices," *BM/E (Broadcast Management/Engineering)*, November 1973, 22–26. On this matter, the FCC issued public notices and memoranda opinion and order in 1962 (23 RR 175), 1965 (6 RR 2d 1540), 1970 (19 RR 2d 1506 and 1507), and 1973 (Seaboard Broadcasting Corp., Docket No. 18814 [1973]), and amended the "double billing" rule to include all forms of fraudulent billing, in section §73.1205 of the Regulations.

32. Norman Pattiz, "Radio's 'Golden Age' Over; This is Its 'Platinum Age' as Webs Take New Tack," *Variety,* 20 January 1988, 176.

33. Data were reported by Advertiser Syndicated Television Association, NAB, *Variety, Television/Radio Age,* and *Broadcasting,* particularly 18 June 1990, 65; 2 March 1992, 42–49; 17 January 1994, 126.

34. Steve McClellan, "Flourishing Barter Poised to Branch Out," *Broadcasting,* 2 March 1992, 42–44.

35. Rich Brown, "Cable Eager to Attract Small Audiences," *Broadcasting,* 31 August 1992, 46–47. About 70% of programs developed for children were on cable, attracting annual ad revenues of $70 million in 1992 and growing 50% to about $115 million in 1993. See also Mike Freeman, "Children's TV: Majors Use Blocks to Build Audience," *Broadcasting,* 31 August 1992, 35–40.

36. Harold Rice, "Monday Memo: The IRS Takes a Dim View of the [Barter] Reporting Used by Much of the Media," *Broadcasting,* 7 December 1992, 18.

37. Estimates by Tom Burkhart, vice-president and general manager, Marketing & Communications, Pittsburgh, in "Monday Memo," *Broadcasting,* 27 May 1991, 22.

38. Cited in "Direct Marketing TV Comes of Age," Broadcast Sales Training's *The Money Maker: Sales & Promotion Idea Newsletter,* September 1993, 1–2. Other data in this paragraph drawn from articles in *Television/Radio Age,* 8 February 1988, 68–70, 97; *Broadcasting & Cable,* 24 May 1993, 69; 25 October 1993, 20–27.

39. This is the term used throughout a substantive, readable, and thought-provoking, analysis of the genre: Craig R. Evans, *Marketing Channels: Infomercials and the Future of Televised Marketing* (Englewood Cliffs, N.J.: Prentice Hall, 1994).

40. Speech to Broadcast Promotion and Marketing Executives and Broadcast Design Associates, Atlanta, June 11, 1987; quoted in *Broadcasting,* 15 June 1987, 39.

41. BPME President Gary Taylor, cited in "BPME 1992," *Broadcasting*, 15 June 1992, 39.

42. Good coverage of the entire subject can be found in Eastman and Klein, *Promotion & Marketing for Broadcasting & Cable,* 2nd ed. The description of promotion in this chapter, and much factual information, was provided by the late E. Boyd Seghers, Jr., vice-president and manager, sales promotion and advertising research, WGN Continental Broadcasting Company, Chicago.

43. Details listed by Jay Hoffer, *Managing Today's Radio Station* Blue Ridge Summit, Pa.: Tab Books, 1971), 277–283.

44. Robert A. Innes, vice-president, Television Division, WGN Continental Sales Company, New York City, in correspondence to author James A. Brown, 26 February 1974; at that time he cited $3,000 as the value, which in the 1994 economy would be more like $28,000–$30,000, according to Harry R. Stecker, executive vice-president of Petry Inc., New York, confirming and updating the 1974 quotation (letter to author Ward L. Quaal, 8 July 1994).

45. Standard Rate & Data Service (SRDS) directory of advertising agencies published in February 1992; outpacing American agencies for media billings were WWP Group with $18.1 billion worldwide and Dentsu Inc. with $8.1 billion; in the early 1990s WPP Group owned both Ogilvy & Mather Worldwide and J. Walter Thompson, but their respective billings in 1991 were reported separately in the SRDS directory.

46. SRDS included as rep firms the national sales offices of group-owned stations, which in addition to retaining a national rep also operate offices in New York or Los Angeles and their other markets, staffed by station personnel to stimulate non-network spot sales; major ones include CBS Radio Representatives, CBS Television Station National Sales; Capital Cities/ABC National Television Sales, NBC Spot Television Sales, and Westinghouse's Group W Television Sales.

47. Practical considerations of the mutual responsibilities of station management and the representative firm are described by Hoffer, *Managing Today's Radio Station*, 266–275. Cf. John B. Sias, "The National Sales Representative," in Yale Roe, ed., *Television Station Management: The Business of Broadcasting* (New York: Hastings House, 1962), 207–216; and 203–206, see Charles Young's discussion of the role of the station rep and self-representation for the independent station. See also Warner, *Broadcast and Cable Selling,* 306–311.

48. Data in this paragraph derived from various sources: Broadcast and Cable Financial Management Association, MediaWatch, BAR, Paul Kagan Associates, Nielsen Media Research Monitor-Plus—cited in *Electronic Media,* 17 August 1992, 25; *Broadcasting,* 15 February 1993, 53; *Cablevision,* 22 November 1993, 39; *Broadcasting & Cable,* 16 May 1994, 6.

49. Don West, "DeSales Street," *Broadcasting,* 13 April 1992, 16.

50. Arbitron's MediaWatch data, commissioned by the Network Television Association, were reported by Brian Lowry, "Network Clean Up Their Primetime Acts," *Variety,* 3 May 1993, 27.

51. Data from the American Association of Advertising Agencies and the Association of National Advertisers reported in "Commercial Clutter Skyrockets, Associations Say," *Broadcasting & Cable,* 14 February 1994, 49; see *Advertising Age,* 7 February 1994, 2.

52. See "Stations Cautioned on Deceptive Ads," *Broadcasting,* 21 December 1970, 30. Cf. "Court's 'Freedom Not to Air,'" *Variety,* 30 May 1973, 1, 70—regarding the BEM case (Business Executives Move for Vietnam Peace) in which the U.S. Supreme Court supported the First Amendment press freedom of editorial judgment against unlimited access to the air by advertisers as proponents of views on issues, or in counter-advertising.

10

Financial Management

Profits are a measure of effective, efficient operation and should be worn as a badge of accomplishment and of honor.
PHILIP REED, GENERAL ELECTRIC COMPANY

Broadcasting in the U.S. is a business enterprise and economic motives are valid if they can be justified in terms of social ethics and technical excellence.
ROBERT H. CODDINGTON, BROADCAST EXECUTIVE AND AUTHOR

My financial strategy is to run up as many bills as possible. Then die.
GARY SHIVERS, PUBLIC RADIO PROGRAM HOST

A manager's activities are diverse, dictated by the complexity of commercial broadcasting. He or she regularly confronts questions involving the programming, engineering, and sales departments. Leadership of these three key areas occupies most of the manager's attention and time.

But behind these operations lies a primary responsibility that affects and often determines the manager's course of action, conditioning every decision. The manager is the representative of owners who judge success principally by the profit produced by the station or cable system.

Typically, the owners of a local broadcast or cable operation are businesspeople successful in fields unrelated to broadcasting; often they represent larger companies doing business elsewhere in the country. Ownership of a radio or TV station or cable system appeals to them for many reasons. It provides some prestige and potential power; it offers an opportunity to become engaged in a community service; and it links them with the exciting and colorful world of "show business." Basically, though, it can earn a substantial profit on an initial investment.

To protect that investment and generate systematic returns requires broadcast-related knowledge and skills that these owners often do not possess. The owners employ a manager experienced in media operation who can produce the profits they seek. How well she can fulfill that expectation, while also exercising her own professional standards, is the measure of her success and contribution to the industry and the community.

While the manager focuses most attention on operational procedures of the station's three major departments, economic considerations govern her *modus vivendi*.

In any business, the product and its distribution and sale are equally important. In broadcasting and cable, programming (with its spot availabilities) may be regarded as the product, and engineering as its distribution system. Sales results, including subscribers' fees, wholly depend on both.

As a catalyst for the programming, engineering, and sales divisions, management supplies the vision and leadership needed for a coordinated effort, as described in chapter 2. As each aspect of the operation improves, station services expand, increasing the opportunity for profits. The manager may delegate many tasks in the three operating departments, but managing the operation's finances is her direct responsibility. She cannot avoid the important decisions affecting profit or loss. Management must take complete charge of approving and administering the operating budget and controlling income and expenses.

Among personal considerations at stake is a manager's liability (best protected by proper insurance) for company losses because of their own "wrongful act." Paul Richard, manager of insurance services for Broadcast Financial Management explained:

> The term 'Wrongful Act' means any actual or alleged error, misleading statement, neglect, breach of duty or any other matter claimed against them by reason of their being an Officer or Director of the corporation. What might some 'Wrongful Acts' be in relation to station directors and officers? Here are a few: misleading representations, anti-trust violations, conflict of interest, improper expenditures, failure to honor employment contracts and finally a catch-all, imprudent management. It is no longer an adequate defense to plead lack of knowledge or inability to understand complex financial data.[1]

The "general and administrative" part of an operational structure includes personnel working directly under the manager's supervision; their work crosses over all three departments, serving each of them. They include employees performing office duties as secretaries, clerks, receptionists, and telephone switchboard operators, but especially the "right arms" of managers in the area of financial administration, the accountants. Accounting furnishes management with accurate data about income and expenses, to help assure fruitful decisions in situations where dollar figures are either red or black with no gray shades.

THE ROLES OF ACCOUNTING VS. ECONOMICS (OR FINANCE)

As broadcast and cable companies over the past decade were swept into the swirl of corporate media mergers and leveraged buyouts, the *business* part of these "show business" entities became more crucial for managers. The growing vortex of station buying and selling and corporate take-overs was described in chapter 1.

A manager spends much of the day with operating budgets, project analyses, and profit statements. Most local radio-TV-cable general managers (GMs) and department heads, and even many national network executives, are by training and aptitude what Finkler calls "nonfinancial managers" who nevertheless must understand and act on financial information, including accounting data. As their broadcast station or cable system has grown, often within a corporate group, and as outside factors of station acquisition, inflation, and competitive marketing forces have compounded, managers have faced increased financial complexity in their job.[2]

Competent financial management serves the twin goals of viability (more than mere survival) and profitability. These relate to the company's return on investment (ROI), return on equity (ROE), and return on assets (ROA) or on net assets (RONA).[3] That return is concerned with liquidity—cash assets or those quickly convertible to cash—and long-term solvency, including interrelated growth among sales, inventory, plant, personnel, and costs.

A manager first needs *accounting* data for a record of *past* transactions in money terms—involving cash accrual and disbursement, government regulations, and tax administration. Those data then provide the basis for managerial *economics* that looks to the *future,* with long-term planning and short-term strategies including marketing analysis. Both accounting and economics (or finance) are keys to successfully managing any business. But they differ in their ideas of costs (as part of either profit or equity), depreciation and capital gains, and valuing assets (exact past charges vs. future purchasing power amid rising price levels). Accounting provides specific data of activity already completed whereas economics or finance involves thinking about future options, which leads to decision making and planning courses of action. Christopher nicely summarizes:

> Managerial economics will help us see operating results and trends from the perspective of the future. It will identify profit opportunity or profit problems before they show up on an accounting income statement. . . . Accounting makes sudden revaluations that have, in reality, been accumulating over time. Economics makes such revaluations currently.[4]

The two disciplines serve different purposes. Accounting provides analysis of past activity through data processing and management information service. Economics (or finance) is speculative, projecting those data in the context of future contingencies in the business environment. Economics helps analyze: future return on investment regarding assets, costs, and income; cash flow projections for planning budgets; and appraisal of new ventures and investment opportunities. Christopher notes that venture analysis is mostly a matter of economics whereas venture achievement is more a matter of motivation—evidenced by such entrepreneurs as David Sarnoff, William Paley, Donald McGannon, Ted Turner, Rupert Murdoch, John Malone, Sumner Redstone, and Stanley Hubbard.

ACCOUNTING

Proper accounting procedures are essential for management's accountability to the broadcast company's private owners, or to the stockholders if it is a publicly owned corporation. Ultimately every management decision is reflected in dollars and cents, either on the company's income ("profit and loss") statement, on its balance sheet, or both. Managers must attempt to quantify in financial terms the effects of decisions before implementing them.

Budgets, cash forecasts, cost justification surveys, and income-and-expense analyses are financial tools that assist management decision making. The manager should work closely with the station's financial officer—the controller, chief accountant, or whatever title—who should develop personal expertise and professionalism to meet the demands of modern business.

THE CONTROLLER

Beyond recording financial facts accurately, the controller must analyze the effects of these historical data and, with management, attempts to forecast the financial impact of prospective actions. Mikita describes this role:

He is responsible for realistically informing the station's management team of the financial result to be achieved in any programming, sales, promotion, publicity, engineering or other operation. To satisfy their own responsibility to the public, broadcasters must do many things which will not immediately enhance the profitability of a given television station. Nevertheless, it is essential that the financial yardstick be applied to all operations if intelligent financial planning is to accomplish an atmosphere of healthy growth.[5]

The controller supervises all accounting, including an office staff of three to fifteen or twenty people. She maintains internal control procedures and regularly evaluates them, especially related to individual department budgets. She is responsible for providing accurate data for union negotiations on contracts, grievances, and arbitration. She oversees salary, hiring and firing of clerical persons, and other personnel matters. The controller sets policies for extending credit to station customers, to achieve maximum sales results while reducing the margin of risk in payment default. Usually she supervises office management of supplies, communications systems, furniture and equipment. She collaborates with the general manager on general financial management to achieve efficient *cost control* (not merely "cost reduction"). She works with the sales department to establish realistic ratecards for airtime and production services which ensure maximum business and profit. She helps the research staff analyze ratings and other data for various department heads. The controller is closely involved with every phase of station operation.

To paraphrase Mikita's analysis, all department heads must be aware of a station's general profitability but they contribute at different levels to an operation's earnings. Particularly in programming and production the profit motive should not undercut creativity and innovation. The program director ensures that his department contributes to the station by creating and scheduling product that attracts maximum audiences each segment of the broadcast day. The controller should work closely with programming staff, evaluating and guiding their efforts so they produce their best creative work at cost levels yielding desired income margins for each project. There are various ways to produce a program, some more cost-effective than others. The controller should also work closely with sales managers, making certain that time rates generate apt profit margins while maintaining ad efficiency for clients and that sales staffers follow ratecard policies. Likewise, the controller should monitor promotion and publicity costs without stifling those key areas. Another challenge is to maintain proper balance of expenditures for improving technical facilities amid constant advances in technology. Profitability demands updated equipment to sustain competitive position in the market, emphasizing "pay-off" impact on viewers' TV screens and speakers. The controller with the chief engineer must forego acquiring expensive hardware and software merely to match rival stations' envied inventories.

BROADCAST ACCOUNTING

Station and cable accounting has become sophisticated over the decades with precise cost analyses and projections. Computer technology permits refined processing of financial data with spreadsheets and programming software. This speeds up handling monthly profit and loss (P&L) and balance sheet data, making possible "what if" scenarios to project effects of alternate spending plans; they also develop graphic presentation materials for management to use with corporate headquarters, investors, and clients.

The Institute of Broadcasting Financial Management (IBFM), organized in 1961, became the Broadcast Financial Management Association (BFM) and expanded in the 1990s to Broadcast & Cable Financial Management (BCFM). It is an association of financial personnel and managers in the broadcast industry devoted to developing progressive concepts of financial management. BFM published *Operational Guidelines and Accounting Manual for Broadcasters,* compiling major analyses and sample forms relating to incomes, expenditures, audits, taxes, personnel, OSHA, EEO, insurance, benefits, payroll, computer processing controls, trade-outs and barter, account collection, record retention, and cable accounting, plus analyses and sample forms for financial statements, lease contracts for films and syndication properties, and other financial procedures.[6] In 1972 the original IBFM established the Broadcast Credit Association to act as a centralized source for agency credit information and related problems such as collection of delinquent billings. Theories and procedures of accounting and economics are discussed in other published works, including some that emphasize accounting principles and practices applied to broadcasting.[7]

The business and accounting department is responsible for bookkeeping and billing. It may also be involved with personnel records including payroll, purchasing, storage of station logs and records, supervision of office personnel, and network accounting. A properly kept set of books records business transactions in a systematic way, objectively itemizing expenses and income. From these data, management can determine the status of business, analyzing measurable reasons for profit and loss, attributing success or failure to respective departments and subareas in the operation.

Different stations require varying kinds of bookkeeping records. Typical accounting operations include: *books of original entry* (or journals) for detailed listing of debits and credits—cash receipts journal, cash disbursements journal, sales journal, voucher journal, and general journal; and *ledgers* for regular summaries of the above according to specific accounts—general ledger, accounts receivable ledger, accounts payable ledger, and plant ledger. These key reports provide data for drawing up the balance sheet and the operating statement.

The accounting department reports on unused facilities, idle time of personnel (technical and general administrative employees), economy of operation of various departments, use of supplies such as television tubes and solid-state components plus audio- and videotapes, budgeted and actual costs of programs (locally originated and leased film or videotape properties), and inventory of airtime hours available and commercial availabilities sold. The accountants keep all records for preparing FCC reports, income tax returns, social security reports, fire and insurance claims, and legal actions. To provide management with data for planning profitable station operation, the accounting department draws up monthly balance sheets and the "profit-and-loss" (operating or income) statement.

Balance Sheet

The balance sheet represents at a given point in time how much a business has, how much it owes, and the investment of owners. A business has *assets,* which are the properties of value such as cash, equipment, buildings, and land. *Liabilities* are what that business owes—claims of outsiders on the business and its assets. The difference between the value of assets and the amount of liabilities is the *owner's equity* or net worth (also called ownership interest). Thus the basic balance sheet formula is ASSETS = LIABILITIES + OWNER'S EQUITY. (A simplified balance sheet form, based on the BFMA model, appears as exhibit 10.1.)[8]

PROFIT & LOSS STATEMENT

The "P&L" is of course *either* profit *or* loss over time (typically, a month or a quarter of the year); the more accurate term is operating statement or income statement. This financial record lists a company's activities through that period of time, expressed in dollars. It reports revenues and expenses incurred, showing profit or loss from those activities during the reported period. Whereas the balance sheet compares a company's position and any differences from similar accounting periods, the income statement indicates how those changes occurred. Management can thus determine major areas within the station where expenses exceed the budgeted amounts, or where patterns of revenue vs. expense are shifting from month to month or between the current year and the past year. Bookkeeping records of areas listed in the income statement provide further details on sources of expenses. (See exhibit 10.2 for a simplified sample of a typical P&L/income statement.)[9]

MEDIA ASSETS

All resources of economic value constitute a company's assets.[10] Assets of radio-TV and cable operations differ from other businesses chiefly regarding: (1) kinds of property, plant and equipment, (2) ownership of certain broadcasting rights, and (3) value of a unique kind of "goodwill."

Property. Amounts expended for land on which to place the building, its transmitter, and the tower and antenna system are considered station assets. Costs for voluntary improvements on that land, including sidewalks, parking lots, roads, landscaping, and so forth, are counted as assets, as are all assessments charged against the property which represent permanent improvements.

Plant. The cost of buildings constructed for studios and offices at a station or cable head-end and transmitting site is classified as an asset, as are charges for permanent additions or alterations to those buildings. Stations renting or leasing buildings may count costs of improvements in land or plant as assets.

Equipment. Broadcast equipment normally considered assets includes the transmitter, tower and antenna systems, satellite facilities, cable lines, studio and mobile equipment, office furniture and fixtures, and vehicles.

Broadcasting Rights. All rights to broadcast/cable program materials already purchased but not yet used (that is, still unamortized) are assets. Rights may include public events where admission is charged, as sports or concerts. They also include such program staples as taped features, syndicated TV packages, and film and record libraries.

"Goodwill." When an established station or cable system is sold, the asking price exceeds the operation's total net assets. The difference between the selling price and net assets represents "goodwill" that it merits from such factors as affiliation with a network, existing contracts for future national spot and local advertising, and its image or reputation in the community and among clients. Sometimes a seller will agree not to compete in the area of coverage for a specified length of time; such an agreement can be advantageous to a new owner and may represent an asset. Possession of patents, copyrights, or leases, may likewise be included

EXHIBIT 10.1
XYZ BROADCASTING COMPANY
BALANCE SHEET
AS OF _____
(DATE)

ASSETS	THIS YEAR	LAST YEAR
Current assets:		
Cash	$....................	$....................
Current investment	....	....
Receivables, less allowance for doubtful accounts	....	....
Program rights	....	....
Prepaid expenses	....	....
TOTAL CURRENT ASSETS	$_____	$_____
Property, plant & equipment at cost	....	....
Less: accrued depreciation	....	....
Net property, plant & equipment	....	....
Deferred charges and other assets	....	....
Programming rights, noncurrent	....	....
Intangibles	....	....
TOTAL ASSETS	$_____	$_____
LIABILITIES & STOCKHOLDER EQUITY		
Current liabilities:		
Accounts and notes payable	$....................	$....................
Accrued expenses	....	....
Income taxes payable	....	....
TOTAL CURRENT LIABILITIES	$_____	$_____
Deferred revenue	....	....
Long-term debt	....	....
Other liabilities	....	....
Stockholders equity:		
Capital stock	$....................	$....................
Additional paid-in capital	....	....
Retained earnings	....	....
Treasury stock	....	....
TOTAL STOCKHOLDERS EQUITY	$_____	$_____
TOTAL LIABILITIES & STOCKHOLDERS EQUITY	$_____	$_____

Source: See note 8.

as a part of a station's "good will." Of course, the FCC license (transfer of which is subject to the Commission's approval) or local cable's municipal franchise represents an enormous asset, obviating expensive, time-consuming, and uncertain procedures of filing against rival applicants for a franchise or new frequency.

OTHER ASSETS

Other assets are standard ones for any businesses. They include: cash on hand or on deposit; investments; accounts and notes receivable; expenses already prepaid (e.g., taxes, insurance, rents and expense-account advances); and unused inventories—primarily programming possessed by the station or cable operation.

～～ **EXHIBIT 10.2** ～～
XYZ BROADCASTING COMPANY
STATEMENT OF INCOME

(DATE)

	CURRENT MONTH				YEAR TO DATE		
	Actual				*Actual*		
	This Year	Prior Year	Budget		This Year	Prior Year	Budget
	$..........	$..........	$..........	Gross Revenues:	$..........	$..........	$..........
	....	....	....	Local Time Sales	....	....	....
	....	....	....	National Rep Time Sales	....	....	....
	....	....	....	Network	....	....	....
	....	....	....	Other	....	....	....
	$_____	$_____	$_____	Total Gross Revenue			
				Direct Expenses:			
	....	....	....	Agency Commissions	....	....	....
	....	....	....	National Rep Commissions	....	....	....
	....	....	....	Local Sales Commissions	....	....	....
	....	....	....	Music License Fees	....	....	....
	_____	_____	_____	Total Direct Expenses			
	$_____	$_____	$_____	Net Revenue	$_____	$_____	$_____
				Operating Expenses:			
	....	....	....	Program	....	....	....
	....	....	....	Outside Production	....	....	....
	....	....	....	News	....	....	....
	....	....	....	Technical	....	....	....
	....	....	....	Selling Expenses	....	....	....
	....	....	....	Research	....	....	....
	....	....	....	Advertising & Promotion	....	....	....
	....	....	....	General & Administrative	....	....	....
	....	....	....	Depreciation	....	....	....
	$_____	$_____	$_____	Total Operating Expenses	$_____	$_____	$_____
	_____	_____	_____	Income before Fed'l. Income Taxes	_____	_____	_____
	....	....	....	Provision for Fed'l. Income Taxes	....	....	....
	$_____	$_____	$_____	NET INCOME	$_____	$_____	$_____

Source: See note 9.

■ In 1988 the editors of *Broadcasting* magazine evaluated the TV-cable industry's assets of hardware, programming, plant, and value of subscribers.[11] They figured the thousand TV stations (apart from networks, program suppliers, or syndicators) were worth some $49 billion, which after outstanding debt could bring in the marketplace some $33 billion. They added an estimated $4 billion as the value of the four networks. They appraised the cable industry (after debt of $25.7 billion) at $80 billion for local systems, plus dozens of cable network program services as worth more than $9 billion. The bottom line: approaching the 1990s, cable's total assets more than doubled those of over-air TV broadcasting. In 1991 former FCC chairman Mark Fowler referred to $100 billion of assets in head-ends and coaxial cable distribution "sunk . . . in the ground" by the cable industry; he contrasted that total to reach 62% of U.S. households to an estimated $2 billion invested by the direct broadcast satellite (DBS) industry to reach virtually all 93 million homes.[12] Foreshadowing the heated contest to come among major players were 1988 valuations: TV broadcasting's assets of $40 billion earned revenues that year estimated at $21 billion, while cable's $90 billion assets earned $17 billion; and massive telephone companies ("telcos") held assets worth $240 billion as they earned $90 billion in revenue. That formidable giant, plus proposed DBS services, appeared over the horizon to challenge both traditional over-air TV and cable. ■

An investment firm in 1995 estimated broadcasting's total assets (for 1993, the latest year data were available).[13] It differed widely from the earlier evaluation: broadcast networks were far higher at $14.2 billion, but assets of radio and TV stations were much lower at $16 billion—perhaps because of group owners' heavy debts in acquiring them.

Amortization and depreciation are accounting procedures that spread out costs of major purchases over a period of time. Those costs can be partly recovered through business tax planning that annually deducts from income a percentage of the costs over the life of each asset. In other words, federal law stimulates companies' investments by permitting them to deduct from income taxes that portion of an asset's cash value which remains yet unused in their inventory. A business *amortizes* intangible assets that retain their value through time, such as employees' contracts, on a "straight-line" basis equally over each year of the item's life (some items from five to eight years, but none to exceed forty years).[14] A business *depreciates* tangible physical assets that wear out over time like plant, buildings, and equipment—either "straight-line" or else "accelerated" whereby more is deducted at the beginning when an item's value is higher, progressively less in each subsequent year.

An early study found average life expectancy for studio equipment and office furniture came to slightly under nine years. Only about one-fifth of radio managers planned to replace studio equipment before it fully depreciated. Depreciation write-offs pose a problem for broadcasters due to rapidly increased prices for broadcast equipment, coupled with IRS revisions of depreciation schedules. Some managers prefer a fast depreciation rate to replace station equipment prior to its obsolescence. Other managers prefer a slower rate, retaining their equipment for the full depreciation schedule or longer.

Media managers need to consult with competent tax advisers to determine lengths currently allowable by the Internal Revenue Code. IRS schedules list depreciation ranges for various items, such as 10 to 30 years for real estate, 20 years for structures, 8 years for furniture and fixtures, 5 years for technical equipment, and 3 years for autos. A heavy expense is long-term licensing rights to syndicated programming, which must be handled properly to comply with the tax code.

Media Liabilities

A firm's liabilities are what it owes to creditors. Ordinary liabilities include: amounts due for prior purchases of goods and services; any collections made in advance for station services not yet delivered; unpaid balances on long-term notes, mortgages, bonds, or other debts; dividends declared but not yet paid; tax monies due and all miscellaneous accrued expenses.

The category of miscellaneous accrued expenses includes various obligations due at some future date. Normally, these encompass such items as: employee salaries, wages, commissions and other benefits; federal and state income taxes; sales and use taxes; interest due on notes, mortgages, and so forth; and materials received and used in trade or barter arrangements for which airtime "payments" have not yet been made.

SOURCES OF INCOME

In the total media mix in 1996 (described in chapter 9), of $102 billion national ad dollars, newspapers and magazines attracted $13 billion; four networks generated the same amount; spot TV $10 billion, with another $5.3 billion going to cable and syndicated barter TV. Of ad expenditures among local media in 1993, $29 million went to local newspapers, while only $8.6 million was spent for advertising on local TV and $7.2 million on local radio spots. Broadcast managers need to compete not only against one another's stations but must also aggressively market themselves against print media (including direct mail) that draws off heavy portions of total advertising dollars.

Broadcasters earn revenue from selling airtime for network, national/regional spot, and local advertising; cable systems acquire almost all their income from subscriber fees, the remainder from advertising. Secondary sources of income include fees for talent services, recorded materials, rental of facilities, and merchandising activities, plus interest and dividends from investments.

Table 9.5 in chapter 9 plotted the growth of various categories of time sales for broadcast and cable television. In the decade and a half up to 1995 the nation's TV stations almost tripled both their local sales to $10 billion and national/regional spot to $9.1 billion. TV networks more than doubled ad revenues from $5.1 billion to $12.4 billion. During that time span, national and local cable advertising revenues (only a fraction of their income from subscriber fees) leaped from $58 million to $5.3 billion. Total revenues in radio almost tripled to $11.5 billion—about half a billion of those ad dollars to radio networks, the rest to local stations.

In 1995 TV station time sales were about evenly split between local advertising and national/regional spot ads, with 3% from network compensation. Those figures differed widely from the early 1980s when national/regional spot dominated with around 55% of all time sales, local about one-third, and network compensation one-tenth of a station's revenue. In 1995 time sales by radio stations were almost 80% local advertising, the rest national/regional spot, except for 1% of revenue from network compensation.

Looking at the television scene in 1993 (table 10.1)—with 221 more TV stations since 1982, most of them independents*—the three networks' $7.8 billion in sales revenue accounted for less than a third of all TV advertising volume, while national/regional spot and local sales were about equal at around $5.5 billion each. Syndication posted massive gains, generating over $4.5 billion during the year—approaching one-fifth (17%) of all TV sales revenue. Cable network ad revenues climbed to almost 10% of the total.

■   Local cable systems in the early 1990s earned about 95% of their revenue from monthly fees paid by cable subscribers. The remainder came mostly from local advertising, plus some national/regional spot time sales. The cable industry's total revenues from advertising in 1991 were $3 billion, having tripled in five years; 70% was sold by cable networks, the rest ($761 million) came from national and local spot sales, including regional sports advertising. That same year consumers paid $11.7 billion in

---

*In 1982 a total of 203 independent stations served 95 markets, reaching 80% of all U.S. households; by 1991 independents had grown to 424 stations in 151 markets, serving 95% of U.S. homes (source: INTS/Association of Independent Television Stations, reported in *Electronic Media*, 14 September 1992, 48).

—ᴡᴡ— TABLE 10.1 —ᴡᴡ—
Pʀᴏᴘᴏʀᴛɪᴏɴ ᴏғ Aᴅᴠᴇʀᴛɪsɪɴɢ Rᴇᴠᴇɴᴜᴇs ʙʏ Tᴇʟᴇᴠɪsɪᴏɴ ᴀɴᴅ Cᴀʙʟᴇ, 1973–1993
($ ɪɴ ᴍɪʟʟɪᴏɴs)

	1973	1983		1993	
Three networks	46%	41.1%	$6,955	29%	$7,880
National spot	31	28.7	4,827	21	5,618
Local spot	21	25.8	4,345	21	5,565
Syndication	NA	1.8	300	17	4,633
Cable networks	—	2.0	331	10	2,670
Local cable	—	0.4	60	2.5	600?

Sources: Data derived from various accounts, not always wholly comparable in line items, by Television Bureau of Advertising, Cable Television Advertising Bureau, Paul Kagan Associates; as reported in *Channels/Field Guide 1989*, 22; *Broadcasting & Cable*, 7 March 1994, 50; 16 May 1994, 6; *Cable World*, 28 March 1994, 3.

subscription fees for basic cable and $5.3 billion for premium-pay cable TV services, plus another $3 billion in fees for installation, remote controls, and second sets; that total of over $20 billion exactly doubled subscription revenues of five years earlier.[15] Average monthly cable rates for basic service more than doubled from 1980 to 1991—from $7.80 up to $17.95—while pay channel fees rose modestly from $8.80 to $10.28 per channel. In 1993 the 21 major basic-cable networks attracted $2.43 billion in ad revenues, up 17% from the previous year; added revenues of 25 smaller program networks brought that total to an estimated $2.67 billion. ∎

The investment firm of Veronis, Suhler & Associates in 1993 projected a half-decade (1992–1997) of compound growth in revenue of almost 12% for cable advertising, 3% for cable subscriber fees, about 5% for radio and TV network revenues, 6%–7% for radio and TV stations, 6% for TV programming, and almost 9% for barter syndication (table 10.2).

Noncommercial radio raised funds increasingly from individual listeners' subscriptions and corporate gifts, and less from state and federal allocations (the latter had begun in 1962). Between 1975 and estimated levels for 1995, state funding for public radio dipped from 43% of all income to 25% while federal funds dropped from 25% to only 9%. Audience contributions became the largest source of financial support (30% of all funds) followed by corporate grants that rose from 7% to 21% of revenue; philanthropic gifts dropped from 8% to 4% of public radio's finances. All public broadcasting—noncommercial radio and TV networks and stations—in 1991 generated revenue almost equally from private contributions (51.4%) and tax-based funding (48.6%, marked by boldface type):[16]

21.0%	Private viewer/listener subscription
16.9	Business underwriting (including "non-ads")
16.2	**State government**
14.0	**Federal government (via Corporation for Public Broadcasting)**
8.4	**State colleges**
5.3	**Federal government (other)**
4.1	Foundation grants
3.6	**Local government**
1.5	Private colleges
1.2	Auctions
1.1	**Other public colleges**
6.6	All other sources

TABLE 10.2
Compound Annual Growth of Media Revenues
Projected 1992–1997
($ in millions)

	1992	CAG*	1997
Cable advertising	$ 2,155	11.7%	$ 3,750
Cable subscribers fees	17,520	2.9%	20,200
TV networks	10,150	5.4%	13,175
TV stations	15,630	5.9%	20,850
Radio networks	424	5.3%	550
Radio stations	8,212	6.8%	11,435
TV programming	7,705	5.9%	10,280
Barter syndication	1,440	8.6%	2,175

*CAG = compound annual growth.

Data reproduced from Veronis, Suhler & Associates "Communications Industry Forecast" released in July 1993 in association with Wilkofsky Gruen Associates, in "Five-year Media Forecast," *Broadcasting & Cable,* 19 July 1993, 46.

In the mid-1990s the federal government budgeted around $300 million in annual funds to the Corporation for Public Broadcasting for distribution to 350 public TV stations and 600 public radio outlets.

More than their sales manager counterparts at commercial operations, directors of development at public broadcast stations must blend subtle appeals with direct marketing savvy to raise funds that meet noncommercial radio-TV's considerable expenses. It was estimated in 1994 that about 1% of public television airtime was given over to fund-raising drives, contrasted with 21% of airtime for ads on commercial television.

CATEGORIES OF EXPENSES

Networks.   Program costs—license fees to outside suppliers plus internal production expenses and salaries—account for almost half of all TV network expenses. By 1987 entertainment, news, and sports programming consumed 40%–50% of all costs by the entire CBS/Broadcast Group—including TV and radio networks, owned and operated (O&O) radio and TV stations, and all other divisions such as marketing and administration.[17] That year the three broadcast networks spent an estimated $5 billion on programming (compared with cable networks' $1.7 billion for program costs). Next highest was network compensation payments to over 200 affiliated stations each. Relatively less went to administrative costs and agency commissions, payments to owned stations, technical and selling costs.

■   In 1980 when each network was spending almost $40 million every week for prime-time programming alone, those three networks paid $310.9 million in compensation to their 600+ affiliates plus $115.5 million to their 15 O&O stations. Compensation rose through that decade, peaking in 1989 at around $475 million, then plummeted back to $415 million in 1991 amid recessionary cut-backs,[18] coupled with diminishing station clearance to carry network programs (down to 90% in the early 1990s, but

Financial Management   395

returning to only occasional pre-emptions by the mid-1990s).* In 1994 Fox raided affiliates from other networks with bountiful affiliation contracts and by outright purchase. That prompted the Big Three to rebuild compensation and offer contracts beyond the usual year or two to as many as ten in some instances; ABC's five-station affiliation contract with Scripps Howard Broadcasting called for $10 million annually for ten years, and NBC wooed Boston's WHDH-TV away from CBS by offering $15 million a year in compensation. Meanwhile, in the 1990s station compensation payments continued to be the largest line item in radio networks' budgets; some top affiliates earned nearly $1 million in annual compensation. ■

Cable's expenditures for programming rose steadily from $1.6 billion in 1984 to over $3.7 billion in 1992. (About 10% of license fees were passed on to actors and creative crews as residuals; that contrasts with up to 20% of off-network syndication rates that go for residual payments.)

As for noncommercial TV expenses, in fiscal 1992, $301 million of public and private funding was earmarked for PBS programs.[19] Public TV programming at the network level cost $115 million in fiscal 1995 (July 1, 1994–June 30, 1995), about one-fourth of its budget; the rest was passed on to local stations for their production and operating costs. But member stations returned a large part of that money to PBS as assessments for receiving national programs, which totaled $87.5 million in 1995. National Public Radio network's budget increased to $47.6 million in 1995, when its staff numbered 471. Two-thirds of revenues came from member station dues—each station paying 10% of its own revenues for NPR's news package and another 2% for cultural programming.

Stations.   Station expenses are usually budgeted under four categories: programming, engineering, sales, and general and administrative.

In the 1990s almost half of TV station expenses were associated with programming and production, including news. Affiliate stations budgeted 24.7% to entertainment programs and production and 20.3% to news, while independent stations allocated 51.5% and 5.3% respectively. General and administrative costs accounted for almost a third of annual expenses; about 8% went to sales plus 5% to advertising and promotion; technical engineering expenses were about 7% of TV station budgets (see table 9.1, chapter 9).

Radio stations budgeted the lion's share (over 40%) of budgets to general and administrative expenses. They devoted around 22% of expenses to programming and production, plus roughly 4% to news. Sales received just under one-fifth of budgets, advertising and promotion one-tenth, and engineering 4%. (Percentages are approximate because in each category they vary widely for AM daytimers, fulltime AM, AM/FM combos, and stand-alone FM operations.)[20]

Local cable systems invest heavily up-front in capital expenditures for laying or stringing cable throughout their distribution area. Costs per mile in a rural area run over $10,000; in urban areas costs can reach $100,000 per mile and more than $300,000 per mile when cable must be buried underground (that means not only dirt trenches but pulling up and replacing sidewalk and street surfaces). Meager if any local origination programming requires only modest outlays for studios and other equipment. As the 11,000 systems laid cable past 90% of all U.S.

---

*Affiliate general managers surveyed across the country said "likely" or "very likely" reasons for pre-empting network fare were primarily to carry special event programming (91.9%) or local programs (78.5%); just under half (48.1%) did so to avoid offensive programs, and only one in five (20.7%) pre-empted to avoid poorly rated programs. Carolyn A. Lin, "Changing Network-Affiliate Relations Amidst a Competitive Video Marketplace," *The Journal of Media Economics* 7:1 (1994): 1–12.

households, they reached apparent saturation level with two-thirds of all homes hooked up. The next phase was to expand the customer base by investing money less in capital plant extension and more in developing expensive new programming to attract and hold subscribers. Most capital-intensive cable-laying was complete (until coaxial cables were replaced by fiber-optic to accommodate hundreds of channels and interactive service). Those massive initial construction costs translated into enormous depreciation that sheltered cable's subsequent annual cash flows. However, some saw the FCC's mandated lowering of rates in 1993–1994 as crimping revenues to a point where some smaller systems might default on their eight-to-nine-year loan payments.

■   Local systems paid national program services according to the number of subscribers. Average cost to operators in 1992 was 50¢ per subscriber for ESPN; 23¢ for CNN; 18¢ each for USA and VH-1; 17¢ each for Nickelodeon/Nick at Nite and MTV; 15¢ for TNN; 12¢ for the Discovery Channel; 10¢ each for Arts & Entertainment/A&E and Financial News Network; 5¢ for superstation [W]TBS; and 4.5¢ for C-SPAN.[21] In the early 1990s program expenses included compulsory license fees for importing distant broadcast signals; by 1991 cable systems had paid over $200 million, estimated at 36¢ monthly per subscriber (although complex formulas determined several strata of royalty fees for various kinds of systems).[22]   ■

All mass media companies are labor intensive; personnel require the heaviest outlay on funds on a continuing basis. At the outset, physical plant and licensing syndicated programming are one-time or sequential costs, whereas salaries are continuous. Executive staff, department heads, and especially on-air newspeople and personalities command high salaries. Managers must often negotiate personnel contracts with agents who represent major talent; and managers in larger markets and networks must also negotiate with some forty different unions representing technical and creative personnel about compensation, including benefits and working conditions.

Normally a station's largest operating cost, salaries and wages consume as much as half the total budget. This includes regular payroll plus all overtime pay and costs of paid vacations and holidays. Added are payments for employee benefits, including social security, unemployment insurance, and workers' compensation. Some stations provide further benefits of profit sharing plans, bonuses, group insurance, hospitalization, and retirement pensions. (Fringe benefit packages can amount to almost another one-third of base salaries budgeted for personnel.) For the convenience of employees, systematic payroll deductions may be processed for purchasing U.S. Savings Bonds or for donations to organizations such as a united community fund. The added cost of administering that service can be regarded as another station expense.

Other major expenses include programming and news costs other than salaries, royalties and license fees, technical equipment, depreciation, sales-related costs, and enormous utility and overhead charges clustered under the general administrative grouping.

PROGRAMMING EXPENSES

The program department usually is the most expensive station unit in terms of salaries, wages, and benefits, especially when news anchors and staffs are included. Program-related expenses include the cost of licensing rights to air syndicated programs and events, music license fees, recordings and transcriptions, news wire services, film and video rental, shipping costs, sets and props, costs of remote origination including line costs and leased satellite time, and other miscellaneous expenses for producing and airing programs.

Theatrical films and syndicated video series leased in package lots are amortized over the period when all episodes in the package are licensed to be shown (usually four to six years with 5 or 6 re-runs contracted for a price set per episode).[*] The Internal Revenue Service's Ruling 62–20 provides that broadcasters must charge film/video contract expenses equally over each year of the contract life, although that is contrary to usual accounting practices in a TV station. The IRS permits sliding-scale amortization of charge-offs in some instances, provided the broadcaster can substantiate clearly the progressively decreasing revenues generated by the program package over several years' exhibition.[23] For example, stations often write off a film with contract rights to four presentations by amortizing 50% for the first airing, 25% for the second, 15% for the third, and 10% for the fourth showing.

Local cable systems pay monthly fees based on number of subscribers to suppliers of national cable program channels (as described in chapter 8 on national programming). Charges may range from a few cents per subscriber to 15¢ or 20¢ or more, as described earlier. Systems with 80,000 subscribers, for example, might pay one channel service some $12,000 (at an average of 15¢ per subscriber) each month, or $144,000 a year. Systems carrying 25 channels, in addition to relatively free retransmission of half a dozen over-air broadcast signals, would need to forward over $3.6 million dollars annually to program suppliers. Of course, those 80,000 subscribers' monthly fees of $20 or so for basic nonpremium service would bring in $1.6 million each month or $19 million a year to the local system.

Music Licensing

A major operating cost in broadcasting is music license fees for performing rights. In 1989 dominant radio stations in largest markets paid an average of over a quarter million dollars each for annual "blanket" music license fees. Stations in other size markets paid from around $5,000 (population 25,000) to $27,000 (250,000–500,000 population) to $70,000 (markets of 1 million–2.5 million).[24]

The business of music licensing is controlled in the United States by three organizations. The American Society of Composers Authors and Publishers (ASCAP) is the largest and oldest, formed in 1914. Broadcast Music, Inc. (BMI) was created in 1940, when broadcasters were restricted from playing any music not in the public domain because they balked at paying ASCAP's increased fees. BMI was created and financed by the broadcast industry as a source for new compositions. Broadcasters and ASCAP resolved their dispute after some ten months, but by then BMI had become a successful enterprise. Each controls far more than a million titles in their repertories. Originally formed in 1931 as the Society of European Stage Authors

---

[*]For example, a package of 120 episodes of an off-network sit-com series might be licensed to a local mid-sized market station at $2,000 per title (x 120 = $240,000 contract, 10% paid up front, the rest in 30% installments over the next three years), with rights to run each episode up to six times within a five-year period. When the contract expires, the package reverts to the syndicator to be renegotiated with the original station for another leasing period or for resale to another station in that market.

and Composers, SESAC is smaller than the other two. Although originally representing European works, SESAC is now based in the United States and its repertory includes American creations.

The licensing organizations have found themselves in conflict with broadcasters down through the years. Per-play fees vs. blanket licenses, rates for performing or playing recordings of music (such as a percentage of a station's net revenue), and other contentions have kept attorneys and courts as well as broadcasters and musicians busy with cyclic negotiations over the last half century. A decade and a half of continuing litigation sought to determine how much TV stations owed ASCAP for eighteen years ending in 1995. If all TV stations used the blanket formula, they would pay an estimated $72 million annually just for ASCAP titles. In recent years some 260 stations saved $6 million a year by using a per-program license instead.

By 1993 all TV stations paid music licensing fees to ASCAP and BMI totaling $110 million. ASCAP charged radio stations a blanket fee equivalent to about 1.6% of their adjusted gross revenue, or tiered flat blanket fees for smaller stations—$450 for stations billing less than $50,000 annually, $800 for those billing $50,000–$75,000, to a top fee of $1,800 for stations billing $125,000–$150,000; an alternate option was a per-program license fee equal to about 2.1% of revenue earned during hours when a station played music. SESAC's fees were based on a station's highest 60-second rate and the size of the city of license.

ENGINEERING EXPENSES

In addition to payroll for supervisory and staff employees of the engineering department, other expenses include major upgrades of technical equipment, and such items as tubes for the transmitter, studio and remote units; audio and video tape; rental charges for transmitter lines; and parts and supplies used to maintain technical facilities.

■ A 1991 survey found that half of all GMs and chief engineers decided together on equipment purchases, while 16% of chief engineers only recommended specific buys for the GM to decide on. Stations in top-10 markets budgeted just over a million dollars each for new equipment, while those in markets 100+ spent a quarter million each. The average for all stations was $470,000; VHFs averaged about $580,000, twice that of UHFs. Affiliates averaged $407,000 for new equipment that year, independents $684,000, and public TV stations $501,000. Despite a weakened economy and sagging revenues for stations that year, U.S. TV stations (not including networks) budgeted more than $689 million in 1991 for replacing and upgrading equipment. More than one-third of chief engineers reported their stations occasionally leased equipment. ■

SALES EXPENSES

Sales commissions and expense accounts are added to the usual costs of salaries, wages, and benefits in this department. Other normal expenses may include: public relations, advertising and promotion directed to the audience; trade advertising and promotion directed to the industry; commissions for the station rep; and the cost of rating services.

In the mid-1980s sales accounted for 10% of TV station budgets, dropping to 8.5% in the early 1990s; another 5% went to advertising and promotion over that time. Radio during that period continued to budget 20% to sales, plus around 10% to advertising and promotion (except AM daytimers budgeted less than 4%).

This is often the largest category in broadcast budgets, 30%–40% or more, because it covers a wide range of expenses within the company, affecting all departments. A movement in the 1990s to consolidate multiple station operations through local management agreements (LMAs) and duopolies brought greater efficiencies by joining two or more local stations' department staffs, eliminating redundant positions. Many jobs in accounting, billing, and other business administration could be eliminated from newly acquired operations. This was a factor in the FCC's loosening restrictions on owning multiple local stations, especially small ones that were highly unprofitable. An internal FCC report in 1992 noted that "if a conservative 10% of general and administrative costs could be eliminated, for example, the savings would raise industry profitability by 30%."[25]

The administrative budget includes salaries and wages of the station manager and all other employees not assigned to sales, engineering, or programming (including news). Additional expenses include some items associated directly with management and other items used by all departments but expenditures for which are controlled by the station's central administration. These include: maintenance and repair of buildings and equipment; heat, air-conditioning, light and power (enormous monthly charges for generating transmission signals); rents; telephone and telegraph; postage; stationery and office supplies and services, including computers; travel and entertainment; membership fees and dues; subscriptions; operating costs of station-owned automobiles and trucks; real and personal property taxes; state and local taxes; insurance; station contributions to charitable, educational, religious, and welfare organizations; and legal, auditing, and other consultant fees. A major component of "G&A" are nonoperating expenses of depreciation and amortization, which are counted as expenses in computing profit before taxes but are excluded when figuring "cash flow" in broadcasting.

See table 10.3 for representative radio and TV station financial figures.

THE MANAGER AND PROFIT

In the early 1990s annual losses were reported by more than 13% of commercial VHF television stations, over half of UHFs, and almost two-thirds of commercial radio stations (especially AM). It is obvious that managing finances, central to running a broadcast or cable operation, must be improved in many instances. One ultimate consequence of neglecting to improve is a reduction in the number of radio and television stations. As noted earlier, hundreds of radio stations went silent while others were submerged into joint-operated LMAs or bought out to become part of duopolies; a number of TV stations (mostly UHF) had also gone dark as the mid-1990s approached.* Problems can arise from an excess of radio stations; it may be true that a larger number has been authorized than can be supported profitably. Any station, however, that has been authorized and in operation deserves

---

*In two years, 1989–1990, the FCC's Mass Media Bureau reported 227 stations went off the air: 175 AM and 17 FM radio stations, plus 32 UHF and 3 VHF television stations. Dun & Bradstreet listed bankruptcies and reorganizations annually from 1990 through 1992 as 50, 88, and 123 for radio and 11, 23, 21 for television stations; in the first half of 1993, 23 radio and 9 TV stations filed for Chapters 7 or 11 bankruptcy.

~~~ **TABLE 10.3** ~~~

REPRESENTATIVE FINANCES AND BUDGETS, RADIO AND TV STATIONS
(AVERAGE OPERATION, MID-SIZE MARKETS: RADIO, 250,000–500,000 POPULATION;
TV, ADI MARKETS 41–50)

SOURCES OF REVENUE	RADIO		TELEVISION	
National/Regional Advertising	$ 235,790	19%	$ 7,120,876	47%
Local Advertising	980,390	79	7,423,892	49
Network Compensation	24,820	2	606,032	4
Total time sales	1,241,000	100%	15,150,800	100%
Agency and Rep Commission	148,920		2,272,620	
Total Net Revenue	$ 1,092,800		$ 12,878,180	
DEPARTMENT EXPENSES				
General and Administrative*	$ 423,550	43%	$ 4,051,048	37%
Sales	216,700	22	985,390	9
Program and Production	197,000	20	3,394,122	31
News	29,550	3	1,313,854	12
Advertising and Promotion	78,800	8	547,439	5
Engineering	39,400	4	656,927	6
Total Expenses*	$ 985,000	100%	$ 10,948,780	100%
Pre-Tax Profit	$ 107,800		$ 1,929,400	
Nonoperating Expenses (included in G&A)				
Depreciation and Amortization	$ 116,000		$ 1,320,250	
Interest	103,500		1,106,500	
Cash Flow**	$ 327,300		$ 4,356,150	

*Includes depreciation, amortization, and interest.

**Computed as difference between revenues and expenses, before taxes, interest, depreciation and amortization expenses.

Composite data based on multiple sources, including NAB/BCFM's annual *Radio Financial Report* and *Television Annual Report,* early 1990s.

a chance to survive. That survival—assuming good programming, technical, and economic practices—depends on effective station management.

Some managers, faced with continued financial loss, show a tendency to panic. They decide to accept any types of commercial accounts, cut their rates to get some semblance of business, program as cheaply as possible without regard to quality standards, cut veteran personnel to a minimum and then hire the cheapest people who can be found, get by with equipment past its lifespan of maximum utility, oppose any improvements in station facilities, cancel their membership in state and national professional associations, and cut all budgetary items to a point that permits only minimal operating standards. Such action may produce some immediate financial improvement but it endangers the reserve potential of the station. Then, when further budgetary cuts may be necessary, no effective retrenchment is possible.

The manager needs to maintain a financial or operating reserve adequate to meet possible future emergencies. That reserve need not necessarily be in cash or securities. It may be created by prepayments of certain station obligations. It may be in the form of manpower reserve. It may consist of some station services that can be eliminated in a crisis. There are various other possibilities. Maintaining a financial reserve is important to any manager (on behalf of the owner-licensee) because increased competition can force him to draw upon it in the future. By way of contrast, there have been managers who, when confronted with almost certain financial losses, have expanded rather than retracted budgets. They increase station expenses in order to become more competitive. Choosing not to make radical overall budgetary cuts, these managers eliminated waste and inefficiencies causing stations to suffer financial losses. They have accepted the possibility of even greater losses over a short period of time so their stations could become effective competitors in the future. These managers have had the courage to defend their strategies to their owners or stockholders. They have been fully prepared to find other employment should station owners disagree with their policies. In most cases, there has been no need for any change in employment. Instead, the vision, integrity, and strength of these managers have been highly regarded by station owners. In almost every case, their stations rebounded to become solid competitors.

Cost Controls for Greater Profits

A station can implement various economies without sacrificing either quality or standards. Constant and careful attention to cost controls can effectuate real savings. Many relatively minor budgetary items can turn into excessive costs due to wastes, inefficiencies, or extravagances. To achieve economy, the administration of expenditures requires constant vigilance. Little is accomplished if the action is sporadic. Supervising cost controls must be a daily process until it receives the cooperation and enthusiastic support of the entire staff. It is said that the test of real executive leadership comes when a company's major objective becomes reducing costs without diminishing quality. An effective manager can produce real savings without losing employees' cooperation or enthusiasm, and can even cause both to increase.

Mikita distinguishes appropriately between "cost reduction" and management's goal to establish an atmosphere of "cost awareness." Cost reduction connotes that reductions are to be made regardless of their impact on the overall operation. But the concept of cost awareness is entirely different; it states merely that at some stage of developing a project or campaign, or in considering new facilities equipment, the measure of profitability must be applied.

When allocating budgets, it is especially important that the manager and the controller possess enough ability and experience so department heads respect their understanding of detailed station operations and appreciation of specific problems in those departments.

Major ways for a manager to strive for more efficient cost control, beyond exercising leadership in motivating all employees: monitor manpower to utilize it effectively; work towards better union agreements; upgrade apt supervision to improve employee performance; encourage employee participation; improve working facilities; reduce costs of services; especially maintain budgetary controls, determining standards of cost operation and analysis; and, of course, increase revenue from sales.

Staff and Working Facilities

Employees can abuse the "coffee break" or use the telephone for personal conversations; but far more important is developing a climate of work where every staff member's full potential is realized. This subject is so important that chapter 5 was devoted to personnel practices and motivating people to accomplish those creative results of which they are capable.

Investments in improved working facilities often produce returns far greater than their cost. Buildings originally designed for other purposes and then transformed to serve as radio or TV stations or cable head-ends often use space inefficiently. Rearranging the physical layout and facilities can result in more productive efforts by the staff. Locating the newsroom and selected office staff closer to studios can improve traffic flow and increase work output. Repositioning filing cases, photocopy machines, computer equipment, or desks can result in greater productivity. Some studio and control room arrangements limit engineers' attention to one activity when they should be able to oversee several. Offices can become so overcrowded they hamper good work. Department heads should regularly assess how space is utilized.

Mechanized/Electronic Systems

Business machines and computers are not restricted to accounting. This technology can handle all routine traffic from the time a spot is sold, through its logging into the daily schedule, its air play, and to its eventual billing.

High-tech equipment is not a panacea in itself. Where control is weak or inadequate, computerization only highlights defects in the existing system. Inefficiencies cannot be eliminated by pushing a button that starts a machine. In a properly organized operation, computers offer swift, accurate processing of highly complex data, including instant interaction with agencies, station reps, and ratings services. Computer programs designed for station and cable operations produce all kinds of reports of basic data analyses, including audience break-outs, avails inventory, and operating logs. Commercial cards can be sorted (a) by agency code to prepare invoices, (b) by salesperson code to prepare commission reports, and (c) by day and time codes to obtain daily or hourly breakdowns of revenue, including break-outs of revenue by products. Computerized spreadsheets and relational databases are particularly valuable for forecasting revenues and expenses.

Sales Revenues

Astute planning and sales staff collaboration can generate increased sales volume, so a radio or TV station or cable operation achieves its fullest potential in a market. Enhanced revenue contributes to upgrading the entire operation. A small-market independent radio station in the Midwest more than doubled its billings by carefully analyzing the "greater market" of its full signal area, by systematic scheduling of fully paid-for spots (no discounts, no rate cutting, and even no free airtime for public service announcements—content of which was covered in local newscasts and commentaries).

While emphasizing service, news, and information programming with community and area involvement, Jerell A. Shepherd, owner-manager of KWIX, Moberly, Missouri, claimed that "we are more of a sales organization than we are a programming operation." He doubled billings by determining concrete, specific, and demanding sales goals to be reached periodically. The initial goal was to make no changes whatever except in the total sales effort. He reversed the normal professional sequence: "We get the sales, and then we get better programming. . . . If you've got the money coming in, programming is all set. You can have the mobile news crews; you can have as many telephones as you want; you can have all the equipment, a capable staff—the works." He summarized his staff's sales success:

It was built by no magic formula nor because the community was any more progressive than any others. It was built by hard work and a determined effort. It was built on the development of a fine sales staff and a fine programming staff. It was built by plowing back much of our earnings into programming after we achieved our earlier sales goals.[26]

While parts of his policy might be debated, success in building sales revenues ultimately enhanced the quality of program service.

A particular concern of the manager is the problem of delinquent accounts—often a small but significant percentage of total business. Bankruptcy proceedings by a few advertising agencies in the early 1970s focused on poorly managed procedures for extending credit, resolving discrepancies, and determining liability for default on payments. Three national organizations able to assist management in shaping policy and procedures in this matter are the Credit Association for Radio and Television, the Broadcast Credit Association, and Media Payment Corporation. For example, a survey by the latter group found that 89% of responding radio-TV stations favored dual liability for advertisers and agencies in payment for broadcast advertising.

BUDGETING

Budgeting annual expenditures is standard practice in any business. Problems of budgetary control usually can be solved by including department heads in budget planning and by using administrative follow-up procedures at regular intervals during the budgeted year.

Budgets should not be prepared entirely by management if they are to be realistic in terms of departments' needs. The manager, of course, gives final approval to all budgets and must exercise the right to cut any requests deemed unnecessary; but every department should be given full opportunity to propose. Each department head should be made well aware of corporate budgetary goals and the importance of departmental cost controls in achieving those goals. Department heads and the station controller or accountant should meet weekly or at least monthly to review actual expenditures and each department's budget balance. Through such conferences, it is possible to curb cases of overspending and to effect some savings.

It is a wise policy for the general manager and department heads to review at regular intervals annual financial reports, breaking out categories of the industry's expenses and sales revenues—formerly provided by the FCC, and then by the National Association of Broadcasters and Broadcast Cable Financial Management Association, and by the Department of Commerce. From these detailed reports, station executives can gain insights into financial practices of similar stations in various categories of transmission power, market size, and competition.

A manager and her staff must make decisions about plant and facilities, such as the advantage of leasing rather than purchasing some kinds of equipment. They must carefully draw up estimates to determine when it might be profitable to expand studio, office, or technical capacity; such capacity must closely relate to programming and engineering costs. The volume of programming or production services (commercials, promotional productions or packaged shows) should relate to the operation's real capacity for output in terms of space, facilities, personnel, and overhead. Hoffer urges managers to analyze carefully such questions as: Will cutting departmental operating expenses necessarily result in a higher net operating income?[27] Will investing more money into programming and facilities be reflected in greater sales (more time sold, at higher rates)? How do current estimates compare with previous years? What rate of return is expected by station owners or company stockholders? Will predicted net operating incomes affect the station's tax status? Have reserves been provided to protect against economic recession? Are there any major projects in the future, such as building new studios or acquiring additional properties, or upgrading technical equipment? Are competitors planning any major expansion? Budgeting necessarily involves projecting estimates of future revenue and expense, sometimes even forecasting for five- and ten-year periods. This sets goals and parameters for business expressed in gross and net figures. Projections must reflect trends in the general economy, fluctuations in the local market's business and population, changes in costs for equipment and salaries and services, expansion of plant and facilities, increases or cut-backs in payroll reflecting cost-of-living scales and merit as well as increased efficiency (particularly if automated equipment is involved). Increases in airtime rate schedules and services must be anticipated to compete with other broadcasters and with print and other advertising media. Management must acquire sound revenue data for these forecasts because business planning can be harmed as much by overestimating as by underestimating.

RADIO INDUSTRY PROFITS

The changing context of radio over the decades caused the industry's profitability to swing widely from highs of 11% of revenues in the late 1960s to lows of 3.5% in 1981, 2.5% in 1985, and less than 1% in 1991—punctuated by intermittent highs of 8%–9% in the late 1970s and early 1980s. Radio profits in 1961 had reached their lowest levels since 1939 in the Depression; almost 40% of all stations lost money, more than in any previous year. But two years later radio stations set new records in sales revenues (totaling $681 million) and profits ($55 million, or 8.1% of net revenues). A decade later revenues had more than doubled to $1.5 billion, with an overall profit ratio of 7.4% (AM and AM-FM combo stations earned $123.3 million, while stand-alone FMs lost $10.8 million). But that was down from the previous year's 9.5% profit ratio because revenue increases failed to keep pace with growing expenses. After 1975, stand-alone FM stations as a group finally began to operate at a profit; then they surged ahead of AMs as audiences with FM receivers grew, tuning in to those stations for high-quality reproduction of music. After March 1986 when the FCC eliminated restrictions on duplicating program services on twin-owned stations, AM-FM combos were able to operate more efficiently by cutting back their double programming staffs (the GM of WMZQ-AM-FM in Washington, D.C., saved $250,000 a year on salaries by simulcasting).[*]

Markets became intensely competitive for the listener's ear as the number of stations multiplied from 3,000 to some 11,000 by the 1990s. The overall average profit margins for stations from 1975 to 1990 were 4%–5%.

[*]Up to that March 1986 ruling, some 192 pairs—384 AM and FM stations—in small markets simulcast their entire schedules, with another 508 pairs (1,016 stations) simulcasting 10%–90% of their programming. Within two years, by mid-1988, 715 pairs (1,430 stations) took advantage of duplicated programming. Arbitron figures cited by *Broadcasting*, 15 August 1988, 59.

By 1990 profit-ratios of fulltime stations doubled to 8%–10%, while day-time-only AMs suffered 5% losses (table 10.4a). That partly positive picture changed drastically as the nation's economy dragged down broadcast sales. Revenues and profits sank by two-thirds or more, with all radio categories reporting average losses between $8,000 and $15,000 (table 10.4b). But cash flows remained positive because added amortization and depreciation figures bolstered final tallies. The FCC estimated that in 1990 the top 50 stations in major markets—only one-half of 1% of all stations on the air—earned 11% of all revenues and half of all the indus-try's profits. Three-quarters of all stations had annual revenues under $1 million and on average lost money that year. The FCC counted 287 radio stations that ceased operating, over half of them during the twelve months of 1991 when 59% of all radio stations lost money. In 1992 and again in 1993 almost two out of three radio stations reported losses. In the mid-1990s marginal stations were made more effi-cient in duopoly or LMA operation or were bought outright by larger groups. Prof-itability rose as stations also began to be hot properties for corporations seeking to enlarge their media empires with deregulated limits on ownership.

Most stations with annual losses reported depreciation expenses that half the time exceeded those losses. Almost one-third of all unprofitable stations made payments to owners, partners, or stockholders that were larger amounts than they lost. A question might be raised why so many radio stations continue in busi-ness when they operate at a loss? A partial answer may be that station owners and investors are given healthy payments before profit/loss or dividends are declared. Another reason—beyond mere prestige or personal satisfaction in owning a pub-lic broadcast property—is the possible tax advantage for a parent corporation sup-porting such a deficit operation. Some stations performing poorly offer potential for sale to aggressive buyers who will pay handsomely for the property with the expectation of turning it into a competitive, profitable operation.

Obviously a station's signal strength (1,000 to 50,000 watts plus clear channel) and operating hours (daytimers vs. twenty-four-hour operation) along with market size are critical factors in audience reach, thus value of airtime. The kind of signal service is also important; from 1987 through 1990 net revenues each year, on average, declined 9.5% for AM fulltime stations (daytime-only suffered even more), and almost 1% for FM stand-alones; AM/FM combos increased 2.1% in net revenues in the same time period. Because of those variables, it is hard to assess which formats will generate highest profitability apart from their relative popularity in general. For example, in 1989 average profits of stations airing country formats differed widely: daytime AMs averaged losses of $5,213 that year while country-format fulltime AM stations earned $22,000 average profits and AM/FM combos drew $156,000 on average; FM stand-alones with coun-try format averaged $70,000 in profits. Looking just at AM/FM combos, those with adult/contemporary (A/C) formats earned average profits of $45,000 in 1989; album oriented rock (AOR) formats reported average losses of $222,000, but contemporary hit records (CHR) earned $846,000 per station whereas mixed news/talk and A/C format lost $47,000 on average.[28]

The key to achieving broadcast profitability is proper managing. Granted radio stations in some markets are oversaturated or too small to support robust broadcasting, but some stations do earn profits even in those circumstances. Yet, when one station generates over a million dollars in revenue but loses money, and another in a small market produces only about $200 a week for all its effort, good business sense demands reassessing those entities as management and business enterprises. Again, when a profitable radio station earns twice the revenues of

━ TABLE 10.4a ━
TYPICAL FINANCIAL PROFILES OF RADIO STATIONS
ANNUAL AVERAGES, 1975–1990
(LOSSES IN PARENTHESES)

	1975	1980	1990 Daytime AM	1990 Fulltime AM	1990 AM/FM	1990 FM
Revenues	$205,000	$376,600	$137,795	$255,417	$554,704	$568,584
Expenses	196,600	361,000	146,256	266,477	564,868	584,299
Pre-tax Profits	8,800	15,600	(8,461)	(11,000)	(10,164)	(15,715)
Profit Margin	*4.3%*	*4.1%*	*(–6.1%)*	*(–4.3%)*	*(–1.8%)*	*(–2.8%)*
Cash Flow	NA	NA	8,027	17,526	54,938	37,103

Sources: Surveys of U.S. TV and radio stations' annual financial reports by National Association of Broadcasters and Broadcast & Cable Financial Management Association, 1984, 1986, 1991; *Broadcasting*, 26 August 1991.

━ TABLE 10.4b ━
FINANCIAL PROFILES OF RADIO STATIONS
ANNUAL AVERAGES, 1990, 1991

	Daytime AM	Fulltime AM	AM/FM	FM
1990				
Net Revenues	$137,795	$255,417	$554,704	$568,584
Expenses	146,256	266,417	564,868	584,299
Pre-tax Profit (Loss)	(8,461)	(11,000)	(10,164)	(15,715)
Profit (Loss) Margin	*(–6.1%)*	*(–4.3%)*	*(–1.8%)*	*(–2.7%)*
Cash Flow	8,027	17,526	54,938	37,103
1991				
Revenues	$155,000	$778,000	$1,260,000	$1,262,000
Expenses	182,000	773,000	1,237,000	1,288,000
Pre-tax Profit (Loss)	(27,000)	5,000	23,000	(26,000)
Profit (Loss) Margin	*(–17.4%)*	*0.06%*	*1.8%*	*(–2.0%)*

Note: Enormous variations between 1990 figures and those in 1991 that more than doubled revenues as well as expenses are partly due to the uneven response rate from radio stations surveyed; depending on which stations chose to return the questionnaires to NAB/BCFM, very high or very low revenue/expense reports could greatly affect averages for various categories.

Source: Survey of U.S. TV and radio stations' annual financial reports by National Association of Broadcasters and Broadcast & Cable Financial Management Association, 1991, 1992; see *Broadcasting*, 26 August 1991, 17; 6 July 1992, 6.

an unprofitable competitor, with expenses only one-fifth higher, the key factor is management.

Radio networks regrouped in the 1970s and 1980s, resulting in annual modest increases in profitability after 1982.

■ In the early 1980s the Mutual Broadcasting System was losing some $5 million a year when Westwood One (a radio programming syndicator) bought it for $30 million; within a year the Mutual division's profits were estimated at about $4 million. In 1985 before acquiring Mutual, Westwood One itself grossed around $19 million, with pre-tax profit margin of 30%, netting over $3 million income after taxes. From 1990 to 1993 Capital Cities/ABC's multiple radio networks regularly reported profit margins of around 25% on $134 million of business, averaging about $36 million profit a year (table 10.5a). But CBS Radio networks earned only half a million to $2 million (1%–3%) annually on revenues of $60 million. The nation's economy and diminishing

TABLE 10.5a
NATIONAL RADIO NETWORK FINANCES, 1990–1993
($ IN MILLIONS)

	CBS RADIO	CC/ABC RADIO	WESTWOOD ONE (NBC/MUTUAL)
1990			
Revenues	$ 60	$ 134	NA
Expenses	58	102	
Income	2	32	
Profit Ratio	*3.3%*	*24%*	
1991			
Revenues	$ 62	$ 149	$ 144
Expenses	61.5	110	160.8
Income (Loss)	0.5	39	(16.8) corporate*
Profit (Loss) Ratio	*1%*	*26%*	*(11.5%)*
1992			
Revenues	$ 57	$ 141.5	$ 137
Expenses	30	117	161.1
Income (Loss)	(5)	24.5	(24.1) corporate
Profit (Loss) Ratio	*(8.7%)*	*17.3%*	*(17.6%)*
1993			
Revenues	$ 60	$ 152	$ 99.6
Expenses	58	114	NA
Income	2	38	(23.9) corporate
Profit Ratio	*3.3%*	*25%*	*(24.0)*

*Westwood One's corporate losses absorbed NBC/MBS networks' positive income of $9 million in 1991; the company's operating loss in 1992 was given elsewhere as $18.7 million, with net loss of $24.1 million; the 1993 figure of $23.9 million was net loss, with operating loss of $2.19 million.

Sources: Head, Sterling, and Schofield, *Broadcasting in America: A Survey of Electronic Media,* 7th ed. (Boston: Houghton Mifflin, 1994), 258–259; annual compilations of network finances in *Broadcasting,* 6 May 1991, 19; 13 April 1992, 4, 16; *Broadcasting & Cable,* 10 May 1993, 5; 7 February 1994, 45; 16 May 1994, 6; Peter Viles, "Westwood Reports $24 Million Loss in 1992," *Broadcasting,* 22 February 1993, 40–41.

ad revenues industry-wide, as well as national radio's volatility along with heavy investment debts, were reflected in Westwood One's corporate net *loss* of $16.8 million in 1991 on over $144 million in total company business. The next year Westwood One's losses increased to $24.1 million on lower revenues of $137 million (corporate losses absorbed the $9 million in operating profits from its NBC and Mutual networks); it sold off two of its three recently acquired radio stations, and reduced news staffs by 20% at both its networks, helping stanch its *negative* cash flow from $11.2 million in 1991 to $2.1 million in 1992. Symptomatic of financial maneuvering widespread in the industry was, despite reporting annual losses, Westwood One's acquiring Unistar Radio Networks for $16.6 million in cash and adding further debt of $84.7 million for the purchase. Motivation was similar to local radio and TV station consolidation: reduce facilities, overhead, and personnel costs of two companies to a single operation while developing new programming formats or niches to draw audiences and ad dollars. (In 1994 Westwood One realigned its multiple networks, renaming them Westwood AC, Westwood CNN+, Westwood Country, Westwood Young Adult, and Westwood Variety.) The company's 1993 net losses totaled $23 million on $99.6 million in revenues.   ∎

Overall network radio revenues dropped 13% from 1991 to 1992. Continued cost-cutting and staff reductions at all radio networks coupled with a strengthened economy in 1993 helped account for CapCities/ABC radio's $38 million profit

⌇⌇ TABLE 10.5b ⌇⌇
FOUR MAJOR TV/RADIO NETWORKS: REVENUES, PROFITS, 1995
($ IN MILLIONS)

	CC/ABC		CBS		NBC		FOX	
	REVENUE	PROFIT	REVENUE	PROFIT	REVENUE	PROFIT	REVENUE	PROFIT
TV network	$ 3,280	$ 374	$ 2,522	$ 24	$ 2,990	$ 330	$ 930	$ 75
O&O TV stations	896	450	500	185	670	360	650	340
Radio networks	165	54	70	NA	—	—	—	—
O&O radio stations	230	77	239	55	—	—	—	—

Profit: Pre-tax operating profit.

CBS stations do not include Group W stations, merged in November 1995.

Source: Estimates by *Broadcasting & Cable,* 25 March 1995, from network executives and securities analysts.

on $152 million revenues (25%) that year, and CBS radio's $2 million income on revenues of $60 million (3.3% profit ratio). In 1995 CapCities/ABC earned $54 million profit before taxes, on revenues of $165 million. Their owned radio stations profited $77 million from $230 million revenues, while CBS's O&O radio outlets grossed about the same, $239 million, but only showed profits of $55 million (table 10.5b).

Radio station groups faced similar financial challenges. For example, Infinity Broadcasting lost $6 million in 1991, and the following year lost $21.8 million on revenues of $150 million; because losses were partly due to common stock public offerings and "extraordinary charges" of $12.3 million, its actual operating income in 1992 was $35.4 million.

Radio managers learned to cope with diminished time sales by finding ways to reduce overhead expenses. Network and station management cut back on news and administrative personnel, partly by offering early retirement bonuses. The latter showed up as balloon outlays in current financial reports, but promised leaner operating costs over the longer run. Broadcasting followed the lead of other industries in "downsizing" staffs, especially by consolidating departments in multiple-station LMAs (local marketing agreements) and legally sanctioned duopolies in many markets. In the recessionary early 1990s Group W cut costs by 20%, including some 250 layoffs—one-tenth of its employees. In seven markets where it owned two or more stations it cut operating expenses by one-fifth.

TELEVISION INDUSTRY PROFITS

The broadcasting industry reflects the nation's economy, with recurring cycles of feast and famine. Radio's general decline in profit in 1958 was experienced by television a year earlier. Where some six years were needed for the radio industry to return to its former profit level, television came back much faster. Within two years, it exceeded its former profit record. TV temporarily slumped again in 1961, then set records for profitability until a 'soft' 1967; it set a new high again in 1968, slumped in 1970 and 1971, then rose to unparalleled heights the next two years.

From 1975 to 1985 TV station profits fluctuated between 20% (1975, 1981) and 26%–28% (1978, 1985), although one out of three stations operated at a loss. Stations owned by network companies (O&Os) and by large

— TABLE 10.6 —
FINANCIAL PROFILES OF TELEVISION STATIONS
ANNUAL AVERAGES, 1975–1991

	1975	1980	1985	1989	1991
AFFILIATES					
Net Revenues	$2,122,400	$4,314,300	$12,974,258	$15,809,909	$14,779,000
Expenses	1,701,700	3,323,300	9,165,547	12,365,172	12,161,000
Pre-tax Profit	420,700	991,000	3,877,800	3,444,738	2,618,000
Profit Margin	*19.8%*	*22.9%*	*29.8%*	*21.8%*	*17.7%*
INDEPENDENTS					
Net Revenues	$3,104,600	5,067,700	13,149,724	14,906,352	$15,084,000
Expenses	2,869,600	4,667,400	11,402,886	14,684,601	14,226,000
Pre-tax Profit	235,000	400,300	1,746,838	221,751	858,000
Profit Margin	*7.6%*	*7.9%*	*13.3%*	*1.5%*	*5.7%*

Source: Surveys of U.S. TV and radio stations for annual financial reports by the National Association of Broadcasters (NAB) and Broadcast Cable Financial Management Association (BCFM), 1984, 1986, 1990, 1992; see also Head, Sterling, and Schoefield, *Broadcasting in America: A Survey of Electronic Media,* 7th ed. (Boston: Houghton Mifflin, 1994), 256–257.

groups typically earned 40% or more profits; of every revenue dollar, less than 60¢ went for expenses and the remaining 40¢ was profit. (Businesses generally look to 3%–5% profit margin—some less, such as grocery supermarkets, others more.)

■ Figures in table 10.6 show affiliates' average revenues doubling the half decade after 1975, then tripling the next half decade. In the latter 1980s the rate of growth slowed down, then reversed in the early 1990s as revenues dropped. Expenses followed a similar pattern, but slower growth in the late 1980s resulted in profit margins dropping from almost 30% in 1985 to 17.7% in 1991, before they began to expand again in subsequent years as the national economy gradually strengthened. During that same decade and a half, independent TV stations reported larger average revenues than affiliates (partly because there were far fewer indies than affiliates in the nation's smaller markets). But indies also bore far higher expenses due to leased programming in lieu of network feeds; their profit margins were only half to one-third as large as affiliates—with a vast gap of 1.5% vs. 21.8% in 1985. ■

Average net revenue for TV stations in the United States was about $15 million in 1991. But market size largely determined level of profitability. Stations in the top-10 ADI markets averaged pre-tax profits of $14 million; in markets 31–40, average profits were $1 million, and in markets 61–70 about $790,000. In markets 111–120 stations averaged losses of $300,000, while stations in markets 131–150 lost more than twice that amount in 1991.[29]

A station's income usually reflects its network's fortunes that wax and wane, reflecting sizes of audiences drawn to competing program schedules. But local news and vigorous market promotion with a strong syndication line-up can help local managers make up for network program shortfalls. In 1993 CBS programming outdrew other networks in ratings numbers; but competitors' younger-skewing demographics, especially of Fox network shows, attracted strong advertising dollars. That plus other factors contributed to affiliates of NBC averaging $15.9 million revenue, while ABC affiliates earned $15.7 million, and CBS

~~~ **TABLE 10.7** ~~~

RATES OF REVENUE GROWTH/DECLINE, TV MARKETS
REPRESENTATIVE ADI MARKETS, 1988–1997
(EACH FIGURE TOTALS NET REVENUES* FOR ALL STATIONS IN MARKET)

ADI MARKETS	1992 NET REVENUES	# STATIONS IN ADI	ANNUAL NET REVENUE GROWTH					AVERAGE GROWTH	1993–1997 PROJECTED GROWTH	
			1988	1989	1990	1991	1992		AVERAGE	1997 NET REVENUE
	(millions)							Annual Rate		(millions)
1–10 Market #5	$ 439	20	7.1%	9.7%	9.9%	(4.9%)	(0.8%)	4%	4.8%	$ 535
11–30 Market #20	145	13	(0.6)	2.9	1.2	(3.9)	5.8	1.1	4.6	182
31–50 Market #40	57	7	1.5	6.2	(3.6)	(4.4)	2.1	0.3	4.6	73
51–80 Market #65	43	5	2.8	6.5	4.0	(4.2)	20.2	5.6	5.0	55
81–100 Market #90	28	5	9.0	12.5	(1.3)	0.0	12.5	6.4	4.6	37
101–140 Market #120	19	4	NA	6.3	7.3	1.1	(1.8)	4.7	4.7	24.5
141–180 Market #160	NA	3	NA	NA	NA	NA	NA	NA	5.7	8.2
181+ Market #195	NA	1	NA	NA	NA	NA	NA	NA	NA	NA

Negative growth (decline) in parentheses.
*"Net revenues" = total time sales (national/regional spot, local, network compensation) minus agency commissions.
Markets selected at midpoint of each ADI market cluster are #5 San Francisco-Oakland-SanJose; #20 Denver; #40 Norfolk-Portsmouth-Newport News-Hampton (VA); #65 Austin (TX); #90 Madison (WI); #120 Lafayette (LA); #160 Clarksburg-Weston (WV); and #195 Anniston (AL).

Source: National Association of Broadcasters, *1994 Television Market-by-Market Review.*

affiliates $15.1 million, with Fox affiliates earning an average $13.1 million in revenues. Local managers cannot merely ride the coattails of their networks but must aggressively compete in their markets with successful syndication and local news programming along with vigorous promotion.

■ Table 10.7 breaks out net revenues in 1992 by market size (taking the midpoint market in each ADI cluster). Stations in ADI market #5 saw drops in revenue in 1991 and 1992, while the 40th ADI market suffered drops in 1990 and 1991; other markets' TV revenues dropped from a previous year at least once each during the half decade 1988–1992. An anomaly was ADI market #65 with a jump of one-fifth in revenues in 1992; that year market #90 bounced back 12.5% in growth after a flat previous year and a loss in 1990. Both those mid-size markets sustained the highest average growth over the half decade, respectively 5.6% and 6.4%; they outpaced the 4.7% of market #120 and the 4% revenue growth of market #5. The 20th market and the 40th market experienced small revenue growth between 1988 and 1992, earning 1.1% and 0.3% respectively. ■

Estimated average growth in revenues from 1993 to 1997 (last two columns of table 10.7) were based on the midpoints of NAB's sets of high- and low-projections for each market over the half decade. Overall TV station net revenues were expected to grow at an annual compound rate of 4.8%, from $11.9 billion in 1993 to $14.4 billion in 1997. That growth was on its way as *Broadcasting* magazine's Steve McClellan reported in mid-1994: "Broadcasting is back. By most accounts the television station business—based on local and national spot advertising sales projections—will have its best year in close to a decade in 1994."[30]

TABLE 10.8
CASH FLOW MARGINS (OPERATING PROFIT MARGINS)
AVERAGES FOR TELEVISION STATIONS, 1992
GROUPED BY ADI MARKET SIZES

Markets	AFFILIATES		INDEPENDENTS	
	VHFs	UHFs	VHFs	UHFs
1–25	40.6%		27.5%@	22.7%
26–50	31.8	} 30.4%*		21.4
51–75	33.0	27.7	18.4#	
76–100	31.9	31.1		} 28.2
101–150		18.6		
150+	} 33.8	31.0		} 15.6

*Markets 21-50 (1-20 NA)
@ Markets 1-16
#Markets 16+
Cash Flow is computed as net revenues less operating expenses, not including interest, income taxes, depreciation, and amortization.

Source: 1993 NAB Television Financial Report, cited in *1994 Television Market-by-Market Review,* B–1–4.

CASH FLOWS

Figured separately from annual profit is the far larger cash flow (net revenues less operating expenses, not including interest, income taxes, depreciation, and amortization). Cash flow margins—or "operating profit margins"—in 1992 averaged about 33% of total revenues for affiliate stations, about 20%–25% for independent stations (table 10.8; compare with previous table 10.3). Affiliates of the Big Three networks in 1993 again averaged cash flow margins of just over 33%, on revenues of $7,980,000 with $5,240,000 in expenses, for a cash flow of $2,700,000.

But each market offered different factors of competing stations and other variables, including UHF or independent status, resulting in wide swings among markets and between stations within markets.

■   The function of cash flow in financial management becomes apparent with data for 1991. That year profit margins for affiliates was 33%, for independents only 18% with one-third of indies reporting pre-tax losses. Independents averaged losses of $315,000, but showed cash flows of $604,000. That million dollar differential came before charging off both depreciation and amortization along with taxes and interest (for indies that year interest payments averaged about 6% of revenues, compared with affiliates' 10% of revenues). A second example comes from group-owned stations: Multimedia for the fourth quarter of 1993 reported a 6% drop in revenue but an 8% increase in cash flow; for the year its total revenue declined 3% to $155.7 million while its cash flow dropped only slightly to $47.8 million.   ■

TV networks rode roller coasters of prosperity and losses over recent decades.

Pre-tax profit margins of networks were 8% in 1965, 12.5% in 1975, and 15.4% in 1979, but began to drop in the next decade as audiences eroded and national inflation compounded escalating production costs and salaries.

TABLE 10.9

TABLE 10.9

NBC, Inc. Pre-Tax Income and Chief Executive Officers, 1977–1987
(Totals include NBC Radio, TV Network, TV O&O Stations)

($ in millions)

NBC CEO	1977	1978	1979	1980	1981	1982	1983	1984	1985	1986	1987
Herb Schlosser	$152										
Fred Silverman		$122	$105	$ 75	$ 48						
Grant Tinker						$108	$156	$218	$333	$430	
Robert Wright											NA

Source: *Television/Radio Age,* 16 February 1987, 35.

Volatile supply/demand forces softened profitability much of the time and even reversed it, but occasionally allowed bubbles of high earnings to one or other network. Executive management often become identified with their company's fortunes, partly because of the stamp of their personality and observable actions and statements, and partly because of internal and external factors not all of which can be directly controlled.

Classic examples include ABC's rise from third place in the mid-1970s to displace CBS from its decades-long #1 spot in prime-time ratings; that coincided with the move of Fred Silverman from head of CBS programming to ABC. Silverman subsequently moved to NBC, Inc. as top corporate executive to bolster its competitive position; but his third tour of duty failed. After four years NBC replaced him with Grant Tinker, respected head of highly successful M-T-M production company. In the mid-1980s with a strong creative team and a string of quality hit programs, he turned NBC-TV around to the top position for several years. He left when General Electric bought RCA-NBC; other executive management and the fickle fate of show business saw NBC gradually drift—then plummet—to third place. "The Fall and Rise of NBC" in the decade from 1977 to 1987 was plotted by *Television/Radio Age* according to annual pre-tax income (table 10.9).[31]

■ Ratings and shares of prime-time audiences track NBC's lock on the top position by as many as 4–5 share points of the national audience 1987–1990 with Tinker's powerful comedy and drama series still hits, then its subsequent slide to a weak third place (table 10.10). Those data also chronicle CBS' unprecedented drop to third place for three seasons, just behind #2 ABC-TV. During those years the young Fox network almost doubled its share of national audiences as it progressively expanded its nightly and weekend schedule along with adding affiliated stations augmented by cable distribution in some markets. The three traditional network's eroding audience leveled off at 60–61 points over four seasons (1990–1994) but then dropped further to a total share of 52 in 1995–1996. Adding the Fox network's growing audience increased those totals to a composite share of 73 in 1991–1993, which then drifted down to 69 and then 64 in two subsequent years as newcomers UPN and WB networks drew off some of the audience. The national audience for cable, independent stations, and PBS grew by 4 share points to 29% of viewing between 1988 and 1991, but then decreased two points 1991–1993. CBS regained the lead for three seasons, 1991–1994, then dropped to third as ABC forged forward in 1994–1995; the next season CBS dug a deeper hole in third place as did ABC in second as NBC regained the top spot. As noted earlier, each national rating point meant multimillions in added sales revenue. In 1991 NBC had lost $100 million and CBS lost $400 million; in 1993

NETWORK PRIME-TIME RATINGS/SHARES, 1987–1996

UNDERLINED MARKS NETWORK WITH HIGHEST RATING/SHARE EACH SEASON

SEASON	ABC	CBS	NBC	FOX	3 NETWORK TOTAL	4 NETWORK TOTAL
1987–1988	12.0/**21**	12.1/**21**	<u>14.4/**25**</u>	3.9/**7**	38.7/**67**	42.6/**74**
1988–1989	11.7/**20**	11.5/**20**	<u>14.3/**25**</u>	5.7/**10**	37.5/**65**	43.2/**75**
1989–1990	11.7/**21**	11.0/**19**	<u>12.9/**23**</u>	6.3/**11**	35.6/**63**	40.2/**74**
1990–1991	11.2/**20**	11.2/**20**	<u>11.5/**20**</u>	6.3/**11**	33.9/**60**	40.2/**71**
1991–1992	11.0/**19**	<u>12.2/**21**</u>	11.6/**20**	7.5/**13**	34.8/**60**	42.3/**73**
1992–1993	12.5/**21**	<u>13.5/**22**</u>	10.9/**18**	7.7/**12**	36.9/**61**	44.6/**73**
1993–1994	12.4/**20**	<u>14.0/**23**</u>	11.0/**18**	7.7/**11**	37.4/**61**	45.1/**72**
1994–1995*	<u>12.2/**20**</u>	11.0/**18**	11.6/**19**	7.3/**12**	34.8/**57**	42.1/**69**
1995–1996*	10.6/**17**	9.6/**16**	<u>11.7/**19**</u>	7.3/**12**	31.9/**52**	39.2/**64**

*The last two seasons, UPN and WB networks each scheduled about two hours of programming two to four evenings a week: 1994–1995 **UPN** 3.7/**6** **WB** 1.8/**3**
1995–1996 **UPN** 3.1/**5** **WB** 2.4/**3.8**

Annual data for 1987–1993 cover twelve months, September to September; for 1993–1996, 30-week seasons, September to April.

Sources: Nielsen Television Index data in *Electronic Media,* 8 March 1991; *Variety,* 25 April-1 May, 1996, 24; other trade magazines and wire service reports.

> ABC's profit was $455 million, and the others had returned to profitability—NBC with $432 million, and CBS $326 million. In 1994 NBC's profit was $700 million on revenues of $4 billion, while CBS TV network and stations profited $426 million on revenues of $3.7 billion, and CapCities/ABC had $1.2 billion profit on revenues of $6.4 billion. ■

Counting audiences became complex because of multiple sets in households, resulting in overlapping data for simultaneous viewing of different programs (and thus total estimates exceeding 100%). Nielsen Media Research confirmed that during the 1993–1994 season the Big Three networks increased total share to 61%, while independent stations (including Fox network outlets) also increased share to 18%; cable remained stable with a 30 share. Together the estimated viewing "shares" total 109%, even without counting public broadcasting's audience.[32]

> ■ Table 10.11 offers a profile of network finances over seven years. In 1989 each of the Big Three generated around $2.5 billion in ad revenues, with about $2.3 billion in expenses, for profit margins ranging from CBS' 4.5% and ABC's 6.3% to NBC's 14%. The difficult year of 1991 found ABC's television network profitable, earning $120 million (4.5% profit margin); so was the recently started Fox network, whose revenues approached its rivals but with far lower expenses to result in $348 million profit (16.5% margin). That contrasted with losses by NBC-TV of $50 million (-2%) and by CBS-TV of $153 million (-6.4%). In 1991 CBS' lackluster single-digit ratings were reflected in the network's mere 1% profit margin, far behind its competitors.
>
> Companies owning TV networks typically profited far more from their clusters of O&O stations. In 1989 only NBC's network outpaced its station division, because high prime-time ratings kept it in first place with top-dollar time sales ($289 million profit from network operations, $256 million from its owned TV stations). Rival ABC's stations were twice as profitable as its network ($417 million vs. $178 million), and CBS' stations earned almost four times more than its faltering TV network (profits of $152 million vs. only $43 million). ■

TELEVISION NETWORK FINANCES, 1989–1994

($ IN MILLIONS)

	ABC TV	CBS TV	NBC TV	FOX TV
1989				
Revenues	$2,424	$2,315	$2,787	NA
Expenses	2,259	2,212	2,394	
Income	165	103	393	
Profit Ratio	*6.8%*	*4.5%*	*14%*	
1990				
Revenues	$2,606	$2,576	$2,638	NA
Expenses	2,367	2,639	2,389	
Income (Loss)	240	(63)	249	
Profit (Loss) Ratio	*9.2%*	*(2.4%)*	*9.4%*	
1991				
Revenues	$2,630	$2,388	$2,531	NA
Expenses	2,510	2,541	2,580	
Income (Loss)	120	(153)	(50)	
Profit (Loss) Ratio	*4.5%*	*(6.4%)*	*(2%)*	
1992				
Revenues	$2,514	$2,736	$2,698	NA
Expenses	2,422	2,731	2,647	
Income (Loss)	92	5	51	
Profit (Loss) Ratio	*(3.7%)*	*0.2%*	*1.9%*	
1993				
Revenues	$2,730	$2,732	$2,421	NA
Expenses	2,546	2,970	2,466	
Income	184	238	45	
Profit Ratio	*6 .7%*	*8.7%*	*1.9%*	
1994				
Revenues	$ 3,060	$2,851	$2,612	$950
Expenses	2,720	2,676	2,522	870
Income	340	175	90	80
Profit Ratio	*11%*	*6%*	*3.4%*	*8.4%*
1995				
Revenues	$ 3,280	$2,522	$2,990	$ 930
Expenses	2,906	2,498	2,660	855
Income	374	24	330	75
Profit Ratio	*11.4%*	*1%*	*11%*	*8%*
1996				
Revenues	$3,125	$2,581	$4,000	$1,700
Expenses	2,715	2,556	3,620	1,610
Income	410	25	380	90
Profit Ratio*	*13.1%*	*1%*	*9.5%*	*5.3%*

*"Paper profit" figures of ABC, CBS, and Fox for 1996 include hundreds of millions of dollars added back in hefty amortization and depreciation accounting benefits; without them, ABC-TV's operating profit would be $210 million instead of $410 million, Fox would have an operating loss of almost $35 million, and CBS-TV a loss of about $106 million.

Sources: Data from network executives and security analysts were basis for estimates by *Broadcasting,* 11 May 1991; 6 May 1991; 13 April 1992; *Broadcasting & Cable,* 10 May 1993; 16 May 1994; 3 April 1995; 25 March 1996; 3 March 1997.

Although networks generated three to six times as much revenue as their owned stations, expenses were also massive, leaving modest profit margins in contrast to high earnings of efficient station operations. For example, in 1993 the #1 ratings network CBS-TV had a profit margin of 8.7% ($237 million on $2.7 billion in sales) while its station division's margin was 35.5% ($180 million on $506 million of business). A close second in audience ratings, ABC-TV network profited 6.7% on revenue equal to CBS; its O&O TV stations generated 35.5% profit ($417 million). Then a distant third in ratings, NBC-TV earned only 1.8% profit on almost as much business as its counterparts, while its owned stations enjoyed 41.2% profitability ($243 million, almost twice the earnings of CBS' station division, but half that of ABC-owned stations).[33]

As for the fledgling fourth network, Fox Broadcasting Corporation (FBC or simply Fox) lost $99 million in 1988, broke even in 1989, earned profits of $33 million in 1990, and expected to earn some $85 million in fiscal 1991. Developing more than its network's profits, Fox's TV station group generated profits of $135 million in 1990, about $150 million the next year, and $180 million in 1992 on revenues of $455 million—a 40% profit ratio—almost five times its network profits of $38 million. Fox's fortunes rose as it added affiliates—plus cable outlets in markets without available TV affiliates—to reach 94% of U.S. households, and as it filled in its seven-night broadcast schedule. Fox's parent News Corporation jolted the broadcast industry when it not only successfully outbid CBS for rights to televise National Football League seasons, but also joined with New World Communications to buy into part ownership of key stations in major markets, abandoning their Big Three affiliations in favor of Fox network programming. CBS lost eight affiliates, ABC three, and NBC one—in markets numbers 8, 9, 11, 12, and the rest from markets 16 to 65.

In 1995 the new UPN network lost $29 million in the second quarter alone, and $35 million in the second quarter of 1996 as it expanded to three nights of programming. Ominously, revenues for the first six months of 1996 dropped to $231 million from the previous year's first half of $234 million.

Network Compensation Payments

The development of new networks that aggressively competed for affiliates, even raiding stations linked to the Big Three traditional networks, reversed a decade-long slide in network compensation payments to affiliates. Instead of being taken for granted as recipients of "free" programming from their networks, stations became sought-after partners to ensure national coverage for each network's programs and advertisers. Successive drops in compensation rates gave way to new largesse as networks reversed themselves and offered long-term affiliation contracts with higher payments.

In the early 1970s, compensation rose an average of 3.2% a year, but in the second half of the decade and into the 1980s—with ABC strongly competitive and a rising number of independent stations siphoning off audiences—compensation payments accelerated to 7.3% annually. In later years, revenues did not keep pace with rising payments to affiliates. As audience shares diminished and advertising dollars dropped in the late-1980s and early 1990s, corporate management sought to cut compensation payments especially to affiliates pre-empting network programs. In 1989 CBS spent $160 million to compensate its stations for carrying network programming; NBC paid $145 million, and ABC $120 million. In 1991 NBC slashed total compensation payments by $15 million, averaging about 10% to affiliates—between 6% and 15% based on their performance in clearing time for the network schedule. CBS announced a 20% cut (one analysis showed that would

The entire debate changed, however, in the mid-1990s when stations gained strong leverage as Fox raided network affiliates for its own line-up. Networks swiftly offered greatly enhanced compensation packages to keep their affiliates or to woo stronger ones in some markets away from competing networks. ABC's $10 million annually to Scripps Howard was far outpaced in 1994 by NBC's $15 million annual compensation to entice Boston's WHDH-TV away from CBS. Now on the defense, the three major networks aggressively extended annual affiliation contracts to five and even ten years while offering increased compensation payments to keep them in the fold or to woo major outlets; CBS forged an alliance with all Group W's TV stations, resulting in two shifting away from traditional NBC ties to ten-year affiliation contracts with CBS.

Further complicating the picture in the mid-1990s, two new networks began to compete for national audiences and ad dollars. By 1994 United/Paramount (merged with Viacom) had contracted with forty-one stations reaching 52% of the country's households; it expected to reach 70% of the country plus another 15% with secondary affiliations when it inaugurated Monday-Tuesday evening programming in January 1995. It would later add Sunday morning programs and weekend movies, then afternoon children's programs, with gradual expansion to ten hours in prime-time. Time Warner's WB network lined up twenty-seven stations, with 43% national coverage, plus Chicago "superstation" WGN-TV that added another 30% of U.S. homes via cable. Executives at both companies acknowledged they needed supplemental cable carriage to succeed as networks.

NEWS BUDGETS

A network's news service must be factored into broadcast budgets. Initially news was considered part of a station's "price of doing business" with a government-granted license to broadcast. By extension, networks contributed to their affiliated stations' public service role by supplementing entertainment programming with news and public affairs. Also, network corporate owners were themselves licensees of half a dozen lucrative stations in the nation's largest markets. Network news and documentaries made those corporate licensees "good stewards" and citizens, justifying heavy expenditures on news.

$23 million in 1980 to $130 million in 1986. In 1989 only ABC's news operation was profitable, but CBS and NBC expected to be in the black the following year.

As broadcast corporations began downsizing staffs and operations in the late 1980s, *Broadcasting* magazine noted that "much of the budget analysis and debate going on at the three broadcast networks is predicated on the assumption that news should be a business, rather than a responsibility of media to inform the voting public."[34] Of course the appropriate course is not either/or, but a synthesis or accommodation between the two: an accurate and timely news service that is also cost-effective. Subsequent success of more efficient news operations was described in previous chapters.

■   In 1988 over 70% of local TV stations reported profitable news operations, 18% broke even, and 11% lost money; 94% of those with news staffs of thirty-six or more were profitable; 96% of affiliates in the top-25 markets and 81% of affiliates in markets 26–100 earned profits for their stations. In 1992 news operations at most TV stations made money in 1992, while at radio stations they broke even or made money. Among news directors responding to an RTNDA survey, 74% at TV stations reported news as profitable, 16% as even, and 10% as losing money; at radio stations 44% said they showed profit, 42% broke even, and 14% operated at a loss.* In 1993 affiliates in the top-20 markets budgeted around $2.5 million annually for their local news operations.   ■

By 1996 *Broadcasting & Cable* magazine identified local news as the major revenue source for most TV affiliates, bringing in 35% to 40% of those stations' advertising dollars. At the same time affiliates typically spent about 25% of their operating budgets on their news departments.

A mark of changing times in the industry was income produced by those broadcast corporations' cable and nondomestic activities.

■   In 1993 NBC's cable division profited $31 million on $94 million in business (a margin of 33%), while CapCities/ABC's cable and international operations earned $138 million on revenues of $781 million (17.8%). But in 1995 that division at NBC reported no profit on $260 million revenue, whereas ABC posted a $360 million profit on $1.156 billion revenues (a 30% profit margin). Income from these and other alternate revenue streams would grow as federal government gradually removed restrictions on cross-media holdings. Rupert Murdoch committed Fox Broadcasting to massive investments beyond the network, including $100 million to start an all-news cable channel with $45 million a year to operate it, $3 billion to buy New World Communications and its ten stations, plus another $1 billion in the late 1990s for a DBS venture with MCI.   ■

Since 1993 the FCC's loosened restrictions on networks' financial participation in program ventures and eventual syndication (fin/syn rules) allowed them to expand activity as program suppliers and distributors, including nonbroadcast avenues such as cable and videotape. In 1995 those restrictions expired entirely; and the prime-time access rule expired in 1996. Management looked to a growing role as programmer and less as merely a distribution network to over-air stations for other studios' productions. The final section of this chapter addresses corporate redefining of increasingly synergistic broadcast/cable/telecommunication media. Diane Mermigas, financial editor of *Electronic Media,* summarized other

---

*The findings may be overly optimistic. The response rate was about 50%; perhaps those losing money were less likely to respond. Further, only stations with newscasts were polled; many radio operators with unprofitable news would have already dropped the service.

analysts' assessments in 1993 that "owning more outlets, controlling production costs, establishing lower overhead and having the ability to sell product to more buyers are the key factors for network profitability in the future."[35]

LOCAL CABLE SYSTEMS

By 1990 cable's revenues from advertising totaled over $2.5 billion, one-fourth of that by individual systems from local and regional spot, the rest by national cable services' network advertising. The cable industry expected to generate $4 billion in 1995, eight times the expenditures a decade earlier. Six years earlier, ad agencies had placed only half a billion dollars for advertising with cable (and over $18 billion for ads on broadcast TV).

■   Because of heavy construction costs with long-term loans coupled with high interest rates in the 1980s, cable's profitability rose and fell. In 1975 it saw a 3.5% profit margin, growing to 11% from 1978 to 1980, then plummeting to 4% loss in 1982; it rebounded to over 6% the next year, and further up to 12% in 1985—sparked in part by federal deregulation of subscriber rates the previous year. In 1990 cable had revenues of over $17 billion, expenses of almost $10 billion, with $7 billion in operating cash flow. In 1992 revenues were affected by the FCC's first round of rate cut-backs (theoretically, 10%) for a majority of systems; the agency's second rollback in 1994 reduced most systems' rates another 5% to 10%. At least two major mergers were called off, attributed in part to decreased cable system earnings due to those government mandated rate rollbacks; TeleCommunications, Inc. and Bell Atlantic backed off, as did Cox Cable Communications and Southwestern Bell Corporation. MSO Cox Cable estimated it would lose $11 million in revenues during 1994 because the FCC reduced local cable rates; Times Mirror looked to $5 million less. Both companies were merging to become the third largest MSO with 3.1 million subscribers and over $1 billion in combined revenues; each had average monthly revenue of over $33 per subscriber (up from $27 five years earlier). Many cable operators who had financed capital building with eight- or nine-year loans based on projected cash flow levels found themselves flirting with default when revenues dipped from restricted subscriber rates; they had to scramble to renegotiate loan extensions with their banks. Although owners denied cable rate reductions as a factor, Maryland Cable Corporation filed for Chapter 11 reorganization in 1994 when they could not pay an $85 million bank loan. The system of 76,000 subscribers had been bought in 1988 for $198 million, or $2,700 per subscriber when cable systems were selling at their highest levels—far beyond the typical high of $2,000 per subscriber.   ■

Table 10.12 breaks out cable's 1993 revenues from advertising by market group mid-point, adding percentage share of market ad dollars, penetration levels, and subscribers. Cable's total revenues from advertising and subscriber fees came to $21 billion in 1994.

National Cable Networks, after initial heavy investments, gradually worked their way towards profitability with twin revenue streams of advertising and of carriage fees from local cable systems.

■   In 1991 cable ad revenues increased half a billion dollars to $3 billion, up 18% from the previous year; that was in addition to massive revenues coming from license fees paid to national cable programmers by local systems out of their subscriber fees (then averaging $31 monthly or $372 a year). That year when CBS-TV's losses were $400 million, fourteen ad-supported cable networks reported profits. Each reached over 47 million subscribers, generating most of the approximately $20 billion in fees going to local systems, most of which was forwarded to cable networks. They included ESPN, USA, CNN, MTV, TBS, Discovery, TNT, Nickelodeon/Nick-At-Nite, Lifetime, A&E,

~~~ **TABLE 10.12** ~~~

CABLE ADVERTISING REVENUES, SELECTED ADI MARKETS, 1993

</div>

ADI MARKETS (at mid-point of each ADI market cluster, table 10.7)	CABLE ADVERTISING REVENUES	% OF TOTAL TV ADVERTISING	PENETRATION	SUBSCRIBERS
#5 San Francisco-Oakland-San Jose	$27,000,000	6.1%	70%	387,500
#20 Denver	7,000,000	4.6	60	287,500
#40 Norfolk-Portsmouth- Newport News-Hampton (VA)	4,700,000	7.7	75	276,200
#65 Austin (TX), #90 Madison (WI)	2,250,000	5.0	68	160,500
#120 Lafayette (LA)	950,000	4.8	72	72,200
#160 Clarksburg-Weston (WV)	300,000	4.5	79	31,400
#195 Anniston (AL)	375,000	NA	81	35,200

Source: National Association of Broadcasters, *1994 Television Market-by-Market Review.*

CNN Headline News, the Nashville Network (TNN), Family Channel, and the Weather Channel.[36] In 1992 seventeen basic cable networks (from among some fifty basic channels) generated almost $2.2 billion in ad revenues, apart from subscribers fees. ESPN, USA, and TBS networks led, each with more than $250 million income from advertising (table 10.13). TNT, CNN, and MTV each carried $163 million to $173 million in ads. Advertising revenues on twenty-one basic cable networks in 1993 totaled $2,436 billion, almost half a billion higher than the previous year. That year basic cable networks were estimated to earn half to two-thirds of revenues from advertisers, the rest from local systems' forwarded subscriber fees. ■

Pioneer national cable networks broke the chicken-and-egg impasse between viewers and programmers. Earliest network cable entrepreneurs risked big capital to attract subscribers to the young cable service. Some major companies' efforts are sketched here as they struggled to turn red ink into black. Most eventually succeeded; a few failed.[*]

■ ESPN started up in 1979, losing $80 million in a single year. Revenues over $20 million in 1982 doubled the next year to $42 million. In 1984 it broke even; its staff of 30 had grown to 400, when it reached 33 million homes as the largest cable network. In 1988 it earned $75 million with 42 million subscribing households. In 1992 its revenues approached $300 million.

CNN paralleled the sports network. It began in 1980 with 2 million homes reached, earning only $1 million in revenue monthly to cover costs of $3 million.[**] It went $250 million into debt before becoming profitable in 1985. By 1987 CNN earned $60 million profit on $205 million in revenues, and in 1988 profits of $85.5 million, when it reached 54 million U.S. homes plus another 6.5 million households abroad and hotels in 89 countries. By then it had a staff of 1,600 and sixteen overseas bureaus; its annual news budget of $100 million was only one-third of each of the Big Three networks but supplied six times as much news programming. Fees to local cable operators rose from 15¢ a subscriber to 21¢ within seven years (if they also took WTBS and

---

[*]See chapter 8, and table 8.10 for complete list of major cable networks and rank by number of subscribers in 1995.

[**]By 1983 CNN and the added Headline News earned revenues of $209 million, compared with the three broadcast networks' revenues of $690 million on expenses of $1 billion for annual combined losses of over $300 million. Turner's cable program network budgeted $30 million at the outset in 1988, then in successive years $40 million, $90 million, and $150 million, offering some 250 films a month (contrasted with 90–100 on premium/pay movie channels or 120 on a strong independent station with an extensive film library), plus originally produced features and mini-series.

17 Basic Cable Networks: Revenues From Advertising, 1992
($ in millions)

$ 298.9	ESPN	$ 98.5	Lifetime
275.4	USA	94.6	Family
263.8	TBS	90.8	A&E
173.6	TNT	65.5	CNN Headline
171.1	CNN	31.2	Weather Channel
163.3	MTV	30.0	VH-1
115.0	Nick/Nite	26.8	CNBC
114.8	TNN	19.8	BET
112.6	Discovery		

Source: *Cablevision,* 10 May 1993, 5A.

Headline News as a package; otherwise 28¢ per subscriber on system). When Turner's largely borrowed $1.5 billion to purchase MGM/UA threatened CNN's stability, fourteen cable operators, including massive MSO TCI, pooled $550 million for one-third stake in the company. In 1990 its 1,800 staff worked in the United States and 27 foreign bureaus including the then-Soviet Union. By 1994 CNN's twenty-four-hour news service reached viewers in 140 countries and territories around the world. Despite enormous audiences for special event coverage such as the Los Angeles earthquake and the "live" highway chase of football star and alleged murderer O.J. Simpson, regular domestic viewing of CNN dropped in 1994, lowering profits by 10% to $200 million. Taken together, Turner's CNN, CNN Headline News, TBS, and TNT exceeded $1 billion in sales in 1989, showing an operating profit of around $224 million with CNN earning almost one-third of that (but because of interest rates on long-term debt and depreciation, the company reported a net loss of $71 million, down from $95 million net loss the previous year). Turner Broadcasting System in its first nine years grew from a staff of 240 in 1980 to 4,500 employees (compared with NBC's single network workforce of 5,700 in 1991 after cuts of 2,300 over previous years). By 1994 CNN was estimated to account for almost 70% of TBS' annual operating profit of some $300 million–$350 million.

Started in 1982, the short-lived CBS Cable culture network and the Entertainment Channel, of Rockefeller Center, Inc. and RCA, faded in less than two years. The latter lost $35 million but reemerged as the Arts & Entertainment network—jointly owned by CapCities/ABC, Hearst, and NBC. A&E's $50 million annual programming costs showed $4 million profit on $50 million revenues by 1988. Its 37 million subscribers, with 350 advertisers, were expected to double in five years (by 1995 it had 62 million subscribers).

In 1983 the Disney Channel introduced family programming with $70 million start-up costs, first going into the black in 1985, and earning $50 million in profit by 1987. By 1990 more than 5 million households paid monthly fees as subscribers.

Viacom's MTV in 1986 had income of $68.8 million on revenues of $523.6 million. By the mid-1990s it was carried by satellite to 240 million homes beyond the United States, operating as MTV Asia, MTV Europe, MTV Brasil, MTV Internacional, MTV Japan, MTV Latino, Nick UK, VH-1 UK, and collaborating with broadcast systems in South Africa and India. In 1994 the combined MTV Networks, including MTV, Nickelodeon/Nick at Nite, VH-1, and International earned revenues of $198 million in the second quarter alone, up 21% from the previous year. ■

By 1987 all national cable programmers were spending an annual $1.7 billion on programming, compared with the Big Three networks' $5 billion. That year subscribers' fees brought in $13 billion, with national and local spot sales bringing in another $1.4 billion, and almost 10% more the next year. In the

1992–1993 season, basic cable networks spent $1.8 billion on original programming, while premium/pay services spent $1.2 billion.

■   Pay-per-view TV, which had drawn $26 million in 1982 revenues, by 1988 had gross revenues of $149 million, around $230 million in 1989 and 1990. In 1992 PPV reached 20 million addressable homes, attracting $327 million, $2 million less than the previous year; in 1993 it brought in about $477 million. Program fare consisted of events and movies; in 1990 each generated almost equal revenue, but in 1991 events (mostly sports) brought in twice as much as films.   ■

Subscriber fees made up the bulk of local and national cable revenue. But local fees did not keep up with the rate of inflation in the early years. From 1972 to 1986 basic cable rates grew by 89%, only about half the consumer price index' 162% rise. When the industry tried to catch up by raising subscription fees 61% between 1986 and 1991, it caused public outcry that mobilized Congress. The FCC was mandated to roll back rates and monitor them. Those actions threatened cable's fiscal growth; Bell Atlantic broke off a planned $33 billion merger with TCI, giant cable MSO. But while the average basic fee of about $12 a month received 27 channels in 1986, the fee of $19 in 1991 brought 35 channels. The per-channel increase was thus nine cents, while cable operators' investment in programming grew from $300 million in 1984 (when the Cable Act was passed) to $1.5 billion in 1991. Complex formulae established by the Commission, were later modified as the 1996 Act further deregulated media industries. Cable fees in 1996 rose between 5% and 10%. TCI announced rate hikes of 15% to 20% for half of its 12 million subscribers in mid-1996 to cover two years of program improvements and three years of modest inflation. Over a dozen years, 1984–1996, subscriber fees more than doubled. But when telephone companies and DBS service began competing head-to-head, cable operators had to keep rates to a buyable level. In addition to advertising, they started to develop adjunct services such as high-speed data and telephony to provide supplemental revenue streams.

Local ads inserted into popular cable network programs soon returned part or more of local operators' payments to program services. In 1996, typical monthly payments per subscriber—ranging between ten to fifty cents depending on the channel service—were offset by ad revenues for each channel. Income from ads equaled from 55% of per-subscriber fees (Nick at Nite) to 105% (USA), 115% (A&E), and as much as 160% (Lifetime) of fees paid to those program suppliers.

EVALUATING MEDIA PROPERTIES

In 1994, 90 mergers and acquisitions of TV and radio properties were worth $2.7 billion; another 111 deals for $15.2 billion involved cable and other broadcasting. In 1995 the 65 radio-TV deals were valued at $26.3 billion; 84 cable and other broadcasting negotiations were worth $9.4 billion. Accurate valuation of media properties is crucial to sustaining and "growing" the business.

A common rule-of-thumb is to evaluate a property based on reported cash flows—taken as initial profit (revenues minus expenses) before subtracting taxes, interest payments, and depreciation and amortization costs. Broadcasters figure on cash flow rather than net profit (before taxes) because "cash flow pays back debt" on interest and depreciation/amortization expensing. Traditionally radio and TV stations were valued in the vicinity of ten times cash flow, depending on current economic conditions and predictors for the future.

■   Those multiples grew in the aggressive mid-1980s, rising to 12 or 14 times cash flow or higher; in 1988 George Gillett paid 17 times cash flow for a station acquisition. After the broadcast market ran into economic recession at the decade's end, multiples plummeted to 6 or 7 times cash flow before rising again in the mid-1990s to 10 to 12 times and higher; one station in San Diego sold for 30 times cash flow (partly because the buyer, the new WB network, needed a key west coast station). In 1995 both radio and TV in large markets sold at 10 to 11 times cash flow, in medium markets at 9 to 10 times annual cash flow; in small markets TV properties sold for 8 to 9.5 times and radio 7.5 to 8.5 times cash flow.[37] In 1996 Rupert Murdoch's News Corporation acquired New World Communications, including ten TV stations, for $3 billion or 15 times the company's cash flow. The same year Westinghouse/CBS bought Infinity radio group for $4.5 billion, or 16.8 times annual cash flow.

Those formulas could be translated into concrete evaluations for 1991. In the top-25 markets, stations affiliated with ABC and NBC averaged $22 million annual cash flow in 1991, while CBS affiliates averaged $15 million, and Fox affils $11 million. Thus the sale price for a Fox station averaged half that of an ABC or NBC affiliate; if multiples then were 10 times cash flow, a Fox outlet might sell for $110 million on average. In markets 26–50, annual cash flows averaged $5.3 million for ABC and NBC affiliates, $4.3 million for CBS, and $2.7 million for Fox stations.[38]

The entire NBC operation (without the radio division which parent GE sold the previous decade) was valued in 1994 at $6 billion; that was based on TV network and cable divisions at $2.6 billion, O&O stations at $2.6 billion, and program syndication operations at $800 million.[39] At that time financial analysts estimated a network ought sell for 7 to 8 times its annual cash flow. But with the three networks' total operating profit approaching the highest level in a decade, one outside estimate reappraised ABC and CBS as able to sell for 11 times cash flow.[40] ■

A variant for computing value is to compute a station's selling price at 2.5 times annual gross (not net) revenue. Benjamin Bates figured typical evaluation multiples in the early 1990s at less than twice revenues for AM stations, around double the revenues for AM-FM combos, and two to three times annual revenues for stand-alone FM stations.[41] The specific multiple chosen will depend on a property's anticipated revenues, affected by macroeconomic conditions such as interest rates, and domestic and even international economic fluctuations, market competition, potential governmental (de)regulation, as well as other external factors outside the industry. Estimates of cash flows are financial experts' best guesses about a property's and industry's stability and predictable return on future business.

Cable system values were also figured on multiples, ranging between 7 times and 14 times cash flow. In 1989 some systems were appraised at 12 to 14 times annual cash flow; that dropped considerably in following years—even to 6 times cash flow in 1996, according to Merrill Lynch analysts. A cable system's value can also be estimated by the number of subscribers. In the mid-1980s a local cable system was priced at some $950 per subscriber, on average. That valuation increased to about $2,000 per subscriber in 1986–1987, rose somewhat briefly (see earlier instance of $2,700), but then remained at the $2,000 level into the late-1990s.

■   In 1994 Comcast Corporation bought Maclean Hunter's U.S. cable systems for $1.27 billion; the 550,000 customers figure to $2,309 per subscriber (ComCast's new total: 3.5 million households). Tele-Communications, Inc. purchased TeleCable Corporation's cable systems with 740,000 subscribers for $1.4 billion, or $1,891 per subscriber (TCI's total: 14.7 million homes). Other acquisitions and joint ventures that year included Time Warner Entertainment's pact with Newhouse (whose 1.4 million subscribers brought Time Warner's total to 8.9 million). Consolidation of systems narrowed the field of MSOs, enabling economies of scale and stronger positioning to compete with telephone companies' impending entry into video delivery.   ■

Broadcast and cable values responded to swift shifts in supply and demand as deregulation re-opened the buying market to major investors. The changing media market's volatility spawned successive turn-arounds as buyers soon became sellers for handsome profits.

Market and investment analysts in 1996 judged that larger groups with multiple stations in various markets would no longer be competing so much with other individual radio outlets; they would put more money and resources into attracting more of the market's total advertising expenditures into radio. That would further enhance the value of radio properties and help make cash flows more predictable.

FORECASTS OF INDUSTRY REVENUES

Shifting national and global economic conditions shape broadcasting and cable revenues from advertising and subscribers' fees. Broadcast managers must track past trends as well as patterns projected for the future, to guide their own short- and long-term budget planning.

■   Growth forecasts by investment firm Veronis, Suhler & Associates projected greater radio and TV growth, but a slower rate for cable, during the five years following 1992 than for the preceding half decade. TV revenue's rate of growth was expected to almost double (from 3% to 5.7%) as would radio's (from 3.7% to 6.8%); but cable TV's revenue growth rate would diminish from a robust 12.2% in 1987–1992, to only 4% during 1992–1997.[42] Gross advertising expenditures going to specific media outlets between 1993 and 1998 were predicted to grow at a compound annual rate of 5.5% for television networks and stations, 3.3% for radio networks, 7.3% for radio stations, and 11.9% for cable; but cable subscribers' fees would grow at a rate of 3.6% over that five-year period (table 10.14). Cable's revenue from subscriber fees roughly equaled all ad dollars spent on TV networks, or about two-thirds of ad sales by all TV stations. The nation's 11,000 AM and FM stations together attract about half the ad revenues earned by 1,100 VHF and UHF TV stations. Growing more slowly (3.6% rate) than cable advertising's 12% clip annually during that half decade, cable subscriber fees were projected to drop to about triple ad revenues, down from four times revenues.   ■

GROUP OWNERSHIP

Management of media enterprises becomes more complex as properties become part of a larger group under one ownership, and then merge with other corporations. Central administrative offices oversee finances, they set budgets with input from scattered lower divisions, and accountability within the company requires detailed communication procedures—corporate meetings, divisional conferences, local planning and interface via correspondence, facsimile, computer, telephone, and face-to-face discussions.

■   The financial statement of a multiple media owner reflects variants of cash flow computations, as in the excerpt from Tribune Company's consolidated statement of income (table 10.15a). Operating revenues of $2.2 billion were offset by operating

TABLE 10.14
PROJECTED GROSS EXPENDITURES ON TV, RADIO, CABLE, 1993–1998
($ IN MILLIONS)

	1993	ANNUAL GROWTH*	1998
TV networks	$ 10,435	5.3%	$ 13,480
TV stations	16,175	5.6	21,230
Radio networks	458	3.3	540
Radio stations	8,940	7.3	12,690
Cable TV advertising	2,510	11.9	4,405
Cable TV subscriber fees	10,256	3.6	12,000
Barter syndication	1,480	7.5	2,120

*Compound annual growth, 1993–1998. Projections are predicated on national economic growth of 4.5% in 1994, then dropping 0.5% each following year.

Source: Veronis, Suhler & Associates; data tabulated in *Broadcasting & Cable,* 25 July 1994, 79.

TABLE 10.15a
SAMPLE CASH FLOW ALTERNATIVES
CONSOLIDATED STATEMENT OF INCOME, 1995: TRIBUNE COMPANY (CHICAGO)
($ IN THOUSANDS; ADD 000)

OPERATING REVENUES		
Publishing	$ 1,312,767	
Broadcasting & Entertainment	828,806	
Education	103,101	
Total Operating Revenue		$ 2,244,674
OPERATING EXPENSES		
Cost of Sales	$ 1,164,609	
Selling, General & Administrative	553,868	
Depreciation & Amortization	120,986	
Total Operating Expenses		$ 1,839,463
Operating Profit [Earnings before Interest & Taxes]		$ 405,211
Operating Cash Flow*	$ 526,197	
Interest Income	$ 14,465	
Interest Expense	(21,814)	
Disposition of Subsidiary Stock and Investment	14,672	
Income Taxes	(167,076)	
Income from Continuing Operations	$ 245,458	
Income (Loss) from Discontinued Operations of QUNO	32,707	
Net Income [Net Earnings]		$ 278,165
Preferred Dividends, Net of Tax	(18,841)	
Net Income Attributable to Common Shares	$ 259,324	
Net Income (Earnings) Per Share	4.00	
Capital Expenditures	117,863	
Net Cash (Flow) Provided by Operations	$ 393,667	

*This "cash flow" computation is the one more commonly used: earnings before interest, taxes, depreciation, and amortization.

Source: *Tribune 1995 Annual Report.*

expenses of $1.8 million, leaving earnings before interest and taxes at $405 million (and net income after taxes, etc. at $278 million). But excluding depreciation and amortization of almost $121 million, that year's *cash flow* (in the common meaning

TABLE 10.15b
BROADCAST AND CABLE OPERATING REVENUES, CASH FLOW, AND PROFIT, 1995:
TRIBUNE COMPANY, CHICAGO
($ IN MILLIONS)

	RADIO	TV	CABLE	TOTAL
Operating Revenues	$ 88	$ 630	$ 7	$ 725
Cash Flow	15	215	(15)	215
Operating Profit	11	186	(16)	181

Radio: Five stations and Tribune Radio Networks (primarily Chicago Cubs baseball).
TV: Ten stations.
Cable: Programming and development, twenty-four-hour cable regional news channel; negative totals for cash flow and operating profit attributed to equity losses from The WB Network and TV Food Network.

Cash Flow = earnings before interest, taxes, depreciation, and amortization.

Source: *Tribune 1995 Annual Report.*

of earnings before interest, taxes, depreciation, and amortization expenses) amounted to $526 million—a more representative operating figure on which to base multiples for purposes of sale or market valuation, than the $393 million reported as net cash provided by operations after other encumbered expenses. Table 10.15b breaks out Tribune's specific broadcast and cable operating revenues, cash flows, and profit. Five radio stations and the Tribune Radio Networks brought in $88 million, with a cash flow of $15 million; revenues from ten TV stations were $630 million, cash flow $215 million; and cable programming and development for a regional news channel, plus equity loss from the WB Network and the Foods Channel, generated revenues of $7 million but a negative cash flow of $15 million. ■

Group owners were stronger financially than individual operators so they could benefit from the FCC's relaxation of regulations against multiple station ownership in the same market.[43] The Commission permitted "duopoly" effective in 1992, partly to keep floundering stations on the air in densely crowded radio markets; it also raised the limit on single ownership of radio stations to 18 AM and 18 FM, raising it further to 20 each in 1994. But duopoly owners generally chose to add stronger stations rather than weak ones. In less than two years, between September 1992 and mid-1994, the FCC received 935 applications for radio duopolies and 335 for LMAs. By 1994 duopoly operations in ten U.S. markets drew half or more of total local radio audiences; in each of twenty markets duopoly stations earned more than half of all ad revenues, with many pulling in two-thirds of market billings (up to 84% in Richmond, Virginia).

■   In Atlanta, according to a critic's analysis, Cox Communications' WSB-TV, WSB-AM/FM, and *Atlanta Journal-Constitution* already attracted half of the market's total media ad revenue (excluding cable) in 1994 as it sought to add yet another local station, WYAI-FM. Duopolies and LMAs (local marketing agreements among multiple stations) permitted staff consolidation and efficiency; a general sales manager oversaw a joint sales agreement with WKBH-AM-FM and WIZM-AM-FM, with a single staff selling all four schedules with accelerated success annually since 1991. Some duopoly operations cornered the market on a single format, as Group W's four signals in Houston, KILT-AM/FM and KIKK-AM/FM, all programmed country format. In 1995 television LMAs, with two separately licensed TV stations in a market managed by one operator, numbered thirty-six—ten of them in the top-30 markets. This raised fear in some quarters that airtime charges for advertising could be driven up by the all-but-controlling set of stations.■

TABLE 10.16
Station Transactions Approved by FCC, 1980–1995

YEAR	RADIO STATIONS	TV STATIONS IN GROUPS*	TV STATIONS	TOTAL VALUE OF TRANSACTIONS
1980	424	35	3	$ 876,084,000
1981	625	24	6	754,188,067
1982	597	30	0	998,398,244
1983	669	61	10	2,854,895,356
1984	782	82	2	2,118,056,053
1985	1,558	99	218	5,668,261,073
1986	959	128	192	6,192,669,871
1987	775	59	132	7,509,154,473
1988	845	70	106	4,947,838,198
1989	663	84	40	3,235,436,376
1990	1,045	75	60	1,976,626,100
1991	793	38	61	1,014,579,000
1992	603	41	24	1,045,373,000
1993	633	101	219 deals**	3,300,883,000
1994	648	NA	89 deals**	4,970,000,000
1995	737	NA	112 deals**	8,320,000,000

*Combined radio and TV stations, or multiple radio or TV stations, except for last two years.
**Transactions each involving multiple stations.
Transactions tabulated in 1994 and 1995 include some pending with FCC.

Source: *Broadcasting,* 4 March 1994, 37; 11 March 1996.

The national economy put brakes on media advertising revenues in the late 1980s. Auto manufacturers and dealers, department stores, and banks cut back severely on ad expenditures in the 1989–1990 period. Many new buyers of stations in the early to mid-1980s had no broadcast experience but were investing heavily, overpaying with inflated multiples. In 1990 it was estimated that almost $1 billion worth of radio properties on the market were in technical default. The total value of station sales annually tracks the state of the economy and the industry (table 10.16). The dollar total of station transactions almost tripled to $2.8 billion in 1983 from the previous year's $1 billion, doubling again within two years to $5.7 billion, then rising to $6.2 billion and peaking in 1987 at $7.5 billion. Dollar volume swiftly dropped more than $1 billion each successive year, leveling off at $1 billion in 1991–1992. Then expanding ad budgets fueled by a growing national economy, plus networks competing for key TV affiliations, as well as radio's duopolies and LMA consolidations, accounted for accelerated station deals, totaling $3.3 billion in 1993. The cycle was clear with successive sales of WHBQ in Memphis; it sold for $40 million in 1990, then for $54 million in 1993, and a year later sold yet again for $80 million to Fox. That year Fox and Savoy Pictures paid three times ($40 million) the 1984 price for WLUK in Green Bay, Wisconsin.

BROADCAST GROUPS

Hundreds of companies had licenses to operate multiple radio and TV stations. Few owned the legal maximum (first 7 each, then 12, among AM, FM, and TV stations, later raised to 18 each in radio in 1992, and then 20), partly because all TV properties were limited to reaching no more than 25%

(later 35%) of the total potential U.S. audience. Then almost unrestricted ownership was introduced with the Telecommunications Act of 1996. Pursuant to that law the FCC further deregulated the industry, permitting wholesale ownership of stations.

Some classic group owners who had pioneered public-spirited program service had already left the scene after decades of profitable service, such as Storer (bought by George Gillett, then New World Communications, and then Fox), Metromedia (purchased by Fox), and Times Mirror; several of them shifted their holdings to cable systems, especially Storer, Times Mirror, and Taft. Others continued to adapt competitively, including Westinghouse's Group W, Cox Enterprises, Tribune Broadcasting, Gannett, Hubbard, Jefferson-Pilot, Multimedia, and Park Communications; some of them also owned cable systems, including Cox and Multimedia. Station group owner Capital Cities Communications acquired ABC, Inc. New groups were created (Fox, from Metromedia and other independents), or were spawned from previous ones like NewCity Communications (formerly Katz), or merged (Shamrock and Malrite in 1993, whose combined 21 stations in 12 major markets made it the fifth largest radio group at that time). Still others were in flux, seeking buyers after the hard-charging 1980s fell prey to fiscal retrenchment in the 1990s; some sold off properties in bankruptcy reorganization, such as Gillett (SCI Television). Then the pendulum swung back again in the mid-1990s, with a growing economy and further deregulation; corporations gobbled up TV and radio stations at prices higher than the previous decade (see table 10.17.). Westinghouse/Group W bought CBS and then Infinity radio group in 1996, merging large clusters of TV and radio stations.

Channels magazine reported in 1989 the ten largest TV group owners ranked according to homes *actually delivered,* measured by Nielsen's average annual ratings for each group's stations (table 10.18). Those rankings of viewership can be compared with *Broadcasting & Cable* magazine's 1996 survey that reported the 25 largest TV station groups in terms of total households—*potential* audience homes—covered by their signals in each group's markets (table 10.19).[44]

■ By 1996 some groups had added or dropped stations; Gillett/SCI, which had bought many of Storer's stations in major markets, went into Chapter 11 bankruptcy reorganization, selling off its stations to New World Communications—linked with Fox Broadcasting; Fox bought New World outright in July 1996, making it the nation's largest group TV owner with twenty-two stations reaching almost 35% of the nation's households (with UHF station markets computed at half their size). Two Spanish-language station groups were among the top 12, at least according to potential audiences in total households in their market coverage areas (table 10.19). At the core of the mix were groups of flagship stations owned and operated by companies that ran major networks.

Personifying the turbulent mid-1990s was the Sinclair Broadcast Group which acquired 31 radio and 10 TV stations from River City Broadcasting in 1996. Sinclair owned more than any other company: 22 TV stations outright (18 of them UHF) plus 7 more UHFs operated through LMAs and time brokerage agreements. President David Smith sought to sustain the company's typical 50% profit margin in those operations, all in medium to small markets.

In radio during that same period, Westinghouse's Group W merged with CBS radio stations, then bought Infinity's group for a total of 82 AM and FM stations. In 1996—by outright ownership, local marketing agreements, joint sales agreements, and pending options to purchase—SFX Broadcasting with Multi-Market Radio operated 59

TOP 25 GROUP OWNERS, RADIO & TV, 1994
RANKED BY POPULATION REACHED BY STATIONS

	TELEVISION STATIONS				RADIO STATIONS		
Rank	Group Owner	Number of Stations	% of U.S. Households	Rank	Group Owner	Number of Stations	U.S. Population Reached
1.	CapCities/ABC	8	23.63%	1.	CBS	22	14,770,000
2.	CBS	7	21.86	2.	Infinity	26	14,170,000
3.	NBC	6	20.23	3.	Group W	18	11,940,000
4.	Tribune	7	19.46	4.	CapCities/ABC	18	11,780,000
5.	Fox	8	19.26	5.	Shamrock	18	7,220,000
6.	Chris Craft/United	8	18.0	6.	Viacom	14	6,540,000
7.	Silver King Communications	11	15.51	7.	Cox	14	6,510,000
8.	Univision	9	10.52	8.	Evergreen	11	5,650,000
9.	Gannett	10	10.4	9.	Bonneville	15	5,580,000
10.	Group W	5	9.72	10.	Emmis	5	4,670,000
11.	Telemundo	6	9.27	11.	Gannett	11	4,520,000
12.	SCI Television	7	8.8	12.	Greater Media	14	4,500,000
13.	Scripps Howard	9	8.43	13.	Clear Channel	26	4,500,000
14.	Cox Enterprises	6	7.62	14.	Susquehanna	16	3,660,000
15.	Post-Newsweek	6	7.03	15.	Summit	7	3,590,000
16.	Hearst	6	6.72	16.	Pyramid	12	3,460,000
17.	A. H. Belo	5	6.0	17.	Jacor	13	3,400,000
18.	Pulitzer	10	5.84	18.	EZ Communications	14	3,340,000
19.	Paramount	7	5.83	19.	Secret Communications	14	3,150,000
20.	Hubbard	9	5.81	20.	Great American	13	3,050,000
21.	Disney	1	5.3	21.	Tribune	6	2,950,000
22.	Great American	6	5.28	22.	Nationwide	11	2,640,000
23.	Providence Journal	9	5.22	23.	Park Communications	22	2,560,000
24.	LIN	7	4.71	24.	Jefferson-Pilot	13	2,550,000
25.	Gaylord	4	4.66	25.	Inner City	5	2,460,000

Source: Julie A. Zier, "Ranking TV and Radio's Top Players," *Broadcasting & Cable*, 21 March 1994, 52–56.

──w── **TABLE 10.18** ──w──

GROUP-OWNED TV STATIONS, 1988
TEN LARGEST GROUPS BY AUDIENCES DELIVERED

GROUP OWNER	NO. OF STATIONS	TOTAL HOMES REACHED	HOMES DELIVERED	AVERAGE RATING (ALL DAY-PARTS)
CapitalCities/ABC	8	22,098,010	1,883,464	8.5
NBC	7	20,199,320	1,436,202	7.1
CBS	5	18,687,960	1,080,469	5.8
Fox Television	7	21,652,750	859,142	4.0
Gillett Communications Co.	12	11,718,520	755,927	6.5
Westinghouse Broadcasting Co.	5	8,928,150	668,349	7.5
Tribune Broadcasting Co.	6	17,810,230	654,624	3.7
Gannett Broadcasting	10	10,219,630	639,196	6.3
Scripps-Howard Broadcasting	9	8,374,530	588,964	7.0
Cox Enterprises, Inc.	7	9,468,430	586,116	6.2

Source: *Channels* magazine, April 1989, citing Nielsen Station Index, November 1988.

stations and operated another 12, plus 32 through Triathlon Broadcasting. Clear Channel Communications and U.S. Radio operated 102 AM and FM stations. Jacor Communications, with Noble Broadcast Group and Citicasters, ran 51 stations. This was a far cry from the decades-old limit of 7+7 AM/FM stations to any owner-licensee. (See table 10.20.) ■

— TABLE 10.19 —
GROUP-OWNED TV STATIONS, 1996:
25 LARGEST GROUPS BY TOTAL HOUSEHOLD COVERAGE
(POTENTIAL AUDIENCE)

GROUP OWNER	NO. OF STATIONS	TOTAL HOMES REACHED*	PERCENTAGE OF TOTAL U.S. HOMES**
1. Westinghouse/CBS	14	30,048,090	31.53%
2. Tribune (including Renaissance)	16	24,501,630	25.71
3. NBC	11	23,491,450	24.65
4. Disney/ABC	10	22,938,710	24.07
5. Fox[†]	12	21,013,650	22.05
6. Silver King	16	20,966,000	22.00
7. Paxson	16	17,125,410	17.97
8. Chris-Craft/BHC/United TV	8	16,868,100	17.70
9. Gannett	15	13,418,240	14.08
10. Univision (Spanish)	11	12,255,580	12.86
11. New World[†]	10	12,179,340	12.78
12. Telemundo Group (Spanish)	8	9,863,550	10.35
13. Viacom (Paramount)	12	9,644,360	10.12
14. Young Broadcasting	13	8,653,240	9.08
15. Sinclair/River City	22	8,491,230	8.91
16. Scripps-Howard	9	7,624,000	8.00
17. A. H. Belo	7	7,614,470	7.99
18. Cox Broadcasting	7	7,376,220	7.74
19. Hearst	7	6,975,960	7.32
20. Post-Newsweek	6	6,651,194	6.98
21. LIN TV	9	6,013,430	6.31
22. Providence Journal	11	5,146,200	5.40
23. Pulitzer	10	4,965,130	5.21
24. Ellis Acq./Raycom	22	3,850,120	4.04
25. Albritton Communications	8	3,754,820	3.94

*Not audiences viewing stations but population in stations' markets; see previous table 10.18 and note 1 for earlier listing of top ten station groups based on actual homes delivered according to average local ratings of each station in group.

**Calculated on base of 95.3 million TV households (approximations due to varied counting options; see source). The Telecommunications Act of 1996 limits TV coverage to 35% of all U.S. households—with UHF stations computed at half their markets' households.

[†]In July 1996 Fox bought New World; the new total of 22 stations reaching 34.83% of U.S. households made it #1.

Source: Data based on survey reported by Broadcasting & Cable, 8 July 1996, 12–20; 22 July 1996, 6.

Concentration of ownership in some markets triggered questions of anti-competitive economic control, such as Jacor's 8 radio stations in Denver and its 6 stations plus another financially related one in Cincinnati. Some radio broadcasters claimed that company controlled 40% to 80% of radio revenues in at least one market. Advertisers raised concerns about forced escalation of airtime rates in such situations. The Justice Department inquired and the FCC reviewed selected markets with large clusters of stations under common ownership and involving LMAs—local marketing agreements (see chapter 11).

In a single week in May 1996, station trading set a record of $1.87 billion worth of deals. This came nine months after the week when corporate media announced $24 billion in mergers: Disney's purchase of CapCities/ABC ($18.5 billion) and Westinghouse's acquisition of CBS ($5.4 billion).

Meanwhile, group-owned Hispanic stations included Spanish Broadcasting System with 6 major-market radio stations; Lotus Communications Corporation's

TABLE 10.20
Top 10 Radio Station Groups, 1995
Ranked by estimated gross revenues

GROUP	# STATIONS	EST. STATION REVENUE
Westinghouse/CBS	39	$ 509,300,000
Infinity Broadcasting	27	361,000,000
CapCities/ABC/Disney	21	273,300,000
Evergreen Media/Pyramid	34	265,200,000
Chancellor Broadcasting/Shamrock	33	185,200,000
Cox Broadcasting	15	145,900,000
Clear Channel Communications	36	136,600,000
Citicasters/OmniAmerica	28	135,000,000
Jacor Communications	26	130,000,000
American Radio Systems	27	122,400,000

Note: Some combined acquisitions were yet to be approved, plus Westinghouse/CBS in 1996 moved to acquire Infinity.

Source: Jim Duncan, *Duncan's Radio Comments* (August 1995), cited in *Broadcasting & Cable,* 18 September 1995, 37.

4 radio outlets, plus 10 others; and Tichenor Media System's 14 stations, including a New York AM, and AM-FM combos in Chicago and Miami.

Group managers at the corporate level oversaw long-range planning and budgeting among their various stations. Some organizations permitted considerable local autonomy within preset parameters, whereas others participated closely in long- and short-term activities of stations to ensure that each carried their weight towards corporate goals of profitability and service. The challenge in complex organizations scattered across several states or in major markets around the country was to ensure that all stations carried out company policy ("macro-intentions") at all levels of local management, while avoiding intrusive micro-managing or second-guessing local managers to whom they delegated authority to make decisions for their own staffs and operations. Local general managers are typically also vice-presidents of their corporations, sharing in the group's deliberations about corporate/divisional and individual stations' revenue projections, budgeting, technical and programming acquisitions, and personnel and regulatory issues.

As noted earlier, when typical affiliates' profitability in the early 1990s was around 25% of revenues, network O&Os were closer to 45% or higher; major-group stations fell somewhere in between. In 1995 in the top-3 markets of New York, Los Angeles, and Chicago, ABC O&Os' profit margins were 62% of revenues, NBC's 53%, and CBS 44%. In markets 4–6 Westinghouse's Group W stations had profit margins of 52%, NBC's 51%, ABC 49%, and CBS 42%.

Cable MSOs

Most local cable systems are owned by corporations that operate or are linked with many other systems in various parts of the country. By far the largest was John Malone's Denver-based Tele-Communications, Inc., whose owned systems in the early 1990s served 5.2 million basic subscribers, plus investments in 12 affiliated systems with 3 million subscribers; by 1996 TCI's cable systems served 12 million households. With that broad base, TCI had the resources to pioneer broadband cable service capable of hundreds of signals to subscribers in Queens

TABLE 10.21
Top 10 Cable MSOS, 1995

RANK	COMPANY	SUBSCRIBERS
1.	Tele-Communications, Inc. (Denver)	14,700,000
2.	Time-Warner Cable (Stamford, Conn.)	11,500,000
3.	Continental Cablevision (Boston)	4,000,000
4.	Comcast Corporation (Philadelphia)	3,400,000
5.	Cox Cable Communications (Atlanta)	3,200,000
6.	Cablevision Systems Corporation (Woodbury, N.Y.)	2,600,000
7.	Adelphia Communications (Coudersport, Pa.)	1,600,000
8.	Jones Intercable (Englewood, Colo.)	1,300,000
9.	Falcon Cable TV (Los Angeles)	1,100,000
10.	Sammons Communications (Dallas)	1,100,000

Source: *Broadcasting & Cable,* 6 March 1995, 53.

on Long Island. Malone announced plans for the late 1990s to establish a $2 billion fiber-optic, 500-channel national cable distribution system.

The clout of MSOs—especially mega-MSOs such as TCI, Time Warner, ATC/ American Television & Communications, and Comcast—lent weight to their strenuous objection to the FCC's rulings in 1992–1993 on must-carry/retransmission-consent, which favored stations and networks over cable operators (see chapter 11 on regulation and law). Obviously large multiple system operators also enjoy leverage when negotiating with cable program suppliers for per-subscriber rates to carry services such as CNN, ESPN, MTV, Nickelodeon, and others.

By 1988 the top-10 MSOs connected almost half of the nation's cabled homes, over 20 million cable subscribers. Leading the ten largest MSOs—including partly owned systems and pending deals—in 1996 were Tele-Communications, Inc. (14.7 million subscribers) and Time Warner Cable (11.5 million).[45]

Tables 10.18–20 list major group owners of radio and television properties, and Table 10.21 lists the top-10 multiple system cable operators, as of 1996.

CORPORATE OWNERSHIP: MERGERS, CONGLOMERATES

"Bigger is better" seemed to be the mantra in the 1980s, as recounted in chapter 1. Media giants merged or bought out major players, sometimes several times their own size (partly because Justice Department concern about megalopoly deterred a colossus from gobbling up a smaller entity). The largest media-related mergers ranked right along with massive conglomerates in oil, food, and tobacco industries (table 10.22).

■ General Electric's acquisition of RCA-NBC in 1986 was the biggest media deal to date, $6.2 billion. That was doubled by Time's merging with Warner Communications for $13.9 billion. The biggest deal of all—a $33 billion union of Tele-Communication, Inc. and Bell Atlantic—was called off, partly because FCC cable rate reregulation and a slow economy dampened cable's cash flows and made the finances less lucrative for Bell Atlantic. That deal would have been half again larger than the country's biggest merger (Kohlberg Kravis Roberts & RJR Nabisco, $24.9 billion) and almost three times larger than Phillip Morris & Kraft ($13.3 billion) or Time/Warner ($13.9 billion).

Some major acquisitions worked despite the costs, while others failed in the slowed economy of the early 1990s. In 1985 Rupert Murdoch's News Corporation bought Twentieth Century-Fox studios and then Metromedia's seven TV stations which reached almost 23% of the national audience. He paid 17 times cash flow or

TABLE 10.22
LARGEST MEDIA-RELATED CORPORATE MERGERS AND ACQUISITIONS
1985–1996

YEAR	ACQUIRER / ACQUIRED	VALUE OF DEAL (BILLIONS)
1985	Capital Cities Communications/ABC	$ 3.5
1986	General Electric / RCA	6.2
1987	Viacom / National Amusements	3.4
	Sony Corporation/CBS Records	2.0
1989	McCaw Cellular/LIN Broadcasting	3.3
	Time, Inc./Warner Communications	13.9
	Sony Corporation/Columbia Pictures	5.0
1990	GTE / Contel	6.2
1991	Matsushita/MCA-Universal	6.5
1993	AT&T/McCaw Cellular (*proposed*)	12.6
	[Bell Atlantic/TCI (*withdrawn*)]	33.0
1994	Viacom/Paramount Communications	9.8
	[CBS Inc./QVC cable net (*withdrawn*)]	7.2
	Comcast/QVC cable net	2.3
1995	Seagram/MCA-Universal (80%)	5.7
1996	Disney/CapCities-ABC	18.5
	Westinghouse/CBS	5.4
	Time Warner/Turner	6.5
	Westinghouse-CBS/Infinity (radio)	4.9
	SBC Communications/Pacific Telesis	17.0
	Bell Atlantic/Nynex (*proposed*)	23.0

Source: *The World Almanac, 1994,* 112; *Securities Data,* 1994; various trade publications, including *Broadcasting & Cable,* 1995, 1996; *Los Angeles Times,* 22 April 1996.

$2 billion for six independent stations in top markets (with total cash flow of $94 million annually), reselling the seventh, WCVB-TV in Boston, for $450 million—double what Metromedia paid for it three years earlier. Murdoch used those six stations as the nucleus for his new Fox network. In 1986 Ted Turner paid $1.5 billion for MGM studios; he financed the enormous debt partly with high-interest "junk bonds" and by selling off part of his holdings, plus the aid of a consortium of cable owners (TCI and Time, Inc. and others who became part owners, dropping Turner's 81% control down to 51%). But Turner maximized the value of his MGM library of 3,301 movies along with those of RKO and Warners, for his WTBS superstation, and for starting his Turner Network Television classic movie channel two years later. Sumner Redstone's National Amusements bought Viacom, including its MTV cable networks, for $3.4 billion in 1987; they assembled a financial package in which their investment was less than $500 million (that investment within four years was valued at over $2.6 billion).[46] In 1994 Redstone's Viacom in heated competition to acquire Paramount Communications successfully outbid Barry Diller's QVC by offering $9.6 billion; that laid the base for a fifth broadcast network (initially called United/Paramount). ■

Others were less successful in leveraged buyouts of other companies. Otherwise competent operators were savaged by the stock drop and economic downturn in the late 1980s and early 1990s. This resulted in a weak advertising market that could not generate adequate revenues to offset enormous burdens of debt to be serviced.

■ George Gillett's purchase of three small UHF stations in 1978 for $6 million led to other buys, and another 12 stations in 1986–1987 with $650 million of high-yield debt. At the peak of station trading in 1987 he paid $385 million for a CBS affiliate in Tampa and then $1.3 billion for six Storer/SCI stations. (The complicated equity/investment deal involved $100 million each from him and Kohlberg Kravis Roberts, the rest financed half by bank loans and half by "junk bonds" and notes.) With stations reaching 13% of the national audience, Gillett was noted by *Variety* reporter Paul Noglows as "the country's largest broadcaster next to the networks." As noted earlier, the crushing debt load could not be met by fading revenues; all station values plummeted. Soon Gillett's properties were in Chapter 11 reorganization with other parts of his operation in bankruptcy; his stations were sold off at enormous loss. Similarly ill-fated were other companies' media properties acquired at the height of easily available funding, which lost heavily as the market plummeted. TVS in Britain bought MTM television production studios in 1988 for $325 million, but could not resell them for $70 million three years later. ◼

In 1989 one-third of all cash flow in the nation's corporations went to pay interest on deep loans, causing some otherwise profitable operations into Chapter 11 bankruptcy. In 1991 that proportion lessened to one-fourth, and by 1994 debt payments dropped to one-fifth of corporate cash flow.[47] (In 1991 over 810,000 commercial and personal bankruptcies far outpaced the 63,000 in Depression year 1930 following the stock market crash.) The bust began to cycle back towards a boom by the mid-1990s, with TV's growth in ad revenues in double digits, the strongest rate of growth since 1984. Media properties again found major players competing for them, once more driving up prices.

◼   Early in 1994 the ComCorp subsidiary of Associated Broadcasters bought WHBQ-TV in Memphis for $54 million, and resold it months later to Fox for $80 million. CBS stock dropped from a high of $221 a share in 1989 to $182 in mid-1990, then to $133 late in 1991; it rebounded prior to Westinghouse's acquiring it in 1995. Through that period, amid complex shared ownership and debt structuring, Ted Turner's total media empire saw revenues rise from $800 million in 1988 to almost $1.8 billion in 1992. When it acquired Turner in 1996, Time Warner was already carrying an enormous debt of $17 billion. At that time TCI's debt was $12 billion, and Viacom's was $10 billion (after its $8.2 billion merger with Paramount—17.5 times 1994 cash flow of filmed entertainment operations). ◼

As a result of mergers, in 1994 *Broadcasting & Cable* magazine revised its ranking of publicly reported media companies based on revenues. CapCities/ABC and CBS ranked behind Time Warner and TCI/Liberty whose revenues from broadcast-related business each exceeded $5.3 billion (table 10.23).[48]

The same trade publication also ranked media companies and electronic communication divisions of other corporations based on revenues for 1993.[49] Table 10.24 lists the top 10 in broadcasting, cable, and program services, according to total revenue from electronic communication operations, each company's electronic revenue as a percentage of total revenues, operating income, and total net earnings. The tabulated listings offer a profile of the continually changing face of broadcast and cable media and related companies. The constant flux would find some moving up, others down or off, the list of leading companies in electronic communications. Despite being outdated because of mergers and acquisitions in the next three years, the list serves as an index to the companies and revenues involved in proliferating deals.

Early in 1986 as merger mania and buyouts first accelerated, CBS/Broadcast Group President Gene Jankowski spoke of the "new marketability of the broadcast license" affected by shifts in regulation and technology; "and out beyond all of this is the circle of investors, arbitrageurs, merger-makers and takeover specialists, a mega-force entirely new to the industry."[50] He noted "this halfway point in the decade was the time Wall Street discovered our industry." Author Christopher Byron

TABLE 10.23
TOP 11 MEDIA COMPANIES WITH PUBLICLY REPORTED REVENUES, 1993

1993 RANKING	COMPANY	REVENUE* (MILLIONS)	1992 RANKING
1.	Time Warner	$ 5,324	1
2.	TCI/Liberty	5,306	3
3.	Capital Cities/ABC	4,663	2
4.	CBS	3,510	4
5.	Sony/Columbia	3,262	6
6.	Viacom/Paramount	3,231	8
7.	General Electric/NBC	3,102	5
8.	Matsushita/MCA	2,572	7
9.	Comcast/QVC	2,528	17
10.	Turner Broadcasting	1,761	9
11.	News Corporation (Fox)	1,720	NA

*Revenues are from corporations' media-related divisions only.

Source: Estimates by *Broadcasting & Cable,* 25 July 1994, 15.

ended his searing account of Time/Life's debacle with short-lived *TV-Cable Week* by reflecting on the changing forces and values in business generally:

> With their rise has come a whole new financial orientation to American business—not the development of new products and processes for the marketplace. As Robert Reich, a one-time policy chief at the Federal Trade Commission [and later Secretary of Labor], says of these 'paper entrepreneurs' of the 1970s and 1980s:
>
> 'They innovate using the system in novel ways: establishing joint ventures, consortiums, holding companies, mutual funds; finding companies to acquire, "white knights" to be acquired by, commodity futures to invest in, tax shelters to hide in; engaging in proxy fights, tender offers, antitrust suits, stock splits, spinoffs, divestitures; buying and selling notes, bonds, convertible debentures, sinking-fund debentures; obtaining government subsidies, loan guarantees, tax breaks, contracts, licenses, quotas, price supports, bail-outs; going private, going public, going bankrupt.'
>
> That is quite a list, but it hardly exhausts the new financial lexicon. Since Reich's comments in 1980, even more jargon has been added to that language of business: leveraged buyouts, corporate greenmail, golden parachute deals, shark repellent strategies, scorched-earth ploys, Pac-man scheme—the list grows and grows, all of it to help business managers who know little about individual products or markets make their firms grow bigger without necessarily helping them to become better. . . .
>
> The root of that struggle traced to yet another expression of financially oriented corporate management: the cult of 'growth by numbers' and the rise of the business conglomerates. In its way, the corporate conglomerate is the final elaboration of professional management theory, a business enterprise in which the focus of decision making is not the customer or the marketplace but Wall Street, the stock market, and earnings per share.[51]

THE MANAGER'S CHALLENGES

The broadcast manager is a person of at least two worlds. She is charged with properly operating a publicly licensed radio or television station to serve the local community. At the same time she is charged by her employers to oversee the entire station operation in a way that produces increasingly greater revenue.

— TABLE 10.24 —

LEADING COMPANIES IN ELECTRONIC MEDIA, 1993 REVENUES
($ IN MILLIONS)

RANK/COMPANY	ELECTRONIC COMMUNICATIONS* REVENUES	% OF TOTAL COMPANY REVENUES	OPERATING INCOME	TOTAL NET EARNINGS
BROADCASTING				
1. CapCities/ABC	$ 4,663.2	82%	$ 778.1	$ 455.3
2. CBS	3,510.1	100	411.2	326.2
3. General Electric	3,102.0	8	264.0	4,315.0
4. News Corporation	1,716.7	23	362.3	605.2
5. Tribune	727.2	37	125.7	188.6
6. Westinghouse	705.0	8	136.0	(326.0)
7. BHC Communications	412.0	100	79.3	224.3
8. Gannett	397.2	11	86.7	397.8
9. Scripps Howard	371.7	100	85.9	100.3
10. New World Communications	369.1	100	38.5	(20.3)
CABLE				
1. TCI	$ 4,153.0	100%	$ 916.0	$ (9.0)
2. Time Warner	3,649.0	44	619.0	(221.0)
3. Continental	1,177.2	100	232.0	(210.8)
4. Comcast	1,095.4	82	311.5	(859.2)
5. Cablevision Systems	666.7	100	57.3	(246.8)
6. Knight-Ridder	491.2	20	242.0	148.1
7. Multimedia	481.9	76	158.7	99.9
8. Times Mirror	470.4	13	106.5	317.2
9. Cablevision Industries	397.0	100	15.2	(101.0)
10. Washington Post	363.1	24	107.0	165.4
PROGRAM SERVICES				
1. Viacom	$ 2,004.9	100%	NA	$ 171.0
2. Turner Broadcasting	1,761.6	92		244.2
3. Liberty Media	1,153.3	100		4.8
4. Paramount	1,407.0	28		197.0
5. QVC Networks	1,222.1	100		59.3
6. Disney	984.2	12		299.8
7. King World Productions	474.3	100		101.9
8. Gaylord	442.1	67		27.6
9. Spelling	216.8	79		17.7
10. International Family Entertainment	208.2	100		17.3

*Revenues from electronic communication operations only.

Source: *Broadcasting & Cable,* 27 June 1994, 40, 42.

Harry Skornia, a critic of American commercial broadcasting, recognized that the inherent organization of modern business as well as supportive legal precedent seem to militate against a full public-interest orientation by commercial broadcasting. He cited the Michigan Supreme Court which supported a Ford Motor Company stockholder who objected to Henry Ford's attempt to reduce the price of that company's automotive products. Skornia noted:

> The court held that the business corporation, created to operate for profit, must serve its stockholders first and the public only secondarily—the corporation interest, rather than the public, must be favored whenever profits may be affected. . . . A corporation's charitable or public-interest expenditures must advance the long-range prospects of profit making. . . . Therefore, a corporation created to operate for profit, regardless of any desire to serve 'public

interest first' by its officials, must serve its owners and stockholders first, returning to them as large a profit as possible. Many corporations have been able to reconcile these two interests satisfactorily; how well broadcast corporations have done so will become clearer as the record is examined.[52]

But at this point the manager is confronted by several kinds of data which suggest that producing revenue and making profits might well contribute to a station's community precisely through its programming. A study of how television station performance relates to revenues cited evidence from FCC files for several hundred television stations. The study documented the predictable conclusion that "stations with higher revenues broadcast more hours of local programming" than do stations with lower revenues, although those extra hours occur outside of large-audience prime-time. Also, more money was spent on local and nonentertainment programming in direct proportion to the additional revenue dollars of a station (a small network affiliate spent 15¢ of each additional revenue dollar on local programming, while a large network affiliate spent 17¢).[53] Similar studies tended to support the finding that multiple-station owners—with large revenues and capital—provide diversity of views in nonentertainment programming, and provide more local and public service programming for their communities than does single-station ownership.[54] A corollary research effort found that commonly-owned media in a single market ("cross-ownership") also provided generally better program service than did stations owned by companies not involved with other media in the market.[55] Some of those earlier issues became less urgent and even moot with the FCC's deregulation in the 1980s, and especially with relaxed rulings permitting local management agreements and duopolies in the 1990s.

The many theoretical and practical challenges to the media manager involve economics, politics, sociology, law, psychology, and most other human, artistic, and scientific disciplines. They relate to political candidates' use of airwaves; crime and violence on the screen; aesthetic, perceptual, and behavioral effects on the audience; "overcommercialism"; and a host of other significant issues in society. A scattered series of court decisions have given precedent for state and local taxation of commercial broadcasting, including local-oriented (intrastate) advertising, and net-income state taxes on broadcast profits, gross receipts taxes, use taxes, and attempts at taxing airtime sales.[56] A station's net income can be directly affected by such legislation. Where instituted, taxing initiatives and statutes have been struck down or reversed by their creators, at least to date.

The station or cable manager is faced with a challenging dilemma. He must provide programming and scheduling for a community with a federally-regulated license or locally sanctioned franchise while at the same time striving for annually increasing profitability to the owners and investors in that station as a business. Les Brown, formerly of *Variety* and then the *New York Times,* noted realistically that "American television is a business before it is anything else, and within the broadcast companies the sales function is pre-eminent." He added the sobering reflection:

> There is no other course but for broadcast managements to dedicate themselves to profit growth; their executive survival depends upon it. They must at the same time convey the impression of being stable and sturdy in the face of the speculative and volatile nature of show business, and so to whatever extent possible they divorce themselves from the impresario risks and behave as companies engaged in the manufacture of goods. They deal, therefore, in programs that will be instantly accepted by the audience, rejecting new and experimental forms that might take weeks or months to catch on, if at all.[57]

That realistic analysis is confirmed by "one who was there," who had to meet payrolls and answer for profit and loss statements to owners. Joseph Mikita, vice-president of finance for Westinghouse Broadcasting's Group W, summarized the manager's multiple responsibilities:

> To satisfy their own responsibility to the public, broadcasters must do many things which will not immediately enhance the profitability of a given television station. Nevertheless, it is essential that the financial yardstick be applied to all operations if intelligent financial planning is to accomplish an atmosphere of healthy growth.
>
> Forecasting has become a necessity in television broadcasting because of the high cost of operation and because the profit margin is highly susceptible to unfavorable variations in an increasing sales pattern. There is a reasonably fixed level of operating costs which bears little or no relation to the amount of commercial business the station enjoys. Beyond this level of fixed costs, or "break-even volume," the television station can convert a substantial part of sales improvements to operating profit. Conversely, a drop-off in business is generally reflected in reduction in operating profit. It is essential, therefore, that for a healthy growth in profitability, the gross business improvement inherent in a young and dynamic industry contain within itself a satisfactory margin of profit. The measure of profitability must therefore be applied to new ventures in programming promotion, facilities improvement and, in fact, to any other area designed to increase the gross revenue of a station.[58]

A final note supporting the traditional, special role of broadcasting in society was sounded by CBS executive Tom Leahy (cited earlier in chapter 1). He cautioned affiliate managers that solely dollar-oriented buying and selling of broadcast properties

> is at great variance with what we have always felt about the essential character of broadcasting. That is not the kind of business that you and I signed on for. It is not the kind of business that we have been in. It is not the kind of business that we believe the public wishes us to be in. . . . The public interest has been the standard by which the industry operated, and we believe that we have thrived as commercial enterprises because we have served the public interest.[59]

He urged media companies not to sacrifice "long-term credibility in the community, and in their business relationships, for short-term transitory gain." Successful veteran broadcasters hope their successors in broadcast and cable management continue to support that viewpoint into the next century.

The next chapter looks to the role of government which monitors how media managers serve society while profiting through competitive free-enterprise.

⁓ CHAPTER 10 NOTES ⁓

1. Paul Richard, "Everything You Always Wanted to Know About Broadcasting/Cable Insurance—But Were Too Busy to Ask" (Chicago: Broadcast Financial Management, 1985), 18; booklet also appeared as an article in *Broadcast Financial Journal*.

2. Media managers will find as readable as it is useful Steven A. Finkler's (of the University of Pennsylvania's Wharton School) *The Complete Guide to Finance & Accounting for Nonfinancial Managers* (Englewood Cliffs, N.J.: Prentice-Hall, 1993).

3. For a readable treatment of terms, concepts, and procedures for processing ratios, including return on investment (ROI) and other ratios, see Finkler, Chapter 10, "Ratio Analysis," 194–211.

4. William F. Christopher, *Management for the 1980's* (New York: AMACOM/American Management Association, 1980), 152–153; see Chapter 10, "Discovering Managerial Economics," 151–163.

5. Joseph K. Mikita, "The Controller's Role in Management," in Yale Roe, ed., *Television Station Management: The Business of Broadcasting* (New York: Hastings House, 1964), 94; see Chapter 6, 91–104, for amplification of these points.

6. Broadcast Financial Management Association, *Operational Guidelines and Accounting Manual for Broadcasters* (BFMA: Des Plaines, Ill.: [various years of booklets published starting in 1978]); booklets often include 70–80 pages each (double-spaced typescript format) and hole-punched for insertion into the jumbo three-ring binder. The earlier compilation was Institute of Broadcasting Financial Management, Inc., *Accounting Manual for Broadcasters* (revised in association with Ernst & Ernst, 1972), 360 North Michigan Avenue, Chicago, Ill. This loose-leaf binder provided a fifty-one-page, single-spaced itemized listing and description of specific account entries together with subentries, and sixteen pages of sample forms for financial statements. Periodically updated 8.5" x 11" booklets for insertion in the binder offer clearly outlined descriptions and sample forms such as "Risk & Insurance Guidelines for Broadcasters & Cable Operators" (79 pages); "Operations Audit Guideline for Broadcasters" (125 pages); "Internal Control Guidelines" (61 pages); "BFM Guidelines for Trade and Barter Transactions," (19 pages); "Discussion of Credit, Collections and Accounts Receivable for Broadcast Entities," (10 pages); "Broadcasters Data Processing Vendor Guide," (21 pages); "Record Retention Guideline for Broadcasters," (23 pages); "Cash Control Accounting Guidelines for Cable Communications Industry," (26 pages); and an "Accounting Manual for Broadcasters" with five sections on account descriptions (25 pages); financial statement forms (7 pages); accounting for film contracts (5 pages); and trade agreements policies and procedures (12 pages).

7. See especially Alison Alexander, James Owers, and Rod Carveth, eds., *Media Economics: Theory and Practice* (Hillsdale, N.J.: Lawrence Erlbaum Associates, 1993) and Bruce M. Owen and Steven S. Wildman, *Video Economics* (Cambridge, Mass.: Harvard University Press, 1992). For earlier analyses see: publications by the Broadcast Management Department of the National Association of Broadcasters (Washington, D.C.): "Accounting Manual for Radio Stations," "Accounting Manual for Television Stations," "Internal Control in Broadcasting Stations"; Warde B. Ogden, *The Television Business: Accounting Problems of a Growth Industry* (New York: Ronald Press, 1961); Jay G. Blumler and T. J. Nossiter, eds., *Broadcasting Finance in Transition* (New York: Oxford University Press, 1991); Alan B. Albarran, *Media Economics: Understanding Markets, Industries, and Concepts* (Ames, Iowa: Iowa State University Press, 1996).

8. Based on the form in BFMA, *Operational Guidelines and Accounting Manual for Broadcasters,* Section II, "Financial Statement Forms," II–1.

9. Ibid., II–3.

10. Helpful explanations of assets and liabilities may be found in Alexander, Owers, and Carveth, "Appendix A: Media Accounting Practices" and "Appendix B: Financial Management," 355–367; Finkler, 12–13; and Erwin G. Krasnow, J. Geoffrey Bentley, and Robin B. Martin, *Buying or Building a Broadcast Station in the 1990s,* 3rd ed. (Washington, D.C.: National Association of Broadcasters, 1991), 22–23.

11. "TV, Cable, Telcos: What Are They Worth?" *Broadcasting,* 26 September 1988, 35–36; see the previous year's "Putting a Price on TV and Cable," *Broadcasting,* 31 August 1987, 31–32.

12. Quoted in *Variety,* 9 September 1991, 27.

13. Veronis, Suhler & Associates, New York, cited by *Broadcasting & Cable,* 5 June 1995, 9.

14. By 1992 the General Accounting Office reported 159 intangible assets were subjects of disputes between the IRS and taxpayers. They included or related to favorable leases, customer bases, FCC licenses and construction permits, income agreements, advertising lists, broadcasting rights, cable franchises, network affiliation agreements, and other intangible assets based on market shares. See John H. Sanders, "Update on Amortization of Intangible Assets," *Broadcast Cable Financial Journal*, January/February 1992, 11.

15. Paul Kagan Associates, Inc., *Cable TV Advertiser,* 23 May 1991, 3; *Cable TV Investor,* 24 December 1991, 5. A 1990 total of $13 billion had been estimated by Richard H. Beahrs, president, The Comedy Channel, in "Monday Memo," *Broadcasting,* 26 February 1990, 25.

16. CPB, *Public Broadcasting Income, Fiscal Year 1991* (Washington, D.C.: Corporation for Public Broadcasting, 1992); Figure 6 reproduced by Head, Sterling, and Shofield, 276. These authors describe details of funding, distribution, and programming policy in Chapter 8, "Noncommercial Services," 264–301.

17. CBS/Broadcast Group President Gene Jankowski's estimate, quoted by *Broadcasting,* 14 September 1987, 109.

18. Estimates were by Television Bureau of Advertising, reported as a graph by *Electronic Media,* 3 August 1992, 20. The previous 1990–1991 season CBS had cut its compensation by 20% to $134 million, and NBC dropped 10% to $132 million; ABC spent $110 million on affiliate compensation that year; *Variety,* 7 January 1991, 1.

19. In fiscal 1992 business corporations contributed $89.5 million to PBS (29% of all funding); public TV stations' funds to national PBS were $84.6 million (28.1%)—including portions of audience subscribers' donations passed on to PBS in the form of program assessments; producers and co-production funding amounted to $47.6 million; the Corporation for Public Broadcasting dispersed $37.7 million of federal funds to PBS; private foundations gave $19.3 million, while federal and state agencies added $18.6 million. Data from PBS report cited by Richard Huff, "PBS: In the Money in '92," *Variety,* 11 January 1993, 51.

20. See NAB and BCFM annual financial reports published in 1992; a well laid out set of summary tables can be found in Sydney W. Head, Christopher H. Sterling, and Lemuel B. Shofield, *Broadcasting in America: A Survey of Electronic Media,* 7th ed. (Boston: Houghton Mifflin, 1994), 256–257. The NAB discontinued surveying finances of radio stations in 1993; it had taken over that task when in the early 1980s the FCC abandoned gathering such data for its own annual reports. The U.S. Department of Commerce began surveying in 1992, reporting the previous year's data. Further reason for NAB's decision to discontinue was weak response from radio stations (fewer than one-fourth of 10,000 stations returned the long-form survey, 77% of the 800 receiving the short-form) coupled with radio's complicated duopolies and local marketing agreements, permitting stations to consolidate sales, engineering, promotion, and other workgroups for multiple stations in the same market. See Peter Viles, "NAB Pulls Plug on Radio Station Financial Survey," *Broadcasting & Cable,* 31 May 1993, 26.

21. Data by Wireless Cable Association, cited in table by *Electronic Media,* 21 September 1992, 4.

22. See Robert G. Picard, "Copyright Royalty Payments Top $200 Million as Payments by Cable Triple," *Broadcasting Cable Financial Journal,* July/August 1992, 8–11.

23. See the presentation by Harold Poole, a director of the Broadcast Credit Association and treasurer of the Institute of Broadcasting Financial Management, at an NAB panel discussion in Houston, Texas, March 20, 1974; published as part of an NAB booklet: "Beating the Profit Squeeze: A Financial Management Workshop," 12–18. He noted that IBFM for several years advocated before the Internal Revenue Service a change of Ruling 62–20. Cf. Ogden, *The Television Business: Accounting Problems of a Growth Industry,* especially 119–137, "Station Accounting"—including amortizing syndicated film packages. The move to accelerated depreciation for syndicated program packages is clearly described with examples illustrating alternate options by Kevin L. Raymond and Michael H. Schwerdtman, "Broadcast Rights Amortization—Emerging Trend," *Broadcast Financial Journal,* March/April 1990, 4–6. The same authors analyzed cash flow requirements for current and future program packages in "Forecasting Cash Flow Needs for Broadcast Rights," *Broadcast Financial Journal,* November/December 1988, 8–9.

24. Data listed by NAB and BCFM in tables throughout *1990 Radio Financial Report* and *1990 Television Financial Report.*

25. "FCC Says Radio is in 'Profound Financial Distress,'" *Broadcasting,* 3 February 1992, 4.

26. From a speech by Jerell A. Shepherd at the 1968 NAB convention in Chicago, quoted in its entirety in "Profile 9, KWIX," by Joseph S. Johnson and Kenneth K. Jones, *Modern Radio Station Practices* (Belmont, Calif.: Wadsworth, 1972), 174–182.

27. Jay Hoffer, *Managing Today's Radio Station,* 50–55. Barry L. Sherman devotes Chapter 10, "Entering the Telecommunications Marketplace," to outlining steps for acquiring media properties by establishing a new station or cable system or else buying an existing operation; *Telecommunication Management: The Broadcast & Cable Industries* (New York: McGraw-Hill, 1987), 190–219.

28. Figures rounded from data in NAB and BCFM, *1990 Radio Financial Report* (Washington, D.C.: NAB, 1990), 85–115. See "Note" at bottom of table 10.2b, cautioning that data were unevenly reported which skewed the averages, especially in some subcategories with fewer stations participating.

29. Data from NAB's *1992 Television Financial Report* cited by Head, Sterling, and Schofield, 258.

30. Steve McClellan, "Station-group Revenue on Comeback Course," *Broadcasting & Cable,* 25 April 1994, 18–19.

31. Alfred J. Jaffe, "Tinker-to-Wright: A Hard Act to Follow as NBC $$ Peak," *Television/Radio Age,* 16 February 1987, 35–37, 101–104.

32. Reported by *Variety,* 25 April–1 May, 1994, 9; cited by "Broadcast Sales Training Executive Summary," 9 May 1994, 1.

33. Data in this paragraph derived from tabulations compiled by Geoffrey Foisie in *Broadcasting,* 1 May 1989, 35; 6 May 1991, 19; 13 April 1992, 4; *Broadcasting & Cable,* 10 May 1993, 5; 16 May 1994, 6. Foisie clarifies how figures were computed in the context of qualifying factors and accounting procedures.

34. "The Shaky Symbiosis of News and Profit," *Broadcasting,* 31 August 1987, 54–56.

35. Diane Mermigas, "Changing Marketplace Forcing Broadcast Networks to Evolve," *Electronic Media,* 1 March 1993, 23.

36. John Dempsey, "Cable in Black, Nets in Blue," *Variety,* 16 December 1991, 25.

37. Estimates were by Brian Cobb of Media Venture Partners, brokers that negotiated radio and TV station sales totaling over $1 billion in 1994, more than any other broker; *Broadcasting & Cable,* 27 February 1995, 47–49.

38. Data from NAB's "Television Financial Report" summarized in "Facts of the Affiliate Business," *Broadcasting,* 21 December 1992, 42.

39. Geoffrey Foisie, "Analysts Ponder NBC Sale Permutations," *Broadcasting & Cable,* 19 September 1994, 10. The *Wall Street Journal,* 14 July 1994, reported Walt Disney Company offered $5 billion to General Electric for the NBC television network and its seven O&O TV stations.

40. Geoffrey Foisie, "Reports of Network Deaths Exaggerated," *Broadcasting & Cable,* 14 March 1994, 6, 14.

41. Benjamin Bates, "Valuation of Media Properties" in Alexander, Owers, and Carveth, *Media Economics,* 108.

42. Data tabulated in *USA Today,* 19 July 1993, 2B.

43. The role of government policy and regulation as a factor in station transactions is treated by H. J. Levin in several publications: *The Invisible Resource: Use and Regulation of the Radio Spectrum* (Baltimore: John Hopkins University Press, 1971); "Franchise Values, Merit Programming and Policy Options in Television Broadcasting," in R. E. Caves and M. J. Roberts, eds., *Regulating the Product: Quality and Variety* (Cambridge, Mass.: Ballinger, 1975); *Fact and Fancy in Television Regulation: An Economic Study of*

Television Alternatives (New York: Russell Sage, 1980). See also P. W. Cherington, L. V. Hirtsch, and R. Brandwein, *Television Station Ownership: A Case Study of Federal Agency Regulation* (New York: Hastings House, 1971); R. T. Blau, R. C. Johnson, and K. J. Ksobeich, "Determinants of TV Station Economic Value," *Journal of Broadcasting* 20:2 (Spring 1976):197–207; B. J. Bates, "The Impact of Deregulation on Television Station Prices," *Journal of Media Economics* 1(1988):5–22 (he found little causal relation); and B. B. Cheen, *Fair Market Value of Radio Stations: A Buyer's Guide* (Washington, D.C.: National Association of Broadcasters, 1986).

44. Jessica Sandin and Harry A. Jessell, "Westinghouse/CBS Tops in TV: Top 25 Television Groups," *Broadcast & Cable,* 8 July 1996, 12–20. "TV Station Groups: Who's on Top," *Channels,* April 1989, 54.

45. Data supplied by MSOs to *Broadcasting & Cable,* 6 March 1996, 53.

46. Paul Noglows provided a detailed summary of major partnerships in "The Best and Worst Entertainment Deals of the Eighties: With Other People's Money," *Variety,* 9 December 1991, 61, 64–65, 68, 71.

47. *U.S. News & World Report,* 8 April 1991, 49; 8 November 1993, 64.

48. "Mergers Reorder Top 10 List," *Broadcasting & Cable,* 25 July 1994, 15.

49. "Top 100 In Electronic Communications," *Broadcasting & Cable,* 27 June 1994, 38–43.

50. Gene Jankowski, "Remarks Before the International Radio & Television Society's Newsmaker Luncheon" (Waldorf-Astoria Hotel, New York, 15 January 1986), p. 10 of speech text.

51. Christopher Byron, *The Fanciest Dive: What Happened When the Media Empire of Time/Life Leaped Without Looking Into the Age of High-Tech* (New York: W. W. Norton, 1986), 271–272, 273.

52. Harry J. Skornia, *Television and Society: An Inquest and Agenda for Improvement* (New York: McGraw-Hill, 1965), 18. Predictably, Dr. Skornia proceeded to analyze what he claimed to be the broadcasters' failure to serve the public interest properly.

53. Rolla Edward Park, "Television Station Performance and Revenues" (based on research funded by the Ford Foundation), *Educational Broadcasting Review* 5:3 (June 1971), 43–49.

54. See Paul W. Cherington, Leon V. Hirsch, and Robert Brandwein, eds., *Television Station Ownership: A Case Study of Federal Agency Regulation* (New York: Hastings House, 1971). The study was sponsored by the Council for Television Development of 42 firms owning 100 television stations and was headed by Ward Quaal, to respond to the FCC's June 1965 notice of proposed rule-making about limiting the number of stations under single ownership.

55. George H. Litwin and William H. Wroth, "The Effects of Common Ownership on Media Content and Influence: A Research Evaluation of Media Ownership and the Public Interest," prepared for the NAB (Washington, D.C.: NAB, 1969).

56. The arguments are briefly outlined by William Joseph Kennedy, "State and Local Taxation of Commercial Broadcasting," *Journal of Broadcasting* 17:1 (Winter 1972–1973), 77–84.

57. Les Brown, *Televi$ion: The Business Behind the Box* (New York: Harcourt Brace Jovanovich, Inc., 1971), 61.

58. Mikita, "The Controller's Role in Management," in Roe, ed., *Television Station Management,* 94–95.

59. Thomas F. Leahy, then executive vice-president of CBS/Broadcast Group, in speech to convention of CBS affiliate managers, Los Angeles, 1986; quoted in "Viewpoints: Transfer of TV Properties Went from $750 Million to $6 Billion in Four Years," *TV/Radio Age,* 9 June 1986, 61.

11

Managers and the Law

*Men are constantly called upon to learn over again how to live
together. It is a hard task. When unprecedented disputes and
difficulties confront them, they repeatedly turn for help to the
government, as the recognized umpire.*

ZECHARIAH CHAFEE, JR.[1]

Throughout this book, we have cited various areas where more attention by
broadcast and cable managers can result in increased efficiency. Managers could
devote more time to those needs if not preoccupied with government regulation.
Some broadcasters assert that continuous requirements—such as record keeping,
balancing political candidates' appearances, and implementing EEO provisions—
and threat of further action by the Federal Communications Commission and other
government agencies, seem to require an excessive amount of time and attention.

In few other areas of American life is a private business owner so constantly
reminded to monitor and modify staff behavior in order to maintain a license. Some
reminders and threats are real; some are imagined. Managers and owner-licensees
keep aware of possible new government encroachments from several sources. The
FCC collectively and individual Commission members regularly issue public statements. The trade press reports governmental activities and recounts speculation
about prospects. Whenever fellow managers or owners assemble, their meeting
agenda and conversation are often dominated by discussions of government regulation, interpreting and applying rules, and legislative threats. Communication
law firms and attorneys retained by stations track laws and regulations affecting
their clients.

Why is so much of a manager's time occupied with these kinds of responsibilities? Why aren't the guidelines clearer? Why do opinions differ on some issues
that have existed through all the decades of broadcasting? What are the reasons
for different interpretations of opinions and why have they not been resolved?

A brief review of the origins of federal regulation provides some context
for these questions. Then follows an analysis of specific practices and issues in
regulation. The final portion of the chapter appraises the role of government regulation and industry self-regulation.

This chapter cannot analyze in detail the many laws, regulations, and court
decisions directly affecting broadcast and cable. Other books are dedicated solely

to that daunting task; managers should go to them for more intensive treatment of issues affecting electronic media today.[*] Some apt sources are cited in endnotes.[2]

Managers must keep abreast of constantly shifting legal factors impinging on how they do business while licensed or franchised by government. In addition to legal counsel needed for specific instances, managers can keep informed by industry publications such as *Broadcasting & Cable* magazine. They should consult their professional associations that regularly publish updates on pending governmental actions. In addition to its annually updated bound volume *NAB Legal Guide to Broadcast Law and Regulation (1996),* the National Association of Broadcasters' Legal Department published its annual *Broadcast Regulation: A Review of [year] and a Preview of [year]* (some 250 spiral-bound large pages), plus many smaller booklets devoted to specific topics such as EEO compliance, political broadcasting, libel and privacy rights.

In the mid-1990s it was "all up for grabs" with some sections of the present chapter subject to modification within months of publication. The Reagan administration's sweeping deregulation in the 1980s removed or reduced a plethora of FCC rules and procedures. The brief swing back in the early 1990s toward tighter oversight of cable and broadcasting was followed in mid-decade by further reduction of federal regulations—such as removing financial interest and syndication rules limiting network operations and twice raising and finally all but abolishing limits on the number of AM, FM, or TV stations licensed to one owner. Further lessening of government constraints was promised with the Republican party's ascendancy in 1995, constituting a majority in Congress for the first time in almost half a century. President Clinton signed into law the Telecommunications Act of 1996 which overhauled many parts of the Communications Act of 1934 to which it was attached as an amendment.[3] The new law opened up the field to cross-ownership and mergers between cable and telephone companies and utilities, extended broadcast license terms to eight years for radio and for TV, removed limits on total number of stations owned by a single licensee, and raised to 35% the national cap on a licensee's multiple TV stations' audience reach.

HISTORICAL BACKGROUND OF FEDERAL REGULATION

After extended hearings Congress passed the federal Radio Act of 1927. The new law created a five-member Federal Radio Commission, each member representing approximately one-fifth of the land area in the United States. The Commission established a pattern of frequencies for all radio service around the country, designated by class of service.

The federal Radio Act of 1927 also set down the underlying philosophy of our American system of broadcasting. The Act established that (1) radio channels on the nation's national resource of the electromagnetic

---

[*]Middleton and Chamberlin's *The Law of Public Communication* (1988) offers a broad perspective of foundations and major topics, with the final hundred pages devoted to broadcast and cable regulation. Bittner's second edition of *Law and Regulation of Electronic Media* (1994) is a readable, descriptive narrative analyzing legal matters specifically in telecommunication. Smith, Meeske, and Wright provide an analytical survey of *Electronic Media and Government: The Regulation of Wireless and Wired Mass Communication* (1995). Other helpful volumes include Francois' sixth edition of *Mass Media Law and Regulation* (1994) and Creech's *Electronic Media Law and Regulation* (1993). Bensman's *Broadcast/Cable Regulation* (1990) briefly digests all topics in alphabetical order—from "advertising" to "trespass-intrusion-consent"—for easy reference.

spectrum of frequencies were a public resource belonging to all the country's citizens, (2) those awarded licenses to broadcast on specific frequencies were authorized to utilize but not own those channels, and (3) licensees would be expected to serve the broad public interest as a condition for using the frequencies. The Act further established that license applicants had to qualify for the privilege of operating a radio station through such tests as the Commission would devise. This Act applied to broadcasting a responsibility to serve "public interest, convenience and necessity" (a phrase taken from an 1892 statute affecting railroads and from the Transportation Act of 1920). This was the broad standard by which every station was expected to operate and which would guide the FRC's regulatory powers. A further major provision of the Radio Act gave broadcasters the right of appeal to the courts for review of Commission decisions.

DEVELOPMENTS TO THE PRESENT

In 1934, President Franklin Roosevelt recommended that Congress expand the FRC's jurisdiction to include all forms of interstate and foreign wire, as well as wireless, communication. A supportive Congress passed the Communications Act of 1934 that incorporated the Radio Act's major provisions and augmented the regulatory agency's jurisdiction; it enlarged the agency to seven members, renaming it the Federal Communications Commission. The Act sustained the philosophy of the 1927 law. It asserted that the FCC granted a license for use, not ownership, of an airwave frequency:

> Sec. 301. It is the purpose of this Act, among other things, to maintain the control of the United States over all the channels of interstate and foreign transmission; and to provide for the use of such channels, but not the ownership thereof, by persons for limited periods of time, under licenses granted by the Federal authority, and no such license shall be construed to create any right beyond the terms, conditions, and period of the license.[4]

The FCC (reduced to five members in 1982) is still governed by the same basic legislative philosophy developed for radio in 1927, reaffirmed in the 1934 Act, and left standing without change in wording despite heavy revision by the Telecommunications Act of 1996. Subsequent decades have seen phenomenal growth of AM and FM broadcasting along with television, plus advancements in radar, microwave relay, cable and fiber optic distribution of signals, digital compression, and space and satellite communication. By the late 1970s some members of Congress, particularly Lionel Van Deerlin (Democratic representative from San Diego), strove to produce wholly new communications legislation. Although unsuccessful, their efforts prepared the way for waves of deregulation the following two decades. The very premises on which American broadcasting had been formally established by the 1934 Act were challenged throughout that period—not only by the media and by critics but by governmental leaders themselves. Successive chairpersons of the Federal Communications Commission, from both major political parties, expressed concern for "reregulating" or "deregulating" or even "unregulating" electronic media. Efforts by consecutive chairmen Richard Wiley and Charles Ferris in the 1970s to untangle bureaucratic paperwork in the licensing and review process were taken to their extreme in the

1980s by chairman Mark Fowler and his successor Dennis Patrick. Fowler vigorously applied the Reagan administration's philosophy of removing governmental intrusion from private business. His crusade for deregulating every aspect of broadcasting, from licensing to program standards to engineering, was based on his bedrock criterion of *marketplace forces* rather than federal fiat. (His marketplace was one of laissez faire economics, not necessarily a marketplace of ideas—the classic meaning of the phrase.) During his tenure the Commission dismantled many criteria, guidelines, and procedures for reviewing broadcast licenses. He based these moves on his assertion that the traditional "public interest" standard had no real foundation in law and so ought not be applied to radio and television; he sought full parity with print media under the First Amendment. He once referred to television as no more than "a toaster with pictures"—just one more household appliance needing no government controls. By the time he left office in 1987, Congress increasingly objected to his campaign—including eliminating the "fairness doctrine" which the Commission claimed was an unwarranted and indefensible intrusion into free enterprise in mass media. Both legislative houses sought to undo what many perceived as the FCC's capitulation to faulty reasoning as much as to forces of big business. Going into the 1990s the pendulum began to swing back from governmental permissiveness to federal oversight, until the Congressional elections in 1994 swept Republicans into office as the majority.

By then many media participants and observers questioned whether the statutory law of 1934, amended many times over the half century, had been stretched to its maximum applicability to modern problems of communications. From time to time appeals emerged from various quarters for an entirely new communications law including overhaul of the regulatory agency. In 1994 Don West, respected editor of influential *Broadcasting & Cable* magazine, spoke out in favor of a wholly new legislative act to govern use of the electromagnetic spectrum for communications. In 1995 the Senate began considering a new omnibus telecommunications bill to supplant the Communications Act of 1934 with its inheritance from the seminal Radio Act of 1927.

Every session of Congress over recent decades turned up a bundle of proposed bills on specific matters affecting broadcast and cable media. Most died in committee, some moved forward only to be snuffed out in subsequent deliberations. A few were enacted, becoming law with the President's endorsement. Key legislation will be discussed in later sections, including laws regarding children's programming, cable subscribers' rates, indecency, and intermedia mergers crossing over between telephone utilities and cable and broadcast sectors.

President Clinton signed into law the Telecommunications Act of 1996, advancing further deregulation with major modifications of the nation's communications law as noted below.

THE REGULATORS: FCC COMMISSIONERS

Legislation mandated oversight of broadcast media by a commission of regulators. How the law is applied through specific regulations is shaped by the successive members of that agency acting as a body. Appointed commissioners bring with them their political and professional expertise and personal viewpoints.[5]

Unless appointed to fill unexpired terms, commissioners serve for five years (formerly seven years, until 1986) and may be reappointed. When they join the agency, they rely heavily on the Commission staff for detailed information and even recommendations about decisions. This middle staff includes not only aides assigned to individual Commission members, but also heads and key personnel of various bureaus that are part of the FCC apparatus. These career staffers long outstay appointed commissioners and thus contribute significantly to the continuity and pattern of the agency's deliberations.

The FCC Workload

Despite an enormous workload, the Commission has conducted its licensing function with relative efficiency and equity. The welfare of the public has been its basic guide in establishing allocation tables, defining technical standards, deciding changes in power, structure, equipment, and ownership, and in screening applications for licenses and renewals. Assisting the commissioners on substantive matters as well as more routine processing details is a staff of some 2,000 fulltime employees, almost 500 of them in the field, in ten categories of offices, five bureaus, and a review board (table 11.1).

While veteran staffers, including department heads, provide factual data and recommendations for action, the commissioners remain responsible for final decisions implementing the congressional statutes in the Communications Act of 1934 as amended, including the major "overhaul" amendment under title of the Telecommunications Act of 1996.

Establishing and supervising technical standards in broadcasting and approving licensees and their facilities to broadcast would seem fulltime work assignments. Yet, that is only part of the Commission's responsibilities. It is also charged with administering the entire electromagnetic spectrum. This includes approving all frequency assignments and supervising their use by telephone and telegraph services, shortwave and amateur radios, remote pickup equipment, relay facilities, facsimile, international wireless activities, experimental research services, various industrial functions, marine, aeronautical, land transportation, cellular phone services, disaster communications, citizens radio, and satellite transponder transmission.

■   Thus radio and television broadcast station authorizations represent but a small part of the FCC's paperwork. By 1995 the FCC oversaw 11,767 commercial and educational AM and FM radio stations, and 1,527 VHF and UHF TV stations. License renewals cycled among them; roughly one-seventh of all radio stations and one-fifth of TV stations were filing for renewal annually when license terms were seven years for radio, five years for TV—until the 1996 Act changed that to eight-year terms for both. Just handling change of ownerships as stations were sold involved 10,494 radio stations between 1987 and 1996—over a thousand transfers of ownership every one of the nine years—with the transactions valued at $21.7 billion.

The Commission also oversees more than 2,000 FM translator and FM booster stations, over 2,500 TV translators, and 1,591 low-power TV stations. In 1993 the FCC issued 1,076 construction permits for low power television service. It processed 746 applications for cable TV relay services; it also monitored rates charged by the nation's 11,000 cable systems. Private radio services (aircraft, maritime, police and fire, businesses, taxi, and many others) include over 3 million stations, with more than 18 million transmitters; in 1993 alone there were 686,046 authorization requests for such services.   ■

— TABLE 11.1 —
FCC Units and Fulltime Staff, 1994

32	Commissioners Offices	6	Office of Inspector General
268	Office of Managing Director	385	Field Operations Bureau
101	Office of Engineering & Technology	330	Common Carrier Bureau
16	Office of Administrative Law Judges	330	Mass Media Bureau
14	Office of Public Affairs	196	Cable Services Bureau
12	Office of International Communications	190	Private Radio Bureau
12	Office of Legislative Affairs	51	Office of General Counsel
12	Office of Plans & Policy	6	Review Board

TOTAL of 1,961 fulltime employees: 1,508 in Washington, D.C. offices plus 453 in the field (most of them in Field Operations Bureau and Private Radio Bureau).

Source: FCC, *60th Annual Report/Fiscal Year 1994* (Washington, D.C.: U.S. Government Printing Office, 1995), 21.

Merely maintaining balanced use of the range of frequencies in the electromagnetic spectrum brings enormous pressures. Requests for increased spectrum space come regularly from agents of public safety such as police or fire departments, the military, the nation's space program, and from industry. The federal agency must protect the total public's interests in administering all these varied uses of the spectrum.

■ The scope of the Commission's work in a given year—in addition to ongoing actions on license applications, renewals, modifications, and transfers—includes major issues such as hearings and rulemaking about political broadcasting, multiple ownership, policies about comparative hearings of competing applicants, limitations on the TV networks' control over programming, subscription television, cable TV rates, UHF service, standards for HDTV, guidelines for automatic monitoring of TV programs and commercials, authorizations for lower-power classifications of FM stations, cellular radio, inquiries into children's television programming, noncommercial broadcasting, excesses in commercial advertising, station identification rules, reregulation of technical requirements pertaining to radio stations, establishing directives about equal opportunity for minorities in employment and in programming, modified allocations and use of frequencies, satellite services, determining station ownership limitations, and the Emergency Broadcast System (EBS) which in 1997 became the digital Emergency Alert System (EAS). Other business before the Commission at any time involves investigations and decisions about pending matters. In fiscal 1993 the Commission testified or commented on legislation pending before Congress, participated in 477 federal appeals and Supreme Court proceedings, and in another 112 federal district court cases. The Commission processed over 4,600 written Congressional inquiries and more than 10,000 telephone inquiries about policies, status of applications, and especially regarding implementing the Cable Act of 1992.[6] ■

With such a volume of work, together with all the attendant pressures, the FCC does not have much opportunity to engage in unhurried philosophical thinking about its future relations with broadcasters. Over thirty years ago former Chairman Newton Minow (a Democrat) appraised the agency:

[A]s we reexamine the status quo, I must confess that I have found the FCC, too, a prisoner of its own procedures. The Commission is a vast and sometime dark forest where we seven FCC hunters are often required to spend weeks of our time shooting down mosquitoes with elephant guns. In the interest of our governmental processes, and of American communications, that forest must be thinned out and wider, better marked roads have to be cut through the jungles of red tape. Though we have made substantial improvements in recent years, the administrative process is still a never-never land.[7]

Those appointed to the Commission have been, for the most part, public servants. They have generally discharged their assignments in a manner that helped develop the finest broadcast service in the world. But in the 1960s and 1970s there grew what many in the industry saw as needless harassment of radio and television licensees, taking their time and energies from more constructive activities. Consumer activists criticized broadcast media for not offering enough "quality" programming and for catering to commercialism; minority advocates sought greater participation in media by African Americans, Hispanics, and others. The deregulatory 1980s, amid proliferating competing media systems, did remove much regulatory "underbrush" and restrictions. Marketplace forces became the criterion under chairman Mark Fowler during the Reagan and Bush administrations; but chairman Reed Hundt again advocated public trusteeship by the mid-1990s. At that time interim chairman James Quello voiced a fair-minded view of the need for some regulatory parameters: "I am for the First Amendment but against anarchy."

The FCC as Protector of Broadcast Interests and the Public Interest

It might seem from broadcasters' complaints over the decades that the Commission is nothing but a hindrance to the progress of professional and commercial media interests. This is not true, despite the distress caused to broadcasters by some commissioners, such as Newton Minow, Kenneth Cox, and Nicholas Johnson during the "activist Commission" years in the 1960s and 1970s.

The FCC has regularly acted to preclude practices within the industry which might constitute restraint of trade. It has refused station licenses to applicants with records of attempts to undercut fair competition or who harmed their competitors in other lines of commerce. It has conducted investigations of applicants suspected of previous antitrust actions.

Establishing and enforcing the duopoly rule (no ownership of two services in the same market) and restricting the number of station licenses any one person or corporation might hold were consistent with the Commission's determination to foster competition within the industry. (Ironically, those limits were dissolved in the 1990s partly to salvage weak stations that could not compete effectively in media-crowded markets.) Early chain broadcasting regulations, elimination of network option time, and limiting networks' service to three of the four prime-time hours of local affiliates' schedules, while not universally popular, were adopted in the interests of individual station ownership and of local management's autonomy in selecting programming. When conducting hearings on competitive applications or on significant policy changes, the FCC has demonstrated a high degree of objectivity and concern for the best interests of the public.

Then in recent years as mass media multiplied with proliferating cable channels, new TV networks, and satellite service, the FCC began to respond to economic realities of the expanding technology mix. It sought to restore a "level playing-field," supporting competitive balance by rescinding many previous regulations, to afford network and local broadcasters flexibility in program strategies to attract audiences and advertisers in an intensely competitive media marketing environment.

Still, the FCC was created by Congress in 1934 to serve as a protector of all the public interest—interests of pluralistic and divergent publics that together make up the nation's citizenry. These publics include not only broadcasters and advertisers, but the general audience and subgroups such as ethnic and cultural minorities. The FCC itself is not wholly autonomous in its deliberations and decision making. Regulatory policy is determined by the FCC; but that policy is also shaped by the broadcasting industry, citizen groups, the courts, the executive branch (White House and Department of Justice), and Congress with its various committees and subcommittees that oversee the FCC's stewardship.[8] Each of these forces mutually interact; they frequently trigger and often significantly affect the outcome of FCC actions.

CONFLICT OF FREEDOM VS. REGULATION: THE COURTS

While imposing regulatory constraints, the 1927 Act nevertheless sought to extend to radio broadcasting freedom of speech and freedom of the press, guaranteed by the First Amendment of the Constitution; the Act specifically denied the FRC any power of censorship over radio. This seeming inconsistency would vex broadcast professionals through the rest of the century (see the later section regarding successive court decisions).

Ambiguity remains about correlative First Amendment rights of the broadcaster on one hand, and of the audience or public on the other. The Supreme Court's "Red Lion" decision in 1969 supported the public's right to limited access to media and thus seemed to abridge somewhat the editorial rights of the broadcasters whose facilities were to be made available to such claimants. In 1971, U.S. Court of Appeals Judge J. Skelly Wright claimed broadcaster's freedom was circumscribed by the public's limited First Amendment right of access to radio and television. Yet his colleague of the same appellate court, Chief Judge David Bazelon, in November of 1972 modified his own previous stance by noting "no factual basis for continuing to distinguish the printed from the electronic press as the true news media." In May 1973, Wright's judgment was contradicted when the U.S. Supreme Court decided 7–2 against demanding that broadcasters sell advertising time to whoever so requested. For the first time Justices of the highest court in the land explicitly endorsed full protection of radio-TV freedoms under the Bill of Rights. Justice William O. Douglas' opinion was particularly outspoken: "My conclusion is that TV and radio stand in the same protected position under the First Amendment as do newspapers and magazines." Justice Potter Stewart concurred: "The First Amendment prohibits the government from imposing controls over the press. Private broadcasters are surely part of the press." Nevertheless, Chief Justice Warren Burger stated in his opinion that "a broadcast licensee has a large measure of journalistic freedom but not as large as that exercised by a newspaper." In 1979 the high court again distinguished broadcast media from print, this time because of broadcast signals' "pervasiveness" which citizen listener-viewers cannot avoid, rather than because of scarcity of frequencies—the original basis for claiming government oversight.[9] That line of reasoning about media's impact on the mass citizenry grew stronger in the wake of widespread criticism of alleged excesses in entertainment content (violence, sex, language) and alleged abuses in news programming (misrepresentation or bias in coverage and sensationalism-oriented, exploitative "tabloid journalism").

PRAGMATICS OF LEGISLATION AND REGULATION

Broadcast and cable managers must be aware of the "political" structure in which they operate—alert to changes in FCC policies or personnel that can affect current decisions as well as plans for coming years. It is important to retain a Washington law firm specializing in communications practice. Also valuable is membership in the National Association of Broadcasters (NAB) or National Cable Television Association (NCTA), headquartered in Washington and closely attuned to governmental political affairs affecting broadcasting. State associations of broadcasters and cable operators assist members with conferences and newsletters summarizing recent rule-makings, clarifying regulatory practices, and reminding about deadlines for filing periodic reports.

While Congress rarely makes major changes in basic broadcast law, its committee rooms often provide a sounding board for regulators and for those with alleged grievances against broadcasters. This is because the FCC, like every government agency, must appear for renewal of its funding each year, and is accountable to Congress annually for its stewardship. Members of Congress, especially members of the House Communications Subcommittees and Senate Commerce Committees, frequently receive complaints against broadcasters. Because members of Congress tend to be responsive to constituents, broadcasters should meet with their elected representatives and senators to discuss media problems. Since the 1970s organized citizen groups have formed a counter-influence among members of Congress and commissioners by participating in the deliberative process leading to regulatory decisions; courts have also become increasingly engaged in broadcast-related issues. Therefore broadcasters and cable operators—individually and through their professional associations—must be alert to present their case effectively in this pluralistic political-social forum.

Because of these contending forces, the FCC tries to compromise conflicting goals. FCC policy-making realistically seeks moderate change that will not generate debilitating opposition. Thus the agency attempts to remain flexible and responsive to reactions by major parties and power groups, even to the point of deleting or changing proposed rulemaking if opposition grows strong. The realities of policy-making also dictate that the FCC use its resources to resolve immediate problems without becoming immersed in massively complicated long-range issues. (This reflects a need for philosophy and broad policy able to withstand future circumstances by offering consistency and predictability instead of a case-by-case approach to rulemaking.)

Krasnow and Longley offer an assessment of the FCC that is reasonably sound as well as pragmatic, when they comment on "Why doesn't the Commission regulate the industry more vigorously?"

Such a question assumes that the public interest will be furthered by greater regulation. However, the history of regulated industries such as transportation and broadcasting has shown that stricter governmental controls may in fact disserve the public interest. Moreover, calls for the imposition of more restrictive regulations by the FCC usually do not take into account the highly complex, politically sensitive, and rapidly changing character of the communications industry. Under the system of regulation established by the Congress, the FCC has operated within a sequential, bargaining, policy-making process. America's stake in broadcasting is too fundamental and precious to be subjected to drastic or politically unpopular policies which do not allow the FCC to modify policies without excessive loss if new information indicates unexpected troubles.[10]

However, those authors perhaps underemphasize the demand for long-term principles and consistent, equitable guidelines by the regulatory agency on behalf of all parties affected, when they conclude their book on a practical note:

> The policies of the FCC are not abstract theories, but political decisions allocating material rewards and deprivations—decisions, in Laswellian terms, concerning who gets what, when, and how. The development of policy in this manner is not easy. Before any proposal can emerge as public policy, it must survive trial after trial, test after test of its vitality. The politics of broadcast regulation offers no escape from that imperative.

Other appraisals of the FCC have been less sanguine. Various broadcasters, commissioners, Congresspersons, researchers, and academicians have lamented the agency's seeming lack of major policy and direction about large issues and problem areas, relying instead on ad hoc decisions about each case before it. But the courts of the land have supported precisely a flexible FCC in its implementing the Communications Act.[11] Although innovative broadcaster Gordon McLendon publicly complained about the ambiguity and inconsistent applications of FCC rules, veteran communications lawyer Marcus Cohn appraised favorably

> the fact that the Commission has no specific and hard and fast rules. I happen to think this is good. I was taught a long time ago that the entire administrative process depends upon a flexibility which is attuned to changing tides. Indeed, in one sense, the Commission ought to reflect the shifting standards, objectives and goals of society. If you ever get rules which are dogmatically specific, you will be assured of an unresponsive Commission.[12]

Demonstrating "Murphy's law," the FCC's rare attempts to take initiative in making policy often resulted in reversals of its actions by higher courts or at least congressional rebuke, such as with the "Family Viewing" embroglio in the mid-1970s, abolishing the "fairness doctrine" in 1985, and its efforts in the 1980s and 1990s to determine clock hours for a nighttime "safe harbor" for adult content considered indecent.

LIMITS ON OWNERSHIP CONCENTRATION

A central policy of the FCC in assessing what best serves "public interest, convenience, and necessity" has been to ensure diversity among broadcast owner-licensees. The intent is to avoid monopolistic control of the media economy and means of disseminating information.

Scarcity of frequencies prompted the first cap on ownership at 7 AM, 7 FM, and 7 TV stations, coupled with prohibitions against "duopoly"—forbidding in any single market one licensee from operating 2 stations in the same service (2 AMs or 2 FMs or 2 TV stations). With stations multiplying over the decades many governmental as well as industry leaders sought to loosen that restriction. In 1984 the Commission voted to raise the limits to 12–12–12 (with a cap of 25% of the nation's homes by any licensee's TV stations). On the heels of the late 1980s' recession, the economic struggle of many radio stations prompted the Commission to expand those limits and to modify the duopoly rule. In 1992 it proposed caps of 30 AM and 30 FM stations to a single licensee and also ownership of three to six stations per market depending on market size; but Congress balked so the FCC revised the limits to 18 AM and 18 FM (after two years the cap rose to 20–20 radio stations), and no more than 2 AMs and 2 FMs in large markets (with 15 or more stations), 3 stations in small markets.

In 1994 the Commission sought to eliminate entirely limits on TV ownership, proposing that national household caps rise triennially by 5% to a ceiling of 50% of all U.S. homes reached by any licensees' multiple stations. In 1995 Senator Larry Pressler, Commerce Committee chair, proposed repealing all or most limits to radio and TV ownership. The industry's National Association of Broadcasters feared small owners would be squeezed out by increasingly larger groups, so it first remained neutral—neither supporting nor opposing the plan—then reluctantly objected to it. Networks and large-group owners predictably endorsed eliminating caps to ownership but affiliate and independent stations opposed outright repeal of all limits.

The Telecommunications Act of 1996 to be implemented by the FCC removed almost all limits for total number of radio and TV stations owned by a single licensee. But total TV stations' "reach" could cover no more than 35% of the U.S. population (UHFs counted only half of their market population); and the number of multiple-owned radio stations in a market were limited according to each market's total number of stations—in the largest markets a maximum of eight stations to a single owner.

In relaxing its one-to-a-market station ownership rules in 1988 and eliminating them in 1992, the Commission looked to joint ownership of stations in the same markets for economies of scale by reducing overhead costs and merging staffs, permitting them to invest in more diverse programming including local news and public affairs and to upgrade technical facilities. For samples of duopoly efficiencies and consolidation of staffs, *Broadcasting & Cable* magazine outlined plans and practices of four companies' duopolies in as many markets in 1993 (table 11.2).

Local marketing agreements (LMAs) and time brokering of one station's schedule/spots by another became widespread among radio stations along with a number of TV stations in the 1990s. Some referred to the LMA as an all-but-duopoly arrangement or "de facto duopolies" (ownership of multiple stations in the same service, such as two AM stations or two FMs or TVs). Similar to newspapers' joint operating agreements, one station took over managing another's operation, in return for a rental fee or a percentage of sales. Such "leasing" instead of outright purchase avoided transfer of license and exceeding limits on owning multiple radio or TV outlets in a market. The procedure also eliminated a competing station's format and sales operation. This arrangement became attractive in smaller markets where local financing of station purchases was unavailable. For some it served as an interim step to buying the other station and transferring license.

EFFICIENCIES, SYNERGIES IN SELECTED DUOPOLIES OF FOUR LICENSEES
1993

LICENSEE:	CLEAR CHANNEL	CAPSTAR	PAXSON	INFINITY
Market/Rank	San Antonio/35	Greenville-Spartenberg/60	Jacksonville/50	Chicago/3
Station Formats *AM or FM*	FM Country AM News/Talk AM News/Talk FM Soft AC FM Urban AC	AM/FM Country FM AC	FM Country FM Classic rock AM Sports AM News	FM Country FM Oldies AM Adult standards -full service
Facilities	1	1	1	2
General Managers	2	1	1	2
Program Directors	2	2	2	4 (2 at Country station)
Sales staffs	3 (2 AM, 1 FM)	2	1	2
National reps	1	1	1	2
Combo buy discounted?	No	Yes	Yes	No
Sales synergies	Increase in national business	Some increase in national business, none in local	More options for clients, attract better salespeople	Stronger pricing, sales managers trade information on pricing and buys
Cost synergies	Combined facilities, personnel, negotiations with vendors	Office, GM, accounting, promotion, traffic	Traffic, sales, promotion, engineering, management, negotiation with vendors, lower sales commissions	"Minor": some business office, engineering

Source: *Broadcasting & Cable,* 19 April 1993, 62.

By late 1994 over 2,000 commercial radio stations were involved in duopolies or LMAs; that number was projected to triple by 1997. Also operating under LMAs were three dozen TV stations in 1995, ten of them in the top-30 markets; a year later there were 100 TV agreements. In 1996 Commissioners considered permitting actual duopolies of TV stations in a market if they were not both VHF. (Late in 1996 the FCC said it would consider counting LMA operations the same as fully owned licensed ones, in computing multiple stations in markets.)

Many broadcasters, especially at small stations, saw duopoly and LMA developments a threat to their competitive independence. They feared affluent corporations that bought clusters of key stations in a market would totally dominate that market in capturing audiences and sales revenues.

■ The Antitrust Division of the Department of Justice began to explore the financial impact of multiple station ownership in markets; in 1996 it required Jacor to dispose of 1 of its 7 Cincinnati radio stations—limiting it to controlling no more than 50% of radio revenue in that market. (Jacor also owned 8 radio outlets in Denver and 6 in Tampa, among its 51 holdings in 15 major markets at the time; when it acquired a TV station in Cincinnati, it further reduced its radio stations to 5.) But the American Association of Advertising Agencies called even that 50% too much control over ad dollars, giving one company power to set (that is, unilaterally raise or lower) ad rates in a market.[13] Justice notified broadcasters it would assess concentrated ownership in other markets as well for their "anti-competitive effects." It also challenged American Radio Systems' control over 8 stations in Rochester, New York, a market with 20 station signals—giving ARS 64% of the market's radio revenue and 49% of the radio audience; the Justice Department required ARS to dispose of 3 stations and a "joint sales agreement" with another station, reducing its sales to 42% of the radio market revenue. But some industry leaders and also Commissioner James Quello suggested that to properly assess the level of real anti-competitive control, multistations' sales revenue should be considered not as a percentage of the market's radio revenue but rather as a percentage of all media ad revenues in that market—radio, TV, cable, print. ▪

The FCC's Mass Media Bureau chief, Roy C. Stewart, cautioned broadcasters engaged in LMAs that each licensee remains fully responsible for maintaining control over their own stations. Despite turning over managing, programming, and selling commercial time to another station's management, the original licensee is obligated to keep up the public file, political records, timely filing with FCC for license renewal, and responsibility for all program content aired by the transmitter.[14] Among complications for multiple operations reporting to the FCC are Annual Employment Reports about EEO staffing (Form 395–B). Licensees with duopolies must file separate reports for each station, listing employees under one or other station according to their primary duties. Where duties are evenly split among stations, the licensee lists employees on one station's Form 395–B, while filing a separate report for each other station noting that employees are listed on the first station's form. With LMAs, each licensee remains responsible for their respective stations, filing separate employment profiles even when an LMA operation has staff working for the dominant station running it.

Clearly, legal counsel must assist owners and managers in preparing documentation to support multistation properties of owner-licensees. Group owners of stations clustered in regions of the United States or scattered across the country are susceptible also to criticism of limiting the public's sources of information and communication. Yet, studies have indicated higher quality and amount of local-oriented community programming, including news, on stations owned by large interests as part of a group than on single stations owned locally.[15] Often, large size and diversity of resources provides economic stability as well as pooled experience and competent personnel. In the early 1970s the FCC emphasized diversity of ownership as a major consideration in license applications. But with the onslaught of competing cable services, the erosion of network affiliates' audiences, and the widespread losses in broadcasting in the late 1980s and early 1990s (two-thirds of all radio stations lost money in 1992 and 1993) coupled with stations going bankrupt and off the air, the FCC relaxed those rules by successive moves, culminating in Congress' highly deregulating communications law in 1996.

License Terms   When original 3-year license terms were still the standard, Senator Barry Goldwater as chairman of the Senate Communications Subcommittee in 1981 favored eliminating all license renewal periods for radio, and expanding TV terms to 10 years. That year radio license terms were expanded to 7 years, TV to 5 years. In 1995 FCC chairman Reed Hundt proposed 7-year terms be granted to TV; meanwhile Congressional committees drafted new telecommunications bills, with Senator Hollings recommending 10 years for both radio and TV while Senator Pressler promoted eliminating all limits and ownership rules. The Telecommunications Act of 1996 finally set 8-year terms for both radio and TV licenses.

REGULATORY PRACTICES AND ISSUES

The Commission has a range of penalties it can impose on broadcast operators who violate its rules. To enforce the Communications Act and its own regulations, the Commission initially inquires by letter about alleged rule violations; it can also take the more serious step of sending investigators to inquire directly about the matter. In 1952 Congress empowered the FCC to issue Cease and Desist Orders for specific violations of federal rules.

FORFEITURES (FINES)

In 1960, the payola scandals and quiz scandals resulted in Congress' amending the Communications Act, authorizing the Commission to penalize infractions by "forfeitures" (or fines) of up to $1,000 per day to a maximum of $10,000.[16] Later, until 1989 the FCC was limited to assessing $2,000 per violation or per day (of a continuing violation) up to a ceiling of $20,000; but in November that year Congress raised the limit to $250,000. In 1991 the Commission announced a new schedule of fines, most of them in five figures for each infraction, up to $25,000 per violation or each day of a continuing violation up to a total of $250,000. The outcry against the fees as excessive prompted the FCC two years later to reduce the "base levels" by about 20%, with the actual amount levied adjusted up or down depending on each case's circumstances. Table 11.3 indicates selected forfeitures or fines as of 1993; most items not listed are technical engineering matters (in 1988 a total of 1,132 technical violations of rules received fines averaging $845).[*] A record fine of $2 million was assessed against a North Carolina cellular company in 1996 for constructing a 187-foot tower without notifying the FAA and getting approval for construction (the tower reached into air safety zones, creating a hazard to air navigation); the FCC discovered discrepancies between data the company filed with it and with the FAA.

In the 1990s the FCC, under chairman Al Sikes' leadership, began to step up the pace of fines for violating regulations. In 1991 FCC field officers issued forfeitures totaling $350,000 for lack of proper painting or lighting on 160 of 1,000 randomly checked broadcast, common carrier, and private radio towers. Violations of EEO rules, alleged indecency, and excessive advertising in children's TV drew increasing fines into the mid-1990s. The Telecommunications Act of 1996 raised the previous $10,000 ceiling on forfeitures for broadcast or cable obscenity to $100,000.

■   For example, the FCC issued notices of apparent liability for forfeitures because of *EEO violations* in 1992 to a station in Avalon, New Jersey ($20,000); in 1993 to stations in Phoenix and in Pittsburg, California (each fined $25,000 plus a short-term renewal of only two years); in 1994 to 22 stations up for renewal in California, Texas, and Seattle (a total of $318,750, ranging from $18,750 to $37,500, plus short-term

---

[*]Most violations (197) related to Emergency Broadcast System rules; others concerned antenna tower painting, lighting, and maintenance (100), incomplete public inspection files (88), deficient station logs (59), failure to designate a chief operator (75), failure to post station or operator licenses and authorizations (67), plus others for operating with incorrect power or frequency, inadequate remote control systems or measurements of equipment performance. *Broadcast Cable Financial Journal* (January/February 1991), 24–25.

~~~ **TABLE 11.3** ~~~

Selected Forfeitures (Fines) Set by FCC in 1993

</div>

$20,000	Misrepresentation/lack of candor	$10,000	Failure to respond to FCC
20,000	Unauthorized transfer of control		communication
20,000	Unauthorized construction or	10,000	Violation of children's TV rules
	operation	10,000	Violation of broadcast hoax rules
20,000	Violation of alien ownership rule	5,000	Failure to maintain public files
18,750	Failure to permit inspection	5,000	Failure to file information
17,500	Malicious interference	5,000	Failure to ID sponsors
12,500	Transmission of indecent material	5,000	Unauthorized broadcast of
12,500	Violation of political rules		telephone calls
12,500	Fraud	1,250	Failure to ID station
10,000	No licensed operator on duty	1,250	Failure to maintain records

The FCC could adjust these base fines *upwards* by 50%–90% for egregious misconduct, intentional violation, or for their financial strength and ability to pay; by 40%–70% for substantial harm or prior violation; and by 20%–50% for substantial economic gain derived from the violation; repeated or continuous violations varied. Adjustments were *downward* from the base by 50%–90% for minor violation, by 30%–90% for good faith or voluntary disclosure, or 20%–50% for history of compliance; lower adjustments also varied according to inability to pay.

Sources: *Broadcasting & Cable,* 16 August 1993, 15; NAB, *Radio TechCheck,* 13 September 1993, 1.

renewals to all but one of them). The Lutheran Church/Missouri Synod's AM/FM station in Clayton, Missouri, was fined $250,000 and a hearing set to consider revoking its license. Cable was not immune to forfeitures; that same year in 1994 a Palm Beach cable system was fined $121,500 for EEO violations.

Alleged *indecency* violations centered on Howard Stern's syndicated scatological radio talk show.[17] Infinity Broadcasting was fined $2,000 for an incident which was multiplied to $6,000 because aired by its three stations (in New York, Washington, Philadelphia); Infinity had been warned in 1987 for material on a previous show by Stern. By 1994 Infinity's cumulative fines totaled $1,278,750, plus another $400,000 for indecency complaints against those three stations and its Manassas, Virginia, station for violations on four days over three months in 1993. The Commission also fined a Las Vegas station $47,500 for carrying those Stern incidents. In 1994 it levied another fine of $200,000 against Infinity for material aired in a Stern broadcast. That year the FCC also issued notices of apparent liability—which stations may appeal within 30 days—to a Dallas station for song lyrics alleged to be indecent and to an FM station in Muscatine, Iowa, for a joke deemed indecent in a phone-in show. Among 1993 fines was a $23,750 penalty against a state college's student-run radio station; the Commission almost doubled the base fine of $12,500 because "the egregious nature of the material exacerbates the violation."[18] The previous year the FCC had reduced the penalty to $3,750 for a South Carolina station—initiated by a complaint that morning DJs used the word "shit" once—because station management disciplined them, and it was a single minor infraction.

Exceeding *limits on advertising in children's TV programs* drew fines from the Commission. The Children's Television Act of 1990 limited commercials to 12 minutes per hour (twenty-four 30-second spots) on weekdays and 10.5 minutes on weekends. In 1992 the FCC audited 141 TV and 27 cable operations, then asked ten who violated limits for explanations before forfeitures were set (up to $25,000 possible for each violation). In 1993 the FCC penalized three TV stations in Michigan, South Dakota, and Montana $15,000 each for multiple violations of the limits (one overshot the mark on 58 occasions the previous year); the Commission noted that one station's response of "weakness in communication and management" did not excuse it from liability.[19] Four more stations that year were fined a total of $80,000; the next year when up for renewal five stations in Texas were fined, three $15,000 each and the other two $10,000 each. Three stations in Phoenix, Tucson, and Albuquerque were fined $10,000, $14,000, and $15,000 respectively by mid-1994. By then the two-dozen fines issued since the law was enacted less than four years earlier totaled almost half a million dollars.

In a single week in 1993 the Commission fined 26 radio and TV licensees in the western U.S. a total of $101,000 (the highest $13,150) for a variety of violations involving public inspection files, EBS, station logs, and unattended stations. ■

Despite sweeping deregulation the FCC went about its business of monitoring departures from federal guidelines. Many station managers found themselves saddled with fines equal to annual salaries of staff members. Those who replied in a timely and convincing way about mitigating circumstances often received lower or waived penalties. But such incidents generate distracting concern and take management's valuable time, plus considerable legal fees in many instances.

License Renewals: Short-Term and Denials

The Commission may also issue short-term renewals to licensees or deny renewal at the time of expiration (§307[d]), and it may revoke licenses (§312[a]).

Of the thousands of stations on the airwaves over the decades, only a few dozen have ever lost licenses (see table 11.4). After 1960, the Commission devoted more attention to short-term license renewals, as a form of probation instead of "capital punishment" by outright revocation. The agency became slower to renew licenses for the full period (then three-year terms) where evidence showed stations lacked efforts to ascertain and meet their community's needs with specific program service. In fiscal year 1994 the agency issued twenty-one short-term renewals of license because of EEO violations.[20]

While many kinds of violations occasion initial action by the FCC, *willful* and *repeated* violations—most commonly technical regulations not related to programming service—are major causes for license deletions, and especially "lack of candor" in reporting to the Commission.

Over the decades, the Commission revoked licenses of fewer than 100 of the 10,000 stations repeatedly filing for renewals. Among multiple reasons for revoking licenses, most were misrepresentation to the Commission and technical violations. In thirty-six years (1934 to 1969), program-related violations were cited against only six stations: one for false advertising, another for indecent program material, a third for overcommercialization, a fourth for departure from promised programming, and against two stations for fraudulent contests.

Although citizen and minority groups have contested renewals of station licenses, revocations almost never occurred because of programming content.

Comparative Renewal Hearings   About 70 challenges were filed annually in the early 1970s when the procedure was common; those proceedings typically lasted more than seven years and averaged 5,000 pages of testimony plus exhibits and other documents. Despite broadcast licensees' usual success against attempts to unseat them, their legal fees averaged $590,000 per radio station and $1.2 million for each TV station to defend against the challenge. Between 1982 and 1987 the Commission designated 40 comparative renewal cases for hearing. In 1989, some of the 4,200 radio stations up for license renewal in sixteen states, plus Washington D.C., Puerto Rico, and

TABLE 11.4
RADIO/TV STATION LICENSES DENIED, DISMISSED, OR REVOKED
1934–1994

DECADE	DENIED/DISMISSED*	REVOKED**
1934–1939	7	2
1940–1949	8	5
1950–1959	3	9[†]
1960–1969	28	17
1970–1979	49	6
1980–1989	11	2
1990–1994...	7+...	12+...

*Applications for license and license renewals denied or dismissed by the FCC.
**Licenses and construction permits revoked by the FCC.
[†]Included are five revocations of construction permits only.
Note: A number of licenses denied or revoked were later reinstated or otherwise permitted to return to operation: 3 in the 1960s, 12 in the 1970s (including 8 noncommercial TV stations in Alabama), and 3 in the 1980s; most in the 1990s were not finalized including 3 under appeal with the Appellate Court.

Source: Federal Communications Commission, *59th Annual Report/Fiscal Year 1993* (Washington, D.C.: U.S. Government Printing Office, n.d.), 38–39; *60th Annual Report/Fiscal Year 1994,* 49–50.

the Virgin Islands were faced with 220 petitions to deny renewal and 22 competing applicants. In latter years some challengers sought an incumbent's license not to serve a community better but rather to maneuver for anticipated settlements by the station or by competing challengers. Often the challenger was not only to be reimbursed for "legitimate and prudent expenses" for withdrawing but also received hefty payoff sums in six figures, tantamount to extortion. (In 1981 Congress had relaxed its 1960 law limiting settlements so compromises could reduce the number of lengthy hearings; but because of abuses the FCC in 1991 limited settlements only to recover documented expenses, with not even that permitted after a comparative hearing began.)

Until the 1980s the Commission did demand some accounting of on-air performance as compared with the promises made in the last application—which was the basis for awarding or renewing the licenses—including program balance, commercial content, and treatment of controversial issues. Broadcasters commonly acknowledged that prudent and competent operators had little problem with the procedure. But most renewal criteria today relate solely to technical matters or EEO provisions. Except for questions of flagrant language and imagery (indecency) or mishandling political candidates' access to airtime, program content is no longer a major factor in renewals or even in most forfeitures imposed. The broad issue put to broadcast licensees has become: Demonstrate that you have assessed your community's major needs along with what ways your program service responds to them. That stood as the basis for "renewal expectancy" by a licensee against any competing applicant.

Proposals and legislative efforts over the decades led to successive lengthenings of the term of license, as described earlier. With outlets multiplying to 11,000 commercial AM and FM stations and 1,200 commercial TV stations by 1996, less frequent renewals seem in the overall public interest and contributes to stability and maturity of the industry. Critics of the media often opposed this extension lest broadcasters become too independent of the public and the regulatory agency. But by reducing by over 40% the workload of the Commission in renewing broadcast licenses, renewal periods of five years (TV stations) and seven years (radio) since

1981—and seven years for both media in the 1990s, lengthened to eight years each by the 1996 Act—permit more detailed qualitative study of licenses when they do come up for review. Further, the longer period, beyond greatly lessening costs for licensees, gives added time to management to pursue its basic responsibility of serving audiences with astute entertainment and news programming strategies.

Deregulation eased the burden of keeping detailed records through the license period. Instead of compiling massive documentation for renewal applications, licensees need submit only a briefly-worded Form 303–S (replacing the postcard form in 1995) to confirm their compliance with federal guidelines. To that form they append their Equal Employment Opportunity Program Report (FCC form 396) with data tabulating the composition of their workforce (FCC form 395–B) and copies of any designated contracts not already filed with the Commission; licensees of TV stations also add evidence of complying with the Children's Television Act of 1990 regarding commercial limits and programming requirements.

Instead of perusing every application as in the past, the FCC randomly selects 5% of radio and 5% of TV stations to audit in detail (so all licensees better have proper documents on file at their stations!). The Commission relies on the short renewal form and accompanying sheets to routinely renew the vast majority of licenses.

Although legal procedures are less complex for the FCC in denials of renewal than with revocation, still at any time during a license period—whether five years or eight or more—the Commission can call a station to account for failure to operate in the public interest. It does so when a formal complaint or challenge has been submitted against the station. Those opposing a station's renewal must submit their competing application or petition to deny renewal within ninety days after the filing dates.

In order to alert members of the public to a station's impending renewal period, starting six months before the license expiration date stations must make on-air announcements on the first and sixteenth days of each month. In effect, the rules were intended to inform local audiences of their right to appraise local broadcast station service, and to comment directly to the broadcaster or to the Commission prior to the FCC's scheduled review of a station's license.

Those rules also instituted procedures for making available to the local public key documentation prepared by the station as part of its application for renewal. Those documents are placed in a Public Inspection File available to the public during business hours throughout the year. The file includes data about policies that identify community needs and interests, together with corresponding programming efforts during each year of operation—in the form of quarterly lists. Also on file are EEO employment data, report of ownership (FCC form 323), renewal filings, requests for political time by candidates for public office, petitions to deny or other formal challenges, and letters from the public.[21]

THE MANAGER AND RENEWALS

The role of the station manager in the renewal process begins the day he or she accepts the position. If she is truly in command of standards of quality for the operation, renewal should be no problem. She should insist that the program director and news director keep her supplied with a monthly performance report. This provides a basis for determining how the station responds to identified community needs and problems.

One function that a manager (with the licensee-owner) should not delegate to another person is preparing all FCC reports including applications for renewal. True, the detail work can be done by others but the manager must personally supervise its preparation. She must also approve and bear full responsibility for everything finally submitted.

An FCC official once noted that the major cause for delayed action on renewal applications is an incomplete filing: Some details are missing on forms and documents or exhibits referred to in the forms are not included. Instead of substantive reasons for delay, mere clerical carelessness at stations (mostly AM) contributes to repetitious refiling that can also be expensive. The manager should double-check renewal filings before they leave the station for the FCC, to be sure the staff and local counsel have included all information and materials needed.

In addition to license renewals, the manager must be sure that other periodic reports required by the Commission are completed properly and submitted on time. Major submissions include the following (those relevant to current circumstances):[22]

Form 159. *Fee Remittance.*
Form 301. *Application for Construction Permit* (CP)
  Also for any changes such as call-letters, location of studios, operating hours, power or antenna system, or equipment.
Form 302. *Application for License*
Form 303–S. *Renewal Application:* Every eight years for both radio and TV stations
Form 323/323-E. *Ownership Report:* Annually, and initially within 30 days of grant of CP (any changes must have prior approval of the Commission)
Form 314. *Assignment of License* (outright sale)
Form 315. *Transfer of Control*
Form 316. *Technical changes in organizational structure*
Form 395. *Employment Report:* Annual (with quarterly tabulations filed)
Form 5072. *Change of Mailing Address*
Proof of performance: Annually
Political Report: By November 30 of primary and general election years
Copies of contracts: Within 30 days of such agreements; these pertain to network services, ownership, sale of time to "brokers" for resale, sharing profit/loss of operations, mortgage and loan agreements, parttime engineer contracts, and management consultants. Not included are contracts with personnel, unions, or music licensing agencies.
Report on advertising to Federal Trade Commission: Annually

Local broadcast managers, including those at noncommercial stations, can find the *NAB Legal Guide to Broadcast Law and Regulation* a very useful source for checking specific regulations. Richard E. Wiley, former FCC chair and founding partner of Wiley, Rein & Fielding communication law firm, commented that "I would recommend that all of my clients keep a copy of the *NAB Legal Guide* at their fingertips. It would make my job and theirs a lot easier." Various sections apply to explicit responsibilities of general managers, sales managers, program directors, news directors, and chief engineers. In addition to topics noted above in this chapter, the volume covers programming and commercial practices, newsgathering, technical operations, cable rules about TV signal carriage, non-commercial rules, taxes, employee benefits, antitrust, copyright, and rules for satellite earth stations.

NEED FOR LEGAL COUNSEL

The manager should consult the Commission's Rules for descriptions of what must be filed in the FCC's Washington offices. For a guide to assist in conforming with federal laws, regulations, and evolving policies as well as with any state and local statutes, the manager needs ready access to an attorney who keeps abreast of cases and criteria and who can competently interpret the legal labyrinth.

Unless a media company has a manager trained in communication and business law generally as well as in other legal areas affecting broadcasting and cable, that operation needs to have local legal counsel available to assist on various problems arising throughout the year. Even if a station does not employ an attorney on a regular retainer basis, it should seek to get the benefit of local legal advice whenever needed. Even though legal services are expensive, the amount is far less than the cost of errors requiring resubmitting documents to the FCC or than complaint suits against the station, or other avoidable difficulties arising from inadequate legal guidance. The risks of operating without at least standby legal counsel are too great.

The process of media legislation and regulation over the decades has been tortuous, with Congress and the FCC often shifting stance on procedures from year to year, while the appellate courts and Supreme Court often undo what the Commission has cobbled out often after years of deliberations. Interpreting and applying those regulations often pit competitors as well as public interest groups against broadcasters and cable operators. Changes and reversals are so common, especially on the swiftly shifting telecommunication scene of media conglomerates, cable MSOs, and giant telephone companies in the 1990s, that no manager actively running a station or cable system can expect to keep up with the swirl of modifications and new rulings that affect him. Veteran lawyer Marcus Cohn, founder of Cohn & Marks communications law firm in Washington, once noted that when writing a book or chapter on broadcast law one should use three-ring notebook paper and write in pencil so pages can be dropped or inserted and lines easily erased from week to week as legislative and regulatory bodies do their work.

In addition to interpreting FCC rules and regulations, legal advice is necessary for such matters as labor problems, taxes, depreciation schedules, and many decisions associated with station promotions and contests. Among recurring instances fraught with legal complexities that can raise havoc with a station's daily operations as well as with management at license renewal time are: equal opportunities ("equal time") for political broadcasts, controversial issues of public importance (the on-again, off-again "fairness doctrine"), editorializing, the right to access to court proceedings, the right of privacy, defamation by libel and slander (in news and documentary as well as entertainment programs), "payola" and "plugola" (sponsor identification), fraudulent and misleading advertising, overcommercialization in children's programming, lottery laws, indecency and obscenity, censorship, and copyright laws.[23]

Program-related issues as the manager's responsibilities will be discussed after the following comments about program accountability required by the federal government.

Programming: Documenting Public Service

This section remains relevant to contemporary broadcasters because managers of leading stations in major markets often follow traditional, but no longer required, guidelines and procedures for confirming their community-oriented program service. This documentation can support their claim of "superior service" with expectancy of renewal before the FCC. Astute, successful managers claim they need to gather similar data anyway, for their business of attracting audiences and advertisers in media intensive markets. They also note that the regulatory pendulum could swing back some day to closer federal oversight of programming and operations; thus, station personnel will already be familiar with procedures while past data provide context and models.

Accountability and Reporting Forms

The practice of applying lofty classifications under public-service categories to make-do or lackadaisical programming is foolish as well as deceptive. (This practice recurred in 1993–1994 when TV stations responded to Congressional and FCC mandates for truly educational children's programming in their weekly schedule; they listed routine animated cartoons and adventure programs as having "elements of instruction" and lost their credibility as a result.)

With 1980s' deregulation came relief from many specific requirements outlined above. In 1984 the FCC reduced paperwork and computation of program-related matters to simply commenting on how a station's program service responded to the community's major problems and needs which management determined by some procedure less strenuous than formal "public ascertainment" techniques. Daily program logs were no longer required to be put into the public file. Instead of the traditional "promise vs. performance" comparative analysis of program percentages or total hours, the FCC's postcard renewal form simply asked for reconfirmation that no major changes from the original license application had occurred. A vestigial remain of the earlier mandates was the "issues/programs lists" to be filed quarterly with the Commission and put in the station's public inspection file. But that "public affairs" orientation no longer required carriage of news programs by every station.

The Supreme Court, overturning the D.C. appeals court, endorsed the FCC's determination that it could not properly order licensees to conform to a given program format, nor could it forbid complete format changes by new owner-licensees (in this instance the buyer of a market's sole classical music station switched to a rock music format).

The Commission also jettisoned its unofficial guidelines of the NAB's former Code that limited radio stations to 18 commercial minutes per hour. (Again, cyclic patterns reemerged when explicit limits to commercial minutes per hour of "kids' TV" were mandated not by FCC regulation but by Congress itself in legislating the Children's Television Act of 1990.)

The Commission's reasoning for deleting program category requirements, especially for radio service in a community, evolved over recent decades. Instead of each licensee being required to offer a full-service program schedule, rather a market's totality of stations together was considered as offering the public a broad range of programming. Individual stations could air a narrow format—such as one music "sound," often with no news or other information, or else all-talk. While most over-air TV stations still program full-service schedules, cable channels specialize in formats like radio stations. In recent years the FCC eventually permitted all-commercial services (shopping channels) as well as "program-length commercials" (demonstration shows solely promoting commercial products throughout). The rationale was that as long as a variety of differing program options were available, the listener-viewer could pick and choose from among the radio-TV-cable specialty or niche services for different kinds of programming. Thus it was no longer necessary to require each station to provide an across-the-board service of news, weather, education, sports, entertainment, and so forth.

Nevertheless, what goes around comes around—at least sometimes. In the 1970s the Commission had authorized its staff to renew TV licenses when applications indicated programming included 10% total nonentertainment, 5% informational, and 5% local. In December 1995 Barry Diller, chairman of the Home Shopping Network and Silver King Communications, wrote an op-ed article for the *New York Times* urging Congress to expand public interest requirements in the pending communications bill; the media executive recommended that the FCC establish minimum guidelines for local, educational, and nonentertainment programming to serve as standards for assessing broadcasters' performance.[24] FCC chairperson Reed Hundt in 1995 said internal staff processing had in the past looked to one half-hour a week of children's educational programming as the minimum to justify routine renewal of TV licenses. The chair pushed for three hours weekly of children's fare to respond to the earlier Congressional mandate for educational and instructional children's television on all stations' schedules (see later section on children's issues). Meanwhile, the U.S. Court of Appeals in Washington upheld the public-interest requirement that DBS operators devote from 4% to 7% of their channel capacity to noncommercial educational or informational programming. (Entering 1997, the FCC had not yet developed rulemaking to implement that provision.)

Program Issues: Politics and Controversy

Several specific issues in programming involve vital First Amendment freedoms pertaining to media: "equal time/opportunities," the "fairness doctrine," editorializing and personal attack, and "editorial advertising." (Topics of indecency, drugs, sex and violence, and children's programming are treated in the next section.) Constitutional guarantees of free speech and press were explicitly applied for the first time to radio and television media by opinions of the Supreme Court justices in mid-1973. But practical application of those rights occasion complex legal analysis and often difficult decisions for broadcast managers.

"Equal Time"

Section §315 of the Communications Act of 1934 regulates the use of broadcast facilities by candidates for public office. Under its provisions, radio and television stations are not obligated to schedule political broadcasts (but a 1971 addition of section §312(a)(7) to the Act requires stations to allow "reasonable access" to station time by candidates for *federal* but not local elective office). When a station or local cable system allows any legally qualified candidate to use its facilities—for a program or a political spot announcement—all other candidates for the same office must be provided equal opportunities to use those facilities.

Although popularly referred to as the "equal time" provision, the law's technical language refers to "equal opportunities"; it demands only that comparable airtime (similar length and schedule placement of approximately equal commercial value or audience availability), rates (lowest unit rate, including discounted ones), and treatment (facilities, services, style of presentation) be afforded to all opposing candidates. Spokespersons for candidates or a political party, or for positions on public issues before the electorate, are not subject to this provision of the law, but rather to applications of the related so-called "fairness doctrine," related to section §315(a)(4), pertaining to controversial issues of public importance. In 1970 the FCC ruled on "quasi-equal opportunities" during political campaigns, extending to political parties' supporters or a candidate's spokesperson those section §315 guidelines that a licensee must afford "equal opportunity" or comparable time to the opponents' supporters or spokespersons if they requested it.[25] Thus, unlike the strict fairness or personal attack doctrines, free time need not be provided; instead, as in strict applications of section §315 to candidates themselves, stations need only provide comparable equality of access (same or similar time, rates, facilities) to spokespersons for opponents. Also the FCC allows the station to make judgments in good faith about which candidates and spokespersons concern the community's interests, thus avoiding the §315 strictures about equal opportunities for all candidates for the same public office. Congress amended the 1934 Act to exempt from equal opportunities provisions any candidates or spokespersons who appear in bona-fide newscasts, news interviews, and documentaries.

It is important for the broadcast manager to anticipate requests for airtime during political campaigns by establishing prior guidelines and policies to be applied to those requests.[26]

All requests by candidates or made in their behalf should be carefully noted and filed, with specific information on the disposition of each candidate's request, including time and length of broadcast and charges paid; this record must be available for public inspection and retained for at least two years.

Who is a legally qualified candidate? The NAB's legal department drafted the following definition:

> In general, a candidate is legally qualified if he can be voted for in the election being held and, if elected, will be eligible to serve in the office in question.
>
> Thus, if the state law permits write-in candidates, an announcement that a person is a candidate for re-election would probably be sufficient to bring him within Section 315. In some states, however, a person is not a legally qualified candidate until he has complied with certain prescribed procedures such as filing a form or paying a fee entitling him to have his name printed on the ballot.

Most station managers, if given the power to decide, would prefer to have Section §315 repealed. In 1984 Senator Robert Packwood proposed a bill (without success) to repeal that portion of the Communications Act. Broadcasters argue that multiple candidates for an office, each having the right to request "equal time," creates problems greatly outweighing its benefits. The assumption at law is that all candidates for the same office are equal. Realistically, however, this is a false assumption.[27] Various fringe-party candidates reported by the press on the back pages can claim broadcast time equal to candidates of the major parties. But in the following elections, candidates of minor parties or of no party at all almost always receive an insignificant percentage of the vote. Legally qualified parties with candidates most often on the ballots in various states include: Republican,

Democrat, American Independent, Prohibition, Socialist-Labor, Socialist, Libertarian, American Labor, Constitution, Christian Nationalist, Liberal, Independent, Socialist Worker's American, America First, American Vegetarian, Greenback, Four Freedoms, and the Poor Man's Party.

The complexity of granting equal time to all candidates for the same office becomes compounded when extended to various positions appearing on a ballot, including state legislators, mayors, councilpersons, and all other state, county, district, and local offices. It can become ludicrous, as in a case once cited by Commissioner Robert E. Lee. A legally qualified candidate of the Progressive Party demanded equal time from a local station in behalf of his candidacy for the office of state governor; the station tried to comply with his request but had difficulty in rigging a line to the candidate's place of residence—a federal penitentiary!

In August, 1960, the equal opportunities requirements were temporarily suspended for the offices of President and Vice-President during the 1960 political campaign. This suspension led to the historic "great debates" between Senator John Kennedy and Vice-President Richard Nixon. The networks offered time for as many as eight debates; they were able to arrange for four. The debates served the American public, which for the first time could see the two major candidates commenting on the same points at the same time and in the same place, although personalities and techniques of public speaking tended to overshadow the substance of issues.[28] Audiences for each debate included over half the country's adult population.

The FCC in later years permitted (in the "Aspen rule") broadcast media to air debates involving only major candidates, without triggering equal opportunities for lesser ones to appear or respond, only if an outside third party "sponsored" or mounted them as a news event. By 1983–1984 the FCC further relaxed the rules: Beyond merely covering them as news, now broadcasters themselves could mount debates or discussions among key candidates. In 1984's local elections one out of three stations broadcast political debates they set up, while another 10% aired debates sponsored by outside groups; over half the stations surveyed said one or more candidates declined invitations to TV debates. Debates in subsequent elections, including those of candidates for President, were assumed to favor the challenger over the incumbent. Maverick millionaire Ross Perot succeeded in pushing himself and his agenda to the forefront in 1992 to the point where he joined rival candidates President George Bush and Arkansas Governor Bill Clinton in TV political discussions during that campaign. But organizers of the 1996 debate between President Clinton and candidate Senator Bob Dole excluded Perot because of his small constituency and diminished impact during that campaign.

Early on, the FCC required equal time for opponents to officeholders running for reelection who were covered in news stories. As a result of broadcasters' complaints, Congress in 1959 amended the Communications Act to exempt from Section §315's "equal opportunities" requirements certain kinds of bona-fide actuality programs where candidates may appear: (1) newscasts, (2) news interviews, (3) news documentaries, and (4) on-the-spot coverage of bona-fide news events, including political conventions.

The Commission interpreted those categories broadly over the years, exempting from "equal time" rules as news programs the CBS newsmagazine *60 Minutes* (1976), syndicated audience-participation talk show *Donohue* (1984), and syndicated *Entertainment Tonight,* among others.

If a Congressperson is a candidate for office, recorded radio or TV reports to his constituency are not exempt from provisions of Section §315; a station is obliged to afford airtime to opposing candidates. If a candidate appears in a network-originated program or announcement broadcast by an affiliated station in the area where the election campaign is being conducted, other candidates for the same office have a right to request equal opportunities to appear on that affiliate station. It is not necessary for a station to advise candidates of "equal time" availability; opponents are responsible for initiating such requests. A candidate must also request airtime opportunity equal to that afforded an opponent within one week of the day when the opponent appeared; after that deadline he or she may no longer seek equal time for that appearance.

The Commission modified how it interpreted the meaning of "use" by candidates for office. It originally called any identifiable voice or image of a candidate as "use" (even an appearance in an old film on TV, as happened when Ronald Reagan was campaigning for office). But in 1992 it narrowed the definition to such appearances controlled, approved, or sponsored by the candidate or the candidate's authorized committee after the candidate becomes legally qualified. Nevertheless, a newscaster or radio personality who is a legal candidate for office triggers "equal opportunities" for opponents whenever he or she appears on the air. The U.S. Court of Appeals in Washington unanimously agreed with the FCC's ruling in 1987 that Section §315's "bona fide news" exemption applied only to the content of newscasts and not to the persons presenting the news if they should be candidates.

The Federal Election Campaign Act of 1971 had amended Sections §312 and §315 of the Communications Act by imposing new requirements that broadcasters provide "reasonable access" to airtime and facilities for candidates to *federal* office, charge political candidates the lowest unit cost for airtime within a period of approximately two months before elections, and obtain certification that the candidates' expenditures are within campaign limits set by law.

Congress added the "reasonable access" provision to the Communications Act in 1971 as a new subsection 312(a)(7). It specified that a station's license could be revoked for "willful or repeated failure to allow reasonable access to or to permit purchase of reasonable amounts of time for the use of a broadcasting station by a legally qualified candidate for federal elective office on behalf of his candidacy." For candidates to other than federal offices, a station applies Communications Act guidelines existing prior to that 1971 amendment, such as "good faith judgments" that the station was serving the public interest in determining which political campaigns were significant and of interest to its service area.

The "lowest unit charge" refers to what a station or cable system charges commercial advertisers for the same class and amount of time for the same period. This rate pertains to a legally qualified candidate for a public office during the forty-five days preceding a primary election and during the sixty days preceding a general or special election, including the election day itself. The rate is the absolutely lowest possible charge (thus local ratecard, if national rates are higher) to the "most favored commercial advertiser" with the advantage of all frequency discounts, and other discounts or bonuses, regardless of the number of announcements by the candidate. Outside those periods of forty-five to sixty days, the "charges made for comparable use" by opposing candidates must be the same for all; no discrimination can be made in charges for time, so that if free time is provided to one candidate, any opponents have a right to the same treatment. Time segments of appearances must have similar commercial value for all candidates for an office;

airtimes scheduled must be equally desirable. Political sponsorship must be identified by visual or aural announcements stating who or what organization paid for the broadcast (under Section §315 a station licensee does not fulfill the obligation by merely announcing "this has been a paid political broadcast"); if the program exceeds five minutes in length, an announcement is required at both the beginning and end.[29]

The "certification" modification in Section §315(c) requires stations to have every federal elective office candidate or his or her representative certify in writing that payment of broadcast charges will not violate the permissible limit to campaign spending according to the provisions of Section §104(a) of the Campaign Communications Reform Act of 1971. Section §315(d) makes the same provision for state and local elections if those states have adopted campaign spending limitations.

A particularly vexing provision of Section §315 was the "no censorship" clause. Under this provision, radio and TV licensees and cable operators have no power to censor material that candidates broadcast. Even if the material is judged by station officials to be libelous, they are prohibited from censoring political candidates. The U.S. Supreme Court resolved the dilemma by holding that if a candidate makes defamatory remarks on the air, because a station is prohibited by federal law from editing political candidates' material it cannot be held liable provided it does not directly participate in that libel.[30] Nevertheless, legal counsel recommends libel and slander insurance to protect from lawsuits. Possibly the simple procedure of pointing out potential defamatory content might prompt the candidate to resolve the problem by personally modifying such portions of prepared text (but a broadcaster cannot demand to see copy in advance). Protection for the station applies only to "legally qualified" candidates and does not apply to supporters of those candidates, or to proponents of ballot issues.

Station managers must ensure that their staff keeps accurate accounts of all requests for political time and how each was handled. A political file must be complete and available for public inspection. Political ads must be checked for proper sponsorship identification. Stations drew heavy fines in the 1990s when they kept poor records of those instances, or violated the guidelines.

To avoid some of the snags and to help reduce the need for enormous financing to conduct campaigns, in 1996 first Fox network then the other major networks including PBS and also some local stations provided modest amounts of free time for major presidential candidates.

THE FAIRNESS DOCTRINE

Closely allied to rules affecting broadcasting by candidates for public office is the now deleted (or temporarily moribund?) "fairness doctrine." A radio or TV station presenting one position on a controversial issue had an affirmative obligation to seek out responsible proponents of opposing or diverse viewpoints and to provide reasonable opportunity to use their facilities. (More stringent guidelines apply to editorializing and personal attack; see later section.)

Unlike the case of political broadcasts by legally qualified candidates, the station licensee may be held liable for defamation, libel, or obscene material in broadcasts of controversial issues. It is therefore necessary for station management to exercise supervisory control over materials aired. The licensee is permitted to edit or censor that content.

The policy first emerged in 1929 when the Federal Radio Commission ruled that radio should present contrasting views about public issues (*Great Lakes Broadcasting, Inc.*).[31] The Federal Communications Commission initially forbade editorializing in 1941 lest a station exercise undue influence by becoming "an advocate"; then in 1949 it reversed itself by permitting editorializing provided stations achieved balance by airing variant viewpoints.[32] In 1959 Congress added to Section §315(a)(4) wording that many consider the legislative basis for the policy; it required licensees to "afford reasonable opportunity for the discussion of conflicting views on issues of public importance." The Supreme Court in 1969 endorsed the doctrine in *Red Lion Broadcasting vs. FCC,* asserting that First Amendment rights are primarily those of the listener-viewers (to have access to a range of views on issues) rather than of the broadcasters (to limit that range): "it is the right of viewers and listeners, not the right of the broadcasters, which is paramount." Yet in 1973 (*CBS v. Democratic National Committee*) the Supreme Court affirmed broadcasters' rights as "editors to edit" in determining what advertisements they would or would not carry—even when the ads treated controversial issues such as the Vietnam war.

In 1986 a federal appeals court undercut the assumed legislative foundation for the doctrine: "We do not believe that the language adopted [by Congress] in 1959 made the fairness doctrine a binding statutory obligation; rather it ratified the Commission's longstanding position."[33] The Commission discarded the doctrine in 1987 as not explicitly legislated by Congress but rather as its own invention which it now claimed was unworkable, unenforceable, and even unconstitutional. In 1990 the Supreme Court declined to review a lower court's decision affirming the FCC's repeal of the doctrine. But influential members of Congress vowed to reinstate it as legislative statute by modifying the 1934 Act. Repeated attempts to codify "fairness" amendments did get through both houses of Congress, but President Ronald Reagan vetoed the legislation. The frequency of the problem was reflected in the 5,000 to 8,000 complaints the FCC received annually in the 1980s claiming unfair broadcast treatment of controversial topics.

But even if not a formal governmental regulation or law, any station serving a community where interests often clash over matters of civic and social concern should strive to serve its audience's need for exposure to balanced presentation of controverted issues. On its face, one-sided treatment of a debated topic is simply bad business; it can alienate much of the community as well as many advertisers.

A problem of offering fair and balanced treatment of a hotly debated issue—or in applying the fairness doctrine if it is reinstated by Congress—is the ambiguity in interpreting or applying what is "fair." What may appear fair to one person may seem incomplete, unfair, or biased to others. As a consequence, treating almost any controversial issue can prompt a number of requests by individuals and groups for opportunities to be heard.

The matter of propaganda heads the list of many problems associated with operating on the principle of fairness. Aware of the broadcaster's obligation to present more than one point of view, and confusing the various forms of "access" and "the right of access" as used in broadcast and First Amendment contexts, individuals and groups of many persuasions regularly requested the opportunity to reply.

■   Major examples include anti-cigarette announcements aired to counter cigarette commercials, as an application of "fairness doctrine" principles; Congress eventually legislated against radio-TV advertising of cigarettes after midnight of New Year's Day, January 1971. Environmentalists sought airtime under "fairness" provisions, to respond to commercials for oil products such as gasoline that contributed to polluting the atmosphere. Similarly, businesspeople opposed to the war in Vietnam tried to purchase airtime to make counter-war announcements. Although the District of Columbia appellate court reversed the FCC and supported that last claim, the U.S. Supreme Court struck down the district court's ruling and reaffirmed the Commission's original decision by a 7–2 vote.   ■

While many requesters gained some sophistication regarding the fairness doctrine, there continues to be general confusion between Section 315's "equal opportunities/time" provisions for *persons* who are political candidates and "fairness doctrine" provisions for *ideas* that are controversial and of public importance.[34] Unfortunately, this same misunderstanding can be found among many broadcast employees as well.

Also problematic is cost for production and airtime when the program first presenting the controversial subject was sponsored. A would-be proponent of another point of view may have no funds to pay for a like broadcast, but still claims the need to air opposing views. The broadcaster may attempt to sell time for a response on an issue but, if unable to find a sponsor, is still obligated to present major dissenting views about the argument.

The complex area, even when under FCC jurisdiction, had ambiguous guidelines and was thus often decided on an ad hoc basis by the Commission. When applying the amended Act of 1934 to "fairness" questions the FCC provided latitude to the broadcaster. For example, not every person or organization claiming to represent a viewpoint about a significant public controversy had to be granted airtime. Managers were free to exercise their judgment "reasonably and in good faith" by selecting apt representatives for various viewpoints on an issue. Nor does mere newsworthiness constitute a controversial issue of significant public concern; again, the broadcaster—often with legal counsel—must make reasonable judgments about what issues seem truly at the level of public controversy. The broadcaster was also free to determine the format and other aspects of the program presentation. What the Commission looked for was the licensee's ability to justify procedures and defend judgment about providing service to the community, by identifying issues of major importance to that community and by affirmatively seeking out proponents of contrasting viewpoints. The problems typically arose with the number of advocates initiating requests for time on these topics or demanding time for response.

The question remains whether society is better served by insisting that all sides be presented, no matter how divergent, or by trusting station managers to use good judgment in presenting significant facets of issues of genuine concern to the community of license. A basic difficulty is that while print media are capable of almost endless expansion, and live speeches to audiences may take place from the streetcorner soapbox to the largest assembly hall, radio and TV transmissions are limited to the number of minutes in the hour and the number of hours in the broadcast day. A pragmatic concern is whether airing successive pro and con arguments really whets the audience's intellectual and judgmental appetite or instead drives them away to competing stations that offer entertainment and news programs.

Because Congress in the mid-1990s continued its effort to reinstate the "fairness doctrine," it is worth considering implications of the policy.[35] Over the decades most licensees, including corporate owners of network stations, vigorously challenged the constitutionality of the "equal time" rule and the "fairness doctrine." CBS President Dr. Frank Stanton referred to it as the "unfairness doctrine" partly because it forced equations between inherently imbalanced controversy (mathematical "fairness" in time and treatment to Fidel Castro's views? to pornographers' demands? to flat-earth proponents?). Further, the government-mandated policy fostered evasive blandness rather than robust debate; media managers avoided topics that might trigger "fairness" mechanisms.

Yet the alleged "chilling effect" of Commission oversight was questionable when over the decades only one station ever lost its license for violating the doctrine.[36] The Commission's usual remedy to a serious fairness complaint was to require the broadcaster to offer opportunities for opposing views. Despite hundreds and even thousands of complaints about fairness annually, the regulatory agency's staff found only a small number apt for serious review, with few going forward to the commissioners for judgment. In the first half-decade of Mark Fowler's chairmanship, the commissioners had to act on only one fairness case (which was appealed to the higher courts). Nevertheless, even the prospect of having to prepare an explanation defending a station's handling of controversial material—with the time and expense along with uncertainty about FCC assessment—can inhibit local managers from offering anything approaching controversy lest they have to devote too much airtime to various other views both substantive and tangential. Since the Commission abrogated the official fairness doctrine in 1987, there was no concerted rush to documentaries and special programs about controversial themes; but the phenomenon of popular interview and talk shows about highly debated social and political issues did begin to flourish when such were freed from accountability to the federal agency.

Respected Westinghouse Broadcasting was joined by another group owner, Fisher Broadcasting in supporting the doctrine. The latter testified in FCC hearings that

> the fairness doctrine is a vital ingredient of the public trusteeship under which Fisher Broadcasting, and all licensees, have operated their stations for so many years. That trusteeship, which is the heart of the legislative scheme for broadcasters, makes broadcasting unique among American industries. The public trusteeship principle in fact enhances the broadcaster's integrity nationwide. In Fisher Broadcasting's opinion, the fairness doctrine, as one aspect of the principle, has added significantly to the institutional credibility of broadcasting.[37]

At its height during the past decade, the heated controversy found major corporations, broadcasters, interest groups, and federal officials either supporting the doctrine or opposing it (table 11.5). Some shifted their stance over time.

Ironically, the fairness doctrine for a long time was favored by conservatives as an instrument to force media they perceived as liberal to permit rejoinders to alleged bias in news coverage of controversial topics. In the 1990s the rise of national and local popular talk and phone-in shows featuring conservative hosts reversed the debate; they accused complaining liberals of attempting to revive the fairness doctrine to stifle those widely viewed and listened to hosts.

—␉— TABLE 11.5 —␉—
Sampling of "Fairness Doctrine" Proponents and Opponents
1985–1995

<u>PRO</u> FAIRNESS DOCTRINE	<u>CONTRA</u> FAIRNESS DOCTRINE
some local station managers	most local station managers
Westinghouse Broadcasting (Group W)	Tribune Broadcasting, and most executives
most Senators and Representatives	of group-owned stations
Sen. Hollings (D) and Danforth (R), Rep.	networks
Dingell (D) and Markey (D): *codify it*	National Association of Broadcasters
into law (1987, 1989, 1991, 1993)	Radio-Television News Directors Association
Supreme Court Chief Justice Warren Burger	Federal Communications Commission
American Legal Foundation	some Senators and Representatives
American Business Media Council	Sen. Proxmire (D), Packwood (R): *abolish it*
Accuracy in Media	*(1977, 1981, 1987)*
The Conservative Caucus	Supreme Court Justice William O. Douglas
Committee for a Free Press	D.C. Appellate Court Chief Judge David
The Leadership Council	Bazelon
Telecommunications Research and	U.S. Department of Justice
Action Center	Association of Trial Lawyers
Media Access Project	National Telecommunications and
Mobil Oil	Information Administration
Accuracy in Media	Sigma Delta Chi: Society of Professional
New York State Consumer Protection Board	Journalists
Democratic National Committee	American Newspapers Publishers
Office of Communication, United Church	Association
of Christ	American Society of Newspaper Editors
Communications Commission, National	Association of American Advertising
Council of Churches	Agencies
Anti-Defamation League of B'nai B'rith	American Advertising Federation
American Civil Liberties Union	Freedom of Expression Foundation

Source: List drawn from various sources, including trade magazines and news articles about the topic.

The whole issue of "fairness" in programming controversial issues had risen in the 1930s because some radio owners editorialized, taking advantage of their favored status as sole or one of few stations in a community.

Editorializing and Personal Attack

The issue of editorializing in broadcasting began with the Mayflower Case, which resulted in a 1941 decision by the FCC. The Commission asserted that "the public interest can never be served by a dedication of any broadcast facility to the support of partisan ends." The FCC claimed that "radio cannot be devoted to the support of principles [a broadcast manager] happens to regard most favorably. In brief, the broadcaster cannot be an advocate."[38]

Reversing itself in June of 1949, the Commission issued a report saying that broadcasters would now be permitted to editorialize provided they sought out and allowed airtime for opposing points of view:

> If, as we believe to be the case, the public interest is best served in a democracy through the ability of the people to hear expositions of the various positions taken by responsible groups and individuals on

particular topics and to choose between them, it is evident that broadcast licensees have an affirmative duty generally to encourage and implement the broadcast of all sides of controversial issues over their facilities.[39]

Despite the fact that some broadcasters had worked diligently through the 1940s to regain the right to editorialize, very few of them exercised their newly won right during the 1950s. The obligation to seek out opposing points of view often seemed burdensome, injudicious, or distasteful. In 1959 the Commission tempered its earlier interpretation and adopted a "fairness" obligation on the part of the licensee, replacing the obligation to "seek out." Then in 1963 and 1964 came FCC policy statements intended to clarify the fairness doctrine but which confused broadcasters yet undecided about editorializing.

The FCC continued to develop its policy, in 1967 adopting rulings about personal attack: Stations must give notice to any persons whose "honesty, character, integrity or like personal qualities" had been criticized and commented on in programs involving controversial issues of public importance—excepting newscasts or live coverage of news events, but including editorials, news commentaries, interviews, and documentaries. The Supreme Court's "Red Lion" decision in 1969 affirmed the constitutionality and legality of the FCC's fairness policies and also its regulations requiring a station to afford opportunity to persons or groups to reply to personal attacks.[40] In effect, a station broadcasting such an "attack" must within one week of that broadcast notify the person or group attacked of the date, time, and identification of the broadcast; it must provide a script or tape (or summary if both are unavailable); and it must offer "reasonable opportunity" to respond through that station's facilities. In the case of political candidates, notice of such attack and the offer for reply time must be given within twenty-four hours after broadcast (but if the program comes within seventy-two hours prior to the election, the station must provide such information and offer in advance of the broadcast).

These considerations apply to three intertwined facets of the "fairness doctrine" that are often confused: discussion of controversial subjects (regardless of format), personal attack, and editorial endorsement of a candidate for public office. Although the fairness doctrine and Section 315 are most often cited when a person or organization has a grievance, state laws concerning libel and slander often can be applied against the person making the allegedly offensive statements and also against the broadcast licensee. In addition, while a broadcaster may not censor the statements of a political candidate (to edit out possibly libelous or slanderous material) and cannot be held liable under state laws, candidates might be well advised that the same protection does not apply to them.[41]

The right to editorialize, and even the obligation to do so, has been confirmed by the Federal Communications Commission and the federal courts. Yet by 1974 almost half of all broadcast stations in the United States never editorialized.[42] The NAB reported that smaller stations, with lesser budgets and staffs, avoided editorializing more than larger stations. By 1983 a survey jointly by NAB, RTNDA, and the National Broadcast Editorial Association found only about 45% of all stations in the U.S. aired editorials. Since then, with deregulation most radio stations eliminated news operations or reduced them to headlines and weather, foregoing traditional public service items such as editorials, for nonstop music.

The editorial is regarded as management's voice (including owner-licensee); it represents station policy or point of view, not merely a newsperson's or staff member's judgment. The choice for air spokesperson should be someone who clearly represents management, embodies authority, is believable, and conveys the management "image." A fulltime editorialist who researches and writes the editorials could be the logical person to broadcast them, if the manager does not.

The responsibility is great. If radio and television editorials are to command the same respect and impact as newspaper editorials, they require investment of time, effort, research, production, and talent. Clearly, only those stations with qualified staff and resources should attempt to editorialize. Neither society nor broadcasting gains from misinformed, superficially biased, or simply fatuous opinions. If a station manager truly wants to be a good influence in her community through editorials with social significance, she is obliged to hire qualified personnel with adequate time for research.

In 1984 the Supreme Court (*FCC v. League of Women Voters*) overturned as unconstitutional restraint the ban by Congress on editorializing by noncommercial public broadcasters. But those radio and TV managers rarely exercised their new-found freedom as they struggled against criticism of "left-leaning bias" in programming and especially under threat of reduction or total elimination of federal funding by the Republican dominated Congress after 1994.

"Editorial Advertising" or "issue advertising" presents a hybrid challenge to broadcast managers by those seeking to promote a viewpoint or action rather than to sell goods and services. Distinctive among corporations was the Mobil Oil company which for decades purchased prominent space in major newspapers to exhort the public about actions involving government, media, or business. But networks and most local stations reject requests to purchase airtime for these "advertorials."

One concern is that those with the most money could dominate public discussion by heavily buying expensive airtime (although Mobil offered to pay time charges for spokespersons with opposing views). Until the FCC shelved the fairness doctrine, such ads would trigger response time for differing views. Further, expensively mounted appeals can exploit the medium's artful production techniques to manipulate audiences. Radio-TV managers prefer to present controverted, complex topics in presumably objective news and public affairs or documentary formats. (But others counter that media news itself recasts reality—not necessarily "the way it is"—and political ads manipulate images, perceptions, and even facts on behalf of a candidate.)

A pragmatic concern is that ads debating issues can also alienate segments of the listening/viewing audience. On the national level, a network is reluctant to send 30- and 60-second "issue ads," embedded in program commercial breaks, down the line to over 200 affiliated stations in communities with diverse views and sensitivities. That would put local station managers in a difficult position when their transmitters become the means to proselytize on behalf of one side of a controverted issue. (However, ABC-TV did accept "issue ads" for late-night airing in the early 1980s.)

Former FCC Commissioner Nicholas Johnson lamented as ironic broadcasters' eager sale of time on the public's airwaves to hawk deodorant and banal trivia while refusing to sell time to promote ideas and to stimulate thinking about society's controversies. Apart from social judgment or even ethics, however, the Supreme Court decided the legal aspects: Broadcasters are free to accept or reject whatever advertising messages they wish, because in planning programming and filling out their daily schedules they act as editors whose proper role is indeed to select and edit.[43] Not operating a common carrier, broadcast management remains responsible for all materials aired, so can and must exercise supervisory control over content.

PROGRAM ISSUES: INDECENCY AND VIOLENCE

Among many legal and ethical problems arising from programming are such issues for management oversight as indecency/obscenity, and censorship. Obviously, individual cases demand competent legal counsel. The NAB and state associations of broadcasters distribute information reports with paralegal opinions and guidelines to assist station managers. Also, management and law texts cited earlier offer commentary about these responsibilities of management.

INDECENCY/OBSCENITY: RADIO

In the spring of 1973 the FCC explored developments in radio telephone-talk formats emphasizing intimate details of sex-related activities.[44] The Commission issued $1,000–$2,000 fines for alleged violations of the obscenity clause in the Communications Act (and §1464 of the U.S. Criminal Code) in radio sex-talk shows, dubbed by some as "topless radio." An effort to bring such matters to court for juridical determination failed when penalized WGLD-FM paid a $2,000 FCC-levied fine rather than contest the decision. But stations throughout the country understood the implied caution by the Commission and most of them modified or canceled their sex-talk oriented radio programs.[45] The Commission subsequently opened a general inquiry into broadcast of obscenity, indecency, and adult "X-rated" movies. The NAB's Television and Radio Code Boards met to consider possible actions.

The Pacifica case involved a noncommercial station in New York City playing a recording of George Carlin's satirical comedy monologue about the seven "Filthy Words" not permitted on the air. The Commission, initially reversed by the appellate court, was supported by the Supreme Court (1978) regarding the tactic of "channeling" border-line program material to hours when children would not normally be in the broadcast audience.[46] Through subsequent years governmental-industry-citizen debate centered on which hours properly constituted a "safe harbor" for adult material. Actions by Congress (mandating bans), appellate courts and Supreme Court (opposed to broad bans as unconstitutional), and Commission (whipsawed back and forth between those two positions) successively set as a "safe harbor" midnight to 6 A.M. (1987, 1992); none—a twenty-four-hour ban (1988, 1990); total safe harbor—no ban (1989, 1992); 8 P.M. to 6 A.M. (1991, 1993); and 10 P.M. to 6 A.M. (1995). The last was supported by the U.S. Supreme Court which refused to hear a challenge to the D.C. Court of Appeals' 1995 ruling in favor of the 10 P.M. to 6 A.M. "safe harbor" period.

Proliferating media, especially permissive cable channels with specialized pay-TV networks, complicated the issue of media "pervasiveness." Media's unavoidable omnipresence, thwarting one's simply turning off the TV or radio, was the Supreme Court's initial basis for ruling in favor of channeling mature radio and TV content to restricted hours to protect youths.

The FCC annually received about 6,000 indecency complaints (one estimate was as high as 20,000). However, when pressure heightened about the issue of indecency between 1987 and 1991, it found only twelve cases actionable during that period.[47] In FCC Chairman Alfred Sikes' first three years (August 1989 to August 1992), the Commission fined 14 stations from $2,000 to $20,000 each, with actions pending against 19 other radio stations late in 1992; all incidents aired prior to 8 P.M.

In 1992 the U.S. Court of Appeals for the District of Columbia struck down the ban on indecency between 6 A.M. and midnight (that is, a "safe harbor" for indecency from midnight to 6 A.M.), claiming neither Congress nor Commission provided adequate evidence about children's listening/viewing patterns. While acknowledging protecting children was a proper concern of government, the court stated "the government has not demonstrated . . . the compelling nature of any interest in suppressing constitutionally protected material in order to protect an abstract privacy of the home at the expense of the First Amendment rights of its inhabitants." Staunch defenders of full First Amendment rights for broadcasters, the editors of *Broadcasting & Cable* magazine nevertheless reminded media readers:

> That does not mean, however, that programmers are relieved of their responsibilities to their audience. Quite the contrary, it is they, in concert with their audience, who must strike the proper programming balance. That is what editors are for, and it is the flip side of the freedom that should be guaranteed by the First Amendment.[48]

That same appellate court also rejected one section of the 1992 Cable Act that had permitted cable operators to refuse indecent program material for public access channels or to place all such programming on a separate channel that subscribers had to specifically request. The Commission, pressured by Congress, in 1993 reinstated the midnight to 6 A.M. "safe harbor" for adult material but the D.C. appellate court in 1995 (7–4) narrowed the more recent "safe harbor" for indecent material, pushing it from 8 P.M. back to 10 P.M. until 6 A.M.[49] Judge James Buckley, writing the majority opinion, stated "it is fanciful to believe that the vast majority of parents who wish to shield their children from indecent material can effectively do so without meaningful restrictions on the airing of broadcast indecency." The court opinion added that "the government's dual interests in assisting parents and protecting minors necessarily extends beyond merely channeling broadcast indecency to those hours when parents can be at home to supervise what their children see and hear." But opponents from civil liberties organizations as well as broadcasters said they would appeal the decision to the Supreme Court.

Many broadcasters were concerned about permissive excesses just as there were defenders on behalf of constitutional rights.

■　Retired broadcaster Joe Meier lamented the debate over legal technicalities in the face of gross affronts to listeners' sensibilities, when he wrote to *Broadcasting:*

> [W]hether or not it is 'legal' may be the most irrelevant question of all. My friend, the lawyer, said with some conviction that it was impossible to arrive at a legal definition of 'indecency.' He is probably right. But that doesn't mean that you and I and Howard Stern don't know what decency is. I spent most of the 35 years I worked in radio as a programmer, DJ, PD, and GM. I knew the first day I went on the air what was decent and what was not. So does every

other jock (shock or not) out there. So do their program directors and managers. That is quite a different thing from fuming over some legal definition that will permit you to do something you know is, if not indecent, certainly beyond the bounds of common decency to begin with. . . .

Your editorial is right. The government has no business whatsoever in determining program content. And you are right again. If the government takes a little bit of your freedom (which they have done for the last 45 years under the guise of the fairness doctrine) you risk losing it all. But then the government should have no reason to be in the discussion at all. If we demand our freedom then we must be willing to pay the price. We must be 'responsible.' If we are irresponsible, whether that irresponsibility takes the form of setting buildings on fire, shooting at Presidents or shoveling verbal excrement into the public ears, the government will surely find a way to respond to the public outrage. . . .

We are truly fools if we think we can park our responsibility at the studio door and rely on the American Civil Liberties Union to write a legal definition of what is decent that will let us sniggeringly thumb our nose at those who find our nasty little intrusions objectionable.

They will surely write our epitaph.[50]   ∎

In the 1990s "shock jock" Howard Stern expanded a metropolitan New York radio audience into 18 million listeners in many radio markets across the nation. His crude, lewd, but witty comments and innovative excesses drew multiple fines by the FCC totaling $1.7 million for the Infinity group of stations carrying the daily shows. More than his rude, obnoxious but tantalizingly outspoken predecessor Andrew Dice Clay, Stern began to build an empire of TV guest appearances and pay-cable extravaganzas—titillating to some and tasteless to many—and added a published book and motion picture. His celebrity encouraged stations to mount local "zoo" morning shows in drive-time, featuring creative but borderline-raunchy comedy DJs.

∎   The FCC continued to impose forfeitures (fines) on stations. In 1993 a Chicago station drew a $33,750 fine for two incidents at midday and afternoon drive time which the Commission cited as "lewd and vulgar and clearly . . . within the definition of broadcast indecency." The largest, noted earlier, was Infinity broadcast group's $1.7 million fine (which it paid rather then contest and jeopardize FCC approval of a large station purchase) in September 1995 for multiple incidents involving Howard Stern, distributed to 353 stations including seven that it owned. Between then and October 1996 the FCC received 90 indecency complaints, some involving Stern. Many complaints it dismissed for lack of documentation of seriousness; otherwise it issued a "notice of apparent liability" and requested a station to respond to a complaint before imposing any penalty. In 1996 it fined a Lansing, Michigan, station $8,000 for an on-air exchange between a phone-in caller and the host; it assessed $10,000 against a Richmond, Virginia, station for a series of comments on two Stern programs that it aired.   ∎

Drug Lyrics

Although their licenses were not necessarily jeopardized, radio stations risked repeated fines because disk jockeys' spontaneous remarks and patterns of comments were not adequately overseen by local management. Compounding the problem were allegedly obscene words or supportive references to drugs in the lyrics of contemporary songs. The Commission called managers' stewardship into question when complaints were brought against stations for programming drug-oriented songs.[51]

The FCC in March of 1971 issued a Public Notice regarding broadcasters' responsibilities for drug lyrics. Noting complaints, it warned licensees of their responsibility to know the content of lyrics they broadcast, just as they are responsible for all other material aired by their stations. Two college FM stations took the matter to the Court of Appeals for the D.C. circuit. That appellate court sustained the FCC's authority to issue both Public Notices.[52] In its decision, the court noted the Commission had provided three options for broadcast licensees: (1) audition each selection before broadcast, (2) monitor each selection as it is broadcast, (3) decide about future broadcast after an initial broadcast—thus leaving broadcasters as much latitude as they enjoyed prior to issuance of the Public Notices. Later in 1973 the U.S. Supreme Court refused to enter the case and let stand the FCC's two Public Notices. Because what is a minority decision today may become a majority decision tomorrow, it is worth noting the statement by Associate Supreme Court Justice William O. Douglas, who with Associate Justice William Brennan dissented from the Supreme Court's action: "For now the regulation is applied to song lyrics; next year it may apply to comedy programs, and the following year to news broadcasts."

Sex and Violence: TV

Television, too, was under increasing scrutiny for alleged permissiveness in drama and comedy programs. The theme of "sex and violence" was sounded regularly in Congressional sessions in the 1970s. By 1974 Congress demanded that the Commission take more specific action about the perennial issues of sex, violence, and crime depicted in television programming. Early in 1975 Chairman Richard E. Wiley reported to Senate and House committees the steps taken in preceding months: discussions with heads of television networks which resulted in (a) a self-declared *"family viewing"* hour in the first hour of network evening prime-time (8:00–9:00, Eastern time); (b) actions by the NAB Television Code review board to expand that "family hour" forward one hour into local station time; (c) program "viewer advisories" about content that might disturb members of the audience, especially younger people; and (d) an effort by the FCC to define what it construed as "indecent" under the law (in a case involving Pacifica's WBAI (FM), New York). The Commission identified the latter as language in "terms patently offensive as measured by contemporary community standards for broadcast media, sexual or excretory activities and organs, at times of the day when there is a reasonable risk that children may be in the audience"; the agency added that when young people are likely to be present, the factor of "redeeming social value" became irrelevant.[53] Meanwhile, the FCC sought to clarify the U.S. Code provision (Title 18, §1464) prohibiting obscene, indecent, or profane language, to extend explicitly to visual depiction of such material.

After extensive testimony by those opposing and defending "family viewing" restrictions for early evening hours, U.S. district court judge Warren Ferguson ruled that, whatever the merit of the concept, the FCC had acted improperly in privately persuading the three network heads to orchestrate the NAB's Code provisions.[54] Voiding the policy satisfied creative producers of programs affected by the schedule restrictions. Syndicators of off-network re-runs also were mollified because the Code limited kinds of shows able to be aired in lucrative early-evening "fringe time" of local stations. Also, network staff personnel were relieved of the complicated task of applying the principle to specific shows and time-slots, directly affecting audience ratings and time-sales for commercial spots. Frustrated with the reversal were many members of Congress, FCC Chairman Wiley, and CBS Inc. President Arthur Taylor—considered the "father of family viewing" who had heralded the policy as the first step in twenty-five years to reduce the level of gratuitous TV violence and sex.

The episode demonstrated the daunting task of guiding a complex mass entertainment medium in a pluralistic society with varied perspectives and values. Through the decades television was under increasing scrutiny for alleged permissiveness in drama and comedy programs. Congress perennially worried about the theme of excessive "sex and violence" on TV, from Senator Estes Kefauver (1950s), to Senator Thomas Dodd (1960s), and Senator John Pastore (1970s). In 1975 House Communications Subcommittee Chairman Torbert MacDonald viewed the "family viewing" plan as a public relations ploy; he threatened to license the commercial networks that distributed national programs.

Deregulated radio and TV in the 1980s and growing permissiveness of cable channels led to widespread calls for "family values" in media. On December 1, 1990, President Bush signed into law the Television Program Improvement Act of 1990 passed unanimously by the Senate and by a vote of 399 to 18 in the House. Engineered by Senator Paul Simon, the law provided a three-year waiver of antitrust provisions to allow major networks and cable companies to collaborate on voluntary self-regulatory practices, to obviate threatened government action. Ironically in light of the scuttled "family viewing" clash almost two decades earlier, the Senator in a speech to West Coast industry leaders in 1993 warned them to take steps within sixty days to form an advisory committee to monitor violence levels on TV: "Either you will initiate the effort for such a monitoring office, or those outside the industry will do it. . . . The surest solution is governmental intervention, but it is also the most dangerous."[55] In the Fall of 1993, eight bills in Congress addressed TV violence.

■ A research study reported in 1994 that first-run syndicated series offered six of the ten most violent inaugural episodes of the 1993–1994 season; a Fox program was fourth highest, an NBC episode was seventh, and two CBS programs were ninth and tenth in the listing.[56] Despite complaints that limited methodology undercut their conclusions, the same organization later in 1994 reported that on ten channels in the Washington, D.C. market, 18 hours of TV and cable programming on a single Thursday night portrayed 2,605 violent scenes (41% more than two years earlier); scenes with "serious violence" totaled 1,411 (67% more than 1992); they claimed violent scenes on network stations rose 72% from 1992 totals to 765 in 1994, while cable violence rose 37% to 1,356 scenes.[57] ■

Historical irony was compounded in 1996 when 74 senators and congresspersons of both parties sent a letter to all six TV networks urging that they create a "safe haven for the family viewing audience" by dedicating one hour of prime-time programming to promoting positive values. The Media Research Center also advocated that networks voluntarily set aside a family viewing hour. (They had tried it in 1975–1976 and aroused an avalanche of protests from both liberal and conservative critics, but was terminated on a legal technicality.)

Congress' cable act in 1984 applied the U.S. Code's provision of fines or imprisonment for transmitting over cable systems constitutionally unprotected material such as obscenity. The 1992 Cable Consumer Protection and Competition Act declared system operators liable for obscene material on leased access channels. Those laws mandated putting indecent material only on a special channel which was to be blocked from reception by all households not requesting it in writing. The courts on four occasions in the mid-1980s struck down state laws totally forbidding indecent material on cable, because there was no scarcity argument nor was cable as "uniquely intrusive" as over-air broadcasting nor as pervasively available to children—customers had to subscribe and pay monthly fees to continue service.[58] In 1991 the Justice Department began to oversee obscenity and indecency on cable services, while the FCC continued to monitor such complaints about radio and television.

The cable and record industries introduced ratings to identify kinds of content in their products. The four TV networks in 1993 and 1994 agreed to provide "parental advisories" ("PA—some violent content") to newspapers and *TV Guide* for program listings. In 1995 broadcasters still struggled with appropriate ways to label without impeding programming. But broadcasters opposed Congressman Edward Markey's proposed computerized "V-chip" for receiving sets which parents could activate to eliminate reception of programs electronically rated by programmers as "V-violent." In mid-1995 the House Commerce Committee Republicans rejected Democrat Markey's proposal 31–15; but a month later after Republican Senator Robert Dole publicly excoriated excesses in movies and TV, Republican Larry Pressler led the Senate to pass the Telecommunications bill which included such a provision. That new law also mandated cable operators to relay sexually explicit programming channels with scrambled audio and video so nonsubscribers would not receive them. The FCC ruled that cable systems unable to scramble signals could deliver those channels only between 10 P.M. and 6 A.M. Further, the new law lifted part of the restriction against cable's editing access channel programs; now they were permitted to exclude access material they deemed indecent or obscene.

The issue, of course, is the broadcaster's freedom to program a station without government censorship, whether by prior restraining action or by ex-post-facto penalty that constitutes implied restraint against subsequent similar actions. The other side is the audience's right to freedom from what some consider offensive program content broadcast over a federally-licensed and pervasive airwave frequency defined by Congress as a natural public resource intended to ensure the "public interest, convenience, and necessity."

The conscientious manager keeps in close touch with programming personnel, not only to oversee them but to get their support in establishing and upholding station policies for properly clearing program content (scripted copy, ad lib comments, and musical lyrics). Management cannot delegate entirely to others the responsibility to monitor a station's daily broadcasts. The themes of responsibility and accountability receive increased attention among the media, including station groups, networks, national associations, and in the trade press.

First the cable industry and then four major broadcast associations agreed to develop a ratings system for programs, similar to the MPAA film ratings introduced in 1969. To be developed "voluntarily" (under threat of government setting up such) by production and distribution groups, the ratings would identify programs activating the "V-chip" also mandated by the new law. All TV receivers were to be manufactured with a computer chip that could be programmed by parents to exclude specifically rated shows ("V" for excessive violence, but also implying sex and language limits).

In 1994 Senator Paul Simon and the four major networks launched a three-year study to monitor violence in TV programming. Initial findings reported late in 1996 that some shows raised concerns about depicting violence; but 1995–1996 levels were reduced on all four networks from the previous season—in regular series, theatrical films, and children's programming.

Noncommercial television at times was hit by criticism that some of its dramas include nudity, strong language, or objectionable themes such as favorable portrayals of homosexuality. For example, on July 16, 1991, the "P.O.V." (Point of View) series offered a documentary-essay about gay African Americans called "Tongues Untied"; 127 large- and small-market stations of PBS' 320 affiliates did not carry the show, while some others rescheduled it after 8 P.M. Partly because of such content and also alleged liberal bias in programming, some in Congress challenged federal funding for the Corporation for Public Broadcasting and its PBS network. Managers of local noncommercial stations appraised their community's potential reaction before carrying some national PBS programs; they also considered possible impact on local fund-raising.

■　In 1993, almost one hundred program managers of PBS stations reported they consulted with other members of station management about potentially objectionable programming—two-thirds always did so, the remaining one-third sometimes.[59] On some occasions most conferred beyond management: Half consulted with their station's board of directors (one-third mostly in extreme cases), and at least in extreme instances 70% conferred with community members and 83% also checked with legal counsel. Almost one in four of responding managers had been threatened with lawsuits when they broadcast such programming; but only a single manager had actually been contacted by the FCC about such, and the Commission was not critical of the program.　■

The noncommercial TV manager's challenge is to offer an outlet for "alternative" kinds of nonmainstream programming while still respecting community standards and sensitivities.

CHILDREN AND BROADCASTING

General programming with "mature" themes of violent or indecent content are not the only concerns related to children. Attention also focuses on programming directed at youths, especially on Saturday mornings. Cartoons and "stylized violence" plus merchandising ploys integrated into animated storylines lures impressionable youngsters.

Chapter 7 already noted the Children's Television Act of 1990 that became law during George Bush's presidency (effective October 1991). The new law limited commercials in kids' shows (aimed primarily at those aged twelve and under) to 10.5 minutes during any hour on weekends and to 12 minutes an hour on weekdays. The law also mandated local, syndicated, and network children's programs (aimed at those 16 and under) to include "educational and informational" content.[60]

In the mid-1990s the FCC levied fines of $10,000–$20,000 up to $80,000 against local stations that exceeded commercial limits in shows aimed at youths, including Ted Turner's WTBS (Atlanta and nationwide by satellite to cable systems). The FCC fined a Tucson station $125,000 and another in Erie, Pennsylvania, $100,000, plus gave them short-term two-year license renewals, for excess commercial time in children's programs.

The FCC also settled on a requirement that every station air three hours weekly of pro-social, constructive programming for youngsters; compliance would become a factor in license renewal. Programs must have a significant purpose of advancing educational and informational needs of children 16 and younger—that is, their intellectual/cognitive or social/emotional needs. Programs must be weekly, 30 minutes or longer, scheduled between 7 A.M. and 10 P.M.

OTHER PROGRAM ISSUES

Broadcast and cable managers must continually watch over specific areas fraught with legal strictures, such as proper sponsor identification and sources of program material (including "payola"), false and deceptive advertising.

"Payola" and "Plugola"   A continuing threat to properly managing a broadcast station is undisclosed payment of money or valuable goods to broadcast employees in return for on-air promotional comment or display or other consideration. The "payola" and "plugola" investigations of the late 1950s resulted in action by Congress and the FCC, strengthening Sections §317 and §508 of the Communications Act, and adding penalties of up to $10,000 and/or one year imprisonment for violations.[61]

Managers must exercise personal vigilance to ensure that station procedures protect against programming content and personalities being used unlawfully.[62] In a word, whenever payment (in money or goods, as in a "trade-out") is made for commercial proposes, the audience should know the source of that payment. Such sponsor identification is intended to inform the public, while also protecting management's control of personnel who might otherwise modify program content because of private advantage from outside the station. These federal regulations also protect the value of commercial time, which erodes if nonclients have access to broadcast promotion without paying commercial rates (while also contributing to "clutter" within programs). The need for sponsor identification is even more apparent for individuals and groups who are not commercial advertisers but who purchase airtime for promotional announcements and for programming related to public affairs and politics. Again, the audience has a right to know the source of such "content with a viewpoint."

■   In the 1996 election, the FCC did not fine Oregon stations but ordered them to change sponsor identification on commercials opposed to a ballot proposition to raise taxes on cigarettes. Placing the ads was the "Fairness Matters to Oregonians Committee" (FMOC) which watchdog Media Access Project complained was actually a front for the Tobacco Institute. The Institute contributed all but $20 of the $2.7 million campaign funding; and both FMOC members, who controlled ad content, were registered in that state as lobbyists for R. J. Reynolds, the largest contributor to the Tobacco Institute.   ■

False and deceptive broadcast advertising is under the jurisdiction of the Federal Trade Commission rather than the FCC. Section 14(c) of the FTC Act protects media that accept advertising from outside sources, through ad agencies or directly from advertising clients. Nevertheless, stations—and networks on behalf of their affiliated and owned stations—have a responsibility to know the material they are broadcasting; stations are still responsible to protect the public from deceptive advertisements. But stations may be liable to FTC action if they contract to write advertising copy or produce radio or TV commercials for clients; they then act as agents of those clients. Further, audience research and market data can qualify as advertising material when, to promote the sale of station time or to elicit audience support, the station misuses them by making deceptive claims of audience coverage, market rank, or competitive status among other local stations.[63]

Aggressive stations that mount **contests and games** to stimulate community awareness and audience participation in their programming activities must be alert to avoid the element of "consideration" which constitutes a lottery and is illegal. Stations must exercise caution with promotional announcements about contests and games, because to broadcast such would be to air information about a lottery which until 1975 was a federal crime.[64] Many Commission actions about license renewals and forfeitures (fines) have resulted from misleading or illegal contest announcements. Three key elements must be present to constitute a lottery: a prize, the element of chance, and the more complex element of "consideration" or what one must do or pay to participate. Even though state laws may permit some kinds of contests (e.g., bingo or state-sponsored lotteries), broadcasters are bound by federal laws affecting contests that are lotteries.

Legal Issues Focused on Persons

In addition to candidates for public office, and those personally attacked in editorials, individual persons have legal right to protection from defamation and right to privacy. Minority persons also have protection and positive support from government in the media industry's hiring practices.

Libel

Inaccurate, misleading, inept, or simply incompetent media coverage, including statements or images in print or on the air, can adversely affect a person's reputation, good name, and career. The courts have granted wider latitude to covering and commenting about public figures (*The New York Times v. Sullivan*), extended to include celebrities and even private individuals swept up into newsworthy events. The vast majority of individuals have strong claim to protection from media intrusion and misstated or erroneous "facts" putting them in a negative light. Those falsely accused or otherwise defamed have recourse to the courts for redress. Most libel suits against media which win jury trials fail on appeal, especially if the plaintiff is widely known.

Some stations in major markets have aggressive news staffs who prepare investigative reports. Managers there say they often have several libel lawsuits pending at any one time. One study of 700 libel suits filed from 1974 to 1984 found 20% of defendants were broadcasters and almost 66% newspapers.[65] Public officials or public figures figured in almost two-thirds of libel lawsuits against news media. Half of all cases arose from news treatment about plaintiffs' business or professional activities. Half of the plaintiffs said they sought an attorney's assistance only after their direct contact with media people did not resolve their complaints.

Researchers concluded that media managers might reduce libel suits by establishing in-house policies for courteous, timely, responsive processing of complaints. Those procedures should be established with the help of the station's attorney.

Lawyers recommend that specific policies not be written lest they be used against a station in a trial. Alert managers quickly advise their libel insurance carrier if litigation appears likely, and consult with them along with legal counsel before airing any correction or retraction. Suits can be prevented by vigilant news producers who check all copy before aired—including proper match between audio and video (identifying correct person by name and image when allegations are voiced); they should confer with top management and even the station attorney in selected instances.

■ In the 1980s, one study found that of dozens of libel suits brought annually by public officials, the print and broadcast media won over 60% of pretrial motions for dismissal or summary judgment; the media lost three-quarters of the rest in trials, but eventually won about 70% of appeals of jury verdicts favoring officials. Plaintiffs ultimately collected damages in about 6% of libel cases against media; average payments were around $40,000 in the 1970s, about $90,000 in the 1980s. Another study of 300 libel trials between 1980 and 1992 reported that print and electronic media won about only one in four cases; but of the 167 judgments against media only 58 were later upheld on appeal and awards paid.[66] In 1990–1992, judgments against media averaged over $8 million. On appeal, damages were reduced dramatically, for example, from $2 million to $100,000. Nevertheless, despite winning or reductions, lawyers' fees can come to 80% of media costs as defendants, and up to half of plaintiffs' awards. As many as nine out of ten libel suits are dropped or settled early on or some time before a courtroom trial; but legal fees mount during those protracted negotiations.

Entertainer Wayne Newton sued NBC for implying he was linked with organized crime in Las Vegas, which he claimed damaged his reputation and diminished future income; the jury awarded him $19.2 million which NBC immediately appealed to set aside or reduce. CBS News President Van Gordon Sauter asserted in 1983 that his division "never lost a libel suit and never settled one out of court."[67] That year CBS News was faced with 57 pending suits. But five years later the Supreme Court refused to block the largest libel award ever against media: $3,050,000 against CBS' WBBM-TV for a news commentary broadcast in 1981, which courts determined included deliberate falsehoods. (The judgment was $1 million against CBS for "presumed damages" to Brown & Williamson Tobacco Company's reputation, plus $2 million for punitive damages, and $50,000 for punitive damages against newsman Walter Jacobson.) To that date the largest judgment against a broadcast company had been $850,000 against Outlet Communications Company's KSAT(TV) in San Antonio.[68] A contemporary classic in libel proceedings was General William Westmoreland's $120 million claim against CBS and *60 Minutes*. From the January 1982 airdate, through public hearings and court sessions amid widespread media commentary, to a stand-off joint statement of a nonmonetary settlement in 1985 (with each side claiming victory), the case distracted corporate and network executives and news personnel, disrupted its business operations, diminished CBS-TV and broadcast journalism in the eyes of many, and cost millions in legal fees. ■

In radio, libel suits have grown since the late 1980s because of comedy material, usually by controversial air personalities. Suits are most prevalent in small markets where stations staffs are generally less experienced and less responsible. "Bits" criticizing local businesses and people, call-in features pretending to be other persons, on-air telephone conversations that insult callers can cause enough distress that aggrieved parties seek legal support. Suits against radio stations average $100,000 or higher. Again, even a dismissed suit costs the station heavy legal fees to process the case. In 1990 WBZZ-FM in Pittsburgh and its two "shock jock"

DJs were hit with a $694,000 judgment in a suit brought by the station's own news director for defamation, invasion of her privacy, and intentional inflicting of emotional distress, plus medical bills and lost wages during a two-week period.

RIGHT TO PRIVACY

Related to the issue of defamation is the problem of broadcast news coverage of individual persons whom media cover in a way that compromises their right to privacy. Discretion must be coupled with good sense and reasonable sensitivity to recognize a person's private activities and feelings where the public has no legitimate concern or where intrusion by media can cause unwarranted distress and problems for the parties involved and for members of their families. Radio-TV-cable remote crews meet complex problems in visually presenting news events that occur in "semiprivate" or quasi-public places (government buildings, sports arenas, controlled admission halls), where permission must be obtained to report events. Cameramen operators must have legal access to a site and, unless the subjects are themselves in public notice, permission must be obtained from persons who will appear in televised reports. This issue is exacerbated in the matter of broadcasters' access to trials and other judicial or legislative functions of governmental groups. For there is the correlative right of the public to know about persons and activities that influence society and affect their own lives at least indirectly, including how government functions and how officials conduct themselves and their business. As in matters of libel, a person's newsworthiness or celebrity status diminishes their claim to normal levels of privacy.

ISSUES OF MINORITIES AND WOMEN

Chapter 5 on personnel discussed this important matter for management. Following the firestorm of civil rights achievements in the 1960s, the federal government sought to advance the role of minority persons in the business of broadcasting. The Kerner Commission Report in 1968 severely criticized media for seriously neglecting the country's minorities. As in other occupations, minorities had been vastly underrepresented among owners and on staffs; one result was the nation's program service reflected almost exclusively mainstream white/Anglo personnel, content, and viewpoints. The FCC vigorously promoted equal employment opportunities for African Americans, Hispanics, Asian Americans, Native Americans, and Aleuts or Alaskans—through annual reporting forms of the composition of each station's employees (beginning in 1971), and by requiring broadcast managers to report in detail their positive "affirmative action" programs to seek out and hire minority persons.

Fulfilling the seemingly mere quantitative quotas sometimes obscured the government's insistence that applicants for hiring and promotion must possess basic abilities needed for positions. The government never intended the hiring of unqualified persons. The problem arose in choosing between similarly qualified applicants, where preference began to go to minority persons over whites in an effort to redress under-hiring of minorities over previous decades.

■   By 1978 studies indicated minorities made up 15% of TV news staffs and 10% of radio news staffs, contrasted with only 3.9% in newsrooms of daily newspapers.[69] By the 1990s, fully 15% of all fulltime employees at the nation's broadcast stations were minority persons, mostly blacks and Hispanics—reflecting respective market populations. Nor were they relegated solely to maintenance and secretarial positions; they rose to become on-air newspersons, news directors, program directors (PDs), and account executives in sales. By 1987 women held 30% of broadcast jobs in the top four categories, and were 37% of the industry's entire workforce.[70] In TV, 9% of general sales managers were women, 2% were minority persons. One out of ten TV news directors was a woman, 3.1% were minorities; 3.4% of radio news directors were minorities. That year the NAB estimated about 500 of radio's 10,000 radio managers were women; general managers (GMs) of almost 1,000 commercial TV stations at that time included fourteen women, nine blacks, and five Hispanics. Although the numbers of women and minorities in broadcasting have risen over the past two decades, there remains much room for further advances—especially in higher positions of management (see chapter 5). ■

See chapter 5 for details of specific FCC actions pursuant to Title VII of the Civil Rights Act of 1964 as amended by the Equal Employment Opportunity Act of 1972.[71] Those actions included hefty fines against stations for EEO violations, such as forfeitures of $10,000 and $18,000 and short-term license renewals for radio stations in the 1990s.

■   Prime Cable of Austin, Texas, also received an $18,000 fine for not complying with the Commission's EEO rules. In February 1994, the FCC issued new guidelines setting a "base forfeiture" of $12,500 per EEO violation, which could be adjusted up or down depending on circumstances of the case. By October the FCC had notified twenty-six broadcasters of penalties totaling $600,000 for EEO violations; but that July the D.C. appellate court ruled against applying the guidelines. Among fines levied in 1996, the Commission fined an Athens, Georgia, station $20,000 for its inability to verify its effort at recruiting minorities because it kept no record of the number, sex, race, or national origin of applicants for jobs. It fined a Greensboro, North Carolina, station $17,500 for similar inadequate record keeping. That same year it renewed a Santa Rosa, California, station's license while assessing $17,500 for deficient minority recruiting efforts. It also penalized an AM-FM station licensed to the Lutheran Church-Missouri Synod in Clayton, Mississippi, for lacking candor in describing EEO efforts; the penalty: $50,000.   ■

The FCC sought to bring minorities into ownership and decision-making positions by granting preferences to minority applicants over nonminority applicants for licenses. Further, tax laws were relaxed to permit more financial flexibility that could attract minority investors to broadcast ownership.

In 1984 the U.S. Court of Appeals in Washington, D.C. held that policies giving preference to minorities and women, intended to promote diversified programming, were constitutional and consistent with administrative and statutory law. The Supreme Court supported that judgment by refusing to review the case. However, in 1985 the D.C. appellate court declared invalid the Commission's policy of preferential treatment of female applicants in comparative license proceedings, distinguishing that from acceptable minority preferences; in 1992 a federal appeals court held gender preferences in awarding licenses as unconstitutional.

The regulatory agency's efforts at structural changes regarding ownership criteria and procedures did aid increased minority participation in broadcasting and cable. The FCC reported that in 1994 almost 40% of the broadcast workforce were women and 18% were minorities; for cable the percentages were about 42% women and 25% minorities (those figures were compared with nationwide workforce totals of 45.6% women and 22.6% minorities).

■   In 1976, minorities owned only one TV station and 30 radio stations. During the Carter administration the number grew to 10 TV and 140 radio stations; then growth slowed during the three Reagan-Bush terms. By 1993 minorities owned 15 TV stations and 182 radio stations. In 1991 black owners who were members of the National Association of Black-Owned Broadcasters (NABOB) numbered 185; of those, 106 owned AM stations, 63 owned FM stations, and 16 owned TV stations.[72] In 1994 FCC Chairman Reed Hundt cited minority ownership figures: 490 of 98,000 telecommunications firms were controlled by minority members, 300 of the nation's 10,000 radio stations and 20 of 1,000 TV stations were minority controlled, while nine of 7,500 cable operators were minority persons. ■

In 1995 the Supreme Court surprised many by seeming to withdraw support from some affirmative action programs, acknowledging the risk of counter-discrimination and preferential treatment based solely on race or ethnic origin. The Republican-dominated Congress had already weighed in with challenges to seeming reverse bias; that and widespread unrest nudged President Clinton to order an evaluation of programs with affirmative policies and their effect on the broader population, especially middle-class white males who increasingly objected to competing for jobs and business on an "uneven playing field."

Personnel with Disabilities

Title I of the Americans with Disabilities Act of 1990 prohibits practices that discriminate against disabled persons regarding hiring, training, promotion, compensation, or termination. This means employers must make "reasonable" efforts to accommodate physical or mental limitations that do not result in "undue hardship" for the employer and business. Job procedures and schedules, workplace access, and equipment may need to be modified. Management must assure that federal guidelines and restrictions are followed regarding job descriptions, application forms, tests, medical examinations, and job interviews.

Legal Issues Involving Economics, Industry Structure and Processes

Other legal issues involve contracts and large sums of money for rights to broadcast, including **music licensing** through the American Society of Composers, Authors and Publishers (ASCAP), Broadcast Music Incorporated (BMI), and SESAC (formerly Society of European Songwriters, Authors and Composers). Stations and networks can hardly afford to do the bookkeeping on "per play" royalty payments for each song used in their daily schedules, so they contract for blanket fees to the respective licensing agencies. Expert legal counsel is required to assist in properly negotiating these important contracts.[73] (See chapters 8 and 10.)

Similarly complex and important is the matter of **copyright** for station originated material that has been broadcast, and for material gathered from other sources (books, magazines, newspapers, stage plays, reports, research study monographs, and so forth). Syndication properties licensed to a station carry copyright protection, complicated by cable's retransmitting of station signals. Legal advice is essential to avoid lawsuits for infringing property rights in print and photographed materials, and in using videotape and film materials including off-air and syndicated properties.

Since the early 1980s, cable and broadcasters engaged in a drawn-out battle marked by shifting Commission rulings and multiple reversals by higher courts. Regulations related to property rights included: syndicated exclusivity to protect local station investments in expensive program packages from the same shows carried on distant signals imported into their markets by cable operators; modest compulsory license fees for cable systems carrying stations' leased programming copyrighted by suppliers; and "must carry" provisions to ensure that over-air signals remained available to local households wired for cable.

By **syndicated exclusivity** ("syndex"), stations paying millions of dollars to lease syndication packages were protected from local cable systems relaying those same shows into their markets from distant station signals, including "superstations" from Atlanta, New York, and Chicago. The FCC discarded that rule in 1980, then readopted it in 1988. The **compulsory license** fee was intended to compensate stations whose signals and programs were carried by subscription-fee-generating cable systems; but the fee was a fraction of the actual value of the program properties, which involved copyrights and licensing revenues for show producers. Total copyright fees paid by cable in 1979 and distributed by the short-lived Copyright Royalty Tribunal amounted to $12.9 million; in 1983 $69.2 million; in 1988 and again in 1989 about $200 million; then dropped to around $175 million by 1991. In thirteen years cable operators paid over $1.1 billion to the nation's thousand TV stations for distant signal cable carriage. Since 1993 when Congress abolished the Copyright Royalty Tribunal, the Copyright Office of the Library of Congress collected cable payments and distributed them to holders of copyright—including imported superstations and other distant broadcast signals.

Introduced as an alternate option to compulsory copyright licensing (must-carry), **retransmission consent** became a knotty debate into the 1990s. Local TV stations wanted equitable compensation from cable systems that relayed their signals while receiving subscribers' monthly fees (two-thirds of cable viewing is the programming on three network-affiliated stations, the remaining one-third divided among dozens of cable's own program suppliers). But cable systems generally refused to pay compensation, instead offering local stations shared or exclusive time on an extra channel, which they could fill with more news department programming that included spots sold to advertisers.

The Commission sought to preserve free, universal TV service to the nation by **must-carry** provisions requiring cable systems to include all local and some regional over-air signals along with their menu of cable program networks. A local station's strength in a market could be greatly affected by whether it was included in a local cable system's service, and where it was placed in their channel listing. A UHF gained by being placed alongside competing VHFs on lower-numbered channel positions—a bonanza for a station previously isolated in the UHF ghetto. In 1985 the D.C. Court of Appeals struck down the "must-carry" provision; its action was sustained on appeal.[74] By 1988, over 20% of 4,000 cable systems reported having dropped or denied carriage to 704 TV stations (because some stations had been carried by two or more cable systems, the total number of "drops" was 1,820). Broadcast complaints to the FCC resulted in revised must-carry rules crafted thirteen months after the earlier ones had been voided on First Amendment grounds; but they, too, were struck down by the D.C. appellate court. Once again in the early 1990s broadcasters prodded the Commission to revisit the issue with must-carry provisions.

To protect the viability of smaller local TV stations, including noncommercial ones, the 1992 Cable Consumer Protection and Competition Act required cable systems to carry all local TV signals. But that "must-carry" provision, although upheld by the District appeals court in 1993 and by the Supreme Court the following year, was reviewed by the U.S. Supreme Court in 1997 and affirmed 5–4.

Cable operators claimed it unconstitutional, infringing on their First Amendment freedom to select programs and program sources without government coercion. When the government offered either "must-carry" (free) or "retransmission consent" for a price, if a cable system wished, most stations worked out arrangements such as to trade their popular program schedule for access to a second cable channel for its own purposes. Networks with O&O stations did just that: NBC developed America Talking program channel that became MSNBC; Disney/ABC added ESPN2; Fox added fX channel; CBS originally held out for cash payment from cable systems with "retransmission consent" and lost out on any fees; reformed as Westinghouse/CBS it worked out an Eye on People cable program channel in exchange for cable's carriage of its programming at no charge. Still, many stations have received compensation from cable systems for rights to carry their popular program line-ups. Until the rules are changed, every three years stations must elect to renegotiate with cable systems for either "must-carry" or retransmission consent for their signals.

The Commission interpreted that law involving "retransmission consent/must-carry" as constituting "a new communications property right, created by Congress, that inheres in the broadcaster's signal; this right is distinct from copyright, which applies to the programming carried on the signal."[75] After the Commission in 1993 gave broadcasters the option of either mandatory carriage on their local cable systems or negotiate a fee to be paid by cable operators, it turned out that only cable operators in small markets were willing to pay cash to carry local and regional TV stations' signals.

Both sides gained and lost in the decade-long struggle. Local stations needed cable relay distribution as much as cable systems needed popular network and local programming.

Cable Subscription Rates

Another continuing struggle through the past decade was what rates a virtually unregulated, uncompetitive cable industry might charge subscribers. The Cable Communications Policy Act of 1984 deregulated that industry, including cable subscriber rates starting in 1987. Attorneys general of five states in 1988 (and in twenty-three states by 1993) monitored cable subscription prices for possible anticompetitive practices. As rates accelerated into the 1990s, nationwide complaints roused Congress to pass the Cable Consumer Protection and Competition Act of 1992; it ordered the Commission to review and reduce excessive charges. After exhaustive study of cable marketing data, in 1993 the FCC announced rollbacks of most rates by 10% wherever cable systems had no "effective competition"—potentially reducing cable operators' revenues by an estimated $1billion–$1.5 billion. But complicated formulas applied to assessing rates later resulted in 4%–8% decreases in most systems, no decreases in many others. The Commission established a new Cable Services Bureau of 240 lawyers, accountants, and economists to administer the complex rate regulations. Responding to widespread complaints and confusion the agency adjusted those regulations in 1994. Beyond immediate cash-flow problems, cable managers put off plans to add further cable channels to their local systems and drew back from ambitious plans to merge MSOs. Bell Atlantic withdrew its plan to merge with TCI cable giant, scrapping the $33 billion deal partly because TCI's cable fees in thousands of systems across the country had to be rolled back.

Regulated regional and national telephone companies ("telcos") also benefited from loosened rules permitting them to provide video signals as a common carrier, initially in other regions and then in their own operating locations. They posed a competitive threat to established cable operations. Soon networks could merge with film studio production companies as well as enter into syndication and cable operations. The continually shifting regulations threatened the stability of established cable operators, especially those with smaller systems.

PTAR, F<small>IN</small>/S<small>YN</small>

From 1970 to 1995, local TV stations, networks, syndication, and cable operations have all been directly reshaped by government regulations affecting industry structure and operations. Congress' ban on advertising cigarette products on radio or television eliminated at one stroke $250 million of billings. Those same activist years prompted the FCC to diminish the power of the Big Three networks by cutting away one-fourth of prime-time available to them on stations in the top-50 markets. That **Prime Time Access Rule (PTAR),** described in chapter 8, opened local station's schedules to their own programs and gave syndicators access to lucrative early evening audiences.

Another structural reshaping came with the Commission's withdrawing the networks from financial involvement in most of their programming—opening the field to outside production companies and independent suppliers—and also from profiting from any kind of syndicating programs which was an alternate form of national distribution.

This **financial interest/syndication** restriction—enacted by consent decrees by networks and the Justice Department—enabled independent stations to blossom, with independent production companies rushing to first-run syndication, now feasible because those precious prime-time hours were open to them. Indy stations doubled from 73 to 149 during the 1970s, and to 320 by 1989; more than 50 production companies supplied over 100 programs in prime-time access alone. By the mid-1980s the syndication business had grown to almost $1 billion, and multiplied to over $3 billion in 1991 ($5.7 billion worldwide).[76] But network audiences shrank from 90% to less than 60% of all prime-time viewing, as new networks arose to compete, and multi-channel cable gradually crept into 65% of the nation's homes; as a result, the government relented and under the banner of deregulation rolled back most of those quarter-century restrictions by the mid-1990s.

Intended in 1970 to ensure fair and open competition in the broadcast marketplace, the fin/syn restrictions were regarded in 1990 as outdated and unnecessary even by the Justice Department which had originally applied them through network consent decrees.

In 1991 the FCC revised fin/syn rules to allow networks to produce in-house 40% of their prime-time schedule and to syndicate it afterwards; but a year later the U.S. Court of Appeals in Chicago struck down the relaxed rules as still too restrictive. As a result, in 1993 the Commission cut back even its relaxed rules, which the same court overturned in 1994. By Fall of 1995, networks were permitted to produce and own whatever they wished and by Fall 1996 they could syndicate their programs in both domestic and international markets.

Likewise the FCC caused the Prime Time Access Rule to expire in Fall 1996, permitting networks to program the entire four-hour evening block in top-50 markets for the first time since 1970. That worked against first-run program syndicators; and until networks actually programmed those hours nationally, smaller stations now had to compete with top affiliates in major markets in bidding for highly popular off-network syndication.

Broadcast and cable managers alike must keep abreast of shifting regulations and franchise requirements by federal and local government. Managers need to lobby their elected representatives on behalf of their own business interests, to affect pending rulings. Managers must also be alert to factors affecting their community's media mix, standing firm where necessary while also becoming informed enough to negotiate skillfully with regulators and competitors.

REGULATORY FEES

In recent decades the FCC repeatedly attempted to recoup some of its heavy annual expenses from the industry it regulated. Federal belt tightening threatened to reduce the multimillion dollar budget supported by the treasury's tax monies. The Omnibus Budget Reconciliation Act of 1993 authorized such fees, in addition to any fees for processing license applications, and the like. In effect the new fees are charges for conducting business. Licensees are responsible for identifying proper fees for their operation and for paying them in a timely way (errors can result in a 25% penalty). The new Form 159 (not automatically provided to stations) must be sent in with fees, subject to annual adjustments. Adjustments are not minor ones, especially for television. VHF stations in the top-10 markets saw fees jump from $18,000 in 1994 to $32,000 two years later; UHF fees increased from $14,400 to $25,000 in 1996 (table 11.6). In markets 11-25, fees for VHFs rose from $16,000 to $26,000, for UHFs from $12,800 to $20,000 in two years.

MEDIA ACCESS TO GOVERNMENT

THE RIGHT TO EQUAL ACCESS

This right differs from access that refers to the general public's limited right to use scarce broadcast frequencies or to air counter-comments under the former fairness doctrine. The issue of "equal access" for broadcasters involves the right to take microphones, cameras, and auxiliary equipment—the tools of the trade—wherever the print press takes its tools of paper, pencil, and laptop computer. Broadcasters seek equality with print media in reporting events of public import, including bringing broadcast equipment into legislative and judicial chambers (most often banned in the latter, while becoming common in the former but without universal rules or application). Broadcasters have argued they are entitled to this privilege under First Amendment guarantees. They claim media have a "right to find out" in their role as surrogate for the general citizenry's "right to know" in a free self-governing society.

ACCESS TO THE COURTS

The American Bar Association's Canons of Judicial Ethics since 1937 restricted broadcasting's access to court proceedings. Although not a law, the ABA's original Canon 35 was long adopted as a rule of court in many jurisdictions.

Various arguments have been advanced against admitting microphones and cameras to the courtroom. Some Supreme Court justices have argued that broadcasting is entitled to the same guarantees as other media. Opponents fear broadcasters would use access to the courts in a spirit not of public service but rather

FCC Annual Regulatory Fees, 1996

TYPE OF LICENSE	FEE	TYPE OF LICENSE	FEE	
AM		**TV**		
Class A fulltime	$ 1,250	Market Rank	VHF	UHF
Class B fulltime	690	1–10	$ 32,000	$ 25,000
Class C fulltime	345	11–25	26,000	20,000
Class D daytime	280	26–50	17,000	13,000
Construction permit	140	51–100	9,000	7,000
FM		Over 100	2,500	2,000
Classes C, C1, C2, B	1,250	Construction permit	5,550	4,425
Classes A, B1, C3	830	**OTHER**		
Construction permit	690	Cable antenna relay service		$ 325
OTHER		Cable TV systems (per subscriber)		0.55
Broadcast auxiliary	35	Satellite TV stations		690
Low power TV, TV/FM		Satellite construction permit		250
translator, booster	$ 190	Earth stations		370
Multipoint distribution service	155	Satellites (including DBS)		70,575

Source: FCC, reprinted in *The Alabama Broadcaster,* August 1996, 5; *Broadcasting & Cable,* 15 July 1996 18.

of entertainment—avoiding cases and litigation with numbing technical detail and complexity, exploiting sensational aspects of dramatic trials. Yet, in most cases where microphones and cameras have been admitted to trials, judges and attorneys have assessed that coverage as handled in the best interest of all directly concerned, as well as the general public.

One concern is that some attorneys and judges as well as witnesses might take advantage of broadcast coverage to engage in "show-off" performances to get publicity for themselves. However, where trials have been presented by the electronic media, judges have demonstrated they were able to keep proper order in their courts. Another claim is that broadcast media would violate the rights of privacy of all people in the courtroom; but legal opinion has held there is no law guaranteeing personal privacy in public matters.

It has been alleged that permission for radio and TV to cover court proceedings would result in confusion of equipment and personnel, prevent orderly legal procedures, and thus jeopardize the defendant's right to a fair trial under the Sixth Amendment. But with advances in miniaturized low-light-level cameras and accessory equipment, plus the demonstrated efficiency of pooled arrangements, facilities for court coverage without distraction are well known. Where exceptions have permitted broadcast coverage of court proceedings, reaction has been largely favorable.

Eventually the radio microphone and television camera became commonplace in many courtrooms. Continued vigilance and efforts by groups such as the Radio and Television News Directors Association (RTNDA), plus the rise of twenty-four-hour cable news and Court TV national program channel along with C-SPAN cable's coverage of Congressional proceedings, have prompted gradual progress in getting microphones and cameras into courts in most states and in many federal courts, but not the U.S. Supreme Court. By 1992, 45 of the 50 states permitted cameras in courts; New York was the only state to drop the practice (in 1991) after an initial probationary period. (That was surprising, because the chief administrator of that state's courts reported 94% of judges said cameras had no adverse effects on trials during the experimental period, with 3% stating that broadcast coverage added to the fairness of trials.)

The Judicial Conference of the U.S. in 1990 approved a three-year test of cameras in federal civil (not criminal) trial and appellate courts. After the test period, in September 1994 the Judicial Conference released the vote of twenty-seven federal judges which banned cameras from federal courts. The major reason given was concern for the effect cameras could have on jurors and witnesses. However, a few months later those same judges (who were permitted only a "yes" or "no" vote previously) sought to review the matter. Most chief judges of the thirteen federal appeals circuits signed a letter urging the Judicial Conference to recommend allowing cameras into appellate courts, letting each circuit decide for itself. At that same time, the American Judicature Society, most of whose 10,000 members are lawyers and judges, issued a policy statement in its journal; it noted that "responsible live or recorded coverage of courtroom proceedings has the unrivaled ability to inform the public about the courts and their processes" and added that "virtually all arguments against TV coverage of court proceedings have been successfully addressed or shown to be unfounded."[77] In 1995 the twenty-seven members of the Judicial Conference voted 14 to 12 (the chairperson only votes in case of a tie) to permit cameras into any of the thirteen federal appellate courts that wished to admit them. This applied to appeals in civil cases, but might possibly also apply to criminal appeals.[78] However, the Conference also recommended that district court judges not admit cameras to their proceedings.

The burden remains on broadcasters to demonstrate that electronic media will in no way disrupt the process of justice.

In court trials especially, but also in other serious procedural activities in society, the presence of broadcast and cable media must be planned and carried out discreetly by broadcasters. For the media do affect in some way the shape of events being reported, by their very presence and by the fact that they are relaying those sights and sounds to the public at large. This calls for deft handling of equipment set up unobtrusively in a timely way before the event. Crews assigned to formal public proceedings should wear proper attire, so as not to distract participants (and to avoid stereotyping "TV types"); blue jeans or cut-offs and tee-shirts might be appropriate for covering an outdoor sporting event, but are entirely out of place where everyone else is dressed formally and the location exhibits dignity.

The notorious trial of athlete-broadcaster O. J. Simpson in a Los Angeles courtroom in 1995 demonstrated both the advantages and the drawbacks of televised coverage, especially of highly visible and volatile cases.

ACCESS TO THE FEDERAL GOVERNMENT

Until recent years, radio and television/cable enjoyed less access to legislative and judicial affairs than to the executive branch.

Sessions of the U.S. Senate are closed to broadcasters except for the carefully controlled pool cameras of C-SPAN cable network. Senate committee hearings can be covered by special permission of the chairperson and ranking minority members. Prior permission, however, is no guarantee of coverage. Access has been granted, refused, limited, and arbitrarily expanded and restricted. The Senate hearings on the "Watergate" disclosures affecting multiple branches and agencies of government (in the Nixon presidency in the 1970s) dramatically demonstrated the impact that access to such sessions can have on the public whose representatives were conducting the meetings and whose elected or appointed officials were being interviewed. The flexibility of media in covering the sessions "live," by "gavel-to-gavel" re-runs the same evening, by abbreviated summaries at the end of a day or week, and by excerpts inserted into the evening newscasts suggested the important role for broadcast media access in a free, democratic, pluralistic society.

Access to Regional and Local Government

Station managers and their news directors need to take advantage of access by covering state legislatures where officially permitted. Except in rare instances, "live" broadcasts may not be as useful for audiences as representative excerpts (more than brief clips and "sound bites") of legislatures, with explanatory comments, in the station's evening news. Similarly, appropriate coverage of city councils can add to better community understanding of issues and actions affecting them. But when attempting to cover city council meetings, broadcasters and other media representatives are sometimes the victims of a convenient maneuver known as the executive session. Many so-called executive sessions are in reality only "closed meetings." The business that may be conducted in an executive session includes deliberation about a public employee's conduct or position or about any person who would be adversely affected by public disclosure. City or state officials who use the "executive session" closed meeting for other purposes may not realize they are in fact denying voter-citizens' rights to know how their elected officials conduct government business on their behalf. This could be a significant matter for investigation and even editorializing by a radio or TV station, to truly serve the "public interest, convenience, and necessity" of a community.

The Challenge

In the fight for equal access, broadcasters do not ask for permission to cover all trials nor do they propose to cover all proceedings of legislative bodies. The commercial schedule would not permit such extensive coverage. What broadcasters do want is the opportunity to cover legislative and courtroom proceedings at times when the event reported is of special interest to the public.

When broadcasters achieve equity with print colleagues in access to courts and to governmental sources, no aspect of their programming will call for greater integrity and professional competence. It is not an assignment for staff announcers and intern videographers with little background in journalism. This coverage is too important to rely solely on microphones, tape recorders, and cameras to transmit the whole story. The trained and experienced reporter's judgment is an absolute necessity. To bring that judgment to bear in one's own radio-television community is the duty and distinction of responsible station management by the GM and the news director.

Appraising the Role of Regulation

The Issue of Regulating Programming and Advertising

From the beginning, there has been continual disagreement between broadcasters and the FCC about program and advertising regulation. A basic point of contention is the phrase in the Communications Act that requires radio and television stations to be operated in the "public interest, convenience, and necessity." Since the Act was established, it has been almost impossible for broadcasters and regulators to agree on just what those words mean. Nor do all broadcasters agree with one another, just as all regulators, attorneys, and judges do not agree about their meaning. This semantic conflict has caused misunderstandings on all sides.

Over the decades some FCC commissioners have interpreted the phrase as a command to act as "watchdogs" over the content of programs and advertising. Congressional committees have generally supported that view. Broadcasters, notably sensitive to criticism, have considered inquires from the Commission in these areas as threats of censorship. Over the years, the "cat and mouse" game has resulted in much wasted energy and time. Commercial broadcasters generally resent any intrusions of the FCC into these areas. Broadcasters feel the proper province of the government regulatory agency is its licensing function and maintaining proper technical standards of broadcast engineering. Broadcasters find nothing specific in the Communications Act giving the FCC the right to intrude into program or commercial issues. But the courts have regularly interpreted the Act to give the Commission some degree of authority over the total program service provided to the public by broadcast licensees.[79] When the Commission outlined policy about programming in 1960, it disavowed any intent to substitute its judgment for that of the licensee in determining an apt program schedule for the audience in a local community. But the Commission did intend to grant and renew licenses based on the licensee's efforts to determine community needs and interests, together with the program service aired to serve those needs and interests. Typically, the Commission depended on comparative appraisals between competing applicants for the same market frequency, instead of attempting to establish any national "standard" or fixed formula to be applied to all licensees.

The knotty issue involves the difficulty of agreeing on qualitative as well as quantitative standards for programming, and then applying them equitably and consistently. But the responsibility for determining quality more properly lies with the professional and dedicated broadcaster, rather than with a national regulatory body of political appointees and career civil servants.

Obviously when relieved of any governmental requirement for balanced programming as a condition for securing and maintaining licenses, most broadcasters concentrate on program types that draw the largest audiences, while eliminating most service-oriented programs. Even so, freedom in programming must rest with the licensee, free of government dictates.

Before the Justice Department occasioned the demise of the NAB Code of practices, stations failing to meet certain standards were moderately disciplined by their own professional associations of colleagues. But more importantly, such stations were often ignored by the more educated, thoughtful, up-scale listening and viewing public. Of course, stations with narrowly defined entertainment programming also affect the quality-conscious station by wooing away large numbers of the audience. Even so, it seems better to have ill-run broadcast properties or even competitively disadvantaged "quality stations" than to entrust to a handful of commissioners and their staffs in Washington, D.C., the decision to tell all the communities in the nation what they should see and hear. Ideally, the most apt arbiter of program acceptability should be the listener-viewer who can freely turn the dial and channel selector.

With deregulation the FCC has given stations most responsibility for determining program balance and quality. Although the Commission has eliminated management's comprehensive surveys of listening/viewing audiences and local leaders to "ascertain community needs," each station must still identify major issues and problems in its community, adding commentary on how some of its specific programming responds to those needs. Many broadcasters had opposed the earlier exhaustive requirements because of the extra effort and expense involved; detailed and periodic surveys cost money and staff time to compile data, along with other data gathering (now abandoned) for license renewals and defenses against challengers filing against those licenses. But truly professional broadcast managers, in smaller as well as major markets, continued many earlier procedures for compiling such data; they reasoned that, after all, a station programs for its audience's needs and interests or else it loses them and goes out of business. They also felt it easier to keep staff members knowledgeable by continued data gathering, than to be faced sometime in the future with renewed federal requirements and having to start over from scratch.

Media professionals at times do not act in what appears to be their own best interests. Then when the public complains, along with critics in print, government representatives take up the cry through legislation and regulation. A compact between broadcaster-cable operators and their audiences is broken when media ignore audience's genuine needs and preferences in favor of pandering to their lesser instincts. In default of mutual respect and support, public pressure builds up, leading to governmental action that is intrusive and often becomes overkill. That was the case with overcommercialization in children's TV programming, overly permissive bad taste in radio programming, and greedy subscription rate increases by cable operations. Key to independent and competitively effective radio-TV-cable management are professional competence and good judgment—discretion mingled with respect for the community that makes up the audiences on which media depend.

INDUSTRY SELF-REGULATION

Most broadcasters maintain that industry self-regulation is preferable to government regulation in all areas except technical standards and the licensing function. Many feel that governmental action would be unnecessary if the industry removed all causes for those actions through effective but nonintrusive self-regulation.

Most managers of broadcast and cable properties are capable of regulating themselves individually or in association with professional colleagues. The standards they adopt and maintain should be in the interests of the listening and viewing public, as well as in their own interest.

The real obstacle to a self-regulatory system has always been the owner or manager not devoted to broadcasting as a profession, but rather as a source of income, with little interest in even minimal standards. A small minority in earlier decades, the number grew with proliferating mergers and acquisitions of media properties solely as vehicles for financial gain. They retard the efforts of the majority to effect changes. In any market, one or two operations that undercut the ratecard, accept questionable advertising, or approve substandard programming can gradually push other stations in that market to compete on the same terms. Before long, the general image of broadcasting in the area can be damaged.

Other businesses have deviant operators, too, as do most social institutions. In most cases, however, the predominant weight of public opinion favors those who maintain readily identifiable high standards. With broadcast and cable media, relatively few people in the audience recognize differences between marginal operators and those who do not approve of loose operational procedures but feel forced to adopt them in order to stay in business. The public is aware, however, of stations that consistently maintain high standards.

The nearest approximation to any sort of official "seal of approval" in radio and television used to be membership in the NAB's Codes of practices for radio and TV. But displaying the Code symbol, indicating station membership, unfortunately meant little to people in the audience. Station membership in the now-defunct Code was fairly high, considering that voluntary membership included annual dues and restrictions on kinds and times of programs and commercials. Fewer than two-thirds of the nation's TV stations and fewer than half of all radio stations subscribed to the Code and paid membership fees.[80] Although sizable numbers of stations showed no interest in an industry code, the fact that a large proportion of stations did subscribe indicated some desire for industry self-regulation—partly to ward off government regulation, partly to set standards and expedite processing of commercial clearance for acceptability. The idea of a self-regulatory code was resurrected in the 1990s, in response to widespread criticism by the public and Congress, to forestall government action.

In June 1995 the Senate voted 73 to 26 to add to its telecommunications reform bill (S.652) a provision creating a Television Ratings Commission (of five members appointed by the President) if the industry failed within one year to develop a ratings system to classify violent and other objectionable programs. But even outspoken critics of program excesses Democratic Senator Paul Simon and Republican Senator Bob Dole opposed mandatory federal ratings as extreme.

MANAGEMENT'S ROLE IN SETTING POLICY

Managers must also realize that their powerful media of communication can perform against the best interests of the public. This results whenever any station lapses into antisocial programming, slanted news, or excessive advertising. As such practices grow, they can lead to more restrictions placed on the industry as a whole. Those restrictions are prompted by citizen groups and other public organizations and individuals critical of media practices—from religion, education, business, labor, and government. While politicians court media for favorable coverage, they still respond to loud public outcry from citizens whose votes they need for reappointment.

While resisting efforts toward regulation, broadcasters and cable operators must thoughtfully analyze the media industry's structure, processes, capacities, and proper relationship to the social fabric—linking research data and judgments with pragmatic understandings that have generally been criteria for successful operation. The NAB and consulting firms as well as academic institutions can contribute to useful research and analysis. Meanwhile, the individual station manager can do much to put his own house in order. American media managers need to get away from operational aspects of radio-TV-cable plants from time to time to develop a consistent philosophy of broadcasting-cable that they put into practice. Such a philosophy of goals or mission, purposefulness and values, as well as strategies and tactics is needed for the industry as a whole as well as for individual stations and cable systems. To minimize the threat of government control, management must define a set of principles for media operation, setting forth freedoms central to the industry while also describing responsibilities joined to those freedoms.

Broadcasters and cable operators are often referred to as large or small based on differences in market size, station power, capital investments, gross sales, net profits, or number of employees. But in one sense the only "small" media managers are those disinterested in or incapable of high standards of performance and who consequently deserve the sort of regulation they get.

Most managers are in broadcasting or cable presumably because they derive a greater sense of satisfaction there than they could gain from other pursuits. They could, with business acumen, make a living in other lines of endeavor—often with less complicating duties and less publicized criticism of their performance—but they could not find any greater opportunity to serve their communities. CBS Vice-President Tom Leahy emphasized in 1993 a key reason for going into broadcasting: "It *matters*!" He felt this career was matched by none other except possibly public service in political life. Modern broadcasters know that the character of their stations as a community force is the greatest asset they possess. The more they do to make their stations a truly vital part of their communities, the more they can rely on the people of their area for support against outside interference. Through their station and cable service and by their personal participation, they should speak not *to* the community but rather as a part of it.

Media managers are in a dynamic industry that does not stand still for very long. Movement and change, however, do not necessarily mean progress unless purposeful; and sustained progress is necessary to maintain leadership.

The Manager's Relationship with Federal, State, and Local Governments

To understand better the meaning of serving the "public interest, convenience and necessity," and to apply this understanding to successful broadcast managing, one must understand the Federal Communications Commission, both structurally and in relation to the problems it deals with (outlined earlier in this chapter). While the broadcast-cable manager and owner-licensee may deal with other agencies and levels of government, relations with the FCC are paramount. In order to insist on their rights, media managers first must clearly understand those rights. They also must learn to accept their responsibilities to the Commission, just as the Commission needs to balance its responsibilities to the public as well as to broadcasters.

Beyond the FCC are other governmental agencies such as the Federal Trade Commission, the Justice Department and its Antitrust Division, the Food and Drug Administration, the Internal Revenue Service, and the two houses of Congress. A manager (and the owner-licensee) serves not only his own operation's best interests but also the needs of fellow broadcasters or cable operators by keeping abreast of and even participating in governmental activities, especially through Washington attorneys, state associations, and national professional organizations. Relationships with federal officials can be mutually beneficial; but they can also possibly hamper proper administration and regulation. Each manager and media company must exercise integrity in communicating and collaborating with federal agency personnel. Krasnow and Longley note the complex web that links the industry and the Commission:

> On a day-to-day basis, Commissioners are forced to immerse themselves in the field they propose to regulate; however, the line between gaining a familiarity with an industry's problems and becoming biased thereby in favor of that industry is perilously thin. . . .
>
> The opinions and demands of the broadcast industry are expressed through consultative groups (such as joint industry-government committees), interchange of personnel, publication of views in the trade press, liaison committees of the Federal Communications Bar Association, social contacts and visits to offices of the Commissioners, informal discussions at state broadcaster and trade association meetings, and the formal submission of pleadings and oral argument. . . .
>
> In the intricate and dynamic relationship between the FCC and the industry, the Washington Communications lawyer plays a special role not only in interpreting FCC policies for broadcast licensees but also in shaping the policy direction of the Commission. In a recent study of Washington lawyers, Joseph Goulden has noted that, while the lawyer's historic role has been to advise clients on how to comply with the law, the Washington lawyer's present role is to advise clients on how to make laws and to make the most of them.[81]

Lawyers attempt precisely those roles on behalf of citizen groups and critics of broadcasting as well as on behalf of clients in radio and television. This is the play of pluralistic forces that constitutes a democratic form of government. Therefore, the broadcast manager has as much a duty as a right to seek to affect favorable legislation and regulation through proper relationships with members of government.

Through the years, many broadcasters have continued to maintain working relationships with members of the FCC and its staff; they found themselves in a position to aid the Commission in its task of regulating broadcasting. They found most commissioners willing to cooperate on all matters affecting broadcasting so long as the practices were honorable in intent and addressed the principles embodied in the Communications Act.

Literally closer to home are other governmental relationships important to station managers and owners. They must maintain regular and effective contacts with representatives of state and local governments. The broadcast or cable manager must be rightly judged a key citizen of the community by the mayor, state legislators, governor, and area Congresspersons. It helps to have a close working relationship with members of the state legislature; here, exchanging information is important to media managers and legislators alike.

Although rarely able to be on hand when elected representatives are in session, a manager can maintain regular communication with them through correspondence, fax, and telephone.

Few persons have more impact on their community and thus greater responsibility than the radio or television broadcaster or, to a lesser extent, the cable operator. Sometimes they fail to realize their own power to influence. Unless they recognize and exercise this power with discretion, they can hardly expect political figures, civic leaders, or the general public to respect them or regard them as community leaders.

The "Public Interest" Standard

The decades have been marked by a continual tug-and-pull between broadcasters and federal government. But the basis for that on-again, off-again struggle has been in broadcasting's perceived successes and failures in serving the public.

In 1966 as appellate court judge, Justice Warren Burger commented on the WLBT case: "A broadcaster seeks and is granted the free and exclusive use of a limited and valuable part of the public domain. When he accepts that franchise it is burdened with enforceable public obligations."[82]

Media managers of substance keep their eye on what the public needs as well as what it wants or is simply interested in. Reacting to accelerated deregulation, in the late 1980s even the National Association of Broadcasters realized dropping the "public interest" standard would remove broadcasting's distinctive claim to support of its somewhat protected status against competing media systems. The Association of Independent TV Stations' President Preston Padden in 1988 told the Federal Communications Bar Association that "unrestrained and critical pursuit [of] the free marketplace and full First Amendment freedoms based on the print model . . . threatens to undermine our special status and the traditional bargain upon which our industry has been built." A year earlier Joel Chaseman, president of Post-Newsweek stations group, claimed that "the public-interest standard [is] what makes broadcasters unique"; his successor Bill Ryan in 1995 echoed that refrain:

> You know, there's a whole bunch of broadcasters out there who take their job very, very seriously—serving the public interest and serving their communities. Radio and television guys who are well known and well thought of in their communities, people who have a high profile and who care. . . . I too have supported deregulation, for the most part. But I should also say that, as a broadcaster, I've never really felt shackled by any of the regulations we had to live under through the years. I've been running stations since I was 28 years old and I've lived through the fairness doctrine and all the NAB Code issues, and I've never felt they really handcuffed us so badly. . . . But if this sort of deregulation [and] these kinds of changes go through and the large companies that emerge from it are allowed to vertically integrate to the extent that I believe they will, the uniqueness of broadcasting—and local broadcasting—and the serving of the public interest will go by the wayside.[83]

In 1988 James Lynagh, president of Multimedia Broadcasting wrote to the editors of *Broadcasting* magazine on behalf of the Television Operators Caucus, noting that "as operators of 70 affiliated and independent television stations in markets that reach over 52 million households" they had published a policy statement "supporting public interest responsibilities for television broadcasters." As chairman of the Caucus he asserted that the anti-regulation trade magazine's "editorial position lacks support among working broadcasters who believe that the public interest standard is a basis for the public and the industry being well-served."[84] Co-signers of that letter were executives of major group-owners: Great American Broadcasting, Post-Newsweek Stations, Hearst Corporation, Tribune Broadcasting, Gillett Holdings, Outlet Communications, Belo Broadcasting, Cox Enterprises, Westinghouse Broadcasting, Taft Broadcasting, and Gannett Broadcasting.

Reflecting the pendulum swing away from his predecessor's equating the "public's interest" with marketplace forces, FCC Chairman Reed Hundt in 1994 announced "the time has come to reexamine, redefine, restate and renew the social compact between the public and the broadcasting industry."[85]

In recent years the "public interest standard" came into question as both misdirected and unfeasible to implement under the First Amendment. FCC Chairman Mark Fowler decried it as anachronistic; he called for marketplace forces instead of governmental fiat. In 1996 Representative Barney Frank (D-Mass.) recommended that broadcasters should be free of any public interest obligation. Senator Larry Pressler (R-S.D.) who guided the Telecommunications Act of 1996 through to law told the FCC the Act did not intend the Commission to add explicit elements beyond a broad public interest review of a station, such as children's programming levels. As presidential candidate Senator Robert Dole claimed "public interest obligations are just liberal code words for federal mandates; I'm the candidate for corporate responsibility, not federal mandates" for broadcasting.[86]

Perhaps a stand can be taken on that middle ground which is not merely a compromise or fashionably realistic, but a serious call to entrepreneurial commitment. *Broadcasting & Cable* magazine supported former CBS President Howard Stringer's remarks on receiving a First Amendment award from the RTNDA in 1996. He called for industry responsibility, coupled with responsibility by parents and viewers generally, as the best way to prevent government intrusion into program content. Still, he added that media leaders

> have a higher responsibility that the government cannot and should not enforce. . . . In the end industry leaders must take personal responsibility for what goes on the screen. If we separate, like church and state, our artistic values from our personal values, then we create programs for others we would not be willing to share with our own family and friends.[87]

The magazine applauded that statement as "a credo worth living by." In other editorials it focused on the issue: "If this is a pivotal moment in history, so is the debate over the public interest standard, or public trusteeship, a pivotal subject of our time. . . . This debate is going to occupy broadcasters and their policymakers for some time." Finally—but not really finally—it editorialized:

> Privilege entails responsibility. Second only to a constant defense of the First Amendment, that might be the other half of a broadcasters' credo. It's been the dilemma that has occupied the industry in recent months, as issues of responsibility competed with issues of press freedom on its public policy agenda.[88]

The debate shifted somewhat with the arrival of multiple media technologies offering competing delivery systems. But the debate continues. And it includes other questions relating to the proper role of media regulation in today's society.

Some call for strengthening professional standards for radio, TV, and cable through guidelines, monitoring, and even self-regulation by professional associations such as the NAB and the National Cable Television Association. Since the 1970s citizen groups have called prominent attention to the complementary role to be played by media consumers, whose actions can affect broadcasters' and advertisers' policies as well as deliberations of the Federal Communications and Trade Commissions. Others call for restructuring the FCC.

REORGANIZATION OF THE FCC

Sometime during every decade, questions rise about reorganizing or doing away with the regulatory agency.

By 1995 some members of Congress, partly piqued by the Commission's stand on several issues, recommended disbanding the agency or else reducing its scope of responsibilities and relocating it under some other division of the federal bureaucracy. Speaker of the House Newt Gingrich advocated phasing out the agency within three to five years; some members of Congress claimed the Commission's anti-competitive duties could be transferred to the Justice Department and its EEO compliance efforts could be taken over by the Labor Department.

One must acknowledge the magnitude of government's regulating an enormous and swiftly expanding field driven by technology, while melding the viewpoints of media operators and consumers into their own as regulator-referees. Just out of sheer feasibility, can an already overburdened FCC be expected to take leadership in framing and enforcing qualitative standards? Is the deregulatory climate conducive to subjective evaluations about program content and quality? Aren't those the areas of broadcasters' particular qualifications? Aren't broadcasters and cable operators in a better position to know the wants and needs of listeners and viewers in their particular communities and to provide the sort of balance in programming best tailored to meet those wants and needs? That, of course, was FCC Chairman Mark Fowler's motivation in reducing significantly the Commission's oversight of day-to-day media operations. He left it to marketplace forces to sort out which stations and cable systems were most attuned to the public's *interest*.

But the challenge remains: what about the public's *needs*? It is up to competent and conscientious media managers to make those decisions. They are now freer to do so with reduced government involvement.

Meanwhile the FCC can serve appropriately by its oversight of such areas as: conflicting interests and needs of telephone companies and their entry into competition with cable systems; satellite use of spectrum space; digital audio radio satellite service; technical engineering including the complex question of digital standards; and high definition television (assigning frequencies, replacing analog and NTSC television signals with digital HDTV standards—over what time frame, should second TV channels be returned to the government for auctioning off, and what impact would this have on dislocating low-power TV operators to make room for the new HDTV assignments?). A technical and social issue forwarded by the 1996 Act was the V-chip technology mandated by Congress, and the ratings or labeling of program content negotiated by the government with media industries. Reemerging was the issue of alcoholic beverage commercials; Seagram's expanding ad campaign on TV drew negative reaction from trade magazines and broadcasters as well as beer and wine companies, and prompted members of Congress to propose bills outlawing all alcohol from broadcasting, not only hard liquor. These onrushing developments along with directives by Congress from time to time—such as reviewing 11,000 cable systems' subscriber rates, and the 1996 Act's calling for FCC oversight of program ratings—put increased demands on the agency, rather than the opposite that permitted phasing out the agency.

—∿∿—

Whatever the current developments in legislation and regulation, station and cable managers tend to rely too heavily on their legal counsels for many important matters related to those laws. We are not suggesting that managers ignore their attorneys (we have already underscored the absolute necessity to retain them). But they should use legal counsel for purposes that management cannot achieve independently. Managers *can* study and become knowledgeable about the Telecommunications Act, Commission regulations, and other laws affecting electronic mass media. This is part of their obligation as licensees and franchise holders and should not be delegated exclusively to attorneys.

Every media manager should know key elements of the law, rules, and regulations, and even major court cases. Owners-licensees and cable franchisees depend on their station executives and system managers to be familiar with relevant documents and keep up to date on continuing developments. A similarly thorough knowledge should be part of a training program within radio-TV-cable companies, and should be required of every student who seeks a career in broadcasting. Close study of this material is essential to understand the many responsibilities of the licensee and top managers and, by extension, of every employee.

～ CHAPTER 11 NOTES ～

1. Zechariah Chafee, Jr., *Government and Mass Communications,* (Chicago: University of Chicago Press, 1947).

2. Books are cited here in reverse chronological order. Some that analyze the whole range of laws and regulations pertinent to mass media include: F. Leslie Smith, Milan Meeske, and John Wright, *Electronic Media and Government: The Regulation of Wireless and Wired Mass Communication* (White Plains, N.Y.: Longman, 1995); John R. Bittner, *Law and Regulation of Electronic Media,* 2nd ed. (Englewood Cliffs, N.J.: Prentice Hall, 1994); William E. Francois, *Mass Media Law and Regulation,* 6th ed. (Columbus, Ohio: Grid, 1994); Kenneth C. Creech, *Electronic Media Law and Regulation* (Boston: Focal Press, 1993); Don R. Pember, *Mass Media Law,* 2nd ed. (Dubuque, Iowa: Wm. C. Brown, 1981); Douglas H. Ginsburg, *Regulation of Broadcasting: Law and Policy Towards Radio, Television, and Cable Communications* (St. Paul, Minn.: West Publishing, 1979); and Harvey L. Zuckman and Martin G. Gaynes' concise compendium *Mass Communications Law in a Nutshell* (St. Paul, Minn.: West Publishing, 1977 [see later editions]); Daniel W. Toohey, Richard D. Marks, and Arnold P. Lutzker, *Legal Problems in Broadcasting* (Lincoln, Nebraska: University of Nebraska, 1974).

For better understanding of legal structures and processes, other commentaries assess legislative and regulatory trends involving media. They include: Donald M. Gillmor and Jerome A. Barron, *Mass Communication Law: Cases and Comment,* 5th ed. (St. Paul, Minn.: West Publishing, 1990); Don R. LeDuc, *Beyond Broadcasting: Patterns in Policy and Law,* (New York: Longman, 1987); Stanley M. Besen et al., *Misregulating Television: Network Dominance and the FCC* (Chicago: University of Chicago Press, 1984); Erwin G. Krasnow and Lawrence D. Longley, *The Politics of Broadcast Regulation,* 3rd ed. (New York: St. Martin's Press, 1982); Barry Cole and Mal Oettinger, *Reluctant Regulators: The FCC and the Broadcast Audience,* revised ed. (Reading, Mass.: Addison-Wesley, 1978).

Practical concerns of creative developers of media, including contracts, releases, royalties and residuals, are explained by Phil Miller in *Media Law for Producers,* 2nd ed. (White Plains, N.Y.: Knowledge Industry Publications, 1993).

Kent R. Middleton and Bill F. Chamberlin, *The Law of Public Communication* (New York: Longman, 1988) also outline legal structures and processes of the larger ambit beyond and including electronic media.

3. House Commerce Committee, *Communications Act of 1934 as Amended by the Telecommunications Act of 1996,* 104th Congress, 2nd Session, 1996, Committee Print 104–L.

4. *Ibid.,* Title III, Part I, *The Communications Act of 1934 as Amended by the Telecommunications Act of 1996* (Washington, D.C. U.S. Government Printing Office, 1996), 276.

5. See Krasnow and Longley, 23–31.

6. Federal Communications Commission, *59th Annual Report/Fiscal Year 1993* (Washington, D.C.: U.S. Government Printing Office, n.d.), 14, 18.

7. Newton Minow, "Address to the National Association of Broadcasters" (speech delivered at annual convention of the NAB, Chicago, April 2, 1963); reprinted in Newton

N. Minow, *Equal Time-The Private Broadcaster and the Public Interest,* Lawrence Laurent, ed. (New York: Antheneum, 1964), 245–262; excerpt pp. 258–259.

8. See Krasnow and Longley, 23–72, for a clear and convincing study of the anatomy of the regulatory process amid pluralistic influences. They outline key Congressional strategies for overseeing broadcast regulation (pp. 56–62); infrequent legislation by statutes and even by inaction on issues; power of appropriating funds for FCC budgets; investigations; advice and consent to the executive branch about FCC appointments; supervision of the FCC by the Senate Committee on Commerce and by the House Committee on Interstate and Foreign Commerce and their respective Subcommittees on Communications and on Communications and Power; and pressures exerted by individual members of Congress and their staffs. Depending on the issue under consideration, FCC commissioners on occasion have been answerable to Senate and House Government Operations Committees, the Science and Astronautics Committees, the Senate Judiciary Committee and its Subcommittee on Antitrust and Monopoly, the House Judiciary Committee and its Antitrust Subcommittee; the Senate Foreign Relations Committee, the Senate Select Committee on Small Business and its Subcommittee on Monopoly, and the Joint Economic Committee.

9. Key court decisions include *National Broadcasting Co. v. United States,* 319 U.S. 190 (1943); *Red Lion Broadcasting Co., Inc. v. FCC,* 381 F. (2d) 908 (1967); *U.S. v. Radio Television News Directors Association,* 395 V, 5, 367 (1969); *Business Executives' Move for Vietnam Peace v. FCC,* 450 F. (2d) 642 (1971); *CBS v. the Democratic National Committee,* 412 U.S. 94 (1973); *FCC v. Pacifica Foundation,* 438 U.S., 98 (1978).

10. Krasnow and Longley, *The Politics of Broadcast Regulation,* 138; the following quotation is from p. 139.

11. John H. Pennybacker and Waldo W. Braden, eds., *Broadcasting and the Public Interest* (New York: Random House, 1969), 11–14, citing appellate courts, the U.S. Supreme Court, Congressional committees, and other governmental reports. See Sydney Head, Chapter 21, "Regulation and the Public Interest: Fact and Fictions," *Broadcasting in America,* 447–465.

12. Marcus Cohn in a personal letter to James A. Brown, 25 October 1973.

13. See Elisabeth A. Rathbun, "Justice Caps Radio Ownership," *Broadcasting & Cable,* 12 August 1996, 9–10.

14. Peter Viles, "Stewart Warns Radio about LMA's," *Broadcasting & Cable,* 26 April 1993, 38.

15. See Paul W. Cherington, Leon V. Hirsch, and Robert Brandwein, *Television Station Ownership: A Case Study of Federal Agency Regulation* (New York: Hastings House, 1971), an analysis published by a grant from WGN Continental Broadcasting, Chicago. Cf. another study underwritten by the NAB: George W. Litwin and William H. Wroth, *The Effects of Common Ownership on Media Content and Influence: A Research Evaluation of Media Ownership and Public Interest* (Washington: NAB, 1969). On the other hand, multiple-ownership effects have been appraised negatively by Harvey J. Levin, *Broadcast Regulation and Joint Ownership of Media* (New York: New York University Press, 1960); but Peter 0. Steiner was not convinced by Levin's data; in *American Economic Review* 51 (June 1961), 472 he concluded that whether or not there was joint ownership of stations or great effort at diversification, the difference to the broadcast service in the United States would not be appreciably altered.

16. See Title V of the Communications Act of 1934, §§501–509, particularly §503(b).

17. For the uninitiated, Stern's witty and articulate but raw, raunchy style is suggested in the Commission's official notice of apparent liability on November 29, 1990, against Infinity stations. It stated that the program dwelled "on sexual matters, including sexual intercourse, orgasm, masturbation, lesbianism, homosexuality, breasts, nudity, and male and female genitalia. . . . [It] contained frequent and explicit verbal references to sexual activities and organs that were lewd and vulgar and that, when taken in context, were made in a pandering and titillating fashion" including talking about a "guy who plays the piano with his penis" and a "big black lesbian . . . out of her mind with lust" (program transcript excerpts). Quoted in HAJ [*sic*], "Infinity to Fight FCC Indecency Fine," *Broadcasting,* 3 December 1990, 38.

18. The FCC excerpt was quoted by Intercollegiate Broadcasting System, *Station Manager's Newsletter* (January/February 1993), 1–2.

19. HAJ, "FCC Fines 3 TV's for Exceeding Kids Ad Limits," *Broadcasting & Cable,* 26 July 1993, 10.

20. Federal Communications Commission, *59th Annual Report/Fiscal Year 1993* (Washington, D.C.: U.S. Government Printing Office, n.d.), 38–39; *60th Annual Report/Fiscal Year 1994,* 49–50. See also John D. Abel, Charles Clift III, and Frederic A. Weiss, "Station License Revocations and Denials of Renewal, 1934–1969," *Journal of Broadcasting,* 14:4 (Fall 1970), 411–421.

21. See Smith, Meeske, and Wright, *Electronic Media and Government,* 204–218.

22. For brief treatment of key details of these forms and reports, see: Bittner, *Law and Regulation of Electronic Media,* Chapter 9, "Broadcast Operations," 190–221; Kenneth C. Creech, *Electronic Media Law and Regulation* (Boston: Focal Press, 1993), Chapter 4, "Licensing Broadcast Stations," 81–107; Smith, Meeske, and Wright, Chapter 8, "Permission to Operate," 192–236. See also Marvin B. Bensman, *Broadcast/Cable Regulation* (Lanham, Md.: University Press of America, 1990), under specific alphabetized listings.

23. Analyses and commentaries about these issues can be found in several useful sources. Rather than repeated citations to them, a composite reference is here given. See recent sources cited in the preceding note, and also traditional treatment by Sydney Head, *Broadcasting in America;* also Daniel W. Toohey, Richard D. Marks, and Arnold P. Lutzker, *Legal Problems in Broadcasting: Identification and Analysis of Selected Issues* (Lincoln, Nebr.: Great Plains National Instructional Television Library, University of Nebraska, 1974). Additional sources are cited later in the text.

24. Chris Stern, "Washington Watch," *Broadcasting & Cable,* 11 December 1995, 26.

25. This became known as the "Zapple doctrine," based on the Commission's letter responding to Nicholas Zapple, aide to Senator John Pastore; *Nicholas Zapple,* 23 FCC 2d 707 (1970); First Report, Docket No. 19260, 24 RR 2d 1917 (1972).

26. Helpful suggestions are provided in the current edition of NAB's regularly updated the *Political Broadcast Catechism* (Washington, D.C.: NAB, 19xx). The Commission published a 100-question primer on Section §315 applications, titled "The Use of Facilities by Candidates for Public Office," (see 31 Fed. Reg., 6660; 7 RR 2d 1901–1930).

27. Perhaps proportionate time relative to the previously measured "impact" of candidates could offer some guidelines. Cf. Branscomb, "Should Political Broadcasting Be Fair or Equal? A Reappraisal of Section 315," *George Washington Law Review* 30 (1961): 63–65. For three differing views of this issue, see Pennybacker and Braden, 129–147, including Nathan Karp's urging that divergent and "unpopular" positions and candidates precisely must be afforded widespread media exposure so that people have access to minority views, 138–147. Thus the First Amendment freedom of speech might be seen to apply here, "guaranteeing that minority views . . . find their place in a free market place of ideas and communications," as noted by Newton Minow, *Equal Time.: the Private Broadcaster and the Public Interest,* Lawrence Laurent, ed. (New York: Atheneum, 1964), 74.

28. See Sidney Kraus, ed., *The Great Debates: Background, Perspective, Effects* (Bloomington, Ind.: Indiana University Press, 1962) for texts of the broadcast and a summary of thirty-one studies.

29. While disruptive of advertising by regular advertisers, political broadcasting is still big business for radio and television.

In 1991 a *Los Angeles Times* survey found that of all politician campaign costs, about one-third (of Senate candidates' spending) to one-fourth (of House of Representatives') went to advertising in all media: radio, TV, newspapers, and billboards; clearly broadcast charges were not the largest outlay in escalating campaign expenses (*Broadcasting,* 25 March 1991, 114).

In 1994 non-presidential political campaigns spent $355 million for TV advertising (*Broadcasting & Cable,* 5 February 1995, 50); projections for 1996 were $500 million for television station airtime. Totals did not include cost of media consultants and production firms to produce spots. Nor did totals include charges for airtime on radio stations or networks. Campaign expenditures in the 1996 elections totaled $1.8 billion; almost one-third of that was spent on television advertising.

30. The Supreme Court decision in 1959 (*Farmer's Educational and Cooperative Union of America v. WDAY, Inc.,* 360 U.S. 252) ruled against previous State courts, striking down their decisions that State libel laws did apply to broadcasters despite federal law forbidding them from censoring candidates' statements.

31. For a readable and documented chronological account of the development of the fairness doctrine, see various editions of Head or Head and Sterling. For detailed citations of documents and for legal analysis, see Gillmor and Barron's current edition of *Mass Communication Law.*

Successive federal actions related to the issue include: FCC, *Public Service Responsibility of Broadcast Licensees* (Washington, D.C.: Government Printing Office, 1946). *In re United Broadcasting Co. (WHKC),* to FCC at 517–518 (1945). FCC, "Editorializing by Broadcast Licensees," 14 Fed. Reg. 3055 (1949). FCC, "Report and Statement of Policy re: Commission en banc Programming Inquiry," 25 Fed. Reg. 10416 (1964). FCC, "Letter to WCBS-TV," 2 June 1967; 67 FCC 641: *Banzhaf v. FCC,* 405 F. 2d 1082–1103 (1968). *RTNDA v. FCC,* 400 F. 2d 1002 (1968). *U.S. v. RTNDA* 395 U.S. 367 (1969). *Red Lion Broadcasting Co., Inc. v. FCC,* 381 F. 2d 908 (1967); 395 U.S. 367 (1969). *Business Executives' Move for Vietnam Peace v. FCC,* 450 F. 2d 642 (C.A.D.C., 1971). *Friends of the Earth v. FCC,* 449 F. 2d 1164 (C.A.D.C:, 1971). *Columbia Broadcasting System, Inc. v. Democratic National Committee,* 412 U.S. 94, 93 S.Ct. 2080, 36 L.Ed.2d 722 (1973).

32. Mayflower Broadcasting Corporation: *In re The Yankee Network, Inc.,* 8 FCC 333, at 340 (1941); FCC, "Editorializing by Broadcast Licensees," 14 Fed. Reg. 3055 (1949).

33. *Telecommunications Research and Action Center v. FCC,* 801 F. 2d 501 (1986); rehearing denied, 806 F. 2d 1115; *certiorari* denied 482 U.S. 919 (1987).

34. Confusion was compounded, for example, by no less than *The New York Times* in a column by Frank J. Prial (26 May 1984, 15): "The Fairness Doctrine mandates that stations broadcasting any program featuring a legally qualified candidate for any public office must provide equal opportunities for all other candidates seeking that office to appear on the air." (Close, but no cigar!)

35. See Donald J. Jung, *The Federal Communications Commission, the Broadcast Industry, and the Fairness Doctrine: 1981–1987* (Lanham, Md.: University Press of America, 1996).

36. At midnight, Thursday, June 5, 1973, WXUR-AM/FM, Media, Pennsylvania, ceased broadcasting; they were the first and only stations to be denied license renewals by the Commission because of violating the Fairness Doctrine by harshly criticizing various religious faiths. Fundamentalist preacher Dr. Carl McIntire and his attorneys had unsuccessfully fought the FCC's original denial of renewal, but both the Commission and the U.S. Court of Appeals in Washington refused to reverse that action closing the seven-year-old stations. *Brandywine-Mainline Radio, Inc. v. Federal Communications Commission,* 473 F. 2d 16 (D.C. Cir. 1972).

37. Quoted in "Fairness Weighed in Comments to FCC," *Broadcasting,* 2 March 1987, 39–40.

38. *In the matter of The Mayflower Broadcasting Corporation and The Yankee Network, Inc. (WAAB),* 8 FCC 333, 338 (1941).

39. *In the Matter of Editorializing by Broadcast Licensees,* Docket No. 8516; 13 FCC 1246; 14 Fed. Reg. 3055; 1 RR Section 91:21 (1941).

40. Although CBS, NBC, and the Radio Television News Directors Association's (RTNDA) appeal to the Seventh Circuit Court of Appeals was supported on September 10, 1968, on the grounds of vagueness in the rule, unduly burdening licensees, and involving possible censorship and violations of the First Amendment, that appellate decision was reversed when the U.S. Supreme Court unanimously upheld the validity of those rules; *United*

States v. Radio Television News Directors Association, et al., coupled with *Red Lion Broadcasting v. FCC,* 395 US 367 (1969).

41. Broadcast media fall under libel (or written defamation) rather than slander (spoken defamation) because of broadcasting's wide dissemination and potential extent of injury. Libel refers to a false and malicious statement that subjects an individual to public hatred, contempt, or ridicule. But communication media may criticize and comment on matters pertaining to public controversy and general public interest. This "fair comment" privilege obtains as long as no actual malice is demonstrated; its protection extends to matters of opinion but not to misrepresentations of fact, and those matters must be of public concern rather than merely private activity of a nonpublic person.

Court cases since 1964 have extended media protection further. "Actual malice" demands proof of knowledge that a statement was false or was made with reckless disregard of whether it was false or not, or even (Supreme Court, 1968) evidence supporting the conclusion that the party charged had serious doubts about the truth of his statement.

The Supreme Court further expanded media's protection against defamation liability by including the "fair comment" constitutional protection not only to high public officials but also to government employees of substantial responsibility, to public figures, and (in June of 1971, *Rosenbloom v. Metromedia*) even to coverage of "matters of public or general concern" regardless of whether the persons involved are famous or anonymous.

In 1975 the U.S. Supreme Court, in an 8–1 decision, supported the appeal of Cox Broadcasting Corporation in an invasion-of-privacy suit. The Justices ruled that an individual's right to privacy must give way when the press—significantly including electronic as well as print news media—made public accurate information based on public court records. Justice Byron R. White noted that "great responsibility is accordingly placed upon the news media to report fully and accurately the proceedings of government, and official records and documents open to the public are the basic data of governmental operations." His conclusion was reasoned "in terms of the First and 14th Amendments and in light of the public interest in a vigorous press." Thus the highest court in the land passed judgment reflecting no distinction between newspaper and broadcast journalism. See "Supreme Court in WSB-TV Case Treats Print and Broadcast Press as One," *Broadcasting,* 10 March 1975, 44.

42. Only 61% of AM stations, 44% of FM stations, and 51% of TV stations presented editorials; the great majority editorialized only occasionally. Fewer than one of five stations that editorialized aired them daily, and even fewer editorialized weekly. "Extent of Broadcast Editorializing," *Broadcasting Yearbook 1974* (Washington, D.C.: Broadcasting Publications, Inc., 1974), 37.

43. *Business Executives' Move for Vietnam Peace [BEM] v. FCC,* 450 F. 2d 642 (C.A.D.C., 1971); the U.S. Supreme Court on May 29, 1973 reversed the D.C. appellate court, reconfirming the FCC's initial judgment against the BEM claim to a right to purchase broadcast time to advertise a viewpoint on the controversy about United States involvement in the Vietnam War: *Columbia Broadcasting System, Inc. v. Democratic National Committee,* 412 US 94.

44. Among others, see Jeremy Lipschultz, *Broadcast Indecency: FCC Regulation and the First Amendment* (Woburn, Mass.: Focal Press, 1996).

45. There has been little court delineation of what constitutes "obscene, indecent, or profane language by means of radio communication" (18 U.S.C. 1464) because of the wide variance of community standards and criteria for decency. However, there have been efforts to make *ad hoc* determinations from the time of the Radio Act: *Duncan v. United States,* 48 F. 2d 128 (1931); WREC Broadcasting Service, 10 RR 1323 (1955); WDKD, Kingstree, South Carolina, 23 RR 483 (1962), and *E.G. Robinson, Jr. t/a Palmetto Broadcasting Company (WDKD) v. Federal Communications Commission,* 334 F. 2d 534, in which Judge Wilbur Miller of the D.C. Court of Appeals noted "I do not think that denying renewal of a license because of the station's broadcast of obscene, indecent or profane language— a serious criminal offense—can properly be called program censorship. But if it can be so denominated, then I think censorship to that extent is not only permissible but required in the public interest. Freedom of speech does not legalize using the public airways to peddle filth." The FCC refused to renew WDKD's license for approximating such program content and for willful misrepresentation to the Commission; the appeals court avoided censorship issues by affirming the FCC's decision on grounds of misrepresenting facts instead of for programming excesses; a petition for writ of certiorari to the Supreme Court was denied in 1964. *In Re Pacifica Foundation* 36 FCC 147, 1 RR (2d) 747 (1964) involved alleged indecency of program content, but the FCC judged the full circumstances of the broadcast and allowed the license to be renewed. A $100 fine was levied against WUHY-FM, educational noncommercial station in Philadelphia, for vulgar vocabulary, 18 RR 2d 860 (1970). Cf. *Apparent Liability of Station WGLD-FM,* 41 FCC 2d 919 (1973); cf. *Sonderlin Broadcasting Corporation,* 27 RR 2d 285 (1973).

46. 56 FCC 2d 98 (1973); *FCC v. Pacifica Foundation,* 438 US 726 (1978).

47. Data cited by FCC staff members and information about Commission and court actions were reported by PJS [*sic*], "Defining the Line Between Regulation and Censorship," *Broadcasting,* 15 April 1991, 74; Harry A. Jessell, "FCC Picks Up Pace on Indecency Enforcement," *Broadcasting,* 31 August 1992, 25; Sean Scully, "Court Throws Out Indecency Ban," *Broadcasting & Cable,* 29 November 1993, 10. Section §16(a) of the Public Telecommunications Act of 1992 had prohibited broadcast of indecent material between 6 A.M. and 10 P.M. on public broadcast stations that went off the air at or before midnight, and between 6 A.M. and midnight for all other stations. The FCC duly adopted regulations to carry out that law, in its Report and Order (FCC 93–42) on January 19. 1993.

48. Editorial, "Square One," *Broadcasting & Cable,* 29 November 1993, 106; the preceding quotation was cited in this editorial. In 1992 Congress had passed the Public Telecommunications Act which led to the FCC's ruling (January 1993) of a safe harbor for adult material between midnight and 6 A.M. The U.S. District Court in Washington first stayed the FCC's midnight-to-6 A.M. ruling (March 1993), then declined to forbid the FCC's enforcing indecency guidelines at other times, thereby reverting the "safe harbor" to the previous 8 P.M. to 6 A.M. period.

49. "Jeannine Aversa, "Federal Ruling Sets Limit on Indecent Broadcasts," Associated Press column in the *Tuscaloosa News,* 1 July 1995, 4A.

50. Joe Meier in "Open Mike," *Broadcasting,* 2 October 1989, 27–30.

51. For a descriptive summary of recent cases, with sample wording and legal references, see Kenneth C. Creech, *Electronic Media Law and Regulation* (Boston: Focal Press, 1993), 120–131.

52. *Yale Broadcasting Co. v. FCC,* 478 F 2d 599 (1973).

53. See "FCC Puts a Chip On Its Shoulder in Declaratory Ruling on Indecency," *Broadcasting,* 17 February 1975, 6; "Wiley Plan to Clean Up Television Goes to Hill," *Broadcasting,* 24 February 1975, 25–26.

54. In 1979 a three-judge panel of the Ninth Circuit Court of Appeals in San Francisco unanimously ruled that Judge Ferguson's Los Angeles District Court was not the proper forum for judgment of the case; they remanded the case, instructing a return to the FCC for review of its own administrative proceedings in mounting "family viewing" with the networks and NAB. Subsequently the full twelve-judge Circuit Court denied the plaintiffs' request for a rehearing, and in October 1980 the Supreme Court refused to review it. In 1981 Judge Ferguson relayed back to the Commission the request to review its previous procedures in the case; *Broadcasting,* 19 November 1979, 28; 6 April 1981, 124.

55. Joe Flint, "Simon Delivers Violence Ultimatum," *Broadcasting & Cable,* 9 August 1993, 18–25.

56. The report by nonprofit Center for Media and Public Affairs listed premiere episodes of first-run syndication and network programs' number of scenes with "serious violence" (defined as armed and unarmed assaults, gunplay, sexual assaults, and suicides); "TV Violence Study Released," *Broadcasting & Cable,* 8 August 1994, 56.

57. Steve McClellan, "Programmers Challenge Violence Survey," *Broadcasting & Cable,* 15 August 1994, 14–16.

58. See 47 U.S. 639; 47 USC 612, 638; 476 U.S. 488; F, 1985c: 989; Charles Ferris et al., *Cable Television Law: A Video Communications Guide* (New York: Matthew Bender, 1983–present), 3 volumes.: 8–38/9. Cited by Head, Sterling, and Schofield, *Broadcasting in America,* 7th ed. (Boston: Houghton Mifflin, 1994), 512.

59. "Tongues Untied: Public TV Managers and Potentially Objectionable Programming," manuscript reviewed (without author identified) for potential publication in *Media Management Review's* first issue in 1995.

60. One industry analysis in 1991 estimated that for a typical independent TV station one less commercial minute in every hour of their children's programming meant an annual loss of $160,000. The Association of Independent Television Stations (INTV) also estimated that the average indy station billed $1,830,000 annually in time sales for kids' shows, of which 84% was through barter.

In 1993 revenue from advertising in children's programs totaled $800 million—about $550 million in national spending and the other $250 million in spot TV. The 40 million children under the age of 12 in the U.S. were estimated to "spend or influence the spending of" $100 billion annually. Those aged 2–5 viewed 28 hours of TV a week (almost four hours daily), while those 6–11 watched 24 hours a week (3.4 hours daily). Steve McClellan, "It's Not Just for Saturday Mornings Anymore," *Broadcasting & Cable,* 26 July 1993, 38. Commercial advertising in programs intended primarily for children accounts for 3%–5% of all ad revenues in TV; *Broadcasting & Cable,* 25 July 1994, 67.

61. The FCC adopted specific regulations on 1 May 1963 (22 RR 1575, 28 Fed. Reg. 4707) to implement Section §317 as revised and the newly added Section §508 of the Communications Act of 1934. Those FCC regulations are numbered 73.119, 73.289, 73.654, and 73.789. Excellent examples of proper and improper relation of station and suppliers are in Robinson, *Broadcast Station Operating Guide* (Blue Ridge Summit, Pa.: Tab Books, 1969), 48–55.

62. Routt provides sample affidavits for announcers to sign as station employees, testifying to their relations to other business enterprises and persons, and to be notarized, in *The Business of Radio Broadcasting* (Blue Ridge Summit, Pa.: Tab Books, 1972), 376–377.

63. See Leon C. Smith, "Local Station Liability for Deceptive Advertising," *Journal of Broadcasting* 15:1 (Winter 1970–1971): 107–112.

64. 18 USC 1304 prohibited broadcast of lotteries or information related thereto, subject to prosecution by the U.S. Department of Justice, and liable to penalties of up to $1,000 and/or up to one year imprisonment. Furthermore, the language of the statute specifically stated that "each day's broadcasting shall constitute a separate offense." The FCC Rules and Regulations banning lottery broadcasts for all AM, FM, and TV licensees were 3.122, 3.292, and 3.656 respectively. Commission penalties ranged from a "cease and desist" order to license revocation. Further guidance can be sought from parallel instances in Post Office Department rulings and from FTC cases. But on December 20, 1974, Congress sent to President Ford new legislation permitting broadcast stations to carry advertising and information about lotteries in their own state and in adjacent states.

65. Iowa Libel Research Project, directed by University of Iowa law professor Randall Bezanson and journalism professors Gilbert Cranberg and John Soloski, was summarized by the National Association of Broadcaster's Legal Department in "How to Reduce Libel Suits," *NAB Management Report,* 3 February 1986, 1–2.

66. Survey by the Libel Defense Resource Center was cited by Head, Sterling, and Schofield, *Broadcasting in America,* 7th ed. (Boston: Houghton Mifflin, 1994), 505.

67. "Sauter Says Libel Suits Aren't Affecting CBS News Operations," *Broadcasting,* 30 May 1983, 49.

68. In 1981 comedienne Carol Burnett's libel suit against the *National Enquirer* brought a judgment of $1.3 million in punitive damages and $300,000 in general damages for her emotional distress. In 1990 the *Philadelphia Inquirer* was assessed $34 million for libeling a prominent lawyer in stories printed seventeen years earlier—the largest judgment against any news medium until then. A month earlier, a heart surgeon won a $28 million libel award against a San Antonio TV station.

69. Frank Newton and Lucienne Lopez Loman, "Hispanics in Broadcasting in 1988," *RTNDA Communicator,* August 1988, 16–18.

70. Carolyn Wean, "Viewpoints: Women and Minorities in Broadcast Management—Room for Improvement," *Television/Radio Age,* 25 May 1987, 57.

71. Useful sources of EEO information, including enormous data reported by formal studies, include: Matilda Butler and William Paisley, *Women and the Mass Media: Sourcebook for Research and Action* (New York: Human Sciences Press, 1980); Gary N. Powell, *Women and Men in Management* (Newbury Park, Calif.: Sage, 1988), especially Chapter 7, "Promoting Equal Opportunity" (pp. 207–235); Alice Sargent, *The Androgynous Manager* (New York: ANACOM, 1981); and Ethel Strainchamps, *Rooms with No View* (New York: Harper, 1974).

72. James Phillip Jeter, "Black Broadcast Station Owners: A Profile," [Broadcast Education Association] *Feedback* (Summer 1991): 14–15.

73. See Sidney Shemel and M. William Krasilovsky, *This Business of Music* (New York: Billboard Publishing Company, 1964), and *More About This Business of Music* (New York: Billboard Publishing Company, 1967).

74. *Quincy Cable TV, Inc. v. F.C.C.,* 768 F.2d 1434.

75. Quoted in Joe Flint, "FCC Lays Down Cable Law," *Broadcasting & Cable,* 15 May 1993, 6.

76. Syndication sales generated $800 million for Hollywood program producers in 1983, according to the *New York Times,* 8 October 1983, 13. *Broadcasting* magazine referred to the "$3-billion-a-year syndication business" (25 February 1991, 19); Prodigy interactive personal service cited "the $5-billion rerun market" (24 October 1991, 7:51 EDT), ©1991 Prodigy Services Company. The "$5.7 billion worldwide rerun business" was cited by *Newsweek,* 22 April 1991, 44–45.

77. The excerpt from the journal *Judicature* was quoted in "Editorials: Amicus Brief," *Broadcasting & Cable,* 2 January 1995, 74.

78. Syndicated news item, "Appeals Courts May Allow Cameras" with a Washington dateline, in the *Tuscaloosa News,* 19 November 1994, 2; see "In Brief," *Broadcasting & Cable,* 21 November 1994, 80; New York Times News Service, "TV Cameras to be Allowed in Federal Appeals Courts," the *Tuscaloosa News,* 13 March 1996, 6C.

79. See the FCC's *Report and Statement of Policy Re: Commission en banc Programming Inquiry,* adopted 27 July 1960 (22 RR 1902 [1960]), which effectively superseded the "Blue Book" of 1946—*Public Service Responsibility of Broadcast Licensees.* For a succinct survey, see Walter B. Emery, *Broadcasting and Government: Responsibilities and Regulations* (East Lansing, Mich.: Michigan State University Press, 1971 edition), 46–50, 318–339. See also Head, *Broadcasting in America,* 2nd ed., 376–378, 420–425. Cf. Pennybacker and Braden, *Broadcasting in the Public Interest,* 11–15; Krasnow and Longley, *The Politics of Broadcast Regulation,* 16–19, 137–139. For one application of the principle of FCC overview of programming by a former Commissioner, see Frederick W. Ford and Lee G. Lovett, "Interpreting the FCC Rules & Regulations: Children's Television Programming," *BM/E,* January 1975, 20–26.

80. By early 1975, a total of 414 of the 703 commercial TV stations (59%) subscribed to the NAB's Television Code, and 2,946 of the 7,005 commercial radio stations (42%) subscribed to the NAB's Radio Code. Those percentages represented a half-decade drop from 1970s' figure of 65% for TV stations and a rise from 34% for radio stations. In addition to stations, Code members included three national television networks and four national radio networks. Code subscribers were not necessarily members of the NAB; but almost half again as many stations were members of the Association as were subscribers to the respective Codes. Sources: FCC figures compiled 31 January 1975, cited in "Summary of Broadcasting," *Broadcasting,* 1 March 1975, 58. Code subscriber totals as of 1 February 1975, in ''Subscriber Status," *Code News,* February 1975, 5. Comparative data from NAB letter of 20 August 1970, cited by Head, 469.

81. Krasnow and Longley, 32–33; reference is to Joseph C. Goulden, *The Superlawyers: The Small and Powerful World of the Great Washington Law Firms* (New York: Weybright & Talley, 1972), 6.

82. Quoted by Vincent L. Hoffart in "Open Mike" letters column, *Broadcasting & Cable,* 17 January 1994, 142.

83. *Broadcasting & Cable,* 22 May 1995, 50–52. The two previous quoted excerpts were from *Broadcasting,* 27 June 1988, 52, and *Channels,* April 1987, 15.

84. [Letters to the editor column] "Open Mike: Separatist Movement," *Broadcasting,* 25 July 1988, 23.

85. *Broadcasting & Cable,* 1 August 1994, 6, 8; the excerpt was also printed on the cover of that week's issue.

86. *Broadcasting & Cable,* 14 October 1996, 29.

87. *Broadcasting & Cable,* 8 April 1996, 90.

88. *Broadcasting & Cable,* 29 July 1996, 86; preceding quote from 11 December 1995, 118.

12

Managers and Technology: Engineering

The world today listens to De Forest and pioneers of his class with an almost boundless faith. From what has been done no man would venture to place a limit upon what may be done in the domain of applied science.

JERSEY CITY JOURNAL, FEBRUARY 20, 1909

Some day the program director will attain the intelligent skill of the engineers who erected his towers and built the marvel. . . . The radio engineer has worked miracles in research, invention, and clever engineering. But a comparable knowledge of the basic factors of the social equations to be solved seldom is found in the executive offices of his employers.

LEE DE FOREST, "FATHER OF RADIO"

Technology doesn't automatically a market make.

JAMES GRAY, WARNER CABLE[1]

Radio and television began as scientific and engineering phenomena. Those who work in broadcast engineering today have inherited the traditions established by great inventors and innovators such as Hertz, Marconi, De Forest, and Zworykin. Members of the engineering fraternity developed the marvels of broadcast and cable communication: the image-orthicon camera, Zoomar lens, audio and video tape, instant replay videodiscs, electronic color, stereo and quadraphonic sound transmission, computerized graphics, digital compression, sophisticated on-line editing equipment, and relay satellites. Technicians applied those developments to storing, retrieving, and transmitting information, entertainment, and commercial messages. Those technical advances were essential for broadcast media to fully utilize performing talent along with sales and business management. The evolution of broadcast technology served as a catalyst to bring together performers, programmers, salespeople, and managers (as well as federal regulators).

THE STATION MANAGER'S
RELATIONSHIP TO ENGINEERING

The engineering world is necessarily technical and apart. Electrical engineers are not subject to the public criticism often directed at programming and advertising. In their unobtrusive way, technicians contribute to the broadcast and cable process without fanfare. Few outside their own department, including the station manager, understand engineering work and motivations. On the other hand, the engineer may be inclined to dismiss programming and sales considerations as having very little personal attraction. Even so, friction between engineering and other departments is due more often to poor communication than to a lack of proper understanding and mutual goodwill.

As radio began to grow in maturity and sophistication in the early 1930s, the close ties between engineering and management that prevailed in the experimental, formative years often eroded. This division between technical and non-technical interests widened at local stations and at networks. The breakdown of close collaboration between the engineering department and administrative projects reflected gradual loss of a common language and a common concern.

It is hard to overstate the perils of estrangement. Most station managers are neophytes in engineering and rely heavily on chief engineers to supervise all technical matters. They risk proper administrative oversight because of their meager technical knowledge about everything from purchasing and operating exotic equipment to the attitudes of engineering personnel. Engineers, too often sketchily or arbitrarily informed (or ignored), sometimes miss opportunities to advance proposals that could significantly benefit the organization. When management plays superior and engineers play canny, both may find themselves overspending on proposals for facilities or equipment and underachieving potential results in broadcasting.

Swift advances in technology require chief engineers to keep abreast of enhanced hardware and software equipment so they can alert management to upgrades needed to keep competitive. Computerized master control operations, digital audio, high performance lightweight ENG cameras, nonlinear computerized disk editing, subcarrier signal service, satellite up- and down-link dishes and transmitters, HDTV compressed digital signals—all contribute to higher quality pictures, sound, and intercutting that enable a station, network, or cable system to match and exceed the competition. Engineers must be able to recommend cost-effective innovations and specific models; they need to estimate the expected operating life of each item for amortizing costs and to budget for subsequent replacements and further upgrades. Sony estimated in 1990 that 25% to 30% of broadcast equipment sales involved meetings with general managers, contrasted with 5% a few years earlier.[2] As noted in chapter 10, almost 50% of general managers (GMs) and chief engineers collaborate on decisions about purchasing equipment; 16% of chief engineers only recommend specific equipment to the GM.

Trade magazines have added equipment details to management-oriented broadcast/cable magazines; other media journals offer language to engineers to help them meet concerns of GMs and financial managers. When discussing budgetary matters, engineers should underplay abstruse technical specifications and stress the results of purchases such as improving the on-air look, greater productivity, and cutting labor costs.

The National Association of Broadcasters publishes regularly updated publications useful to technical engineers.[3] The subjects reflect the range of responsibilities for chief engineers and their staffs. Titles available in 1997 included:

Practical Tips for Choosing and Using Consulting and Contract Engineers
 (40 pages)
The Nonlinear Video Buyers Guide, 3rd edition (1997, 105 pages)
The Tapeless Audio Directory, 4th edition (1995)

NAB Engineering Handbook, 8th edition (1992, 1,300 pages)
The NAB Guide for Broadcast Station Chief Operators, 2nd edition (1996)
Ghost Canceling: An Implementation Guide (128 pages)
NAB Broadcast and Audio System Test CD, vol. I (1988) and vol. II (1992)
NAB Guide to AM and FM Audio Performance Measurements (81 pages)
NAB Guide to Aural and Vidual Performance Measurements (246 pages)
NAB Guide to Unattended Station Operation (1996, 200 pages)
Antenna and Tower Regulation Handbook (1996)
Radio and Television Towers: Maintaining, Modifying and Leasing, 2nd edition (54 pages)
A Broadcaster's Guide to FCC RF Radiation Regulation Compliance, 3rd edition (1997, 85 pages)
NAB Technology Reference Guide (55 pages)
A Radio Broadcaster's Guide to Successful Station Self-Inspection (1995)
Advanced Television Transmission: Planning Your Station's Transition (1995, 250 pages)
Record Retention Guide for Radio and Television Stations (60 pages)
Perspectives on Widescreen and HDTV Production (85 pages)
1997 NAB Proceedings: 51st Annual Broadcast Engineering Conference
Understanding DAB: A Guide for Broadcast Managers and Engineers, 2nd edition (140 pages)
NAB Guide to HDTV Implementation Costs (85 pages)
Advanced Broadcast/Media Technologies: Market Developments and Impacts in the '90s and Beyond (147 pages)
IMA Digital Audio-Pac (1994)
IMA Set Top System (1994)
Television Data Broadcasting (1997)

The NAB also distributes selected books from other publishers such as Focal Press:

Digital Compression in Video and Audio (1995, 256 pages)
Digital Audio and Compact Disc Technology, 3rd edition (1995, 320 pages)
Digital Multimedia Cross-Industry Guide (1995, 321 pages)

The chief engineer can keep abreast with these and other books, technical journals, and electronics magazines; he should make them available to staff members to keep them current.

FUNCTIONS OF THE CHIEF ENGINEER

Among a chief engineer's many functions—described in detail in chapter 4—is helping the station manager better understand this department's work and keeping him informed about its problems and achievements.[*] In addition to scheduling manpower with efficiency and cost control, the chief engineer seeks every possible means of improving the broadcast or cable operation's technical performance. Engineering and operations contribute significantly to the difference in

---

[*]In small operations, especially radio stations, an operations manager or similarly titled middle manager sometimes oversees these areas when an engineer is hired only parttime or on a retainer basis ("on call" as needed), as permitted by FCC regulations.

the *sound* and in the *look* of well or poorly supervised broadcast properties. Although that difference is sometimes achieved by superior equipment and facilities, more often it is due to well trained and committed personnel. Even in large markets, some allegedly "major" TV stations and cable systems repeatedly mismatch slides and video (IDs, PSAs, and even commercials) with voice-over announcements; display badly-timed or poorly-cued video and film inserts; and lapse into extended moments of black during network and local program breaks. Some radio stations display inept operations by repeated interruptions and failed transmission signal or by garbled audiotape or miscued satellite service from unattended, malfunctioning automated equipment. This reflects careless staff and inadequate technical supervision.

As is every other department head, the chief engineer is responsible for allocating work to staff technicians, coordinating and assessing their performance. He achieves maximum efficiency along with satisfaction of personnel by assigning work to staff members based on their special interests and skills. Some engineering staffs increase efficiency by rotating assignments among studio operations, transmitter duty, maintenance, and remote work. Other stations prefer to assign staff to specific functions, because some prefer to work in studio production or master control while others are better suited to technical assignments such as ongoing maintenance away from programming personalities and the tensions of on-air production. The chief engineer serves as his staff's representative to higher management regarding compensation, promotion, and working conditions.

Unions

As noted in chapter 4, a key role of the chief engineer is cooperating with unions—their representatives and staff stewards—with thorough knowledge of negotiated contract details. This ensures uninterrupted operations by productive, satisfied crew members. The chief engineer should not wait until just before union negotiations start or are under way to explain to his staff why certain steps must be taken or why particular approaches cannot be implemented. Those explanations should be an ongoing process, taking place every day as a part of interpreting company policy to them (in terms of the "V Theory": top management's directives, along with their and the company's macro- and micro-intentions behind policies).

Equipment Purchase and Maintenance

Purchase

Every station manager and chief engineer should take time to become familiar with electronics gear on the market or even those still on the drawing boards of the world's major suppliers. Both must seriously investigate equipment exhibits at the National Association of Broadcasters' annual spring conventions if they want to avoid getting caught in the quicksand of surging expenses. Management's knowledge of the field of engineering and its equipment can make a difference in future profit and loss. A most useful pursuit for station managers is to spend an hour or two each week over several months becoming familiar with engineering equipment, procedures, and attendant problems.

Managers of established stations must purchase new equipment in planned cycles. Replacement is necessary for equipment that is becoming outworn or outmoded. Sometimes, however, there is a tendency to postpone equipment purchases or the remodeling of plant or building in order to improve the condition of the annual financial report. This policy may have certain immediate gains, but it only postpones the day of reckoning. A manager who moves wisely and constantly in these matters strengthens his station's advantages over his competition and thereby improves his station's profit position. Justification for buying the latest hardware is not merely to keep up with others, but as a calculated investment; purchase of a new piece of equipment should lead to an eventual increase of profit because of impressed viewer-listeners and satisfied advertisers. Acting on their chief engineers' recommendations, many managers have continued to show strong profit margins while engaging in regular expansion of facilities and equipment. They have proceeded on the sound management belief that one has to spend money to make money. This contrasts with some businesspersons who believe they can be successful by keeping costs down and operating on a narrow margin between income and outgo. But that margin soon becomes only enough for survival; and eventually, if their narrow view persists, even survival becomes questionable in hotly competitive markets.

Studio and field cameras built on computer chip technology can offer high quality, dependability, and low maintenance; but they come with high price tags. The same goes for sophisticated digital editing equipment.

■ Cosmos Broadcasting planned to spend $2.5 million on new equipment in 1992 for maintaining facilities at its seven TV stations, down about 23% from the previous year; they replaced three analog tube studio cameras with CCD units at each of two stations, projecting savings of $17,000 annually at each by eliminating tube replacement. Over two years the company bought Ampex ACR 225 digital editing decks for all its stations, costing $300,000 each; each had five recording decks and unattended storage and playback capability for entire spot inventories—permitting redeploying technical personnel.[4] The company expected to invest in HDTV transmission equipment after the mid-1990s. ■

News equipment is a high-cost area that is constantly evolving; for example, ENG cameras, relay transmitters, remote vehicles with microwave, and even satellite capability (SNG). Equipment can ensure quality images and sound of late-breaking events and "live" coverage, which can become a hallmark for a broadcast operation. But sometimes an elaborate item might serve a crew's curiosity or ego more than result in significant improvement in coverage. A well informed engineer along with news department personnel can assess what "bells and whistles" can truly add to the finished on-air product. There is also sometimes the paradox of fancy equipment being used for stories just because the station owns it, whether it really contributes or not. Such can occur with late-night "live" pickups from darkened locations long after the action has ceased, just because a remote unit can transmit a live picture at the news hour; it tells the audience little while keeping the newsperson far from the studio where they might better spend time carefully editing a substantive piece.

In broadcasting, the rate of technological change is rapid and constant. To maintain an improved profit position, and to achieve regularly improved standing among those with whom radio and television do business, managers must realize that facilities and equipment of many years' service are not assets if they are outmoded. On the other hand, purchasing additional or replacement equipment and improving building or plant should not be determined by mere physical obsolescence. Outlays must always be weighed against how they contribute to profitability. Decisions about large expenditures are not easy ones; only the manager with an alert and informed engineering department can make them wisely.

Sometimes a decision is forced by circumstances, as when satellite feeds from network suppliers shifted from C-band (transponders in the 4 to 6 gigahertz area of the spectrum) to high-power transmissions via the Ku-band (located from 12 to 14 gHz); this involved modification in up-link and down-link transmitters/receivers and dishes.

Cost savings can be calculated between buying new or used equipment, especially by smaller operations, and also by considering leasing rather than buying. Even networks increasingly look to leasing big-ticket items.

Among the most important cost-effective purchasing procedures reported by broadcasters have been: (a) automation (for programming, control-room operations, logging, traffic, billing, and bookkeeping), if its use is carefully adjusted to the station's actual and anticipated needs; (b) equipment surveys before buying, to analyze the range of alternative choices and to discover lesser-name equipment that costs less yet does the job as well or better than the "standard"; and (c) exploring reconditioned equipment that can serve as well as new yet costs far less. The key for management is the twin concern of being both cost-conscious and quality-conscious.

MAINTENANCE

The engineering administrator must maintain electronic equipment in top operating condition. Even the best technical staff at an affluent operation cannot keep up with all forms of repairs and still perform their other assignments effectively. Often it is more economical to send malfunctioning equipment to the manufacturer for repair and updating instead of investing crew time in-house on those repairs.

A well-managed engineering department gets the maximum life out of all major equipment. Engineering personnel run equipment checks more often than minimally required and take special pride in how all equipment performs. They use various functional charts for troubleshooting studio, transmitter, and remote gear. They submit discrepancy reports for every malfunction, no matter how small. The efficient engineering department prepares an operations manual and keeps it up to date. All station equipment should be checked thoroughly at least four times a year in addition to regular measurements for proof of performance.

Some broadcast stations lack adequate test equipment to check their facilities. But proper maintenance depends on adequate electronic testing as a major way to protect costly investments in sophisticated equipment. Apparently to save the relatively small expense of proper test and monitoring equipment, some stations run the risk of inadequate signal transmission and programming standards and of violating FCC technical regulations, as well as not protecting the longevity of capital facilities.

Maintaining a strong, clear signal is paramount to building an audience that enjoys fine reception. WZZK-AM-FM in Birmingham, Alabama, prides itself on sustaining a lofty first place for more than a decade partly due to the top-quality signal its transmitter and antenna generate, coupled with carefully matched audio equipment in studios and control rooms. A dedicated engineer, supported by management, monitors and maintains those facilities to ensure optimum reception on increasingly sophisticated auto and home receivers and stereo systems. In radio that becomes critical as more stations start up and others receive authorization for power or location changes. Probably the greatest competition for a radio station is the other radio operations in a market, especially in nonmetropolitan areas. Referring to program content but probably also to the technical operations and signal quality of radio stations, NAB President Eddie Fritts noted in 1993 that "while the universe is losing money, a lot of it is because a lot of small markets, and rural areas, have gone from good two-station markets to very lousy eight-station markets."[5]

A never-ending vexing problem for cable operators is maintaining proper signals from the head-end through many miles of feeder cables—strung from utility poles and buried underground, with amplifiers all along the lines—and through multi-taps down to individual drop lines to each subscribing household. Subscribers' major complaint, next to level of monthly fees, has been deteriorating or interrupted cable service. When any point on the distribution cable is disturbed, all feeds below that point are affected by "snow" or picture break-up or total loss of signals for all channels. Technicians must respond swiftly to requests for restoring service or risk losing subscribers. Cable managers struggle with the continuing challenge of "churn": as marketing succeeds in attracting new subscribers, many current ones cancel. Disconnects sever the economic life-line of a cable system. Engineering costs are heavy for repairing and replacing aging capital equipment sprawling in a maze of miles of wires spread across the service area; those costs include heavy expenditure for a small army of technical staff and a fleet of equipment trucks.

Automation/Computerized Technology

The increased cost of doing business in broadcasting is a constant concern to management. But that very economic fact has been the catalyst bringing engineering and management closer together, including introducing automated procedures. Managers must clarify for engineering personnel what is needed for the immediate and long-range future. The road to management-labor relationships can be rocky unless all labor-saving possibilities and their effects on operations, personnel, and economics are explored fully and honestly. This does not mean slashing the engineering payroll out of hand. But in this ever-developing industry every operation, regardless of size, must continually plan for the future by taking automation into account. The economic facts of life make this inevitable. Managers must keep technical personnel apprised of the situation and the reasons for it.

With proper planning, automation not only may bring about sizable savings, but it can gradually replace costly manpower without undue hardships for individuals involved. Well operated stations with forward-looking management can, with advanced planning, absorb into other phases of their broadcast and cable operations most employees affected by introducing computerized operations. Technical personnel should be told exactly why automation gear—well conceived, well installed, and knowledgeably utilized—can eliminate many of the human errors that constantly recur and pose so many sizable losses over a period of time. Also, automated equipment can release highly qualified technical people from merely routine functions to pursue more productive and creative duties, including careful preventive maintenance and supervision of the complex, highly expensive equipment.

In the 1960s, radio stations continually increased their automated technical and programming operation; larger stations also automated bookkeeping and paperwork. Television gradually followed suit. By the mid-1970s trade journal articles about automated equipment and procedures were commonplace. In 1977 *BM/E* magazine headlined a four-page report "KCBS All-News Radio Enters the Computer Age with Broadcasting's First Electronic News Room" in San Francisco.[6] In 1991 about half of all TV station newsrooms had computers; by 1996 virtually all TV newsrooms used them for: writing and editing stories, managing and filing wire service material, program rundown and timing logs, production notes, assignment sheets and story logs, archiving and retrieving (scripts, rundowns, story contacts, wire copy), electronic mail, and remote bureau operations.[7] Units most generally used included cuing and teleprompting, character generators, electronic still-storage of images, and closed captioning. Alabama-based Datacount installed its first DARTS radio traffic and billing system in 1979, and by 1995 had its computer system in 130 Alabama stations and more than 1,500 radio stations across the United States, Canada, the Pacific Rim, and Europe. Through every decade the NAB's spring conventions have annually showcased the latest high-tech applications for broadcasting and cable.

Broadcast and cable operations really had no choice but to adopt digital. Digital was the standard in computer, telephone, and fax applications, and increasingly in electronics equipment—even apart from HDTV. Cable was moving to digital, DBS used it exclusively, and telcos developed high-quality fiber-optic transmission paths. Manufacturers no longer invested in designing new analog equipment.

Computerized post-production linked creative digital workstations for nonlinear editing of text and image, for graphics, character generators, and special effects. Servers expedited inserting commercials and instant updating of news program copy and video. Completely outfitting a radio station for digital technology cost some $1.5 million, and up to $15 million for a major-market TV station. Cox Broadcasting spent $13 million in 1996 for Sony to design and supply a complete integrated digital operation.

Broadcast towers and new transmitters needed to move to the digital standard. In the interim they could be used with compression for ancillary services, developing a supplemental revenue stream. Costs were $1 million–$3 million for a second transmitter and antenna, $8 million–$12 million to equip a station for complete HDTV origination and network relay, or about $1 million if it only rebroadcast network feeds. Radio and TV networks in 1996 and 1997 installed digital compression production and transmission equipment, for satellite digital distribution to their stations.

In 1996 half the nation's cable subscribers were already served by some form of fiber-enhanced line. The cable industry looked to investing $14 billion in new technology in the half decade to 2000. Cable and telephone companies estimated the cost for wiring homes with fiber optic lines by cable or telephone companies at around $5,000–10,000 per home, or $250 billion–$450 billion for the entire United States.

■   One digital VTR (around $75,000) for post-production editing can do the work of four or five conventional one-inch VTRS ($20,000 each). On- or off-line nonlinear random access video editing, such as the Avid system, permits almost infinite flexibility in manipulating images, colors, and fonts down to the most minute element through computerized digital storage/retrieval of content transferred from videotape. Similar units capable of going from nonlinear editing directly to air—manufactured by Avid, Data Translation, Grass Valley, ImMIX, Chyron, Lightworks, EMC, or D-Vision—cost in mid-range between $30,000 and $250,000. The Digital Commercial Systems (DCS) unit has cart or reel-to-reel tape commercials transferred to it where they can be edited digitally and also cued for airplay by the DJ directly from a computer in the control room. Digital Generation Systems (DGS) records spots from satellite feeds. A digital computer "server" will gradually replace cart machines for replaying spot commercials. Automation with open platform systems, rather than "dedicated boxes" for single applications, can centrally control character generator, still-store archiving and retrieval, "paint box" functions, and random-access editing. CBS installed disk-based TV broadcasting equipment and in 1995 added computerized audio editing to network radio operations in New York and Washington, and eventually in all its owned and operated (O&O) radio stations. Audio compression saves disk space; the system permits nontechnical news reporters to control editing and switching, including routing audio signals and transmitting signals by satellite. Digital disk cameras will speed up transfer of images and editing, replacing real-time tape-to-disk duplication.

Robotic camera equipment for newscasts cost Westinghouse's KPIX (TV) in San Francisco $250,000 in 1988. Operated from the control room, they supplemented the one manned camera in the studio. Of concern to unions is the downside: considered the leader in introducing technology, NBC-TV in 1989 introduced robotic cameras, automatic cart machines, and other systems that eliminated 186 broadcast engineering jobs (but those remaining were paid at a higher level and no one was laid off, rather choosing to accept a generous buy-out plan negotiated with the National Association of Broadcast Employees and Technicians, 32 NABET Local 11).[8] ABC rebuilt a New York news studio with robotic cameras late in 1990. CBS brought robotic cameras to its network and O&O stations in 1991. CNN early on installed digital decks to edit field tape shot in digital, providing top-quality images that do not degrade when digital tape is reused over and over or dubbed through various generations—important for its multiple news services on five national and international feeds plus material for fifty-seven cable systems, national Fox news, and twenty-six foreign language newscasts (1993).　■

A result of introducing digital equipment including robotic cameras is fewer and better paid workers who are highly skilled and versatile, rather than larger numbers of single-skill technicians. Loren Tobia, news director at KMTV in Omaha, estimated that savings began two years after installing robotic cameras in the news studio; a five-person crew was reduced to three and eventually two. He put the decision in these terms: "As a news director, do you prefer losing members of a studio crew or reporters on the street?"[9] The chief engineer might appraise the equation differently, more in terms of releasing technicians to other tasks of maintenance and repair. That news director acknowledged that "everyone is affected—anchors, producers, reporters, production personnel and engineering. . . . The effects on morale remain and must continually be dealt with." Bob Seidel, CBS engineering vice-president, cautioned in 1995:

> You can be led down the road where you think one person can do everything, and the danger there is that one person has to be an artist and a typist and have multifaceted skills. So it's not just a question of converting the hardware. You have to look very closely at the technical pool of people who will be operating it, and invest in training to make sure they have these interdiscipline skills. You have to be careful—if you have a producer doing everything, it can limit creativity.[10]

Careful planning and astute use can make digital equipment cost effective as well as extremely functional for master control, production control room, central equipment area, news, program delay systems, post-production suites, and graphics and effects. In 1995 Jim Waterbury, chairman of NBC's affiliate board, estimated it would cost stations an estimated $12 million–$15 million each to convert to digital operations. That year's NAB convention featured digital optical disk storage, disk-based video camcorders (including digital Betacams by Sony at $62,000 each), nonlinear editors, and transmitters.

Some predict that videotape will be gone by 2000, in large part because digital servers for multiple tasking of recording, storage, editing, retrieval, and playback offer eventual savings as well. Digital disks reduce costs from that of space-consuming reels of videotape, fewer breakdowns and less maintenance for disk drives than for VTRs, and lower cost to operate because less human handling and intervention is needed to run a disk-based operation.

■   A survey of affiliate, independent, and public broadcasting engineers in 1995 reported two-thirds of them did not expect to purchase any more studio videotape machines, but rather disk-based digital-compression "tapeless" recorders; almost half of the final one-third said the next set of tape machines would be the last they would buy.[11] Half of them planned to buy tape machines in 1995; half also planned new ENG equipment; one-third looked to automation equipment, and one-fourth planned to buy SNG equipment. One-third of the engineers said their stations had satellite newsgathering trucks; 22% of the others were considering investing in one. One out of five had to buy new satellite down-link facilities because of shifting affiliation in the previous year. One-third of their stations used nonlinear editing equipment (half of those were Avid units), and 38% of the others planned to buy such that year. Over half (56%) were looking into buying disk-based servers, mostly for storing and playing back commercial spots (89%); half claimed servers would be useful for workstation graphical interface, 44% for time shifting, and 31% for storing and playing back long-form programs. One out of three engineers reported their station was planning budgets for purchasing high definition equipment for local production. Half of those 125 engineers surveyed estimated station budgets for new broadcast equipment that year at over $500,000; 7% said between that level and $300,000; 11% said below that figure and $200,000; while 9% said their 1995 budget for buying new equipment was less than $100,000 (8% did not respond). Decisions about purchasing equipment were made in collaboration, but primarily by chief engineers (71%), and/or by general managers (39%) and the corporate broadcast group (19%); 24% noted that others also participated in final decisions.   ■

The chief engineer must figure how to integrate these complex new devices and systems into current operations. Continued analysis and methodical planning include retraining current staff, upgrading their skills, and astute hiring of technicians capable of exploiting the full capabilities of new technologies.

PLANNING FOR THE FUTURE

A good engineering executive keeps up with developments in the industry and plans future steps to provide better performance and, where possible, at less cost than in the past. He must plan sufficiently in advance so he can arrange to relocate within his department or elsewhere in the station's operation those personnel to be relieved of their duties due to the change.

He must know what variables can affect major equipment purchases, such as technical design advances. The forty-year evolution of video recording devices (table 12.1) demonstrates progressive developments—just as every home PC owner experiences with hardware and software purchases. The challenge is which format to invest in, for how long; when to buy; when to start replacing—gradually or all at once?

TABLE 12.1
Broadcast Videotape Formats
(Year of first sale)

ANALOG VIDEOTAPE FORMATS		DIGITAL VIDEOTAPE FORMATS	
1956	2-inch quad	1986	D1, D2
1964	2-inch high band quad	1990	D3
1971	Super high band quad	1993	Digital Betacam
1972	3/4" U Matic	1994	D5
1976	1" B format (helical)		
1978	1" C format (helical)		
1981	Betacam		
1982	1/2" M, 1/2" L		
1984	8 millimeter		
1985	MII		

Source: National TeleConsultants, "The Migration to Digital: A Practical Guide to TV Managers," *Broadcasting & Cable,* 1 November 1993, D1-D16.

The chief engineer must also check carefully when selecting vendors of newly introduced high-tech equipment. National TeleConsultants urge broadcast buyers of digital facilities to require "specific, real answers" to "lots of questions," including explicit commitments put in writing by suppliers. Beyond basic points of technical specifications, costs, and delivery time-line, these questions should be asked:

> What is the manufacturer going to do about catastrophic failure? What emergency support services are offered by the vendor? Is there a 24-hour hotline? Is there a 'next day delivery' policy for spare parts?
>
> Should I invest in phone lines and modems for current/future diagnostic capabilities?
>
> What kind of training are my people going to get? Is it on-site or do I have to send my staff? How much is it going to cost me? Are refresher programs available?
>
> What kind of operations manuals are provided? Written? CD-ROM?
>
> What emergency spare parts should I store? Should I get them from the supplier? (Here, you need to compare the cost of storing spare parts vs. the manufacturer's standby repair service.)
>
> With software, do I get de-bugged for free? Are software upgrades free? For how long? Is it compatible with what I've got?
>
> How will software updates get to me? Will spare parts be upgraded at the same time or will they be made obsolete?[12]

An example of pragmatic, cost-conscious planning came when NBC-affiliate WVTM in Birmingham, Alabama, approached the mid-1990s with aging studio cameras. The general manager and the chief engineer chose to replace them with high-quality but only moderately expensive compact field cameras (looking almost stunted when mounted on massive studio dollies) for several years' use. The money saved they put aside towards purchasing studio high-definition cameras down the line when HDTV came on the scene; at that time they would move the interim cameras over to the news department for remote field work.

The typical shelf-life of electronics means continued turnover in state-of-the-art equipment. Veteran CBS executive engineer Joseph Flaherty suggested that in perhaps thirty years, digital itself might be absorbed into further technological systems—just as NTSC color in the 1960s was to be replaced in the 1990s. A new generation of computers occurs every two to three years and interconnected computer systems may last five to ten years.

Parallel Issues: Color Television and HDTV

All around us the real world is a sea of color. But television, the medium most suited to use this visual dimension effectively, was exceedingly slow to adopt it. NBC's owned stations introduced full-schedule color in the late 1950s, to stimulate sales of color TV equipment and receivers patented and manufactured by parent company RCA. But left to stand alone as the sole color programmer, the network suspended the effort a year or two later.

The issue of synergy and critical mass for HDTV was foreshadowed by the shift from black-and-white to color TV. Without going into complex legal and technical as well as psychological reasons, in 1965 a large part of the American public finally became excited over color television. NBC presented in color almost 96% of its entire evening schedule during the 1965–1966 season. CBS followed with half of its evening programming and ABC telecast more than a third in color. The following year, all prime-time programs on all three networks were in color. It was estimated that up to January 1, 1966, some $80 million had been spent on color by the networks, another $55 million by television stations and around $25 million by production companies, for a total of approximately $160 million. But the heaviest expenditures were made by the general public, which by that time had invested around $2.8 billion in receiving sets.[13] The number of color sets increased from a total of 7 million by 1965 to 60 million by 1973. In 1995 virtually all 95 million U.S. households had color TV receivers; over half of them had two or more sets.

Several interrelated factors account for that slow early development: (a) reluctance of set manufacturers to enter the field and consequent dealer passivity; (b) the public's indifferent attitude because of TV receiver costs and relatively few programs transmitted in color and thus weak support by advertisers. Those same factors loom as the industry approaches the next plateau of picture enhancement: high definition television (HDTV) with wide-screen aspect ratio of 9 units high by 16 units wide (or about 3x5 compared with the traditional 3x4 picture frame ratio) and double the quality of detailed image (1,125 lines rather than 525). The Japanese paced the electronic world by introducing high-definition TV using analog signals in 1981; the first feature-length program was produced in 1987, and HDTV programs were transmitted to home receivers by satellite in 1990. But after eight years of study and experimentation by the U.S. electronics industry, the consortium in November 1995 finalized a universal *digital* standard for HDTV transmission to replace the original NTSC standard and supplanting the lesser-quality Japanese *analog* HDTV system. Manufacturers, networks, stations, cable operators, and satellite home-signal distributors circled nervously, poised until HDTV approached a critical mass of high-definition technical standards, production equipment, program supply, audiences, advertisers, and subscriber fees.

■   The nation's first two HDTV stations, albeit experimental, began transmitting signals within a week of each other in July 1996. So-called WRAL-HD in Raleigh,

North Carolina, used prerecorded videotapes and pictures from its single HDTV camera; they added a converter to upgrade analog videotapes. WHD-TV, in the basement of NBC's WRC-TV in Washington, D.C., followed in days with eight hours daily on air, with plans for twenty-four-hour transmission the next year; it was to serve as a model operation for the industry (two receivers picked up its signal—one in WRC's basement, the other in a truck in the station's parking lot). Noncommercial WETA-TV in Washington, D.C., announced plans to build a complete HDTV station in 1998–1999 for $10 million–$14 million; it bought its HDTV transmitter in 1996 for around $450,000. Maryland Public Television planned gradual upgrading of its six stations; in 1996 it modified transmitters at three stations for twin analog and digital service, doubling power to over 3 megawatts—to introduce digital-quality service even prior to HDTV transmission. ■

The FCC proposed a plan for introducing HDTV over a fifteen-year cycle: Stations must install HDTV transmission facilities within five years to send out HDTV-standard digitally compressed transmission on a second channel; simultaneously they would continue to transmit NTSC signals on their current channel for fifteen years (to around 2010). At the end of that period, stations would discontinue the older analog service and turn back their original channel for government reassignment to other services. That time-frame allowed broadcasters to phase in capital purchases of HDTV equipment incrementally over a decade, while the public gradually began to buy HDTV receivers to replace their aging NTSC sets.

The FCC hesitated in the mid-1990s to fully endorse the Grand Alliance's recommendation for a single HDTV system, because computer interests argued it was not adequate to interface with computer information transmission and imaging. But the Commission did set a September 30, 1996, deadline for filing applications for any analog NTSC stations (300 applications were then pending); thereafter all assignments would be for digital TV transmission.

The National Television and Information Administration in 1988 projected that, after introduced, high definition TV would reach only 1% household penetration by the first ten years (market valuation of HDTV products that decade worth $1.7 billion to $3.5 billion); in fifteen years HDTV would reach 25% of homes ($26 billion to $52 billion in costs), and by twenty years 70% penetration ($72 to $144 billion).[14] At the same time the Electronics Industry Association projected HDTV would be in 10% of U.S. homes within four years of becoming available, and 25% penetration another four years after that.[15] Most of the public would wait for program offerings in HDTV to expand and for prices on receivers to come down.

■   The very first, almost experimental, HDTV receivers in Japan cost $27,000 (3.5 million Yen when $1 equals 131 Yen); in 1992 the cost of a 36-inch set had plummeted 70% but was still $7,955 (1 million Yen), while a 20-inch set cost $4,500 in 1995. European 36-inch receivers sold for $6,000 in 1992 but later dropped to $2,000–$3,000. (In mid-1995 Japanese viewers owned 60,000 HDTV sets and 140,000 NTSC widescreen receivers; within five years, by 2000, they were expected to own 4.2 million widescreen sets.) U.S. prices in the mid- to late-1990s were expected to be $3,500–$5,000 at first, then gradually lower to $2,000–$3,000 after five years, and further to $900–$2,000 by year 2010. ■

Broadcasting magazine estimated in 1992 that the total investment to change over the U.S. system to high definition television would cost local and national broadcasters $10 billion and American consumers $100 billion.[16] (This recalls the perennial question of "who owns the airwaves?" even if considering only the amounts invested in making them operative; still, broadcasters must put up most of their money first before consumers begin to spend theirs, and if consumers fail to follow then the broadcasters' enormous risk capital is doomed.)

Just as audiences in the 1970s came to expect all networks and stations to provide programming in color, audiences in the early 2000s will come to assume widescreen, high definition pictures available on all channels. Back then, a station that carried network color but aired local programs and news in monochrome lost sizable numbers of viewers to competing stations that provided color throughout their schedules. The same for HDTV as coaxial/fiber-optic cable and over-air and direct broadcast satellite (DBS) delivery systems compete for audiences. Just as color adds to the visual and emotional impact of entertainment programming and commercials, so HDTV will enhance the viewing experience. (Immediate beneficiary of HDTV will be the vast storehouses of motion pictures on high-quality 35mm and 70mm film in wide aspect ratio such as Panavision, with Dolby surround-sound; for the first time homebound TV audiences will be able to view classic and popular films in all their breadth and fine detail along with CD-quality audio.) But as with the introduction of color TV, after the novelty factor diminishes when all services offer superior images and sound, neither HDTV nor color alone will save a second-rate program. Broadcasters and producers must continually search for new program ideas and fresh approaches.

How long will it take to introduce HDTV to the consumer mass market? Some said fifteen to twenty years if the government did not help expedite the transition by supporting guidelines (as with manufacturing deadlines for all-channel receivers mandated to help UHF stations compete in the VHF world). One proposal by the White House in 1995 was ten years for transition; but even if first digital broadcasts could start in 1998, only seven more years would not be enough for complete national transition to digital.

COMPETING DELIVERY SYSTEMS

Managers must keep alert along with their engineering colleagues to assess shifting factors in the mass media mix, both local and national, as competition proliferates for audience attention and subscribers' and advertisers' dollars. Terrestrial broadcast stations face satellite TV (DBS) and radio (DAB) delivery systems which compete for audiences and for national network and spot advertising (but not for local ad dollars). Local stations do compete with cable systems for local advertising and audiences. Along with cable operators, stations also face entry of affluent telephone companies into media delivery by wire or fiber optic cable. Meanwhile media conglomerates such as Disney and Time Warner and Tele-Communication, Inc. merge with or take over broadcast networks, local stations, and cable systems, creating multiple-threat competitors for the public's leisure time and discretionary dollars.

■   Those companies need to be affluent. The National Cable TV Association in 1990 estimated the cost of wiring each home with fiber optic at $5,000–10,000, for a total of $450 billion to bring fiber to all U.S. households; Northern Telecom estimated less: $3,875–$5,650 per home for a national total of $200 billion–250 billion. Another projection in 1991 was that telcos could rebuild the nation with fiber optic by 2015, the same year the Japanese set for their total fiber telecommunications infrastructure. United States Telephone Association President John Sodolski in 1993 estimated fifteen years and $150 billion to wire half of U.S. homes with fiber optics for a complete video delivery service. Those figures contrast with $420 per household for a start-up cable system, and an estimated $3 per household for Comcast's early-1980 C-band DBS system which it abandoned after losing some $200 million. MMDS systems installed "wireless cable" at $300–400 per home in 1991. In 1993 TCI announced plans to spend $2 billion over four years to build a fiber optic network in thirty-seven states, while Pacific Bell announced $16 billion for a fiber optic and coaxial cable network in California.[17] ■

Potential rewards are matched by high-stake risks in committing multimillions to new technologies. CBS and Paramount both withdrew from a co-venture with Comsat to develop a DBS service; it was judged too costly over too long a time before generating adequate revenues. NBC, Cablevision Systems, Hughes Communications, and Rupert Murdoch's News Corporation never developed their Sky Cable 108-channel DBS service, first announced in the early 1990s (Hughes in 1994 launched DirecTV on its own). CBS Cable's cultural programming channel was aborted after 330 days and $30 million dollars; Rockefeller Center, Inc. and RCA spent $35 million on cultural programming, then bowed out after 300 days (this was later resurrected as the Arts & Entertainment cable network). ABC and Westinghouse together forged Satellite News Channels, spending $35 million–$40 million a year, until they abandoned the effort after 480 days. Success-prone entrepreneur Ted Turner followed his CNN and CNN Headline News (then losing $9 million per half year) with a Cable Music Channel, which lasted just 36 days.[18]

Back in 1982 Julius Barnathan, president of ABC Broadcast Operations and Engineering, cautioned "I think the name of the game is programming. I wish they'd get off this technology kick, because nothing has changed. All you've got now are more ways of failing than you had before."[19] But irrepressible Stanley Hubbard compared the $250 million his company needed to establish USSB's satellite coverage of the entire nation's 95 million households to the $510 million paid by Tribune Broadcasting for a single independent TV station, KTLA in Los Angeles reaching fewer than 3 million households.

Coaxial Cable and Fiber Optics

Established cable operators began to adapt distribution to optical conduits and to signal compression over standard coaxial wire. Optical fiber and compression technologies can expand cable capacities tenfold or more.

■ In 1993 of the 75% of all cable operators planning to expand channel capacity, most expected to do so by fiber optics.[20] Almost 40% of all cable systems intended to install fiber technology, while over 20% expected to introduce digital compression on current cables within three to five years, and between 16% (offering 12–36 channels) and 7.2% (with more than 36 channels) planned continued use of conventional coaxial cable. About 25% foresaw no expansion; one-half percent said they planned to cooperate with a regional Bell operating company.

The cable industry expected compression technology to be ready in 1994 with the proposed MPEG-2 standard (Moving Picture Experts Group). Five of the largest cable MSOs had placed orders in mid-1993, including TCI's million units to serve almost 10% of their subscribers. ■

On the other hand, the Telecommunications Act of 1996 opened the door for telephone companies to establish competing video delivery systems. For example, regional Bell company Ameritech set up its own competing cable service in Detroit suburbs, and held cable franchises it could activate in areas around

Chicago, Cleveland, Columbus, and Milwaukee—competing with established local cable systems of Comcast, Time Warner, Cablevision, Continental Cablevision, and TCI. In 1996 AT&T entered into marketing and programming agreements along with financial investments with DBS operators DirecTV and USSB.

SATELLITES

Networks first leased time on satellites in the late 1960s for news feeds from Europe to New York headquarters. In 1975 Home Box Office (HBO) pioneered nonstop daily satellite relay to syndicate its movies to the country's cable systems. The following year Ted Turner's WTCG-TV (later renamed WTBS) in Atlanta was the first station relayed to the nation's cable systems via satellite. The first TV network to shift from microwave relay and long-lines cable to satellite distribution to their affiliated stations was noncommercial PBS in 1978. Its main transmit/receive earth station outside Washington and four other regional earth stations fed signals to 162 PBS stations with receive-only earth stations. The commercial networks followed suit when their contracts for telco long-lines/microwave links expired. Turner launched twenty-four-hour CNN in 1980; ABC and Group W sought to compete two years later with their Satellite News Channel, but capitulated in sixteen months by selling it to Turner for $25 million (they had lost $35 million in the first year of operation). Mutual radio network (MBS) by 1982 used its up-link and 650 earth terminals to deliver programming to 900 affiliates. By 1983 almost 85% of TV stations in the top-50 markets had satellite capacity; a total of 375 stations in the United States were so equipped, plus some 6,500 cable systems. That year more than twenty syndicated program series were delivered by satellite.

In 1988 12 orbiting carrier satellites with multiple transponders relayed Ku-band 120-watt signals to four-foot receiving dishes, and 19 fed lower-power C-band 20-watt transmissions to eight- to ten-foot dishes; in 1993 there were 14 Ku-band and 19 C-band satellites, most with 24 transponders but many with 16 or 18 and a few with only 6 transponders. For direct transmission to homes, DBS powered 240-watt signals to eighteen-inch dishes.

Satellites spawned regional, national, and global news distribution by operations such as Hubbard Broadcasting's Conus (Continental U.S.) serving 160 stations and Turner Broadcasting's CNN Newsource. Other organizations included The Associated Press' TVDirect, Visnews International, Group W's Newsfeed, and major network news interchanges—ABC's Absat/NewsOne, CBS' Newsnet/NewsPath, and NBC's Skycom/News Channel.

In 1990 Unistar Radio Networks and Satellite Music Networks each offered ten different formats of high quality stereo via satellite to their 1,100 affiliated radio stations. By 1991 NBC-TV, the first all-Ku-band network (1983), had 116 up-links and 800 down-links; it used only one-third of its satellite time to distribute network programming, devoting the other 70% to news and other "backhaul" program material and data from its affiliated stations.

■ In 1990 a motor-driven, steerable up-link/down-link earth station cost $28,000 to $44,000, while a nonsteerable unit cost $7,000–$19,000; a top-of-the-line mobile satellite up-link truck cost up to half a million dollars. Local TV stations owned 300 up-link satellite trucks, with prices starting at $230,000, primarily for "satellite news gathering" (SNG)—for remote pickups for itself and for interchange with other stations in the region. ■

TVRO *and* DBS

As for TVRO (TV receive-only) satellite earth stations or "backyard dishes" that pick signals off satellites, they grew from 4,000 in 1980 to more than 3.6 million by 1994; the seven- to ten-foot wide dish, preamplifier, and receiver cost about $2,000 to receive some 75 unscrambled channels, plus additional costs for descramblers and fees for other services.[*]

In the early 1980s Comsat (Communications Satellite Corporation) announced it would develop Direct Broadcast Satellite (DBS) service. RCA/NBC, CBS, and Western Union indicated they were exploring the venture. Those plans faded in the face of enormous costs and risk of failure. From 1983 until 1985 United Satellite Communications, Inc. had offered Ku-band DBS service; the $30 monthly fee for four channels attracted only 20,000 in the Northeast and Midwest before the USCI terminated service.

Stanley S. Hubbard championed DBS throughout the decade and collaborated with Hughes Electronics in jointly launching DBS satellite VSS #1 in December 1993; six months later Hughes launched VSS #2. In 1994 Hughes' DirecTV and Hubbard's USSB (United States Satellite Broadcasting) introduced high-powered DBS programming service to the United States; they transmitted at 120 watts per transponder (capable of going to 240 watts) on the high power Ku-band of the spectrum, enabling reception by eighteen-inch antenna dishes—contrasted with earlier C-band's 40 watt signals requiring broad receiving dishes. Satellite coverage area reached all forty-eight states.

■   Their two satellites each had 16 transponders with a 4:1 digital compression ratio; USSB offered 20 video signals on five channels, while DirectTV relayed almost 200 signals, including 40 program channels of pay-per-view (PPV) movies and 30 PPV sports channels plus three dozen digital audio channels. By mid-1995, 600,000 subscribers had bought receiver hardware. The initial $800 cost (plus $200 installation) for the eighteen-inch antenna dish and "blackbox" converter was followed by monthly subscription fees of $20 to $50 or more, depending on the program packages ordered; entry of competing hardware suppliers and satellite programmers in 1996 slashed consumers' start-up costs to $200 for reception hardware. DBS converter boxes were fitted on back with a multi-pin plug and switch, already equipped for future HDTV reception from satellite feeds.   ■

A coalition of TCI, Time Warner, Comcast, Continental, Cox, and General Electric Americom launched Primestar in 1994 using lower-power Ku-band that required a thirty-six-inch dish; it lowered initial costs to subscribers by renting rather than selling the receiving apparatus.

USSB and DirecTV plus Prime Star served 1.3 million households going into 1996; they expected to attain their break-even point of 3 million subscribers in several more years, and hoped to reach 10 million within ten years. USSB conservatively estimated 1.8 million to 2 million subscribers as the point when its venture produced profits; DirecTV's break-even point, with more channels and programming costs, was 3 million subscribers, while Primestar's profit threshold was 2.5 million. In mid-1996 all three had reached half their goals.

---

[*]In 1990, an estimated 80% of the country's 6,404 local dealers of home satellite equipment were involved directly or indirectly in piracy—assisting dish owners to illegally intercept satellite signals without paying required fees. Wholly unrelated to that issue was the anomaly that among the fourteen earliest applicants to the FCC in 1981 for DBS authorization, representing Home Broadcasting Television Partners and Unitel were two inmates at a federal prison in Terre Haute, Indiana.

Seven other FCC-licensed DBS companies intended to launch and become operational during the 1990s, including EchoStar's DISH Network, and Alpha-Star (requiring a thirty-inch dish receiver) in 1996. News Corporation (Fox)/MCI's American Sky Broadcasting (ASkyB) was to launch in 1997 or 1998 and planned to incorporate transmission of local TV stations along with the array of cable/DBS program services. Subscribers had to purchase separate reception equipment for each of the services, except for DirecTV and USSB that shared satellite transmission.

By mid-1995, 1.14 million subscribers had bought receiver hardware; twelve months later households receiving one or more DBS services were estimated at 2.95 million—far ahead of projections. Total households receiving those services and other direct satellite systems (DSS) plus over 2 million homes receiving pioneer C-band transmissions (not counting another 2 million nonsubscribers to unscrambled signals), were estimated at 8 million going into 1997. (Those totals contrast with 62 million subscribers to the nation's 11,000 cable systems carrying national program services, and 97 million households reached by broadcast networks via the nation's over-air TV stations.) The many uncertain factors influencing media growth—such as telephone companies' fiber optic video services—are evident in USSB President Stanley E. Hubbard, II's broad projection that satellite penetration in five to ten years (2000–2005) would lie somewhere between 10 million–15 million to 25 million–30 million subscribers.

LPTV, MDS, MMDS, SMATV

Other delivery systems competing for fragmenting audiences include low-power TV (LPTV), translators, multipoint distribution service (MDS), multichannel multipoint distribution service (MMDS), and satellite master antenna TV (SMATV).[21] Most are not major threats to traditional broadcast.

■   LPTV is limited to 10 watts on VHF and 1,000 watts on UHF to reach a five- to twenty-mile radius. Yet in 1993 the country's largest LPTV station, WFBT in Chicago, reportedly could reach three-quarters of the 8 million viewers in that ADI market. LPTV stations on the air grew from 427 in 1988 to 1,771 by 1996 (561 VHF and 1,211 UHF). In 1996 translators totaled 2,263 VHF with another 2,562 UHF.   ■

Fledgling networks UPN and WB both served some of their markets via LPTV stations where no VHF or UHF outlets were available. In 1995, of UPN's 150 affiliates 11 were low-power stations; WB's 100 affiliates included 14 LPTVs plus cable carriage in some small markets; Fox network had low-power TV affiliates in 4 markets—including 117th-ranked Fort Smith-Fayetteville, Arkansas, which the network reached through a cluster of 7 LPTV stations, all with call letters KPBI-LP.

■   From their start in 1973 single-channel MDS services had grown to half a million subscribers in the early 1980s until cable penetration drew away audiences. Some 140 multiple-channel MMDS (or wireless cable) systems in 1993 served 400,000 homes, most in rural areas; each system averaged 1,400 subscribers (who typically paid about $30 a month for thirty channels). The largest number of subscribers to MMDS in individual markets were 42,000 in Riverside/San Bernardino and 35,000 in New York City. By 1996 MMDS systems had grown to 170, with over 800,000 subscribers. The seven major publicly traded MMDS companies together add about 175,000 new customers annually.[22]   ■

In these technologies, too, the telephone companies began to invest and build. Pacific Telesis, joined by Nynex and Bell Atlantic, set up its digital "wireless cable"

MMDS in Los Angeles, to compete for 6 million homes with fourteen cable MSOs covering that area. The telco consortium had similar plans for San Diego, San Francisco, and possibly in Washington state, Florida, and South Carolina.

NONBROADCAST TV REVENUES

The dimensions of the competitive arena are reflected in estimates of cable, DBS, and telephone company TV operations in the year 2000 (table 12.2). The dollar figures include capital and operating expenses, with revenues from subscriptions, fees, and advertising. Cable subscribers were expected to drop to 59 million, while 10 million subscribed to telco video services and another 5.6 million to DBS.

STEREO

In 1961 the FCC propelled FM radio's success by endorsing a common standard for transmitting stereo separation of high fidelity reproduction. That standard was set by industry collaboration with the Electronic Industry Association's (EIA) independent National Stereophonic Radio Committee. Two decades later, to strengthen AM radio's attraction, the FCC deliberated about which of four competing technical systems to endorse: Motorola, Kahn/Hazeltine, Harris, or Belar Electronics. By default, it was left to the marketplace to determine, leaving engineers and managers to make their best guess if they wanted to embark on AM stereo. Regrettably, the incompatible systems created mixed-up markets for enterprising broadcasters and potential audiences alike. Like the Beta/VHS battle for VCR dominance, multiple conflicting stereo transmission systems turned up in the same market, or auto and home stereo receivers built by one or more of the manufacturers could not receive the differing system used by local AM stations in that area. That makes investing in stereo AM equipment a capricious roulette game for stations. In 1993 the Commission did finally choose Motorola's C-Quam system, but the forty stations using Kahn's "Powerside" system could continue to use that technology at least to reduce signal interference from other stations.

Engineers must carefully analyze technical specifications and their expected long-term usefulness before recommending capital outlay from management.

By 1985 TV with stereo sound reached half of all U.S. homes; although concentrated in the top-40 markets, stereo TV was available in thirty-four states. In 1987 a third (500) of all TV stations transmitted stereo audio, while one out of four new color TV sets sold were equipped for stereo.

DIGITAL AUDIO BROADCASTING (DAB)

Broadcasting magazine in 1980 trumpeted "The Dawning of Digital in Radio Broadcasting" as microprocessors and computers were installed not only for bookkeeping and programming but in technical areas including digitized audio. By the 1990s the issue turned to satellite transmitted CD-quality audio signals—an attractive dream for national quality radio service,

TABLE 12.2

ESTIMATES OF SCOPE OF TV OPERATIONS IN THE YEAR 2000

	CABLE	TELCO	DBS
Homes Passed	98,940,000	31,000,000	21,010,000*
Subscribers	59,050,000	10,089,312	5,622,191
Total revenue	$ 38.47 billion	$ 4.84 billion	$ 3.09 billion
Total cash flow	$ 17.72 billion	$ 1.97 billion	$ 1.29 billion

*Assumes DBS actively marketed only in limited areas.

Source: *Broadcasting & Cable,* 22 March 1993, 34; citing projections by Grantchester Securities and Wasserstein Parella Securities.

but a potential nightmare for terrestrial radio stations. On the other hand, although satellite DAB could not displace local advertising (almost 80% of radio stations' revenue), it could siphon off national/regional spot from local stations and would undercut network and syndicated radio operations. An alternate revenue option was a $5–$10 monthly subscription fee to receive the multi-channel audio signal. Apart from sheer time-selling questions, the quality and unlimited reach of digital signals would draw off masses of radio listeners in all markets, except for those interested in community-oriented traditional stations with localized personalities, news, and sports.

To forestall satellite DAB competition, a consortium of CBS Radio, Gannett Broadcasting, and Group W transmitted a digital stereo audio signal over a standard AM radio channel in Cincinnati (August 26, 1992) to demonstrate the feasibility of digital service from existing terrestrial stations on present frequencies.

■ As with every innovative technology, varied systems can only develop when technical standards have been established. Broadcast engineers from radio networks and station groups formed the Committee for Digital Radio Broadcasting (CDRB) in 1990. A reluctant NAB, concerned about the fate of terrestrial radio stations, formed a task force to study DAB spectrum matters. The radio networks had already been distributing by satellite for a number of years with Scientific-American's Digital Audio Technology (DAT), upgraded in 1990 to its Spectrum-Efficient Digital Audio Technology (SEDAT); it transmitted by digital compression up to 80 radio signals per transponder with CD quality audio. The NAB proposed a terrestrial DAB radio service to gradually replace the AM and FM service now in place, with DAB channels going to current radio operators. ■

The chicken-and-egg dilemma recurred: How much would DAB radio receivers cost the public and would they buy into a service if it charged subscriber fees? Radio group owner Mel Karmazin in 1991 felt that "digital radio probably will be rejected by the marketplace."[23]

In 1992 General Electric proposed a 500-digital channel satellite DAB service, costing $600 million, offering ad-free quality audio for $5 a month (plus $200 for a DAB satellite receiver). Sixteen channels would carry national programming, the others would be divided for service to thirty-one regions. The next year a half dozen other proponents offered variations on that plan for harnessing DAB—or satellite-delivered Digital Audio Radio Service (DARS) as it came to be called. Into 1996 the FCC continued to wrestle with the technical and economic implications for potential regulatory guidance, as with the emerging HDTV standard.

Managers and Technology: Engineering

531

Pragmatic Strategies for DBS and HDTV

As technology innovates varied delivery systems, media managers must determine with their engineers how to compete effectively, avoiding redundancy by focusing on what is distinctive to themselves. For local radio and TV stations that can mean substantive local news service for the community of license, unavailable from DBS; for DBS that can mean swift introduction of HDTV enhanced images and sound, far ahead of the networks of stations. Each of those stations must spend up to $3 million for receiving/transmitting just to pass through HDTV program signals from their networks, and perhaps three times that amount each to originate HDTV programs locally.

■ Two studies, by CBS and by PBS, estimated conversion to HDTV by a local station would cost about $12 million spent in steps over five to nine years; small-market stations might be converted for $5 million. An early estimate contrasted the cost of building a new TV station from scratch: $14 million for a traditional NTSC signal station, compared with over $38 million for an HDTV station—with HDTV cameras costing up to $350,000 each in 1990, vs. $140,000 for a standard NTSC camera. NBC estimated that to install HDTV operations would cost the network $400 million for equipment and $500 million for labor; that nearly $1 billion did not include costs for O&O stations. For a local cable operator, it would cost almost $900,000 to equip a system to deliver a dozen HDTV channels to 100,000 subscribers; total cost including subscribers' "black box" receiving units would mount to $1.2 million for that size local cable operation.[24] ■

A variant strategy may call for collaborating with new media services. Local TV and radio already offer their programming (especially news) to cable and MMDS companies or negotiate for a channel dedicated to that broadcast station. To compete with telco fiber optics, local and MSO cable operators need to consider digital compression for cable to multiply channels on already installed coax lines and for top quality pictures and sound. Expert media analysts in 1996 predicted that during the next five years basic cable revenues would drop 20% and premium services 8%, while pay-per-view would grow from 1995's $590 million to $2.35 billion in 2000 and to $4.76 billion by 2005 with enhanced technology permitting video on demand. For that same period analysts projected $8 billion in new revenue supported by new technologies, including cable telephony, video games, home shopping, and Internet access.

Over the two decades needed to develop a critical mass of households with HDTV receivers, the FCC expects local stations to broadcast both traditional NTSC and the new HDTV signals; that means maintaining dual transmitters. The extra 6 mHz spectrum space assigned to each over-air station, along with digital capability, offers opportunity for packaging other kinds of services. Corporate management looks to such alternate revenue streams that creative engineering makes possible, such as data information services on FM, TV signal subcarriers, or on the added second channels for HDTV. (In the early 1990s one small-market TV station earned $200,000 a year from its data broadcasting service which involved no added expenses.) Engineers must help management decide whether to develop other ventures including interactive programming video and data services, and multimedia mixes of computer, video, sound, graphics, and text.

To ensure informed decision making, strategic planning demands that top management confer closely with middle managers—chief engineers along with heads of programming and sales departments and financial administrators.

FCC-Related Responsibilities

The chief engineer must keep alert to federal regulations involving technical operations. That responsibility, crucial to guiding top management, was outlined earlier in chapter 4; already described in chapter 11 were details about typical fines (forfeitures) and patterns of penalties including short-term license renewals.

FCC Rules Violations

Historically, the most common reasons for FCC penalties, including license revocations, have been violations in technical operations—far more than for programming matters. Typical failings involved maintenance of proper operating power or modulation, transmitter "proof of performance" reports, Emergency Broadcast Service (EBS), logging daily check of tower lights and quarterly inspections, entries in operating log, and operator requirements for engineers on duty. Forfeitures were usually imposed after FCC field inspectors confirmed a station did not comply with specific rulings. Inspections were either routine technical inspections by the agency's Field Operations Bureau, or those triggered by complaints handled by the Enforcement Division of the FCC's Mass Media Bureau in Washington.

When visiting stations, FCC field inspectors are particularly interested in checking the points listed in table 12.3. During the 1980s the Commission issued modest fines averaging $845 for violating a technical rule. But Chairman Al Sikes strengthened enforcement for the 1990s, significantly increasing the number and level of penalties. The FCC delegated to staff the authority to assess forfeitures up to $10,000, while the commissioners authorized fines up to $25,000 per violation or for each day of a continuing violation and up to a total of $250,000.

Technical violations in the 1990s were assessed with base fines (table 12.4), which could be adjusted up or down according to circumstances, including "the degree of culpability, any history of prior offenses, ability to pay, and such other matters as justice may require."[25]

Smaller market stations without fulltime engineers might engage the services of a certified senior broadcast engineer to visit and inspect their operation on a confidential basis, to identify any problem area that the station can then correct to avoid potential penalties levied by an official FCC inspection.[26]

Federal as well as state and local regulations also affect aspects of the physical plant which are often under the purview of the chief engineer. Building codes underwent significant changes in recent decades; they apply to major renovations and new construction alike. The chief engineer and general manager must work closely with architects and construction companies to ensure compliance.

■   The Energy Policy Act of 1992 pulls together and extends states' varied standards for energy conservation, especially for air conditioning and lighting. The National Electrical Code sets safety requirements for installing electrical equipment; all cables must be listed and marked with fire rating codes (most broadcast cable made prior to 1987, while not in compliance, were grandfathered in already constructed facilities). The Environmental Protection Agency mandates kinds of chemicals permissible such as for film cleaners and for asbestos and PCBs found in older buildings. The Americans with Disabilities Act in 1990 mandated access for disabled persons in work spaces as well as visitors' areas (affecting plans for locating doors and light switches, ramps, bathroom layouts, and appropriate signs).[27]   ■

TABLE 12.3
Technical Items Given Most Attention by FCC Field Inspectors

TECHNICAL INSPECTIONS

Frequency
Modulation
Emissions
Power (under or over authorized strength)
AM directional parameters
AM monitoring points

INSIDE INSPECTIONS

On-duty operator
Adequate meters and warning devices
Proper transmitter control
Complete public inspection file
License and authorizations posted
Station operating as authorized

OUTSIDE INSPECTIONS

Proper tower lighting and painting
Proper AM antenna grounding
Proper fencing around transmitter

EBS INSPECTIONS

Current checklists and authenticators
EBS monitor (missing, malfunctioning, tuned to wrong station)
EBS generator (missing, malfunctioning)
EBS tests received
EBS generator tests conducted
EBS generator tests on log

Source: Adapted from "An FCC Inspection Checklist," *Broadcast Cable Financial Journal,* January/February 1991, 26.

Deregulating "regulatory underbrush" and paperwork minutiae began under Richard Wiley's chairmanship at the FCC, and continued with successive chairs, especially Mark Fowler. Aided by dependable automated equipment, technical housekeeping chores have greatly diminished. A chief engineer must keep alert to periodic changes; the Commission not only has rescinded operational procedures but has modified them or introduced variations or entirely new requirements from time to time. These typically involve remote control transmitter operation, operating logs, directional antennae, and equipment maintenance logs. For instance, early on the Commission relaxed rules requiring measurement of the transmitter frequency at specified intervals by a frequency monitor located at the station. Since modern transmission equipment demonstrates extremely reliable capability to maintain frequency control within limits specified by the Commission, it is no longer mandatory that a broadcaster has equipment to measure the frequency of emission. The licensee is required, however, to ensure that deviation of frequency of emission beyond FCC specified tolerances will be detected and corrected within a reasonable time period. The Commission began to permit stations to operate transmitters remotely, meter readings could be clocked automatically without detailed personal monitoring; operating (transmitter) logs and programming logs could be kept by automatic logging devices.

The regulatory agency relaxed operator requirements. The designated first-class radio engineer may be employed on a parttime, contract basis. (Nondirectional AM stations operating with power in excess of 10 kw and FM stations operating with power in excess of 25 kw may employ an operator on duty who holds a third-class operator's license, provided the station employs on a fulltime basis at least one operator holding a first-class license.) A first-class engineer employed on a parttime contract should be regarded as an independent contractor, which should be designated in writing. Other contractual items include a statement of the engineer's specific duties, the rate of payment, a specification that he will supply his own tools and testing equipment, the policy for purchasing supplies

TABLE 12.4
FORFEITURES IMPOSED BY FCC FOR SELECTED TECHNICAL VIOLATIONS
(EFFECTIVE: 1990s)

BASE FINE	VIOLATION
$ 20,000	Construction or operation without instrument of authorization for service
15,000	Exceeding authorized antenna height
10,000	Exceeding power limits
10,000	Unauthorized emissions
10,000	EBS equipment not installed or operational
10,000	Violation of main studio rule
10,000	Construction or operation at unauthorized location
8,000	Failure to comply with prescribed lighting and marking
5,000	Failure to file required forms or information
5,000	Violation of file requirements for technical logs
2,500	Failure to make required measurements or to conduct required monitoring
1,250	Failure to maintain required records

Source: *The Alabama Broadcaster,* September/October 1993, 12.

and equipment, a clause about nondisclosure of station information, and a statement about the length of the contract with termination and renewal conditions.

PERMANENT DEFENSE SYSTEM

The Emergency Broadcast System (EBS), designed specifically for use in case of a war emergency, went into effect in January, 1964. It replaced the former Conelrad system in use since 1951, and the EBS itself was to be fully replaced in 1997 by the digital Emergency Alert System (EAS).

EBS used AM stations as its base, with FM and TV aural channels assigned to relay and network assistance when necessary. Under the alert system, a federal government message would be issued to the Associated Press and to United Press International who would flash the alert to stations. Immediately on receiving the alert, stations must cease all regular programming to read prearranged announcements. Stations not holding National Defense Emergency Authorizations must advise listeners to tune to an authorized station and then must leave the air. (Any station may apply for NDEA authorization; they have a right of appeal to the FCC if they are refused authorization.) Stations authorized to remain on the air may not broadcast any commercials. They may program music, when there is no information to be carried, and they also may broadcast information to specific individuals. In the event of an emergency necessitating the use of EBS, daytime-only AM stations could operate beyond their normally licensed hours if no other service is available in the area. Unlimited-time AM stations could maintain their daytime facilities at night. The United States Weather Bureau makes use of EBS for warnings of hurricanes, tornadoes, and other physical catastrophes.

The core consideration by the government, in the words of FCC Chairman Al Sikes, is that unlike selective cable systems and other media, over-the-air "broadcasting constitutes the only truly effective means of providing emergency information to the public very quickly."[28] The new digital emergency service—its first stage mounted in 1995 and scheduled to be completed in 1997—includes multiple source monitoring, an alerting tone reduced to 8 seconds (vs. the previous 22 seconds), and automation control. This more sophisticated system can interface with an automated satellite station, to alert automatically for dangerous weather or other emergencies. By July 1, 1996 all stations were required to have new EAS equipment installed; they were to continue traditional weekly tests but the digital signals and very brief "tone burst" would not be obtrusive to listeners. After July 1, 1997 the two-tone test signal would be used only during an emergency or during monthly tests.

In the mid-1990s various state associations of broadcasters filed comments with the FCC about further reducing regulation. They supported the Commission's proposal to permit unattended operation of stations; they urged eliminating the requirement for licensed chief or duty operators, noting that new technology makes feasible a designated person easily contacted when warranted. The broadcast organizations argued against daily physical inspections of antenna tower lighting because such can be monitored electronically. They also asked the FCC to clarify language—replacing vague guidelines such as "periodically" and "occasionally"—in regulations about technical logs, power determining values, directional antenna parameters, monitoring point values, and checks of tower lighting circuits and controls.

These and the earlier topics of technical engineering are regulated by the federal government on behalf of the public whom the airwaves are intended to serve. The previous chapter explored in detail major regulatory matters involving broader social, legal, and economic aspects of radio-television and cable operations which are a primary responsibility of broadcast management.

—∿∿—

We conclude this chapter and the book with a group of comments about pragmatic priorities and a philosophical perspective for managers in the broadcast/cable media mix.

In 1993 the editors of *Broadcasting & Cable* noted the instability caused by so many changes in electronics technology over such a short period of time. They editorialized:

> The question is, when will it all settle down? In this page's view, not soon. Technology is so driving all the Fifth Estate industries—and technology itself is accelerating in such quantum steps—that there's no time to amortize one development before the next is upon us.
>
> And the debt goes on.
>
> If there's an anchor out there, it's the realization that media distinctions mean less and less and programming means all. As Roger King puts it in this week's cover story, 'It always gets back to software.' The sooner the better.[29]

Gene Jankowski, then president of CBS/Broadcast Group, offered a similar assessment in 1979 (when his network held first place in the ratings):

> The viewer is concerned with what the system delivers, not with the delivery system. The public is interested in the message, not the technology. It matters little whether the picture on the screen comes over the air, by landline or

from outer space. And that's what we really must consider in the decade ahead—how to continue to satisfy the audience through programming, not through technology.[30]

Finally, amplifying an aphorism cited at the start of this chapter, NAB executive Rick Ducey summarized the pragmatic issue for radio-TV-cable management:

> While technology is certainly a driver in changing the circumstances of the media marketplace, consumers do not buy technology, they buy services—either with their attention or with their dollars. Even as technological capabilities expand dramatically in the electronic media, new consumer needs and demands develop at a much slower rate. It is very challenging to build a new market. That means that more competitors will be fragmenting a relatively constant market.[31]

The continuing challenge for media managers is to sustain and increase profitability for owners and investors in stations, networks, and cable systems—through astute programming that not only attracts audience attention by responding to their wants but also earns public support by serving their genuine needs.

Yet beyond economics and technical efficiencies, a larger issue remains central. Largely due to technological achievements, mass media may grow faster than the apparent ability of their leaders to master them. The many changes are often accepted as phenomena of science rather than as challenges for equivalent human achievement. It now has become imperative that broadcast, cable, and satellite managers understand the full import of what is happening before they can make intelligent plans for using those technical advances.

Robert W. Sarnoff, when chairman of the board of the National Broadcasting Company, expanded on the philosophy of his father, David Sarnoff who built RCA and NBC:

> We are entering an era when the most progressive enterprises will possess built-in electronic systems to gather, digest, measure, correlate and analyze all information relevant to management decisions. They will link every corner of an enterprise, no matter how far-flung. And they will centralize management control by endowing the manager of the future with a vast new capacity for swift and efficient response to the most intricate and varied requirements of his own operations and to the most subtle trends of the marketplace. Thus, the new generation executive must not only have a basic understanding of these powerful aids but must also realize that they are, after all, only machines, and the answers they provide can only be as good as the questions that men ask them.
>
> More than ever before, success in business and industry will require of managers that they be both specialists and generalists. They must be knowledgeable in detail of their day-to-day operations, yet steeped sufficiently in science and technology to relate research and engineering developments to the specifics of their enterprises. And they must be equipped with knowledge of the economic, social, political and cultural forces that shape their environment.
>
> To the individual this signifies a longer period of schooling and broader preparation for his career—a trend already evident in the growing number of men and women who enter their careers with graduate degrees.[32]

The professional broadcaster and the broadcast educator who collaborated on this book are pleased to close it on that note.

1. Cited in *Broadcasting,* 4 July 1988, 33. The previous quotations are from: Lee De Forest, *The Father of Radio: The Autobiography of Lee De Forest* (Chicago: Wilcox & Follett, 1950), 444, 446; the *Jersey City Journal* cited on p. 255.

2. E. M. Rosenthal, "GMs Make $ the Issue in New Equipment Buys," *Television Broadcast,* August 1990, 1, 54. Data and sources in these two paragraphs were drawn partly from a graduate term paper by Michael Ian Silbergleid, "Evaluating Capital Equipment Expenditures in Broadcast Television" (Telecommunication and Film Department, The University of Alabama, April 1991).

3. See National Association of Broadcasters, *NAB Publications and Resources Catalog, 1997* (Washington, D.C.: NAB, 1997). Items may be ordered directly from the NAB at 1771 N St. N.W., Washington, D.C. 20036-2891; by phone at 202/429–5373, or by toll-free number 800–368–5644. Listed prices generally ranged from $30 to $50, with some costing $75 up to $400; costs to NAB members were usually half the list price.

4. "Group Owners Buying Equipment, Cautiously," *Broadcasting,* 28 October 1991, 40.

5. Don West and Kim McAvoy, "Staking a Claim on the Future," *Broadcasting & Cable,* 19 April 1993, 20–26; Fritts quoted on p. 26.

6. *BM/E* [Broadcast Management/Engineering], November 1977, 60–63.

7. Sethu Sekhar and Edward Fischer, "The Electronic Newsroom: What It Means to Your Station," *Broadcast Cable Financial Journal,* July/August 1991, 33–35.

8. David Bollier, "Unions Learn to Go with the Flow," *Channels,* May 1989, 44–45.

9. Loren Tobia, "Robots in the Studio: Payback in 2 Years," *RTNDA Communicator,* June 1991, 16.

10. Quoted in an interview by Glen Dickson, "CBS Looks Bravely Into the Digital Future," *Broadcasting & Cable,* 4 September 1995, 60–62.

11. "Engineers Make Vegas Wish Lists," *Broadcasting & Cable,* 10 April 1995, 40–43. Sixty-eight percent of the 125 station engineers surveyed were at affiliates of the four major networks, 11% were at PBS outlets, 4% were at independent stations, 3% at UPN affiliates, 2% at Warner affiliates, plus 11% at other operations.

12. National TeleConsultants, "The Migration to Digital: A Practical Guide to TV Managers," *Broadcasting & Cable,* 1 November 1993, D1–D16; cited on p. D16.

13. Electronics Industry Association figures, tabulated in *Broadcasting Yearbook 1974,* 73. These were cumulative totals, representing all U.S. sales of domestic and imported color receivers through two decades—not the number of color sets actually in service in the final year 1972.

14. Al Jaffe, "Final Edition: Future of Advanced TV Technology?" *Television/Radio Age,* 18 April 1988, 16.

15. "Consumers Will Go for HDTV, Says EIA Study," *Broadcasting,* 5 December 1988. Based on HDTV's introduction in 1992, the two benchmark years were 1996 and 2000. But it took eight years rather than only several for the "Grand Alliance" of manufacturers and research institutes to arrive at a single HDTV standard, in November 1995; that set back by half a decade such prognoses. Late in 1996 a compromise resolved objections: the computer industry criticized those digital standards as not adequate to support computer input for proper image interfacing; and the Hollywood creative community urged a slightly wider 2:1 aspect ratio for HDTV screens, to accommodate complete images of wide-screen motion pictures.

16. Peter Lambert, "HDTV: A Game of Take and Give," *Broadcasting,* 20 April 1992, 6–10.

17. Sources for consecutive data in this paragraph: *Variety,* 15 August 1990, 37; *Broadcasting,* 7 October 1991, 30; *Broadcasting & Cable,* 1 May 1993, 6; *Broadcasting,* 18 February 1991, 49; 15 July 1991, 17; *Electronic Media,* 15 November 1993.

18. *Broadcasting,* 20 July 1984, 64; 10 December 1984, 54; 25 June 1990, 67.

19. "Julie Barnathan and the No-frills Approach to Broadcast Engineering," *Broadcasting,* 19 April 1982, 42–56; quotation on p. 50.

20. Rich Brown, "Operators Plan Growth Along Fiber Lines," *Broadcasting,* 1 February 1993, 29, reported a survey at the end of 1992 by Myers Reports.

21. For further details, including technologies and economic projections, see Thomas F. Baldwin and D. Stevens McVoy, *Cable Communications,* 2nd ed. (Englewood Cliffs, N.J.: Prentice Hall, 1988), Chapter 18, "Impact of New Communication Technologies."

22. Rich Brown, "MMDS (Wireless Cable): A Capital Ideal," *Broadcasting & Cable,* 1 May 1995, 16–18. The largest group owner of wireless cable was American Telecasting, with 152,600 subscribers in a service area of 9.7 million households; Cross Country had the highest penetration, serving 42,000 households out of 6 million its area; the eleven largest groups averaged 2.2% penetration in their market areas; *Broadcasting & Cable,* 18 September 1995, 41.

23. Mel Karmazin, quoted in "Fifth Estater," *Broadcasting,* 11 November 1991, 91.

24. Estimates were by Ronald Horchler, Warner Communications, engineering member of the FCC's advisory committee on advanced television service; NAB newsletter *TV Today,* 12 June 1989, 4.

25. Cited in *The Alabama Broadcaster,* September/October 1993, 12.

26. The Alabama Broadcasters Association in 1993 mounted a pilot project offering to member stations the services of a certified Senior Broadcast Engineer (with 21 years in broadcasting, including 18 years of engineering experience and 12 years as owner/operator) who had an FCC ID permitting access for inspecting EBS equipment and operation records. He used the FCC's Radio Broadcasters Inspection Checklist to examine and make recommendations to forty commercial and noncommercial stations in EBS District 8 in the Mobile area. This was to identify and correct inadvertent technical noncompliance before Commission field engineers inspected facilities. "ABA Initiates Technical Audit Pilot Program in South Alabama," *The Alabama Broadcaster,* September/October 1993, 1.

27. "Codes that Affect Station Design," *Broadcasting & Cable,* 16 May 1994, S–14.

28. Cited by Peter Lambert, "New EBS Technology Vying for Adoption," *Broadcasting,* 9 December 1991, 48–49.

29. "The Great Stabilizer," *Broadcasting & Cable,* 13 September 1993, 66.

30. Gene Jankowski, "Remarks before the Academy of Television Arts and Sciences" (Los Angeles, Calif., November 29, 1979), p. 4 of text.

31. Rick Ducey wrote these comments when he was senior vice-president, NAB Research & Information Group, in *NAB News,* Spring 1992, 17.

32. Robert W. Sarnoff, (speech delivered at Bryant College in Providence, R.I., July 25, 1964).

INDEX

——◈——

Note: References to material in end-of-chapter notes are indicated by page number, abbreviation *n.,* and note number.

Employees. *See* Personnel; Staff
Engineering
  and automation, 518–21
  chief engineer's role, 99–103, 514–15
  equipment purchase and maintenance,
    515–18
  expenses for, 399
  FCC inspections and, 533–35
  HDTV and, 523–25
  relationship with management, 512–14
Entertainment programs
  costs of, 278–81
  developing, 275–77
  popular genres, 283–87
  programming issues, 274–75
  sports, 287–92
  See also Programming
Equal access, 491–94
Equal Employment Opportunity guidelines,
    118, 119, 456–57, 486
Equal time rule, 464–68
Equipment
  as media asset, 389
  purchase and maintenance, 101, 515–18,
    521–23
  testing, 517
ESPN, 287–88, 290, 420
Ethics in broadcasting, 51–53, 66–67 n. 61,
    68 n. 63
Ethnic populations. *See* Minorities
Ethnic programming, 227–29
Executives, 36–37. *See also* General
    manager; Station manager
Executive sessions, 494
Expenses, 395–400
Exploitative climate, 143

F

Face the Nation, 296
Fair comment privilege, 507 n. 41
Fair Labor Standards Act, 128–33
Fairness doctrine, 468–72
Family viewing restrictions, 478–79
Farm stations, 229–30
Fayol, Henri, 37
Feature films, 240–41
Federal Communications Commission
  and children's programming, 248, 379
  and community service, 204
  creation of, 445
  and deceptive advertising, 379
  and drug oriented lyrics, 477–78
  and duopoly, 265, 426
  and editorials, 469, 472–73
  and EEO rules, 118, 119, 120, 485–87
  fees for, 491, 492 (table)
  fin-syn rules, 313, 490–91
  and HDTV, 524
  and indecency, 475–77

Federal Communications
    Commission—*Cont.*
  liaison with, 101
  members of, 82, 446–47
  and ownership concentration, 452–55
  prime-time access rule, 236, 237,
    490–91
  reorganization of, 501–2
  responsibilities of, 10–11, 447–51,
    498–99
  rules violations, 456–58, 533–35
  station licensing, 4, 458–60
  See also Regulation
Federal Election Campaign Act of 1971, 467
Federal Radio Commission, 444
Federal regulation. *See* Regulation
Federal Trade Commission, 379
Ferris, Charles, 10, 445
Fiber-optic cable, 519, 525–26
Financial interest/syndication restriction, 490
Financial issues, 47, 385–86, 401–2
Fines, 456–58, 533–35
The Firestone Hour, 275
Firing employees, 146–49
First Amendment rights of broadcasters,
    450–51
Fish, Marjorie J., 88, 92, 95–96
Fisher Broadcasting, 471
Floor plans for stations, 79–80
FM stations
  advent of, 4–5, 160, 201
  vs. AM, 233–34
  number of, 162
  profitability of, 160, 234, 406,
    407 (table)
  and stereo broadcasting, 530
Food service, 132
Football, 288–90
Forfeitures, 456–58, 533–35
Formats
  determining, 215–21
  dominant, 216, 217–18 (illus.), 219
    (table), 220 (table), 222–26
  emerging, 227 (table)
  in news programming, 209–10
  specialized, 227–32
For Our Times, 296
Fowler, Mark, 10–11, 236, 305, 446, 501
Fox Broadcasting, 428
Fox network
  development of, 270
  football coverage, 289
  growth in popularity, 174, 201–2
  profitability of, 409 (table), 416
  provocative programming of, 278
Franchising, 254
Frank, George, 65 n. 45
Freedom of speech, 450–51
Friedman, Milton, 68 n. 67
Friendly, Fred, 141, 299

Marketing, vs. sales, 337
Markey, Edward, 480
M*A*S*H, 237, 285
Masterpiece Theater, 310
Matsushita, 15
Mayflower case, 472
MCA, 15
McClendon, Gordon, 142
McGregor, Douglas, 21, 23–25, 38–39
McIntire, Carl, 506 n. 36
McLarney, William J., 39, 42, 43, 44,
        69 n. 77
McLean, Austin, 173
McLuhan, Marshall, 209–10
Media assets (accounting), 389–92
Media companies, largest, 435 (table),
        436 (table)
Media courses, 116
Media liabilities (accounting), 392
Median family income, changes in, 170
Media personalities, 186, 194 n. 68
Media properties, 422–24
Meet the Press, 296
Meier, Joe, 476–77
Merchandising, 361–62
Mergers, 14–15, 302, 432–35
Mermigas, Diane, 418–19
Metropolitan Opera broadcasts, 266
MGM/United Artists, 15
Micro-intention, 26
Microsoft, 296
Middle managers, 36–37, 71–103
Middle-of-the-road stations, 224
Mikita, Joseph K., 387, 438
Military service for employees, 132–33
Miller, Karen, 243
Minimum wage, 129
Mini-series, 284
Minorities
    in audience, 171–73
    as general managers, 38, 86
    hiring, 117–24, 127, 133–34
    as news directors, 94
    number employed in broadcasting, 111,
        124 (table)
    owning stations, 124, 486–87
    percentage in U. S. work force,
        118–19, 120
    programming for, 248–49, 270–71,
        307–8
    promoting, 143
    radio networks for, 266–67
    regulations about, 485–87
    in sales, 341, 342 (table)
Minow, Newton, 7, 448–49
Mintzberg, Henry, 61–62
Mobile units, 212, 244, 245
Mohn, Reinhard, 29–30
Morning Edition, 215, 231, 267, 268
Morning news, 295–96

Morse, Robert, 85, 207
Motivation, 41–42
Moyers, Bill, 300
MSNBC, 16, 296, 322
MTV, 421
Multi-channel multipoint distribution
        service, 254, 529
Multiple system operators, 13–14, 322–23,
        431–32
Multipoint distribution service, 529
Murdoch, Rupert, 15, 296, 432–33
Murrow, Edward R., 141, 298, 299
Music
    changes in, 215–16
    drug oriented, 477–78
    licensing, 398–99, 487
Musical-variety shows, 286
Must-carry provisions, 488–89
Mutual Black Network, 266–67
Mutual Broadcasting System, 263, 407

N

*NAB Legal Guide to Broadcast Law and
        Regulation,* 461–62
National Association of Black-Owned
        Broadcasters, 82
National Association of Broadcast
        Employees and Technicians, 133
National Association of Broadcasters
    code for children's programming, 305
    code of practices, 379, 387, 497
    continuing education offerings, 145
    engineering publications, 513–14
    leadership guidelines, 41
    legal publications, 444, 461–62
    salary range chart, 128
National Association of Television Program
        Executives, 89, 237
National Black Network, 267
National Football League, 289
National Labor Relations Board, 134–35
National Public Radio, 215, 267
National sales, 354 (table)
Native Americans, programming for, 229
NBC
    CEOs and income, 413 (table)
    compensation payments to affiliates,
        416–17
    and development of radio networks, 263
    management structure of, 77
    morning news programming, 295–96
    profitability of, 409 (table)
    sports coverage, 288, 289, 290, 291
    Wayne Newton libel suit against, 484
    See also Big Three networks; Television
        networks
NBC News Channel, 274
NCAA, 288–89
Network affiliation changes, 273

Tribune Company, 425 (table), 426 (table)
Trout, Jack, 337
Turner, Ted, 13, 14, 142, 296, 433
Turner Broadcasting, 15, 291, 420–21
Turnover (staff), 112–13, 146
TVRO equipment, 528
TV stations
    capital investments of, 10
    group owned, 76–77, 430 (table)
    image with ad agencies, 366
    major expenses, 249–52, 339, 396
    management structure, 73
    minority owned, 124
    as network affiliates, 272–74
    number of, 163
    physical layout of, 79–80
    profitability of, 6–7, 409–10
    research by, 181–83
    scheduling for, 249
    staff size of, 77–79, 124–25
    typical budgets, 36
    valuation of, 422–24

U

UHF stations, 5, 201–2
Umansky, Martin, 8, 206
Underground stations, 230–31
Unions, 101, 133–36, 515, 520
Unistar Radio Networks, 408
United/Paramount network, 416, 417
United Satellite Communications, Inc., 528
U.S. Catholic bishops, 50
U.S. Satellite Broadcasting, 528
U.S. Supreme Court, 450, 473, 507 n. 41
Univision, 270, 308, 321
The Untouchables, 318
Upfront sales, 373–75
Urban contemporary stations, 227–28

V

Vacation pay, 132
Van Deerlin, Lionel, 445
V-chip, 480, 481
Viacom, 14, 271, 316
Victory at Sea, 298
Videocassette recorders (VCRs), 5, 16, 253
Video editing, 519–20
Videotape formats, 521, 522 (table)
Violence
    effect on audience, 185–86
    news coverage of, 246–47
    in TV programming, 478–81
V Theory of management, 25–29, 37, 44

W

Wage-Hour Division, U. S. Department of
    Labor, 131
Wages, 127–31

The Waltons, 283
War and Remembrance, 15
Warner, Charles, 348
Warner Brothers, 271
Warner Communications, 15
Watergate proceedings, 493–94
WBBM-TV, 484
WBZZ-FM, 484–85
WCCO-TV, 245
WCVB-TV, 204–5, 242
WCVX-TV, 247
Weaver, Sylvester, 141–42
Webs of inclusion, 80
Weekend Edition, 215, 231, 267
Wells, Robert, 55
Wells, Theodora, 49
West, Don, 378, 446
Westinghouse Electric, 14–15
Westmoreland, William, 484
Westwood One, 268, 407, 408
WGN-TV, 137, 247
WHBQ-TV, 427
White, Barton C., 348
WHLL-TV, 247
Widmann, Nancy, 81, 105 n. 15
Wiley, Richard, 10, 445
Wilson, Joseph C., 54
Wire services, 211
Wirth, Timothy, 305
Women
    in broadcast management, 38, 80–82
    as general managers, 86
    hiring, 117–24, 133–34
    as news directors, 94
    number employed in broadcasting, 111
    as percentage of staff, 123 (table), 124
        (table)
    as program directors, 91
    promoting, 143
    regulations about, 485–87
    in sales, 341, 342 (table)
    as sales managers, 98
    and sex discrimination, 155 n. 47
Wood, Robert D., 144
Work attitude, 55–56
Working conditions, 131–32
WOR-TV, 316
WPIX-TV, 238
WQOW-TV, 246
Wright, J. Skelly, 450
Wrongful acts, 385
WSB-TV, 246
Wussler, Bob, 359
WVOX-AM/FM, 204
WVTM-TV, 522
WXUR-AM/FM, 506 n. 36
WZZK-AM-FM, 517